Exploring Chemistry with Electronic Structure Methods

Third Edition

James B. Foresman

Æleen Frisch

Gaussian, Inc.
Wallingford, CT USA

For more information about the Gaussian and GaussView programs, visit *www.gaussian.com*.
For information about WebMO, visit *www.webmo.net*. We used WebMO 14.
For more information about Pcmodel, visit *www.serenasoft.com*. We used Pcmodel 10.

Photo and figure credits follow the Acknowledgments on p. xiv.

Publication History:

		Corresponds to:
April 1993	First edition	Gaussian 92 Revision D.1
April 1996	Second edition	Gaussian 94 Revision D.2
October 2015	Third edition	Gaussian 09 Revision E.1 GaussView Version 5

ISBN: 978-1-935522-03-4

Printed in the U.S.A.

Contents

Preface to the Third Edition

It has been many, many years since the second edition of this book was published. The landscape of computational chemistry has changed considerably in the interim. Consequently, we have produced what is essentially a new book. Today, we tackle problems that couldn't even be considered in the late 1990s, using methods that were far too expensive to use back then.

Like its previous two editions, this book serves as an introduction to modeling chemical compounds and reactions using electronic structure methods. Several different types of chemists will benefit from reading this work:

- Experimental research chemists with minimal or no experience with computational chemistry may use this work as an introduction to electronic structure calculations. They will discover how electronic structure theory can be used as an adjunct to their experimental research to provide new insights into chemical problems.
- Students of physical chemistry, at the advanced undergraduate or beginning graduate level, will find this work a useful complement to standard texts, enabling them to experiment with the theoretical constructs discussed there.
- Experienced Gaussian users may use the discussions of advanced features found in each chapter to acquaint themselves with program capabilities with which they are unfamiliar.

Our goal in this work is to show readers how to use electronic structure calculations to address their chemical problems. All calculation types and Gaussian features are discussed in the context of real world chemistry. At all times, we strive to present the "best practices" with respect to calculation validity, research accuracy and interpretation of results.

Organization

The book's 10 chapters cover a broad range of topics and techniques:

CHAPTER	PRINCIPAL TOPICS
1	Using calculations within a chemistry research project; overview of model chemistries.
2	Basic calculation types & results: energy, geometry optimization, frequency; ONIOM calculations; wavefunction stability.
3	Potential energy surfaces; transition structure optimizations; handling challenging optimization cases.
4	IR and Raman spectra; thermochemistry; NMR; conformational searching/averaging; anharmonic frequency analysis.
5	Modeling compounds/reactions in solution; free energies in solution (SMD model); explicit solvent molecules.
6	Interpreting volumetric data; IRC calculations; PES scans.
7	NMR shielding tensors & spin-spin coupling constants; VCD, ROA & ORD spectra; hyperfine coupling constants.
8	Excited states (TD-DFT) and UV/Visible spectra; modeling fluorescence & emission; Franck-Condon analysis; CASSCF.
9	ONIOM input for large biomolecules; relativistic effects; weakly bound complexes; electronic spin organization.
10	Background discussions of the major theoretical methods.

About the Examples and Exercises

The text in each chapter includes many fully-worked examples highlighting various aspects of the topic under discussion. Exercises are also provided, again with full solutions. Rather than simply recapitulating procedures demonstrated in the worked examples, most exercises include new material that expands on themes first introduced in the text. Accordingly, you will find it beneficial to read through each problem and solution even if you do not choose to complete every exercise.

We have also included an advanced section in each chapter containing more complex exercises and sometimes discussions of additional topics. Experienced researchers may wish to examine the advanced track even in the earlier, more elementary chapters where the basic concepts are very familiar.

Given the power and low cost of computers at the present time, we were able to study substantially larger molecular systems than were practical in previous editions of this book. With only a very few exceptions, the calculations for this work were completed on a computing resource consisting of two quad core processor computers whose total cost was less than $800 US. Jobs were run in parallel across the two systems using the Linda facility.

Input files are available for all jobs on the book's website (see below).

About the References

References: *Method*
Definition: [Ref1]
Review: [Ref2]
Implementation: [Ref3]

References are given for all results reported in this book. In addition, boxes in the margin (like the one here) provide fundamental references for all of the major methods introduced in the text, including the fundamental definition of the method, any available review articles and references for the implementation in Gaussian 09 as of this writing. See the *Gaussian User's Reference* for the full reference list for the program.

Graphical User Interfaces

This book discusses using both GaussView and WebMO as graphical interfaces to Gaussian. Although some procedures for each program as discussed in detail, we generally assume that you have a basic understanding of how to use each program. For this information, consult the documentation for the relevant software package.

The Book's Website: expchem3.com

New in the third edition is the website companion to the text. Throughout the text, you will be directed to go to the website for further details and information, via the following icons:

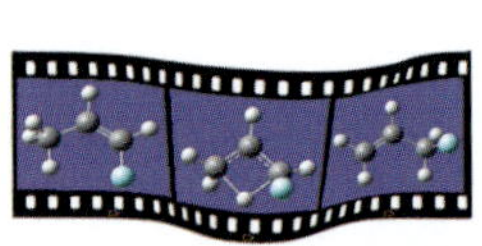

An animation related to the text is available on the website.

Additional information related to the theoretical background is available on the website. This icon is used only in chapter 10.

The site also contains:

- Input files for the examples and exercises
- Links to all references
- A sortable list of all molecules studied in the book
- Larger versions of the illustrations for visually impaired readers
- Alternate color versions of the illustrations for readers with color vision deficiencies

Finally, the website will also be the source of updated information as it becomes available after the printing of the book, so be sure to check it periodically.

Information for Teachers

We have included information for teachers and others wishing to pursue topics in more detail at appropriate points throughout this work. They are enclosed in a purple shaded box generally titled "Going Deeper."

To facilitate assigning reasonable amounts of work to individual students, many examples and exercises study a series of related compounds. The individual molecules can be divided up for study among the various students are appropriate for the needs of the class.

Acknowledgments

We are extremely grateful to the many people who helped us with this book at various points in its creation:

- Prof. Benedetta Mennucci (Univ. Pisa), Prof. Donald Truhlar (Univ. Minnesota) and Dr. Alek Marenich (formerly Univ. Minnesota; Gaussian, Inc.) read an earlier version of chapter 5 and made very helpful comments.
- Dr. Lee Thompson (Univ. CA Merced) and Greta Donati (Univ. Federico II, Napoli) gave very generously of their time in helping us prepare the ONIOM input section of chapter 9.
- Prof. Will Polik (Hope College) and Prof. Berny Schlegel and Dr. Shivnath Mazumder (Wayne State Univ.) read the opening chapters and used them in courses, providing excellent feedback from a classroom setting. The students in Will Polik's Spring 2011 Chem 422 class at Hope College gave us many helpful suggestions: Mr. Banks, Joshua Borycz, Kristian Cunningham, XiSen Hou, Mr. Kammermeier, Bruce Kraay, Jeff Largent, Elizabeth Miller and Suzie Stevenson.
- Jessica Freeze (Univ. Rochester) read the entire manuscript very carefully and made many useful suggestions.
- Kyle Throssel (Wesleyan Univ.), Patrick Lestrange (Univ. WA) and Mark Maturo (So. CT State Univ.; now Wesleyan Univ.) provided helpful comments on several individual chapters.

We were fortunate in having an excellent set of technical reviewers:

- Prof. Vincenzo Barone and Dr. Julien Bloino (Scuola Normale Sup., Pisa)
- Prof. Ed Brothers (Texas A&M Qatar)
- Prof. Hrant Hratchian (Univ. CA Merced)
- Prof. Nadia Rega (Univ. Federico II, Napoli)
- Prof. Mike Robb and Prof. Mike Bearpark (Imperial College, London)
- Prof. George Petersson (Wesleyan Univ.)
- Prof. Gustavo Scuseria (Rice Univ.)
- Prof. Jason Sonnenberg (formerly Stevenson Univ.; Gaussian, Inc.)
- Gaokeng "Daniel" Xiao (Molcalx, Guangzhou, China)
- Dr. Jim Cheeseman, Dr. Fernando Clementi, Dr. Doug Fox, Dr. Mike Frisch, Dr. Alek Marenich, Dr. Giovanni Scalmani, Dr. Gary Trucks and Dr. Lufeng Zhou (Gaussian, Inc.)

Any errors that remain are our own.

Joanne Kurnick, Joanne Hiscocks, Patti Flint, Joan Cie and ÆF's fabulous summer interns—Jessica LeClerc, Alex Smith, Brett Kurpit and John McDonald—all made important contributions to the production process.

Finally, JF would like to thank Lisa and Grace for their constant love and understanding during a period when many nights were spent monitoring calculations. ÆF thanks her partner Mike for his patience and love and all of her friends for their longtime support.

Photo and Figure Credits

Page 3: Photo of William Hartree is from the AIP Emilio Segrè Visual Archives, Hartree Collection.

Pages 24-25: The orbitals figure and the λ_{max}/energy/dipole moment graph are adapted with permission from W.H. Steel, J.B. Foresman, D. K. Burden, Y. Y. Lau and R. A. Walker, "Solvation of Nitrophenol Isomers: Consequences for Solute Electronic Structure and Alkane/Water Partitioning," *J. Phys. Chem. B* **113** (2009) 759-66 (pp. 761 and 762 respectively). Copyright © 2009 American Chemical Society.

Page 30: The non-negative data for the sulfur orbitals figure is republished with permission of John Wiley and Sons from D. Porezag, M.R. Pederson and A.Y. Liu, "The Accuracy of the Pseudopotential Approximation within Density-Functional Theory," *physica status solidi (b) [basic solid state physics]* **217** (2000) p. 222. The original article is © 2000 WILEY-VCH Verlag Berlin GmbH, Fed. Rep. of Germany.

Page 69: Photo of liquid oxygen poured between the poles of a magnet is courtesy of Prof. Robert Burk of Carlton University (Ottawa, Canada). This still photo is taken from his video: *www.youtube.com/watch?v=Isd9IEnR4bw.* The color has been slightly enhanced for print clarity.

Page 88: Saddle point figure courtesy Prof. John Lee, Mathematics Dept., Oregon State University.

Page 139: The photo of NCG 7023 was taken by the Hubble space telescope (courtesy ESA/Hubble). It is part of image *heic0915b*, available at *www.spacetelescope.org/images/heic0915b.*

Page 148: The photo of cosmic dust in the northwest region of NCG 7023 (the Iris Nebula) was taken by the Hubble space telescope (courtesy ESA/Hubble). It is part of image *heic0915a*, available at *www.spacetelescope.org/images/heic0915a.*

Page 149: The figure of the Raman spectrum is reproduced by permission of the AAS from Kris Sellgren, Michael W. Werner, James G. Ingalls, J. D. T. Smith, T. M. Carleton and Christine Joblin, "C_{60} in Reflection Nebulae," *The Astrophysical Journal Letters* **722** (2010), p. L55.

Page 152: The two Raman spectra of benzocaine are republished with permission of the Society for Applied Spectroscopy from Kathryn Y. Noonan, Melissa Beshire, Jason Darnell and Kimberley A. Frederick, "Qualitative and Quantitative Analysis of Illicit Drug Mixtures on Paper Currency Using Raman Microspectroscopy," *Applied Spectroscopy* **59:12** (2005) p. 1495; permission conveyed through Copyright Clearance Center, Inc.

Page 180: The $C_{60}O$ IR spectrum reprinted with permission from Cardini, G.; Bini, R.; Salvi, P. R.; Schettino, V., "Infrared spectrum of two fullerene derivatives: $C_{60}O$ and $C_{61}H_2$," *JCP* 1994, **98**, 9966–9971 (p. 9968). Copyright © 1994 American Chemical Society.

Pages 284 and 301: The IR and VCD spectra of fenchone and camphor are adapted with permission from F. J. Devlin, P. J. Stephens, J. R. Cheeseman and M. J. Frisch, "Ab Initio Prediction of Vibrational Absorption and Circular Dichroism Spectra of Chiral Natural Products Using Density Functional Theory: Camphor and Fenchone," *J. Phys. Chem. A*, **101** (1997) 6322–6333 (pp. 6324, 6326, 6327, 6328). Copyright © 1997 American Chemical Society.

Pages 285 and 289: The Newman projection diagrams and VCD spectrum for desflurane are adapted with permission from Prasad L. Polavarapu, Chunxia Zhao, Ashok L. Cholli and Gerald G. Vernice, "Vibrational Circular Dichroism, Absolute Configuration, and Predominant Conformations of Volatile Anesthetics: Desflurane," *J. Phys. Chem. B* **103** (1999) 6127-6132 (p. 6128). Copyright © 1999 American Chemical Society.

Pages 290 and 305: The ROA spectra for alpha-pinene, alpha-pinene oxide and 2,3-pinanediol are adapted from S. Qiu, G. Li, P. Liu, C. Wang, Z. Feng and C. Li, "Chirality transition in the epoxidation of (−)-α-pinene and successive hydrolysis studied by Raman optical activity and DFT," *Phys. Chem. Chem. Phys.* **12** (2010) 3005-3013 (p. 3008) with permission of the PCCP Owner Societies.

Page 294: The ROA spectra of methyl-β-D-glucose is adapted from J. R. Cheeseman, M. S. Shaik, P. L. A. Popelier and E. E. Blanch, "Calculation of Raman Optical Activity Spectra of Methyl-β-d-Glucose Incorporating a Full Molecular Dynamics Simulation of Hydration Effects," *JACS* **133** (2011) 4991–4997. Copyright © 2011 American Chemical Society.

Page 297: The ORD plot for methyloxirane in various solvents is adapted from S. M. Wilson, K. B. Wiberg, J. R. Cheeseman, M. J. Frisch and P. H. Vaccaro, "Nonresonant optical activity of isolated organic molecules," *J. Phys. Chem. A* **109** (2005) 11752–11764 (p. 11753). Copyright © 2005 American Chemical Society.

Page 302: The VCD spectrum for the Baeyer-Villiger oxidation product of (+)-(1R,5S)-bicyclo[3.3.1]nonane-2,7-dione is adapted with permission from P. J. Stephens, D. M. McCann, F. J. Devlin, T. C. Flood, E. Butkus, S. Stončius and J. R. Cheeseman, "Determination of Molecular Structure Using Vibrational Circular Dichroism Spectroscopy: The Keto-lactone Product of Baeyer-Villiger Oxidation of (+)-(1R,5S)-Bicyclo[3.3.1]nonane-2,7-dione," *J. Org. Chem.* **70** (2005) 3903–3913 (p. 3906). Copyright © 2005 American Chemical Society.

1

Using Computations in Chemical Research

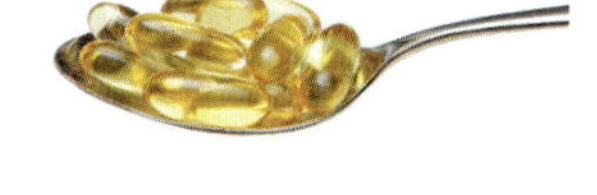

In This Chapter:

Model Chemistries: Theoretical Methods & Basis Sets

Chemical Models and Real Molecules

Predicting Properties of Substances

Simulating the Chemical Environment

Combining Calculations with Experiment

Advanced Topics:

Studying Compounds Containing Post-Third Row Elements

Grids: Numerical Integration Considerations

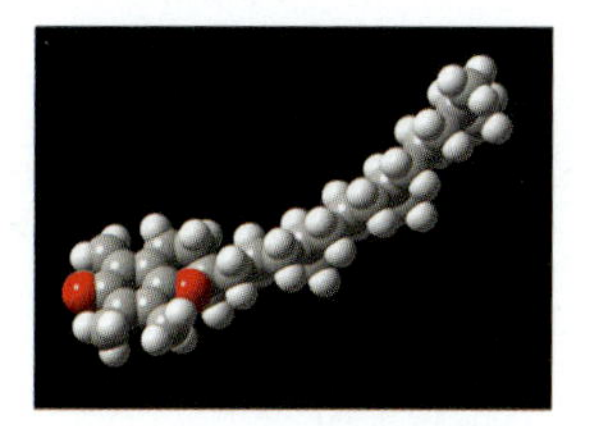

1 ◆ Using Computations in Chemical Research

This chapter provides an introduction to designing and running calculations on molecules and their reactions. It begins with a concise overview of electronic structure theory and its relation to the broader field of computational chemistry. Subsequent sections discuss the key modeling decisions that must be made for any computational investigation, in the context of a real-world research problem: the antioxidant behavior of vitamin E. The final section of the chapter describes an example use of calculations in close conjunction with experimental studies.

What is Electronic Structure Theory?

Computational chemistry is a general term covering any use of computing in the direct study of chemical problems. As such, it includes the entire range of computational techniques that are applied chemistry, whether their roots lie in physics—e.g., quantum mechanics, statistical mechanics—mathematics, informatics and/or other underlying scientific disciplines. Computational chemistry techniques can predict molecular properties for comparison with experiment, to elucidate ambiguous or otherwise unclear experimental data, and to model short-lived, unstable intermediates and transition states which are impossible to observe directly.

Quantum chemistry is a more specific term which refers to methods that were derived, in whole or in part, from the basic laws of quantum mechanics, most directly, the Schrödinger equation (see the sidebar). Quantum chemistry can be conceptually divided into two broad areas:

- Approaches which treat the nuclei within a molecule as essentially stationary particles surrounded by moving electrons. This approximation is reasonable since the electrons move much, much faster than the nuclei. These methods are called *electronic structure* methods due to their focus on the electrons.
- Approaches which model molecular behavior over time. These *chemical dynamics* methods study the detailed motion of the electrons and the nuclei, using the laws of quantum physics (*quantum dynamics*), classical Newtonian mechanics (*molecular dynamics*) or a combination (*semiclassical dynamics*).

$$\mathbf{H}\Psi(\vec{\mathbf{r}},\vec{\mathbf{R}}) = E\Psi(\vec{\mathbf{r}},\vec{\mathbf{R}})$$

*The Schrödinger equation underlies common electronic structure methods. In this time-independent version, Ψ is the wavefunction of a molecule. It is a function of the positions of the electrons and of the nuclei of its component atoms ($\vec{\mathbf{r}}$ and $\vec{\mathbf{R}}$ respectively). The equation states that applying the Hamiltonian operator **H** to the wavefunction yields the wavefunction multiplied by E, the energy of the molecule (see p. 467 for details).*

Source: AIP Emilio Segrè Visual Archives, Hartree Collection

This photograph from the late 1930s shows William Hartree, an engineer and mathematician and a lecturer in Engineering at Cambridge University [Gavroglu12], *as well as the father of the early computational chemist* Douglas Hartree: the Hartree of "Hartree-Fock."*

Beginning in 1935, when William retired, he performed calculations for his son, using linear and spiral slide rules as well as pencil and paper (these tools can be spotted in the foreground of the photograph). In this way, Douglas Hartree became the first scientist to use an external "computer" for theoretical calculations of molecules.

* *Although Douglas Hartree himself would never have used that term. He is traditionally described as a mathematician and physicist.*

There are many different families of electronic structure methods. Some of them base their computations solely on the laws of quantum mechanics and the values of fundamental constants—such as the speed of light and the masses and charges of electrons and nuclei. Others combine versions of the same approach with parameters derived from experimental data to simplify the computations. Still others combine parts of several methods to model molecular systems.

The various methods differ in the trade-offs made between computational cost and accuracy. For example, semi-empirical calculations are relatively inexpensive and provide reasonable qualitative descriptions of molecular systems and fairly accurate quantitative predictions of energies and structures for systems where good parameter sets exist. In contrast, DFT computations provide high quality quantitative predictions for a broad range of systems; they are not limited to any specific set of elements, class of system or chemical environment.

Electronic structure software often provide an additional class of methods, known as *molecular mechanics* (MM) methods. Based on classical mechanics, MM methods are extremely inexpensive and can be applied to systems as large as millions of atoms. MM methods employ *force fields* comprised of parameter sets and specific functional forms to model the molecular potential energy as the interactions between pairs of atoms. In the context of electronic structure calculations, these methods are used primarily for modeling very large molecules using hybrid quantum mechanics-molecular mechanics (QM:MM) approaches (see p. 61).

For more details on the physics underlying electronic structure methods, see Chapter 10.

Example Research Problem: Vitamin E

Vitamin E is a fat soluble antioxidant. It was discovered in 1922, isolated in 1936 and first synthesized in 1938. The latter year also saw the first study testing its potential health benefits. It was generally accepted as an essential nutrient for human health in the late 1960s. More recently, it has been widely touted in the popular press for potential/ostensible anti-aging benefits, and the substance is in wide use in pharmaceutical, over-the-counter health care and cosmetic products.

Technically, vitamin E is a generic term for the set of tocol and tocotrienol derivatives that perform the specific biological activity associated with it, comprising the four tocopherol compounds and the four tocotrienol compounds, denoted as α-, β-, γ- and δ-homologues.

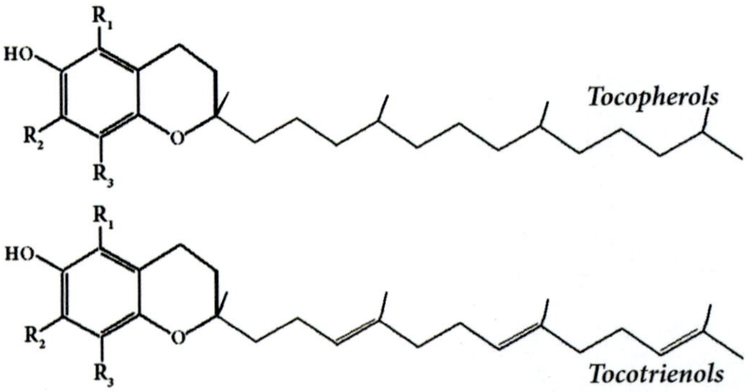

Tocopherols

Tocotrienols

α: $R_1,R_2,R_3=CH_3$
β: $R_1,R_3=CH_3$; $R_2=H$
γ: $R_2,R_3=CH_3$; $R_1=H$
δ: $R_3=CH_3$; $R_1,R_2=H$

Of these, α-tocopherol is the form of relevance to human physiology; accordingly the term "vitamin E" is synonymous with this compound in common usage. α-tocopherol* is found

* IUPAC: (2R)-2,5,7,8-tetramethyl-2-[(4R,8R)-(4,8,12-trimethyltridecyl)]-6-chromanol.

naturally in many vegetables including leafy greens such as spinach, avocados, asparagus and broccoli, in some fruits, including kiwis, mangoes and tomatoes, and, most abundantly, in nuts and various nut- and vegetable-based oils, especially wheat germ oil, sunflower oil and safflower oil.

Vitamin E's physiological role has traditionally been seen as an antioxidant: a compound that reacts with so-called free radicals—reactive oxygen species (ROS)—thereby preventing their harmful effects, such as the deterioration of cell membranes (i.e., lipid peroxidation). Vitamin E is thought to work as a sacrificial compound: by participating in the oxidation reaction, it prevents the ROS from reacting with an oxidized site in the cell membrane. More recently, theories have emerged which describe vitamin E participation in cellular signaling mechanisms related to hormonal responses, cell growth and even memory, by interacting with specific proteins or enzymes. For example, it can react with protein kinase C (PKC), which would otherwise cause vascular smooth muscle cell proliferation; the reaction with α-tocopherol's presence arrests this undesirable effect.

It is not currently known with certainty whether vitamin E plays one or both of these proposed roles in the bodies of human beings and other mammals. However, both of them involve the oxidized form of α-tocopherol. Accordingly, there is great research interest in the chemistry related to the oxidation of this compound, including the various associated intermediates. See [Webster07, Rosenau07, Yao09] and the references therein for details on recent experimental and computational studies of vitamin E chemistry.

Both thermodynamic and kinetic factors are of potential relevance to the oxidation reactions of α-tocopherol. We will focus on the former in our modeling.

Reactions involving vitamin E are challenging. α-tocopherol has the potential for losing two electrons and two hydrogens (protons) in order to reach a stable intermediate: an ortho-quinone methide (OQM) [Rosenau02], structure 7 in the figure below. This (-2p-2e) loss corresponds to a hydrogen molecule. These items can be removed in a variety of ways (orderings), in several combinations; that is, the proton and electron losses can be strictly consecutive or combined into one or more concerted processes. The following figure illustrates several possible pathways, as described by [Webster07]:

R=[(4R,8R)-(4,8,12-trimethyltridecyl)]
$=((CH_2)_3CH(CH_3))_3CH_3$

In the presence of an acid or an acidic environment, the compound initially loses an electron to form the radical chromanolate cation, followed by the loss of a second electron, yielding the +2 cation, followed by the loss of two protons, resulting in the chromanoxlyium cation and then the OQM intermediate (i.e., 2↔3↔6↔7). A base/basic environment takes a different route, first losing a proton, yielding the chromanolate anion, then losing an electron, resulting in the neutral chromanoxyl radical, then losing another electron to produce the chromanoxlyium cation, which loses the second proton to arrive at the intermediate (4↔5↔6↔7). In the absence of an added acid or base, there is a third path, beginning with an electron loss, followed by the loss of a proton,* then the second electron, and finally the second proton (2↔5↔6↔7), with the latter two steps requiring only a non-acid environment. There are also other routes to the intermediate. Note also that each of the reactions is reversible.

Selecting the Initial Calculations

In general, different pathways are taken, depending on the exact conditions of the chemical environment and other species that are present. However, these differing routes often lead to the same OQM intermediate, making it a plausible place to begin our investigations.

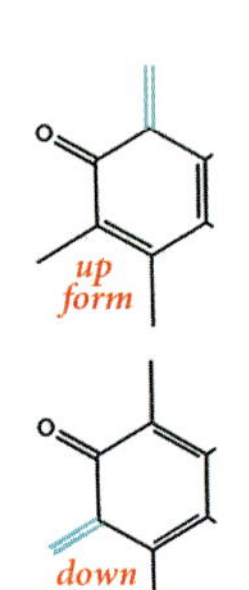

The OQM intermediate has two possible forms, differing in the location of the carbon atom double bonded to the ring (as illustrated in the margin). The one with the double bond pointing "up" is what is observed to form. This regioselectivity is based on the preferential reactivity of the corresponding carbon atom over the one at the alternate location.

Calculations can be used to confirm the thermodynamic preference of the up form of the OQM intermediate in α-tocopherol. Furthermore, calculations can predict how changes to the structure of the molecule might alter this preference. One modification to consider is breaking or altering the six-membered aliphatic ring to discern how important this component is in controlling the reactivity of the aromatic ring. Therefore, we will perform side-by-side calculations using the corresponding compound modified to include a five-membered aliphatic ring instead. Since the structure is based on 2,3-dihydrobenzofuran, we refer to this modified version as the "DHBF-based" structure.†

We could also consider how important the long chain is in determining the preference by replacing it with a methyl or ethyl group. If we did so, we would find that there is no change in preference. This hydrophobic part of the molecule may have some other important biological purpose, but it does not control the reactivity of the aromatic ring.

TO THE TEACHER: MOTIVATING VITAMIN E RESEARCH FOR STUDENTS

Undergraduate students might find the preceding discussion of vitamin E chemistry too detached from everyday experience. It might be helpful for them to take the discussion down a level of abstraction. They might consider the sort of questions we might want to ask about vitamin E:

- Is vitamin E important for nutrition? That question involves human physiology and would be at an operational level beyond what can be approached by using electronic structure theory.
- What products can form when vitamin E is digested? This is still a very broad question and would involve knowing how the substance interacts with other molecular structures in a cellular environment.
- What happens when vitamin E is oxidized? This question is within the realm that we can investigate directly with calculations.

It is known that the oxidation of vitamin E results is highly regioselective: of several possible products, one dominates. We will define our initial problem as investigating the thermodynamic stability of these products to see if theory can predict the observed preference. In addition, we will experiment with the molecular structure to see if a modification will alter this preference.

* These reactions have been observed to occur consecutively [Costentin06, Rhile06].

† IUPAC: DHBF derivation of α-tocopherol: (2R)-2,4,6,7-tetramethyl-3-hydro-2-[(4R,8R)-(4,8,12-trimethyltridecyl)]-5-benzofuranol.

Planning and Carrying Out Calculations

Every computational study involves certain operational considerations: a series of choices as to exactly what model(s) and techniques will be used to investigate the species and reactions of interest. These decisions must be made each time a research project begins, and they must be revisited—and possibly revised—periodically throughout its course. This problem analysis and decision-making process is applicable to any researcher, from beginners running their first calculation to experienced Gaussian users. While there will be variations according to the specific research project, we find it helpful to think about five principal aspects of calculation design:

- *Problem definition*: Narrowing the larger problem down to the first/next species/phenomenon for study.
- *Theoretical model progression*: Selecting an appropriate model chemistry (defined in the next section) or series of model chemistries. We often use a less expensive/accurate model chemistry for the initial calculations and then proceed to higher cost models in order to achieve the desired accuracy level.
- *Chemical representation progression*: Deciding how to represent the real molecular system(s) within the calculations. For example, initial calculations may focus on the key area within a large molecule, while subsequent ones will include more/all of the atoms in some fashion.
- *Calculation series progression*: Planning the series of calculations necessary to predict the desired properties. Most computational studies begin with geometry optimizations in order to locate the ground state equilibrium geometry. Later calculations will begin with that structure when predicting molecular properties.
- *Environmental modeling progression*: Ensuring that the relevant chemical environment is treated accurately. For example, in order to study chemical phenomena that occur in solution, the solvent environment must be taken into account in the calculations.

There are three important things to emphasize about these steps. First, with the exception of the first one, they need not be tackled in this particular order. In fact, the planning phase for most computational studies will involve moving back and forth among them until the best approach is found.

Second, these steps are not really completely separable and independent. Rather, decisions made in one area will affect those in another: deciding to study certain classes of problems will limit model chemistry choices to those capable of handling them accurately, which in turn will limit the potential calculations that can be run to ones that are supported by the relevant model chemistries. For example, if you are interested in a problem involving excited states, you will need to select a model chemistry which can treat excited states, and you will be limited to the calculation types that the one you choose is able to run. Nevertheless, we find it conceptually helpful to consider them as separate items.

Finally, each one of these items typically involves a progression from simple to complex. Making decisions in each area inherently involves trade-offs between ideal computational accuracy and the exigencies of real-world time requirements and computing resource availability. When we begin a study, we tend to use models which are faster and less expensive. As we make progress and understand our problem and its features in more detail, it is natural to move to more resource intensive models and calculation types in order to achieve the accuracy required for understanding the chemistry.

We will see examples of each of these progressions throughout this book and highlight how they affect the decision-making process in the following sections devoted to individual steps.

For our study of vitamin E and its antioxidant properties, we completed the first item when we decided on the initial focus in the previous section. We will perform calculations to verify the regioselective preference for the up form of the OQM intermediate and to contrast this with the five-membered ring correlative compound where the preference should be reversed.

The Theoretical Model Progression: Model Chemistries

A model chemistry is defined most simply as the combination of a theoretical method and a basis set. If you want to compare properties of different molecules/reactions, they must be predicted using the identical model chemistry.

The full definition of a model chemistry includes all calculation specifications which will affect the predicted energy value: theoretical method, basis set, any density fitting set, solvation model, numerical integration grid (if applicable).

These descriptions are designed to be concise. Detailed descriptions for each of these methods, including the relevant theoretical formulations, are given in Chapter 10.

Theoretical Methods

Different theoretical methods correspond to different approximations to the Schrödinger equation, each with their own accuracy, computation cost and resource requirements. Theoretical methods are often referred to as *levels of theory*, where higher levels are the more accurate ones.

Gaussian includes a variety of theoretical methods which can treat systems at varying levels of accuracy and in varying molecular states and chemical environments. The following are some important families of methods available in Gaussian:

Method Family	Examples
Semi-empirical methods are low-cost qualitative methods. They are useful for molecules that are too large to be practical with more accurate methods, for initial modeling of large molecules and as components of ONIOM MO:MO and MO:MM calculations.	**PM6** **AM1**
Hartree-Fock: The least expensive/accurate ab initio method. It was the initial method of choice for many years. Hartree-Fock suffers from the fact that it does not include the effects of electron correlation: the energy contributions arising from electrons interacting with one another.	**HF**
Møller-Plesset perturbation theory: Based on the subdiscipline of mathematical physics known as many-body perturbation theory, these methods add corrections to Hartree-Fock theory to achieve much better accuracy for systems where electron correlation effects are important, at correspondingly higher cost.	**MP2** **MP4**
DFT: These methods include many of the effects of electron correlation with only a modest increase in cost over our Hartree-Fock theory, leading to an almost universal supplanting of it. A plethora of DFT methods exist. One major distinction is between *pure* and *hybrid* methods. A pure method consists of a pair of functionals describing two different terms in the energy expression, known as *exchange* and *correlation*. Hybrid functionals also include some amount of the Hartree-Fock exchange term in their formulation. Recent DFT advances include treatments of *dispersion effects* (described in the margin) and of *long-range interactions* (the electron-electron interactions present at large distances from the nucleus in processes such as electron excitations to high orbitals).	**B3LYP** **APFD** **CAM-B3LYP**
Coupled cluster methods add consideration of various levels of electron excitations—swappings of various numbers of real and virtual orbitals within the Hartree-Fock wavefunction—yielding very high accuracy (albeit at great cost).	**CCSD(T)**

Dispersion refers to the weak interaction arising from the fact that the charge distribution around a molecular system isn't constant; rather, it fluctuates due to the movement of electrons. Consider two argon atoms. There is no electrostatic interaction between them since they are both uncharged (cf., e.g., Li^+ and F^-). However, the movement of electrons within one argon atom can create a temporary, *instantaneous dipole* in some region, which will in turn induce a dipole in the other argon atom, resulting in a weak interaction between them. Many commonly-used functionals were designed without taking these effects into account.

Method Family	Examples
Compound models approximate very high accuracy methods for thermochemistry predictions by performing a series of less expensive calculations. Such models include the *Complete Basis Set* (CBS) and the *Gaussian-n* (also known as G*n*) families of methods.	**CBS-QB3** **G3, G4** **W1BD**
Specialized methods have been developed for modeling the excited states of molecules. Excited states are the subject of Chapter 8.	**TD** **EOM-CCSD**

Note that all methods have limitations as to the type of problems and/or range of elements to which they are applicable. For example, even the very accurate CCSD(T) method is less accurate for transition metal compounds than for ordinary organic molecules.

In this book, our primary method will be the APF-D hybrid DFT method including dispersion [Austin12] (keyword **APFD**). It was selected because at present it represents the best trade-off between accuracy and computational cost for the largest range of molecular systems and chemical problems. In general, it provides good accuracy while being feasible for use on typical computer workstations. The book will also use other methods when specific chemical problems require them.

Why Not B3LYP?

The B3LYP hybrid functional [Becke93a] has been the most-used functional for many years and is widely accepted in the literature. Its one lack is a treatment of dispersion. For this reason, we have chosen APF-D for this book. Nevertheless, if you choose to run the calculations presented here with B3LYP instead, in most cases the results will be comparable.

See [Becke14] for a review of DFT methods in computational chemistry.

Basis Sets

The second component of a model chemistry is the basis set. A *basis set* is a collection of mathematical functions used to build the quantum mechanical wavefunction for a molecular system. Specifying a basis set can be interpreted as restricting each electron to a particular region of space. Larger basis sets impose fewer constraints on electrons and more accurately approximate exact molecular wavefunctions, and they require correspondingly more resources. In general, calculations should be performed with the largest basis set that is practical. Different methods and calculation types have different basis set needs, as we'll see throughout this book.

In the quantum mechanical picture, electrons have a finite probability of being found anywhere in space. Theoretical methods using basis sets converge to an energy limit as more and more basis functions are added; this value corresponds to an infinite, complete set of basis functions and is known as the *basis set limit*.

Gaussian offers a wide range of pre-defined basis sets, which may be classified by the number and types of basis functions that they contain. Basis sets assign a group of *basis functions* to each atom within a molecule. In basis sets currently in use, each individual basis function is typically composed of a linear combination of several gaussian functions, known as *primitives*. The collection of basis functions on an atom mathematically approximates its orbitals.

Early developers of electronic structure methods quickly realized that using simply the minimal set of atomic functions on each atom would be inadequate since atoms within molecules have their electron probability distributions significantly distorted from what they would be in the isolated atom. Defining multiple functions with different gaussian exponents—known as *zeta values*—for the orbital types that would be occupied in the atom is the first way to accomplish this. Later, it was also realized that to represent the polarization of the orbitals (which are important in the formation of molecular bonds), one needed higher angular momentum types than are present in the isolated atom. Modern basis sets accomplish both of these objectives, using different numbers of basis functions of each type based on different philosophies of how to best determine the zeta values.

The Pople Criteria: Essential and Desirable Features of Model Chemistries

The theoretical philosophy that underlies the Gaussian program is based on the concept of model chemistries, formulated by John Pople [Pople73]. It is characterized by the following principle:

> A model chemistry should be *uniformly applicable* to molecular systems of any size and type, up to a maximum size determined only by the practical availability of computer resources.

This is in contrast to the view which holds that the most accurate modeling method which is practical ought to be used for any given molecular system. However, using different model chemistries for different sized molecules makes comparing results among systems unreliable.

This principle has several implications:

- A model chemistry should be *uniquely defined* for any given configuration of nuclei and electrons. This means that specifying a molecular structure is all that is required to produce an approximate solution to the Schrödinger equation; no other parameters are needed to specify the problem or its solution.
- A model chemistry ought to be *unbiased*. It should rely on no presuppositions about molecular structure or chemical processes which would make it inapplicable to classes of systems or phenomena where these assumptions did not apply. It should not in general invoke special procedures for specific types of molecules.

Once a model chemistry has been defined and implemented, it should be systematically tested on a variety of chemical systems, and its results should be compared to known experimental values. Once it demonstrates that it can reproduce experimental results, it can be used to predict properties of systems for which no data exist.

Other desirable features of a model chemistry include:

- *Size consistency/extensivity*: The results given for a system of molecules infinitely separated from one another ought to equal the sum of the results obtained for each individual molecule calculated separately. Another way of describing this requirement is that the error in the predictions of any method should scale roughly in proportion to the size of the molecule. When size consistency does not hold, comparing the properties of molecules of different sizes will not result in quantitatively meaningful differences.
- Reproducing the *exact solution* for the relevant *n*-electron problem: a method ought to yield the same results as the exact solution to the Schrödinger equation to the greatest extent possible. What this means specifically depends on the theory underlying the method. For example, Hartree-Fock theory should be—and is—able to reproduce the exact solution to the one electron problem; it can treat cases like H_2^+ and HeH^+ essentially exactly.

  Higher order methods similarly ought to reproduce the exact solution to their corresponding problem. Methods including double excitations (see Chapter 10) ought to reproduce the exact solution to the two-electron problem, methods including triple excitations, like CCSD(T), ought to reproduce the exact solution to the three-electron problem, and so on.
- *Variational*: The energies predicted by a method ought to be an upper bound to the real energy resulting from the exact solution of the Schrödinger equation.
- *Efficient*: Calculations with a method ought to be practical with existing computer technology.
- *Accurate*: Ideally, a method ought to produce highly accurate quantitative results. Minimally, a method should predict qualitative trends in molecular properties for groups of molecular systems.

Not every model can completely achieve all of these ideals.

The following table describes the various kinds of basis functions/basis sets:

Type	Description
split-valence	Basis sets in which valence orbitals are represented by two or more basis functions of different size. For example, carbon would be represented as: 1s, 2s, 2s', $2p_x$, $2p_y$, $2p_z$, $2p_x'$, $2p_y'$, $2p_z'$—where the primed and unprimed orbitals differ in size.
polarized	Split valence basis sets allow orbitals to change size, but not to change shape. Polarized basis sets remove this limitation by adding orbitals with angular momentum beyond what is required for the ground state to the description of each atom. For example, polarized basis sets add d functions to carbon atoms. Current basis sets often add multiple polarization functions to each atom.
diffuse functions	Diffuse functions are large-size versions of s- and p-type functions (compared to the standard valence functions). They allow orbitals to occupy a larger region of space. Basis sets with added diffuse functions are vital for systems where electrons are relatively far from the nucleus: molecules with lone pairs, anions and other systems with significant negative charge, excited states, systems with low ionization potentials, descriptions of absolute acidities, and so on.

Note that the relative orbital sizes are not to scale. Diffuse functions are much larger than the other two types.

Some basis sets allow you to specify the number and types of desired polarization and diffuse functions explicitly while others provide preset combinations. Basis set names encode information about their contents in various manners. Here are two examples:

- **6-311+G(2d,p)**: The **6-311G** **split-valence** basis set with **2 additional polarization functions on heavy atoms** (non-hydrogen atoms), **1 additional polarization function on hydrogen atoms**, and **diffuse functions on heavy atoms**. If you wanted diffuse functions on hydrogen atoms as well, then the + would be replaced by ++.
- **aug-cc-pVTZ**: The **split-valence triple-zeta** correlation-consistent basis set of Dunning and coworkers with added **diffuse functions**. The parent cc-pVTZ basis set is defined to include additional polarization functions. The term *triple-zeta* specifies that there are three basis functions for each valence orbital; *double-zeta*, *quadruple-zeta*, etc. are defined analogously. These designations are more loosely descriptive than rigorous in that not every atom will necessarily have exactly three times the minimal number of valence basis functions included in its definition. References: [Dunning89, Kendall92, Woon93, Davidson96].

6-311 = 6 primitives/core orbital basis function; 3 basis functions per valence orbital, consisting of 3, 1 and 1 primitive(s), respectively.

In this book, our standard basis set will be 6-311+G(2d,p). We will also use other basis sets from time to time as appropriate.

In Advanced Exercise 2.13 (p. 79), you will examine the basis functions within a basis set in detail, using Gaussian's **GFOldPrint** keyword.

About 6-311+G(2d,p)

This basis set's origins are diverse and eclectic. It is the union of high quality basis sets for different parts of the periodic table, constructed to provide consistent quality descriptions of the elements in the first three rows of the periodic table:

- First row, most non-transition metal third row elements: the 6-311G basis set of McGrath and coworkers [Binning90, McGrath91, Curtiss95].
- Second row atoms: the McLean-Chandler [McLean80] basis set, as modified by Raghavachari and coworkers [Raghavachari80b].
- Ca, K: the basis set of Blaudeau and coworkers [Blaudeau97].
- Third row transition metals: the Wachters-Hay basis set [Wachters70, Hay77], as modified by Raghavachari and Trucks [Raghavachari89].

Model Chemistry Names

We will designate model chemistries using the following naming convention:

method/basis_set
method/basis_set/fitting_set
energy_method/energy_basis_set // geometry_method/geometry_basis_set

The first form is what is specified most often: the theoretical method and basis set, separated by a forward slash. The second form is used only for calculations involving pure DFT functionals for which a fitting set is used in order to speed up the numerical integration (see p. 31).

In the final form, the model to the left of the double slash is the one at which the energy is computed, and the model to the right of the double slash is the one at which the molecular geometry was optimized. Thus, MP4/6-311+G(2d,p)//APFD/6-311+G(2d,p) specifies an energy computed with the MP4 method and the 6-311+G(2d,p) basis set, using the geometry optimized with the APFD/6-311+G(2d,p) model chemistry. When this form is not used in this book, it means that the geometry was optimized using the same model chemistry as was used to predict the molecular properties under discussion.

EXAMPLE 1.1 MOLECULAR STRUCTURE OF FOOF

Historically, predicting the structure of F_2O_2 has been a challenge for electronic structure methods; it was a longstanding "pathological case" until Gustavo Scuseria demonstrated the efficacy of CCSD(T) for this problem [Scuseria91]. The FOOF molecule* has an unusually long O-F bond length, indicating a fairly weak interaction. FOOF is a case where MP2 does rather poorly, even with a large basis set. The following table lists the predicted structures for a variety of levels of theory and a large basis set:

PARAMETER	HF/ 6-311+G(2d)	MP2/ 6-311+G(2d)	CCSD(T)/ 6-311+G(2d)	APFD/ 6-311+G(2d)	EXP.†
R(O-O) (Å)	1.301	1.145	1.220	1.228	1.217±0.001
R(O-F) (Å)	1.357	1.716	1.595	1.499	1.575±0.001
∠F-O-O (°)	106.3	112.4	109.3	109.3	109.5±1.0
∠F-O-O-F (°)	85.2	89.8	88.4	87.7	87.5±1.0

†[Scuseria91]

Hartree-Fock theory and second order Møller-Plesset perturbation theory cannot model this molecule with high accuracy, even with the large basis set we selected. In fact, enlarging the basis set with even more polarization functions, to the 6-311+G(3df) basis set, does not improve the results for these methods. This problem requires coupled cluster theory with single, double and triple substitutions for very accurate results. The CCSD(T) method is very accurate, almost to within the experimental error.

This problem also illustrates the advantages of density functional theory. The hybrid APFD method is almost as accurate as the drastically more expensive CCSD(T) coupled cluster method using the same basis set (which happens to be the one we recommend for general use). Be aware, however, that DFT methods are not always in alignment with coupled cluster results as they are in this case.

The following table shows the effect of basis set size on the results for this problem. It focuses on the O-O bond distance and F-O-O bond angle, whose experimental values are 1.217±0.001 Å and 109.5±1.0°, respectively [Scuseria91]:

	BASIS SET (# basis functions, # primitive gaussians)						
METHOD	6-31G(d) (60,112)	6-31+G(d) (76,128)	6-311G(d) (72,128)	6-311+G(d) (88,144)	6-311+G(2d) (108,168)	6-311+G(3d) (128,192)	6-311+G(3df) (156,232)
HF	1.311 Å 105.8°	1.305 Å 106.1°	1.301 Å 106.4°	1.300 Å 106.5°	1.301 Å 106.3°	1.296 Å 106.2°	1.296 Å 106.3°
MP2	1.293 Å 107.0°	1.167 Å 111.7°	1.155 Å 111.5°	1.130 Å 114.8°	1.145 Å 112.4°	1.132 Å 112.6°	1.142 Å 111.4°
APFD	1.260 Å 108.4°	1.236 Å 109.2°	1.228 Å 109.3°	1.216 Å 109.8°	1.228 Å 109.3°	1.219 Å 109.2°	1.224 Å 109.1°
CCSD(T)	1.276 Å 107.5°	1.233 Å 109.3°	1.204 Å 109.8°	1.193 Å 110.6°	1.220 Å 109.3°		

* IUPAC: dioxygen difluoride.

As we mentioned earlier, even the largest basis sets do not result in accurate predictions for the Hartree-Fock or MP2 methods. The results with the 6-31G(d) basis set demonstrate how a small basis set inhibits high accuracy even for methods like coupled cluster theory including triples (although the addition of a diffuse function improves on those results significantly for the bond angle). CCSD(T) produces excellent results—within 0.1 Å and 1° of experiment—for both parameters only with the larger 6-311G(2d) basis set including a diffuse function. The results for the APFD method also illustrate the effect of the diffuse function: cf. 6-31G(d) and 6-31+G(d). The method also achieves excellent results with the larger basis sets.

This study illustrates how both the theoretical method and the basis set—in other words, both components of the model chemistry—are significant in determining the accuracy of the results. Both a sophisticated method and a large basis set are required for this molecule.

Other Model Chemistry Components

The theoretical method and basis set are the primary and most fundamental components of a model chemistry. However, other variations in calculation procedures can also be part of it. Understanding what is and isn't part of the model chemistry is important because the same model chemistry—in every detail—must be used for every calculation on every molecule within a computational study.

The following list briefly describes other items that are part of the model chemistry when relevant to particular calculations:

- Many theoretical methods include different variations for closed shell and open shell molecules. Open shell systems can be studied with *spin restricted* or *spin unrestricted* methods (see p. 53). When it differs from the default, the choice of a restricted or unrestricted method becomes part of the model chemistry specification.
- The solvent environment employed in the calculation (see Chapter 5).
- For DFT methods, the same integration grid must be used for all calculations whose results will be compared or combined (see p. 30).
- When a fitting basis set is used in combination with a density functional method, it too becomes part of the model chemistry (see p. 31).

Calibration

In order to successfully carry out a computational study, you need to determine the appropriate model chemistry for your calculations. This will often involve a preliminary calibration process where you test various candidates on one species, often the smallest molecule in the set in which you are interested or a related compound for which you have experimental data with which you can compare. Thus, in the preceding example, our results demonstrate that APFD/6-311+G(2d) is an appropriate model chemistry for determining structures of various FOOF-based compounds. Once you have determined the appropriate model chemistry, you can use it to carry out the calculations necessary for your investigation.

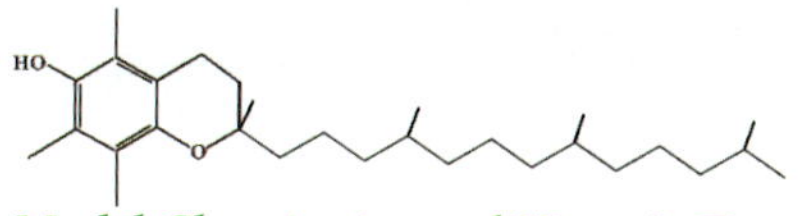

Model Chemistries and Vitamin E

Our first goal in our study of vitamin E chemistry is to compute the free energy difference between the two regioisomers (up and down) for a-tocopherol OQM and the DHBF-based variation we are using as a control. In order to do so, we need to optimize the molecular geometries of each compound and then compute their energy and the thermochemical corrections to it required to predict the Gibbs free energy. We want to do so with the most accurate method that is practical for these molecules.

A plausible way to begin a study of relatively large molecules like these is to optimize the structures with an inexpensive model chemistry, such as a semi-empirical method. Molecular structures built in chemical graphics programs or imported from PDB files typically do not correspond to the lowest energy geometry. Even small changes in the structural parameters of a molecule can produce significant changes in the predicted energy and other molecular properties. For this reason, finding the minimum energy molecular structure for the system of interest—a process known as a geometry optimization—is the first step in any modeling study. Using a relatively low cost model chemistry allows the later more accurate calculations to begin from a reasonable starting point.

Historically, the next step would have been to model the system with a Hartree-Fock-based model chemistry, reoptimizing the geometries and then predicting the thermochemical corrections needed for the Gibbs free energy at the minimum energy structures. The process might have concluded with high accuracy predictions of the energies, using the best affordable method. In summary, the minimum energy geometries would be located with Hartree-Fock theory, which would also be used to compute thermochemistry corrections. The latter would be applied to energies predicted using a model chemistry with higher accuracy, employing a more accurate theoretical method and/or a larger basis set.

Today, however, computer workstations are capable of running quite large calculations in reasonable periods of time. Accordingly, this process of running a series of calculations with gradually increasingly accurate model chemistries is no longer the norm. Instead, in many cases, it is possible to use the desired model chemistry immediately.

For illustration purposes, we ran a series of calculations on each compound using the following model chemistries:

- PM6: We used this semi-empirical model to produce reasonable initial geometries, and we also predicted the Gibbs free energies for comparison purposes.

- HF/6-31G(d): We next modeled these systems using Hartree-Fock theory with a modest-sized basis set.We reoptimized the PM6 geometries and then computed the energy and thermochemistry corrections, producing the predicted Gibbs free energies.

- APFD/6-311+G(2d,p) approx.: Our final calculations approximated the results of our standard model chemistry. They were achieved using the ONIOM method in Gaussian, which we will discuss in the next subsection.

The following table summarizes the free energy differences between the two forms of each compound predicted by each model chemistry. Note that a negative value for ΔG indicates that the up form is the lower in energy.

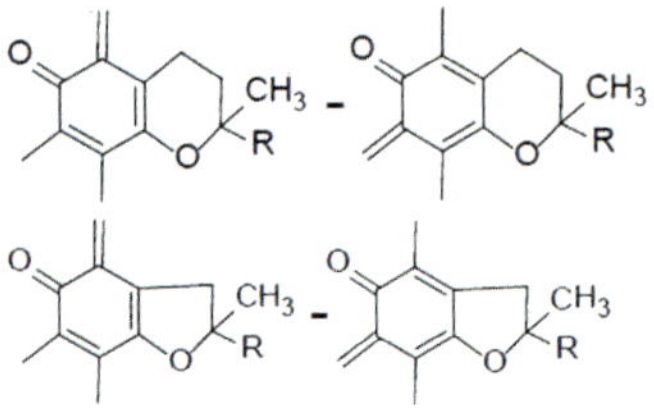

	$\Delta G^{up\text{-}down}$ (kJ/mol)	
MODEL CHEMISTRY	α-TOH OQM (1)	DHBF-BASED OQM (2)
PM6	-9.0	8.9
HF/6-31G(d)	-15.1	8.2
ONIOM APFD/ 6-311+G(2d,p):PM6	-9.9	9.9

All three of these model chemistries predict the regioselectivity that we expect, and the quantitative predictions have the same general magnitude. Fortuitously, PM6 produces results very close to the DFT-based model chemistry. The former has been parameterized for organic compounds like the ones under consideration, and predicting relative energies of organic isomers is a task semi-empirical methods often do well, so this agreement is not surprising. This does not mean, however, that PM6 would do as well studying other aspects of our problem.

Hartree-Fock theory with a modest-sized basis set predicts larger energy differences between the up-and-down forms than the APFD-based model chemistry using ONIOM. The former's only advantage is that it is inexpensive enough to be able to model the entire molecule directly, something that is not possible with our desired, standard model chemistry. The ONIOM model allows us to approximate our desired model chemistry at greatly reduced cost. It is the subject of the next section of this chapter.

The data in the last row of this table represents our final results for this problem. It originates from four separate frequency calculations (we discuss these in the next chapter). However, there were many more calculations required before we got to this point. We began with simpler models and smaller basis sets (see the table) to get preliminary results to use as starting points for the final calculations. We also tried various parameter and layer definitions for the ONIOM model until we settled on the one we used for the final calculation. Finally, we explored several modifications in the geometry of the DHBF-based control compound until we located a conformation which exhibits the observed switch in isomer stability. All of this experimentation was necessary in answering the initial question.

Chemical Models and Real Molecules

Selecting the method and basis set to use for a computational study almost always involves finding an effective trade-off between desired accuracy—as high as possible—and the (finite) available computing resources. Unfortunately, as we've noted, high accuracy model chemistries scale unfavorably with the size of the molecule, resulting in a practical limit on how large a system can be studied, thereby placing many systems of chemical and/or biological interest out of reach of traditional approaches.

Historically, there were limited options available to you for studying systems that were too large for the most desirable methods. You could study the systems with the most accurate practical method and hope that it would be sufficient. Alternatively, you could focus your calculations on only a subset of the compound—for example, the active site or a few amino acid base pairs—and again hope that doing so would enable you to draw conclusions about the molecule as a whole. In either case, preliminary calibration studies on well-understood compounds were essential in order to determine the level of accuracy that could be reasonably expected from studying the real systems of interest in the same manner.

Beginning in the late 1990s, Morokuma and coworkers developed an approach which separated large molecules into computational layers which could be treated at different levels of accuracy [Dapprich99].* This computational technique is known as ONIOM, which stands for **O**ur own **N**-layered **I**ntegrated molecular **O**rbital and molecular **M**echanics (the onion metaphor is more descriptive than the name). In a sense, ONIOM combines both previous strategies, focusing the computationally-intensive, higher accuracy modeling on the most important portion of the molecule while still including the effects of the entire molecule as a whole within the calculation.

* The ONIOM method is a novel alternative to the merged Hamiltonian of traditional QM/MM methods (see [Warshel76, Singh86, Field90]).

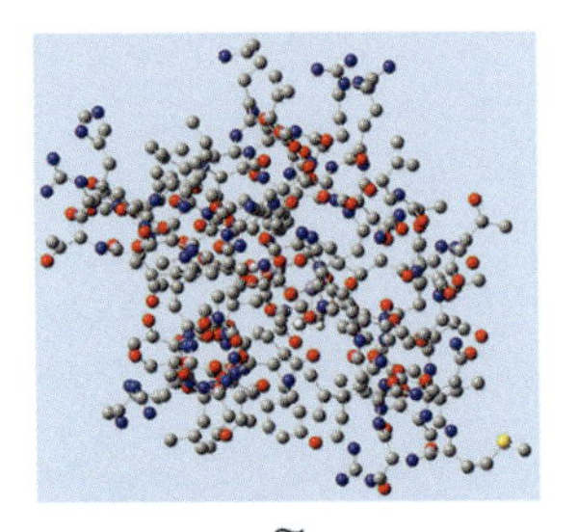

≈

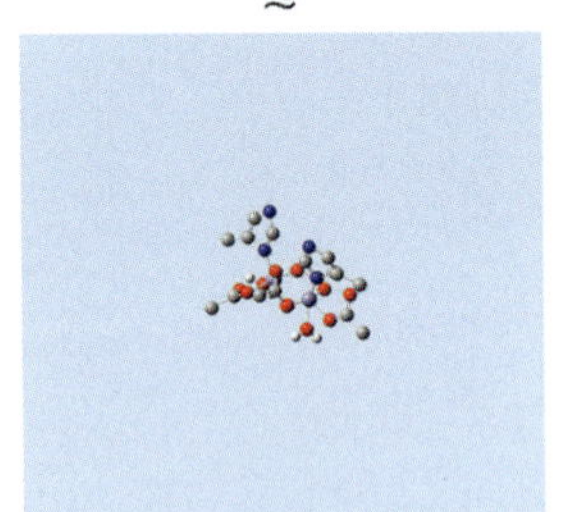

High layer: DFT/large basis set

\+

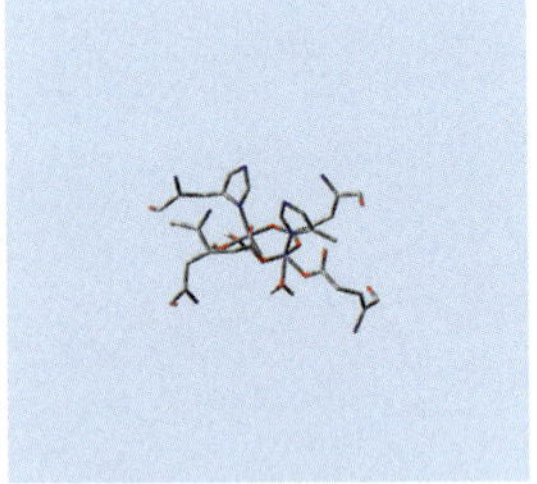

Middle layer: semi-empirical

\+

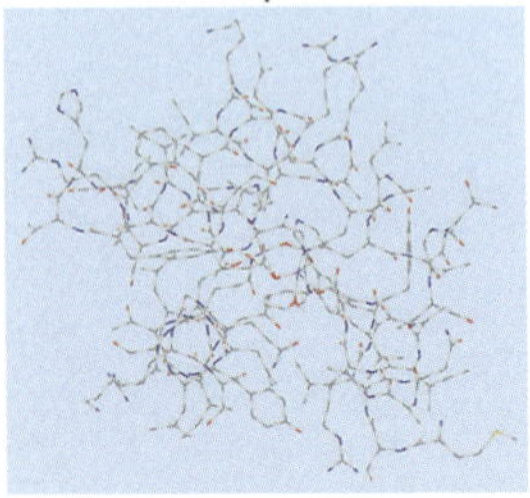

Low layer: molecular mechanics

In an ONIOM calculation, two or three regions within the molecular system are defined. The smallest one, known as the high layer, is treated with the most accurate model chemistry. Bond formation and breaking take place in this region.

The low layer is the largest one, and it consists of the entire molecule. It is typically treated with an inexpensive model chemistry such as a molecular mechanics or a semi-empirical method. This region corresponds to the environmental effects of the molecular environment on the site of interest (the high layer). Defining these two regions corresponds to a two-layer ONIOM model.

In some calculations—known as three-layer ONIOM calculations—a middle region, comprising all of the high layer as well as some portion of the remainder of the molecule, is also defined. It is treated with a model chemistry corresponding to an intermediate level of accuracy.

The protein in the margin illustrates ONIOM layers. Note how each successively higher layer is a subset of the layer below it. These figures also illustrate GaussView's default method of distinguishing ONIOM layers visually: ball-and-stick, tubes, and wireframe representations for the high, medium (if defined) and low layers, respectively.

ONIOM models using molecular mechanics to treat the low layer are known as MO:MM models: for "molecular orbital–molecular mechanics." If electronic structure methods are used for all layers, then the ONIOM model is referred to as MO:MO. MO is sometimes replaced by QM, for "quantum mechanics."

In the case of MO:MM calculations, the ONIOM technique can be optionally elaborated further with a more sophisticated treatment of the interaction between the MM and MO regions. *Electronic embedding* incorporates the partial charges of the molecular mechanics region into the quantum mechanical part of the calculation. This technique provides a better description of the electrostatic interaction between the QM and MM regions and allows the QM wavefunction to be polarized.

The ONIOM technique is designed to reproduce the results that you would get if you used the high layer's model chemistry for the entire molecule. ONIOM presents its own calibration challenges: you need to be sure that its treatment of the molecule in fact is a reasonable approximation of a traditional calculation using the specified model chemistry. This can be accomplished by comparing the results of an ONIOM treatment with those of the actual calculation which approximates for a smaller related or partial molecular system similar to those you want to study. The researchers who developed the ONIOM method have also performed a variety of calibration studies which have been published in the literature.

ONIOM calculations are discussed in chapters 2 and 5 (see pp. 61-64, 69 and 206).

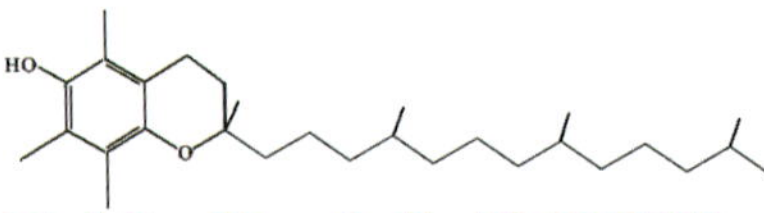

Modeling Vitamin E with ONIOM

The vitamin E-related compounds we are investigating are natural candidates for a 2-layer ONIOM treatment. First, it is too large to treat directly with our standard model chemistry, APFD/6-311+G(2d,p). Second, the long phytyl tail probably does not affect the chemistry of the ring atoms to a considerable degree, but including it to some extent is better than limiting the calculation to the two fused rings using an accurate model chemistry or modeling the entire molecule with a lower accuracy model chemistry.

As we noted before, the researchers who developed the ONIOM method performed many calibration studies. Many of the molecules used for this purpose were biological species, similar enough to vitamin E and its related compounds that we do not need to perform separate ONIOM calibration studies before proceeding with our own investigations.

The ONIOM partitioning is straightforward for this molecule. The fused rings and their attached substituents will become the high layer, and the substituents at position 2 in the righthand ring will be the low layer. In other words, the atoms in the *phytyl tail** that lie outside of the ring wil comprise the low layer.

These layers are illustrated in the following figure:

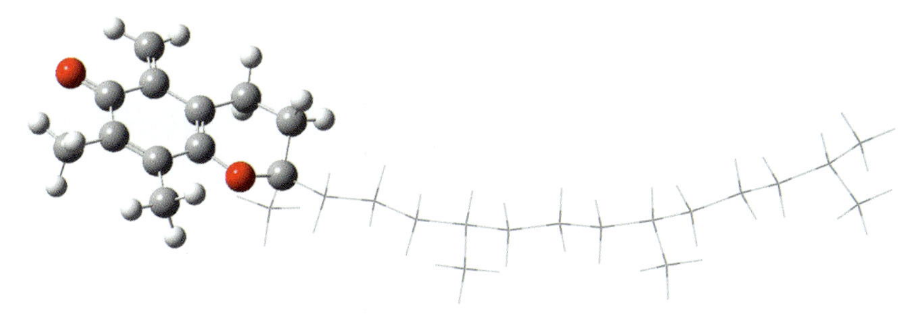

We will use our standard model chemistry for the high layer and the PM6 semi-empirical method for the low layer. We've already seen the results of the up/down form's relative energy calculation in the preceding subsection of this chapter.

Predicting Properties of Interest

The next investigative dimension we will consider is the progression of successive calculation types that are run during the course of a study. Few research questions can be addressed with just a single calculation, even if only one model chemistry is used. Exactly what calculations will need to be run depends on the specific chemical problems.

In general, molecular properties vary with even small changes in the molecular structure. Accordingly, it is necessary to locate the equilibrium geometry for the molecular systems under investigation before proceeding with properties predictions. Therefore, you generally begin by optimizing the geometries of the relevant compounds.

A geometry optimization locates a point on a potential energy surface where the forces on the nuclei are essentially zero, located near the structure given as the optimization's starting point. Such points can be either equilibrium geometries—minima—or transition states. A geometry optimization must be followed by another calculation—a frequency calculation—which determines which of these two possibilities has been found. Since this calculation sequence is so common, Gaussian allows you to specify it via a single job (see p. 57). The technical details of geometry optimizations are discussed in Chapter 3 (p. 87).

Once you have optimized the structures of your compounds, locating the equilibrium geometries, the calculation you run next depends on exactly what you are studying. For example, if you are studying a reaction, the next step after optimizing the reactants and products is to locate the transition structure connecting them. Once you have done so, it is often useful to follow the reaction path from the reactants through the transition structure to the products.

* The phytyl tail is the carbon chain that begins in the six-membered aliphatic ring and contains methyl groups at its positions 2, 6, 10 and 14.

In contrast, if you are studying spectroscopic properties, the next step is to predict the spectrum in question, and then go on to compare it with any experimental data that is available. Once this is done, you may choose to repeat the entire process for a series of molecules with varying substituents in order to compare trends and differences across a range of molecular systems.

If you are interested in thermochemistry problems, you may choose to calculate the energies and thermochemical corrections of the compounds of interest using an even more accurate model chemistry than was used for the geometry optimization. This process may also need to be carried out after locating the relevant transition structure if you are studying, for example, reaction kinetics.

The following table summarizes the most common molecular properties which can be predicted with Gaussian and indicates the type of calculation that produces them. Many of them will be discussed in the course of this book.

MOLECULAR PROPERTY	JOB TYPE
Antiferromagnetic coupling	**SP**‡ **Guess=Fragment**
Atomic charges	**SP**
ΔG of solvation	**SP SCRF=SMD**
Dipole moment	**SP**
Electron affinities	**CBS-QB3**
Electron density†	**SP**§
Electronic circular dichroism	**TD, EOM**
Electrostatic potential†	**SP**
Electrostatic potential-derived charges	**SP Pop=Chelp**, **ChelpG** or **MK**
Electronic transition band shape	**Freq=FCHT**
Polarizabilities/hyperpolarizabilities	**Polar**
High accuracy energies	**CBS-QB3**
Hyperfine coupling constants (isotropic) including spin-dipole terms (anisotropic)	**SP** **SP Prop=EPR**
Hyperfine spectra tensors (including g tensors)	**Freq=VCD** or **NMR**
Ionization potentials	**CBS-QB3**
IR spectra	**Freq**
Pre-resonance Raman spectra	**Freq CPHF=RdFreq**
Molecular orbitals†	**SP**
Multipole moments	**SP**
NMR shielding and chemical shifts	**NMR**
NMR spin-spin coupling constants	**NMR=Mixed**
Optical rotations	**Polar=OptRot**
Raman optical activity	**Freq=ROA**
Raman spectra	**Freq=Raman**
Reaction path	**IRC**
Thermochemical analysis	**Freq**
UV/Visible spectra	**TD, EOM**
Vibration-rotation coupling	**Freq=VibRot**
Vibrational circular dichroism	**Freq=VCD**

†These items can generally be computed on demand by visualization software.

‡**SP** is the default job type, and it refers to a *single point energy* calculation: the prediction of the total energy of the molecule at the specified molecular structure. Items indicating **SP** (alone) as the relevant job type are calculated during every Hartree-Fock/DFT Gaussian calculation unless they are explicitly suppressed with the **Population=None** option.

§**Density=Current** is also needed to select the density of interest for post-SCF methods.

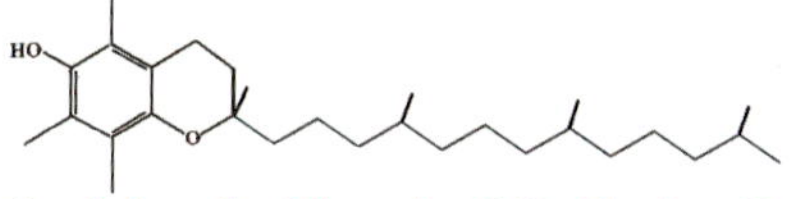

Studying the Vitamin E Oxidation Reactions

Our study of vitamin E oxidation reactions will require a great many calculations over its course. Once we have compared the up and down forms by optimizing their structures and computing the corresponding thermochemical corrections to the energy, a logical next step would be to locate the transition structures connecting α-TOH para quinone and the OQM intermediate for each configuration. This will enable us to assess the effect of kinetic factors on the regioselectivity that has been observed.

Before we can proceed, we need to determine how we will model the receiving molecule for the departing hydrogen atom. This selection is important since choosing an appropriate substance to participate in the reaction is vital to sensible and accurate modeling of the underlying physiological process. It is also a difficult and complex question as the chemistry of the cell membrane and the surrounding environment is only partially understood and is the subject of significant ongoing research.

While we will eventually want to use an ONIOM model for studying the transition structure and the associated reaction path, we will begin with a somewhat simpler approach which can be refined in subsequent calculations. Since transition structures are sometimes tricky to locate, especially for large molecules with many attached groups, it is often good practice to perform the initial transition structure optimization with smaller molecular structures focused on the site of interest and with a less expensive model chemistry.

In our initial calculation, we will simplify the α-TOH para quinone reactant by replacing the substituent in the phytyl tail with a methyl group and make the same replacement in the product OQM intermediate. We will use ethoxy radical as the receiving group for the hydrogen transfer reaction. We will perform the transition structure optimizations using the APFD method with the 6-31G(d) basis set; the latter is significantly smaller than the basis set we used for our previous energy comparisons, and the calculation should accordingly run quite a bit faster.

The following illustration displays the reactants (upper left), products (lower left) and transition structure (right) for our model reaction in the up configuration:

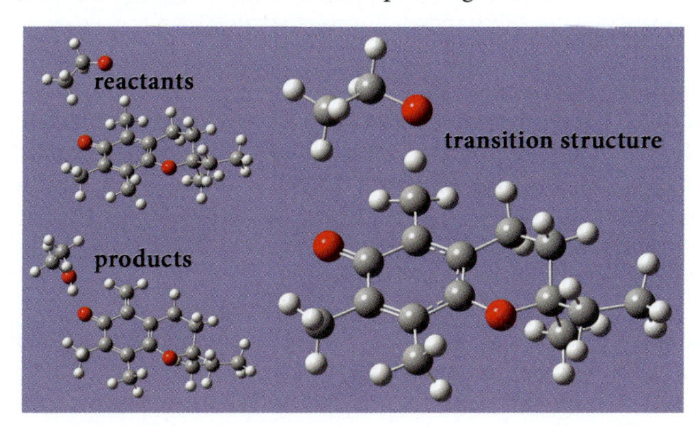

Reactants: (R)-2-ethyl-2,5,7,8-tetramethyl-6-chromanoxy radical + ethoxy radical
Products: (R)-2-ethyl-2,7,8-trimethyl-5-methylene-6-chromanone + ethanol

In the transition structure we located for the up form, the hydrogen atom has moved toward the receiving oxygen atom in the ethoxy radical molecule. Its distance from the carbon atom

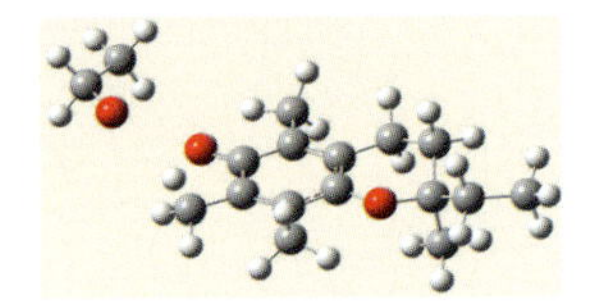

has increased by about 0.1 Å. The C-C distance has also decreased, tending toward a double bond. Finally, the repositioning of the ethoxy radical molecule toward its final position in the products has begun. Similar changes take place in the transition structure for the down form, which is illustrated in the margin.

For the up transition structure, the predicted reaction barriers are 8.1 kJ/mol (~1.9 kcal/mol) for the forward reaction, and 142.3 kJ/mol (~34.0 kcal/mol) for the reverse reaction. For the down transition structure, the values are 2.6 kJ/mol (~0.63 kcal/mol) forward and 121.1 kJ/mol (~28.9 kcal/mol) reverse. These results indicate that the up regioisomer is thermodynamically favored while the down regioisomer is kinetically favored. Thus, regioselectivity for the up product is observed when the reaction occurs under thermodynamic control.

There are several possible ways to proceed at this point in our investigation.

- It is likely that the basis set we chose for our initial calculations was insufficient for describing these molecules accurately. We could begin from the transition structures we located and reoptimize them using a more accurate model chemistry and/or an ONIOM model.
- As we noted earlier, the choice of reacting species is both critical and challenging. In fact, we have investigated a number of candidates, and this question remains a topic of ongoing research.
- All of our calculations up to this point have been performed in the gas phase. However, the environment in the human body is in solution, and these effects are probably crucial to accurate modeling of vitamin E and its physiological activity. Modeling the effects of a solvent environment is considered in the next subsection.

Once we have completed this part of the study, we can go on to model other aspects of the relevant substances and reactions related to vitamin E activity.

Modeling the Chemical Environment

So far, we have focused on simple reactions involving one or two molecules in isolation, and we have paid little attention to the chemical environment in which the reaction takes place. Doing so may be a reasonable first step in a chemical study, but eventually the full complexity must be taken into account in our modeling. We use the term "chemical environment" to refer to any of the aspects of the molecular systems and/or reactions which differ from the ground state in the gas phase. We will briefly consider three such departures here.

Chemistry in Solution

It is well known that a solution environment has a significant effect on chemical properties and reactions. The structures of molecules can change significantly from the gas phase to solution and between solvents with different properties. In the same way, the characteristics of chemical reactions can be substantially altered according to the presence/absence of a solvent and the specific solvent environment: different products may be favored, reaction barriers may be increased/decreased, and the like.

Gaussian provides sophisticated features for modeling chemistry in solution. It offers self-consistent reaction field (SCRF) models of solvation in which the solute is placed inside an empty cavity within the solvent, and the solvent itself is treated as a continuous, uniform dielectric medium. These models take into account the interactions between the solute and solvent and their mutual polarization.

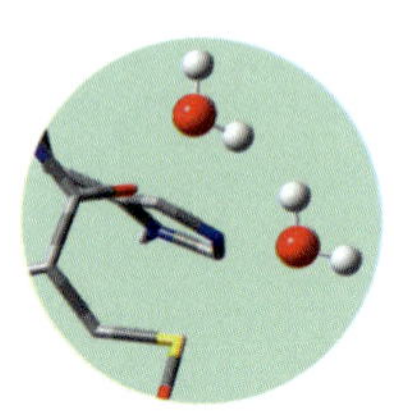

The SCRF models in Gaussian are very effective at modeling chemistry in solution for a wide range of molecular systems. Nevertheless, some solution chemistry—especially that involving aqueous solution—requires including one or more explicit solvent molecules as part of the molecular system under investigation, which is treated using an SCRF model.

Chapter 5 discusses modeling systems in solution in detail.

Excited States

Excited states are stable, higher energy electronic configurations of molecular systems. For example, such states are produced when a sample is exposed to the light source in a UV/visible spectrophotometer. Excited states are relevant to many areas of chemistry, including photochemistry and electronic spectroscopy and can be used to study phenomena such as the absorption properties of dyes and other chromophores, the photodecomposition rates of pesticides and other compounds and the properties of materials of potential use for solar energy.

Special techniques are required to model systems in their excited states. Gaussian provides a range of methods for doing so:

Method (Keyword)	**Description**
Time-dependent DFT (**TD**)	A time-dependent formulation of density functional theory for modeling excited states. Due to its good accuracy and reasonable computational cost, it is the standard method for predicting excitation energies in medium to large molecules.
Equation of Motion Coupled Cluster (**EOMCCSD**)	An extension of CCSD for modeling excited states. It provides CCSD-level high accuracy for excited state calculations and requires comparable computational cost (which is substantial).
Complete Active Space Multiconfiguration SCF (**CASSCF**)	By default, electronic structure calculations employ a single electronic configuration when modeling a molecular system. In contrast, multi-configuration approaches describe the electronic wavefunction as a linear combination of multiple electronic configurations (arrangements of electrons within the molecular orbitals). Such an approach is required for accurately modeling certain types of ground states: e.g., ones which are quasi-degenerate with low-lying excited states. This approach can also be beneficial when modeling excited states. In a full MCSCF treatment of a molecule, both the set of coefficients of the various electronic configurations and the molecular orbitals are varied in order to obtain electronic wavefunction with the lowest energy. The CASSCF method limits the multiple electronic configurations to all configurations within a subset of the orbitals known as the *active space*, thereby reducing the complexity and the attendant computational requirements. It can be used to predict excitation energies as well as to model reactions involving transitions from excited to ground states (conical intersections). In practice, CASSCF calculations are limited to 12-16 electrons.

Modeling excited states is discussed in detail in Chapter 8.

Polymers, Surfaces and Crystals

Periodic systems, characterized by repeating units in one, two or three dimensions, can also be treated with electronic structure methods. Such substances can be modeled as a finite number of the fundamental unit. For example, a surface might be approximated as some number of the component atoms arranged in proper configuration, and reactions with the surface can be studied by placing the relevant molecule in proximity to it. Similarly, polymers may be approximated by truncating their structures to a small number of component units, and higher order periodic systems may be treated as small clusters of the relevant group. Gaussian also provides a special facility for treating periodic systems via a technique known as periodic boundary conditions (PBC).

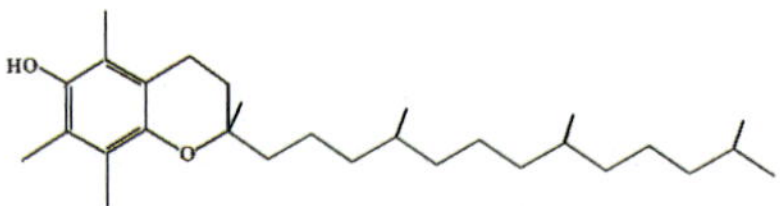

Modeling Vitamin E in the Body

As we noted in the previous subsection, the gas phase studies of the vitamin E-related compounds neglect the chemical environment in which these reactions take place. An organic solvent is a much more appropriate approximation of the environment within the body where vitamin E chemistry takes place. The cell membranes where vitamin E exists are hydrophobic environments (although they are themselves quite inhomogenous).

We revisited our comparison of the favored regioisomers of α-TOH OQM and the DHBF-based compound by modeling the various compounds in solution with acetonitrile. We found that the lower energy regioisomer for both compounds was the same as that in the gas phase. However, the energy difference between the two forms increased in solution for both compounds by about 4 kJ/mol (~1 kcal/mol).

Summary: Studying Vitamin E Chemistry

Obviously, we've taken only the very first steps in our study of vitamin E chemistry. We've isolated four different progressions of calculations in separate discussions, but in reality the work in these areas would overlap and interact from beginning. For example, we would perform every calculation in a solution environment as we know that the gas phase is a poor approximation of environment in the body.

We've also seen how the computational research process does not always proceed directly forward but rather sometimes loops back on itself. In our case, we cannot proceed to study the reaction path connecting the reactants and products based on the initial transition structure that we located as we need to rethink how we are modeling the cell membrane.

A topic such as this one is obviously extremely complex, and studying it thoroughly will take a substantial amount of time. Nevertheless, these first calculations discussed here provide a model for approaching an example problem with electronic structure methods.

Combining Calculations with Experiments

The vitamin E study we considered was a solely computational investigation. However, calculations are often used in conjunction with experiments. This can be done for various reasons, including as an aid to interpreting the experimental data or to explain the experimental results on the basis of molecular structure and properties. For some types of experiments, calculations are required in order to make use of their results. For example, electronic structure calculations are necessary to determine absolute configurations using vibrational circular dichroism (VCD) or Raman optical activity (ROA) spectroscopy. We will look at an example of each type in this section.

Alkane/Water Partitioning

Here we consider an example of using calculations to explain experimental results based on molecular structural features. The researchers in this study [Steel09] are interested in physiological processes that depend on the equilibrium distribution of the solute between two bulk phases. Such considerations are relevant, for example, to the activity of medications taken orally as they move from the intestines through the blood to the region of the cell membrane, three very different environments in terms of solubility. Similarly, studies are relevant to investigating how pollutants that are more soluble in hydrophobic environments can accumulate in the body and become health risks. The goal of the study is to contribute to understanding how the properties of the solute promote or inhibit solute partitioning.

Some alkane/aqueous systems are known to be a good approximation to the hydrophilic/hydrophobic boundaries found in biological systems. The researchers measured the alkane/aqueous partitioning of three solutes in four solvents:

Solutes	**Solvents**
p-nitrophenol* (PNP)	cyclohexane
3,5-dimethyl-*p*-nitrophenol (3,5-DMPNP)	methylcyclohexane
2,6-dimethyl-*p*-nitrophenol (2,6-DMPNP)	octane
	iso-octane (2,2,4-trimethylpentane)

In each case, the compounds were mixed and agitated and then allowed to equilibrate. The researchers then obtained aliquots from each layer, and analyzed them in order to calculate partition coefficients: the ratio of the solute concentration in each alkane relative to that in water (determined by a Beer's Law analysis).

The results from these experiments are given in the table below. PNP has the smallest partition coefficients in all cases, an expected result given this compound's highly hydrophilic nature, arising from the polar O-H and NO_2 groups. 2,6-DMPNP and 3,5-DMPNP have much higher partition coefficients than PNP (two orders of magnitude), with the former being three to four times more soluble in the alkane phase relative to water compared to the latter. In terms of solvents, the largest partition coefficients were observed for methylcyclohexane/water, followed by cyclohexane/water, octane/water and iso-octane/water; methylcyclohexane is thus most accommodating to the nitrophenol-based solutes, and iso-octane is the least able to accommodate them. The difference in magnitude for the partition coefficients between their highest and lowest values is a factor of about 1.5.

In order to understand these results at a deeper level, beyond the easily identifiable qualitative trends, the researchers performed a series of calculations designed to study the substances' molecular structures and properties in detail.

The first set of calculations located the minimum energy structures for the various compounds, predicting their dipole moments, total energies and other properties, and also considered solvent effects in two different ways:

- The molecular structures are optimized using the MP2/6-311++G(2d,p) model chemistry. These calculations were performed in the gas phase. The MP2 method includes the effects of electron correlation within the molecular wavefunction, and the basis set is large and includes additional polarization and diffuse functions on all atoms. This model chemistry was chosen on the basis of preliminary calibration calculations.
- Single point energy calculations were performed for the optimized structures using the SCRF solvation model for cyclohexane in order to compute $\Delta G^{solvation}$ in that solvent.

* The IUPAC name of *p*-nitrophenol is 4-nitrophenol.

- The geometries of the various solute-water complexes (hydroxy hydrogen to water oxygen) were optimized using the same MP2/6-311++G(2d,p) model chemistry. Each complex was constructed using the solute and that explicit water molecule, with a hydrogen bond from the latter nitrophenol hydroxyl group. The predicted energies of the optimized structures were used to estimate hydrogen bond binding enthalpies.
- Thermal energy corrections were computed via B3LYP/6-311++G(2d,p) frequency calculations (frequency calculations using the same model chemistry as the optimizations were prohibitively expensive). These values were used in the computations of the free energies of solvation and the solute-water binding energies; the latter also incorporated the results of basis set superposition error (BSSE) calculations.

The following table summarizes the major results of the experiments and the preceding calculations:

	CALC. PROPERTIES		EXPERIMENTAL $K_{alkane/water}$			
SOLUTE	μ (debye)	SOLUTE-WATER BINDING E† (kcal/mol)	CYCLOHEXANE	METHYL-CYCLOHEXANE	OCTANE	ISO-OCTANE
PNP	4.82	5.87	0.009±0.001	0.010±0.001	0.008±0.0008	0.006±0.0002
2,6-DMPNP	5.43	4.33	1.47±0.01	1.63±0.02	1.31±0.05	0.86±0.002
3,5-DMPNP	3.78	5.35	0.38±0.03	0.47±0.02	0.37±0.03	0.34±0.01

†This value is $-\Delta H^{complex-(solute+water)}$ estimated at 298 K.

The binding energy results indicate that the water molecule is most weakly bound to the hydroxyl group in 2,6-DMPNP. There is thus a lower barrier to migration from the aqueous phase to the organic phase. The researchers explain this result by concluding that the methyl groups in that compound weaken solvation interactions with water and accordingly promote them in the nonpolar alkane solvent environment.

Structural features of the equilibrium geometries are also relevant to understanding the chemistry here. In both PNP and 2,6-DMPNP, the nitro group is in the same plane as the aromatic ring. In contrast, the methyl groups adjacent to the nitro group in 3,5-DMPNP cause the latter to be twisted about 55° with respect to the ring plane. This separates the nitro group from the ring's resonance structure, an effect which can be observed by examining the LUMO, as illustrated in the following figure:

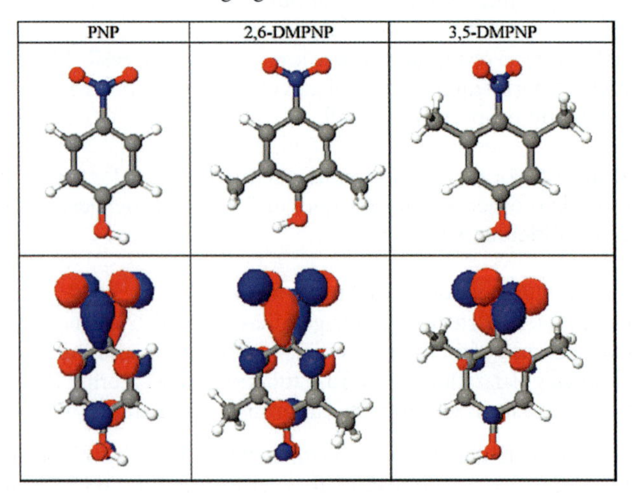

The LUMOs for the planar molecules reflect charge transfer between the nitro group and the ring which contributes to the S_1 excited state. In 3,5-DMPNP, however, electron densities of the nitro group and the aromatic ring are separate and decoupled.

In order to study the interactions between the nitro group and the methyl groups in 3,5-DMPNP further, the researchers performed the calculation known as a relaxed potential energy surface scan for this compound and for PNP. This analysis varied the dihedral angle between the nitro group and the ring between 0° and 90° in 10° increments. At each point, this orientation was held fixed while the molecule's other structural parameters were optimized, a technique known as a *constrained optimization.*

The following graphs plot the dipole moment, energy relative to the PNP ground state and maximum absorption for PNP and 3,5-DMPNP as the angle of the NO_2 group rotates from 0° to 90° (bottom to top). The green points in the graphs indicate the equilibrium geometry values. Note that the barrier to rotation—the middle graph—in 3,5-DMPNP is asymmetric, and these values are quite small (less than about 1 kcal/mol) except when the nitro group is nearly coplanar with the ring. In contrast, the barrier in PNP rises steadily as the nitro group moves out of the plane, reaching a value of nearly 6 kcal/mol at 90°. As we saw in the previous table, the dipole moment of the ground state for 3,5-DNPMP is significantly smaller than for the other two compounds.

The upper plot relates to the solvatochromic behavior of the solutes, and we will discuss it below. Solvatochromism is the ability of the solvent environment to alter the relative energies of ground and excited electronic states. This effect is typically seen in the absorbance spectrum of solutes whose excited states involve movement of electrons across the molecule. As such, it can be used to elucidate solute/solvent interactions. Solvatochromism can be measured experimentally and predicted computationally. Previous experimental work has shown that the maximum absorbance wavelength for PNP and 2,6-DMPNP significantly shifts location as the polarity of the solvent increases: by about 30 nm between the alkane and aqueous solutions. Once again, 3,5-DMPNP exhibits different characteristics, and its maximum absorbance shifts only about 10 nm between the two environments.

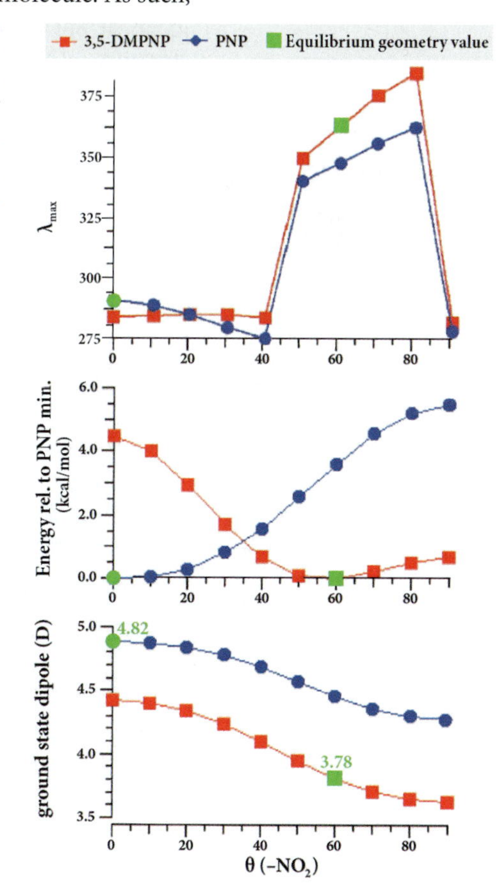

Solvatochromism is studied computationally by modeling the first excited state of the relevant molecules. The researchers in this study performed TD-DFT calculations for the compounds under consideration and located the three lowest excited states. A larger dipole moment in the excited state than in the ground state will enable the electronic excitation to shift to lower energy, resulting in a longer wavelength maximum absorption as the solvent polarity increases. The differences in dipole moment between the ground state and the first excited state are 6.2, 6.4 and 4.2 for PNP, 2,6-DMPNP and 3,5-DMPNP, respectively (values in debye). These differences are consistent with the experimental observations, and the relatively larger differences in the cases of the first two compounds compared to the third provides insight into their differing behavior.

The upper plot in the figure shows λ_{max} for the two compounds as a function of the angle of the NO_2 group with respect to the plane of the ring. This is computed as the vertical excitation energy of the lowest excited state as predicted by the TD-DFT calculations. The two curves have the same shape and are nearly coincident, indicating that the differences in solvatochromism are entirely due to the position of the nitro group altering the electronic structure of the ground and excited states.

In summary, this study provides an excellent example of how real research problems require a variety of techniques and a series of calculations in order to study the chemical phenomena in which they are interested. It also illustrates how experimental data and calculation predictions can be combined to reach significant chemical conclusions.

Determining Absolute Configurations in VCD Spectroscopy

Chirality is a chemical property that is very important to many aspects of current chemical research. For example, consideration of the different enantiomers of a compound is often a key part of pharmaceutical development processes. Chirality can be studied using the *vibrational optical activity* (VOA) of molecules. VOA is the differential response to left vs. right circularly polarized infrared radiation during a vibrational transition. The IR absorption form of VOA is known as *vibrational circular dichroism* (VCD). VCD can be used for many types of analysis related to the structure and conformations of molecules of biological interest:

- Determining the enantiomeric purity of a sample relative to a known standard.
- Determination of absolute configurations.
- Determination of the solution conformations of large and small biological molecules.

In most cases, VCD studies consist of a combination of spectral measurement and electronic structure calculations, with the latter playing a vital role in interpreting the former.

The VCD spectrum of a chiral molecule is dependent on its three-dimensional structure: more specifically, its conformation and absolute configuration. Thus, VCD can allow you to determine the structure of a chiral molecule. Both the IR and VCD spectra of diasteriomers—e.g., (R,R) vs. (R,S)—differ. However, for enantiomers—(R,R) vs. (S,S)—the IR spectra are again identical, but the VCD spectra have opposite sign. The VCD results provided by electronic structure calculations can be used to identify the absolute configuration producing a given observed spectrum, and they can also elucidate the contributions of different conformations to a spectrum.

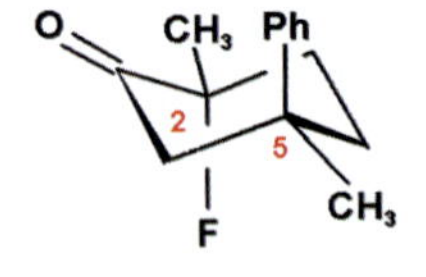

Consider the IR and VCD spectra for the fluoroketone compound in the margin. We've numbered the chiral centers. The following figure displays the IR and VCD spectra for the lowest energy conformers of the (R,R) and (R,S) enantiomers:

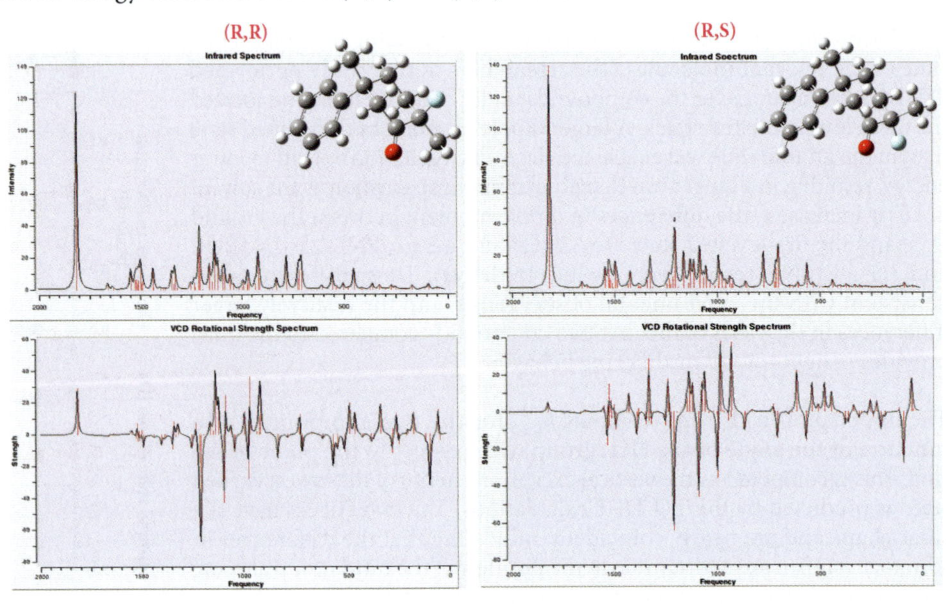

Although the IR spectra are similar, there are differences between them. The VCD spectra are clearly different. All spectra are truncated at 2100 cm^{-1} (there are additional peaks in the 3000-3300 range).

In contrast, the IR spectra of the (R,R) and (S,S) enantiomers are identical. However, the VCD spectra have opposite sign (the (S,S) spectrum is in green):

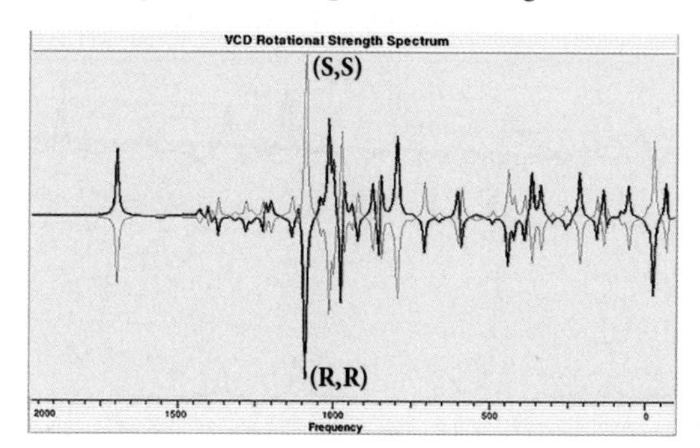

Thus, computed VCD spectra can easily distinguish enantiomers. Here, we have compared only the lowest energy conformations. In reality, however, several conformations can contribute to the experimental spectra of molecules. The plot on the left in the following figure compares the final predicted VCD spectrum for the (R,R) enantiomer, averaged over several conformers, to the observed spectrum. Comparing the two spectra, in terms of the sign and relative intensity for the majority of the bands, makes it clear that experimental spectrum corresponds to the (R,R) enantiomer. In this case, the conformational averaging (see p. 164) also reproduces the observed spectrum very well:

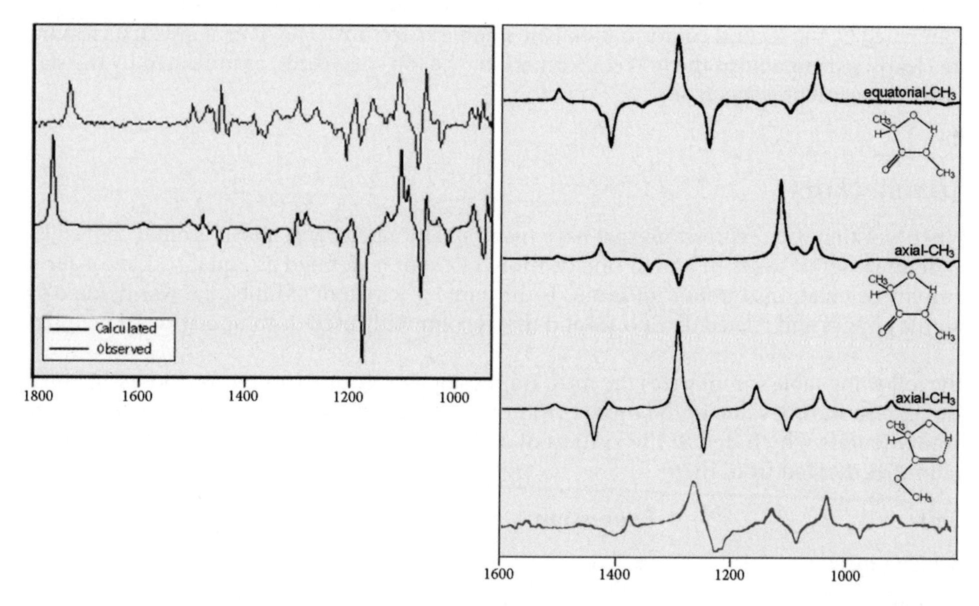

VCD can not only determine the absolute configuration, but can also determine solution conformation. As an example, we examine the VCD spectra for three conformations of (+)-(R)-methyllactate. Internal hydrogen bonding within the ring can take place either to the C=O group or to the OCH_3 group. The former is planar, while the latter is puckered and has two conformations corresponding to equatorial and axial placement of the methyl substituent.

The predicted VCD spectra for the three conformations are illustrated on the right in the preceding figure. When we compare them to the observed spectrum (in blue), it becomes evident that the conformer having internal hydrogen bonding to the O=C group dominates it.

Sometimes, a VCD spectrum exhibits features from more than one conformation. The following diagram shows the predicted VCD spectra for two conformations of the (+)-(R) enantiomer of 3-methylcyclohexanone along with the observed spectrum (bottom):

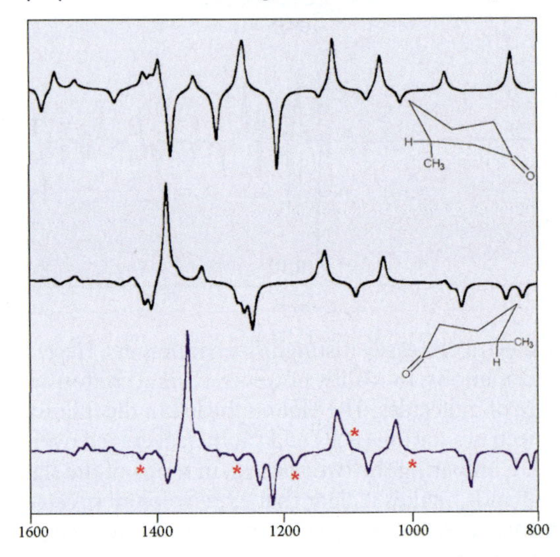

In this case, comparing the calculated and experimental spectra indicates that the latter is dominated by the second conformation (the middle spectrum). However, some of its features are clearly attributable to the first conformation (the top spectrum), as indicated by the stars in the experimental spectrum.

Atomic Units

When working with expressions that have many small valued constants, it is often convenient to define a set of units in which one or more of them is defined as equal to 1 in order to simplify calculations. *Atomic units* (au) is the name of a system of units that was defined for atomic physics and related disciplines and that is commonly used in computational chemistry.

The following table summarizes the most important atomic units for our purposes and gives the corresponding value in the more common SI units. The first three items are among the fundamentals which define the system of atomic units, and the final two are important quantities derived from them:

Unit	Expression	SI value
electron mass	$m_e = 1$	9.109382×10^{-31} kg
elementary charge	$e = 1$	$1.6021765 \times 10^{-19}$ C
angular momentum (reduced Planck's constant)	$\hbar = h/2\pi = 1$	$1.0545716 \times 10^{-34}$ J·sec
energy (hartree)	$E_h = m_e e^4/(4\pi\varepsilon_0\hbar)^2 = 1$	4.359744×10^{-18} J
length (bohr)	$a_0 = 4\pi\varepsilon_0\hbar^2/(m_e e^2) = 1$	$5.2917721 \times 10^{-11}$ m

1 hartree is the amount of energy equal to twice the ionization energy of the hydrogen atom.

The distance which is the most probable radius of the 1s electron in the hydrogen atom is 1 bohr.

Gaussian reports energy values in hartrees. The following conversion factors will be useful in comparing with experiment:

1 hartree = 2625.4996 kJ/mol
= 627.5095 kcal/mol
= 27.2116 eV
1 kcal/mol = 4.184 kJ/mol

Advanced Topics

In this final section of our introductory chapter, we provide brief explanations of three more fundamental concepts related to electronic structure modeling: techniques for handling the heavy atoms found in the fourth and higher rows of the periodic table, numerical integration techniques employed in electronic structure calculations and density fitting basis sets used in conjunction with pure DFT functionals.

Treating Post-Third Row Elements

In general, the usual approximations used by electronic structure theory neglect relativistic effects. The most important of these arises from the large velocities of electrons near the nucleus. Treating all electrons as nonrelativistic is valid for lighter elements, but it becomes more and more problematic the farther one travels down the periodic table. Relativistic effects have a modest effect on elements such as iron (i.e., through the third row of the periodic table). Neglecting these effects gives rise to substantial inaccuracy for elements in the fourth row and higher.

Chemical activity arises primarily from the valence electrons within an atom. The core electrons are essentially inactive. When effects of relativity are ignored in large atoms, the core electrons are predicted to be farther out than they in fact are. This causes the valence electrons to also be pushed out too far, giving rise to a distorted description of their interactions and reactivity.

Effective core potentials are used as a zeroth order approximation to these relativistic effects for heavier atoms. In the potential function describing the atom, the core electrons as a whole are replaced with a single, simpler potential function representing their collective repulsion on the valence electrons. Such a function is known as an effective core potential or a pseudopotential.* This potential results in a simplified term within the wavefunction, sometimes referred to as a pseudo-wavefunction.

The following figure (adapted from [Porezag00]) illustrates the valence orbitals in the sulfur atom as described by an all-electron basis set and by an effective core potential (the latter is composed of separate components for each orbital type). The thick line in each case corresponds to the all-electron basis set, and the thin line to the ECP.

* The latter term is favored by solid-state physicists deploying plane wave methodologies, where pseudopotentials are used to simplify the computational problem. Many online descriptions of pseudopotentials are written from this point of view.

Notice that at a certain point in each orbital, the two curves coincide. The rapid oscillations of the all-electron curve in the region of the core electrons are replaced by a simpler, smoother one anchored at 0. The ECP thus has the correct behavior in the valence region while eliminating the "wiggles" in the core region.

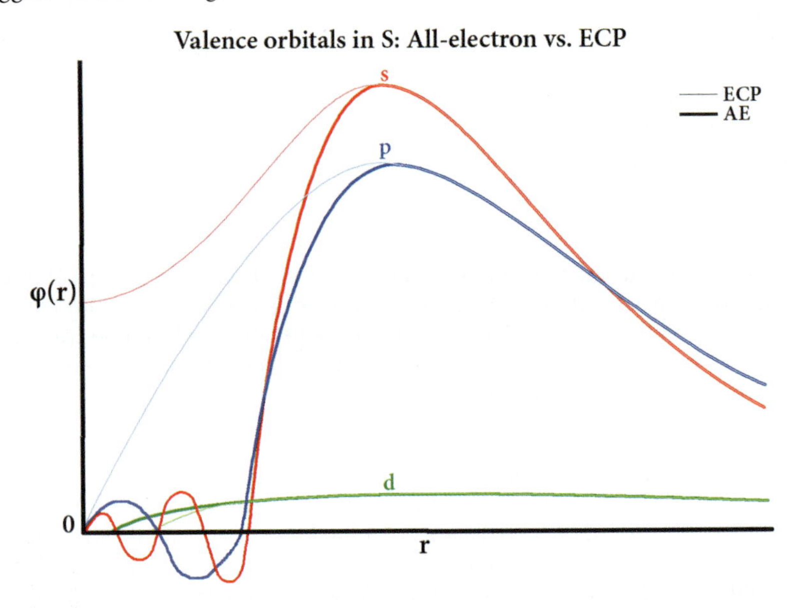

ECPs were first proposed by Hans Hellmann in the 1930s [Hellmann35]. In the late 1950s, Phillips and Kleinman built upon this and other early work to develop the present form of ECPs [Phillips59]. [Schwerdtfeger11a] provides an excellent recent review of ECPs in the context of electronic structure theory.

ECPs are developed in combination with a specific ordinary basis set to be used for the lighter elements and for the valence region for the heavier atoms. In general, different ECPs support different subsets of the periodic table. Using ECPs is discussed in detail in Chapter 9 (see p. 437).

Numerical Integration & Grids

Integral evaluation is a central component of the electronic structure calculations. Integrals may be evaluated analytically, using the exact form of the antiderivative function, or they may be computed approximately using a mathematical technique known as *numerical integration*.*
The general technique involves approximating the integral's value as the weighted sum of the values obtained by evaluating the integrand at a finite set of points. For example, density functionals are evaluated via numerical integration over a three-dimensional grid in space.

Integration grids are used whenever integrals are evaluated numerically, most commonly in calculations using DFT methods. They are one of the largest sources of numerical error in current calculations. The number of points contained in the grid will affect both the accuracy and the stability of the results. As you might expect, larger grids are more stable and accurate but also require longer computation times. Selecting a grid which is the best trade-off between these competing factors is a complicated question, and different choices may be best for different problems. These choices may also change over time as computers become faster, enabling larger grids to be practical for more problems.

* Numerically computing the value of a one-dimensional (single) integral is also called *quadrature*.

In Gaussian 09, the default numerical integration grid is the **FineGrid**. At that program version's release time, this grid was selected as the most reasonable trade-off between cost and accuracy for a large range of molecular systems and calculation types. Subsequent major versions of Gaussian may have different defaults.

You can also set the default grid by including the relevant keyword and option (e.g., **Int=UltraFine**) within the *Default.Route* file (*Default.Rou* under Windows). See the *Gaussian Users' Reference* for details.

For many molecules, especially larger ones, a larger grid provides better stability. This is especially true for molecules containing many tetrahedral centers, for locating minima and transition structures in very flat regions of a potential energy surface, for computing very low-frequency modes of systems, for modeling systems in solution and several other cases. Some DFT functionals also require larger grids for their proper functioning [Johnson04].

For such calculations, the recommended grid is the **UltraFine** grid. It is specified with the **Integral=UltraFine** keyword. In a few rare cases, the additional keyword **CPHF=UltraFine** will also be beneficial. Consult the *Gaussian Users' Reference* for full details on the available integration grids and their characteristics.

Using a larger integration grid can help overcome various calculation issues that may arise for some systems. For example, if a calculation has SCF convergence problems, a larger integration grid may eliminate them. Similarly, some geometry optimizations which have difficulty converging will behave better when computed with a larger integration grid.

Be aware that the integration grid used for calculation is an essential component of the model chemistry. This means that the same integration grid must be used for all calculations whose results you want to use in making comparisons. Thus, if you find that increasing the integration grid is necessary to successfully model one compound in a series which you are studying, it will be necessary to use that larger grid for *all* calculations on *every* compound in the study.

Density Fitting Basis Sets

Gaussian provides the density fitting approximation for pure DFT calculations [Dunlap83, Dunlap00]. This approach expands the density in a set of atom-centered functions when computing the Coulomb interaction instead of computing all of the two-electron integrals. It provides significant performance gains for pure DFT calculations on medium-sized systems too small to take advantage of the linear scaling algorithms, without a significant degradation in the accuracy of predicted structures, relative energies and molecular properties.

The desired fitting basis set is specified as a third component of the model chemistry, as in this example: **BLYP/TZVP/TZVPFit**. Note that slashes must be used as separator characters between the method, basis set, and fitting set when a density fitting basis set is specified.

An example of the use of density fitting basis sets is given in Exercise 3.2 (p. 112).

2

Getting Started with Calculations

In This Chapter:

Fundamental Job Types: Energies, Optimizations & Frequencies

Predicting Dipole Moments and Atomic Charges

Visualizing Molecular Orbitals

Biorthogonalized Orbitals

Thermal Energy Corrections

Modeling Large Molecules: An Introduction to ONIOM

◆

Advanced Topics:

Wavefunction Stability

Comparing Model Chemistries for Accuracy

Exploring Basis Sets in Depth

Understanding CPU Resource Requirements

◆

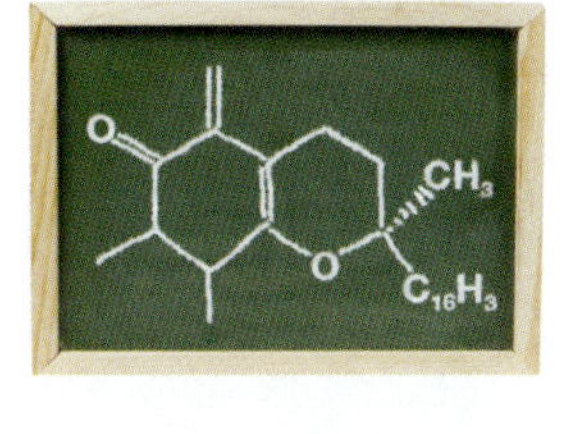

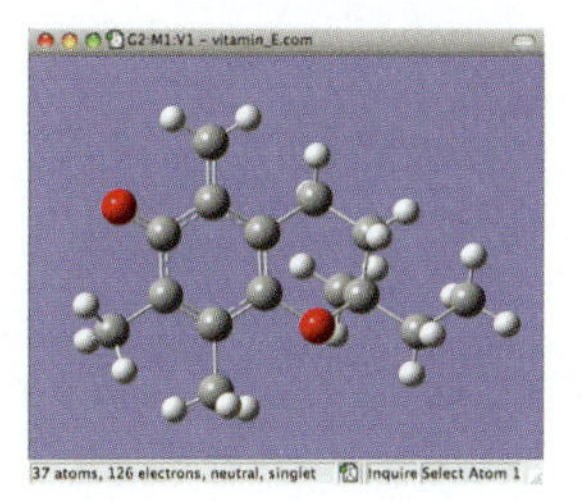

2 ◆ Getting Started with Calculations

In the previous chapter, we considered the general process of using computational studies to conduct research. In this chapter, we will look at the fundamental building blocks of that endeavor. We will introduce the three primary calculation types, consider the process of setting up and running a calculation in detail, and discuss the most important results and predictions that they provide.

First Steps: An Energy Calculation on Formaldehyde

Our first calculation will be an energy calculation on formaldehyde. We have chosen this simple molecule so that this first job can be run quickly and easily.

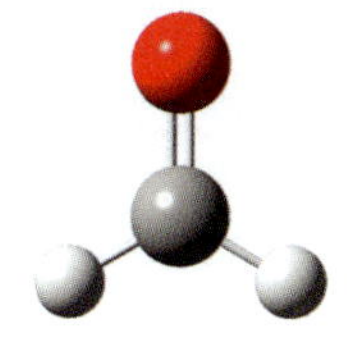

A single point energy calculation is a prediction of the energy and related properties for a molecule with a specified geometric structure. The phrase *single point* is key, since this kind of calculation is performed at a single, fixed point on the potential energy surface for the molecule, meaning that the geometry of the molecule does not change during the course of the calculation. In order to obtain valid results, single point energy calculations require a reasonable structure for the molecule under investigation.

Single point energy calculations are performed for many purposes, including the following:

- To obtain basic information about a molecule.
- As a consistency check on a molecular geometry that you plan to use as the starting point for an optimization.
- To compute very accurate values for the energy and other properties for a molecular structure optimized at a lower level of theory.
- When it is the only affordable calculation for a system of interest.

In addition to the total energy of the molecule, energy calculations with Hartree-Fock and DFT methods* also predict several other properties, including:

- Dipole and higher multipole moments
- Atomic charge distribution
- Molecular orbitals and orbital energies
- Electron density and electrostatic potential

Single point energy calculations can be performed with any theoretical method, with small or large basis sets, and for systems in any chemical state and/or environment for which Gaussian offers modeling capabilities.

What we are referring to as "energy" here is the *electronic energy* of the molecule along with the *nuclear-nuclear repulsion energy*: that is, the energy due to the electrons moving and being attracted to the nuclei and repelled from one another added to energy due to the nuclei repelling one another. This quantity is sometimes referred to as the *total energy*. This energy is an entirely theoretical quantity, an estimate of the molecular energy as if the atoms were

* In order to compute these properties with electron correlation methods such as MP2 or CCSD or with excited state methods like TD or EOM-CC, you will need to specify additional keywords: **Pop=Full** to include the orbital information in the output file, and **Density=Current** to request that all of the properties be computed with the specified method's density rather than the default of the SCF density.

all stationary. Real molecules possess additional energy—called *thermal energy*—arising from the translational, rotational and vibrational motion of the nuclei. We will discuss how to add in the thermal components of the energy later (see p. 59).

Setting Up and Running Calculations

A Gaussian single point energy calculation requires the following information in order to proceed:

References for our Standard Model Chemistry

APFD functional: [Austin12]

6-311+G basis set: [Binning90, McGrath91, Curtiss95, McLean80, Raghavachari80b, Blaudeau97, Wachters70, Hay77, Raghavachari89]

- The model chemistry to be used for the calculation. Our standard model chemistry will generally be APFD/6-311+G(2d,p).
- A brief description of the job. This is referred to as the *job title*.
- The structure of the molecule: its charge and spin multiplicity and the locations of the nuclei in space. This data is typically generated automatically from a sketched-in or imported structure by the graphical environment (i.e., GaussView, WebMO, and so on), but it can also be retrieved from the results of a previous calculation.

  The *spin multiplicity* value specified to Gaussian is a measure of the unpaired spin within a molecule. Technically, it is the number of possible orientations of spin angular momenta. It can be computed for a molecule as the number of electrons with parallel spin which are unpaired with ones of opposite spin plus 1. See p. 45 for examples and p. 478 for mathematical details.

This fundamental information is required by every Gaussian calculation. Many calculation types also require additional information.

EXAMPLE 2.1 FORMALDEHYDE ENERGY, MOLECULAR ORBITALS & ATOMIC CHARGES

We want to perform a single point energy calculation on formaldehyde.* We will do so by:

1. Building the molecule in our visualization program.
2. Setting up the Gaussian job using the program's menus and dialogs.
3. Submitting the job to Gaussian.
4. Reviewing the results once the job is completed.

For this first example, we will examine the building and job setup processes in detail for both GaussView and WebMO. For future examples, however, we will discuss building structures and setting up calculations only when we need to highlight specific structural features and/or software capabilities. Consult your package's documentation for full details on all of its building and job setup features. We will consider GaussView first; skip ahead to page 41 for WebMO.

Building Formaldehyde in GaussView

GaussView

When you initially begin a GaussView session by opening the program, you will see two windows similar to those in the illustration following. The top portion of the GaussView main window contains the program's toolbars, and the icons on the left side of the top row control atom/fragment selection for the building process. The default item is a tetrahedral carbon atom, as indicated in the popup menu in the first toolbar row.

* IUPAC: methanal.

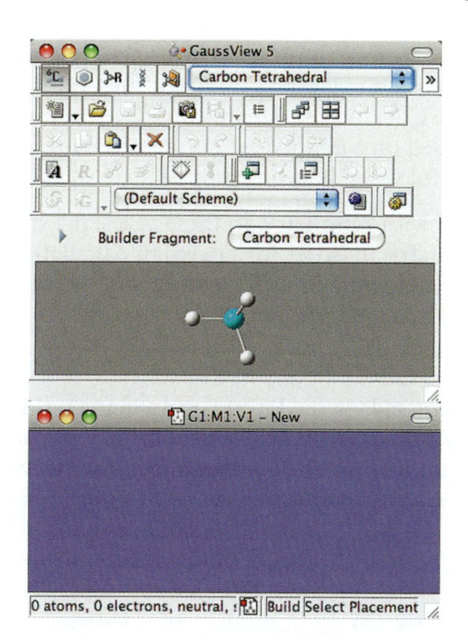

GaussView toolbars: the layout can be customized and may differ on your computer.

This area displays the Builder's current fragment: what will be added with the next mouse click. The connecting atom is highlighted.

View windows contain molecules being built and studied.

We will begin the build process by adding a carbon atom with sp^2 hybridization to the **View** window:

1-1 Select **Carbon Trivalent (S-S-D)** from the hybridization popup menu (corresponding to sp^2 hybridization):

Carbon Trivalent (S-S-D)

The letters in parenthesis refer to the bond types: single, single and double.

1-2 Click anywhere within the **View** window. The resulting structure appears at the right.

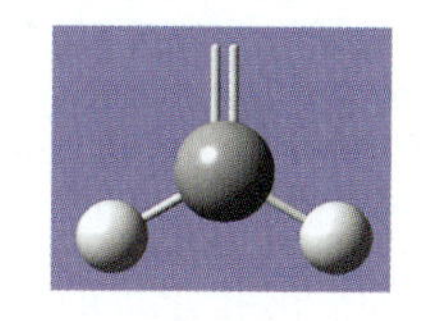

1-3 Next, we add the oxygen atom. Click on the **Select Atom** icon in the upper left corner of the toolbar: . A window containing the periodic table will appear; use it to select the element oxygen and the bare atom fragment form:

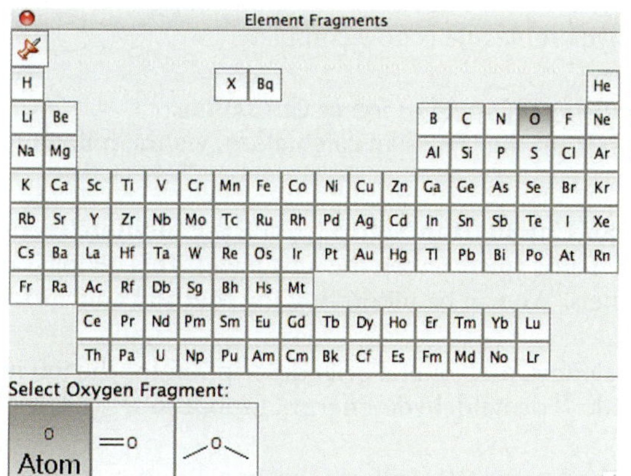

Select the element type from the periodic table.

Select the desired hybridization from these selections.

1-4 Click on the open double bond in the **View** window. An oxygen atom will be added at that location within the molecule.

15 The final step is to regularize the structure before running the Gaussian calculation. There are two options for doing so. If the structure has no symmetry, then you can use the GaussView structure cleanup feature. Click on the **Clean** icon. This function adjusts the geometry of the molecule, based on a defined set of rules, to more closely match chemical intuition. It is often helpful when building ground state structures.

The second method applies to molecules with symmetry. In these cases, the starting structure should be of the proper symmetry in order for Gaussian calculations to recognize and take advantage of it. Formaldehyde has C_{2v} symmetry, and we will ensure that our structure corresponds to this point group.

Imposing Symmetry in GaussView

The GaussView symmetry features are accessed via the **Point Group** item on the **Edit** menu. Once the **Point Group Symmetry** dialog opens, you must activate it by clicking **Enable Point Group Symmetry** in the upper left. The current molecular symmetry will then be displayed at the upper right: see the annotations in purple in the following figure:

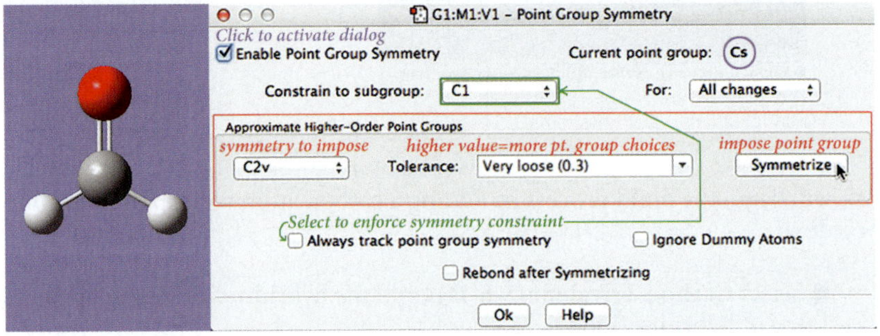

You can instruct GaussView to maintain symmetry across build operations by checking **Always track point group symmetry** *and then selecting the desired point group in the* **Constrain to subgroup** *popup (these controls are labeled in green above). However, there is no need to use this feature now.*

You can impose symmetry on the molecule using the controls in the area designated in red. The **Tolerance** popup specifies how close the structure must be to symmetric before a point group can be applied. As you loosen the tolerance, additional point groups will appear in the popup menu to the left. You can select the desired popup there, and then click the **Symmetrize** button to modify the structure so that it attains that symmetry.

In this case, we impose C_{2v} symmetry on the formaldehyde molecule (it originally had C_s symmetry).

16 The formaldehyde molecule is now complete.

Preparing and Submitting a Gaussian Job in GaussView

We are now ready to set up the Gaussian calculation, via the following sequence of steps:

21 Select **Gaussian Calculation Setup** from the **Calculate** menu. This will open the corresponding dialog (pictured on the next page). This dialog is divided into a number of separate panels. We will be modifying the **Method** and **Title** panels.

22 Select the **Title** panel, and enter a brief description for the job into the **Job Title** area. We used the title "Formaldehyde Energy" for our job.

23 Select the **Method** panel. We will use this panel to specify the model chemistry: the theoretical method and basis set that we want to use. Our calculation will use the APF-D DFT method and the 6-311+G(2d,p) basis set. The items that have been changed from their defaults are highlighted in red in the illustration following.

The first row of popup menus specifies the model chemistry via the following selections:

- The desired electronic state: the ground state or an excited state method.
- The theoretical method or method family.
- Whether a closed shell or an open shell calculation should be performed.* The default choice is to perform the calculation type that is appropriate to the molecule's spin state.
- The specific theoretical method to be used. This control is present when a method family (rather than a specific method) is selected in the second popup menu.

24 Specify the **DFT** method family and the **APFD** method in the **Gaussian Calculation Setup** dialog.

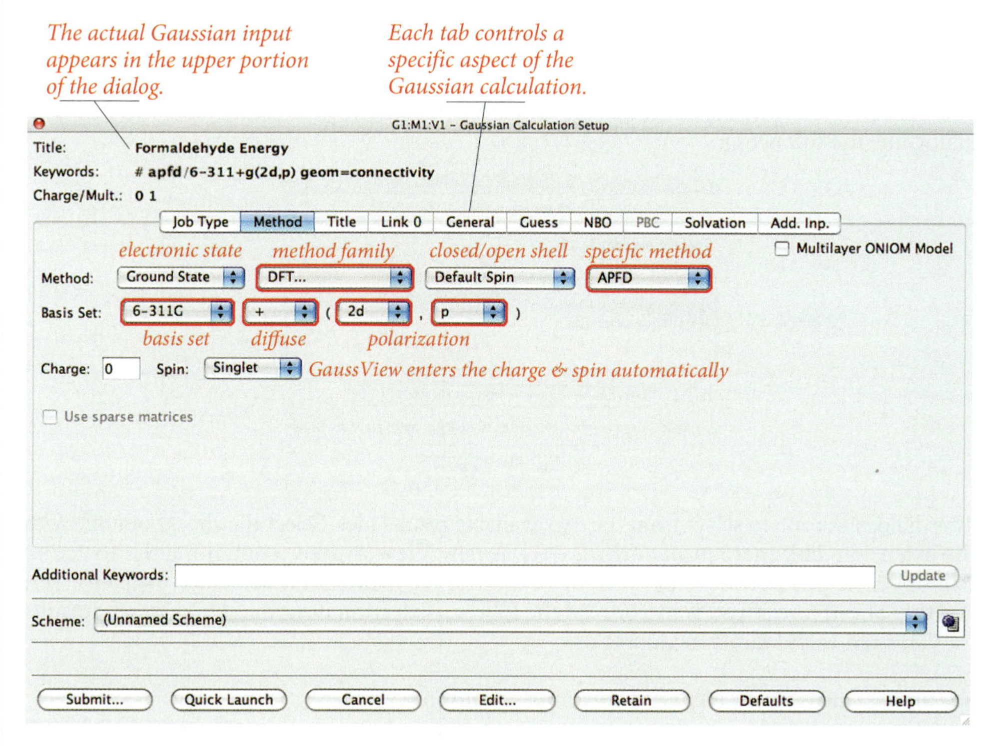

If your version of GaussView does not include **APFD** on the specific method popup menu, then you can specify it in the following way. Select **Custom** from the main method menu (second from the left), and then enter "APFD" into the field that appears to its right (one or both of the other two popup menus may disappear):

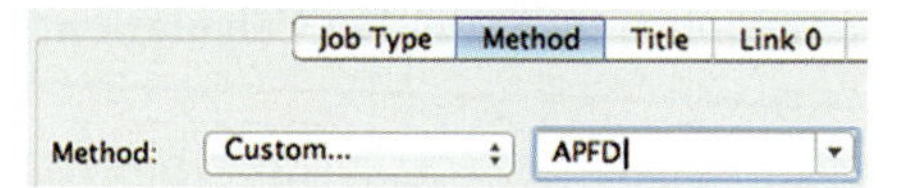

The second row of popup menus specifies the desired basis set for the calculation: the basis set name along with any additional diffuse functions (second popup) and/or polarization functions: on heavy atoms (third popup), on hydrogen atoms (fourth popup).

25 Specify the 6-311+G(2d,p) basis set by selecting **6-311G**, **+**, **2d** and **p** from the various popup menus in the second row.

* More specifically, this item allows you to select from among the restricted, unrestricted or restricted open shell calculation types. These concepts are discussed on page 53.

⑥ After verifying that the charge and spin multiplicity selected by GaussView are correct, we are ready to run the job. Begin the process by clicking on the **Submit** button. GaussView will first ask whether you want to save the Gaussian input file it has generated. Answer **Save** to this prompt, and then select a name and location for the file. Once the file has been saved, GaussView will ask you to confirm that you want to submit the file to Gaussian; select **Ok** to begin the calculation.

In many cases, you will also use the dialog's **Job Type** panel to specify the type of calculation that you want to run. In this case, however, we do not need to do so as a single point energy calculation is the default type.

You can examine the Gaussian input file that GaussView created using this dialog's **Edit** button prior to submission or by opening the external file. The Gaussian input file for this job is discussed on page 44.

The Gaussian job will run in the background. When it is finished, GaussView will display a dialog like the following:

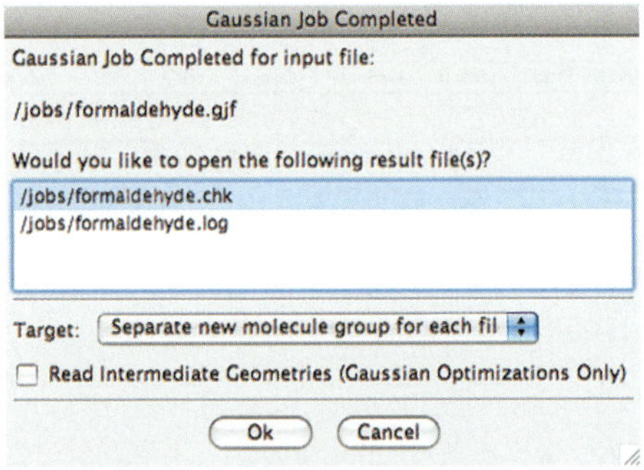

Gaussian Job Completed
Gaussian Job Completed for input file:
/jobs/formaldehyde.gjf
Would you like to open the following result file(s)?
/jobs/formaldehyde.chk
/jobs/formaldehyde.log
Target: Separate new molecule group for each fil
Read Intermediate Geometries (Gaussian Optimizations Only)
Ok Cancel

This dialog asks you to select from the two available results files. Select the checkpoint file with the extension **.chk** (here *formaldehyde.chk*). A new **View** window containing the structure will open. At this point, we are ready to examine the Gaussian results (see page 46). You can skip ahead now if you want, or read the follow subsection describing how to set defaults for Gaussian calculations in GaussView.

Specifying GaussView's Default Gaussian Calculation

You can save a lot of time when setting up calculations by configuring GaussView's default calculation type. Doing so will allow you to apply a group of common settings to a calculation with one mouse click.

To begin the process, access the GaussView **Preferences**; its dialog is on the left in the figure below. Select the **Gaussian Setup** preference (1 below), and then click on the **Calculation** button (2). This will open a version of the **Gaussian Calculation Setup** dialog (on the right in the figure following).

Use this dialog to specify as many of the job settings as you like, using the usual dialog panels (3). You can also include additional keywords in the **Additional Keywords** field (4). In this example, we have specified the job type, model chemistry and the **Int=Ultrafine** option.

When you have finished making selections from the various panels in the dialog, ensure that the **Scheme** popup menu is set to **(Default Scheme)** (5), and then click the **Retain** button (6).* This will close the dialog and return you to the **Preferences**. Click **Ok** (7) to save your settings.

* If you have previously saved a default scheme, the button will be labeled **Update**, and clicking it will save the changes as the new default.

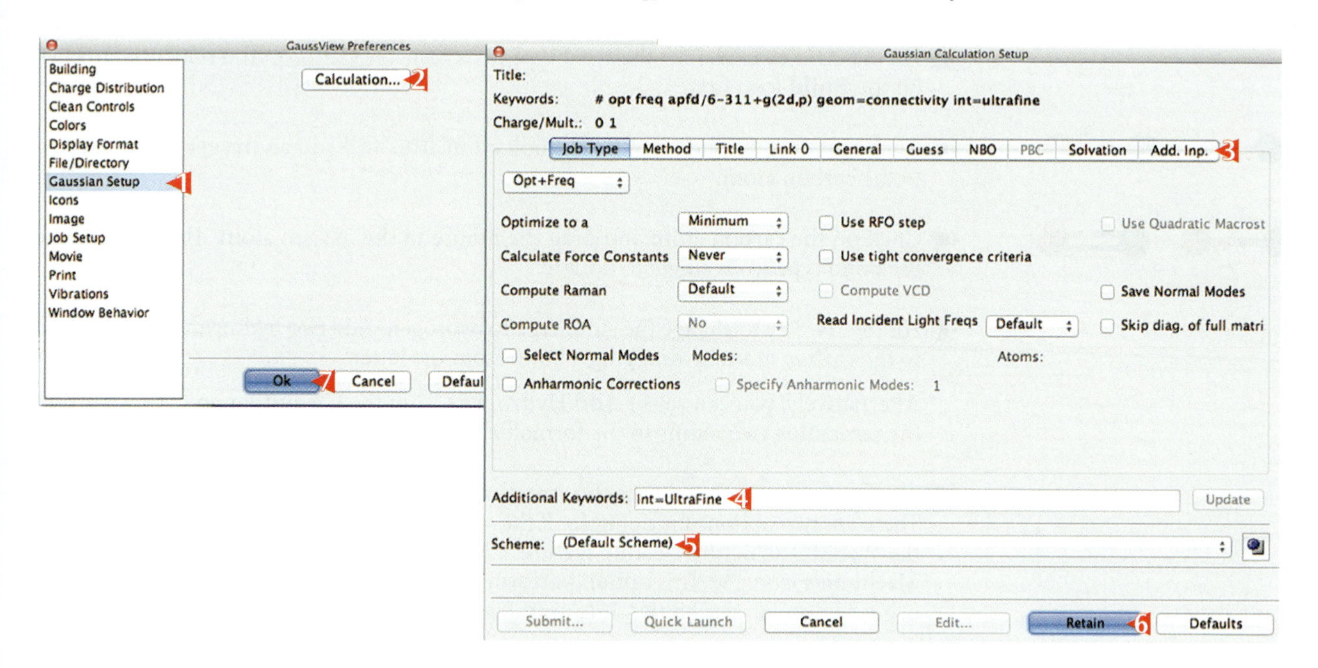

After you have set up the default settings you want, you can apply them quickly using the **Defaults** button in the **Gaussian Calculation Setup** dialog:

You can define other common calculation types using the GaussView calculations schemes feature, which you can access via the **Scheme** popup menu. Consult the GaussView manual for details.

Building Formaldehyde in WebMO

WebMO After opening your browser, navigating to the URL for WebMO and logging in, you will see the WebMO **Job Manager** page. In order to build a new molecule, select **Create New Job** from the **New Job** menu:

This will open WebMO's **Build Molecule** page. The toolbar along the left side of the build area contains icons which invoke the builder's various features.

1. Select the **Build** icon: . The currently selected element will be displayed in the status area at the bottom of the window, along with several hints for various build scenarios.

2. Click anywhere within the build area. This will add a carbon atom.

13 Hit the "O" key to change the atom type to oxygen: Build Mode - O. You may need to click on the **Build** icon first.

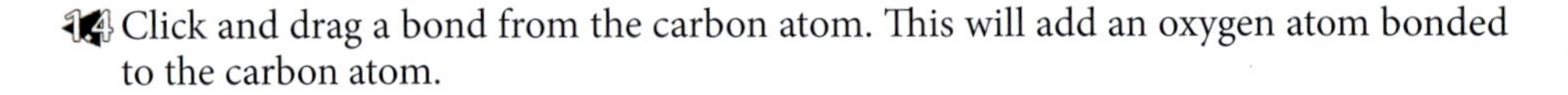

14 Click and drag a bond from the carbon atom. This will add an oxygen atom bonded to the carbon atom.

15 Click on the carbon atom and drag the mouse to the oxygen atom. This will change the bond type from single to double.

16 Hit the "H" key to change the atom type to hydrogen. Add two hydrogen atoms bonded to the carbon atom by dragging a bond from the latter.

Alternatively, you can select **Add Hydrogens** from the **Cleanup** menu. This will add the remaining two atoms to the formaldehyde structure.

17 The final step is to regularize the structure before running the Gaussian calculation. There are two options for doing so. If the structure has no symmetry, then you can use the WebMO structure cleanup feature. Click on the **Comprehensive Cleanup using Mechanics** icon: . This option performs a simple, one-step geometry optimization using Molecular Mechanics. It is useful for regularizing the structure.

The second method applies to molecules with symmetry. In these cases, the starting structure should be of the proper symmetry in order for Gaussian calculations to recognize and take advantage of it. Formaldehyde has C_{2v} symmetry, and we will ensure that our structure corresponds to this point group.

Imposing Symmetry in WebMO

WebMO's symmetry feature is accessed via the **Symmetrize Molecule** item on the **Symmetry** menu in the molecule builder. It opens the following dialog:

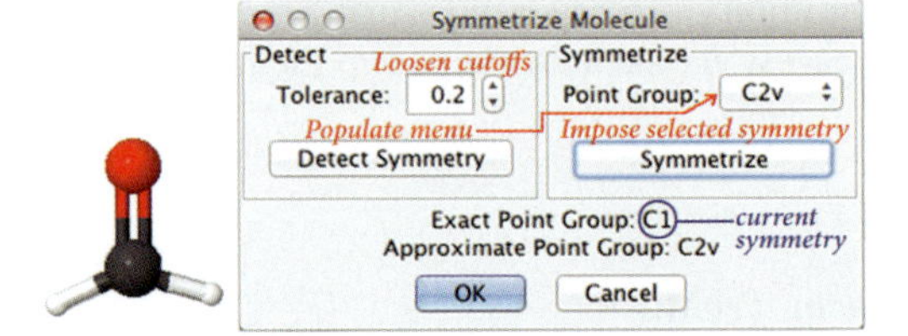

The **Detect** area is used to specify the desired tolerance and to search for higher symmetry point groups. Increasing the **Tolerance** value will make it more likely that a higher symmetry point group will be applicable to the molecular structure. The **Symmetrize** area is used to select and impose higher symmetry than currently present in the molecular structure.

In order to impose higher symmetry on a molecule, increase the **Tolerance** and then click **Detect Symmetry**. Available point groups at that tolerance level will appear in the **Point Group** popup menu, with the highest one displayed. Repeat this process until the desired point group becomes available. Then select it and click **Symmetrize**.

The symmetry of the molecular structure is displayed in the symmetry icons in the building toolbar (upper icon at left). When the molecular structure is close to a higher symmetry point group—in other words, within the default tolerance value—then the upper icon will turn red and indicate the available point group: . In this case, clicking on the icon will apply the indicated point group to the molecular structure in a single step.

In our case, we use the symmetry feature to impose C_{2v} symmetry on the molecular structure.

18 This completes the formaldehyde molecule.

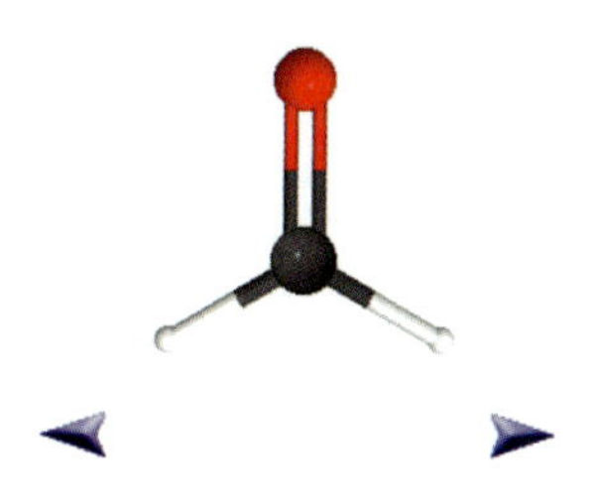

Setting up and Running a Gaussian Job in WebMO

The arrows at the bottom of the various WebMO pages allow you to move back and forth among the various steps involved in setting up and running a calculation and examining calculation results.

21 Click on the right arrow to leave the builder and move on to the calculation setup module. If you are presented with a choice of computational engines, choose **Gaussian** and then click on the right arrow. The controls for a Gaussian job are illustrated in the figure following.

WebMO's **Calculation** *menu has several choices which correspond to Gaussian single point energy calculations, including* **Molecular Energy**, **Molecular Orbitals**, **Bond Orders** *and more. They differ in the post-calculation options which are offered.*

22 We begin with the **Job Options** tab. We need to specify a brief description for the job in the **Job Name** field. We also specify the calculation type as **Molecular Orbitals** in the **Calculation** popup and the theoretical method as **APFD** in the **Theory** popup.

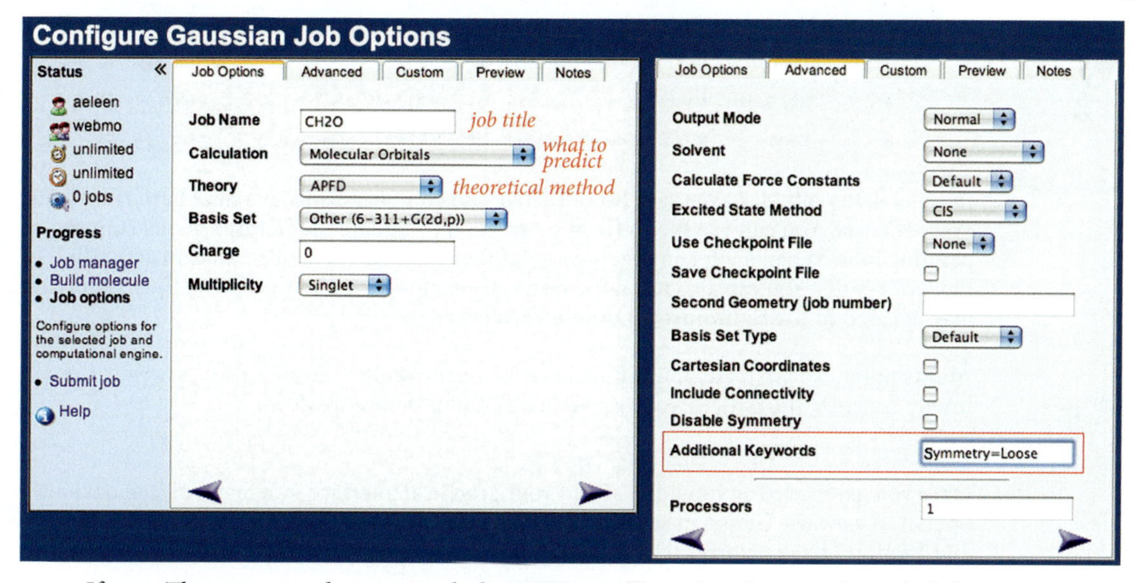

If your **Theory** menu does not include **APFD**, you'll need to do something slightly different. You can always specify any method not found on the **Theory** menu by selecting the **Other** option and then entering the desired method into the subsequent dialog:

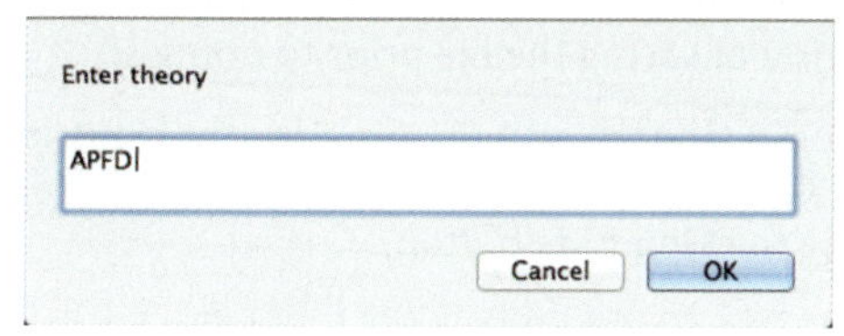

You can add an item to the **Theory** menu by adding a line to the *gaussian.html* file in WebMO's main web server folder (named *webmo*):

```
<TD>Theory</TD>
<TD>
    <SELECT NAME="theory" ...>
        <OPTION VALUE="HF" SELECTED>Hartree-Fock
        <OPTION VALUE="APFD">APFD
        <OPTION VALUE="B3LYP">B3LYP
```

Add this line.

Items can be added to the basis set menu in an analogous manner. If you are unsure of the location of this file or do not have access to it, then contact your system administrator.

23 We will be using the 6-311+G(2d,p) basis set. If it is present in the **Basis Set** menu, you can simply select it. Otherwise, select the **Other** option and then type this basis set name into the resulting dialog. The result of doing the latter is illustrated above.

24 We are now ready to begin the calculation. You can do so by clicking on the right arrow icon in the lower right corner of the page. This will bring you back to the **Job Manager** page:

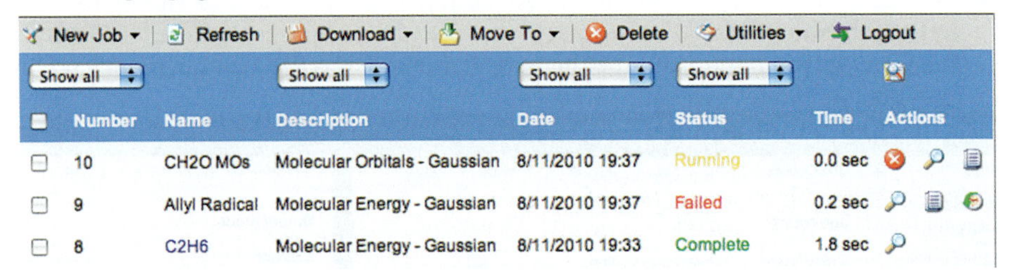

In general, the **Job Manager** page lists all active and completed jobs that have been run by this WebMO user. You can use the **Refresh** menu item to update the statuses of any running or pending jobs. When a job completes successfully, WebMO makes its name an active link to the job's results. You can also reach the results by clicking the **View Job** icon: the magnifying glass located in the rightmost column of each entry.

At this point, we are ready to examine the Gaussian results (see page 46). However, we will briefly consider the structure of Gaussian input files before we do so.

Understanding the Gaussian Input File

Once you have saved an input file within your graphical interface program, you can obviously open it to view the Gaussian keywords the package generated. In addition, both GaussView and WebMO have features that allow you to see the actual input file it is producing from within the package:

- *GaussView*: Clicking the **Edit** button in the **Gaussian Calculation Setup** dialog will allow you to view and/or modify the current input file using an external text editor. You will be given the option of saving the file prior to doing so.

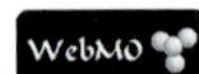

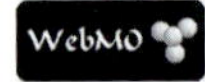

- *WebMO*: Prior to running the job, you can examine the Gaussian input file that WebMO is creating within the **Configure Gaussian Job Options** page by navigating to the **Preview** tab and then pressing the **Generate** button.

If you examine the Gaussian input file created by your graphics program, you will see a file that is similar to this one:

```
%chk=form.chk
# APFD/6-311+g(2d,p) geom=connectivity

Formaldehyde Energy

```

1: link 0 command section
2: route section
required blank line
3: job title section
required blank line

```
0 1     4: charge & spin multiplicity
C   -1.78001774    -0.58048206     0.00000000     5: molecular structure
H   -1.24484917    -1.50703189     0.00000000
H   -2.85001773    -0.58067672     0.00000000
O   -1.15101602     0.50943876     0.00000000
                                                  required blank line
1 2 1.0 3 1.0 4 2.0    6: atom connectivity information
2
3
4
                                                  required blank line
```

A Gaussian input file is comprised of several sections:

Link 0 Command Section (1): This optional section specifies the location and handling of Gaussian's scratch files (and optionally additional directives related to executing calculations in parallel). In our example, Gaussian's primary scratch file, known as the checkpoint file, has been given a name so that it will be saved at the conclusion of the calculation. This is typically done so that molecular orbitals and other spatial data will be available for fast visualization. You will see lines like these in Gaussian input files created by GaussView and the examples provided for use with this book. This section is not used by default by WebMO.

Route Section (2): The route section of a Gaussian input file indicates the precise calculation that you want to run. The first line of the route section begins with a number sign (#), optionally followed by a letter indicates how detailed you want the output to be: # and **#N** request normal output, **#T** requests terse output, and **#P** requests detailed output.

The remainder of the route section tells Gaussian what calculation to perform and how to perform it. It typically includes keywords specifying the kind of calculation you want to run and the specific theoretical method and basis set which should be used; other keywords specifying other aspects of the calculation may also be present.

A single point energy is the default calculation type in Gaussian, so no keyword is required in our example route section (although the **SP** keyword can be used if desired). The route section contains the **APFD** and **6-311+G(2d,p)** keywords, which are conventionally separated by a forward slash (/). The example route section also includes the **Geom** keyword with the option **Connectivity**, which specifies the format of the molecular structure to follow (see below). The route section for your job may contain different additional keywords.

The route section can go on for as many lines as necessary. It is terminated by a blank line.

Title Section (3): This section of the input file consists of one or more lines describing the calculation in any way that the user desires. It often consists of just one line, and the section ends with a blank line.

Molecule Specification Section (4 & 5): Information about the molecule(s) to be modeled comes next in the input file. The first line of this section (4) contains the charge and spin multiplicity. The charge is the positive or negative integer specifying the total charge on the molecule: 0 represents a neutral molecule, +1 or 1 is used for a singly-charged cation, and -1 designates a singly-charged anion.

The spin multiplicity for a molecule (defined above, p. 36) is determined by counting the number of spin-unpaired electrons with parallel (like) spin and adding 1. All electrons in doubly occupied orbitals are of opposite spin and so can be ignored. What must be examined are the spins of the electrons in singly occupied orbitals. For example, a closed shell singlet—a system with no unpaired electrons—has a spin multiplicity of 1. A doublet, which has only a single unpaired electron, has a spin multiplicity of 2. A triplet system with two unpaired electrons of like spin has a spin multiplicity of 3. A biradical, an open shell system with two unpaired electrons of opposite spin, is also a singlet, since all of the electrons are spin-paired.

The structure of the molecular system to be investigated follows the initial charge and spin multiplicity line in the molecule specification section (5). Visualization programs typically express the positions of the nuclei in Cartesian coordinates (GaussView) or as a Z-matrix (WebMO). In the latter case, the molecular structure will often consist of two subsections (separated by a blank line).

The molecule specification section ends with a blank line.

Additional Input Sections (6): Many calculation types require additional input, which will appear in one or more additional sections within the input file. In our example, the input required by the **Geom=Connectivity** option appears as the final section of the file. It specifies atom bonding within the molecule. This option is used by default by GaussView but not by WebMO.

The input file as a whole ends with a blank line.

Although we will not look at Gaussian input files as a matter of course in this book, we will mention the keywords corresponding to the Gaussian job types and other features that we use.

Gaussian Calculation Results

In this section, we'll examine the major results predicted by a single point energy calculation. We will begin with the results summary displays, which you can access as follows:

- *GaussView*: Select **Summary** from the **Results** menu. This will open a window containing the most important quantitative results from the calculation.

- *WebMO*: On the **Job Manager** page, click the magnifying glass icon for the completed job (located in the **Actions** column at the right). This takes us to the **View Job** page whose top portion contains a modified version of the builder page (it includes only the controls related to moving and examining the molecular structure). The page also contains numeric results from the job. In order to view them, you must scroll down the page quite a bit.

Different graphics programs display quantities predicted by the calculation somewhat differently. The figure following illustrates the results from our calculation as they appear in WebMO (left) and GaussView (right).

The summary displays begin by repeating the basic specifications for the job: the job type, model chemistry, stoichiometry and other basic data. They then report a variety of basic quantitative results, including:

- The total energy of the molecular system, as predicted by the specified model chemistry. This value is in atomic units: in this case, in hartrees.
- The magnitude of the dipole moment, in debye (see below for more on the dipole moment).

total energy *dipole moment*

Calculated Quantities

Quantity	Value
Route	#N APFD/6-311+G(2d,p) SP
Stoichiometry	CH_2O
Symmetry	C2V
Basis	6-311+G(2d,p)
RAPFD Energy	-114.543994561 Hartree
Rotational Constants	
Dipole Moment	2.4555 Debye

Constant	Frequency (GHz)	Frequency (cm^{-1})
a	273.7377299	9.13090781957
b	38.9345943	1.29871827196
c	34.0863794	1.13699923031

G1:M1:V1 – Gaussian Calculation Su...

Formaldehyde Energy

File Name	formaldehyde	
File Type	.chk	
Calculation Type	SP	
Calculation Method	RAPFD	
Basis Set	6-311+G(2D,P)	
Charge	0	
Spin	Singlet	
Total Energy	-114.53860583	a.u.
RMS Gradient Norm	0.00000000	a.u.
Imaginary Freq		
Dipole Moment	2.7208	Debye
Point Group		

Ok View File Save Data

When we discuss energies in this work, we will generally use hartrees; when we discuss energy differences, kilojoules per mole (kJ/mol) and kilocalories per mole (kcal/mol) will often be more intuitive and convenient units, especially when comparing calculation predictions to experimental results. Accordingly, the following conversion factors are useful:

1.0 hartree ≈ 2625.5 kJ/mol ≈ 627.51 kcal/mol ≈ 27.21 eV

Interpreting Energies & Energy Differences

Notice that the energies reported by the two programs are different. This is because the structures for formaldehyde that we created in the two programs are not identical: they have slightly different default values for the bond lengths, bond angles and dihedral angles within the molecule. If you decide to try this example for yourself, your energy result may not match either of them.

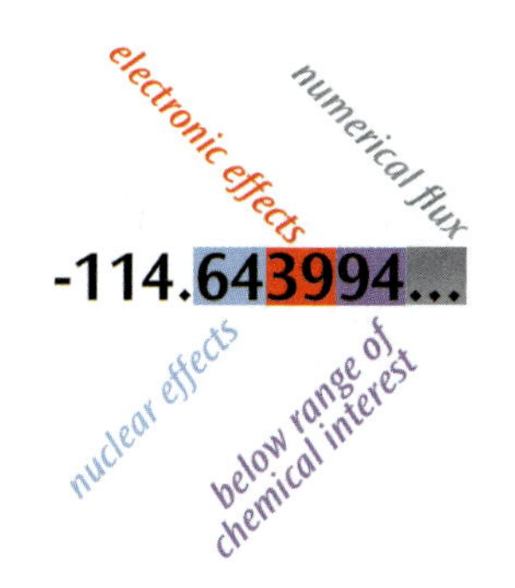

In general, energy differences arising from electronic characteristics are typically in the millihartree range: in other words, the third and fourth digits following the decimal point.* Energy differences of one or more orders of magnitude are the result of nuclear positions, and energy differences more than an order of magnitude smaller are too small to be the result of bonding. When you compute the energy of the same structure using a specific model chemistry on different computer systems, you will typically see energy differences even smaller than the latter, which are insignificant chemically and arise from the slight numerical differences between processors and/or operating systems.

The energy difference between the two structures we built is 0.00539 hartrees, corresponding to about 14.2 kJ/mol (≈3.4 kcal/mol). This results illustrates the fact that structures built with visualization software are somewhat arbitrary. In general, when we compare energies in this book, we will do so only for structures that we have optimized. When we run a geometry optimization starting with the two systems we built and then compare their energies, the difference is 5.0×10^{-7} hartrees, a minute difference of no chemical significance.

* This statement assumes the calculation uses reasonable values for the various convergence criteria (SCF, geometry optimizations), an adequate integration grid for DFT calculations and similar considerations for other program factors. Overly loose convergence criteria, inadequately-sized integration grids and other dubious choices can introduce errors as large as electronic effects. Gaussian default values are selected to avoid such problems and so should be modified downward only with great care.

Dipole and Higher Multipole Moments

The dipole moment is the first derivative of the energy with respect to an applied electric field. It is a measure of the asymmetry in the molecular charge distribution, and is expressed as a vector in three dimensions.

The summary output reports include the magnitude of the dipole moment vector. However, this number does not tell you anything about the direction of the dipole moment. However, the vector itself can also be displayed graphically.

In GaussView, you can display it by selecting **Charge Distribution** from the **Results** menu and then checking **Show Vector** in the **Dipole Moment** section of the dialog (located in its lower portion). You can use other controls to change the length of the displayed vector and its origin; we've selected the molecule's **Center of Electronic Charge** for the latter in the illustration The dipole moment vector is aligned with the O=C double bond. It points away from the oxygen atom (the negatively charged part of the molecule), (following the physics convention). GaussView displays the dipole moment vector with an arrow pointing in the positive direction: in other words, as a mathematical vector.

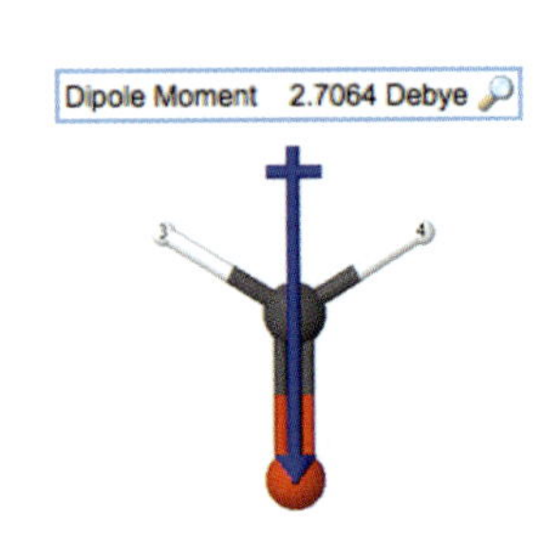

In WebMO, you can display the dipole moment vector by clicking on the magnifying glass icon to the right of the dipole moment value in the **Calculated Quantities** section. The dipole moment vector will then be added to the molecule display, as illustrated in the margin. Its display convention uses an arrow with a plus sign at the tail pointing toward the location of negative charge in the molecule (following chemical convention).

Gaussian also predicts quadrupole moments and higher multipole moments (through hexadecapole). Here is the portion of the Gaussian output file reporting the dipole and quadrupole moments for formaldehyde:

```
Dipole moment (field-independent basis, Debye):
  - - - - - - - - - - - - vector components - - - - - - - - - - - - -        magnitude
  X=     0.0000    Y=     0.0000   Z=     -2.4223   Tot= 2.4223
Quadrupole moment (field-independent basis, Debye-Ang):
 XX= -11.7060   YY= -11.5827   ZZ= -12.1807
 XY=     0.0000   XZ=     0.0000   YZ=     0.0000
```

The quadrupole moment is the second order term in the expansion of the total electron distribution, providing at least a crude insight into its overall shape. For example, equal XX, YY, and ZZ components indicate a spherical distribution. This is approximately the case for formaldehyde. One of these components being significantly larger than the others would represent an elongation of the sphere along that axis. If nonzero, the off-axis components—XY, XZ and YZ—represent trans-axial distortion (stretching or compressing) of the ellipsoid. Quadrupole (and higher) moments are generally of significance only when the dipole moment is 0.

Another way of obtaining information about the distribution of electrons is by considering how it changes in the presence of an external field. *Polarizabilities* are the second and higher derivatives of the energy with respect to an electric field. Higher order polarizabilities—the third and subsequent derivatives—are also of interest. See "Materials Properties: Polarizability, Hyperpolarizability & Gamma" on page 190 for an introduction to these properties.

Molecular Orbitals

A molecular orbital is a mathematical function that describes the behavior of an electron or pair of electrons within a molecule. These functions are typically plotted as surfaces around the molecular structure.

In GaussView, you can view them with **MOs** on the **Edit** menu or by clicking the corresponding icon (pictured in the margin). This will open the **MOs** dialog.

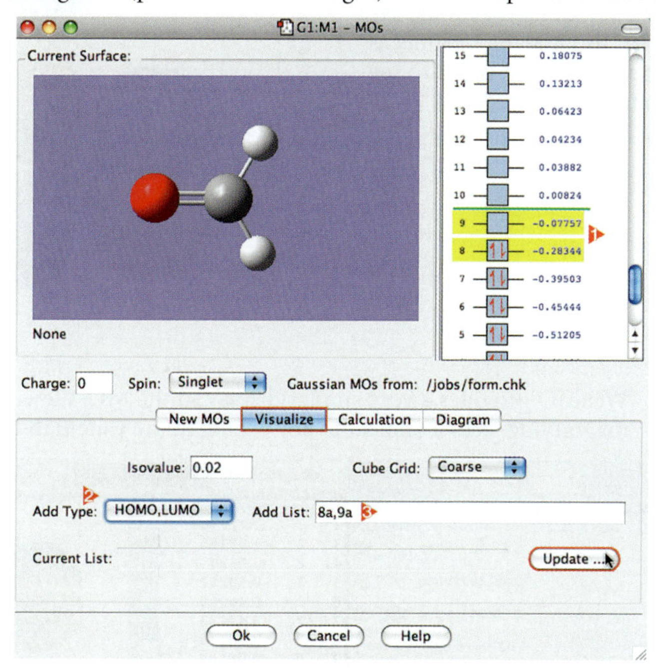

Molecular orbital surfaces must be generated before they can be viewed. Navigate to the **Visualize** tab within the dialog, and then select the desired orbitals in one of the following ways:

- Click on the desired orbitals within the list to select them (1).
- Select one of the common choices from the **Add Type** popup menu (2).
- Enter the desired orbitals in the **Add List** field (3).

Finally, click the **Update** button to compute the orbitals. Once the brief calculation has finished, small squares will appear to the right of the orbitals within the orbital list, indicating that they are available for viewing. Clicking in the square next to an orbital will cause it to appear in the window's display area.

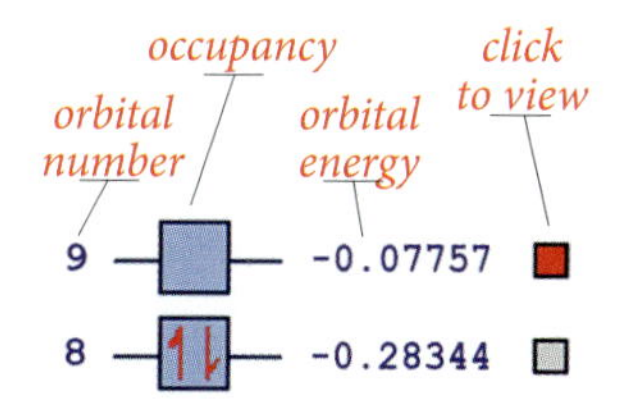

The highest occupied molecular orbital (HOMO) and lowest unoccupied molecular orbital (LUMO) may be identified by finding the point where the orbital occupancy changes from occupied to unoccupied. This is often, but not always, the point where the orbital energies change from positive to negative. In the case of formaldehyde, for example, the LUMO has a very small negative orbital energy.

Transparent Surfaces

GaussView: In the context (right click) menu, select **View→Display Format**; *Go to the* **Surface** *panel; Choose* **Transparent** *from the* **Format** *popup; Adjust* **Transparent–Opaque** *slider as desired.*

WebMO: Right click in the MO display; Select **Opacity→Transparent**.

Here are the HOMO and LUMO as computed by our calculation:

The HOMO is orbital 8 (an *n* orbital), and the LUMO is orbital 9 (an antibonding π^* orbital). The positive and negative orbital lobes are displayed in different colors. We have chosen to view the orbitals as transparent surfaces in order to see their orientation more easily.

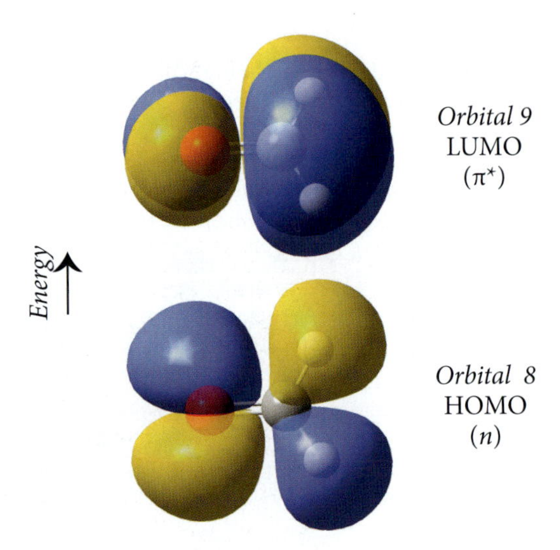

WebMO provides a very similar display in the **MO Viewer** when you click the magnifying glass icon next to any orbital in the **Calculated Quantities** section:

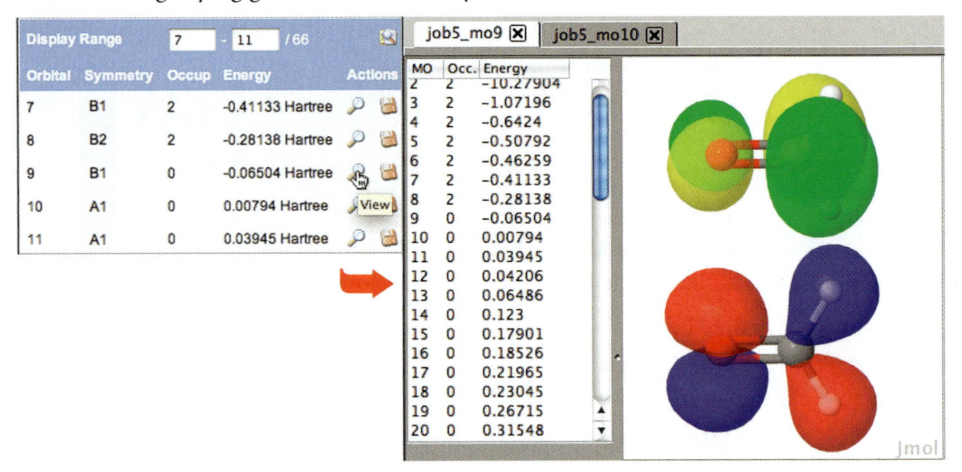

The display again shows the HOMO and LUMO. Notice that WebMO uses different coloring schemes for occupied and virtual orbitals. Within the MO Viewer, clicking on any item in the orbital list will create a new tab displaying that orbital.

GOING DEEPER: MOLECULAR ORBITALS

It is important to emphasize that orbitals are actually mathematical conveniences and not physical quantities, despite how real visualization may make them seem. While the energy, electron density, and optimized geometry are physical observables, the orbitals are not. In fact, several different sets of orbitals can lead to the same energy. Nevertheless, orbitals are very useful for qualitative descriptions of bonding and reactivity.

Atomic Charge Distribution

By default, Gaussian jobs perform a Mulliken population analysis [Mulliken55], which partitions the total charge among the atoms in the molecule.

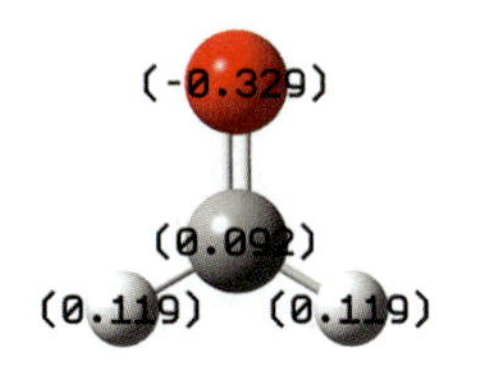

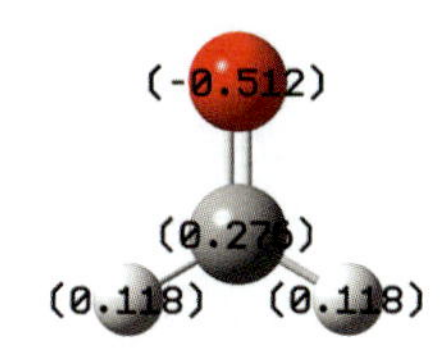

Predicted atomic charges from Mulliken population analysis (top) and Natural Population Analysis (NPA, bottom).

WebMO Results of the population analysis are included in the **Calculated Quantities** section in WebMO, and you can view the molecule with each atom labeled with its charge by clicking on the magnifying glass icon in the upper right corner of one of the atomic charges table (the predicted Mulliken charges are in the table labeled **Partial Charges**).

In GaussView, you can view the predicted numerical charge distribution by selecting **Charge Distribution** from the **Results** menu and checking **Show Numbers** (see below).

For formaldehyde, the Mulliken population analysis places a slight negative charge on the oxygen atom and divides the balancing positive charge between the remaining three atoms (see the upper figure in the margin). Mulliken population analysis is a widely-used but quite approximate charge distribution scheme which computes charges by dividing orbital overlap evenly between the two atoms involved.

GaussView also offers a color-based charge distribution display, controlled by the **Display Charge Distribution** dialog's **Color Atoms by Charge** check box, as in the following example. Positive charges are green, and negative charges are red, with brighter hues indicating larger charge magnitudes:

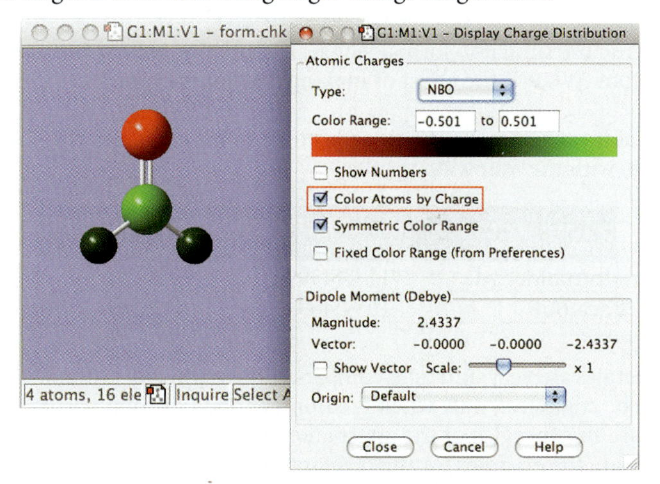

Gaussian provides other population analysis methods as well. For example, Natural Population Analysis (NPA) [Reed85] is carried out in terms of localized electron-pair "bonding" units. To compute charges using this method, we ran a single point energy calculation on formaldehyde as above and included **Pop=NPA** in the **Additional Keywords** field. When an alternate population analysis method is specified with the **Population** keyword, Gaussian still performs a Mulliken analysis as well.

The table below lists the controls for specifying and viewing the computed atomic charges:

CONTROL	GAUSSVIEW	WEBMO
Location of **Additional Keywords** field	Lower part of the **Gaussian Calculation Setup** dialog	**Advanced** tab of **Configure Gaussian Job Options** page
Select which population analysis to view	**Type** popup in **Display Charge Distribution** dialog	All are displayed in the charges table in **Calculated Quantities** section (view with mag. glass icon)

NPA assigns charges somewhat differently, placing more of the negative charge on the oxygen atom. The predicted charges are displayed in the lower illustration in the margin on page 51. More detailed analysis is available in the output file for the Gaussian job, including the number of core electrons, valence electrons, Rydberg electrons, located in diffuse orbitals, and the charge partitioning among atomic orbitals for each atom.*

Gaussian also provides several methods for computing electrostatic potential-derived charges. Electrostatic potential-derived charges assign point charges to fit the computed electrostatic potential at a number of points on or near the van der Waals surface. This sort of analysis is commonly used to create input charges for molecular mechanics calculation.

More about Energies and Orbitals

When Energies Can & Cannot be Compared

Every Gaussian calculation will include the predicted energy for the specified molecular system. However, the energies of different systems cannot always be directly compared to one another. In general, in order for an energy comparison of two or more molecules (or groups of molecules) to be valid, the following conditions must be met:

- The molecules/groups of molecules must contain the same number and types of atoms.
- The calculations must be performed using the same model chemistry.

EXAMPLE 2.2 COMPARING FORMALDEHYDE AND ACETONE

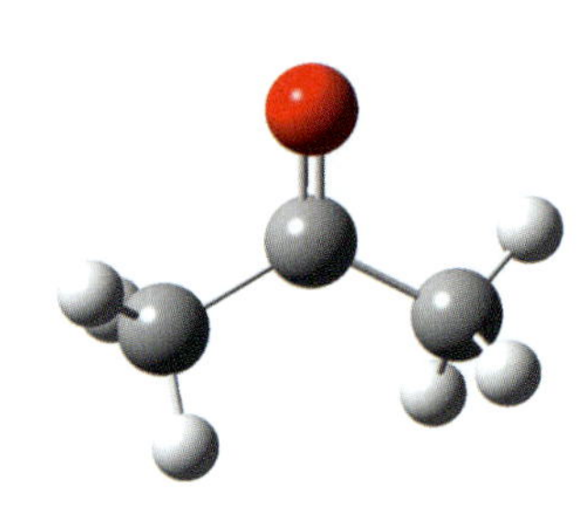

Acetone† has a structure similar to formaldehyde, with methyl groups replacing the hydrogens on the carbon atom. What is the effect of making this substitution?

We ran single point energy calculations on both systems, using the APFD/6-311+G(2d,p) model chemistry, with the following results:

MOLECULE	TOTAL ENERGY (hartrees)	DIPOLE MOMENT (debye)
formaldehyde	-114.44399	2.42
acetone	-193.04694	3.23

Although the energies are very different, comparing them directly is of no value. Energies for two systems can be compared only when the number and type of nuclei are the same. Thus, we could compare the energies of the alternate forms of 1,2-dichloro-1,2-difluoroethane, and we can compare the energies for the reactants and products of reactions when the total number of nuclei of each type are the same. But we cannot make any meaningful statement about formaldehyde versus acetone based upon comparing their energies.

We can compare their dipole moments, however. In this case, we note that the methyl groups in acetone have the effect of increasing the magnitude of the dipole moment, which points away from the oxygen along the double bond in both cases. This means that the centers of positive and negative charge are farther apart in acetone than they are in formaldehyde. In addition, there are negative charges on the two methyl carbon atoms in acetone, as indicated by the NPA charge distribution (red indicates negative charge):

* The NPA facility is part of the version of the NBO program which is integrated within Gaussian (as link 607). If it is used to produce published results, then it should be cited as well as Gaussian itself: [NBO3.1]. Gaussian can also prepare input for recent standalone versions of NBO. See the discussion of the **Population** keyword in the *Gaussian User's Reference*.

† AKA propan-2-one.

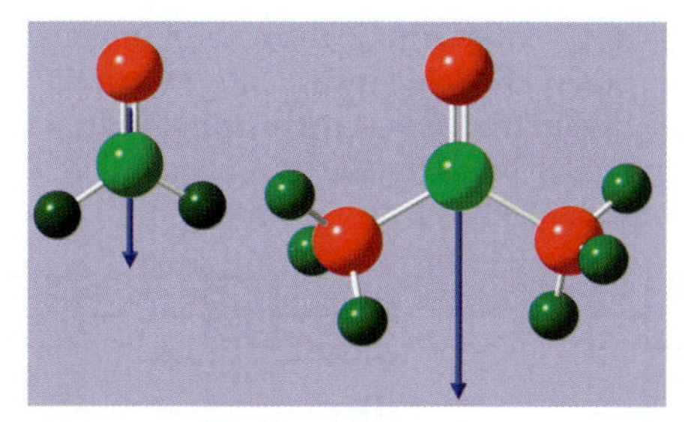

The preceding illustration also includes the dipole moment vectors.■

EXAMPLE 2.3 1,2-DICHLORO-1,2-DIFLUOROETHANE CONFORMER ENERGIES

The three stereoisomers of 1,2-dichloro-1,2-difluoroethane (stoichiometry: CHFCl–CHFCl) provide an example where molecular energies can be directly compared. We will run single point energy calculations using the APFD/6-311+G(2d,p) model chemistry for the three molecules and compare their energies and dipole moments.

Here are the three conformers of the molecule:

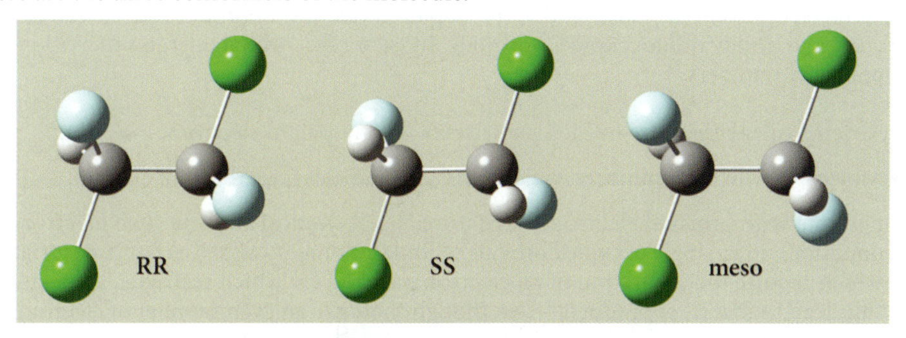

Running all three jobs yields the following results:

FORM	TOTAL ENERGY (hartrees)	μ (debye)
RR	-1197.19290	2.3
SS	-1197.19240	2.5
meso	-1197.19400	0.0

The RR and SS forms have very similar values for the energy and the dipole moment. The energy difference between the SS and the meso forms is about 1.1 millihartrees, which corresponds to about 2.9 kJ/mol or about 0.7 kcal/mol. This is a small but significant difference in energy.

The RR and SS forms have dipole moments of 2.3 and 2.5 debye (respectively), and the dipole moment vector is perpendicular to the Cl–C–C–Cl plane in both molecules. The meso form has no dipole moment.■

Modeling Open Shell Systems

The simple molecules we've considered so far are all uncharged, ground state, singlet species. Closed shell molecules like these have an even number of electrons divided into pairs of opposite spin. Calculations on closed shell systems use a *spin restricted* model which assumes that all molecular orbitals are doubly occupied, each containing two electrons of opposite spin.

In contrast, open shell systems—for example, those with unequal numbers of spin up and spin down electrons—are usually modeled by a *spin unrestricted* model. Restricted, closed shell calculations force each electron pair into a single spatial orbital, while open shell calculations use separate spatial orbitals for the spin up and spin down electrons, known as α and β orbitals, respectively:

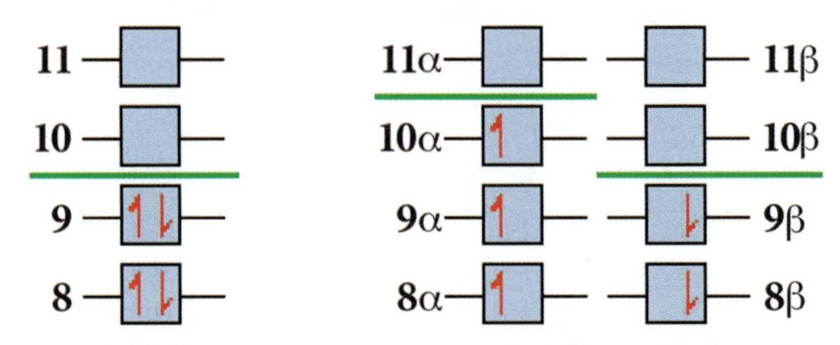

The illustration on the left depicts four orbitals from a closed shell system: the two highest energy doubly occupied orbitals—orbitals 8 and 9—and the two lowest energy unoccupied orbitals (orbitals 10 and 11). The illustration on the right depicts orbitals from an open shell system, in this case, a doublet. Separate α and β orbitals are displayed. The diagram shows the paired electrons in orbitals 8 and 9, the unpaired electron in orbital 10α, as well as one unoccupied α and two unoccupied β orbitals. In each case, orbital energies increase with higher orbital numbers.

Unrestricted calculations are needed for systems with unpaired electrons, including:

- Molecules with odd numbers of electrons: e.g., radicals (regardless of charge).
- Systems with unusual electronic structure: e.g., biradical singlets (two electrons of opposite spin in singly occupied orbitals). Similarly, processes such as bond dissociation which require the separation of an electron pair and for which restricted calculations thus lead to incorrect products (even though there are an even number of electrons).

By default, Gaussian performs a restricted calculation for singlet spin multiplicity systems and an unrestricted calculation for all other systems. However, open shell calculations may be explicitly requested by prepending a **U** to the method keyword (for unrestricted): e.g., **UAPFD**. You can modify the Gaussian input file manually before submitting the job if necessary.

GaussView

In GaussView, you can also select **Unrestricted** from the second-to-rightmost popup menu on the **Method** line in the **Method** panel of the **Gaussian Calculation Setup** dialog.

Be aware, however, that merely requesting an unrestricted method will not result in an open shell wavefunction for closed shell systems. Modifying the initial guess will also be necessary. This topic is discussed in detail in chapter 9 (p. 444ff).

Biorthogonalized Orbitals

The orbitals normally produced by a Gaussian calculation are known as the *canonical* orbitals. For spin polarized systems, these orbitals can be difficult to interpret. Gaussian now has a feature which will transform the canonical orbitals via an energy-invariant rotation which attempts to maximize the overlap between the alpha and beta orbitals, producing what are referred to as the *corresponding* orbitals. The process of producing this representation is known as *biorthogonalization*. Examining biorthogonalized orbitals can make it easier to determine the locations of the unpaired electrons in molecules with high spin multiplicity.

EXAMPLE 2.4 COMPARING CANONICAL & BIORTHOGONALIZED ORBITALS

The following figure shows two sets of orbitals for FeO^+. This system is a quartet, and it is accordingly modeled using an unrestricted method (using our standard theoretical method and basis set):

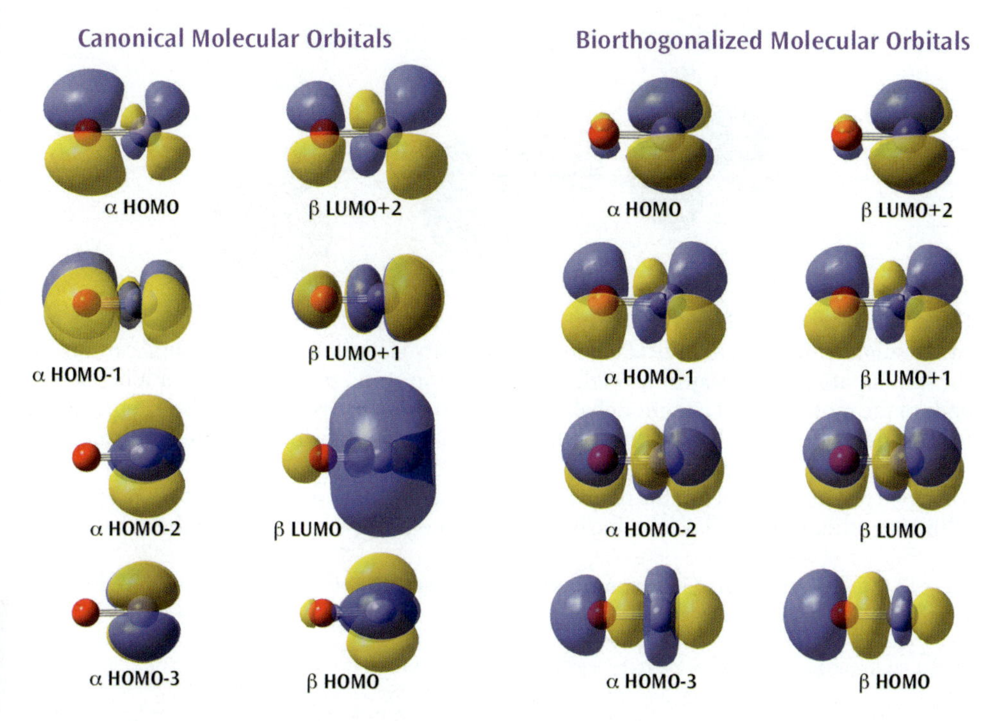

The set of orbitals at the left are the canonical orbitals. Understanding the biorthogonalized orbitals on the right is much more straightforward. With this representation, it is clear that the molecule has five singly occupied orbitals consisting of four α unpaired electrons localized on the iron atom, and one β unpaired electron localized on the oxygen atom.

EXAMPLE 2.5 SPIN POLARIZATION IN HETEROSUBSTITUTED ETHENE RADICALS

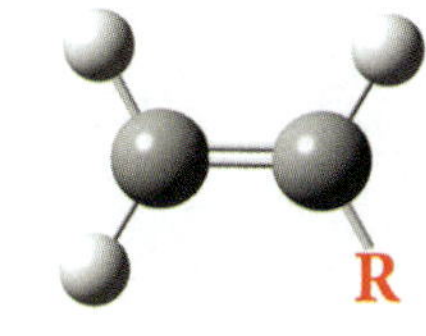

In this example, we will be studying *spin polarization* in a series of molecules of the form CH_2=CH–R, starting with allyl radical:* CH_2=CH–CH_2. Spin polarization is the degree to which spin within the molecule is aligned in a particular direction. We will compute the molecular orbitals and spin polarizations for the carbon and oxygen atoms and compare the latter to the experimental values. We will then go on to predict these same items for two other substituents: S and Be. You will later explore two other substituents—Mg and O—on your own in Exercise 2.5 on page 67.

All of these systems are doublets: spin multiplicity 2. We will request biorthogonalized orbitals for all calculations with the **Population=(BiOrtho,SaveBioOrtho)** options.

Spin densities are computed as part of the population analysis. The *spin density* is the electron density in a radical species, taking into account spin polarization. As the difference of the α and β electron densities, it indicates likely regions for unpaired electrons. When the spin density is partitioned among the nuclei in the molecule, it is known as the *atomic spin density*. This quantity provides the same information for each nucleus. The predicted values are found

* IUPAC: 1-propen-3-yl radical.

in the Gaussian output file following a header line that ends with the phrase "spin densities." We'll see an example shortly.

Here are the singly occupied molecular orbitals for the three molecules:

For allyl radical (left), the orbital is equally divided between the two terminal carbon atoms. In the sulphur substituent radical (center), the density is located on the terminal carbon and the sulphur atom, with very little on the central carbon atom, resulting in larger lobes at both ends of the molecule and a slight imbalance between them in favor of the substituent. Finally, for the radical with the Be substituent, the orbital is essentially localized to the Be atom.

The bond orders also differ in the three compounds: allyl radical exhibits a resonance structure, while the other radicals have double bonded carbons and a single bond to the substituent.

Here is the section of the Gaussian output that reports the predicted atomic spin densities for the Be substituent radical:

```
Mulliken charges and spin densities:              Output section heading
               charges           atomic spin densities
               1                 2     Atom order is same as input file:
    1  C    -0.272735    0.037643  Central carbon
    2  C    -0.353876   -0.010187  Terminal carbon
    3  H     0.107186    0.003957
    4  H     0.113024    0.002711
    5  H     0.109328    0.003940
    6  Be    0.297074    0.961937  Substituent
                            total charge      # unpaired electrons
Sum of Mulliken charges =    0.00000      1.00000
```

The following table gives the computed atomic spin densities for each atom:

R	TERMINAL C	CENTRAL C	SUBSTITUENT
CH_2	0.73	-0.30	0.73
S	0.46	-0.21	0.80
Be	-0.01	0.04	0.96

In the allyl radical, the atomic spin density is evenly divided between the two terminal carbon atoms. In contrast, the atomic spin density is drawn toward the substituent for the Be compound. For the S compound, the atomic spin density is divided between the two terminal atoms, somewhat more drawn toward the substituent than the terminal carbon atom.

See the original paper by Wiberg and coworkers [Wiberg95] for a more detailed discussion of the chemistry of these compounds.

GOING DEEPER: WAVEFUNCTION STABILITY & ORBITAL MIXING

The calculation for the sulphur radical in the preceding example provides an opportunity to discuss wavefunction stability should you want to introduce it. The input file provided for the example uses the **Stable=Opt** and **Guess=Mix** keywords to locate a lower energy wavefunction than is computed by default with our standard model chemistry. Thus, this example can be used to teach more about the mathematical characteristics of orbitals and wavefunctions to students who are prepared for such material. Stability calculations are introduced in the *Advanced Topics* section of this chapter (p. 71).

Minimizing Molecular Structures: Opt+Freq Calculations

So far, we haven't been concerned with the precise molecular structure used for our Gaussian calculations. However, as we've seen, structural changes within a molecule produce differences in its energy and other properties.

From this point forward, all of our calculations will start from optimized molecular structures. Most geometry optimizations attempt to locate minima on the potential energy surface for a molecule, thereby predicting equilibrium structures of molecular systems. Sometimes optimizations are performed in order to locate transition structures. In this chapter, however, we will focus on optimizing to minima, which are also known as *minimizations*.

Once an optimization has completed successfully, however, we must ensure that the predicted structure is in fact a minimum and not a transition structure. This can be accomplished by running a frequency calculation at the optimized geometry.

Frequency calculations take into account the nuclear vibrations in molecular systems in their equilibrium states. They predict a variety of molecular properties, including the following:

- Vibrational spectra for the molecule: IR and optionally Raman and VCD.
- The *zero point energy*: a correction to the electronic energy that allows us to estimate the energy of the molecule in its lowest vibrational state at 0 K, where it does not possess translational or rotational energy yet.
- The *thermal energy* (E^{therm}), *enthalpy* (H=E^{therm}+PV), *entropy* (S) and *Gibbs free energy* (G=H-TS), computed at the default conditions of 298.15 K and 1 atmosphere known as *standard temperature and pressure* (STP).
- The *constant volume heat capacity* (Cv) and *molecular partition functions* (Q) for each thermal motion (normal mode). Cv is the derivative of E^{therm} with respect to temperature.

By default, all of the thermodynamic quantities are estimated by assuming that the molecule is in the gas phase, that it behaves ideally and that the harmonic oscillator and rigid-rotor quantum mechanical models can be used to describe the motions of the atoms. Conditions other than STP can be easily specified with the **Temperature** and **Pressure** keywords.

The partition function Q contains the information about the probability that the system is at a particular energy level under the specified conditions. This is used to evaluate E^{therm} and S, since these are weighted averages of the contributions from each energy level. The other quantities can then be defined from these two. For more details about these relationships, please see the white paper "Thermochemistry in Gaussian" (*www.gaussian.com/g_whitepap/thermo.htm*).

The geometry optimization+frequency calculations must be performed using the same model chemistry. However, since both calculations can be performed by a single Gaussian job, this is easy to ensure. For GaussView, select the **Opt+Freq** job type, and in WebMO, select the **Optimize + Vib Freq** job type.

See Chapter 3 for more about geometry optimizations, potential energy surfaces and transition structures. Frequency calculations and their results are discussed in detail in Chapter 4.

EXAMPLE 2.6 FORMALDEHYDE OPTIMIZATION & FREQUENCY CALCULATION

We performed a geometry optimization and a frequency calculation for formaldehyde using the APFD/6-311+G(2d,p) model chemistry, beginning from a structure that we built. The resulting optimized structure is shown at the right below:

Icons for Examining Bonds & Angles

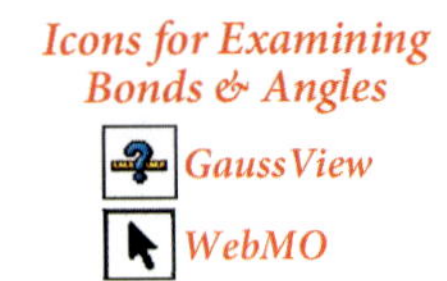

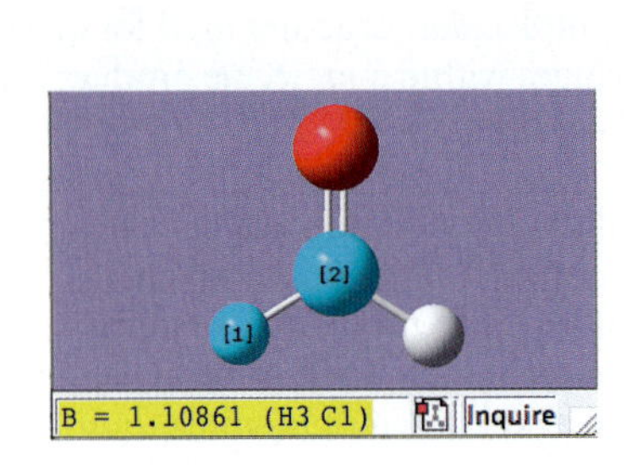

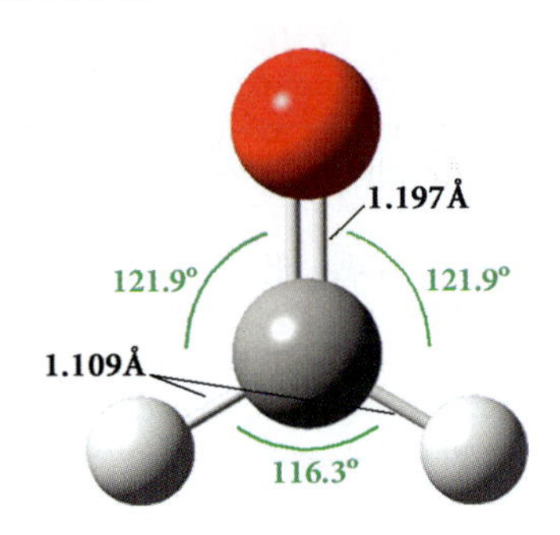

You can view the bond lengths, bond angles and dihedral angles for the optimized molecule using the visualization software's appropriate tool (see the icons in the margin above). GaussView's **Inquire** mode is illustrated at the left in the preceding illustration. We have selected the carbon atom and one of the hydrogen atoms, and the length of the C–H bond is displayed in the bottom left corner of the window.

A saddle point is a location on a surface that is a maximum in one direction and a minimum in another. On a real saddle, the rider sits at a minimum with respect to the cantle and horn (blue line), but at a maximum with respect to the horse's girth (green line).

We could also examine the predicted vibrational frequencies at this point, but we will postpone that until Chapter 4. Our next task is to verify that this structure is a minimum and not a transition structure. This is done by determining the number of *imaginary frequencies:* frequencies with negative values. The presence of exactly one imaginary frequency indicates that the structure is a transition state and not a minimum. More technically, it means that the structure corresponds to a *saddle point* on the potential energy surface.

You should examine the frequency values to determine the number of imaginary frequencies:

- In GaussView, the **Calculation Summary** window includes the number of imaginary frequencies for frequency calculations (see the usage note on the next page).
- In WebMO, the **Calculated Quantities** section includes a table of predicted frequencies, labeled **Vibrational Modes**.

GaussView's imaginary frequency display behavior differs depending on whether you open an output (log) file or a checkpoint file. In the latter case, you must select **Results→Vibrations**, and then click **Ok** in the next dialog (close it afterwards). When you next examine the summary results, the imaginary frequencies field will be filled in.

In this case, there are no imaginary frequencies for our optimized structure, so we have verified that it is a minimum. Chapter 3 discusses this topic in detail (beginning on p. 102).

The dipole moment of our minimized structure is 2.41 debye, slightly different from the value calculated previously for the unoptimized structure.

Thermal Energy Corrections

Our final task is to determine the energy for the molecule including the thermal energy correction. This data is included in the Gaussian output file in the section titled "Thermochemistry." The corrections to the total energy are listed first, followed by computed quantities including the correction. For formaldehyde, the thermal energy correction is about 26.6 millihartrees, and the E^{therm} is -114.41452 hartrees.

```
-------------------
- Thermochemistry -
-------------------
Temperature   298.150 Kelvin.  Pressure   1.00000 Atm.
...
Zero-point correction=             0.026624 Hartree/Particle
Thermal correction to Energy=  0.029493  Add to predicted total energy
...
Sum of electronic and zero-point Energies=  -114.471385
Sum of electronic and thermal Energies=     -114.414517
```

WebMO includes the thermal-corrected energy in its **Calculated Quantities** section, labeled **Internal Energy**:

RAPFD Energy	-114.4440
ZPE	0.026624
Conditions	298.150K,
Internal Energy	-114.4145

Note that we do not use this terminology in this book.

In general, when an energy value is required (rather than an enthalpy or free energy), the thermal-corrected energy is the value that should be used. From this point forward in this book, numerical values for the energy will always include thermal energy corrections unless explicitly stated otherwise, and E will refer to the same energy (which we denoted E^{therm} in the opening section of this chapter).

Clean vs. Optimize

Our next example will illustrate the difference between a full geometry optimization and molecular building software's clean features.

EXAMPLE 2.7 CLEANING VS. OPTIMIZING ANILINE

The aniline molecule* consists of a benzene ring with NH_2 substituting for one of the hydrogen atoms. It is a simple molecule which illustrates some of the pitfalls associated with the structure regularization functionality—**Clean** features—in molecule building programs and why geometry optimizations are necessary.

If you begin building this structure with a benzene molecule, replace one of the hydrogens with an NH_2 group, and then execute the **Clean** function, the resulting structure may be a good starting point for an optimization of the ground state of aniline. However, depending on the specific building process used, it may also be distorted in ways which will ultimately be optimized to a transition state that is nearby on the molecule's potential energy surface.

* IUPAC: phenylamine.

The following illustration shows the structure of the ground state of aniline. We've identified the values of several key structural parameters:

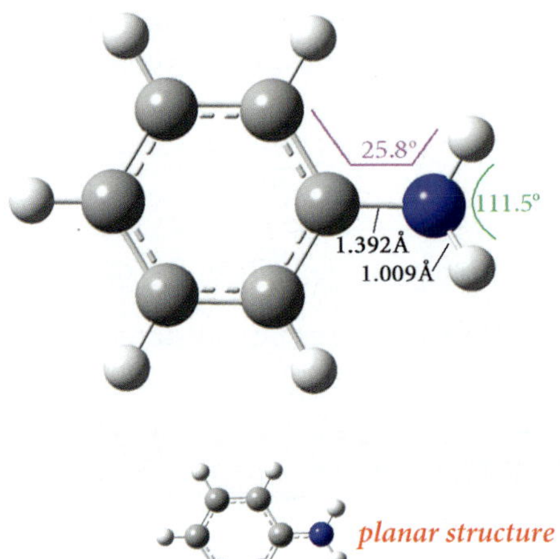

The values in black, green and purple are bond lengths, bond angles and dihedral angles, respectively. Note that the two hydrogen atoms bonded to the nitrogen lie above the plane of the benzene ring.

In order for the default optimization procedure to succeed, the NH_2 group in the starting structure needs to be in the same approximate orientation as the preceding structure. However, two distorted molecular structures commonly result from attempts to build this molecule:

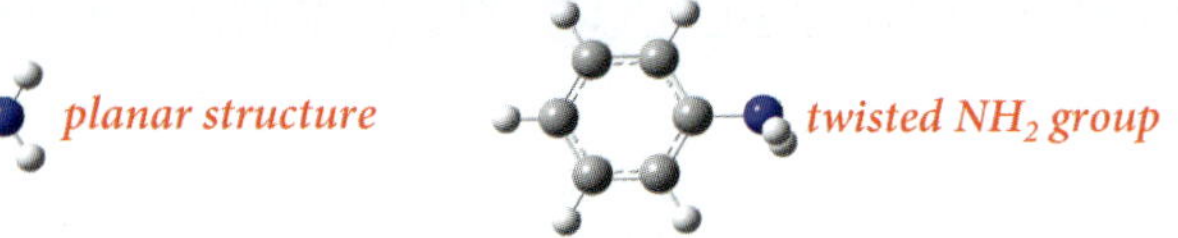

For example, in WebMO, using the **Idealized Clean** function will result in a structure that is planar. Similarly, in GaussView, replacing a hydrogen atom with a tetravalent nitrogen and then cleaning will result in a structure in which the NH_2 group is twisted.

If you perform an ordinary optimization+frequency job for these structures, you will achieve results like the following:

STARTING STRUCTURE	OPTIMIZATION RESULTS		
	# IMAG. FREQS.	STRUCTURE TYPE	C–C–N–H DIHEDRAL
planar	1	transition state	~0°
twisted	1	transition state	~60°
C_s symmetry	0	minimum	25.8°

In two cases, the stationary point located by the optimization will be a transition structure rather than the minimum we wanted. In fact, this is a common occurrence in general.

For this simple molecule, preparing a starting structure with the appropriate features—a non-planar, symmetric geometry—is quite straightforward:

- *WebMO*: Using the **Comprehensive Cleanup Using Mechanics** function will result in an excellent starting structure.
- *GaussView*: You can modify the orientation of the NH_2 group by adjusting the dihedral angle (illustrated in the margin).

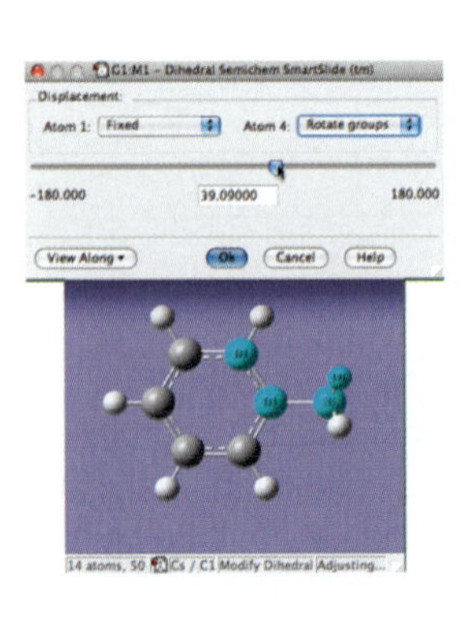

In addition, both distorted starting geometries can be successfully optimized to the minimum by computing the force constants explicitly prior to the optimization (at the cost of some additional CPU time):

- *WebMO*: Check **Calculate Force Constants** on the **Advanced** panel of the **Configure Gaussian Job Options** page.
- *GaussView*: Set **Calculate Force Constants** to **Once** on the **Job Type** panel of the **Gaussian Calculation Setup** dialog.

This feature will be discussed in more detail in Chapter 3.■

EXAMPLE 2.8 ACETALDEHYDE/OXIRANE ISOMERIZATION ENERGY

Our task is to compute the isomerization energy between acetaldehyde and oxirane* at STP with the APFD/6-311+G(2d,p) model chemistry. We will perform optimization+frequency calculations for both systems and then compare their predicted energies, incorporating thermal corrections.

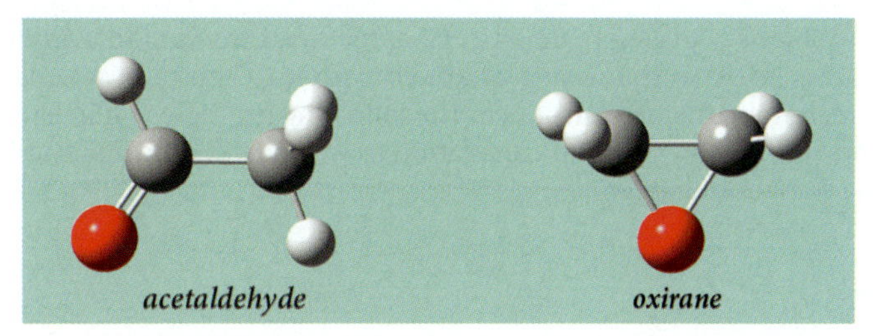

The experimental isomerization energy is 118.07 kJ/mol, equivalently 28.22 kcal/mol (oxirane minus acetaldehyde) [NIST08, Pell65, Wiberg91].

We optimized the two structures with the specified model chemistry, and the two frequency calculations confirmed that the optimized structures are minima. The following table provides the thermal-corrected energy results:

COMPOUND	THERMAL ENERGY (STP)	UNITS
oxirane	-153.64969	hartrees
acetaldehyde	-153.69221	hartrees
ΔE^{STP}	0.04252	hartrees
	26.7	kcal/mol
	111.6	kJ/mol
Experiment	118.07	kJ/mol

Our results are in very good agreement with the observed isomerization energy.

Modeling Large Systems: Gaussian's ONIOM Facility

The molecules we have considered so far have been very small. However, the real systems of research and commercial interest can be much, much larger, especially for ones related to biological substances and processes. Such molecules give rise to additional modeling and job setup considerations:

- The structures for these systems are typically obtained from X-ray diffraction or NMR experiments and are stored in Protein Data Bank (PDB) format files. For structures resulting from X-ray diffraction experiments, this format includes only the heavy atoms within the molecular system, typically organized and tagged by residue. Both GaussView and WebMO can read PDB files, and both have facilities for adding hydrogen atoms to the retrieved structures (GaussView does so automatically by default).
- Most biochemical reactions occur in solution, rather than in the gas phase, making it necessary to include the solvent within the model chemistry. Gaussian provides features for doing so (as we will see in the next example).
- Calculations on very large molecules may not be practical at the desired model chemistry with the available computing resources or within the desired time constraints.

ONIOM References
Definition: [Dapprich99]
Review articles: [Clemente10, Chung15]
Algorithms/Implementation: [Vreven06, Vreven06b, Vreven08, Lipparini11a, Thompson14]

* Also known as ethylene oxide (common name) .

Fortunately, Gaussian's ONIOM facility is available for modeling very large systems at substantially lower computational cost, making them practical for computational study. ONIOM works by dividing the molecule into distinct regions—known as *layers*—which are treated with different levels of accuracy. Accordingly, ONIOM has somewhat more complex job setup requirements than ordinary calculations.

Setting up ONIOM jobs is treated in detail in Chapter 9, but we want to introduce the facility to you early in this book so that it is available to you for each new calculation type and computational technique that we introduce. The following example will illustrate the process of setting up and running an ONIOM calculation as we consider the vitamin E molecule we discussed in the previous chapter.

EXAMPLE 2.9 QM:QM CALCULATIONS ON TWO VITAMIN E STRUCTURES

We examined part of the posited oxidation process of vitamin E in Chapter 1. In this exercise, we will compute the structures for two of the structures in that study: α-tocopherol ortho quinone, with the methide group in the "up" and "down" orientations (see the illustration in the margin).

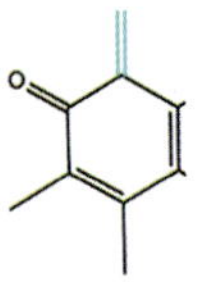

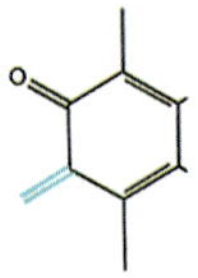

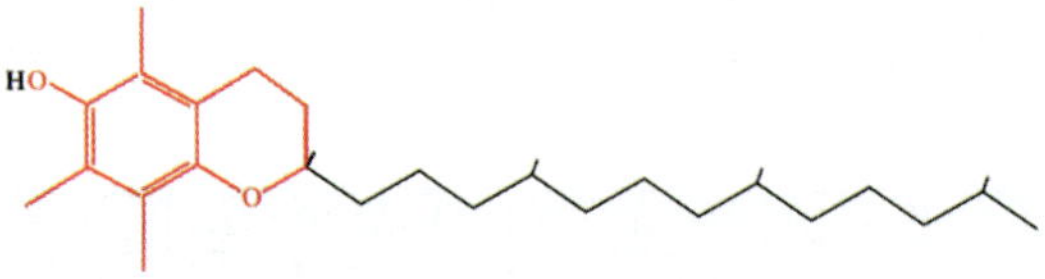

The canonical structure for α-tocopherol—the form of vitamin E that functions in the human body—appears above. We are interested in the -2p-2e product of vitamin E, and specifically in structures where one of the methyl groups adjacent to the oxygen atom within the outermost ring becomes a double-bonded methide group. We will compute the energy difference between these two conformers. The atoms in red above will become the high accuracy layer in the ONIOM calculation.

We will be defining two ONIOM layers for this molecule. The high accuracy layer, which will be modeled using our standard methods: APFD/6-311+G(2d,p). This high layer will be much smaller than the entire molecule in order to reduce the computational requirements for our geometry optimization+frequency calculations. It will consist of all atoms within the two fused rings. This region is displayed in red in the preceding structure.

The low accuracy layer will consist of the remaining atoms in the molecule. We will model it using the PM6 semi-empirical method. Since we are using quantum mechanical methods for both layers in our calculation, it will be a QM:QM calculation. Calculations which use a molecular mechanics method for the low layer are referred to as QM:MM calculations; we will consider such calculations in Chapter 9.

In GaussView, all atoms are initially defined as being in the high ONIOM layer. In order to define our low layer, we select the portion of the tail that we want there. There are many ways to do so; in the illustration following, we use the marquee selection mode to drag a rectangle around some of the desired atoms (the methyl group behind the ring is already selected).

The selected atoms will be highlighted in yellow when we release the mouse button to complete the selection.

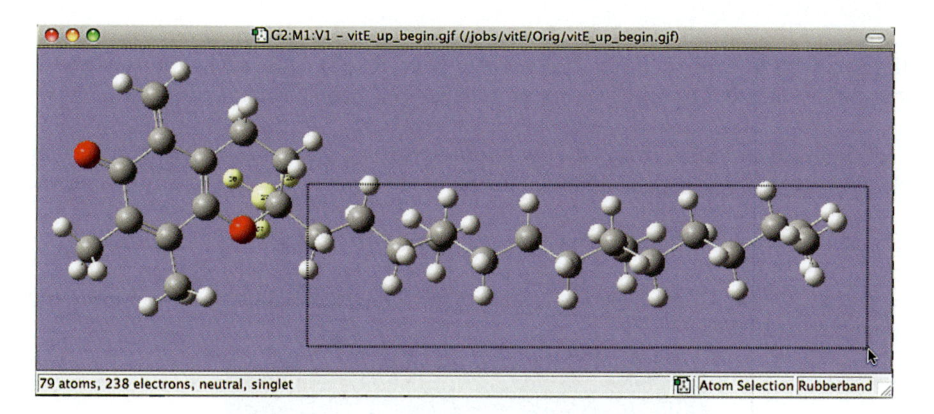

Once we have selected all of the atoms that we want for the low layer, we select **Atom Groups** from the **Edit** menu. We change the **Atom Group Class** to **ONIOM Layer** and then select the **ONIOM Layer (Low)** row by clicking in the corresponding **Group ID** cell. Finally, we use the **Add Selected Atoms to "ONIOM Layer (Low)"** on the **Group Actions** menu to complete the layer assignment:

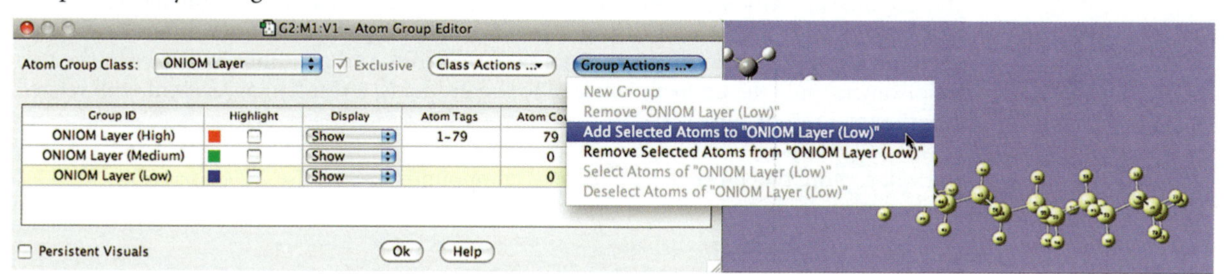

The appearance of the key columns in the **Atom Group Editor** dialog after this operation is illustrated below:

Group ID	Atom Tags	Atom Count
ONIOM Layer (High)	1-26	26
ONIOM Layer (Medium)		0
ONIOM Layer (Low)	27-79	53

We are now ready to set up the Gaussian calculation. We need to select the optimization+frequency job type and specify the desired methods for the ONIOM calculation.

When we open the **Gaussian Calculation Setup** dialog, our first step is to check the **Multilayer ONIOM Model** box (highlighted in red at the upper right in the following illustration).

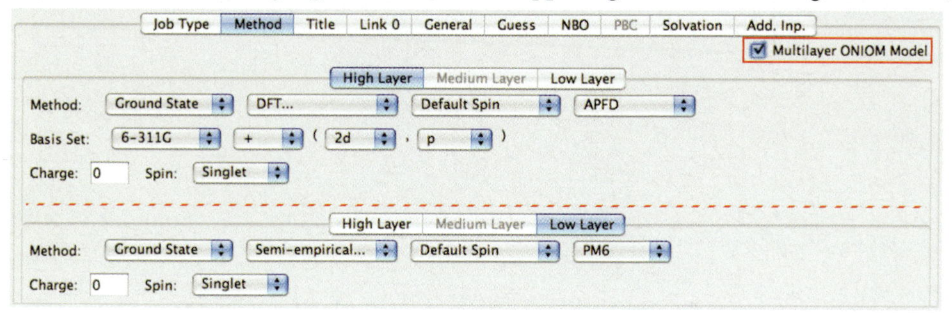

This box must be checked in order to define the multiple methods required by the facility. The **Method** panel is now divided into two subpanels, accessed via the row of buttons above the various popup menus. We select the appropriate model chemistry for the high and low layers.

The following figure superimposes the subpanels for the two active layers into a single illustration, but only a single subpanel will appear in the actual dialog at any given time.

We repeat this process for the "down" structure, and then run the two calculations.

The Gibbs Free Energy is labeled as follows in Gaussian output files: "Sum of electronic and thermal Free Energies"

We will compare the Gibbs Free Energies (G) for the two molecules at room temperature and standard pressure:

FORM	GIBBS FREE ENERGY (hartrees)
up	-615.35442
down	-615.35066
ΔG^{STP}(up-down)	-0.00376

The "methide up" structure is predicted to be lower in energy by about -3.7 millihartrees (about -2.5 kcal/mol or -9.9 kJ/mol) using this model chemistry. An energy difference of this size indicates that, given the experimental conditions and equilibrium between the two regioisomers, only the up form will lead to the products, a result in agreement with related experimental results [Rosenau04]. ■

You will explore two more pairs of related compounds in Exercise 2.7 on page 69.

GOING DEEPER: BOLTZMANN'S FORMULA

Boltzmann's formula gives the ratio of populations of two states *i* and *j*:

$$\frac{N_j}{N_i} = e^{-(\varepsilon_j - \varepsilon_i)/kT}$$

The value of kT is roughly equal to 1 millihartree at room temperature, so under these conditions we can estimate the relative abundance of two isomers by simply raising e to the negative power of the energy difference (in millihartrees). In the case just discussed, this difference is 3.7 mHa: $e^{-3.7}=0.023$. This indicates that the lower energy (methide up) state is >40 times as populated as the higher energy (methide down) state. This demonstrates the convenience of discussing results using units of millihartrees.

Exercises

Unless stated otherwise, use the APFD/6-311+G(2d,p) model chemistry for all calculations. Optimized structures should be located for all molecules, and a frequency calculation should verify that they are minima.

WebMO

WebMO Technique:

If you will want to examine the MOs from a geometry optimization, you must include **Pop=Full GFInput** as **Additional Keywords** on the **Advanced** panel when setting up the job.

EXERCISE 2.1 COMPARING ETHYLENE AND FORMALDEHYDE

Ethylene* is a molecule that is similar to formaldehyde. In fact, the two compounds are isoelectronic. In the case of ethylene, the oxygen in formaldehyde is replaced by a carbon with two additional hydrogens attached to it.

Optimize the structure of ethylene. Then, compare the dipole moments of ethylene and formaldehyde. Examine the HOMO and LUMO in both molecules. How do you explain the differences you see?

SOLUTION

We'll begin with the HOMO and LUMO for the two molecules:

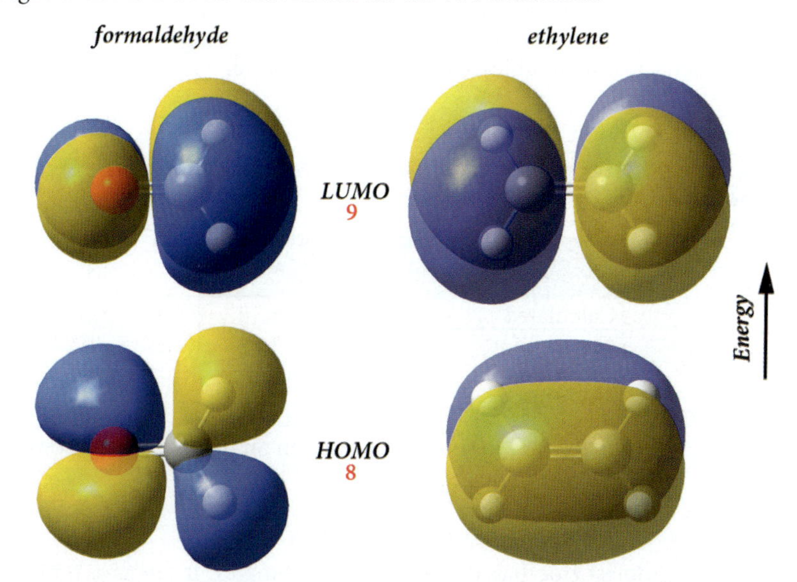

Since both molecules have the same number of electrons, the orbital numbered 8 is the HOMO, and the one numbered 9 is the LUMO in both cases. However, they are not the same type orbitals. In ethylene, both the HOMO and LUMO are formed primarily from p_x orbitals from the two carbon atoms, lying above and below the C-C bond. In the HOMO, the orbitals have like signs, and so they combine to form a bonding π molecular orbital. In contrast, in the LUMO, they have opposite signs, indicating that they combine to form an antibonding π* molecular orbital.

In formaldehyde, orbitals 7 and 9 (HOMO-1 and LUMO) comprise a similar pair. Orbital 7 is a bonding π orbital (see the illustration in the margin), and orbital 9 is a π*.

The HOMO in formaldehyde is an orbital formed from p_y orbitals from the carbon atom and the oxygen atom and from the s orbitals on the hydrogen atoms. The contributions from the carbon atom and the oxygen atom are situated along the double bond. In contrast, the HOMO in ethylene was perpendicular to this bond.

* IUPAC: ethene.

This difference is due to the two lone pairs on the oxygen. Of the six valence electrons on the oxygen atom, two are involved in the double bond with the carbon, and the other four exist as two lone pairs. In Chapter 4, we'll examine the IR spectra for these two molecules. The orbitals suggest that we'll find very different frequencies for the two systems.

Here are the dipole moments for the two molecules:

COMPOUND	DIPOLE MOMENT (debye)
formaldehyde	2.41
ethylene	0.0

While the oxygen atom induces a dipole moment in formaldehyde, the center of inversion in ethylene results in no dipole moment.■

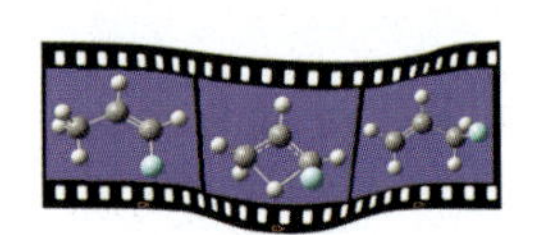

EXERCISE 2.2 OPTIMIZING CHROMIUM HEXACARBONYL

Optimize the structure of chromium hexacarbonyl.* Be sure that you begin with a structure having the appropriate symmetry. See the website for a video showing how to use GaussView's point group symmetry feature to quickly build this compound.

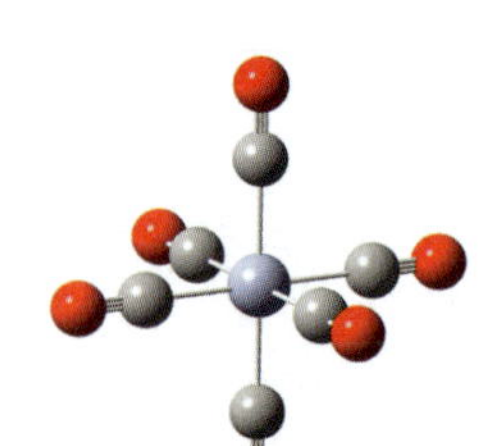

SOLUTION

The following table presents the results of our optimization as well as the experimental electron diffraction results:

	BOND LENGTHS (Å) Cr–C	C–O
Calculated	1.89	1.14
Experiment	1.91	1.14

The APFD/6-311+G(2d,p) model chemistry predicts the structure of this molecule with high accuracy. See [Jost75] and [Ishikawa07] for more information on this interesting compound.■

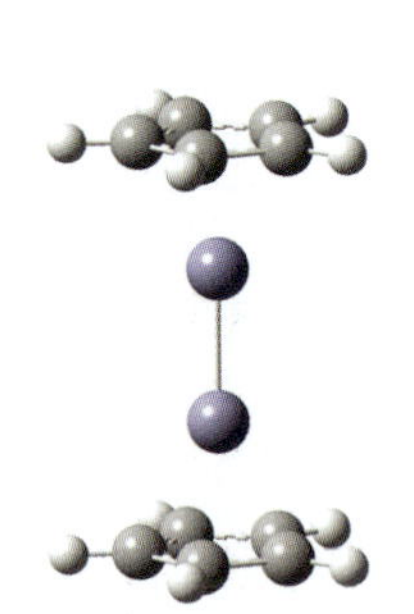

EXERCISE 2.3 ATOMIC CHARGE ANALYSIS FOR DIZINCOCENE

Dizincocene† is the simplest zinc-based dimetallocene complex. It serves as an interesting starting point for investigating the many similar species involving metal-metal bonded Zn_2^{+2} (e.g., decamethyldizincocene).

Predict the atomic charges for dizincocene using Natural Bond Orbital analysis (**Population=NBO**). Compare the predicted charges to the Mulliken population analysis results.

Building and optimizing dizincocene can be quite challenging. See the website for a video illustrating the building process and for hints on optimizing this structure. In addition, we have provided an optimized structure for this molecule in the file *s2_03_molecule.gjf* which you can use if desired.

* IUPAC: hexacarbonylchromium.

† IUPAC: bis(η5-2,4-cyclopentadien-1-yl])di-, (Zn-Zn) Zinc.

SOLUTION

The predicted atomic charges for dizincocene are listed in the following table:

ATOM	MULLIKEN	NBO
Zn	0.55	0.57
C	-0.22	-0.35
H	0.11	0.24

Generally, one expects the zinc atoms in this system to have a charge of 1. However, the NBO analysis shows the net charge to be about 0.6. This differentiation from the expected charge of zinc is most likely a result of the covalent bonding in the system. The Mulliken method underestimates the charges on the carbon and hydrogen atoms.■

EXERCISE 2.4 COMPARING ETHYLENE AND FLUOROETHYLENE

Replacing one of the hydrogens in ethylene with a fluorine atom produces fluoroethylene.* Determine the effect of this substituent on the molecule's structure.

SOLUTION

The following table compares the structures of the two molecules:

COORDINATE	ETHYLENE	FLUOROETHYLENE
C-C bond length	1.33 Å	1.32 Å
C-R bond length	1.09 Å	1.34 Å
C-C-R bond angle	121.7°	122.0°
C-C-H bond angle	121.7°	125.8°

Substituting the fluorine for a hydrogen results in a longer bond length for that substituent with the carbon. It also produces a slight shortening of the C-C bond, resulting in a stronger bond, and larger bond angles for both atoms with the adjacent carbon. The latter has the effect of bringing the atoms closer together on this end of the molecule.■

EXERCISE 2.5 TWO MORE HETEROSUBSTITUTED ETHENE RADICALS

Continue the study of spin polarization in heterosubstituted allyl radicals that we began earlier (see page 55). Perform an Opt+Freq calculation for the substituents O (vinoxy radical†) and Mg. Compute the atomic spin densities for the heavy atoms and compare them to the results for the previous three substituents. What trends do you observe? How do you account for them? How well do the predicted atomic spin density values for vinoxy radical agree with the experimental values of 0.7 for the terminal carbon atom and 0.2 for the oxygen atom?

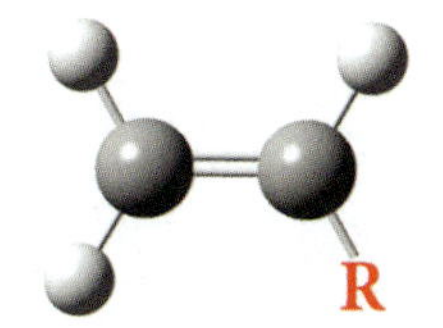

Visualize the singly occupied MO for these two compounds, comparing both of them to ones for the substituents. Remember to request biorthogonalized orbitals in order to make identifying the singly occupied orbitals straightforward. How do the orbital forms relate to the trends you have observed in the atomic spin density?

* IUPAC: fluoroethene.

† IUPAC: oxidanyl ethene.

SOLUTION

The following table gives the computed atomic spin densities for each atom:

R	ELECTRONEGATIVITY	TERMINAL C	CENTRAL C	SUBSTITUENT
CH_2	2.5	0.73	-0.30	0.73
Be	1.5	-0.01	0.04	0.96
S	2.5	0.46	-0.21	0.80
Mg	1.2	0.18	-0.05	0.84
O	3.5	0.89 exp. 0.7	-0.15	0.35 exp. 0.2

In the magnesium substituent, both the Mg and the terminal carbon atoms have an unpaired electron, as was the case for Be. However, a bit more of the atomic spin density remains near the central carbon atom in the case of Mg. In vinoxy radical, the atomic spin density is pushed away from the substituent. These trends in the atomic spin density values follow the varying electronegativity of the substituent heavy atom. Finally, the predicted atomic spin density values for vinoxy radical are in good agreement with experiment.

Here are the singly occupied molecular orbitals for the five compounds:

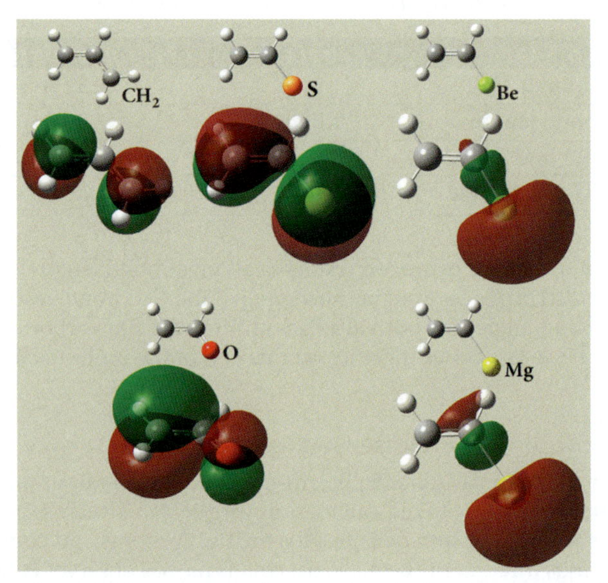

The orbitals in the Be and Mg compounds are quite similar. The orbital for vinoxy radical resembles that of allyl radical and the S compound, but with less evenly balanced lobes: the larger lobe is on the terminal carbon atom.■

EXERCISE 2.6 THE GROUND STATE OF O_2

The photo below depicts the well-known experiment of pouring liquid oxygen between the poles of a powerful magnet. Since oxygen is paramagnetic, the liquid is attracted by the magnetic field at the poles and "sticks" to them. In contrast, if you pour, say, liquid nitrogen at the same spot, it simply falls through.

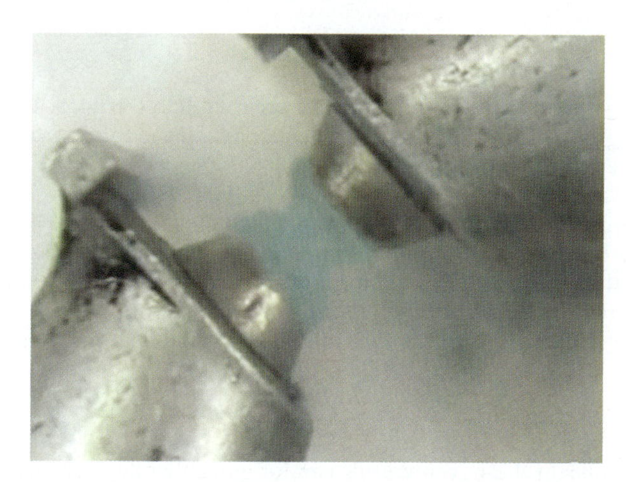

Photo courtesy Prof. Robert Burk of Carleton University (Ottawa, Canada).

You can view the entire experiment on YouTube at *www.youtube.com/watch?v=Isd9IEnR4bw* [Burk07].

Oxygen's paramagnetic nature results from the electronic state of the molecule's ground state. Determine the spin multiplicity of the ground state of oxygen by optimizing the singlet and triplet forms in order to determine which one is lower in energy. Perform both calculations using our standard model: APFD/6-311+G(2d,p), and predict the energy at STP.

SOLUTION

The triplet state is the ground state. The following table summarizes our energy results (which include the thermal correction):

FORM	THERMAL ENERGY (hartrees)
singlet	-150.18769
triplet	-150.25219
ΔE^{STP}(triplet-singlet)	0.06151

The triplet form is significantly lower in energy than the singlet form.

We'll have more to say about this molecule in Advanced Example 2.10 (p. 72).■

EXERCISE 2.7 ONIOM CALCULATIONS ON VITAMIN E-RELATED MOLECULES

In this exercise, you will examine the -2p-2e products from two modified versions of α-tocopherol. First we will change the 6-membered aliphatic ring to a 5-membered aliphatic ring (modification I). This is now a substituted 2,3-dihydrobenzofuran (DHBF). Next, we will replace the hydrogen atoms at carbon 3 with methyl groups (modification II).

Here are the structures of the ortho-quinone methide (OQM) intermediates of these two compounds as compared to the original vitamin E structure.

Original vitamin E structure *Modification I* *Modification II*

In the center structure, a five-membered ring replaces the six-membered ring, with the topmost carbon in the furan group bonded to two hydrogen atoms. In the third structure (on the right), the hydrogens are replaced with methyl groups.

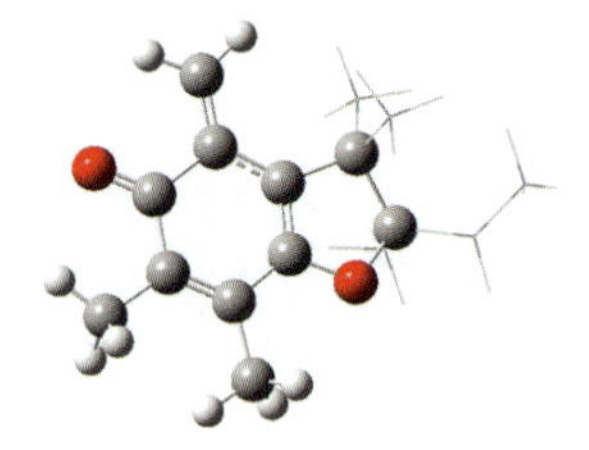

Compare the Gibbs Free Energies at STP of these systems with the methide group (=CH_2) placed at carbon 4 (up) and carbon 6 (down). The up positions are shown in the figure above. Define the ONIOM layers in the same way as we did for α-tocopherol in Example 2.9, with the high layer consisting of only the ring atoms. Place the two methyl groups attached to the ring atom in modification II into the low layer (see illustration in the margin, which is truncated for clarity).

SOLUTION

The optimization+frequency jobs on these compounds run for quite a while. Here are the predicted Gibbs Free Energies (G) of the various molecules at room temperature:

	G (hartrees)		
MOLECULE	UP	DOWN	ΔG^{STP}(UP-DOWN)
α-tocopherol	-615.35442	-615.35066	-3.7 millihartrees ≈ -2.4 kcal/mol ≈ -9.9 kJ/mol
modification I	-576.08791	-576.09168	3.8 millihartrees ≈ 2.3 kcal/mol ≈ 9.9 kJ/mol
modification II	-576.05416	-576.05653	2.4 millihartrees ≈ 1.5 kcal/mol ≈ 6.2 kJ/mol

Both modifications result in the down products being favored, reversing the results we had found for vitamin E itself. Modification 1 results in the down form being favored by ~3 millihartrees. An energy difference of this magnitude suggests that only the down product will form (see the "Going Deeper" box on p. 64). Modification 2 results in the down product being favored by ~2.4 millihartrees. Therefore, under these experimental conditions with two regioisomers in equilibrium, both up and down products would be expected to form, although the down form will still dominate the population (~92% to 8%). These predictions are in full agreement with experimental studies of model compounds that have the same structural features [Rosenau04].*

Finally, notice that the essential difference between modification I and II is the presence of two methyl groups. Treating these groups at the less expensive, low level in the ONIOM calculation succeeds in calculating a significant energy difference between the two forms. Thus, we have verified that this ONIOM layer partitioning is a valid model for the real compound.■

* While this analysis ignores kinetic factors, they are negligible for this case.

Advanced Topics & Exercises

The information in this section is designed to take you beyond the basic level represented by the basic chapter and its examples and exercises. You will find a section like this one at the end of most chapters of this book. Some advanced topics are introduced by brief sections of explanation (with examples), but a good deal of the material is contained within the exercises.

Wavefunction Stability

In this section, we introduce stability calculations. A stability calculation determines whether the wavefunction computed for the molecular system is stable or not: in other words, whether there exists a lower energy wavefunction corresponding to a different solution of the SCF equations (see Chapter 10). This analysis involves relaxing the constraints placed on the wavefunciton by default: for example, allowing the wavefunction to become open shell or reducing the symmetry of the orbitals.

> **Stability References**
> Definition/Implementation: [Seeger77]
> Review articles: [Schlegel91a]

If the wavefunction turns out to be unstable, then the calculation you are performing is fundamentally flawed, and any energy comparisons made or other conclusions drawn from its results will be invalid.

Hartree-Fock and DFT wavefunctions can be tested for stability. Instability types are labeled using the terms "RHF" and "UHF," but they apply to density functional methods as well.

Gaussian reports two types of instabilities. The following output line indicates an "RHF-to-UHF" instability:

```
The wavefunction has an RHF -> UHF instability
```

This output indicates that the current closed-shell wavefunction is unstable and that there is a lower energy open-shell wavefunction. Note that the form of the lower energy wavefunction is not known. It may be an open-shell singlet (e.g., a biradical), it may be a triplet, etc. The system must be studied further in order to determine the proper wavefunction.

The following output line indicates that the current wavefunction is unstable:

```
The wavefunction has an internal instability
```

This result can occur for closed-shell or open-shell wavefunctions. It indicates that there is more than one solution to the SCF equations for the system, and that the calculation procedure converged to a solution which is not the minimum (often a saddle point in wavefunction space). This type of instability is sometimes referred to as an RHF-to-RHF instability or a UHF-to-UHF instability, depending on whether the wavefunction is closed shell or open shell.

Gaussian stability calculations can test the stability of a wavefunction and optionally reoptimize the wavefunction to the lower energy solution if any instability is found. When we speak of *reoptimizing* the wavefunction, we are not referring to a geometry optimization, which locates the lowest energy conformation near a specified starting molecular structure. Predicting an energy with Hartree-Fock or DFT theoretical methods involves finding the lowest energy solution to the SCF equations. Stability calculations ensure that this optimized electronic wavefunction is a minimum in wavefunction space—and not a saddle point—and it is an entirely separate process from locating minima or saddle points on a nuclear potential energy surface.

In general, stability calculations tell you *only about the wavefunction under consideration*. Identifying an instability may provide a direction for further study, but having a stable wavefunction does not guarantee that it is the electronic state that you want. For example, you can locate a stable open shell singlet state for a biradical species using stability calculations. Nevertheless, it will not be the lowest energy electronic state when there is a triplet state with lower energy.

When Should Wavefunction Stability Be Tested?

The wavefunction stability for unknown systems should always be tested. For more information on stability calculations, see [Schlegel91].

ADVANCED EXAMPLE 2.10 OXYGEN STABILITY CALCULATIONS

In order to illustrate how stability calculations work, we'll run a stability calculation on singlet oxygen, which we already know is not the ground state.

You can specify a stability calculation directly via the **Job Type** panel in GaussView (at the left in the illustration below). In WebMO, you must specify the **Stable** keyword to the prompt presented by the **Other** item on the **Calculation** menu. The final appearance of the **Job Options** panel is illustrated on the right below.

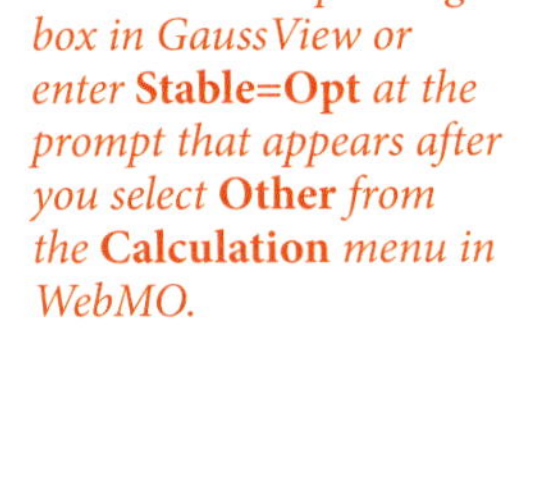
If you want to reoptimize the wavefunction if there is an instability, then check the corresponding box in GaussView or enter **Stable=Opt** *at the prompt that appears after you select* **Other** *from the* **Calculation** *menu in WebMO.*

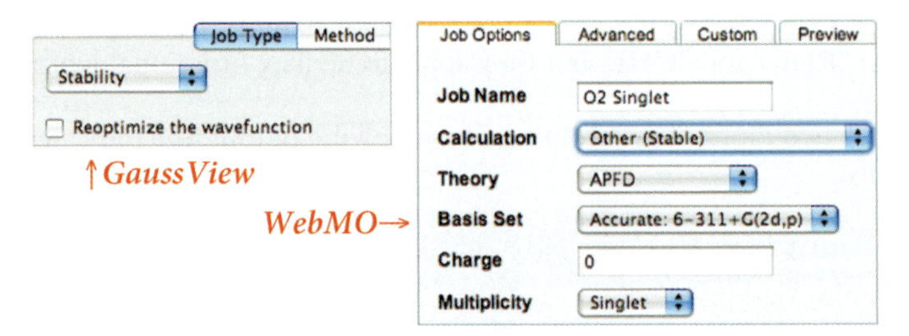

When we run the calculation and examine the output file, we find an "RHF-to-UHF" instability. This indicates that there is an unrestricted wavefunction which is lower in energy than the restricted wavefunction. We are not surprised by this result given that the ground state of the molecule is a triplet.

When we run a stability calculation on the true, triplet ground state of molecular oxygen, we find that the wavefunction is stable:

```
The wavefunction is stable under the perturbations considered.
```

We know that we have the ground state wavefunction for the molecule with this model chemistry.■

ADVANCED EXERCISE 2.8 OZONE WAVEFUNCTION STABILITY

Ozone* is a singlet, but it has an unusual electronic structure and is thus often difficult to model. Devise and run calculations which will determine the lowest energy electronic state for ozone. Use our standard model chemistry and the experimental geometry:

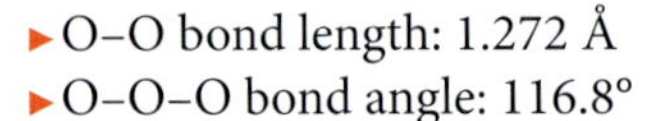

- O–O bond length: 1.272 Å
- O–O–O bond angle: 116.8°

* IUPAC: trioxygen.

SOLUTION

A stability calculation using the default method finds an "RHF-to-UHF" instability, and the reoptimization of the wavefunction leads to a solution with an energy of -225.31729 hartrees.

At this point, we might expect that running an unrestricted calculation in the first place would have been sufficient. However, when we perform a **Stable=Opt** calculation with an unrestricted wavefunction, the predicted wavefunction is again found to be unstable, and reoptimization of the wavefunction leads to the same lower-energy electronic state as was found by the restricted calculation. Thus, it is necessary to modify the default electronic configuration in order to specify the proper ground state of ozone even for an unrestricted calculation. This is not surprising given the known significantly biradical character of ozone resulting from the coupling of the singly-occupied p orbitals on the terminal oxygen atoms.

This example illustrates the general necessity of modifying the default electronic configuration for open shell singlets to avoid the wavefunction converging to the closed shell solution.■

Comparing Model Chemistries for Accuracy

ADVANCED EXERCISE 2.9 BOND ENTHALPIES OF SECOND & THIRD ROW HYDRIDES

In this exercise, we compute the bond enthalpies for the hydrides of the atoms in the second and third rows of the periodic table (excluding the noble gases). We will be comparing the values predicted for each of the molecules in the hydride series. We will also use this exercise as an opportunity to compare two different model chemistries.

The enthalpy is one of the properties predicted by a frequency calculation, as in the following Gaussian output file excerpt:

```
Temperature   298.150 Kelvin.  Pressure    1.00000 Atm.
Zero-point correction=                     0.006441
Thermal correction to Energy=              0.008801
Thermal correction to Enthalpy=            0.009746
Thermal correction to Gibbs Free Energy= -0.010357
Sum of electronic and zero-point Energies=    -38.488560
Sum of electronic and thermal Energies=       -38.486199
Sum of electronic and thermal Enthalpies=     -38.485255
Sum of electronic and thermal Free Energies= -38.505357
```

WebMO includes this information within the **Computed Quantities** section of the results page (as illustrated in the margin).

APFD Energy	-38.4950005709 Hartree
ZPE	0.006441 Hartree
Conditions	298.150K, 1.00000 atm
Internal Energy	-38.486199 Hartree
Enthalpy	-38.485255 Hartree
Free Energy	-38.505357 Hartree
C_v	4.968 cal/mol-K
Entropy	42.309 cal/mol-K

The bond enthalpy is computed as the difference of the hydride enthalpy and the sum of the atoms' enthalpies:

$$\text{Enthalpy}^{\text{bond}} = \text{Enthalpy}^{\text{hydride}} - (\text{Enthalpy}^{\text{atom}} + \text{Enthalpy}^{\text{H}})$$

In order to complete this exercise, you will need to do the following:

- Determine the optimized geometry and frequencies for the hydride compounds H_2, LiH, ..., SH and HCl (excluding He and Ne).
- Run a frequency calculation for the atoms from hydrogen through chlorine (again excluding helium and neon).
- Run each calculation using the following two model chemistries: HF/6-31G(d) and APFD/6-311+G(2d,p).

Some of these atoms and molecules have unpaired electrons, so you will need to specify the proper spin multiplicity for each calculation.

Compare your results to the experimental enthalpies. Remember these are not typical bond energies (C-H does not represent one of the bonds in methane). The data with which to compare comes from spectroscopic measurements. A nice online source for the data has been created by Prof. Mark Winter of Sheffield University as the WebElements database:

www.webelements.com/hydrogen/bond_enthalpies.html

The data itself comes from [CRC00].

Present your results as a plot and comment on the trends that you observe, both in terms of element progression across the periodic table and model chemistry accuracy.

SOLUTION

The following plot reports the experimental (red diamonds) and predicted (lines) bond enthalpies. The grey vertical lines in the plot indicate the row boundaries within the periodic table. Note that the graph's vertical axis begins at around -600 kJ/mol and increases to 0.

The experimental (and predicted) bond enthalpies exhibit similar trends as we move across the second and third rows of the periodic table. The highest value is at the second element in the row, with an increase from the first element. The middle of the row has a series of three values that are similar, making the plot rather flat in these regions. The end of each row sees two additional decreases in the bond enthalpy, with the lowest value occurring for the element just before the noble gas.

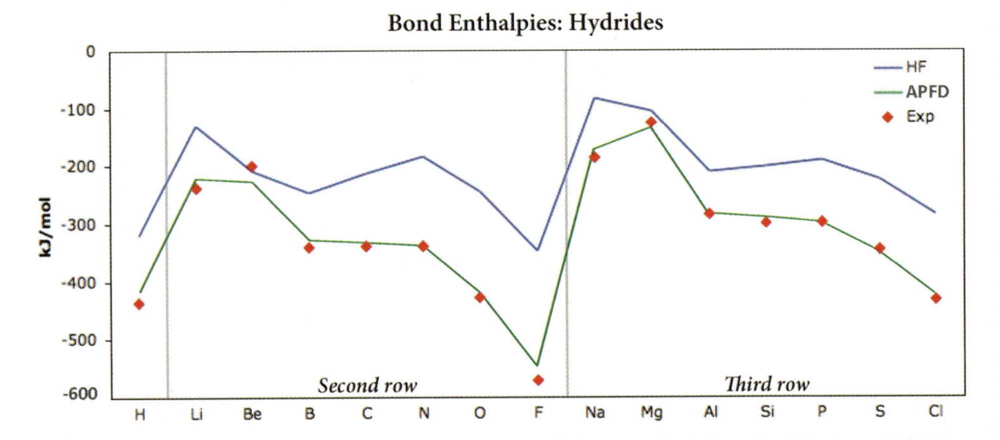

Thus, after the decrease from the first to second elements, the bond strength in the hydride compounds generally increases as we move across the periodic table. The weakest bond is in the second compound in each series, and the strongest is in the final compounds. There is also a dramatic decrease in bond strength as we move from one row to the next—hydrogen to lithium and fluorine to sodium—corresponding to starting to fill a new electron shell. The dip in bond enthalpy in the middle of each row corresponds to beginning a new subshell.

The plot of the bond enthalpies has a similar overall shape for both rows, but the values for the third row are higher than the corresponding ones for the second row, often significantly so. This effect arises from the additional shielding from the nucleus by the filled second shell for the third row compounds.

The APFD/6-311+G(2d,p) results are generally in excellent agreement with the experimental data except for the Be compound (the coincidence of the experimental value with the Hartree-

Fock prediction is merely fortuitous). While the Hartree-Fock results are quantitatively quite inaccurate, they nevertheless reproduce the general shape of the actual results with the exception of the higher value for the second element in each row.

Why is the beryllium compound difficult for our standard DFT method with a large basis set? The electronic configuration for this element is unusual in that there is significant mixing between the 2s and 2p orbitals in the ground state (although it is formally 1s2 2s2). This means that it is poorly described by single determinant methods like Hartree-Fock and Density Functional Theory (see Chapter 10 for details about this topic).

These trends can be illustrated in another manner via mapped electron density plots. For surfaces, the shape of the object is determined from the *isodensity* of the electron density. An isosurface is one where the plotted property has the same value at all points on the surface defined by equal values of the electron density; an isodensity surface is the surface consisting of the points where the electron density has some specific value (e.g., 0.002 electrons/bohr^3). The colors on the surface are determined from the value of the electrostatic potential at that point (where red is the most negative, and dark blue is the most positive).*

Here are the electrostatic potential-mapped isodensities for the compounds we are considering:

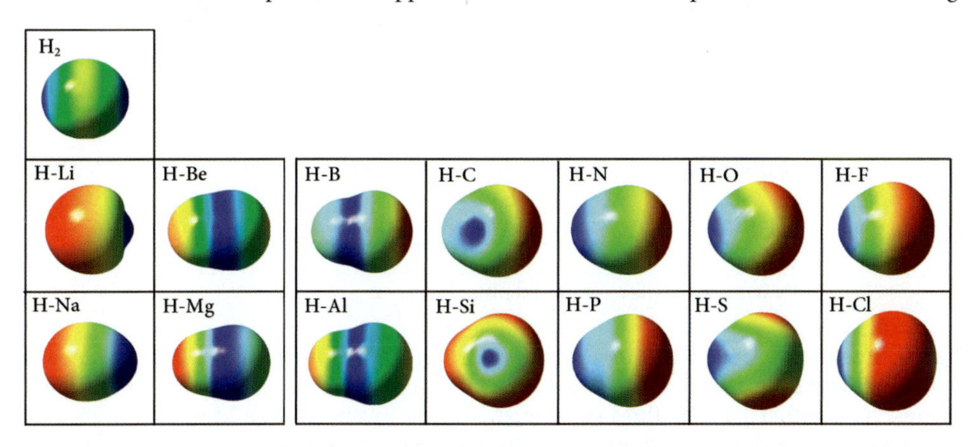

The hydrogen atom is on the left in each surface. The H_2 surface illustrates the covalent nature of the bond in this compound; there is relatively little variation in electrostatic potential across the surface. In contrast, the bonds in the other hydride compounds are ionic, with the negative electrostatic potential localized on the hydrogen atom and the positive electrostatic potential on the substituent for the elements at the beginning of each row. As we move across each row, the positive and negative electrostatic potential gradually swap locations, culminating in a large negative potential on the heavy atom for the final compounds in each row.■

ADVANCED EXERCISE 2.10 BUTANE ENTHALPY OF ISOMERIZATION

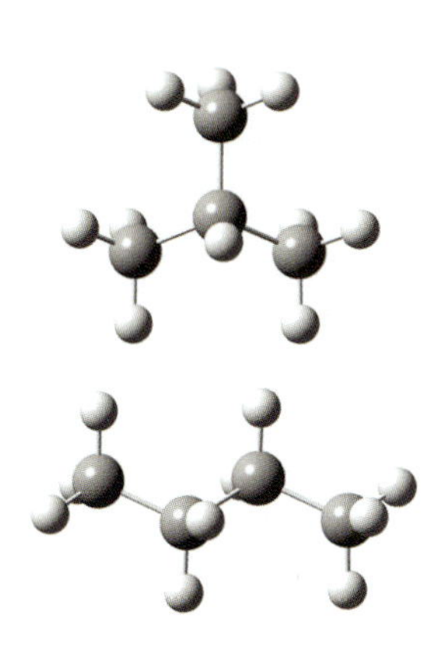

In this exercise, we will compare the accuracies of several theoretical methods. We will ignore basis set effects for this exercise and use the same modest sized basis set with all methods.

Compute the enthalpy of isomerization between two forms of C_4H_{10}: iso-butane and n-butane using the following theoretical methods:

- Hartree-Fock
- B3LYP

* Techniques for producing such visualizations for both GaussView and WebMO are described in Chapter 6 (page 231).

- APFD
- MP2
- CCSD [Bartlett78, Cizek69, Purvis82, Scuseria88]
- CBS-QB3 [Montgomery99, Montgomery00]

Use the 6-31G(d) basis set [Ditchfield71, Hehre72, Hariharan73] for all calculations except for CBS-QB3; the latter is a compound method that does not require a basis set specification.

How does each model chemistry compare with the observed enthalpy difference (iso-butane minus n-butane) of -8.6 kJ/mol [NIST08, Pittam72]? How computationally expensive are the various calculations?

SOLUTION

We optimized the two structures at each model chemistry and then performed a frequency calculation. The results of our calculations are given in the table below, along with an MP4 calculation for reference:

	$\Delta H^{ISO\text{-}N}$		RELATIVE
METHOD	**kJ/mol**	**kcal/mol**	**TIME**
HF/6-31G(d)	-2.84	-0.68	1
B3LYP/6-31G(d)	-3.39	-0.81	3.7
APFD/6-31G(d)	-5.11	-1.22	3.7
MP2/6-31G(d)	-8.51	-2.04	5.7
MP4(SDQ)	-6.84	-1.63	133.1†
CCSD	-6.68	-1.60	788.5†
CBS-QB3	-8.08	-1.93	22.1
Experiment	-8.60	-2.06	

†Frequencies were computed numerically.

Neither Hartree-Fock nor B3LYP does particularly well on this problem. APFD does somewhat better, but still differs from experiment by >3 kJ/mol. A larger basis set does not improve the results of either DFT method very much; e.g., B3LYP/6-311+G(2d,p) predicts a value of -3.68 kJ/mol, and APFD with the same basis set predicts a value of -5.44 kJ/mol. Clearly, electron correlation is important for this problem.

The MP2 electron correlation method predicts a value that is very close to experiment, but this turns out to be fortuitous. The MP2/6-311+G(2d,p) model chemistry predicts an isomerization energy of -9.94 kJ/mol. The larger basis set again lowers the predicted enthalpy difference, causing MP2 to overshoot experiment.

The high accuracy methods including electron correlation, MP4(SDQ) and CCSD, are fairly accurate even with the small basis set (more so than MP2 with the larger basis set). A larger basis set would probably improve these values, but such calculations would be quite expensive.

The CBS-QB3 high accuracy compound method seeks to extrapolate a coupled cluster calculation to the basis set limit (at significantly lower cost). For this problem, the model produces excellent agreement with experiment.

The three smallest calculations run quite rapidly. On our single processor reference system, the MP2 calculation requires a little over three minutes of CPU time. The MP4(SDQ) and CCSD calculations take much, much longer because these methods must perform numerical frequency calculations; their relative times represent worst case performance. Finally, the CBS-QB3 method has significantly lower CPU requirements than MP4(SDQ) and CCSD.■

ADVANCED EXERCISE 2.11 MALONALDEHYDE OPTIMIZATION

Malonaldehyde* provides one of the simplest examples of intramolecular hydrogen bonding. Its experimental structure is given in the following illustration [Baughcum81, Baughcum84]. Bond lengths are in angstroms (black), and bond angles are in degrees (green):

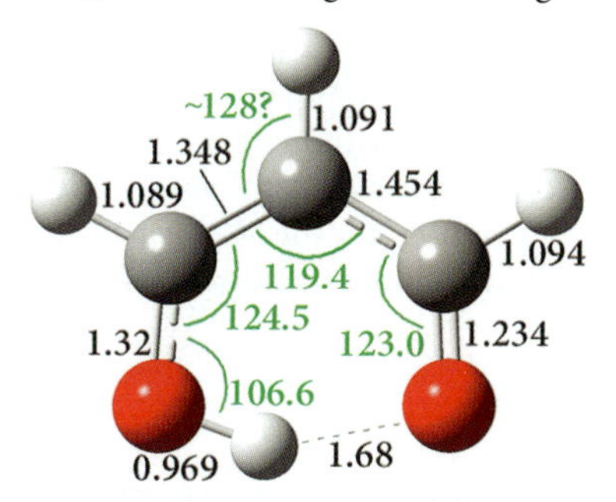

What level of theory is required to produce an accurate structure for this molecule? When we speak of "accurate" geometries, we generally refer to bond lengths that are within about 0.01-0.02 Å of experiment and bond and dihedral angles that are within about 1-2° of experiment (with the lower ends of these ranges being more desirable).

Use the 6-311+G(2d,p) basis set for all your calculations.

SOLUTION

We ran optimization+frequency calculations using the PM3 and PM6 semi-empirical methods and the HF, APFD and MP2 methods with the specified basis set. All of the optimized structures were verified as minima by the frequency calculations.

The following table summarizes the results of the five calculations we ran (bond lengths are in angstroms and bond angles are in degrees). Values outside of the desired agreement with experiment are indicated in red:

PARAMETER	PM3	PM6	HF	APFD	MP2	EXPERIMENT
B1 (H–O)	0.97	1.10	0.96	1.01	1.00	0.969
B2 (O–C)	1.34	1.33	1.31	1.31	1.32	1.320
B3 (C–C)	1.36	1.36	1.34	1.36	1.36	1.348
B4 (C–C)	1.46	1.45	1.45	1.43	1.44	1.454
B5 (C–O)	1.22	1.24	1.20	1.24	1.24	1.234
B6 (O···H)	1.83	1.68	1.89	1.61	1.67	1.680
B7 (C–H)	1.09	1.09	1.08	1.09	1.09	1.089
B8 (C–H)	1.09	1.09	1.07	1.08	1.08	1.091
B9 (C–H)	1.10	1.10	1.09	1.10	1.10	1.094
A1 (H–O–C)	109.3	112.1	109.6	105.4	105.4	106.3
A2 (O–C–C)	124.8	123.9	126.2	123.6	123.9	124.5
A3 (C–C–C)	121.8	120.4	121.2	118.8	119.6	119.4
A4 (C–C–O)	124.6	121.0	124.2	123.2	123.3	123.0
A5 (C–C–H)	121.2	121.1	119.3	120.4	119.9	128.1?

The PM3 semi-empirical method [Stewart89] differs significantly from experiment for several structural parameters. The newer PM6 method [Stewart07] considerably improves on these results, but still predicts an angle value that deviates quite a bit from experiment.

* IUPAC: propanedial.

The Hartree-Fock geometry also differs significantly from the experimental structure. This is most noticeable in the long hydrogen bonding distance and the corresponding errors in the ring's internal bond angles.

In contrast, the APFD and MP2 structures are in excellent agreement with experiment. The only major discrepancy comes with the C–C–H bond angle involving the hydrogen atom on the central carbon atom. However, the reported experimental value is known to be quite uncertain. The APFD method successfully models the effects of electron correlation for this system at substantially lower computational cost than MP2 (the APFD calculations require only about 20% of the CPU resources used for the MP2 calculations). See the original reference for more information about this interesting compound [Frisch85c].■

Exploring Basis Sets in Depth

ADVANCED EXERCISE 2.12 **THE PO BOND LENGTH: THE BASIS SET LIMIT**

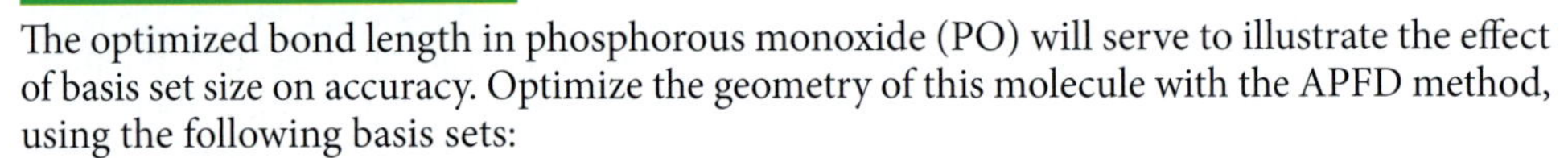

The optimized bond length in phosphorous monoxide (PO) will serve to illustrate the effect of basis set size on accuracy. Optimize the geometry of this molecule with the APFD method, using the following basis sets:

- 6-31G(d)
- 6-311G(d)
- 6-311G(2d)
- 6-311G(2df)
- 6-311G(3df)

The experimental bond length is 1.476 Å.

Calculation Setup Tip: You can place several calculations within the same input file by separating them with a --**Link1**-- line, as in the following example:

```
# APFD/6-31G(d) Opt                          Input file for first calculation

PO Optimization: APFD/6-31G(d)

0 2
molecule specification
                                             Required blank line
--Link1--                                    Indicates a new "job step," in this case, a separate job
# APFD/6-311(d) Opt                          Input file for the second calculation

PO Optimization: APFD/6-311G(d)

0 2
molecule specification
                                             Required blank line
```

The first input section ends with a blank line as usual, and it is followed by the --**Link1**-- line. The latter indicates the beginning of a new calculation. In this case, all of the calculations are independent. We'll see how to link jobs to one another later (see p. 82).

SOLUTION

The following table summarizes the results of our calculations:

BASIS SET	BASIS FUNCTIONS	BOND LENGTH (Å)
6-31G(d)	34	1.493
6-311G(d)	44	1.486
6-311G(2d)	54	1.477
6-311G(2df)	68	1.475
6-311G(3df)	78	1.472
Experiment		1.476

Both the larger 6-311G basis set and multiple polarization functions (≥2d) are needed to produce a very accurate structure for this molecule (<0.01 Å).

ADVANCED EXERCISE 2.13 BASIS SET DEFINITIONS

In this exercise, we examine how a basis set is defined in more detail. Gaussian and other ab initio electronic structure programs use gaussian-type atomic functions as basis functions. Gaussian functions have the general form:

$$g(\alpha,\vec{\mathbf{r}}) = cx^n y^m z^l e^{-\alpha r^2}$$

In this equation, α is a constant determining the size (radial extent) of the function. The exponential function is multiplied by powers (possibly 0) of x, y, and z, and a constant for normalization so that the integral of $\mathbf{g}^2$ over all space is 1 (note that therefore c must also be a function of α).

The actual basis functions are formed as linear combinations of such primitive gaussians:

$$\chi_\mu = \sum_p d_{\mu p} g_p$$

where the coefficients—the $\mathbf{d}_{\mu p}$s—are fixed constants within a given basis set.

Run a single-point energy calculation on methanol using the HF/6-31++G(d,p) model chemistry, including the **GFOldPrint** and **GFInput** keywords in the route section, which request that the basis set information be included in the output file (in tabular and input format, respectively). Examine the basis set output and identify its main components.

SOLUTION

The **GFOldPrint** output is best viewed in a window that is wide enough for 132 characters. The output for the carbon atom is shown in the listing following. We have reformatted it somewhat and reduced the number of digits for clarity, as well as eliminating zero entries in the coefficient section at the right.

Here we have a carbon atom described by 19 basis functions. The left section of the table lists the type and coordinates for the atom in question, along with the orbital type and orbital scaling factor for each basis function on this atom. Details about the composition of each basis function are given in the right half of the table, in the section labeled "Gaussian Functions." The columns in this section hold the exponents (α in the preceding equations) and the coefficients (the $\mathbf{d}_{\mu p}$'s) for each primitive gaussian.

ATOMIC CENTER				ATOMIC ORBITAL			GAUSSIAN FUNCTIONS				
				FUNCTION	SHELL	SCALE	α	$d_{\mu p}$'s →			
ATOM	X-COORD	Y-COORD	Z-COORD	NUMBER	TYPE	FACTOR	EXPONENT	S-COEF	P-COEF	D-COEF	F-COEF
C	-0.08715	1.23644	0.00000	1	S	1.00	0.3047D+04	0.5363D+00			
							0.4573D+03	0.9894D+00			
							0.1039D+03	0.1597D+01			
							0.2921D+02	0.2079D+01			
							0.9286D+01	0.1774D+01			
							0.3163D+01	0.6125D+00	0.0000D+00		
				2- 5	SP	1.00	0.7868D+01	-.3995D+00	0.1296D+01		
							0.1881D+01	-.1841D+00	0.9937D+00		
							0.5442D+00	0.5163D+00	0.4959D+00		
				6- 9	SP	1.00	0.1687D+00	0.1876D+00	0.1541D+00		
				10- 13	SP	1.00	0.4380D-01	0.6823D-01	0.2856D-01		
				14- 19	D	1.00	0.8000D+00			0.1113D+01	

For example, basis function 1, an s function, is a linear combination of six primitives, constructed with the exponents and coefficients (the latter are in the column labeled "S-COEF") listed in the table. Basis function 2 is another s function, comprised of three primitives using the exponents and S-COEF coefficients from the section of the table corresponding to functions 2-5. Basis function 3 is a p_x function also made up of three primitives constructed from the exponents and P-COEF coefficients in the same section of the table:

$$p_x \approx 1.30 g_x(7.87,\vec{\mathbf{r}}) + 0.99 g_x(1.88,\vec{\mathbf{r}}) + 0.50 g_x(0.54,\vec{\mathbf{r}})$$
$$\approx 1.30 e^{-7.87 r^2} + 0.99 e^{-1.88 r^2} + 0.50 e^{-0.54 r^2}$$

In the first equation, the function $\mathbf{g}_x(\alpha,\vec{\mathbf{r}})$ is of the form $c\mathbf{x}e^{-\alpha r^2}$ where c is a normalization constant also depending on α. It is expanded in the second equation assuming c=α (no normalization required).

Basis functions 4 and 5 are p_y and p_z functions constructed in the same manner using the same exponents and coefficients.

Functions 6 through 9 are another set of uncontracted s and p functions, each composed of a single primitive, formed in the same way as functions 5 through 9. Note that functions 1 through 9 form the heart of the 6-31G basis set: three sets of functions formed from six, three and one primitive gaussian.

Functions 10 through 13 comprise the diffuse s and p functions (note the small value for the exponent α, which will fall off to zero at a much greater distance than the earlier gaussian functions).

Functions 14 through 19 are d functions. This basis set uses six-component d functions: $\mathbf{d}_{x^2}$, $\mathbf{d}_{y^2}$, $\mathbf{d}_{z^2}$, $\mathbf{d}_{xy}$, $\mathbf{d}_{xz}$ and $\mathbf{d}_{yz}$. They are constructed using the exponent and D-COEF coefficient from the final section of the table.

Typically, the contracted functions themselves are also normalized. This has two consequences:

- The coefficients specified for the component primitive gaussians are chosen so that the resulting constructed basis functions are normalized. This means that one coefficient in each set is effectively constrained so that this condition is fulfilled.

- The coefficients for uncontracted basis functions—consisting of only a single primitive—are usually 1.0. This is the case for functions 6 through 19 in the 6-31++G(d,p) example we have been examining.

The **GFInput** output gives exactly the same information in the format required for Gaussian general basis set input (see the discussion of the **Gen** keyword in the *Gaussian User's Reference* for more information).■

Understanding CPU Resource Requirements

ADVANCED EXERCISE 2.14 **CPU USAGE BY PROBLEM SIZE**

This exercise is concerned with the way that CPU resources increase with system size. Run APFD/3-21G and/or APFD/6-31G(d) single point energy calculations for various lengths of alanine* chains and compare the CPU resources that are required, starting with just a few residues and stepping up to 25 or 50 (if feasible on your computer system).

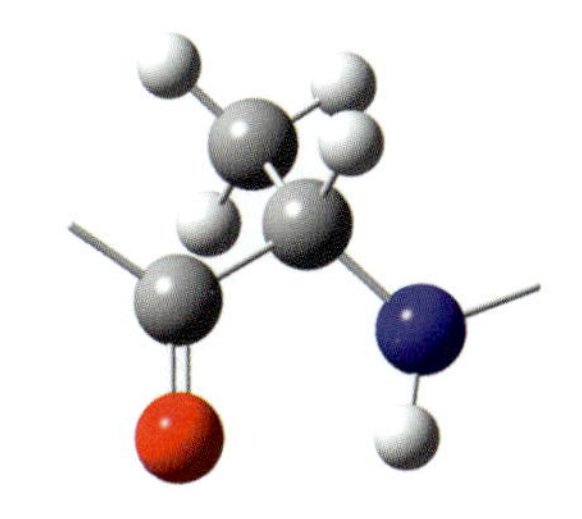

Be sure to terminate your alanine chains. We chose to terminate our chains with an NH_2 group bonded to the C=O carbon on one end and an acetyl group bonded to the nitrogen on the other end (in order to compare with some specific experimental results in a different study). Here is an example, $Ac(Ala)_2NH_2$ (which we will refer to in Gaussian input filenames and plots as "Ala2"):

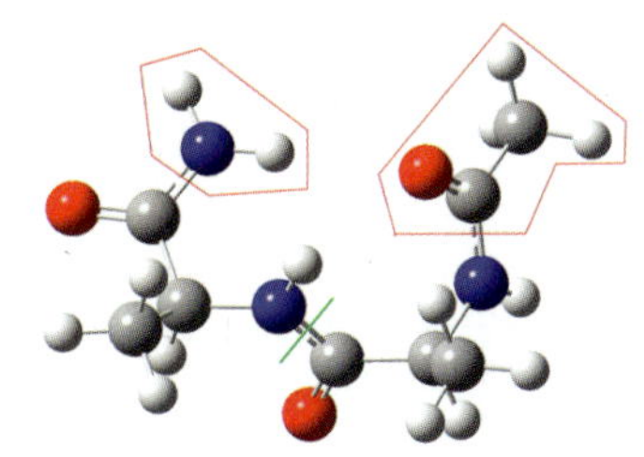

The terminating groups are outlined in red, and the green line indicates the boundary between the two alanine residues.

When your calculations have completed, plot the data as number of atoms vs. relative CPU time (i.e., normalize the data with respect to the shortest job). How does CPU usage scale with problem size (the number of atoms)? If you use a log-log plot type, the scaling with N_{atoms} corresponds to the slope of the line between the two corresponding points on the graph.

Job Preparation Hints:

- For all jobs, include the **SCF=NoVarAcc** additional keyword. This will force full integral accuracy for all cycles of the iterative solution of the energy so that a more useful comparison can be made.

- You will find it helpful to turn on additional print in the Gaussian output (with **#P** in the route section or via the corresponding control in your graphical interface) and also to include the extra keywords **IOp1=Timestamp**, which provides detailed time stamp information at various points in the calculation. We'll explain how to interpret the output in the Solution section below.

- Whenever you are concerned about comparing calculation times, you should run all the jobs on the same computer with the same parameters (i.e., number of processors,

* IUPAC: phenylamine.

amount of memory, etc.). You should also ensure that nothing else is running on the system at the same time.

- Be aware that the ease of difficulty in converging the SCF wavefunction can affect your results. Pay attention to how many SCF cycles are required for each job. If these numbers are not very similar, then it will be more helpful to plot the time required per SCF cycle rather than the entire job type.

- For smaller molecules (under about 8-9 residues), also include the **SCF=NoInCore** option: **SCF=(NoVarAcc,NoInCore)**. This prevents these jobs from using the much faster algorithms that are available when the problem fits entirely in memory, something which is only true for modest size calculations.

- One method for aiding SCF convergence for large systems like these is to perform a preliminary job with a less expensive model chemistry and then use that job's results as the initial guess for the larger model chemistry. This involves setting up a multistep job according to the following scheme:

```
%Chk=ala9
#P APFD/3-21G IOp1=Timestamp                 Input file for first calculation

Alanine 9

0 1
molecule specification
                                             Required blank line
--Link1--                                    Indicates a new "job step," in this case, a separate job
%Chk=ala9                                    Specifying the same checkpoint file lets us retrieve its results
# APFD/6-31G(d) IOp1=TimeStamp               Input for 2nd calculation
  Geom=AllCheck Guess=Read                   Read in molecule spec. & initial guess
                                             Required blank line
```

The second line in the route section tells Gaussian to read in the title, charge and spin multiplicity and molecule specification from the checkpoint file (**Geom=AllCheck**), and also to retrieve the converged wavefunction for use as the initial guess for the job's SCF procedure (**Guess=Read**).

SOLUTION

We set up our calculations as described above and ran jobs on 7 alanine structures. We ran slightly more complex calculations in order to gather more data. We performed **Force** calculations using both model chemistries.* This calculation type is equivalent to the combination of a single point energy plus a single step of a geometry optimization. We did so in order to determine the CPU requirements for more than one kind of job.

* We performed the calculations with the B3LYP DFT method. Results for other functionals, including APFD, will present the same characteristics.

These are the compounds that we modeled:

SYSTEM	# ATOMS
Ala-2	29
Ala-3	39
Ala-6	59
Ala-9	99
Ala-17	177
Ala-25	259
Ala-25 dimer	518

The best timing data to compare is generally elapsed time when jobs are run on an otherwise idle system. Comparing CPU times can be problematic on computers with multiple cores or processors.

The following commands illustrate how to extract the desired timing information from the Gaussian output file on a Linux, Mac OS X or UNIX system:

```
$ grep -i link x2_14_ala2.log | egrep '  1 |502|9999'
 Leave Link    1 at Tue Oct 12 11:14:10 2010 ...
 Top of link  502.    Tue Oct 12 11:14:12 2010 ...
 Leave Link  502 at Tue Oct 12 11:14:31 2010 ...
 Top of link 9999.    Tue Oct 12 11:14:31 2010 ...
```

These output lines are the timestamps we requested. The first search (**grep**) command finds all lines that contain the word "link" in upper or lowercase (**-i**); the second command (**egrep**) further narrows the output to lines for the specific links in which we are interested: Link 1 (matches " 1 ", a one with two spaces before it and one after it), Link 502 and Link 9999. The search expression in this command uses vertical bars to separate the various search strings which we want to match ("|" denotes logical OR).

The corresponding command on a Windows system would be:

```
findstr /I link file | findstr /C:"  1 " /C:502 /C:9999
```

The time for the SCF portion of the job is the amount spent in Link 502, which can be computed by subtracting the second timestamp value from the third. The total time for the job is similarly computed by subtracting the first timestamp value from the fourth. When a **Force** calculation is run, then the time for the gradient can be estimated by computing the time spent in Link 703.

The following plots present the results of our calculations. We have plotted the relative job time vs. the number of atoms (problem size) for both model chemistries. Here is the plot for the SCF portion of the calculation:

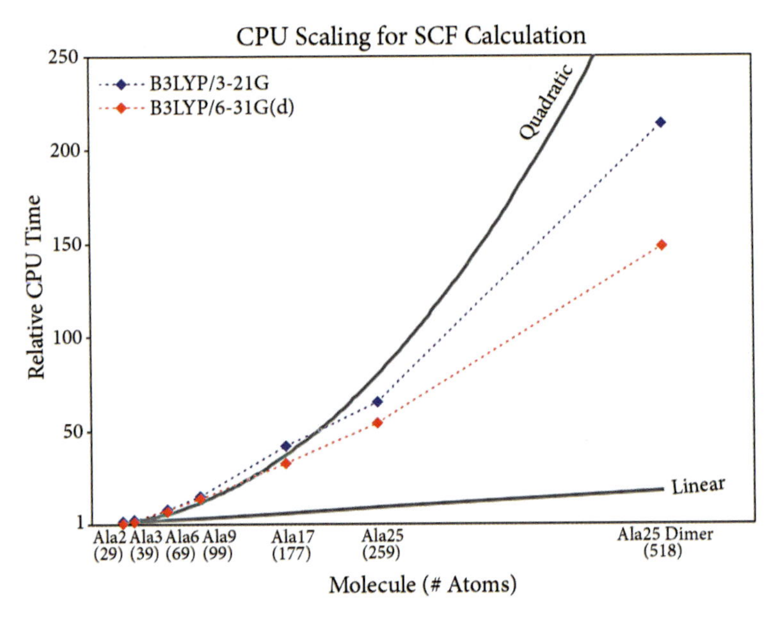

The plot also includes two reference lines (grey) indicating perfect linear and quadratic (N^2) scaling with the number of atoms. SCF calculations using the 3-21G basis set remain quadratic until about 180 atoms. In contrast, SCF calculations with the larger 6-31G(d) basis set are no longer quadratic after about 100 atoms.

Gradient calculations scale much better and are less dependent on the basis set, with near linear behavior after 100 atoms for both basis sets, as demonstrated by the following graph:

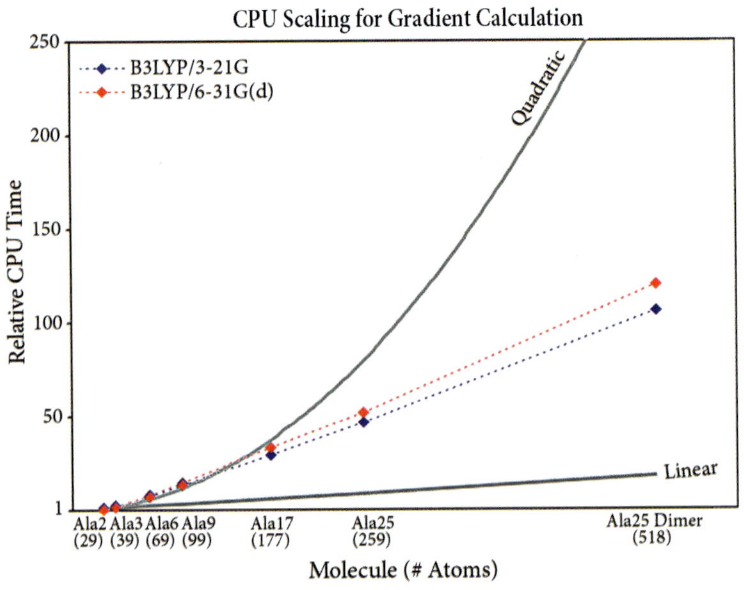

Overall, CPU requirements increase quadratically for the small problem sizes, and improve for larger problem sizes. For very large problems—larger than what we performed here—the diagonalization portion of the calculation becomes more significant and dominates the performance, so CPU performance falls off from linearity somewhat.■

3

Geometry Optimizations

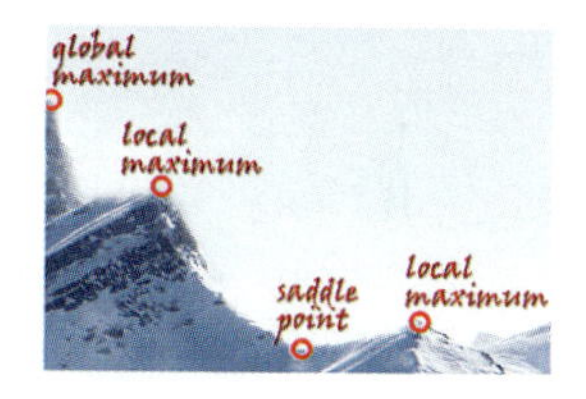

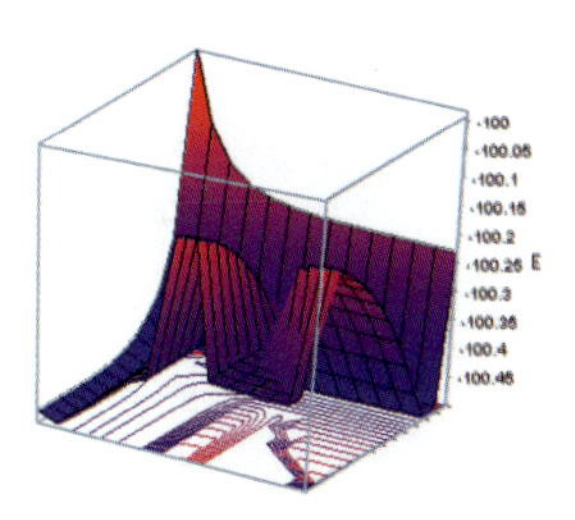

In This Chapter:

Potential Energy Surfaces

Understanding Optimizations & Convergence Criteria

Locating Transition Structures

Advanced Topics:

Techniques for Optimizing Large Molecules

Handling Difficult Cases

3 ◆ Optimizing Molecular Structures

We performed several geometry optimizations in the course of the previous chapter. In this chapter, we look at optimizations in detail, including ones for both minima and transition states.

Potential Energy Surfaces

The way the total energy* of a molecular system varies with small changes in its structure is described by its *potential energy surface*. A potential energy surface is a mathematical relationship linking molecular structure and the resultant energy. For a diatomic molecule, it is a two-dimensional plot with the internuclear separation on the X-axis (the only way that the structure of such a molecule can vary), and the energy at that bond distance on the Y-axis, resulting in a curve. For larger systems, the surface has as many dimensions as there are degrees of freedom within the molecule. Each point on a potential energy surface corresponds to different values for the various bond distances, bond angles and dihedral angles within the molecule.

A potential energy surface (PES) is often represented by illustrations like the one below. This sort of drawing considers only two of the degrees of freedom within the molecule, and plots the energy above the plane defined by them, creating a literal surface. Each point corresponds to the specific values of the two structural variables—and thus represents a particular molecular structure—with the height of the surface at that point corresponding to the energy of that structure.

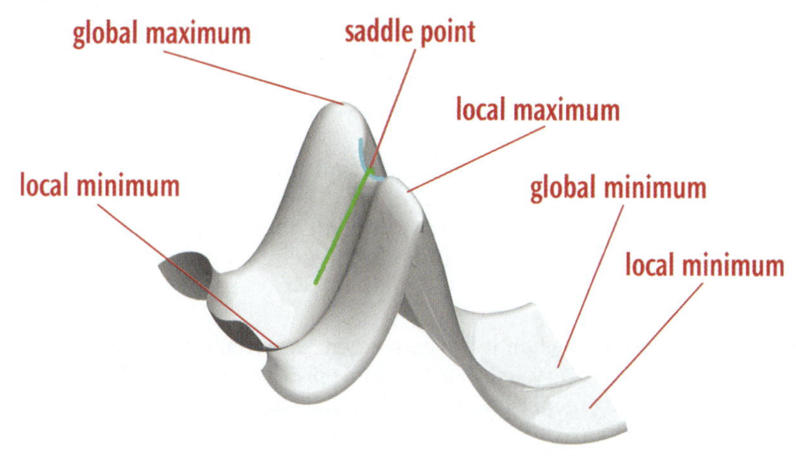

There are three minima on this potential surface. A *minimum* is the bottom of a valley on the PES. From such a point, motion in any direction—a physical metaphor corresponding to changing the structure slightly—leads to a higher energy. A minimum can be a local minimum, meaning that it is the lowest point in some limited region of the potential surface, or it can be the global minimum: the lowest energy point anywhere on the potential surface. Minima occur at equilibrium structures for the system, with different minima corresponding to different conformations or structural isomers in the case of single molecules, or reactant and product molecules in the case of multicomponent systems.

* Precisely, the sum of the electronic and nuclear repulsion energies (see p. 35). This quantity is often referred to simply as the "energy."

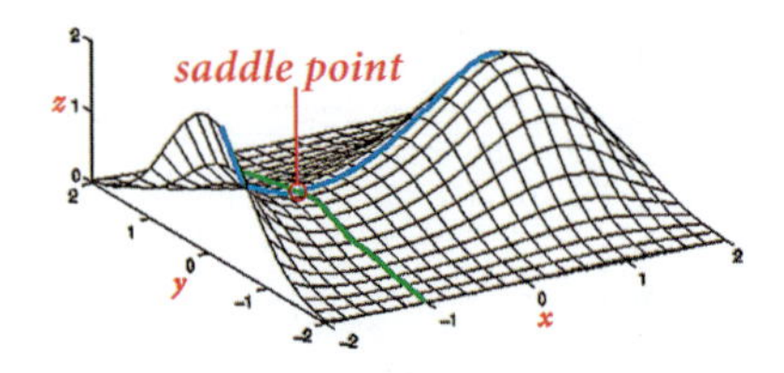

Figure courtesy Prof. John Lee, Mathematics Dept., Oregon State University

Peaks and ridges correspond to maxima on the potential energy surface. A peak is a maximum in all directions (i.e., both along and across the ridge). A low point along a ridge—a mountain pass in our topographical metaphor—is a local minimum in one direction (along the ridge, marked in blue above), and a local maximum in the perpendicular direction (climbing up to the ridge, crossing over, and then going back down; marked in green above). Mathematically, a point like this—a maximum in one direction and a minimum in the other (or in all others in the case of a larger dimensional potential surface)—is called a *saddle point* (based on the shape of the surrounding surface). For example, the saddle point in the diagram is a minimum along its ridge and a maximum along the path connecting minima on either side of the ridge. In chemical terms, a saddle point on a potential energy surface corresponds to a transition structure connecting the two equilibrium structures.

Locating Stationary Points on the PES

References: Optimizations to minima in redundant internal coordinates

Definition: [Pulay69, Pulay79, Schlegel82, Fogarasi92, Pulay92]

Review articles: [Hratchian05a, Hratchian12]

Algorithms/Implementation: [Schlegel84a, Peng96, Li06]

Geometry optimizations usually attempt to locate minima on the potential energy surface, thereby predicting equilibrium structures of molecular systems. Optimizations can also locate transition structures. Optimizations to minima are also called *minimizations*.

At both minima and saddle points, the first derivative of the energy, known as the *gradient*, is zero. The forces on the nuclei are also zero at such a point (since the gradient is the negative of the forces). In general, a point on the potential energy surface where the forces are zero is called a *stationary point*. All successful optimizations locate a stationary point—although not always the one that was intended.

Gaussian geometry optimizations begin at the molecular structure specified as input, and step along the potential energy surface. The energy and the gradient are computed at each point. The gradient indicates the direction along the surface in which the energy decreases most rapidly from the current point, as well as the steepness of that slope, and this data is used to determine the direction and size of the next step.

Most optimization algorithms also estimate or compute the value of the second derivative of the energy with respect to the molecular coordinates, updating the matrix of force constants (known as the *Hessian*). These force constants specify the curvature of the surface at that point, which provides additional information that is useful for computing the next step.

Optimization Convergence Criteria

An optimization is complete when it has converged: essentially, when the forces are near zero, making the next step is very small (below some preset value defined by the algorithm). These are the specific convergence criteria used by Gaussian:

- The forces must be essentially 0. Specifically, the maximum component of the force must be below the cutoff value of 0.00045.*
- The root-mean-square of the forces must be below the defined tolerance of 0.0003.
- The calculated maximum displacement for the next step must be smaller than the defined threshold value of 0.0018 (i.e., essentially 0). The maximum displacement is the largest change in any coordinate in the molecular structure.
- The root-mean-square of the displacement for the next step must be below its cutoff value of 0.0012.

* The unit for the forces and RMS forces is hartrees/bohr or hartrees/radian (depending on the coordinate). The unit for the displacement and RMS displacement is bohr.

Note that the change in energy between the current and next points is not an explicit criterion for convergence. It is reflected in the tests of the size of the next step, since small steps near a minimum will usually result in small changes in the energy.

The presence of four distinct convergence criteria prevents a premature identification of the minimum. For example, in a broad, nearly flat valley on the potential energy surface, the forces may be near zero (within the tolerance) while the computed steps remain quite large as the optimization moves toward the very bottom of the valley. Or, in extremely steep regions, the step size may become very small while the forces remain quite large. Checking the root-mean-squares of the items of interest also help prevent premature acceptance of convergence.

Modified Criteria for Large, Floppy Molecules

There is one exception to the criteria we just looked at, designed to aid in the optimization of large molecules. When the forces are two orders of magnitude smaller than the cutoff value (i.e., 1/100th of the limiting value), then the geometry is considered converged even if the displacement is larger than the cutoff value. This criteria comes into play on very, very flat potential energy surfaces near the minimum, a common condition with large, floppy molecules.

Examining Optimization Results

Optimization calculations produce the optimized molecular structure as well as the energy and related properties. The structural changes that the molecule undergoes in the course of the optimization can also be animated in both GaussView and WebMO:

In GaussView, check the **Read Intermediate Geometries** box at the bottom of the **Open File** dialog. This will cause all available structures found in the file to be loaded into a single molecule group. You can use the green button at the left edge of the toolbar to start animation as well as the **Multiview** icon (highlighted in red in the margin illustration) to display multiple structures at the same time. Use the context (right click) menu in **Multiview** mode to control the placement and sizing of the various subwindows. You can set the animation speed using the **Animation Delay** field on the **Molecule** panel of the **Display Format** dialog (or the corresponding preference option); a higher value corresponds to slower animation.

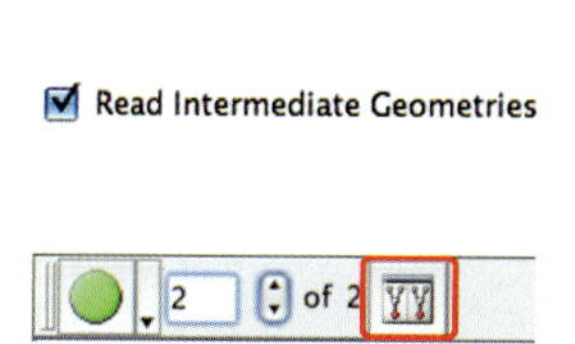

You can view graphs of various quantities at each step in the optimization using the **Optimization** item on the **Results** menu. By default, the total energy and RMS gradient norm are plotted. You can create additional plots using the **Plot Molecular Property** item on the plot window's **Plot** menu. We'll see some examples of the latter below (p. 92).

WebMO In WebMO, you can animate an optimization using the filmstrip icon in the **Geometry Sequence** section of the **View Job** page (a truncated version of the section appears in the margin). The **Loop** popup menu at the bottom of the section controls whether the geometry animation sequence repeats once it finishes. The choices are **None** (no repeat), **Loop** (begin again with the first structure) and **Cycle** (display the structures first in normal sequence and then in reverse). You can set the animation speed via the corresponding field; lower numbers correspond to slower animation.

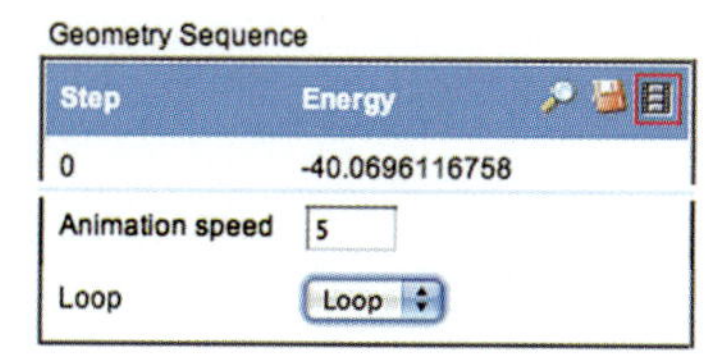

Clicking on the magnifying glass icon will display a plot of the energy at each step in the optimization in the **Data Viewer** panel of the **View Job** page.

You can also view an animation for an optimization in progress in either visualization program by opening the checkpoint file or the log file corresponding to the running job.

EXAMPLE 3.1 OPTIMIZING DECAMETHYLZINCOCENE

Metallocene compounds are an interesting area of organometallic chemistry. Ferrocene* was first synthesized in the early 1950s [Kealy51, Miller52], and these compounds continue to be studied by both experimental and computational methods today. Metallocenes can be formed with a variety of metal atoms bonded to the two rings. We will consider a zinc-based compound in this example.

Zincocene is the simplest zinc-based metallocene, consisting of a zinc atom sandwiched between two cyclopentane rings. Although they are superficially similar, ferrocene and zincocene have quite different structures. The ferrocene molecule has D_{5d} symmetry and a hapticity† of $\eta^5\eta^5$, while zincocene has C_s symmetry and hapticity $\eta^1\eta^5$ [Haaland03].

The most common oxidation state for both iron and zinc is +2. Zinc was long considered to have only a single oxidation state, prior to the unexpected synthesis of decamethyldizincocene* from a reaction of decamethylzincocene* and diethyl zinc in diethyl ether [Resa04]. The analogous reaction with zincocene does not lead to this product. A new synthesis of decamethyldizincocene using a reduction process has also recently been reported [Grirrane08]. Decamethyldizincocene has hapticity $\eta^5\eta^5$.

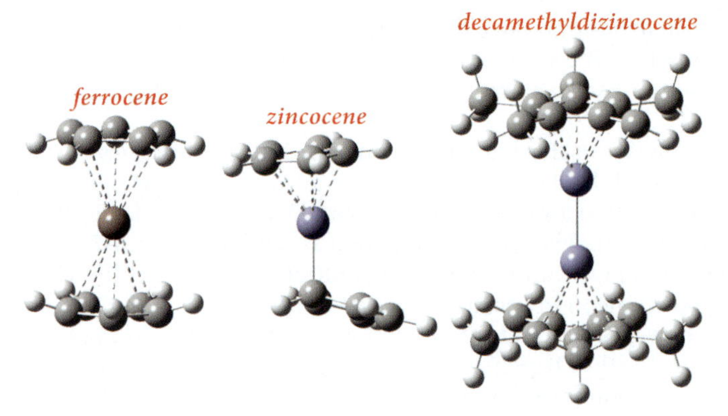

What would you predict for the structure of decamethylzincocene? How will the methyl groups affect the structure of zincocene?

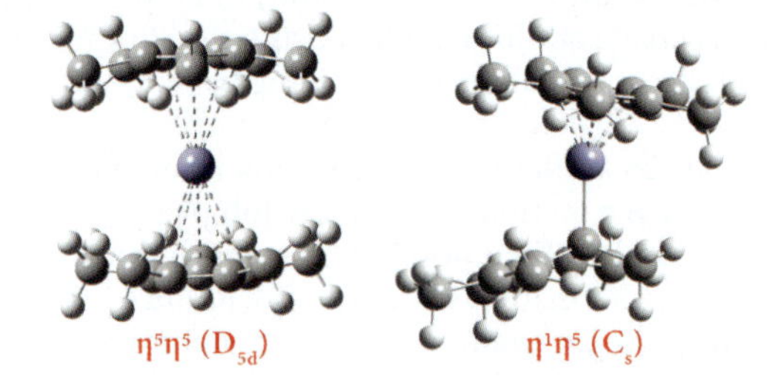

We optimized the structure of decamethylzincocene beginning from three starting structures:

- An $\eta^1\eta^5$ structure with C_s symmetry
- An $\eta^5\eta^5$ structure with D_{5d} symmetry
- An $\eta^5\eta^5$ structure with D_{5h} symmetry

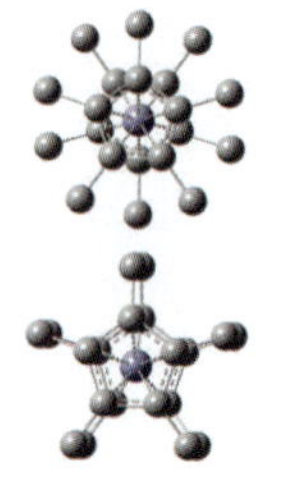

* IUPAC: ferrocene: bis[(η^5)-cyclopentadienyl]iron; decamethylzincocene: [(η^1)-pentamethyl-2,4-cyclopentadien-1-yl][(η^5)-pentamethyl-2,4-cyclopentadien-1-yl]zinc; decamethyldizincocene: bis[(η^5)-1,2,3,4,5-pentamethyl-2,4-cyclopentadien-1-yl]di-(Zn-Zn)zinc.

† *hapticity*: The number of electrons in a ligand that are directly coordinated to a central atom (typically a metal), denoted by the symbol η.

The second and third structures differ as to whether the two rings are staggered (upper illustration in the margin) or eclipsed (aligned: lower illustration in the margin).

We optimized all three starting structures with the B3LYP/6-31G(d) model chemistry, which was chosen so that we could compare directly to results from the literature. Each optimization computed force constants at the first point to aid the optimizer by including the **Opt=CalcFC** option (this technique is discussed later in this chapter).

All three optimizations lead to very similar structures with $\eta^1\eta^5$ hapticity and nearly identical total energies (within 0.000064 hartrees); frequency calculations confirm that it is a minimum. We then reoptimized the structure with the B3LYP/6-311G(2d,p) model chemistry.

The results of our optimizations as well as two related calculations from the literature are presented in the following table:

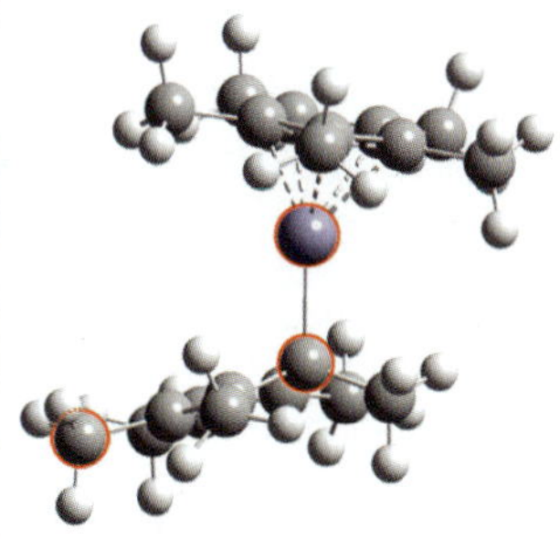

	METHOD: B3LYP/			
PARAMETER	6-31G(d)	6-311+G(d,p)†	6-311G(2d,p)	6-311+G(2d,p)‡
Zn–C η^1 (Å)	1.98	2.02	1.97	2.02
Zn···C η^5 (Å)	2.22-2.26	2.26-2.32	2.22-2.27	2.28-2.29
Zn–C–CH_3 (°)	102.9	101.4	106.2	104.1

†[DelRio08] ‡[Ludlow05]

The third structural parameter reported in the preceding table is the angle formed by the Zn–C bond and the carbon in one of the two equivalent methyl groups on the opposite side of the ring (see the illustration in the margin). This angle is a measure of the ring slippage for the η^1 bonded ring. The angle was measured as 106.7° in the X-ray diffraction experiments of [Fischer89] for the related compound $(C_5Me_4Ph)_2Zn$ (i.e., decamethylzincocene with one of the methyl groups replaced by a phenyl ring).

We can make several observations on the basis of these results:

- The low level B3LYP/6-31G(d) geometry is reasonably good for this molecule. This model chemistry is a good starting point for geometry optimizations and generally produces reasonable geometries for compounds for which it is the only practical method.
- The diffuse function (+) has the effect of lengthening the bonds involving zinc. Calculations on transition metals compounds often benefit from including diffuse functions.
- The second d polarization function is required for an accurate ring slippage angle for this molecule.
- The 6-311+G(2d,p) basis set achieves the best symmetry in the predicted structure, with the Zn-C bond distances in the η^5 bonded ring varying by only 0.01 Å.

The first $\eta^1\eta^5$ starting structure converges in 46 steps. The following illustration displays some representative steps within the optimization. From step 1 until about step 18, the Zn–cyclopentane distance increases and the positions of the methyl groups in the η^1 bonded ring are adjusted.

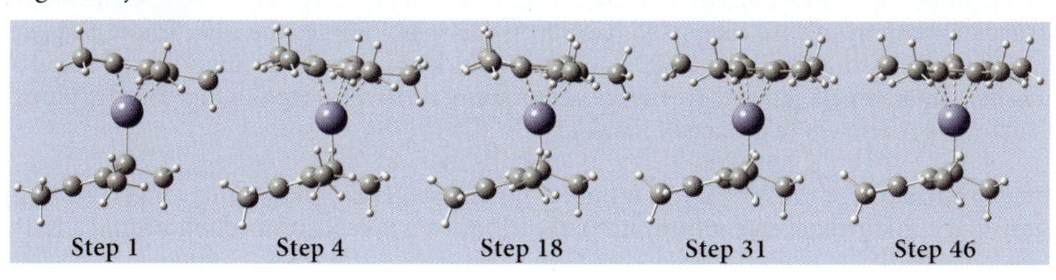

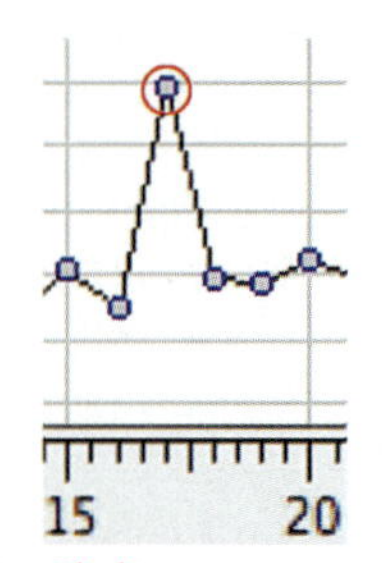

Tip: Clicking on any point within a plot will display the corresponding structure in the view window.

Next, the angle between the two rings changes (around steps 19-30). The latter part of the optimization consists of rotating the η^5 bonded ring (through step 39) and making small adjustments to that ring and its component groups.

The following plots from GaussView further elucidate the optimization's progress. They plot the total energy, Zn–C bond length, one of the Zn···C bond lengths and the slippage angle of the η^1 bonded ring:

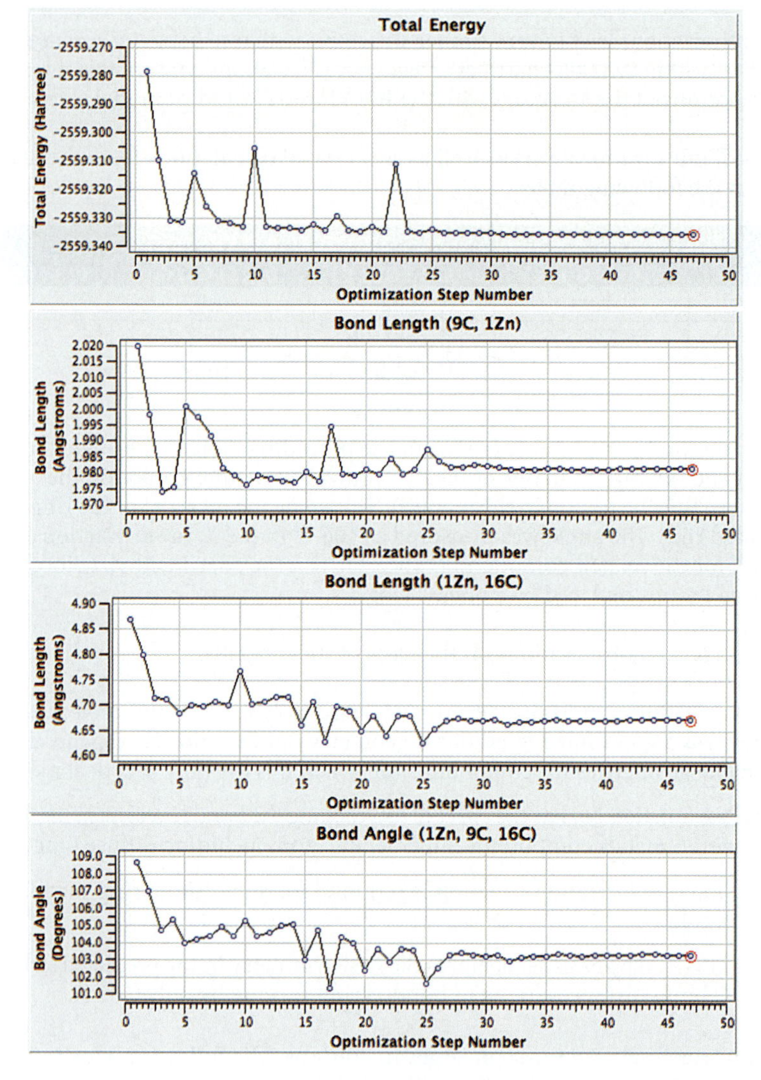

The iterative nature of the optimization process is clear from these plots as the energy and the values of the various parameters oscillate above and below their final values in the course of the optimization. Many times, large changes in energy and in the values of the structural parameters are correlated. However, this is not always the case. Interestingly, at times relatively small changes in structure result in large changes in energy (e.g., step 10); at other times, small changes in energy accompany relatively large changes in the plotted structural parameters (e.g., steps 15 and 25).

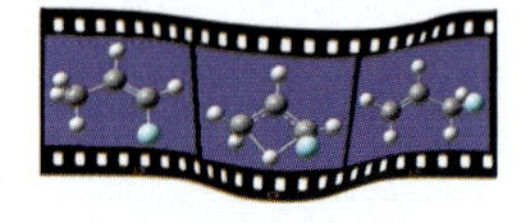

The optimization of the $\eta^5\eta^5$ D_{5d} symmetry starting structure takes much longer: 86 steps. During the first part of the optimization, the zinc atom is repositioned with respect to the two rings, and the structure changes from $\eta^5\eta^5$ to something much nearer to the final $\eta^1\eta^5$ form. Initially, the zinc atom moves from the center to the edge of the lower (η^1) cyclopentane ring (steps 1-11), and then it moves to the center of the upper (η^5) ring (starting at step 12).

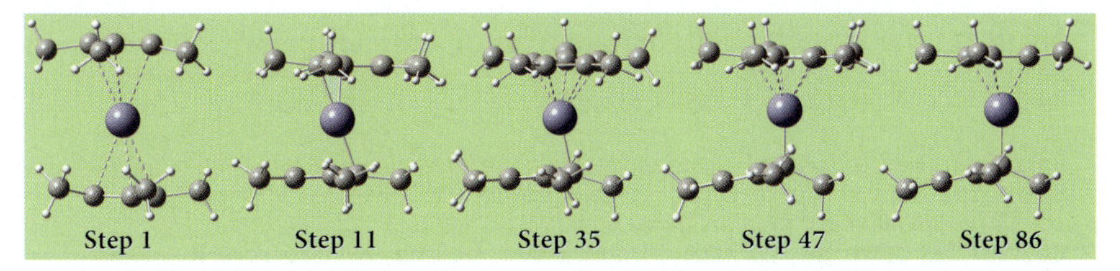

Beginning at about step 35, the major structural changes involve the angle between the two rings, which continues until about step 47. During this portion of the optimization, the Zn–C–C bond in the η^1 ring changes from ~94° to ~100°, and the Zn–C–Me bond changes from ~86° to ~103°. After step 47, only minor adjustments are made to the structure, first in the methyl groups in both rings, and later (after step 60) only to those in the η^5 ring. Very small rotations in the η^5 ring also occur during this final period.

The following plots show the total energy, Zn–C bond length and Zn–C–Me bond angle throughout the course of the optimization:

Experienced Gaussian users monitor their geometry optimizations as they run. In cases like this one, often the best strategy is to stop the calculation when behavior like this becomes apparent, and then start another optimization from the last point using the Opt=CalcFC option to improve the Hessian (discussed later in this chapter).

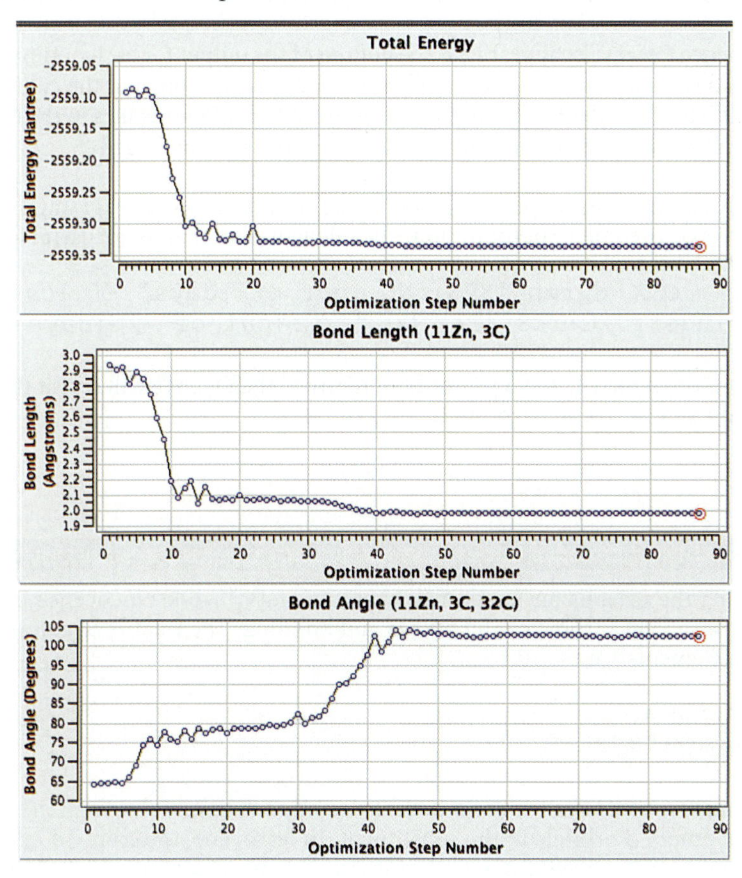

The energy gets very near its final value relatively early in the optimization, and almost half of the optimization consists of relatively small adjustments to the geometry in order to locate the minimum on this molecule's rather flat potential energy surface.

We will now look at a part of the output from the decamethylzincocene optimization. After some initial output from the setup portion of the optimization job, Gaussian displays a section like the following for each step:

```
GradGradGradGradGradGradGradGradGradGradGradGradGradGrad
Berny optimization.                      Optimization algorithm details
Using GEDIIS/GDIIS optimizer.
...
Search for a local minimum.     Optimization goal: minimum or saddle pt.
Step number  36 out of a maximum of  306      Current step number
...  Old & new values for structural parameters:
Variable Old X -DE/DX  Delta X   Delta X   Delta X   New
                        (DIIS)    (GDIIS)   (Total)
   R1    4.2684 0.00004 0.00037  0.00095  0.00131   4.2697
   R2    4.1929 0.00005 0.00026 -0.00036 -0.00011   4.1928
...  Results of convergence tests after this step:
        Item                  Value     Threshold  Converged?
Maximum Force              0.000143     0.000450     YES
RMS     Force              0.000039     0.000300     YES
Maximum Displacement       0.033997     0.001800     NO
RMS     Displacement       0.009219     0.001200     NO
Predicted change in Energy=-2.37667D-06 Not a convergence criterion
GradGradGradGradGradGradGradGradGradGradGradGradGradGrad
```

The convergence test results appear near the bottom of the output for each optimization step. The threshold column indicates the cutoff value for each criterion, and the column labeled "Converged?" indicates the results for that criterion. When all four values in the Converged? column are YES, then the optimization is completed and has converged.

Periodically examining the output file for a running optimization is an easy way to keep track of the progress of a geometry optimization. The following commands are useful for this task:

You may want to pipe the output of these commands to **more** since it can be quite lengthy.

UNIX/Linux/Mac OS X: `egrep 'YES| NO |out of |days' job.log`
Windows: `findstr /C:YES /C:" NO " /C:"out of" /C:days job.log`

The optimization of the D_{5h} form proceeds similarly to the D_{5d}. We will revisit this example in Advanced Exercise 3.9 on page 136. ■

GOING DEEPER: UNDERSTANDING FERROCENE AND ZINCOCENE

Considering the reasons for ferrocene's hapticity of $\eta^5\eta^5$ while zincocene's is $\eta^1\eta^5$ is a useful exercise in electronic configuration and bonding. The electronic configurations of the two elements are:

- Fe: [Ar] $3d^6$ $4s^2$
- Zn: [Ar] $3d^{10}$ $4s^2$

Very different types of bonding occur in the two compounds, resulting from the presence/absence of empty d orbitals in the substituent. In ferrocene, the empty d orbitals can form a bond with the π system in each ring. In contrast, in zincocene, the bond to the ring must use the s orbital, accordingly forming a bond to a single carbon atom. The connection to the second ring in zincocene is electrostatic, involving the higher energy 4p orbitals on the zinc atom.

Examining the molecular orbitals for these elements bonded to a single cyclopentane ring and comparing with those for the components alone can be helpful. The following illustration shows the relevant orbitals from the various metal ions and compounds. They are plotted from the results of **Population=SaveMix** calculations, which generates the natural bond orbitals (NBOs) for occupied orbitals and the natural localized molecular orbitals (NLMOs) for virtual orbitals and saves them to the checkpoint file for later visualization. We used our standard model chemistry for these calculations.

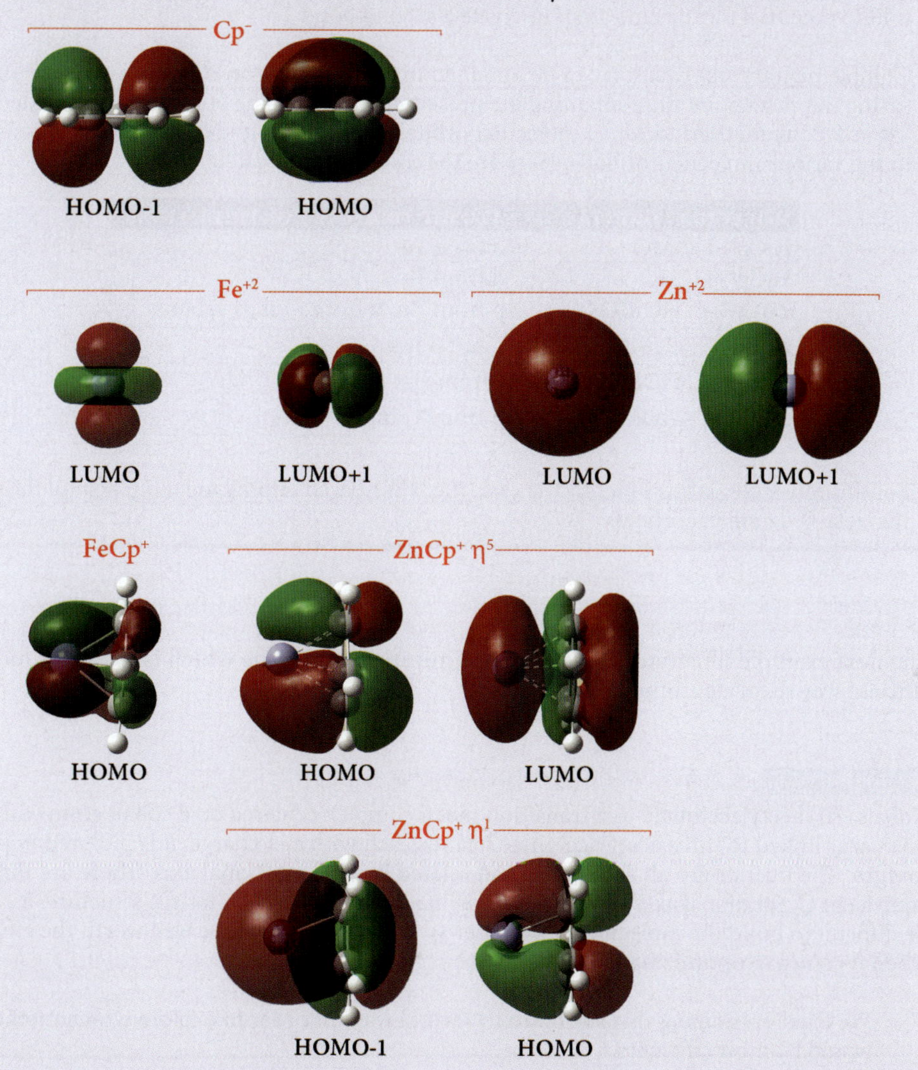

The two highest occupied MOs in cyclopentane are π^* orbitals, well suited for combining with the unoccupied d orbitals in Fe^{+2}. In contrast, the lowest unoccupied orbitals in zinc are an s orbital and a p orbital.

When we consider the three complexes, we see further divergence. The $FeCp^+$ HOMO indicates bonding between the iron atom and the ring. The iron atom still has empty d orbitals with which to make a similar bond to another cyclopentane ring. The superficially similar η^5 $ZnCp^+$ molecule has vastly different orbitals. In the HOMO, the π system from the ring is merely distorted by the zinc atom without forming a bond. In the LUMO, we can see the s orbital of the zinc atom and the π system on the ring; the former is clearly available to create a bond to a second ring.

Finally, in the η^1 configuration of $ZnCp^+$, we see the bond between the s orbital on the zinc and the ring in the occupied orbital below the HOMO. The HOMO is similar to the LUMO in the η^5 configuration in that the zinc atom distorted the MO of the ring without forming a bond (note how the zinc atom remains outside of the electron density). There is little potential for the zinc atom to create a second bond.

Complementary information can be obtained from a **Population=Orbitals** calculation (using the checkpoint file containing the preceding orbitals). The output section headed "Atomic contributions to Alpha molecular orbitals" lists the atomic orbital contributions to the various molecular orbitals. Here are the results for our calculations:

MOLECULE: ORBITAL	ATOMIC ORBITAL CONTRIBUTIONS
$FeCp^+$: HOMO	d from Fe
$FeCp^+$: LUMO	d from Fe
$ZnCp^+ \eta^5$: HOMO	p from Zn, p from 3 ring carbons
$ZnCp^+\eta^5$: LUMO	s and p from Zn
$ZnCp^+ \eta^1$: HOMO-1	s from Zn, p from C
$ZnCp^+ \eta^1$: HOMO	p from 4 ring carbons

For this molecule, as is frequently the case, the NBO-based orbitals are more helpful than the default canonical orbitals.

The next example illustrates a two-step optimization technique which is often the most efficient way to locate a minimum.

EXAMPLE 3.2 OPTIMIZING COBALT(III) ACETYLACETONATE

Cobalt (III) acetylacetonate* is a transition metal complex centered on a cobalt atom with a +3 charge linked to three acetylacetonate ligands each with a -1 charge. It is pictured in the margin. The ligands are all equivalent and related by a C_3 rotational axis. There are three equivalent C_2 rotational axes as well, resulting in a point group of D_3 for the structure. It can be difficult to build the molecule with proper symmetry, but it will be well worth the effort when it comes to optimizing the geometry.

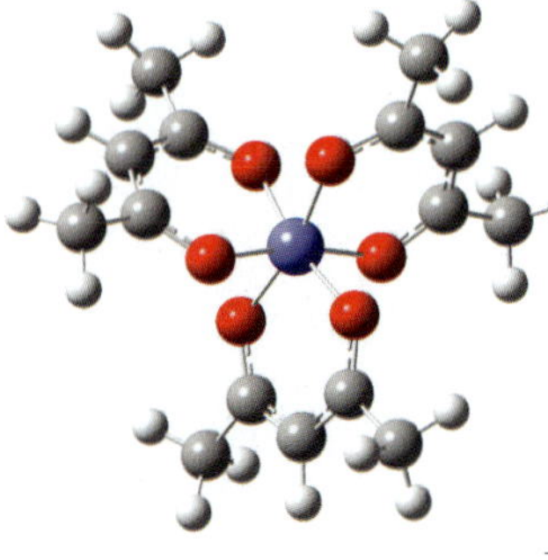

We tried optimizing this structure in several ways in order to explore which method would be most efficient:

- Beginning from a structure built without imposing symmetry and then optimizing with our standard model chemistry modified to use the default† integral options.
- Beginning from a structure with D_3 symmetry and then optimizing with our standard model chemistry.

* IUPAC: tris(2,4-dioxo-3-pentanyl)cobalt.

† In other words, without **Int=(Ultrafine,Acc2E=12)**.

- Beginning from the D_3 structure and then computing the force constants at the first point with **Opt=CalcFC** (see p. 116).

- Beginning from the D_3 structure and then performing a preliminary optimization+frequency calculation with the smaller 6-31G(d) basis set and the default integral options.This was followed by an **Opt=RCFC** optimization using our standard model chemistry, beginning at the optimized structure. The **RCFC** option retrieves the force constants computed by an earlier frequency calculation from the checkpoint file for use in the current optimizaiton.

The following table summarizes the results of our four optimization attempts:

METHOD SUMMARY	RESULT	TIME RATIO	TOTAL STEPS
C_s starting structure (terminated after 50 steps)	failure	40.4	?
D_3 starting structure	success	2.4	12
D_3 starting structure, **Opt=CalcFC**	success	6.9	9
D_3 starting structure, 2 step optimization	success	1.0	12+2

The fastest method is the two step optimization technique, followed by the normal optimization with our standard model chemistry. The use of **CalcFC** marginally reduces the number of steps taken, but the cost of the second derivatives results in a much longer time to solution. For this structure, the **CalcFC** option provides no real benefit. The two step process takes the greatest number of steps, but the vast majority of those are with the inexpensive model chemistry.

Note that the optimization starting from our hastily built C_s structure never finds a minimum. While a different starting point might converge, it would certainly take longer than using the properly built structure, emphasizing the importance of starting from a structure with the proper symmetry. ■

Locating Transition Structures

The optimization facility can be used to locate transition structures as well as ground state structures since both correspond to stationary points on the potential energy surface. Gaussian provides two methods for locating a transition structure.

- By specifying a reasonable guess for the transition state geometry and directing the optimizer to locate a first order saddle point. However, this can be challenging in many cases.

- By automatically generating a starting structure for a transition state optimization based upon the reactants and products that the transition structure connects. This technique is known as a QST2 optimization, and it uses the STQN method (which employs a quadratic synchronous transit approach to get closer to the quadratic region of the transition state and then uses a quasi-Newton or eigenvector-following algorithm to complete the optimization).

  Input files using this option must include two title and molecule specification sections. Graphical interface programs have features which automate this process for you. The facility generates a guess for the transition structure which is midway between the reactants and products, in terms of redundant internal coordinates.

References: Transition structure optimizations
Definition: [Schlegel82]
STQN Definition: [Peng93]
Review article: [Hratchian05a]
Algorithms/Implementation: [Peng96, Li06]

A variation of the QST2 approach allows you to specify a starting structure for the transition state in addition to reactants and products. It is known as QST3.

We will generally perform QST2 and QST3 optimizations for transition structures in this book.

EXAMPLE 3.3 LOCATING A TRANSITION STRUCTURE WITH QST2

Our first transition structure optimization will be a simple reaction: $SiH_2+H_2 \rightarrow SiH_4$*. We will do so via a QST2 calculation.

In GaussView, you can set up a QST2 calculation in the following way (assuming that you have already optimized the reactants and the products):

1. Open the optimized reactant, creating a new molecule group.

2. Open the optimized product, adding the molecule to the molecule group containing the reactant by selecting **Add all files to active molecule group** from the **Target** popup menu at the bottom of the **Open Files** dialog. You can also cut and paste the second structure from a different molecule group via the menu path illustrated in the margin.

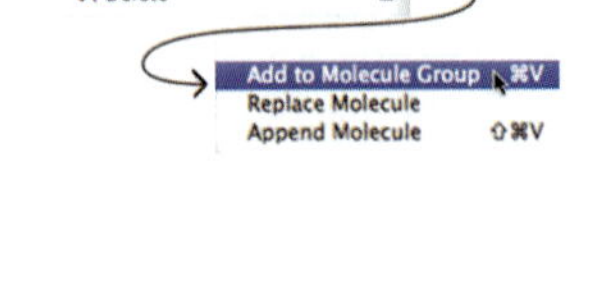

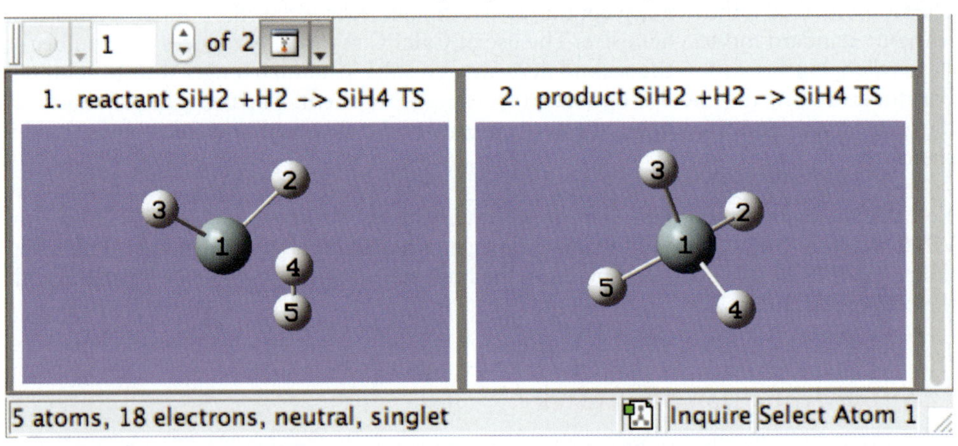

The molecule group will now contain two structures. Note that it does not matter whether the reactant is structure 1 or structure 2.

3. Use the **Gaussian Calculation Setup** dialog to specify the transition structure optimization. Set the **Optimize to a** field to **TS (QST2)**.

The job is now ready to be submitted to Gaussian in the usual manner.

WebMO In WebMO, the process is similar:

1. After optimizing both the product and the reactant, select the reactant from the completed jobs from the list on the **Job Manager** page. Note the job number for the product as well. Once the structure has loaded, click the **New Job Using This Geometry** button at the bottom of the **Molecule Viewer**.

2. Navigate to the **Configure Gaussian Job Options** page. Select **Saddle Calculation** from the **Calculation** popup menu as well as appropriate selections from the other menus.

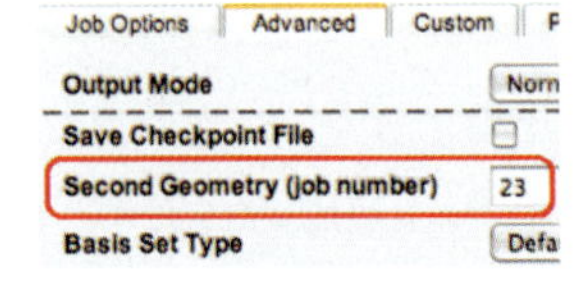

* IUPAC: silylene, dihydrogen and silane (respectively).

On the **Advanced** tab, specify the job number for the product in the **Second Geometry (job number)** field.

The QST2 feature additionally requires that the atom numbering be consistent between the two structures. In this case, we verify that the silicon atom has the same atom number in the reactants and in the product. If necessary, the numbering of the hydrogen atoms can be brought into correspondance by rotating the product and/or the hydrogen molecule in this simple structure.

The job is now ready for submission.

If we examine the input file for this QST2 optimization, we find it has the following format:

`# APFD/6-311+G(2d,p) Opt=QST2 Freq`	
`SiH2+H2 --> SiH4 Reactant`	*First title section*
`0 1` *structure for* SiH_2+H_2	*First molecule specification*
`SiH2+H2 --> SiH4 Product`	*Second title section*
`0 1` *structure for* SiH_4	*Second molecule specification*
	Required final blank line

The input file contains two titles and two complete molecule specifications, with blank lines terminating all four sections.

The QST2 calculation begins from the structure on the left in the following figure. The optimized transition structure is shown on the right:

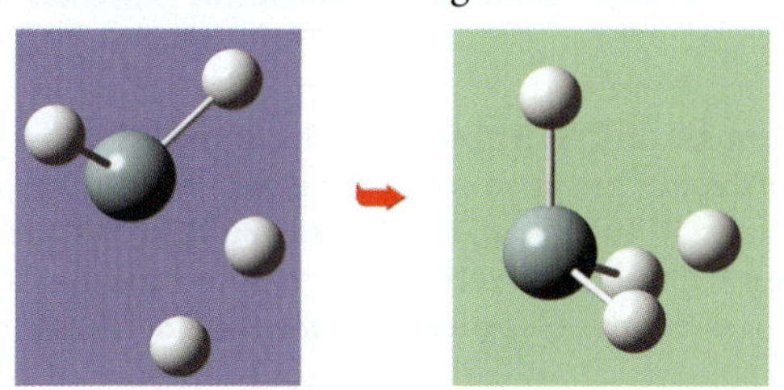

One of the hydrogen atoms from the hydrogen molecule has bonded to the silicon atom. The normal mode corresponding to the imaginary frequency shows motion in all four hydrogen atoms.

EXAMPLE 3.4 TRANSITION STRUCTURE FOR VINYL AZIDE DECOMPOSITION

Studies of thermal rearrangement of various C_2H_3N compounds have concluded that vinyl azide decomposes at 650 K to form 2H-azirine and N_2*, with the azirine converting to acetonitrile above 750 K [Bock81a, Bock83] (for a theoretical study, see [Lohr83]). We will consider the second phase of this process. We want to find the transition structure for the rearrangement of azirine into acetonitrile: a 1,2 hydrogen shift reaction. These are the steps we will take to accomplish this:

* IUPAC: vinyl azide: azidoethene; N_2: dinitrogen.

Rebuilding Azirine into Acetonitrile

- Remove the double bond.
- Change the N-C-C bond angle to 180°.
- Switch the bond for the single hydrogen atom to the other carbon atom.
- Reposition the hydrogen atom so that it is near the carbon atom to which it is now bonded.
- Clean the structure.

1. For all but the simplest cases, accurate starting structures are needed for the QST procedures to function well. Accordingly, the first step here is to build and optimize the azirine reactant. Verify that the resulting structure is a minimum with a frequency calculation.
2. Beginning from the optimized reactant geometry, transform the reactant into the acetonitrile product (see the hints in the margin). Optimize the product, and verify that the resulting structure is a minimum.
3. Set up a QST3 optimization+frequency calculation.
4. Submit the job to Gaussian.
5. Examine the predicted transition structure once the job is completed.

Creating the product structure by starting with the reactant structure is important because the STQN facility in Gaussian requires that corresponding atoms appear in the same order within the two molecule specifications (although it does not matter whether the reactants or the products appear first). The bonding in the two structures need not be the same, but ensuring that it is by adding half bonds where necessary may avoid some coordinate errors in difficult cases. GaussView provides features for specifying corresponding atoms within the two structures. We will see an example of its use later in this chapter (see page 115).

If we try to run a QST2 calculation for this transition state, Gaussian will usually fail to be able to produce a starting structure from which to begin the optimization process.* This is not surprising in that the reactant and product differ in several structural features. For such cases, a QST3 calculation is called for.†

QST3 calculations are similar to QST2 calculations, but they require three molecular structures as input: the reactants, the products and a starting guess for the transition structure (the order of the first two does not matter, but the starting TS guess must be the third molecule specification). As with QST2, each molecule specification is preceded in the input file by its own title section.

GaussView can set up QST3 optimizations as well as QST2 jobs. These calculations require a molecule group containing three structures. After selecting the **TS (QST3)** optimization type in the **Optimize to a** field on the **Job Type** panel of the **Gaussian Calculation Setup** dialog, you must also specify which structure is the transition state guess via the **Geometry to use for TS Guess** field, the final setting in the list.

Geometry to use for TS Guess = 3

WebMO does not support QST3 calculations directly, but you can trick it into doing so:

1. Build and optimize the reactant and the product structures.
2. Build the TS initial structure. Run an energy calculation on it using a fast method (e.g., **UFF mechanics**).
3. Examine the job list in the **Job Manager** page. Note the job number of the TS guess.
4. We will now retrieve and copy the coordinates for the optimized product molecule. Click the magnifying glass icon corresponding to the product optimization job. Then

* Some program versions and input structures may succeed, but in general the QST2 procedure fails for these reactants and products.

† In fact, fully-optimized geometries are not always the optimal choice for QST2. Sometimes, structures which are distorted from the minimum to be closer to the expected TS work better with QST2.

click on the **Export Molecule** button in the bottom center of the page (below the build area). The dialog on the left in the following figure will open:

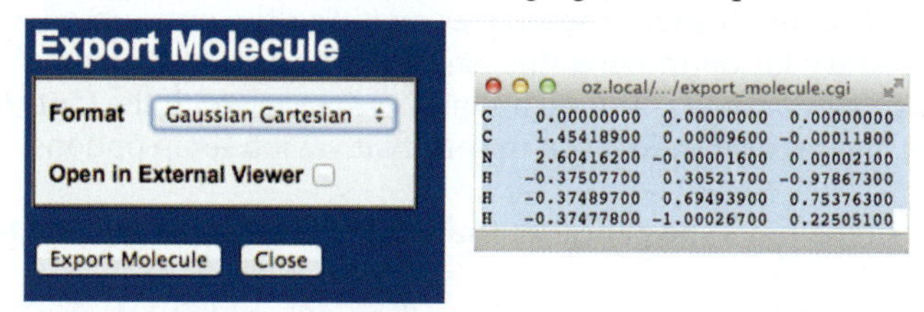

Select **Gaussian Cartesian** as the **Format** and then click the **Export Molecule** button. A windows like the one on the right in the preceding figure will appear. Select and copy the coordinates from the window.

5. Return to the **Job Manager** page without running a job. Once there, click on the job title link of the reactant optimization calculation. Once the molecule appears, click the **New Job Using This Geometry** button. Then click the right arrow icon to continue to the **Configure Gaussian Job Options** page.

6. Select a **Saddle Calculation** and an appropriate method and basis set.

7. In the **Advanced** panel, specify the job number of the TS guess energy calculation as the **Second Geometry**. Check the **Cartesian Coordinates** box, and uncheck the **Include Connectivity** box. Set the **Calculate Force Constants** field as appropriate (discussed below).

8. Move to the **Preview** panel, and click the **Generate** button. When the input file appears in the window, change the **QST2** option to the **Opt** keyword to **QST3**. Place the cursor at the end of the final line of the first molecule specification. Enter two carriage returns, followed by a title line for the products, two more carriage returns and finally the charge and spin multiplicity for the products.

9. Paste the copied structure directly below the lines you just added. Make sure that there is a blank line separating the second molecule specification from the third title section. Verify also that there are a couple of blank lines at the end of the file.

The items in red in the following illustration were typed in or edited manually, while the green lines were pasted in:

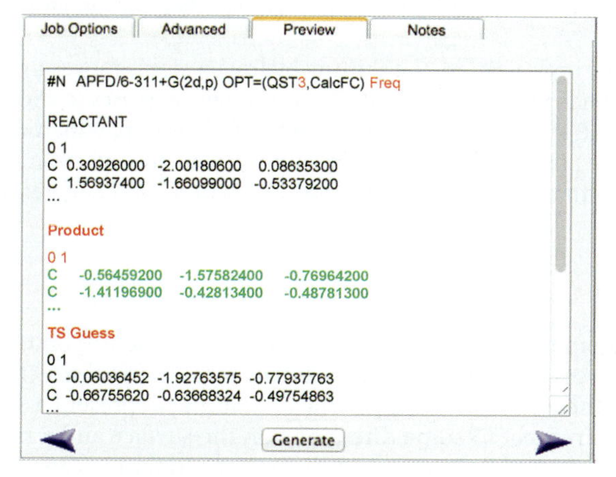

The job is now ready to run. You can initiate it in the usual way using the right arrow icon in the bottom right corner of the page.

Challenging geometry optimizations—of which transition structure searches are one example—often benefit from an extra step prior to the optimization procedure. This consists of computing the force constants at the initial geometry as an aid to the optimizer (see page 116 for details). It is necessary for optimizing this particular transition structure but need not be specified for most optimizations. This technique is requested with the **Opt=CalcFC** option. Both GaussView and WebMO allow you to specify it via job setup options:

- *GaussView:* In the **Job Type** panel, set **Calculate Force Constants** to **Once**.

- *WebMO:* On the **Advanced** tab of the **Configure Gaussian Job Options** page, select **Once** from the **Calculate Force Constants** popup menu.

  If your version of WebMO has a checkbox instead of a popup menu for this item, then you will have to specify this option manually. Navigate to the **Preview** tab, and click the **Generate** button. Edit the route line in the resulting input file, changing **OPT(EstmFC)** to **OPT(CalcFC)**.

In this particular case, these are the structures we used for the QST3 optimization:

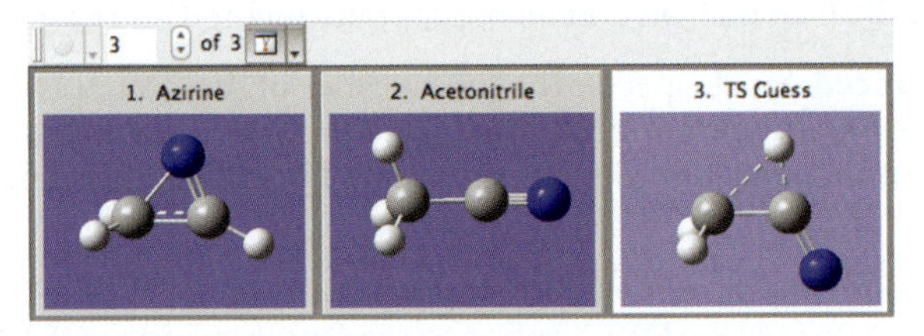

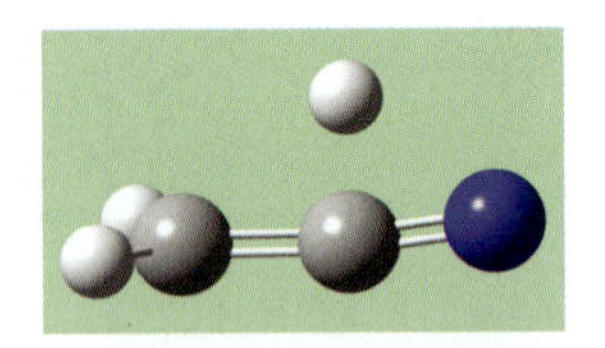

The QST3 optimization using our standard model chemistry completes in 20 steps. The subsequent frequency calculation finds one imaginary frequency, indicating that the structure is a transition state. The optimized transition structure is illustrated in the margin. The normal mode corresponding to the imaginary frequency indicates movement of the hydrogen atom between the two carbon atoms.

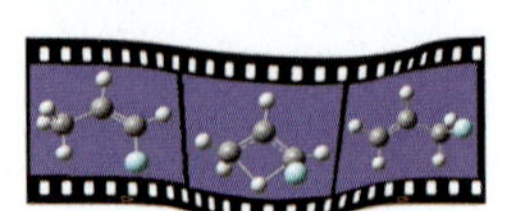

The optimization begins from a structure which is a much deformed version of the reactant, with a C-C-N bond angle of about 132°. The single hydrogen atom bonded to the carbon atom in the CN group in the reactant is a larger distance away from the methyl carbon atom (1.47 Å) than a normal C-H bond. Its distance to the other carbon atom is 1.62 Å.

In the first half of the optimization, the shifting hydrogen atoms moves toward the CN group's carbon atom and bonds to it while the C-C-N bond angle increases to about 160°. The second half of the optimization sees the hydrogens in the CH_2 group, which were initially nearly perpendicular to the plane of the two carbon atoms, rotate to a position where both are only a few degrees out of that same plane while the C-C-N bond angle increases to about 173°. ■

Characterizing Stationary Points

As we've noted, geometry optimizations converge to a structure on the potential energy surface where the forces on the system are essentially zero. The final structure may correspond to a minimum on the potential energy surface, or it may represent a saddle point, which is a minimum with respect to some directions on the surface and a maximum in one or more others. First order saddle points—which are a maximum in exactly one direction and a minimum in all other orthogonal directions—correspond to transition state structures linking two minima.

There are two pieces of information that are critical to characterizing a stationary point:

- ►The number of imaginary frequencies.
- ►The normal mode corresponding to the imaginary frequency.

Imaginary frequencies are listed in the output of a frequency calculation as negative numbers. By definition, a structure which has *n* imaginary frequencies is an *n*th order saddle point. Thus, ordinary transition structures are characterized by one imaginary frequency since they are first-order saddle points.

If applicable, Gaussian notes that there is an imaginary frequency present in the Gaussian output file, just prior to the frequency and normal modes output. The first frequency value(s) will also be less than zero. Log files may be searched for this line as a quick check for imaginary frequencies, as in this example (Linux/UNIX):

```
$ grep imaginary job.log
****** 1 imaginary frequencies (negative signs) ******
```

Under Windows, use the **findstr** command to locate this line if present. As we've seen, you can also view the number of imaginary frequencies in GaussView and WebMO (see page 58).

It is important to keep in mind that finding exactly one imaginary frequency does not guarantee that you have found the transition structure in which you are interested. Saddle points always connect two minima on the potential energy surface, but these minima may not be the reactants and products of interest. Whenever a structure yields an imaginary frequency, it means that there is some geometric distortion for which the energy of the system is lower than it is at the current structure (indicating a more stable structure). In order to fully understand the nature of a saddle point, you must determine the nature of this deformation.

CHARACTERIZING STATIONARY POINTS

If you were looking for ...	And the frequency calculation found ...	It means ...	So you should ...
A minimum	0 imaginary frequencies	The structure is a minimum.	Compare the energy to that of other isomers if you are looking for the global minimum.
A minimum	≥ 1 imaginary frequencies	The structure is a saddle point, not a minimum.	Continue searching for a minimum: try distorting the molecular structure along the normal mode corresponding to the imaginary frequency. Often, this involves modifying the structure to lower the molecular symmetry.
A transition state	0 imaginary frequencies	The structure is a minimum, not a saddle point.	Try performing a QST2 or QST3 optimization to find the TS.
A transition state	1 imaginary frequency	The structure is a true transition state.	Determine if the structure connects the correct reactants and products by examining the imaginary frequency's normal mode or by performing an IRC calculation.
A transition state	> 1 imaginary frequency	The structure is a higher-order saddle point, not a transition structure connecting two minima.	Examine the normal modes corresponding to the imaginary frequencies. One of them will generally point toward the reactants and products. Modify the geometry based on the displacements in the other mode(s), and rerun the optimization. QST3 may again be of use.

One way to do so is to examine at the normal mode corresponding to the imaginary frequency and determine whether the displacements that compose it tend to lead in the directions of the structures that you think the transition structure connects. The symmetry of the normal mode is also relevant in some cases (see the following example). Animating the vibrations with a chemical visualization package is often very useful. Another, more accurate way to determine what reactants and products the transition structure connects is to perform an IRC calculation to follow the reaction path and thereby determine the reactants and products explicitly; this technique is discussed in Chapter 6 (see page 244).

The table on the preceding page summarizes the most important cases you will encounter when attempting to characterize stationary points.

EXAMPLE 3.5 EXPLORING THE C_3H_5F POTENTIAL ENERGY SURFACE

We are interested in exploring the C_3H_5F* potential energy surface. We will begin by running optimization and frequency jobs on these three isomers of 1-fluoropropene:

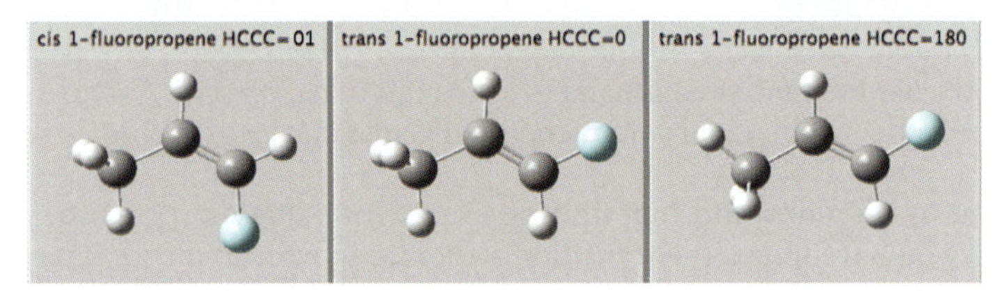

All of the optimizations are successful. The frequency jobs for the two forms where the H-C-C-H dihedral angle is 0° produce no imaginary frequencies, and the cis form is lower in energy than the trans form by about 2.8 kJ/mol (~0.7 kcal/mol).

The frequency job on the rightmost structure produces one imaginary frequency, indicating that this conformation is a transition structure and not a minimum. But what two minima does it connect? Is it the transition structure for the cis-to-trans conversion reaction (i.e. rotation about the C=C bond)?

We look first at the energies of the three compounds:

CONFORMER	THERMAL ENERGY (hartrees)	RELATIVE E (kJ/mol)
cis (0°)	-216.96055	0.0
trans (0°)	-216.95950	2.8
trans (180°)	-216.95696	9.4

The 180° trans structure is only about 6.7 kJ/mol (~1.6 kcal/mol) higher in energy than the 0° trans conformation, a barrier which is quite a bit less than one would expect for rotation about the double bond. We note that this structure is a member of the C_s point group (as are all three minima). Its normal modes of vibration, therefore, will be of two types: the symmetrical A' and the non-symmetrical A" (since point-group symmetry is maintained in the course of symmetrical vibrations).

To investigate the status of this structure further, we next examine the frequency data and normal mode corresponding to the imaginary frequency. Note that the magnitude of the imaginary frequency is not very large (-223), indicating that the geometric distortion desired by the molecule is modest.

* IUPAC: 1-fluoropropene.

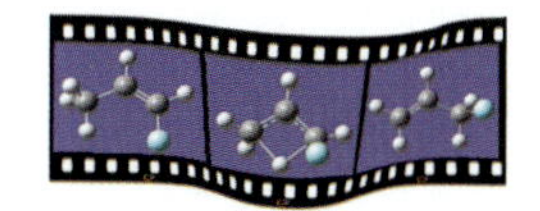

When we animate this normal mode, we see that the largest motion is the rotation of the three hydrogen atoms in the methyl group, accompanied by some lesser motion in the other two hydrogen atoms:

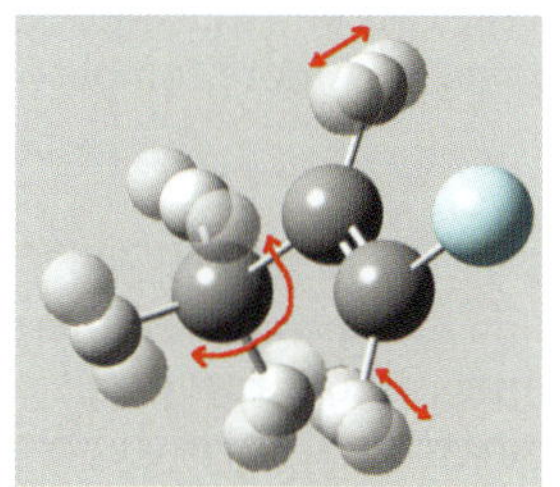

If we examine the output file for this job, we can determine the symmetry of this mode:

```
****** 1 imaginary frequencies (negative Signs) ******
Harmonic frequencies (cm**-1), IR intensities (KM/Mole),
Raman scattering activities (A**4/AMU), depolarization
ratios for plane and unpolarized incident light, reduced
masses (AMU), force constants (mDyne/A), and normal
coordinates:
                      1            2            3
                      A"           A"           A'
Frequencies --   -220.7242     282.378     288.0130
```

Its symmetry is A", indicating that this is a symmetry-breaking mode. The molecular structure has C_s symmetry, indicating that there is a single plane of symmetry (in this case, the plane of the carbon atoms). The structure wants to move down the PES to a lower-energy structure of equal or lower symmetry.

From all of this, we can deduce that this transition structure connects two structurally-equivalent minima, and that the path between them corresponds to a methyl rotation. This is not a very interesting transition structure.

We must look further in order to locate the transition structure linking the cis and trans forms of 1-fluoropropene. Since we are looking for a normal mode which suggests rotation about the C=C bond, then we can expect that its major motion will be in the dihedral angles involving those carbon atoms and the fluorine and hydrogen atoms.

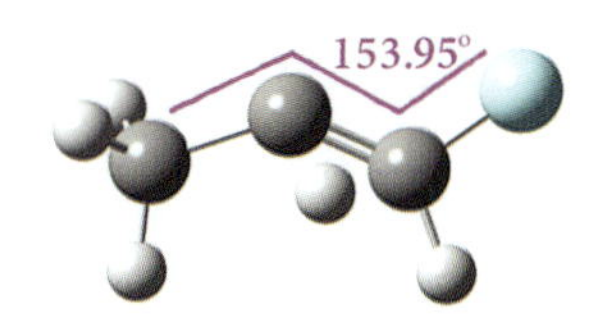

Another transition structure that we located is illustrated in the margin. An optimization+frequency calculation on it reveals that it too has one imaginary frequency, of significantly larger magnitude (-896 vs. -221). Examining the normal mode reveals displacements in the dihedral angles involving the two carbons of interest, strongly suggesting that this is the transition structure that we seek.

The predicted energy of this structure is approximately -216.82927 hartrees, yielding a reaction barrier of about 341.9 kJ/mol (~81.7 kcal/mol) for the 0°-180° transformation. This value is more in line with expectations, although it is on the high side.

We will continue exploring this potential energy surface in Exercise 3.4, p. 115.

Interpreting Normal Mode Information

The preceding example introduced using normal mode vibrations as an aid to interpreting optimized molecular structures. In the case of a transition structure, this information suggests which reactants and products it connects. A more complex example of this technique follows.

EXAMPLE 3.6 AZIDE DECOMPOSITION: CONCERTED VS. STEPWISE MECHANISMS

A Curtius rearrangement is a reaction that involves the thermal decomposition of a carboxylic azide to an isocyanate. For example, in the reaction below, isopropylazide converts to dimethylimine plus molecular nitrogen (N_2), which involves a hydrogen migration from the central carbon atom to the adjacent nitrogen atom as well as N_2 elimination. Two possible mechanisms are illustrated in the diagram below: concerted elimination and migration proceeding through a transition structure (upper pathway) or a two-step process involving a nitrene intermediate (lower pathway):

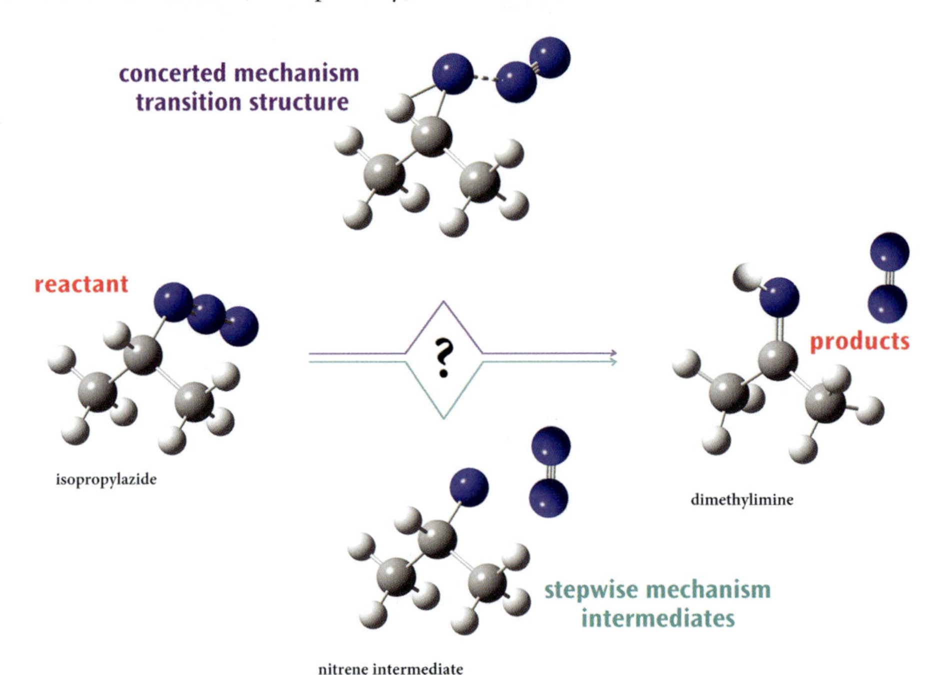

In order to investigate this reaction, we ran the following calculations:

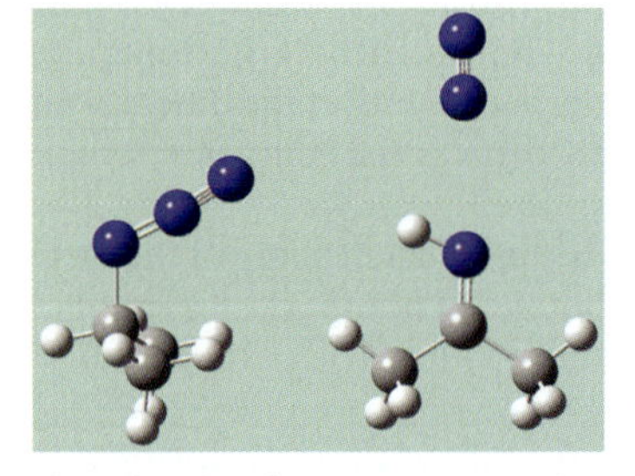

- An Opt+Freq on isopropylazide*
- An Opt+Freq on dimethylimine
- An Opt+Freq on N_2
- An Opt+Freq on the nitrene intermediate
- A QST2 transition structure search, using the reactant and product structures at the right. The distance between the nitrogen atom in dimethylimine and the N_2 molecule is about 2.8 **Å**.

The optimizations on the reactant and two product molecules lead to minima. The optimizations on the nitrene intermediate and the QST2 transition state optimization both produce structures with one imaginary frequency: transition structures.

* IUPAC: isopropylazide: 2-azido-propane; dimethylimine: 2-propanimine.

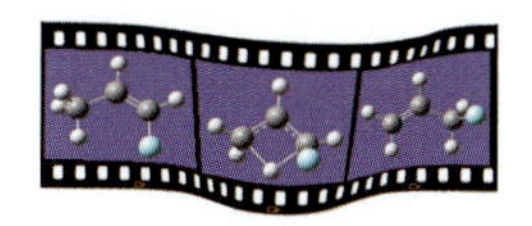

When we animate the normal mode corresponding to the imaginary frequency for the structure resulting from the QST2 calculation, we see that it connects the reactants and products. The vibration consists of the unbonded hydrogen atom moving between the central carbon atom and the nitrogen atom. As this hydrogen approaches the carbon atom, the N_2 molecule moves and twists away from the dimethylimine molecules, and the remaining nitrogen atom bends out of the C-C-C plane away from the hydrogen atom.

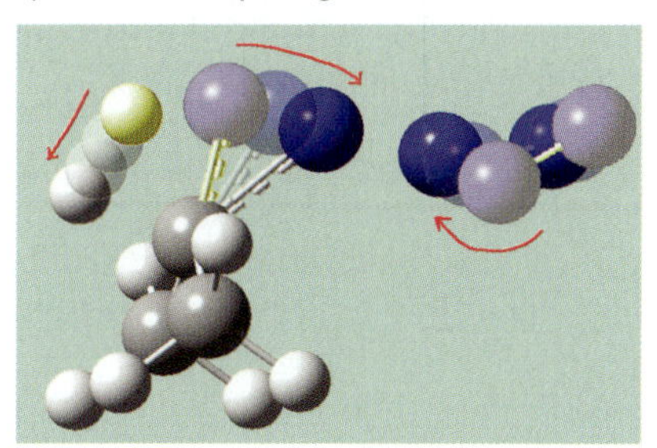

For nitrene intermediate optimization we also found a transition structure (illustrated in the margin). The normal mode corresponding to the imaginary frequency consists of minor displacement of the hydrogen atom bonded to the nitrogen. This transition structure clearly does not correspond to the hydrogen shift.

These results suggest that the reaction proceeds via a concerted mechanism. Although the mechanism was originally proposed as stepwise [Stieglitz1896], modern literature supports the concerted mechanism. The first experimental evidence of a concerted mechanism was observed for related compounds [Linke67], and [Bock88] provides experimental evidence for this reaction and related reactions.

The following table summarizes the results for this reaction (all values are in hartrees):

MOLECULE/JOB	OPTIMIZED TO ...	THERMAL ENERGY (E)	ENTHALPY (H)	FREE ENERGY (G)
isopropylazide	a minimum	-282.46362	-282.46268	-282.50090
dimethylimine	a minimum	-173.07782	-173.07687	-173.11062
N_2	a minimum	-109.46018	-109.45923	-109.48096
QST2 TS search	a transition state	-282.39411	-282.39317	-282.43354

The following figure plots the energetics of this reaction. Note that the symbol E in the table and the graph refers to the thermal energy (E^{therm}).

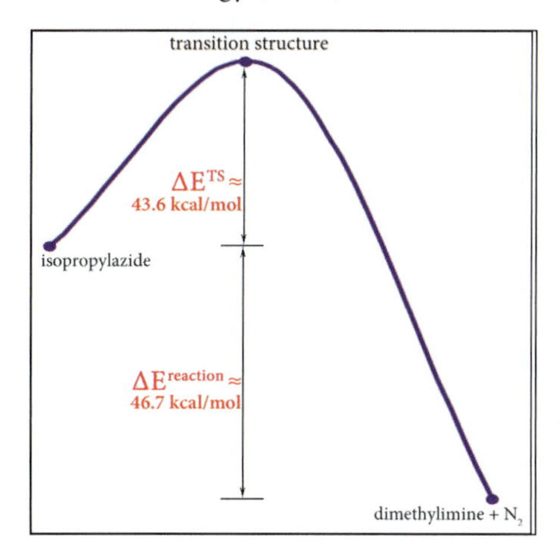

The predicted value of $\Delta G^{\ddagger}$ is 0.067 hartrees or ~42.3 kcal/mol (176.8 kJ/mol). This compares well with the observed value of 40.5 kcal/mol for the analogous reaction for the reactant methyl azide [Odell70]. The predicted value of ΔE for the reaction* is 0.074 hartrees (~46.7 kcal/mol, 195.3 kJ/mol), which is also in line with the published B3LYP/6-311+G(d) and MP2/6-31G(d) values of 53.25 and 41.8 kcal/mol (respectively) [Zeng04], although the values from the literature include no ZPE/thermal correction.

* $\Delta G^{\ddagger}=G^{\text{transition structure}}-G^{\text{reactant}}$; $\Delta E^{\text{reaction}}=E^{\text{reactant}}-E^{\text{products}}$

GOING DEEPER: WHAT IS THE GIBBS FREE ENERGY CHANGE FOR A REACTION?

The previous example introduced the concept of a reaction activation energy: the free energy difference between the transition structure and the reactants ($\Delta G^{\ddagger}$). A related quantity is the reaction free energy change: the difference in free energies between the products and the reactants (ΔG). Here, we consider the Gibbs free energy in more detail.

Absolute values of the Gibbs free energy (G) and free energy changes for reactions are made complicated by the fact that a reference state must be specified. Several different conventions are in common use:

- $\Delta G^{\text{o-atm}}$: Standard reference of 298.15 K and 1 atmosphere.
- $\Delta G^{\text{o-bar}}$: Standard reference of 298.15 K and 1 bar.
- ΔG^{*}: Standard reference of 298.15 K and 1 mole/liter.

These conventions dictate the conditions that apply to the species present in a balanced chemical reaction, and the ΔG is then the G for the products minus the G for the reactants, each multiplied by their stoichiometric coefficient.

This free energy change is not observed directly in the same sense that the heat of a reaction (ΔH) is measured. A system containing reactants and products will, given enough time, reach a state of equilibrium in which the reactant free energies and the product free energies of the mixed system are equal and ΔG is zero. The difference between this zero free energy change and the standard reference free energy change is captured in the equilibrium constant (K) for the reaction, whose value then also depends on the chosen reference state:

$$\Delta G = -RT\ln(K)$$

As an example, let's consider the famous Haber Process for the synthesis of ammonia:

$$N_2\,(g) + 3H_2\,(g) \rightarrow 2NH_3\,(g)$$

We can use our standard model to calculate the Gibbs Free Energies (G) for the species present in this equation using the various reference states and then compute both ΔG and K for each one:

REF	T (K)	P (atm)	$G(N_2)$ (au)	$G(H_2)$ (au)	$G(NH_3)$ (au)	ΔG (kJ/mol)	K
ATM	298.15	1.0000	-109.48096	-1.17388	-56.51520	-73.03	6.24×10^{12}
	800	1.0000	-109.52075	-1.20198	-56.55623	+37.38	0.00363
BAR	298.15	0.9870	-109.48097	-1.17389	-56.51522	-72.97	6.08×10^{12}
	800	0.9870	-109.52079	-1.20201	-56.55626	+37.55	0.00353
MOL/L	298.15	24.465	-109.47794	-1.17086	-56.51218	-88.88	3.74×10^{15}
	800	49.234	-109.51015	-1.19138	-56.54563	-18.29	15.639

The pressures associated with the unitary molar reference are derived from the ideal gas equation:

$$P = \frac{nRT}{V} = \frac{(1.0mol)\left(0.08206\frac{L \cdot atm}{K \cdot mol}\right)T}{1.0L}$$

Since T is in K, this yields P (atm) = 0.08206 T.

In Gaussian, the default values for the temperature and pressure are 298.15 K and 1.0 atm, respectively. To specify other values, use the **Pressure** and **Temperature** keywords. Accordingly, the first row of the preceding table contains the values obtained from Gaussian by default.

Notice how ΔG has both positive and negative values depending on the reference temperature. The correction in going from atm to mol/L reference is a factor of 600 in the equilibrium constant! Clearly we must take care when calculating and discussing free energy.

The effect due to the reference pressure is easily calculated:

$G^{1\ bar} = G^{1\ atm} + RT\ \ln(0.987)$
$G^{1\ mol} = G^{1\ atm} + RT\ \ln(0.08206T)$

So to convert the default Free Energy Changes computed in Gaussian, we can use:

$\Delta G^{1\ bar} = \Delta G^{1\ atm} + \Delta nRT\ \ln(0.987)$
$\Delta G^{1\ mol} = \Delta G^{1\ atm} + \Delta nRT\ \ln(0.08206T)$

Here Δn is the change in the number of moles going from reactants to products. In our example above, Δn equals -2. There are many examples where the change in the number of moles is zero for a given reaction stoichiometry; in such cases, the ΔG values are all the same regardless of the reference pressure.

Finally, assuming we are interested in the 298.15 K reference state, our conversion formulas will simplify to:

$\Delta G^{1\ bar}$ (kJ/mol) = $\Delta G^{1\ atm}$ (kJ/mol) – Δn*0.03244
$\Delta G^{1\ mol}$ (kJ/mol) = $\Delta G^{1\ atm}$ (kJ/mol) + Δn*7.9255

$\Delta G^{1\ bar}$ (kcal/mol) = $\Delta G^{1\ atm}$ (kcal/mol) – Δn*0.007753
$\Delta G^{1\ mol}$ (kcal/mol) = $\Delta G^{1\ atm}$ (kcal/mol) + Δn*1.894

We will look at the additional care required when computing free energy changes in solution in chapter 5.

Exercises

EXERCISE 3.1 COMPARING STRUCTURES IN THE VINYL SERIES

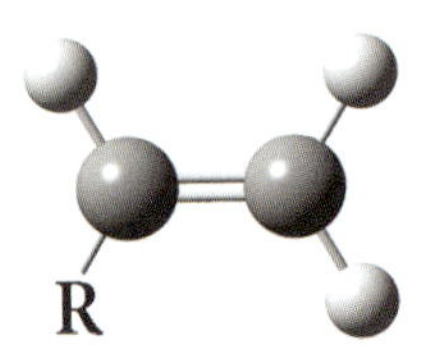

The vinyl group is characterized by its two double-bonded carbon atoms: CH_2=CHR. The name purportedly derives from the Latin word for wine. Vinyl chloride†, the compound with Cl as the substituent, is the precursor to PVA, the substance commonly known as "vinyl."

We will consider several compounds in this series, where R is H, F, CH_3, NH_2 and OH*:

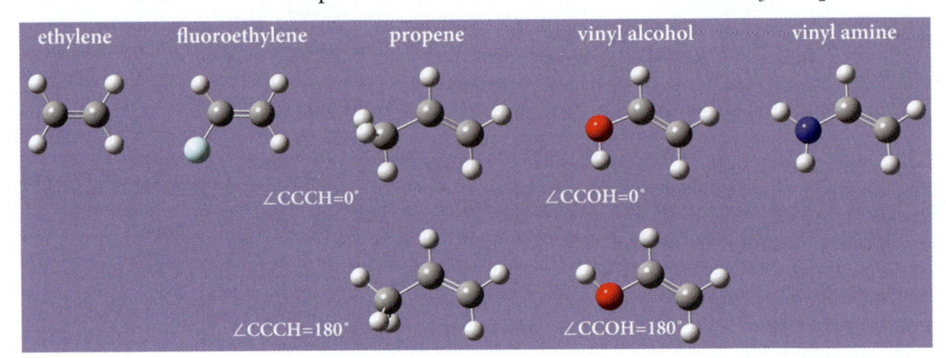

Note that propene and vinyl alcohol each have two conformers corresponding to 0° and 180° values for the C-C-C-H dihedral angle.

Find the minima for each of these compounds, and then compare the key structural parameters and predicted quantities for them. Model all compounds using our standard model chemistry: APFD/6-311+G(2d,p). What trends can you discern among these compounds in the vinyl series?

SOLUTION

Propene & Vinyl Alcohol Conformer Optimizations

All four of these optimizations converge to minima. The jobs for the propene conformers converge in 4 steps, and those for the vinyl alcohol conformers converge in 5 steps. For both pairs, each optimization leads to a different structure (compare the optimized geometries to verify this), and therefore to a different point on the potential energy surface. This means that the same stoichiometry corresponds to two different stationary points on the potential energy surface. If this were not the case, then both of the input structures would converge to the same optimized geometry, although one would probably take much longer to do so than the other.

The following table presents and compares the energies for these optimized structures:

MOLECULE	CONFORMER	STRUCTURE TYPE	THERMAL ENERGY (hartrees)	$\Delta E^{180°\text{-}0°}$
propene	0°	minimum	-117.86526	2.20 millihartrees ≈
	180°	minimum	-117.86307	1.38 kcal/mol ≈ 5.8 kJ/mol
vinyl alcohol	0°	minimum	-153.80925	2.28 millihartrees ≈
	180°	minimum	-153.80736	1.43 kcal/mol ≈ 6.0 kJ/mol

* IUPAC: vinyl chloride: chloroethene; vinyl alcohol: ethenol; vinyl amine: ethenamine.

In both cases, the 0° is slightly lower in energy than the 180° structure. The partial ring-like arrangement of the three heavy atoms and the planar hydrogen from the substituent is slightly preferred over the other form. However, the small energy difference suggests that that converting from one to the other happens very easily.

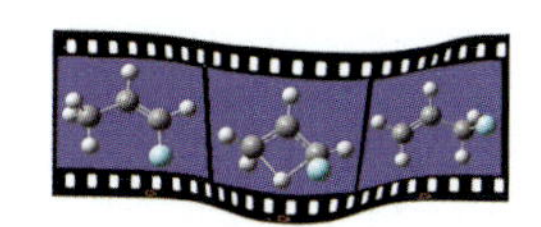

Characterizing Planar Vinyl Amine

Our optimization of vinyl amine begins from a planar structure, and it converges to a planar structure in four steps. However, the frequency calculation on this structure reveals that it is a transition state, not a minimum. When we animate the normal mode corresponding to the imaginary frequency, there is motion in the nitrogen atom and its hydrogens are out of the plane of the molecule:

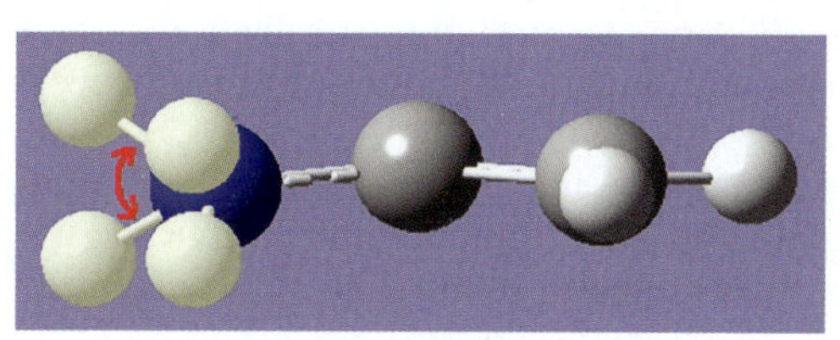

This suggests that if we vary the structure of the NH_2 group, we will be able to locate the minimum. We modify the structure so that the hydrogen atoms attached to the nitrogen atom are positioned out of the C=C plane and perform another geometry optimization.

This optimization succeeds in locating a minimum in which the nitrogen atom exhibits pyramidalization in the optimized structure. The results of these two optimizations are given in the table below:

STRUCTURE	TYPE	H-C=C-H (°)	H-C=C-N (°)	THERMAL ENERGY (hartrees)
planar	transition state	0.0	0.0	-133.80279
out-of-plane	minimum	-32.8	-0.8	-133.80294

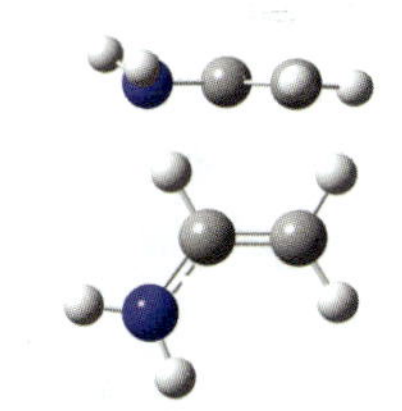

Note that the minimum is also lower in energy than the transition structure, by 0.16 millihartrees (~0.1 kcal/mol, ~0.4 kJ/mol).

Vinyl Series Results

The following table summarizes the results of all of our calculations on these vinyl compounds, including the ones on ethylene and fluoroethylene from the previous chapter (see Exercise 2.4, p. 67):

MOLECULE/CONFORMER	R	C=C (Å)	C-R (Å)	C-C-R (°)	C-C-H (°)†	μ (debye)
ethylene	H	1.33	1.09	121.7	121.7	0.0
fluoroethylene	F	1.32	1.34	122.0	125.8	1.44
propene	CH_3					
0° conformer		1.33	1.50	125.0	118.8	0.43
180° conformer		1.33	1.50	124.6	118.5	0.40
vinyl alcohol	OH					
0° conformer		1.33	1.35	127.3	122.3	1.02
180° conformer		1.32	1.36	122.4	122.0	1.92
vinyl amine minimum	NH_2	1.33	1.39	126.5	120.0	1.60

†This bond angle is to the hydrogen atom on the central carbon atom.

We can draw the following conclusions based on these results:

- Fluorine is the only substituent which shortens the C-C bond. Fluorine is highly electronegative and wishes to obtain additional electron density. It attempts to draw it from the two carbons, which move closer together in order to share the remaining electrons more easily.

- The C-C-R bond angle changes significantly in the 0° conformation of vinyl alcohol and in vinyl amine and propene.

- The C-C-H bond angle also varies with substituent (the hydrogen attached to the same carbon atom as the substituent), most dramatically in the case of the substituents CH_3 and F, which decrease and increase the bond angle, respectively.

- Ethylene has no dipole moment, and propene has only a very small one. The other systems have nontrivial dipole moments. Thus, the more electronegative substituents produce nontrivial dipole moments, in contrast to a single hydrogen atom or a methyl group.■

TO THE TEACHER: FURTHER EXPLORATION

The keto-enol tautomerization of acetaldehyde and vinyl alcohol is a proton migration reaction that complements this examination of the vinyl series.

EXERCISE 3.2 COMPARING $C_{60}O$ ISOMERS

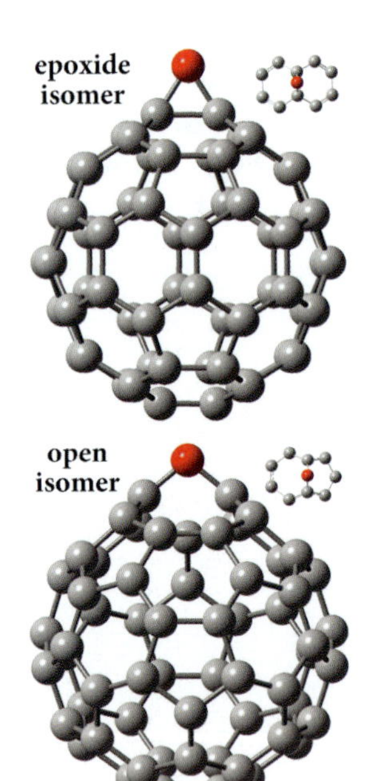

There has been considerable interest in fullerenes and fullerene derivatives since their initial discovery in the mid-1980s. Finding the lowest energy isomer among a variety of choices of attachment is always an interesting and important question, and doing so remains an open issue for many of these compounds.

In C_{60}, all carbons are equivalent, but there are two types of C–C bonds:

- A bond joining two six-membered rings: a 6-6 bond
- A bond joining a five-membered ring to a six-membered ring: a 5-6 bond

An oxygen atom can bind to either of these two sites, potentially forming a three-member (C-O-C) ring. Thus, there are two possible forms for $C_{60}O$* (see the illustrations in the margin). The isomer with the 6-6 bond is known as the epoxide isomer. The other (5-6 bond) isomer is characterized by a very long bridging carbon distance, indicating that the two carbons are no longer bonded (which is why we refer to it as the open form).

Perform an optimization of these two derivatives in order to discern which is the more favorable isomer. What are the most dramatic structural features that characterize these two isomers? Do the bridging carbons remain bonded in the derivative? Experimentalists have proposed that oxygen should bind to the 6-6 bond. Does your calculation support or refute this? Can you justify any inconsistencies?

Select one or more of the following model chemistries for your calculations, depending on your available computing resources. If you run more than one job, begin each successive optimization from the optimized geometry from the lower level of theory. Note that this is often a good way to proceed when you are optimizing a large molecule.

* IUPAC: [5,6]fullereno-C60-Ih-oxiren.

- BLYP/6-31G(d)/Def2SV: This uses the BLYP pure functional in conjunction with a modest size basis set and a density fitting set for enhanced performance.
- APFD/6-31G(d): Our usual hybrid functional with a modest size basis set.
- APFD/6-311G(2d,p): This is our standard model chemistry omitting the diffuse functions which should make little difference for these molecules.

SOLUTION

The optimized structures are quite similar for all of the model chemistries we are considering. The second isomer is characterized by a very long bridging carbon distance in all of them. The geometries obtained with the hybrid density functionals are all in very good agreement with one another, regardless of the basis set. ΔE and ΔZPE differ more substantially for different model chemistries.

The following table summarizes our calculations as well as some recently published work:

MODEL CHEMISTRY	ISOMER	C-O (Å)	C-C (Å)	C-O-C (°)	ENERGY [hartrees]	$\Delta E^{epoxide-open}$ (millihartrees)
BLYP/6-31G(d)/Def2SV	epoxide	1.45	1.57	65.2	-2377.96923	4.13
	open	1.41	2.17	100.2	-2377.97336	
B3LYP/6-31G(d)	epoxide	1.42	1.54	65.5	-2361.35611	3.36
	open	1.39	2.15	101.8	-2361.35947	
B3LYP/6-311G(2d,p)	epoxide	1.42	1.54	65.5	-2361.86695	4.88
	open	1.38	2.15	101.8	-2361.87183	
APFD/6-31G(d)	epoxide	1.41	1.53	65.5	-2359.66801	-3.81
	open	1.38	2.13	100.9	-2359.66419	
APFD/6-311G(2d,p)	epoxide	1.41	1.54	65.6	-2360.15295	-3.15
	open	1.37	2.12	100.8	-2360.14980	
B3PW91/cc-pVDZ [Sohn10]	epoxide	1.41	1.59	65.6		-0.16
	open	1.38	2.13	101.1		
B3PW91/6-311G(2d,p)	epoxide	1.41	1.53	65.6	-2361.01333	0.16
	open	1.38	2.13	101.0	-2361.01349	

In most cases, the epoxide isomer is slightly higher in energy than the open isomer. However, for APFD, which includes treatment of dispersion, the epoxide isomer is the lower in energy at both a small and large basis set. Sohn and coworkers [Sohn10] obtained the same result,* albeit with a very small magnitude, although increasing the basis set size with the functional that they chose reverses the isomers' energy ordering, by the same small magnitude.

Here are the predicted energy differences between the two forms ($\Delta E^{epoxide-open}$) at many levels of theory (the number of basis functions for this molecule is indicated in parentheses below each basis set name):

	$\Delta E^{epoxide-open}$ (kJ/mol)				
MODEL CHEMISTRY	3-21G (549)	6-31G (549)	cc-pVDZ (854)	6-311G(d) (915)	6-311G(2d,p) (1403)
B3LYP/	33.9	19.2	9.6	8.8	13.0
APFD/	16.8			-10.0	-8.4
B3PW91/	22.6[c]	8.4[c]	-0.4[c]	-2.9[c]	0.4

* [Sohn10]'s discussion centers on the B3PW91/cc-pVDZ result, and this is the model that they use for their frequency calculation. As the following table on this page will indicate, they obtained other results which also suggest the epoxide isomer is the ground state, as well as some which suggest the opposite.

MODEL CHEMISTRY	ΔE epoxide-open (kJ/mol) 3-21G (549)	6-31G (549)	cc-pVDZ (854)	6-311G(d) (915)	6-311G(2d,p) (1403)
MP2/	24.3[c]	5.4[c]	3.3[c]	-0.4[c]	
HF/	37.7[a]				
MNDO[d]	24.3[a]				
PM3[d]	27.2[b]				
PM6[d]	37.9				

[a][Raghavachari92]; [b][Foresman96]; [c][Sohn10]
[d]No basis set is specified with these models.

These results cover a very wide range of model chemistry accuracies. Raghavachari's early results reflect the very small basis set, which was all that was computationally practical at the time; the higher accuracy methods perform comparably using the very small 3-21G basis set.

Basis set size increases from left to right within the table. For DFT and MP2, the extent to which the open form is favored decreases as they move from 3-21G to the 6-31G through 6-31G(d) basis sets. However, for the larger 6-311G(2d,p) basis set, the results vary greatly by functional. B3LYP continues to find the open form to be lower in energy, by a greater amount than with the smaller basis sets. B3PW91 finds near isoenergeticity between the two forms with cc-PVDZ and larger basis sets.

In contrast, APFD favors the epoxide form. The energy difference is slightly smaller with the 6-311+G(2d,p) basis set than for the smaller basis sets. These results led us to model these systems with another functional incorporating dispersion. The wB97xD functional [Chai08, Grimme06] with the cc-pVTZ basis set (comparable in size to our standard basis set) again found the epoxide form to be lower in energy, by -3.3 kJ/mol.

These varied results make us reluctant to draw a definitive conclusion as to which isomer is the ground state. It seems clear that the two compounds are very close in energy. We will continue to consider these molecules in the next chapter (see Exercise 4.9, p. 180) where a comparison of the predicted IR spectrum with experiment will shed additional light on this question.■

EXERCISE 3.3 LOCATING A TRANSITION STRUCTURE ON THE GeH_4 PES

Predict the structure of the transition state for the following reaction* involving germane:

$$GeH_4 \rightarrow GeH_2 + H_2$$

What are the predicted Ge-H and H-H bond lengths for the departing H atoms?

Hint: Set up this calculation by building the reactant, copying it, and then modifying the copy into the products (or vice-versa). If you are not sure of the structure of the products, then optimize that complex first.

* IUPAC: germane → λ2-germane + dihydrogen.

SOLUTION

We ran a QST2 optimization, using the reactant and products below:

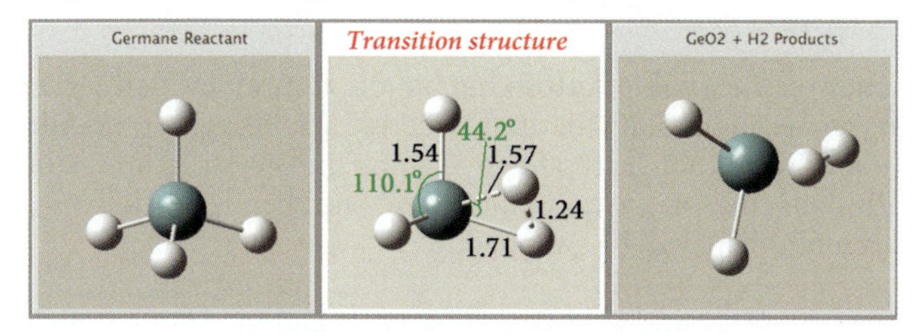

The QST2 geometry optimization, which started from the transition structure guess in the margin, converged in 23 steps, resulting in the structure in the center. The subsequent frequency job confirmed that it is a transition structure.

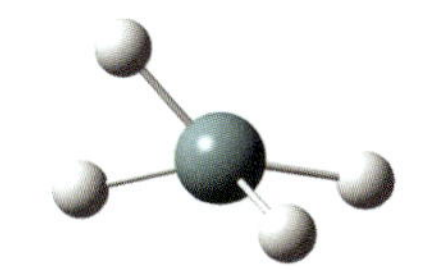

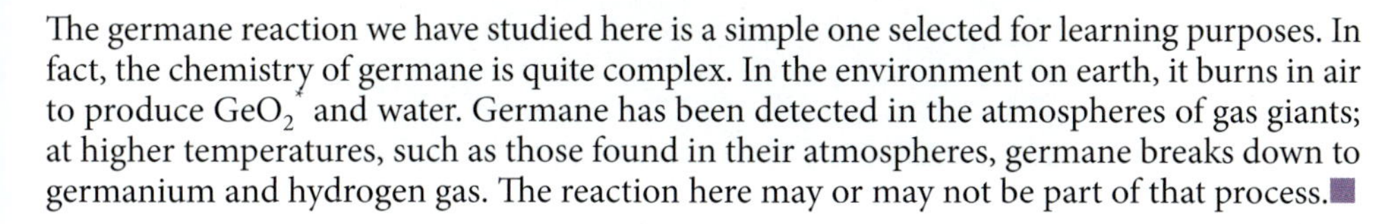

The germane reaction we have studied here is a simple one selected for learning purposes. In fact, the chemistry of germane is quite complex. In the environment on earth, it burns in air to produce GeO_2* and water. Germane has been detected in the atmospheres of gas giants; at higher temperatures, such as those found in their atmospheres, germane breaks down to germanium and hydrogen gas. The reaction here may or may not be part of that process.■

EXERCISE 3.4 MODELING HYDROGEN SHIFTS IN C_3H_5F

In this exercise, we continue our study of C_3H_5F potential energy surface begun in Example 3.5, p. 104. Another sort of transformation that cis 1-fluoropropene can undergo is a 1,3 hydrogen shift, resulting in 3-fluoropropene† (right):

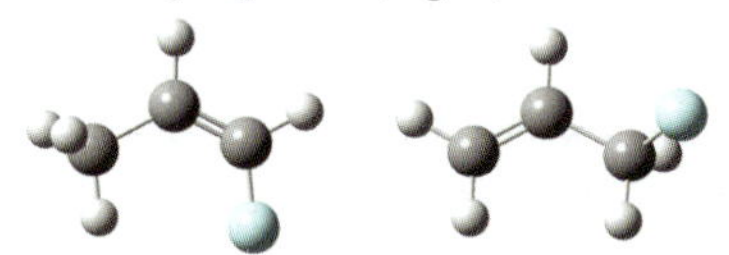

Locate the transition structure for this reaction. You will need to optimize 3-fluoropropene, and then set up a QST3 calculation for the reaction.

It can be challenging to create your starting product structure by copying and modifying the optimized reactant, and it will require many steps. Another alternative is to build and optimize the product separately, and then specify the corresponding atoms using the GaussView's **Connection Editor**. In order to make this tool work optimally, we add bonds between the migrating hydrogen atom and the target carbon atom in all structures before opening the **Connection Editor**. The following illustration shows the results of doing so for the reactant and product structures:

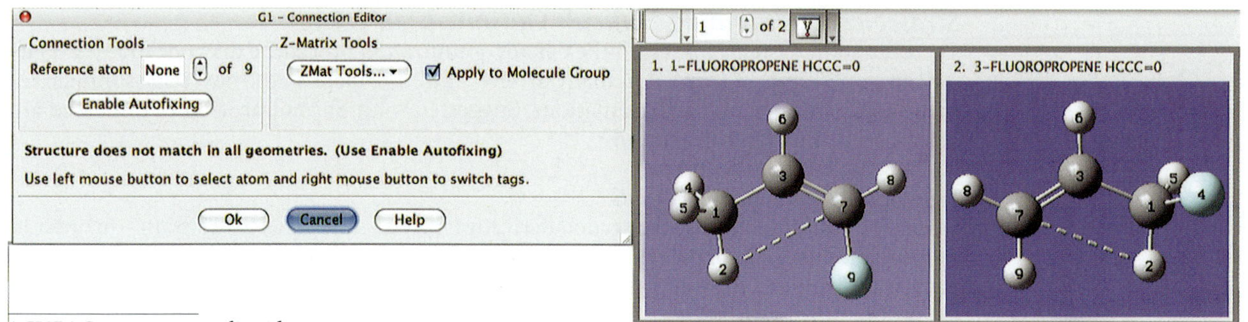

* IUPAC: germanium dioxide.

† IUPAC: 3-fluoro-1-propene.

Notice that the atom numbering is quite different. For example, the fluorine atom is atom 9 in the first structure and atom 4 in the second. The STQN facility in Gaussian will not be able to locate the proper transition state given these input structures.

However, if we click on the **Enable Autofixing** button, GaussView will successfully identify all corresponding atoms and modify the atom ordering accordingly. If the **Autofixing** feature in the **Connection Editor** is not successful in aligning the two structures, atom ordering can be done manually. Left click on an atom to select it, and then right click on a second atom, and their two atom numbers will be swapped.

Here are the three structures we used in our QST3 calculation. After autofixing the reactant and product, we copied the latter into the same molecule group and modified it into the transition structure guess:

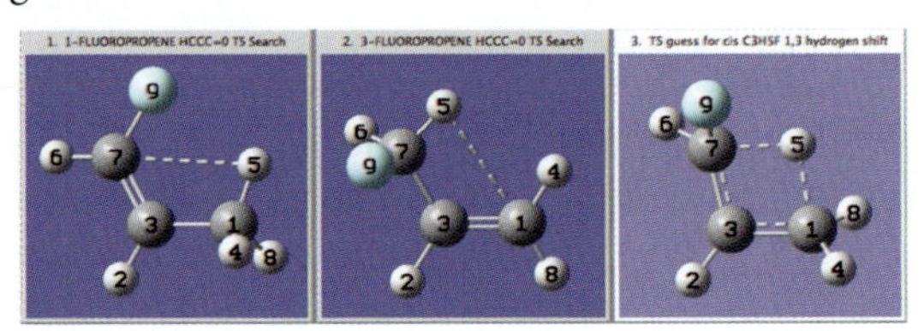

Our Gaussian calculation can now be set up and run.

SOLUTION

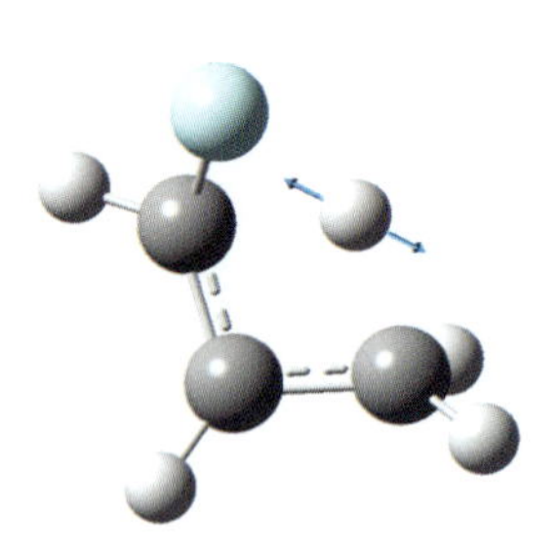

The QST3 transition state optimization converges in 11 steps, and the frequency calculation finds one imaginary frequency. When we examine the associated normal mode, we observe that the majority of the motion in this mode involves the shifting hydrogen atom, so it appears that this is the correct transition structure (see the illustration in the margin, where the normal mode motion is indicated by displacement vectors). This could be confirmed with an IRC calculation, as we'll discuss in Chapter 8. The large magnitude of the frequency (about -1875 cm^{-1}) also indicates a substantial change in structure. Finally, the predicted energy barrier of 0.14648 hartrees (~91.9 kcal/mol, ~384.6 kJ/mol) is of the right order of magnitude.■

GOING DEEPER: ADDITIONAL REACTIONS

If you would like to explore this potential energy surface further, the 1,3 fluorine shift provides an appropriate extension to our studies so far.

Advanced Topics & Exercises

Strategies for Handling Difficult Optimization Cases

There are some systems for which the default optimization procedure may not succeed on its own. An optimization may fail in many diverse ways. The table on the next page summarizes many commonly-encountered situations. Suggested solutions not previously discussed are described in detail in this section.

Providing Initial Force Constants

Geometry optimizations use the second derivative matrix—known as the Hessian—in order to estimate the curvature of the potential energy surface and compute the next point. A common problem with many difficult cases is that the force constants estimated by the optimization procedure differ substantially from the actual values. By default, a geometry optimization starts with an initial guess for the second derivative matrix, which is updated by numerically

WHEN SEEKING A ...	THE JOB ...	POTENTIAL SOLUTIONS
Minimum	Locates wrong minimum	▸ Improve the starting structure. ▸ Run a scan calculation along the structural parameter(s) of interest (see p. 238ff). ▸ Optimize the structure using a less expensive model chemistry. ▸ Perform a constrained optimization in order to get closer to the minimum.
Minimum	Locates a transition structure	▸ Distort the structure in along mode corresponding to the imaginary frequency and reoptimize. ▸ Alternatively, restart the optimization using the force constants from the checkpoint file: **Opt=RCFC**.
Transition structure	Locates a minimum	▸ Perform a QST3 optimization. ▸ Perform a relaxed PES scan along the coordinate that should be a maximum (see p. 238ff).
Either	Runs out of optimization steps	▸ In most cases, provide initial force constants, and restart the optimization from the last point where it seemed to be making progress. ▸ For TS optimizations, try QST2 or QST3 if you did not do so.
Transition structure	Fails: too many negative eigenvalues	▸ The optimization has proceeded into a region on the PES having incorrect curvature. Examine the vibrational mode(s) corresponding to the extra negative eigenvalue(s) and modify the structure accordingly. ▸ Perform a constrained optimization first, and then allow the problematic coordinates to relax in a full optimization.
Either	Fails: coordinate system error	▸ Start a new optimization from the last point and allow the program to regenerate the redundant internal coordinates. if you are retrieving the geometry from the checkpoint file, you must include the **Opt=NewDefinition** option or the old redundant internal coordinates will be retained. ▸ Add redundant internal coordinates for linear bends if applicable.
Either	Is very expensive using the desired model chemistry	▸ Optimize the molecule at a lower, less expensive level of theory and then begin the high accuracy optimization from the resulting structure using the force constants from the inexpensive frequency calculation (**Opt=ReadFC**).

differentiating the forces computed at subsequent steps. However, for some cases the initial guess may be poor, and the optimization will fail to make progress. The optimization will take many steps where the energy rises instead of falling (although some steps like this at the beginning of the optimization are normal), or it may have trouble finding the direction in which the energy decreases (giving a message stating that there is no minimum in the search direction).

For such cases, you need a more sophisticated—albeit more expensive—means of generating the force constants. This is especially important for transition state optimizations.

Gaussian provides a variety of ways of obtaining force constants. Here are some of the most useful alternatives, along with their corresponding Gaussian options and visualization program controls:

- ▸ Compute the force constants at the initial point using the method and basis set as the optimization. This is more expensive than the default of merely estimating the force constants, costing the same as a frequency calculation. Gaussian option: **Opt=CalcFC**.

GaussView: Set **Calculate Force Constants** to **Once** on the **Job Type** panel of the **Gaussian Calculation Setup** dialog.

WebMO: Set **Calculate Force Constants** to **Once** on the **Advanced** tab of the **Configure Gaussian Job Options** page (or check the box in older versions of WebMO).

Note that using the **Opt=CalcFC** option as a matter of course is very likely to be wasteful of computing resources since the initial estimated Hessian is often sufficient. In fact, in many cases, better force constants do not save many optimization steps, and these extra steps are usually much faster than computing second derivatives. In other words, for large systems you can afford to allow the optimizer to take many additional steps before it equals the time needed to evaluate the second derivatives.

- Perform an optimization+frequency calculation at a lower cost model chemistry, and then read in the force constants from that job for use with the desired model chemistry. Gaussian option: **Opt=ReadFC**.

GaussView: Set **Calculate Force Constants** to **Read** on the **Job Type** panel of the **Gaussian Calculation Setup** dialog. Specify the checkpoint file explicitly in the **Link 0** panel.

In WebMO, check the **Save Checkpoint File** box on the **Advanced** panel of the **Configure Gaussian Job Options** page to retain the checkpoint file after a calculation completes.

WebMO: Navigate to the **Preview** tab of the **Configure Gaussian Job Options** page, click the **Generate** button, then change **OPT(EstmFC)** in the route section of the resulting file to **OPT(ReadFC)**. You will also need to specify the saved checkpoint file from the previous job in the **Use Checkpoint File** field on the **Advanced** tab.

- In extreme cases, you can compute the force constants at every step of a geometry optimization. This is a very expensive procedure and is only helpful in drastic situations (and not always even then). Gaussian option: **Opt=CalcAll**.

GaussView: Set **Calculate Force Constants** to **Always** on the **Job Type** panel of the **Gaussian Calculation Setup** dialog.

WebMO: Set **Calculate Force Constants** to **Always** on the **Advanced** tab of the **Configure Gaussian Job Options** page. If your version of WebMO has a checkbox rather than a menu for this item, then edit the input file manually from the **Preview** tab as in the previous bullet, changing **OPT(EstmFC)** to **OPT(CalcAll)**.

ADVANCED EXAMPLE 3.7 APPROACHES TO THE ACETALDEHYDE-VINYL ALCOHOL TS

We ran several QST2 transition structure searches for the transition state for the acetaldehyde-vinyl alcohol isomerization, using a variety of optimization strategies. Our final target model chemistry was our usual APFD/6-311+G(2d,p) in all cases.

All of our calculations located the same transition structure (pictured to the right of the table below). The following table summarizes our results:

CALCULATION(S)	# STEPS	RELATIVE TIME
APFD/6-31G(d) Opt=QST2 Freq; APFD/6-311+G(2d,p) Opt=(TS,ReadFC) Freq	12 3	1.0
APFD/6-31G(d) Opt=QST2 Freq; APFD/6-311+G(2d,p) Opt=(TS,CalcFC) Freq	13 2	1.3
APFD/6-311+G(2d,p) Opt=QST2 Freq	13	1.2
APFD/6-311+G(2d,p) Opt=(QST2,CalcFC) Freq	11	1.5
APFD/6-311+G(2d,p) Opt=(QST2,CalcAll)	7	3.1

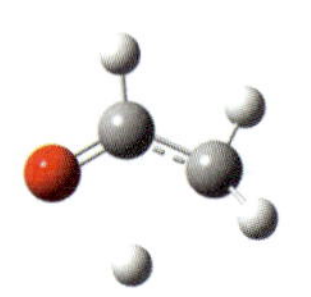

In this case, the fastest approach was to optimize the structure with a smaller basis set first, compute the frequencies, and then use that geometry and those force constants for the optimization with the target basis set. Note, however, that simply running the optimization with the desired model chemistry is not prohibitively more expensive. For this molecule, computing force constants at the first point of the optimization with the **CalcFC** option provided no benefit.■

We'll look at an example comparing optimization strategies later in this chapter (see page 133).

When a Minimization Finds a Transition Structure

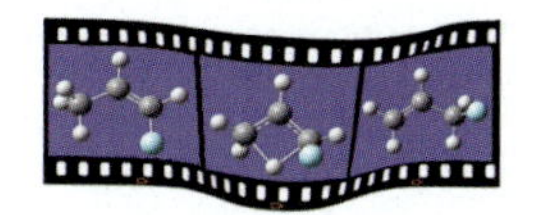

As an example, consider an optimization of acetic acid which initially locates a transition structure rather than a minimum (indicated by the presence of an imaginary frequency). The methyl group in this molecule moves quite freely.

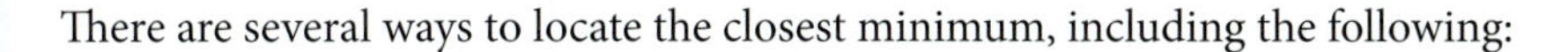

There are several ways to locate the closest minimum, including the following:

- Distort the geometry in the direction of the normal mode corresponding to the imaginary frequency. In this case, the mode corresponds to a rotation of the methyl group, and it indicates that the molecule wants to break C_s symmetry. Modifying the geometry is easily accomplished in GaussView, as illustrated in the dialog following.

  We use the **Manual Displacement** feature to break the symmetry of the current structure. We save our modified molecule by clicking the **Save Structure** button. We then set up an optimization calculation; when run, it succeeds in finding a minimum.

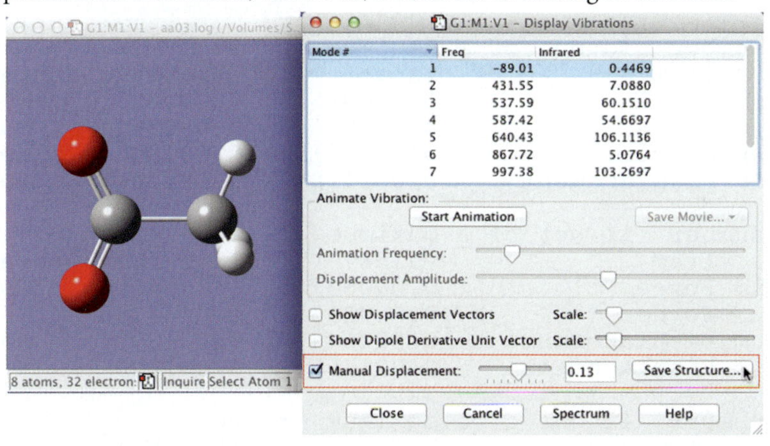

- Alternatively, since the frequency calculation has indicated that the molecule needs to distort to lower symmetry, we can specify the **Opt=RCFC** option to restart the optimization using the Cartesian coordinate force constants saved in the checkpoint file, via a route section like the following:

```
%Chk=name
# APFD/6-311+G(2d,p) Opt=RCFC Freq
  SCRF(Solvent=Chloroform) Guess=Read Geom=AllCheck
```

  This technique has the effect of removing the symmetry constraint of the previous optimization. Running this job also locates the desired minimum.

This technique is useful any time your optimization to a minimum yields a structure with one or more imaginary frequencies.

Getting an Optimization Back on Track

Not every optimization will proceed easily and directly to the desired minimum or transition structure. A certain amount of wandering around on the potential energy surface is normal, especially as molecular size increases. However, some optimizations simply fail to make progress toward a stationary point, and they exceed the allotted number of steps.

If an optimization runs out of steps, it is seldom the case that increasing the number of steps will fix the problem. Examine the output and determine whether the optimization was making progress or not. A graphical interface like GaussView will allow you to "play back" and observe the sequence of geometry optimization steps. Alternatively, the following commands will provide a quick summary of an optimization's progress:

```
egrep 'out of| NO | YES |exceeded' opt.log                  UNIX
findstr "out of" " NO " "YES" "exceeded" opt.log            Windows
```

Here is some output from an optimization that goes off track (blank lines have been added for readability):

```
Step number   1 out of a maximum of  20
        Item               Value      Threshold Converged?
Maximum Force              0.133925   0.000450     NO
RMS     Force              0.044866   0.000300     NO
Maximum Displacement       0.085381   0.001800     NO
RMS     Displacement       0.022118   0.001200     NO

...
Step number  10 out of a maximum of  20
        Item               Value      Threshold Converged?
Maximum Force              0.000925   0.000450     NO
RMS     Force              0.000866   0.000300     NO
Maximum Displacement       0.005381   0.001800     NO
RMS     Displacement       0.002118   0.001200     NO

Step number  11 out of a maximum of  20
        Item               Value      Threshold Converged?
Maximum Force              0.000532   0.000450     NO
RMS     Force              0.000295   0.000300     YES
Maximum Displacement       0.002544   0.001800     NO
RMS     Displacement       0.001755   0.001200     NO

Step number  12 out of a maximum of  20
        Item               Value      Threshold Converged?
Maximum Force              0.003417   0.000450     NO
RMS     Force              0.000447   0.000300     NO
Maximum Displacement       0.006199   0.001800     NO
RMS     Displacement       0.008898   0.001200     NO

...
Step number   20 out of a maximum of  20
        Item               Value      Threshold Converged?
Maximum Force              0.093772   0.000450     NO
RMS     Force              0.029250   0.000300     NO
Maximum Displacement       0.250102   0.001800     NO
RMS     Displacement       0.069076   0.001200     NO
Maximum number of steps exceeded.
```

This optimization seemed close to a minimum at step 11, but then it moved away from it again in subsequent steps. Merely increasing the number of steps will not fix the problem. A better approach is to start a new optimization, beginning with the structure corresponding to step 11 and including the **Opt=CalcFC** option.

Here is an example Gaussian input file which will begin an optimization from the eleventh step of a previous optimization which used the checkpoint file named *opt1.chk*:

```
%Chk=opt1.chk
# APFD/6-311+G(2d,p) Opt=CalcFC Freq
  Geom=(Check,Step=11) Guess=Read

Opt restarted from step 11 with CalcFC

0 2
```

Required blank line

If you don't mind having the same title for the second job as for the first, you can use the **AllCheck** option with the **Geom** keyword and omit the title line and charge and spin multiplicity. Note that if the **Step** option is not included, it defaults to the final geometry present in the specified checkpoint file.

You can set up and run the same job from GaussView and WebMO. In both cases, you will need to have retained the checkpoint file from the optimization (see p. 118 above).

There are two ways to set up an optimization starting from a point in an earlier calculation in GaussView. The first is to open the previous job's checkpoint file or log file, selecting the **Read Intermediate Geometries** box at the bottom of the **Open** dialog. Navigate to the structure corresponding to the desired step in the multiple view window that results. Select that structure, and then copy and paste it into a new molecule group. You can then go on to set up the desired Gaussian calculation starting from that geometry.

If you are restarting a failed QST2 optimization, you should specify an optimization to a transition structure on the **Job Type** panel of the **Gaussian Calculation Setup** dialog (set the **Optimize to a** popup menu to **TS (Berny)**) and specify that the force constants should be calculated at the first point (set **Calculate Force Constants** to **Once**).

The second method for restarting an optimization from an intermediate structure involves setting up a job which retrieves the desired geometry directly from the checkpoint file in the following manner:

1. Open the **Gaussian Calculation Setup** dialog. Select the desired job type and model chemistry from the relevant panels. Remember to handle a restarted QST2 optimization as noted above.

2. Navigate to the **Link 0** panel. Set the **Chkpoint File** popup to **Specify**, and then enter the name of the desired checkpoint file. You can also use the **...** button at the right of the checkpoint file name field to navigate to the desired file.

3. On the **Guess** panel, set the **Guess Method** menu to **Read checkpoint file**. On the **General** panel, uncheck **Write Connectivity**. Finally, enter **Geom=(Check,Step=*n*)** in the **Additional Keywords** field (where *n* is the desired step number).

   These settings are illustrated in the dialog at the right.

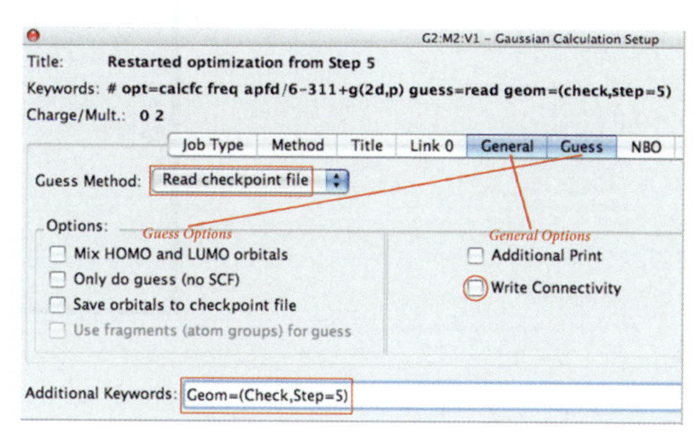

4. Click the **Edit** button, saving the job to a new file when asked to do so. When the editor opens, switch to that application and modify the generated input file to remove any atom lines in the molecule specification. Be sure to retain the charge and spin multiplicity line and the following blank line.

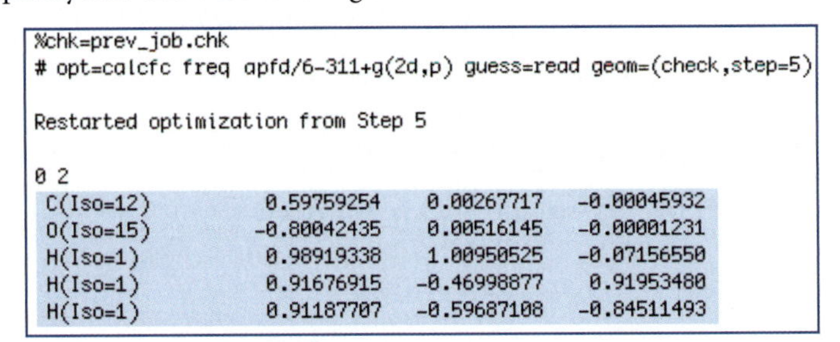

```
%chk=prev_job.chk
# opt=calcfc freq apfd/6-311+g(2d,p) guess=read geom=(check,step=5)

Restarted optimization from Step 5

0 2
 C(Iso=12)          0.59759254     0.00267717    -0.00045932
 O(Iso=15)         -0.80042435     0.00516145    -0.00001231
 H(Iso=1)           0.98919338     1.00950525    -0.07156550
 H(Iso=1)           0.91676915    -0.46998877     0.91953480
 H(Iso=1)           0.91187707    -0.59687108    -0.84511493
```

5. Save the modified input file. Close the editor, return to GaussView, and submit the job to Gaussian by answering **Ok** to the prompt.

In WebMO, the following steps will restart a geometry optimization from a checkpoint file:

1. Be sure that the previous optimization job appears in the **Job Manager**. Select it and then click **New Job Using This Geometry** in the bottom center of the **Builder** page. Then click the right arrow icon to move to the **Configure Gaussian Job Options** page.

2. Select the desired settings on the **Job Options** panel. If the job you are restarting was a QST2 optimization, select **Transition State Optimization** from the **Calculation** menu. Failed QST2 optimizations must be restarted as regular transition structure optimizations (**Opt=TS**).

   After selecting the calculation type, navigate to the **Advanced** panel.

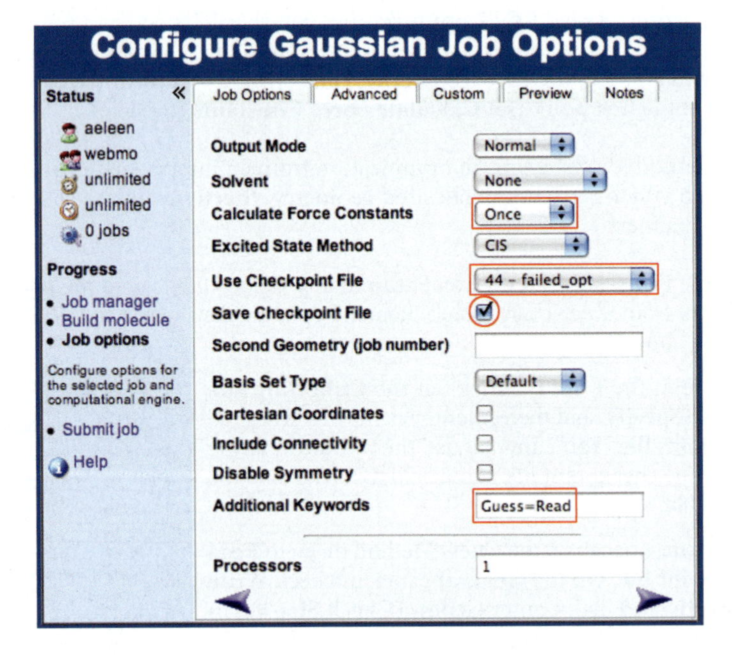

3. Select the checkpoint file from the previous job in the **Use Checkpoint File** popup menu. Next, set **Calculate Force Constants** to **Once**,* and check the **Save Checkpoint File** box. Finally, enter **Guess=Read** into the **Additional Keywords** field.

4. Navigate to the **Preview** panel and click the **Generate** button. This will create and load the actual Gaussian input file. Edit the file by adding the **Step** option to the **Geom** keyword, along with parentheses to enclose the option list:

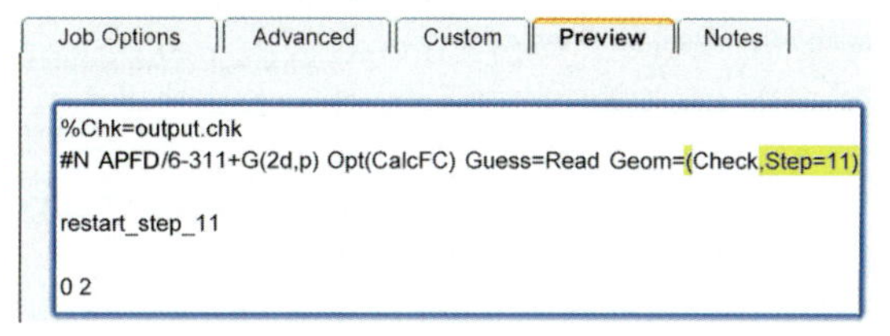

5. Click on the right arrow icon to submit the edited job to Gaussian.

Reducing the Step Size to Eliminate Thrashing

Sometimes the technique described above will still fail to find an optimized structure. This can occur when an optimization is thrashing badly: oscillating among a series of structures without making progress. This can be seen for very flat potential energy surfaces even when the **CalcFC** or **CalcAll** option has been used.

The following plots display the total energy computed for the various successive structures in a geometry optimization. In the optimization on the left, the optimizer takes a bad step at step 11, resulting in a structure with a much higher energy. However, it quickly recovers and continues on to eventual success. Such missteps are common and normal.

In contrast, the plot on the right illustrates an optimization which is thrashing and which will never converge.

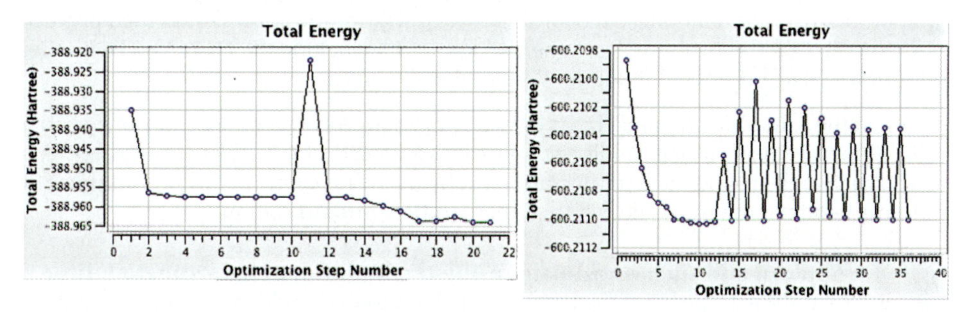

In this particular case, the optimization had already specified the **CalcAll** option.

The following figure plots the values of two key structural parameters over the course of the latter optimization:

* Or check the box if you are using an older version of WebMO.

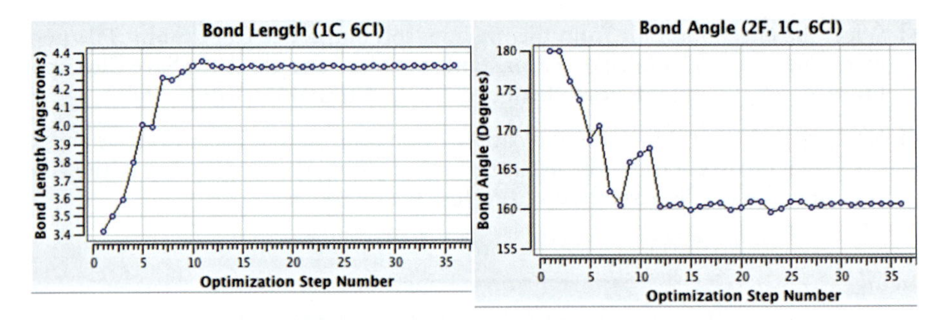

Both of these values have essentially converged. The structural changes that are occurring in the optimization at this point are limited to methyl rotations.

In some cases like this, reducing the optimization step size will enable it to converge. This is accomplished using the **MaxStep** option to the **Opt** keyword, which accepts a numeric value as its argument. The default step size is 30 (the units are unimportant), so setting it to 15 will reduce the step size by half. In a case such as this one, the optimization should be restarted from one of the lower energy points with a reduced step size, and the force constants can be retrieved from the checkpoint file along with the desired geometry: e.g., **Opt=(RCFC,MaxStep=15) Geom=(Check,Step=30) Guess=Read**.

Coordinate System Errors

An optimization will fail with a coordinate system error if the redundant internal coordinates it has generated become invalid due to changes in the geometry. This is a fairly rare occurrence, and it occurs primarily in two situations:

- When the optimization correctly enters into a region of the potential energy surface that is no longer described well by the generated redundant internal coordinates. In such cases, simply starting a new optimization from the last point will generate new, appropriate redundant internal coordinates, and the optimization should then proceed. If you retrieve the structure from the checkpoint file, include the **Opt=NewDefinition** option to discard the old redundant internal coordinates.
- When the molecule contains linear or near-linear angles. Recent revisions of Gaussian have improvements which handle such situations automatically. In older versions, however, you can work around such optimization problems by explicitly adding coordinates for linear bends in the input file, via the **Opt=ModRedundant** option.

You can add such coordinates using the **Redundant Coordinate Editor** in GaussView. The list at the top of the dialog shows the added coordinates, and you can click the **Add** button there to add a new one. The popup menus and fields in the middle of the dialog are used to specify the type of coordinate, the atoms associated with it, the action to be taken with the coordinate (adding it, removing it, and so on), and the coordinate's value or value constraints (as desired). You can select atoms for a coordinate by clicking on them in the View window.

For example, in the following dialog, we are adding a linear bend coordinate for the near linear angle in the molecule. In order to do so, we select the three linearly bonded atoms and then any other atom that is not colinear with them. Note that we do *not* modify the setting of the value popup. These steps would be repeated for each linear angle within the molecule.

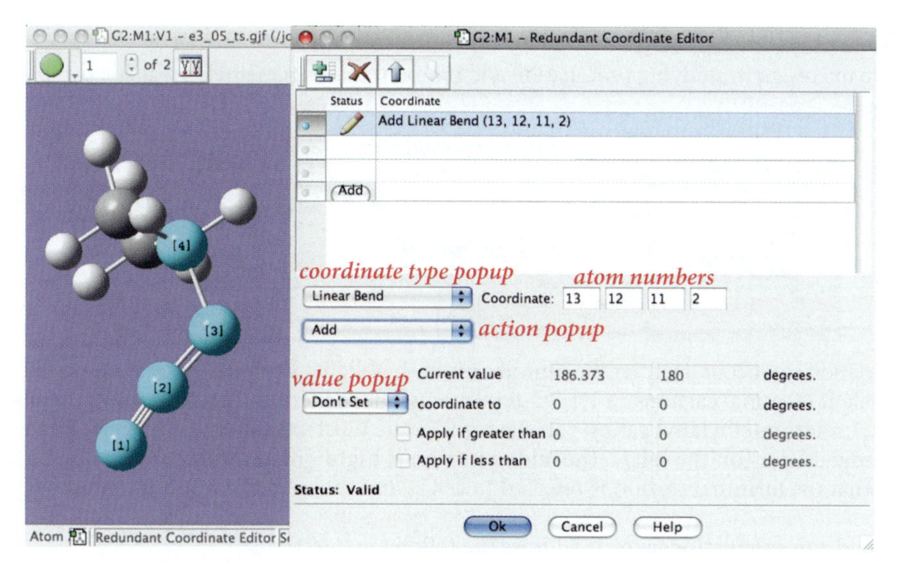

The redundant coordinate modification input can also be included directly within a Gaussian input file, as in this example:

```
# ... Opt Freq Geom=(Connectivity,ModRedundant)
```

title section and molecule specification & connectivity section

```
L 13 12 11 2
```

ModRedundant input: linear bend definition
blank line

The coordinate modification section terminates with a blank line as usual. See the *Gaussian User's Reference* for more details about modifying redundant internal coordinates.

Higher Accuracy Optimizations

The default optimization methods produce accurate structures in general. Sometimes, however, even higher accuracy optimized structures are required. The following features are available for such needs:

- Use a more accurate integration grid: **Int=UltraFineGrid**. In GaussView or WebMO, enter this option into the **Additional Keywords** field.
- Apply stricter convergence criteria: **Opt=Tight**. In GaussView, check **Use tight convergence criteria** on the **Job** panel. In WebMO, select **Other** from the **Calculation** menu, and enter **Opt=Tight** into the resulting dialog.

Both of these features will increase the computational requirements for the optimization.

ADVANCED EXERCISE 3.5 **PROTONATION & PROTON TRANSFERS IN ALLENES**

We are interested in studying protonation and proton transfer in methylated allene compounds [Gronert06, Gronert07]:

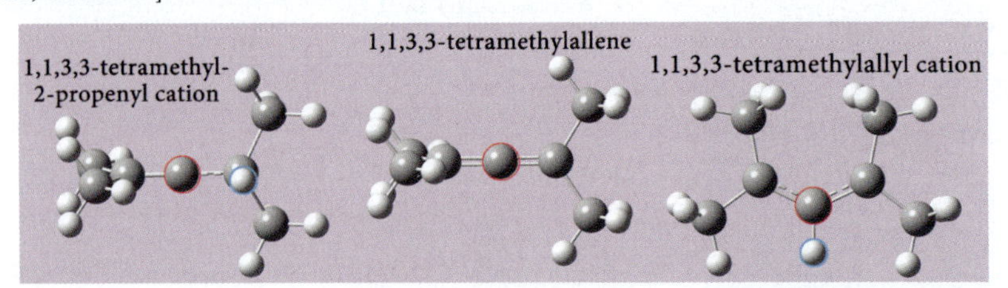

Protonation can occur at either the central carbon (highlighted in red above) or one of the two equivalent terminal carbons in 1,1,3,3-tetramethylallene (center structure above), resulting in 1,1,3,3-tetramethylallyl cation (on the right in the illustration) or 1,1,3,3-tetramethyl-2-propenyl cation (on the left)*. The added proton is highlighted in blue in the illustration. Note that the terminal carbon is referred to as C_1, and the central carbon is denoted C_2.

Plan and run calculations which address the following questions:

- Which protonation site is thermodynamically favored? This can be determined by comparing the predicted energies of the two species.
- What is the barrier to protonation via identity reactions where the proton migrates intermolecularly from the C_1 to C_1 or C_2 to C_2?
- What is the barrier for intramolecular proton transfer from C_1 to C_2 and vice versa?

An *identity reaction* is one in which the reactants and products are the same. In this case, the reactions in which we are interested are ones where a proton is passed from a cation to the same carbon atom in allene, resulting in the same compounds as products. You will need to locate the transition structures for the two reactions.

This exercise is quite challenging, both in setting up some of the calculations and successfully running the optimizations to completion. An in-depth discussion of setting up the transition structure optimizations follows.

Required Calculations

6 optimization calculations will be required:

- 1,1,3,3-tetramethylallene
- 1,1,3,3-tetramethylallyl cation
- 1,1,3,3-tetramethyl-2-propenyl cation
- Transition structure for the identity reaction with the proton at the terminal carbon: C_{2h} symmetry
- Transition structure for the identity reaction with the proton at the central carbon: S_4 symmetry
- Transition structure for the 1,2 proton shift: C_s symmetry

Optimizing the three minima is generally straightforward. The same cannot be said for the transition structures. The easiest way to perform the transition state optimizations is to build a reasonable starting guess for the structure and then optimize it with **Opt=TS** (i.e., without

* IUPAC: 1,1,3,3-tetramethyl-2-propenyl cation: 2,4-dimethyl-2-penten-3-ylium; 1,1,3,3-tetramethylallene: 2,4-dimethyl-2,3-pentadiene; 1,1,3,3-tetramethylallyl cation: 2,4-dimethyl-2-penten-4-ylium.

using the QST2 facility). We will also need to approach each transition structure gradually via a series of progressively larger optimization calculations.

Starting structures which do not have the proper symmetry will encounter problems as the optimization proceeds. We will use GaussView's molecular symmetry features extensively throughout the following discussion.

Optimizing the Transition Structure for Intermolecular Protonation (C_1 to C_1)

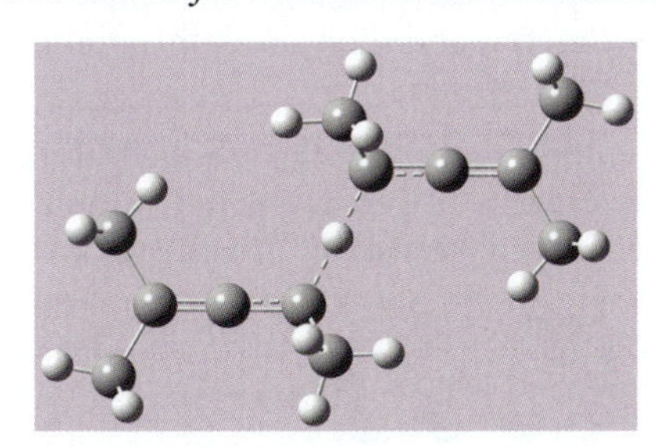

This structure consists of a proton—built as a hydrogen atom—midway between the terminal carbon atoms from two methylated allene molecules. The 6 atoms in the carbon backbones of the two molecules are all in the same plane, and the molecules are positioned so that their lengths stretch in opposite directions.

We will describe the way that we built this molecule in GaussView. A similar process can be followed in WebMO. In either case, we make extensive use of the program's ability to impose a specific point group symmetry on a structure.

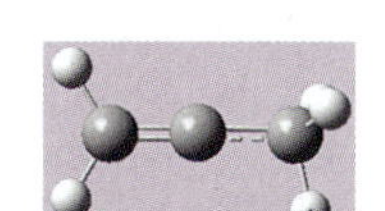

1. Build the 2-propenyl cation, and then remove a hydrogen atom from one methyl group.
2. Copy the resulting molecule and paste it into the same **View** (**Edit→Paste→Append Molecule**).* Position it as illustrated below, and then add a bond between the two carbon atoms. Clean the resulting structure.

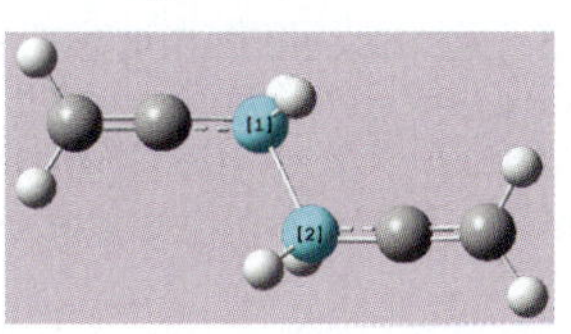

3. Ensure that the carbon backbone atoms are all in a plane by adjusting the dihedral angle involving the highlighted atoms above and the carbon adjacent to each of them to be 180°.
4. Impose C_{2h} symmetry on the structure using the **Point Group Symmetry** feature. You may need to increase (loosen) the **Tolerance** setting in order for this symmetry group to appear in the popup menu in the **Approximate higher-order point groups** area.
5. Increase the C-C distance between the two fragments to about 2.8 **Å** and remove the bond. Place a hydrogen atom midway between them. One method is to select the hydrogen atom as the current fragment, and then right click in the View window to select the **Builder→Place Fragment at Centroid of Selected atoms.** Finally, attach the hydrogen atom to the two allyl molecules with half bonds.

* *GaussView Hint*: A copied molecule can also be added to a **View** by simply clicking within it.

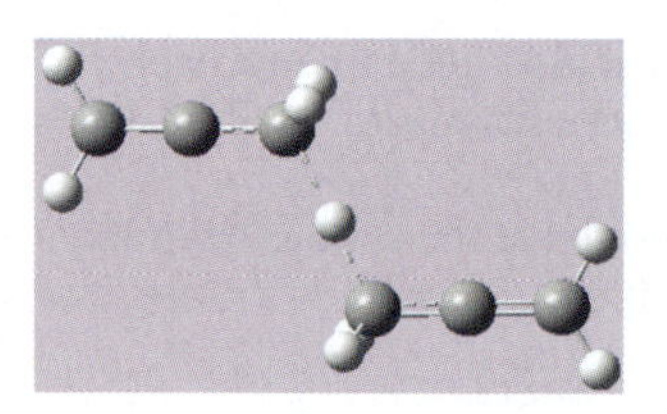

6. Return to the **Point Group Symmetry** dialog and reimpose C_{2h} symmetry. Click **Always track point group symmetry,** and then select this point group in the **Constrain to subgroup** popup.

7. Return to the **View** window and select the tetrahedral carbon atom in the **Builder**. Click on one of the *exterior* hydrogen atoms in the molecule; this atom, along with the three equivalent hydrogen atoms, will become methyl groups:

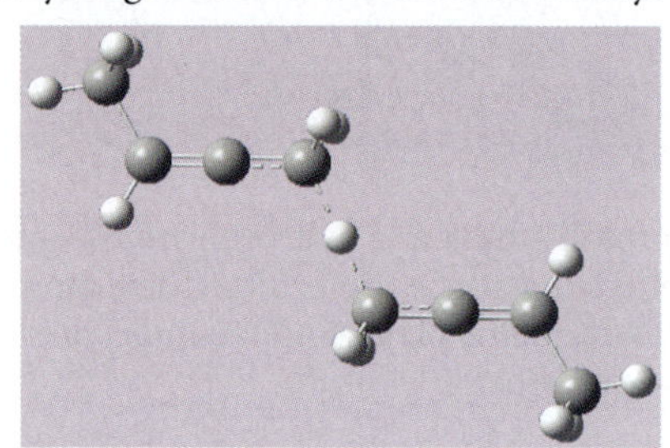

8. Once again, use the **Point Group Symmetry** dialog to reimpose and constrain C_{2h} symmetry. Click on another of the remaining exterior hydrogen atoms to replace it and its equivalent on the other end with methyl groups. Reimpose and constrain to C_{2h} symmetry once again, and then replace one of the remaining hydrogens of the allyl moiety with a methyl group. Keep repeating these steps until all hydrogens other than the central one have been replaced with methyl groups.

The starting structure for the optimization is now complete.

9. An efficient way to proceed with the calculation is to first optimize the structure using the inexpensive HF/6-31G(d) model chemistry. Set up the optimization+frequency job to do so, selecting **TS (Berny)** from the **Optimize to a** popup menu, and **Once** from the **Calculate Force Constants** popup menu. Submit the job to Gaussian.

This will optimize to a transition structure that we can use as a starting point for the TS search using our standard model chemistry. Our strategy will be to approach the TS in two phases:

- First, we will perform a constrained optimization to a minimum, freezing the coordinates associated with the imaginary frequency: the two C-H bonds (indicated in the illustration in the margin).
- Once the first optimization has converged, we will activate those coordinates and perform the final TS optimization.

10. Open the optimized structure in GaussView. Use the **Redundant Coordinate Editor** to freeze the two C-H bonds involved in the intermolecular protonation. Click the **Add** button to add a coordinate to the list, and then click on the atoms for that coordinate within the **View** window. Change the coordinate type to **Bond** and the action to **Freeze Coordinate**. Finally, repeat the process for the second bond to be frozen. The completed dialog is shown in the illustration following.

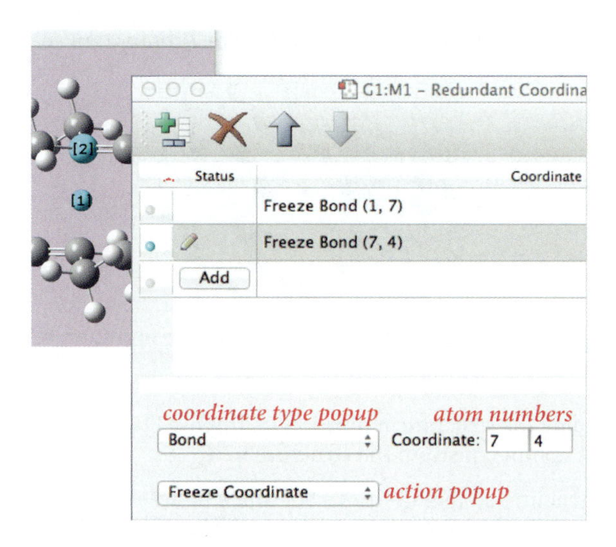

Set up an optimization+frequency calculation using our standard model chemistry, specifying an optimization to a minimum. Set the **Force Constants** popup to **Read**. This adds the **Opt=ReadFC** option to the input section, which says to read the force constants from the previous frequency calculation in the specified checkpoint file: in this case, the HF/6-31G(d) TS optimization+frequency job. Add the **Opt=(NoDownhill)** option in the additional input section; this option prevents the optimization from taking an energy-lowering step that would break symmetry, and it can be helpful in scenarios such as this one.

Once all of these settings are made, click **Edit** in the **Gaussian Calculation Setup** dialog. Save the file when asked to do so. When the editor opens, modify the input file to add a second job step:

```
%Chk=name                                         HF Opt checkpoint file
# APFD/6-311+G(2d,p) Opt=(RCFC,NoDownhill) Freq
  Guess=ModRedundant ...
```

title & molecule specification sections

```
B 1 7 F                                ModRedundant input: freeze 2 bonds
B 7 4 F

--Link1--                                              Add these lines.
%Chk=name                                          Same file as above.
# APFD/6-311+G(2d,p) Opt=(TS,RCFC) Freq
  Geom=AllCheck Guess=(Read,ModRedundant) ...

* * A                            ModRedundant input: activate all bonds.
                                                      Final blank line.
```

These additions add a second transition structure optimization+frequency calculation (to a transition structure) using our standard model chemistry. Force constants will be taken from the previous frequency calculation. The final input section contains the **ModRedundant** option input, and it uses wildcards to activate all bond coordinates for the optimization.

This two-step job is now ready to run and should succeed in locating the desired transition structure.

Optimizing the Transition Structure for Intermolecular Protonation (C_2 to C_2)

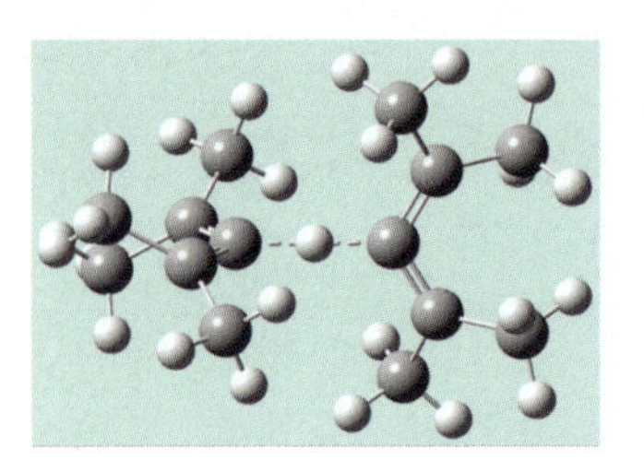

We begin by building the transition state starting structure and then run the optimization in three phases as we did with the preceding molecule.

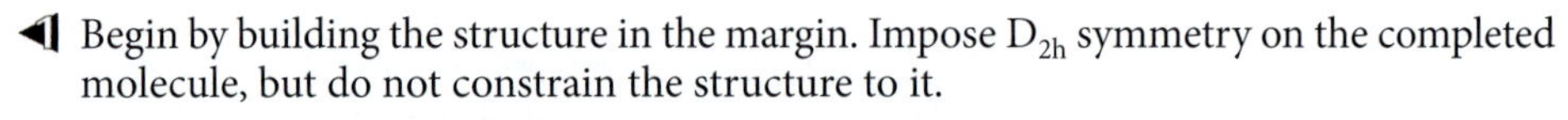

1. Begin by building the structure in the margin. Impose D_{2h} symmetry on the completed molecule, but do not constrain the structure to it.

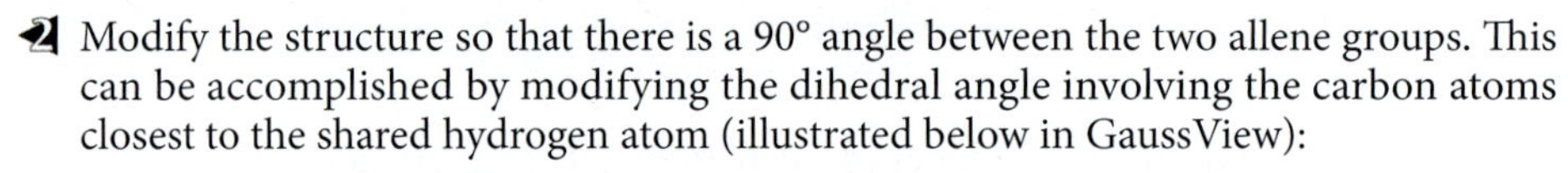

2. Modify the structure so that there is a 90° angle between the two allene groups. This can be accomplished by modifying the dihedral angle involving the carbon atoms closest to the shared hydrogen atom (illustrated below in GaussView):

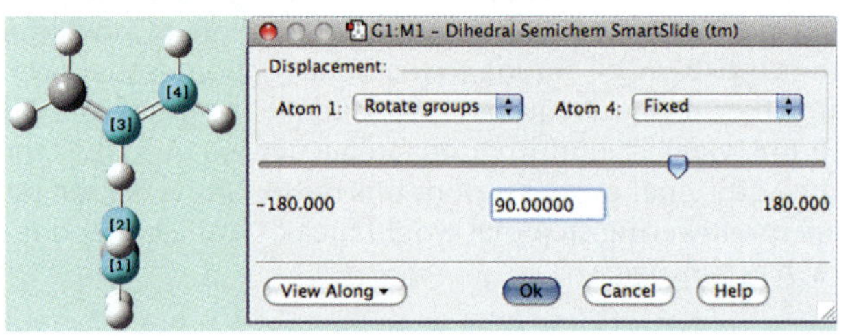

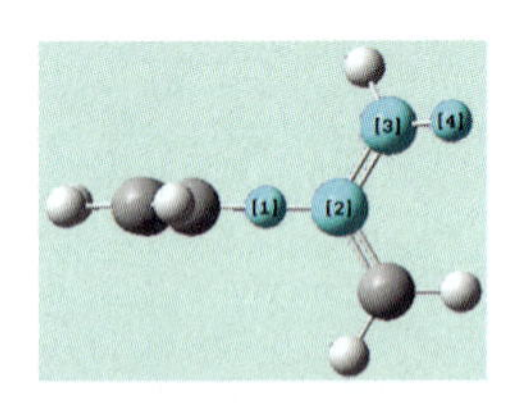

3. Impose S_4 symmetry on the molecule, and constrain the structure to this point group as well. Then, twist one of the CH_2 groups so that it is out of plane. You can do so by setting the indicated dihedral angle to about 144°. The symmetry constraints will cause all four groups to twist.
4. Reimpose S_4 symmetry and symmetry constraints. Then add methyl groups to the molecule using the same technique as previously. Remember to reimpose symmetry constraints on the molecule after adding the first set of methyl groups.
5. Set up and run the HF/6-31G(d) transition structure optimization+frequency as for the previous molecule.

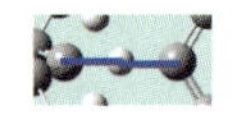

6. Beginning from this TS, freeze the bonds involved in the normal mode corresponding to the desired imaginary frequency, and perform a constrained optimization and frequency calculation using the same options as before. Then perform a TS optimization+frequency calculation with all coordinates active to find the actual transition structure, using the force constants computed by the previous frequency calculation.

Optimizing the Proton Shift Transition Structure

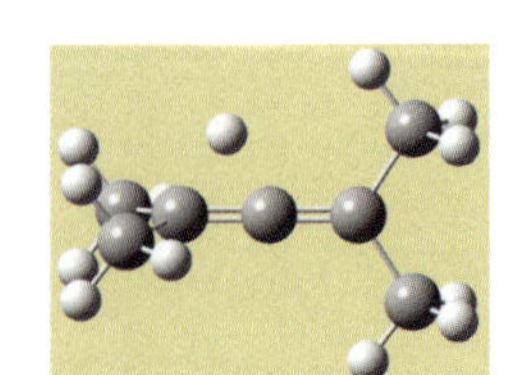

The final remaining structure is the proton shift transition structure. Building it is straightforward:

1. Begin by building allene. Ensure that the molecule has D_{2d} symmetry using the **Point Group Symmetry** feature (as before), and constrain the symmetry before closing the dialog.
2. Add the methyl groups: adding one of them should place all four if the molecular symmetry is correct.

◀3 Next, impose C_s symmetry. You may need to rotate one or more methyl groups until a plane of symmetry exists. If you start from the structure on the left, you will need to modify the two indicated dihedral angles until the atoms outlined in orange all lie within the same plane:

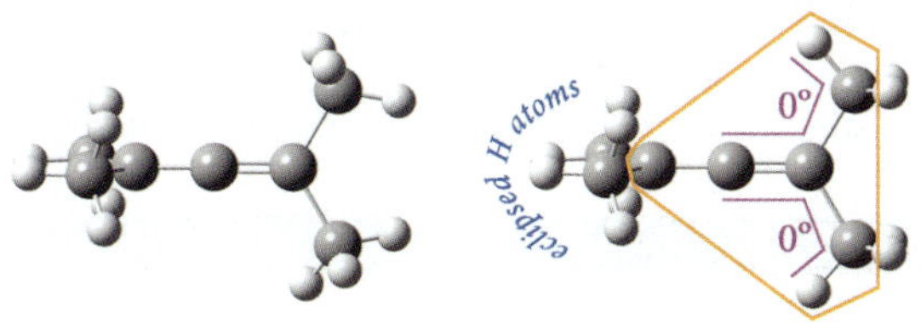

Be sure that the two methyl groups above/below the plane of symmetry (on the right above) are eclipsed.

◀4 Add a hydrogen atom halfway between the two carbon atoms. Ensure that the final structure has C_s symmetry.

◀5 Set up an HF/6-31G(d) transition structure optimization+frequency job. Run this job, which will successfully locate a TS.

◀6 Finding the transition structure beginning from the results of the preliminary low-level optimization is challenging. We use the same general approach as for the other two transition states: first optimizing the TS with an inexpensive model chemistry and then freezing the two relevant C-H bond coordinates in a constrained optimization using our standard model chemistry. However, when we reactivated those coordinates, the optimization entered a region of the PES with incorrect curvature at step 13—indicated by two negative eigenvalues instead of only one—and the optimization terminated.

We started the new optimization from the last point where it had seemed to be making progress (step 8), calculating the force constants at that point. This calculation drew closer to the TS, but again entered a region of incorrect curvature after six steps. Repeating this tactic of starting a new optimization with **Opt=CalcFC** from the final point of the second failed optimization succeeded in reaching the transition structure in six additional steps.

SOLUTION

The following table lists the results from our APFD/6-311+G(2d,p) geometry optimizations and frequency calculations:

MOLECULE	STRUCTURE TYPE	THERMAL ENERGY (hartrees)
tetramethylallene (1▶)	minimum	-273.58977
tetramethylallyl cation (2▶)	minimum	-273.95187
tetramethyl-2-propenyl cation (3▶)	minimum	-273.89600
C_1 protonation TS (4▶)	transition structure	-547.51083
C_2 protonation TS (5▶)	transition structure	-547.53187
proton transfer TS (6▶)	transition structure	-273.89752

We also optimized complexes of 1,1,3,3-tetramethylallene with the two cations in order to compute the actual activation barriers for the protonation reactions, with the following results:

MOLECULE	STRUCTURE TYPE	THERMAL ENERGY (hartrees)
Me_4allene···Me_4allyl cation complex (7)	minimum	-547.55613
Me_4allene···Me_4-2-propenyl cation complex (8)	minimum	-547.50991

From these predicted quantities, we compute the required energy differences (following [Gronert07]):

	COMPUTED AS ...	ΔE (hartrees)
thermodynamic preference (C_2 vs. C_1)	2 – 3	-0.05587
identity intermolecular trans. barrier (C_1 to C_1)	4 – 3 – 1	-0.02506
identity intermolecular trans. barrier (C_2 to C_2)	5 – 2 – 1	0.00977
intramolecular proton shift barrier: $C_1 \rightarrow C_2$	6 – 3	-0.00151
intramolecular proton shift barrier: $C_2 \rightarrow C_1$	6 – 2	0.05436

The protonation of 1,1,3,3-tetramethylallene is thermodynamically favored at the central position by ~56 millihartrees. The identity protonation is barrierless at the terminal position, and so is the intramolecular proton shift from C_1 to C_2. These results suggest that 1,1,3,3-tetramethyl-2-propenyl cation is in a shallow minimum where the proton at C_1 can easily move from one molecule to the next and also intramolecularly relocate to C_2. However, the proton at C_2 is in a deep well and will not easily migrate to another molecule or to C_1.

The barriers for the identity protonation at the central position and for the C_1 to C_1 proton shift are ~24 and ~54 milihartrees, respectively.

We can now examine the energy profile for these processes. The figure following plots the barriers to protonation/proton migration for these three scenarios. The solid lines plot the predicted electronic energies, and the dotted lines plot the predicted thermal energies.

The y-axis value of each structure corresponds to its relative energy with respect to the electronic energy of the tetramethylallene-tetramethylallyl cation complex in order to bring all of the values into the same scale. Note that the difference between the electronic and thermal energies for the reference compound is 0.37011 hartrees.

Identity protonation at the central carbon has a high barrier. In contrast, the intermolecular protonation (C_1 to C_1) is barrierless when we include the thermal corrections. Similar features are seen in the barrier to proton migration from the terminal to the central carbon atoms, while the reverse proton shift has the largest barrier of all.

Thus, this plot visually depicts the likely protonation and proton shift paths that we deduced from our numeric results.■

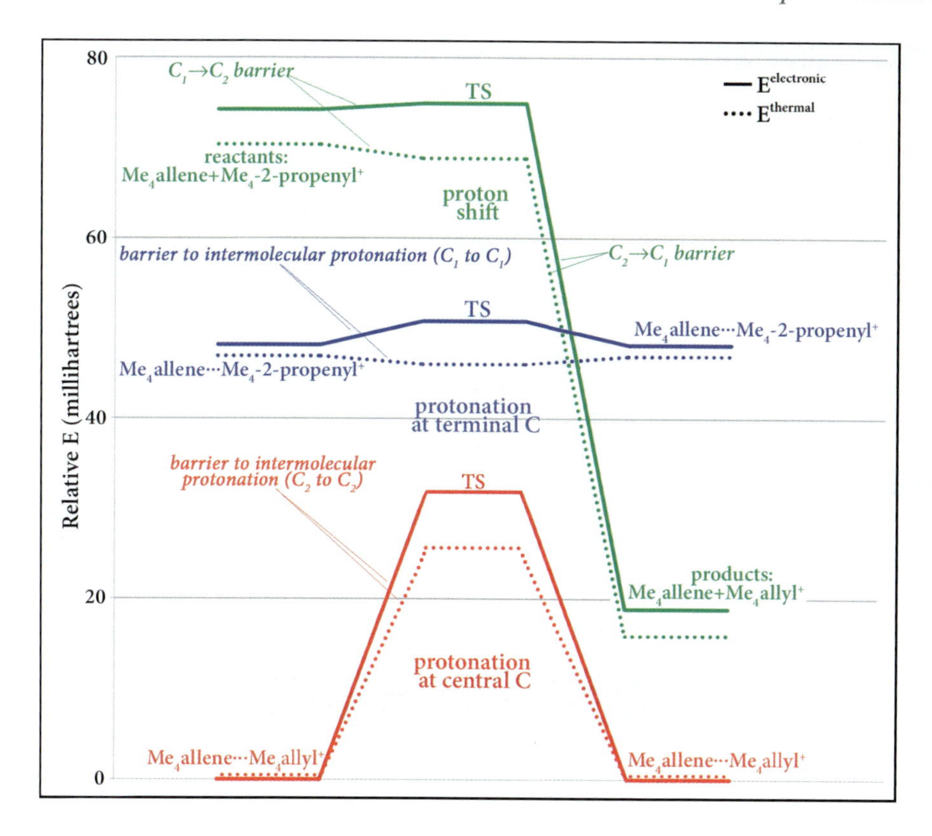

1,1,3,3-Tetramethylallyl Cation Optimization Variations

We ran the optimization of 1,1,3,3-tetramethylallyl cation in several ways to provide another example of how different optimization techniques can vary in performance. The following table presents our results:

BASIS SET & OPT OPTIONS	# STEPS	RELATIVE TIME
6-311+G(2d,p) Opt	21	1.3
6-311+G(2d,p) Opt=CalcFC	21	1.9
6-31G(d) Opt Freq 6-311+G(2d,p) Opt=ReadFC	21 3	1.0

All of the optimizations located the same structure. For this molecular system, the fastest method was an optimization with a small basis set followed by one with the desired model chemistry. Computing the force constants using the large basis set was unproductive.

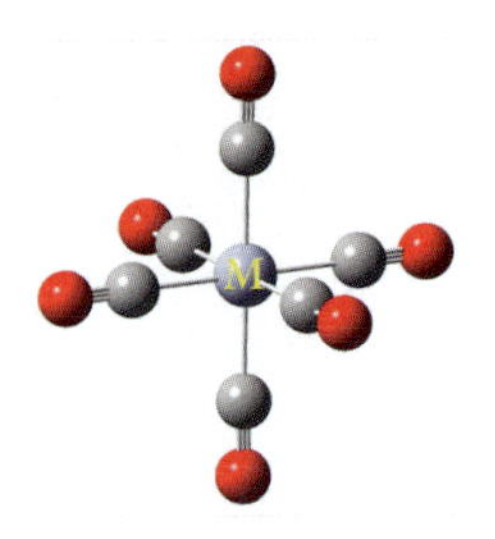

ADVANCED EXERCISE 3.6 **PERIODIC TRENDS IN TRANSITION METAL COMPLEXES**

In the previous chapter, we optimized the structure of chromium hexacarbonyl (see page 66). Now we will investigate the structures of $M(CO)_6$, where M is chromium, molybdenum and tungsten.*

Optimize these three molecules using the B3LYP density functional with the LANL2DZ basis set. LANL2DZ is a basis set containing effective core potential (ECP) representations of electrons near the nuclei for post-third row atoms. While it is a smaller basis set than our standard one, it is applicable to all three elements that we wish to study.

Compare the $Cr(CO)_6$ results with those we obtained previously. Then compare the structures of the three systems to one another, and characterize the effect of changing the central atom on the overall molecular structure.

SOLUTION

The $Cr(CO)_6$ geometry computed with the LANL2DZ basis set is similar to the optimized geometry we obtained previously with the 6-311+G(2d,p) basis set, and accordingly both produce good agreement with experiment.

The LANL2DZ results for the three compounds are presented in the table following.

M	SYMMETRY	M-C (Å)	C-O (Å)
Cr	O_h	1.86	1.17
Mo	O_h	2.04	1.17
W	O_h	2.03	1.17

All three structures have O_h symmetry and are very similar. The bond length from the central atom to the carbonyl group is slightly different in each compound, and it is longest for the molybdenum substituent. The internal structure of the carbonyl groups is essentially unchanged by substitution.■

ADVANCED EXERCISE 3.7 **$HCo(CO)_4$ ISOMERS**

$HCo(CO)_4$† is a frequently studied organometallic compound (see [Torrent00] for an excellent review of the literature on this compound). There are two possible trigonal bipyramidal structures for this compound, with differing symmetries:

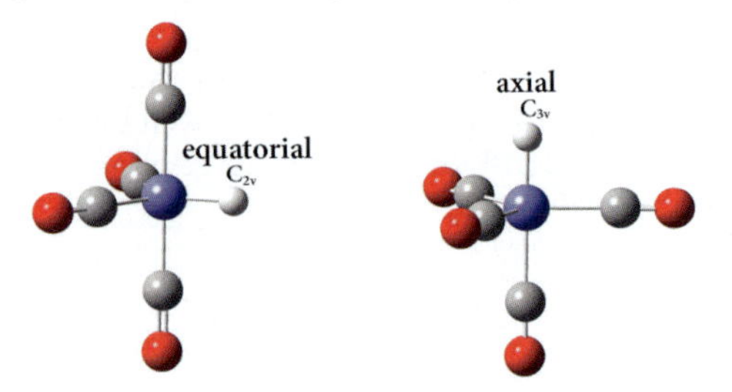

Determine which of these structures is the lowest energy conformation for $HCo(CO)_4$.

* IUPAC: $Mo(CO)_6$: hexacarbonylmolybdenum; $W(CO)_6$: hexacarbonyltungsten.

† IUPAC: tetracarbonyl(hydrido)cobalt.

SOLUTION

We performed geometry optimizations for the three structures using our standard model chemistry, with the following results:

- We successfully located a minimum for the C_{3v} symmetry isomer using the default optimization procedure.
- The optimization on the C_{2v} symmetry isomer using the default estimated force constants lead to a transition structure. When we repeated the optimization, this time calculating the initial force constants, we located a minimum. However, it contains distorted equatorial Co-C-O bond angles (by about 4.5°).

The following table summarizes our optimization results:

ISOMER	OPTIMIZATION RESULT	# STEPS	REL. E (millihartrees)
equatorial (C_{2v})	transition structure with C_{2v} symmetry	11	8.9
	minimum, ~C_{3v} (second opt. from displaced TS)	15	0.0
	minimum, ~C_{3v} (**Opt=CalcFC**)	29	0.0
axial (C_{3v})	minimum with C_{3v} symmetry	7	0.0

The optimizations of the equatorial form reach the same minimum by two paths:

- An optimization with the default options (which finds a transition state) followed by a second optimization from a structure displaced in the direction of the normal mode corresponding to the imaginary frequency (which finds the minimum), taking a total of 26 steps.
- An optimization with the **CalcFC** option, which located the minimum in 29 steps.

Thus, our calculations indicate that the ground state is the axial isomer. This is in agreement with electron diffraction studies [McNeill77]. Understanding the nature of the TS we encountered would required additional tools (discussed later in this book).■

ADVANCED EXERCISE 3.8 **OPTIMIZING THE BOND LENGTH OF HF**

The experimental bond length for the hydrogen fluoride molecule is 0.917 Å. Determine the basis set required to predict this structure accurately. Perform your optimizations using the APFD functional as usual.

SOLUTION

Here are the predicted values for some small basis sets, our standard basis set (in green below), and several larger ones:

BASIS SET	BOND LENGTH (Å)
6-31G(d)	0.930
6-31G(d,p)	0.921
6-31+G(d,p)	0.924
6-31++G(d,p)	0.924
6-311G(d,p)	0.917
6-311G(2d,p)	0.918

BASIS SET	BOND LENGTH (Å)
6-311G(3df,3pd)	0.916
6-311+G(2d,p)	0.920
6-311++G(2d,p)	0.920
6-311+G(3df,2p)	0.919
6-311+G(3df,2pd)	0.919
6-311+G(3df,3pd)	0.918

All of the values except that for the smallest basis set are within 0.01 Å of the experimental value. All of the geometries predicted with the 6-311G basis set are quite accurate. For this molecule, diffuse functions provide little benefit (e.g., compare the 6-311G(3dp,3pd) results with and without the diffuse function), and adding them to the smaller basis set actually leads to a poorer prediction. However, adding additional polarization functions on the hydrogen does generally improve the results. These calculations demonstrate that when working with density functional theory, there is no guarantee that adding functions to a basis set will systematically improve its results.■

ADVANCED EXERCISE 3.9 SEARCHING FOR A SYMMETRIC MINIMUM

In Example 3.1 (p. 90), we investigated the structure of decamethylzincocene, concluding that the $\eta^1\eta^5$ structure is the lowest energy conformation. We did so by optimizing structures exhibiting the various possible symmetries and hapticities for this molecule. However, it is possible that our procedure failed to locate a higher symmetry minimum because our starting D_{5d} and D_{5h} structures happened to be closer to the C_s minimum than to another minimum of higher symmetry.

In order to investigate this possiblity, perform the following procedure for one or both $\eta^5\eta^5$ isomers, using the B3LYP/6-31G(d) model chemistry as we did previously for this system:

1. Build and optimize a zincocene molecule with the same symmetry.

2. Add methyl groups to the optimized structure, taking care to maintain the symmetry. Perform a constrained optimization on this structure: this kind of optimization holds some atoms fixed and optimizes only the remaining atoms. In this case, fix the zincocene structure and optimize only the positions of the methyl groups.

   The **Geom=ReadOpt** option is used to specify which atoms to freeze and which to optimize. These specifications are read from a separate section in the input file (which is terminated by a blank line). There are many ways to specify how the various atoms are to be treated. In this case, use the **notatoms** keyword to exclude the atoms in the zincocene molecule from the optimization:

   ```
   notatoms=1-11
   ```
   blank line

3. Run a frequency calculation on the optimized geometry to determine the structure's status on the unconstrained PES. Then run an **Opt Freq** calculation using our standard model chemistry.

   If you set up these calculations in GaussView by starting from the results of the previous constrained optimization, be sure to unfreeze all of the atoms in the final input file. In order to activate them, change the -1 in the second field of the molecule specification lines to 0:

   ```
   C    -1    0.37479400    1.15351700    1.95533600
   ```

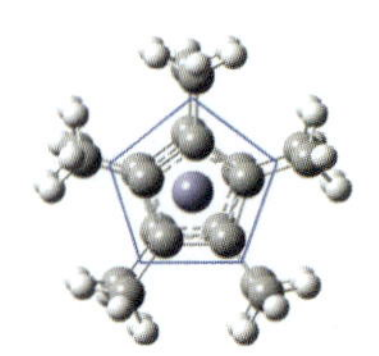

The following figure indicates the atoms to be optimized in each phase of this study by cordoning them in green. Another view of the structure for the second optimization is given in the margin where the blue line indicates the boundary between the frozen and active atoms.

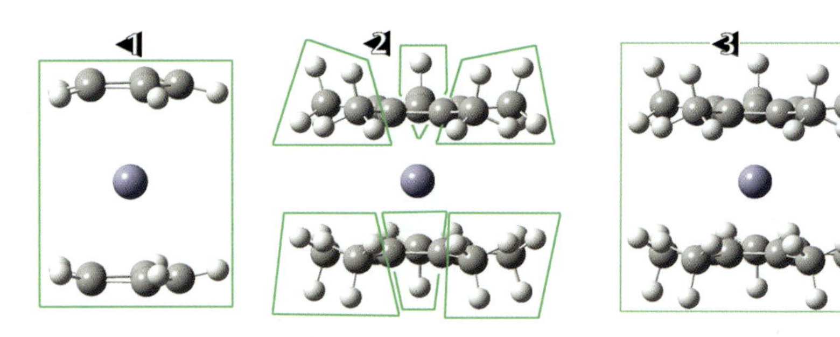

SOLUTION

The following table presents the results of our calculations for the two $\eta^5\eta^5$ isomers:

	OPTIMIZATION RESULTS		
ISOMER	**1: ZINCOCENE**	**2: CONSTRAINED**	**3: FULL OPT.**
D_{5d}	Minimum in 10 steps with D_{5d} symmetry.	Minimum in 6 steps with D_5 symmetry (D_{5d} with loose criteria).	▸ Initial structure has 13 imaginary freqs. ▸ Minimum in 80 steps: $\eta^1\eta^5$ C_1 structure. ▸ E=-2559.33532 hartrees.
D_{5h}	Minimum in 12 steps with C_{2v} symmetry (D_{5h} with default criteria).	Minimum in 5 steps with D_{5h} symmetry.	▸ Initial structure has 3 imaginary freqs. ▸ Minimum in 126 steps: $\eta^1\eta^5$ C_1 structure (C_s with loose criteria). ▸ E=-2559.33539 hartrees.

Both optimization procedures ultimately produce the same structure which is also identical to the one we found earlier by optimizing the $\eta^1\eta^5$ structure. The predicted energy of the latter is -2559.33532 hartrees. Thus, we can conclude that there is no higher symmetry, lower energy minimum for this compound.■

Unfreezing Coordinates in GaussView

In GaussView, you can reactivate frozen coordinates for structures retrieved from a checkpoint file using the **Atom Group Editor** (accessed via the **Edit** menu). First, select **Freeze** as the **Atom Group Class**. Then move all atoms to the **Freeze (No)** group. One way is to click on the **Freeze (Yes)** item, select **Select atoms of Freeze (Yes)** from the **Group Actions** menu (upper right), and finally click the plus sign icon in the **Freeze (No)** row. Close the dialog and save a new input file.

4

Predicting Chemical Properties

In This Chapter:

IR and Raman Spectra

Modeling Thermochemistry

High Accuracy Compound Methods

NMR Spectra

Conformational Searching and Boltzmann Averaging

Advanced Topics:

Frequency-Dependent Raman Spectra

Anharmonic Frequencies

Hindered Rotations

Materials Properties: Polarizabilities, Hyperpolarizabilities, Gamma

Wavefunction Stability Revisited

Region around NGC 7023. Photo credit: ESA/Hubble

4 ◆ Predicting Chemical Properties

As we have seen, Gaussian's predictive capabilities all begin with the computation of the electronic energy of a molecular system, given the coordinates of the nuclei within it and the number of electrons; that is, the energy of the electrons existing in the static field of the nuclei. From the analysis of a molecule's vibrational frequencies, the energy due to the motions of the nuclei can be added to the electronic energy to predict the total energy, enthalpy and free energy for the molecule under specific physical conditions.

Some molecular properties are simply derivatives of the electronic energy with respect to a perturbation. For example, the dipole moment is the derivative of the energy with respect to an applied electric field. Geometry optimizations locate a stationary point on the potential energy surface using the first derivative of the energy with respect to nuclear positions (forces on the atoms). IR and Raman frequencies and intensities are properties which are computed from the second derivatives of the electronic energy.

Some molecular properties cannot be directly calculated by Gaussian, but they can be easily computed from Gaussian results. For instance, a chemical shift is a property directly observed in an NMR experiment* but not directly computed by Gaussian. In an NMR calculation, Gaussian computes the second derivatives of the energy, once with respect to an applied magnetic field and once with respect to the magnetic moment of each nucleus, producing the absolute shielding tensor at each atom, which is used to obtain the isotropic shielding value. To predict the chemical shifts, you subtract the predicted shielding values for the molecule of interest from the one for a corresponding nucleus in the reference compound used in the NMR experiment (typically TMS).

There are also properties that depend on multiple conformations, and hence multiple calculations. For example, predicting optical rotations for a molecule with many rotatable bonds requires a Gaussian calculation for each rotomer, followed by a Boltzmann weighting of the results to compute a composite final prediction. This latter manipulation is typically performed by the user after obtaining the results for each individual compound (e.g., in a spreadsheet application).

Some chemical properties are not experimental observables. For instance, you may wish to discuss the charge present on a specific atom within a molecule. Because atomic charge is not a physical observable, various schemes for partitioning the total electron density among the atoms of a molecule can be chosen. Different researchers have different preferences among them, depending on what chemical features interest them.

Finally, there are properties that are too complex for Gaussian to predict. Macroscopic properties such as elasticity and bulk dielectric constants of liquids are beyond the scope of the Gaussian software.

For some properties that Gaussian does not currently compute directly, there are third-party programs that can use the output from Gaussian to obtain these. For example, Pickett's program [Pickett91] simulates hyperfine spectra from the data produced by a Gaussian calculation. Likewise, Serena Software's Pcmodel program [Gilbert14] performs a conformational search for a specified compound, prepares Gaussian input for each located conformation, and finally

* When the zero point on the scale is set to the signal for the reference nucleus, the results for the nuclei on the molecule under investigation appear on a scale relative with respect to the references: i.e. as chemical shifts.

computes the Boltzmann-averaged VCD spectrum based on the completed set of Gaussian VCD calculations.

The following are the major chemical properties predicted by Gaussian that we will consider in this book:

CHAPTER 2
- Atomic charges
- Multipole moments through hexadecapole
- Population analysis
- Biorthogonalized molecular orbitals (corresponding orbitals)
- Natural Bond Orbital (NBO) analysis

CHAPTER 3
- Vibrational frequencies and normal modes

CHAPTER 4
- IR and Raman spectra
- NMR shielding tensors
- Thermochemical properties: atomization energies, ionization energies, electron & proton affinities, heats of formation
- Anharmonic frequencies
- Frequency-dependent Raman intensities
- Polarizability, hyperpolarizability and gamma (second hyperpolarizabilities)

CHAPTER 5
- Free energy in solution

CHAPTER 6
- Electrostatic potential
- Enthalpies of reaction

CHAPTER 7
- NMR spin-spin coupling
- Electronic circular dichroism (ECD) rotational strengths
- Vibrational circular dichroism (VCD) rotational strengths
- Raman optical activity (ROA) intensities
- Hyperfine spectra tensors (microwave spectroscopy)

CHAPTER 8
- UV/Visible spectra
- Absorption and emission spectra via Franck-Condon, Herzberg-Teller and Franck-Condon/Herzberg-Teller analyses
- Photochemical properties via conical intersections

Frequency Calculation Results: IR and Raman Spectra

Energy calculations and geometry optimizations ignore the vibrations in molecular systems, using a static view of nuclear position. In reality, the nuclei in molecules are constantly in motion. In equilibrium states, these vibrations are regular and predictable, resulting in discrete (i.e., quantized) vibrational energy states. Transitions between vibrational energy states occur by the absorption of infrared radiation, specifically photons of the appropriate energy. Frequency calculations model this behavior and can be used to predict IR and Raman spectra, normal modes and related properties.

IR spectroscopy stems from vibrational transitions and consists of measuring the absorptions resulting from exposing a substance to a range of IR frequencies. The amount of absorbed energy is known as the *intensity*, and it is plotted as the height of the peak at that frequency.* The resulting spectrum is thus uniquely indicative of the molecule's component atoms and bonding (i.e., its structure). Note that not all vibrational motions will produce lines in the IR spectrum. Only vibrational modes which modify the molecular dipole produce non-zero IR intensities; they are said to be *IR active*.

Raman spectroscopy is another spectroscopic technique used to study vibrational modes in a molecular system. It relies on the inelastic scattering of light within or near the visible range that occurs when it interacts with molecular vibrations, resulting in the energies of the applied photons being shifted up or down in characteristic ways. The amount of scattering is known as the Raman *activity*. Raman spectrometers typically employ a laser as the light source. This technique's name comes from C. Venkata Raman, who discovered this radiation effect, also known as Raman scattering, in the 1920s. Vibrational modes are said to be Raman active or Raman inactive, depending on whether their Raman activity is non-zero.

Some modes are both IR inactive and Raman inactive. Such modes are termed *forbidden*.

Because of the nature of the computations involved, frequency calculations are valid only at stationary points on the potential energy surface. Thus, frequency calculations must be performed on optimized structures, and the easiest way to do so is to use the built-in optimization+frequency feature: the **Opt Freq** keyword combination. However, you can provide an optimized geometry as the molecule specification for a stand-alone frequency job. A frequency job must use the same theoretical model and basis set as were used for locating the optimized geometry. Frequencies computed with a different basis set or method have no validity.

Gaussian can compute the vibrational spectra of molecules, predicting the frequencies and intensities of spectral lines corresponding to the nuclear vibrations comprising each of the molecule's normal modes. The IR spectrum is computed by default. You can request the Raman spectrum as well if desired:

- *GaussView*: Select **Yes** from the **Compute Raman** popup menu on the **Job Type** panel.
- *WebMO*: Select **Opt** from the **Calculation** menu and then enter **Freq=Raman** into the resulting popup window (you should select **Other** instead of **Opt** if you do not need to optimize the structure).

* Intensity can be described in other terms: for example, as the charge on the dipole moment. Viewed this way, a carbonyl stretch has a high intensity while a C=C stretch has a weak to medium intensity.

GaussView Tip:

If you are examining the results from a checkpoint or formatted checkpoint file, the **Run FreqChk** *dialog will display the first time you select the* **Results→Vibrations** *menu path. Simply click* **Ok** *to reach the* **Display Vibrations** *dialog.*

The various spectral data are accessed in separate operations in our graphics programs. The various operations are summarized in the following table:

Property	GaussView	WebMO
View frequency results	The **Results→Vibrations** menu path opens the **Display Vibrations** dialog.	Access them from the **View Job** page.
Plotted spectra	Click the **Spectrum** button in the **Display Vibrations** dialog. This will open a new scrollable window containing all predicted spectra.	Click the magnifying glass icon labeled **IR spectrum** in the **Vibrations** section of the **Calculated Quantities** portion of the page. The IR spectrum will be displayed in the **Data Viewer** panel on the page. WebMO cannot currently display Raman spectra.
Table of frequency data	The list of predicted frequencies, along with corresponding IR intensities and Raman activities, appears in the upper portion of the **Display Vibrations** dialog.	The list of predicted frequencies, along with their symmetries, appears in the **Vibrations** section of the **Calculated Quantities**.
Animation of normal modes	Click the **Start Animation** button in the **Display Vibrations** dialog to begin normal mode animation. Select the mode to be animated by clicking on its row in the frequency table or on its peak in one of the plotted spectra.	Click on the filmstrip icon in the frequency list for the mode you want to animate.

Scaling Frequencies

Conventionally, raw frequency values are multiplied by a standard scale factor prior to comparison with experiment. This practice arose from the fact that the values computed at the Hartree-Fock level contain known systematic errors due to the neglect of electron correlation, resulting in overestimates of about 10%-12%. Therefore, it became common practice to scale frequencies predicted at the Hartree-Fock level by an empirical factor of 0.8929 [Pople93]. Use of this factor has been demonstrated to produce very good agreement with experiment for a wide range of systems. The same scale factor was also used for zero-point energies and thermal corrections.

With the introduction of DFT methods, frequency scale factors were also computed for a variety of model chemistries (see e.g. [Wong96]). However, the DFT-predicted frequencies were generally significantly more accurate than for Hartree-Fock, and the scale factors were accordingly larger, an effect which increased with basis set size. For example, Bauschlicher and Partridge computed the B3LYP/6-311+G(3df,2p) scaling factor to be 0.989 [Bauschlicher95], a value close enough to 1.0 to call the necessity of scaling into question.

Nevertheless, in this book, we will scale frequencies computed using our standard APFD/6-311+G(2d,p) model chemistry by this value. Scaling raw frequencies typically has the effect of shifting the entire predicted spectrum downward, often producing better alignment with the experimental spectrum. We will not, however, scale zero point energies or thermal energy corrections.

Frequency data can be automatically scaled by Gaussian by including the **Scale** keyword within the job's route section: e.g., **Scale=0.989**. Be aware that this scales the frequency locations and the zero point and thermal energy corrections. You can also scale the frequencies for a completed job within your visualization program:

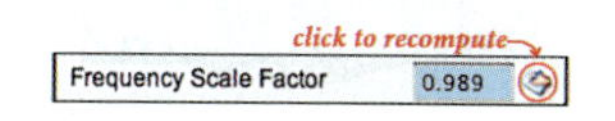

- *WebMO*: The **Vibration Modes** section in the **Calculated Quantities** portion of the **View Job** page contains a scale factor field. After entering a value into the field, click on the icon to its right to recompute the frequency data. The table will now display both the raw and scaled values. You will also need to regenerate the spectrum if you have already opened it.

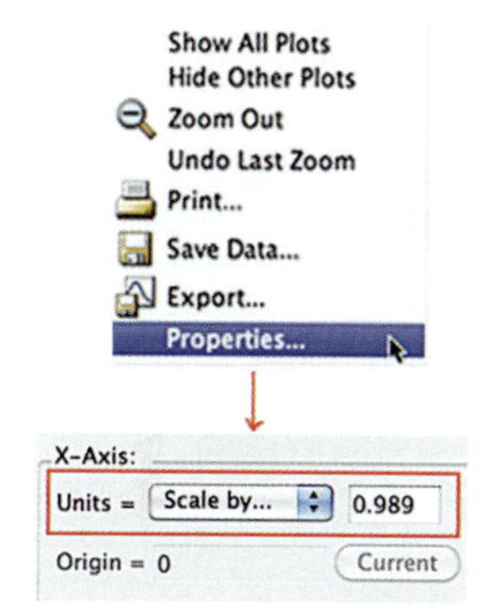

- *GaussView*: Right click in the plot area of the spectrum to open the context menu, and then select **Properties**. Choose **Scale by** from the **Units** menu in the **X-Axis** area of the **Plot Properties** dialog, and then specify the desired scale factor in the field that appears. This will cause the plotted spectrum to be scaled by the specified amount. Clicking on any peak will display the scaled value in the status area of the plot window.

  Note that this scale factor *does not affect* the frequencies values listed in GaussView's **Display Vibrations** table.

EXAMPLE 4.1 IR SPECTRUM OF FORMALDEHYDE; RAMAN SPECTRUM OF BENZENE

Previously, we ran an optimization+frequency calculation on formaldehyde (see page 58). We will now examine the IR spectrum predicted by that calculation.

WebMO Here is the frequency table and IR spectrum as displayed by WebMO (we've scaled the frequencies and narrowed the peak width from the default):

Vibrational Modes

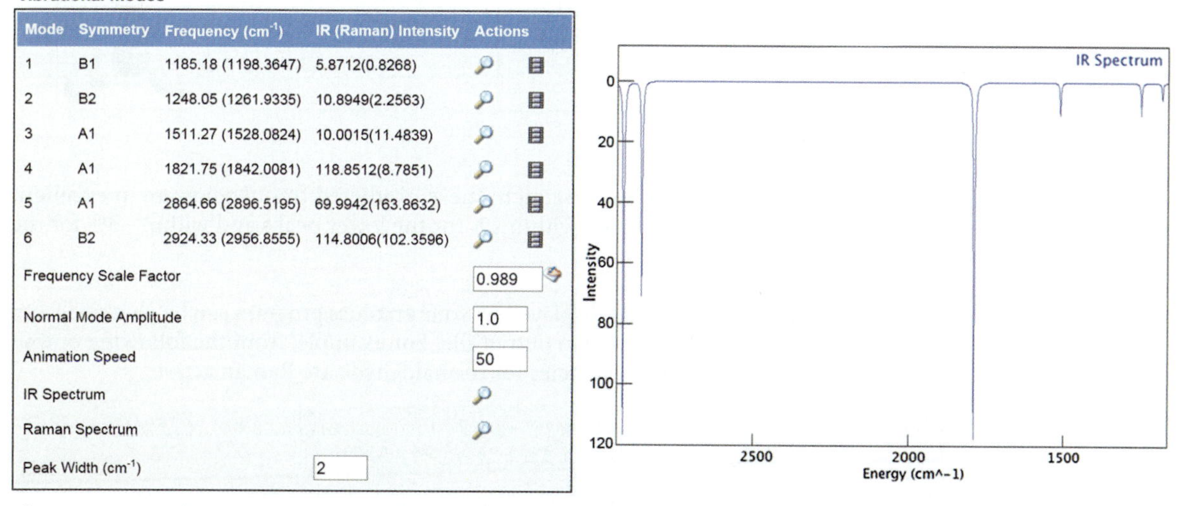

Mode	Symmetry	Frequency (cm^{-1})	IR (Raman) Intensity	Actions
1	B1	1185.18 (1198.3647)	5.8712(0.8268)	
2	B2	1248.05 (1261.9335)	10.8949(2.2563)	
3	A1	1511.27 (1528.0824)	10.0015(11.4839)	
4	A1	1821.75 (1842.0081)	118.8512(8.7851)	
5	A1	2864.66 (2896.5195)	69.9942(163.8632)	
6	B2	2924.33 (2956.8555)	114.8006(102.3596)	

The spectrum shows three strong peaks and three weak ones (the latter are also at the lowest frequencies). The following table summarizes our results and compares them with experimental data [Herzberg91 p. 300]:

	CALCULATED		NORMAL MODE	
OBSERVED PEAK (cm^{-1})	SCALED FREQ. (cm^{-1})	INTENSITY (km/mol)	MOTION	SYMM.
1167.26	1185	5.9	CH_2 wag (out-of-plane)	B1
1249.09	1248	10.9	CH_2 rock (in-plane)	B2
1500.17	1511	10.0	CH_2 scissors (in-plane)	A1
1746.01	1822	118.9	C=O stretch	A1
2782.46	2865	70.0	C-H symmetric stretch	A1
2843.33	2924	114.8	C-H asymmetric stretch	B2

With the exception of the carbonyl stretch, these predicted frequencies are in excellent agreement with the observed peaks: within 1% for the lower peaks and within ~3% for the higher ones.

Any desired quantities that are not displayed by your graphics program can be obtained from the frequency portion of the Gaussian output file. For example, from the following output we can see that the first three frequencies for formaldehyde are Raman active:

Frequencies are unscaled in the Gaussian output unless the **Scale** *keyword was specified for the job.*

```
Harmonic frequencies (cm**-1), IR intensities (KM/Mole),
Raman scattering activities (A**4/AMU), ...

                          1              2              3
                         B1             B2             A1
Frequencies --    1198.3647      1261.9335      1528.0824
Red. masses --       1.3686         1.3447         1.0956
Frc consts  --       1.1580         1.2617         1.5073
IR Inten    --       5.8712        10.8949        10.0015
Raman Activ --       0.8268         2.2563        11.4839
Depolar (P) --       0.7500         0.7500         0.5267
```

Here are the GaussView results displays for benzene:

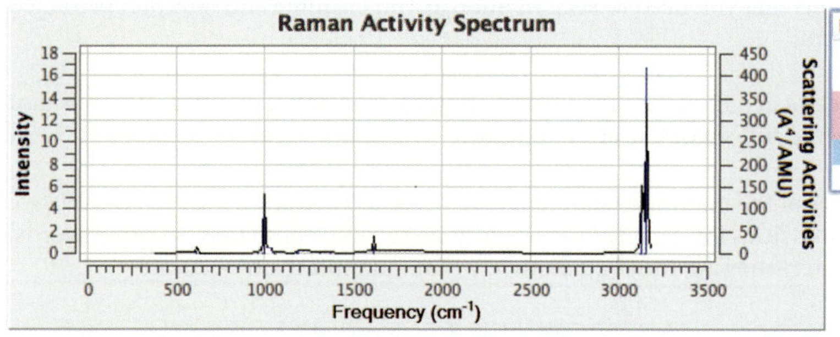

Mode #	Freq	Infrared	Raman Activity
1	408.57	0.0000	0.0000
2	408.57	0.0000	0.0000
3	624.36	0.0000	3.8821
4	624.36	0.0000	3.8811
5	685.04	117.7599	0.0000
6	713.94	0.0000	0.0001

The plotted spectrum includes only peaks corresponding to vibrational modes that are Raman active, and it includes one strong peak in the upper frequency range, one medium strength peak around 1000 cm^{-1}, as well as several very weak peaks. We've shaded these modes light red in the peak list in the illustration above (which we've truncated after the first 6 modes). The nearby IR active mode is shaded blue, and the forbidden modes are left white.

GaussView plots spectra using a different orientation than WebMO. If you think GaussView's plot is upside down and backwards, you can right click anywhere within the plot, select **Properties** from the resulting context menu, and then click one or both of the **Invert Axis** checkboxes in the resulting dialog.

The following table summarizes the predicted frequencies and Raman activities and compares these results with experiment [Herzberg91, p. 364]:

	CALCULATED	
OBSERVED PEAK (cm^{-1})	SCALED FREQ. (MODE) (cm^{-1})	RAMAN ACTIVITY (a^4/amu)
608	617 (3,4)	3.9
845	852 (7,8)	1.0
993	1014 (13)	74.9
1178	1193 (17,18)	5.3
1591	1632 (23,24)	11.8
3057	3142 (26,27)	127.7
3074	3168 (30)	413.5

Some observed peaks correspond to two predicted frequencies whose location and Raman activities are identical. Such doubly degenerate modes correspond to distinct vibrational motions that are equivalent by symmetry. Such modes always appear with the label E in character tables.

The predicted Raman spectrum is in good agreement with experiment. Interestingly, in this molecule, there are no modes which are both IR active and Raman active. Such results are common for molecules with high symmetry and could be predicted by consulting a character table for the D_{6h} point group. Those modes that transform as x, y or z translations will be IR active: A_{2u} and E_{2u}. Those modes which transform as quadratic functions of the Cartesian displacements will be Raman active: A_{1g}, E_{1g} and E_{2g}. All other modes are neither IR nor Raman active.

Understanding & Interpreting Frequency Data

There are several points that you need to keep in mind as you examine and interpret predicted IR and Raman spectra:

- Only active peaks are included in plotted spectra. You must examine the table of frequencies for the full set of vibrational modes.
- Computed values of the intensities should not be taken too literally. However, the relative values of the intensities for each frequency may be reliably compared. In other words, the predicted intensities are more qualitatively than quantitatively accurate.
- The vibrational frequency analysis performed by default uses the harmonic approximation which assumes that there are no interactions between vibrational modes.* Accordingly, only fundamental bands are predicted. Observed spectra also may include overtone and combination bands (which are treated as simple factors and sums of fundamentals in the harmonic approximation).

  Anharmonic frequency analysis must be used to predict the latter band types. However, it is much more computationally expensive and is not always needed. We will explore this topic in Advanced Exercise 4.12 (p. 187).

TO THE TEACHER: THE HARMONIC OSCILLATOR

This chapter offers a number of opportunities for relating the text to topics in an elementary quantum mechanics course, including the following:

- Discuss how to compute vibrational frequencies using a simple harmonic oscillator model of nuclear motion.
- Present formal definitions of intensities and Raman depolarization ratios.
- Rationalize zero-point energies by reference to the harmonic oscillator model once again, and its energy: $h\nu(n+\frac{1}{2})$. The ground state corresponds to $n=0$, yielding a non-zero energy.
- Relate characterization of stationary points via the eigenvalues of the Hessian to the corresponding matrix under the harmonic oscillator problem.

Applications of IR and Raman Spectroscopy

IR and Raman spectroscopy is often used for substance identification. The next two examples provide illustrations of such real world problems.

EXAMPLE 4.2 **DETECTING C_{60} IN INTERSTELLAR SPACE**

The IR spectrum of C_{60}† was reported soon after the compound's initial discovery. Only four vibrational modes are IR active, resulting in spectra lines at 527.1, 570.3, 1169.1 and 1406.9 cm^{-1} [Frum91].

From the beginning, scientists wondered whether fullerenes might occur naturally. In 2010, an article in *Science* generated a great deal of excitement when it reported the presence of C_{60} and C_{70} in Tc 1, a planetary nebula about 6000 light years away [Cami10]. In fact, C_{60} had been

* More specifically, it solves the harmonic approximation of the vibrational nuclear Schrödinger equation, treating vibration motion like springs around a minimum. This approximation neglects terms in the expansion beyond second derivative (which are included in anharmonic frequency analyses).

† IUPAC: [5,6]fullerene-C60-Ih.

detected before, in IR observations of NGC 7023 (the Iris nebula) taken using the Spitzer Space Telescope in 2004 [Werner04]. The Hubble telescope website poetically describes the Iris nebula as a "blushing dusty nebula" (*www.spacetelescope.org/news/heic0915*); it is a vast cloud of dust in interstellar space, parts of which appear unusually red in the visible range. NGC 7023 is categorized as a reflection nebula, meaning that it scatters light from a nearby star. Detecting and analyzing that light allows astronomers to determine the composition of the dust. The majority of such dust is typically composed of polycyclic aromatic hydrocarbons (PAHs).

Closeup of an area in the northwest region of NGC 7023 illuminated by the bright star HD 200775 (top). Photo credit: ESA/Hubble

The 2004 paper's identification of C_{60} was highly tentative and speculative. However, recent follow-up observations by Sellgren and coworkers have confirmed the initial spectral identification [Sellgren10].

Here is their observed spectrum:

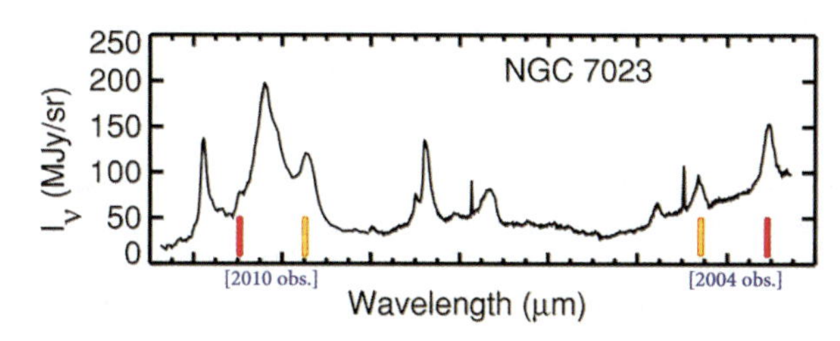

The 2010 observations extended the range of the 2004 results, and the spectrum above is a composite of the two runs. The peaks corresponding to C_{60} are indicated by the solid lines under the plotted spectrum.

The following table summarizes the experimental IR spectroscopy results, the IR observations of NGC 7023 and our own APFD/6-311+G(2d,p) frequency calculation. Note that astronomers typically report spectral observations in micrometers rather than inverse centimeters, so we have converted all data to these units:

1 micron = 10000/cm^{-1}

OBSERVED PEAKS (μm)		CALC.	ASTRONOMY	
C_{60} IR	NCG 7023	(scaled)	INTERPRETATION	REF.
18.97	19.0	18.6	C_{60}	[Werner04]
17.53	17.4	16.9	PAH/C_{60} blend	[Werner04]
8.50	8.5	8.2	PAH/C_{60} blend	[Sellgren10]
7.10	7.04	6.8	C_{60}	[Sellgren10]

We can make several observations about these results:

- The majority of the nebula's observed spectrum comes from PAHs, and C_{60} adds localized perturbations at four points.
- Only two of the C_{60} peaks can be completely isolated (corresponding to the red lines below the plotted spectrum). The other two are blended with the PAH lines (colored orange). For example, the width of the peak at 8.5 μm comes from the combination of that C_{60} peak and a PAH peak at 8.6 μm [Sellgren10].
- There is near-perfect agreement between the IR spectrum and the astronomical observations, within experimental error of the latter, which is ~0.05 μm.

- The results of our frequency calculation agree well with the experimental IR spectrum. The predicted frequencies are all lower than experiment (expressed in μm, corresponding to systematically higher frequency values in cm^{-1}), deviating by 2-4%.

The predicted C_{60} spectrum is sufficiently accurate to allow you to interpret the IR spectrum (assuming the compound's structure was unknown), and it might be suggestive in identifying the presence of C_{60} in the interstellar dust cloud, but it is probably not sufficiently accurate to be definitive alone in the latter case.■

IR and Raman spectroscopy is frequently used in compound identification and differentiation in many contexts. Our next example will consider one from the criminal justice realm.

EXAMPLE 4.3 RAMAN CRIME SOLVING: IDENTIFYING SUBSTANCES ON CURRENCY

Identifying controlled substances in a variety of environments is an important task for forensic chemists. Traditionally, gas chromatography mass spectrometry has been used, but this technique has the disadvantage that it consumes the sample that it analyzes. In recent years, there has been considerable interest in Raman spectroscopy for such purposes, one advantage of which is that it does not render the evidence unfit for future reanalysis.

Interestingly, virtually all US currency contains trace but detectable quantities of cocaine [Oyler96].

Using Raman spectroscopy to identify cocaine* and distinguish it from similar compounds, such as the local anesthetic benzocaine,* has been studied extensively. The Raman spectrum of cocaine was obtained by Gamot in 1985 [Gamot85], and the technique has been employed in the context of cocaine on raw textile fibers [Cho07], on clothing [Ali08], on human fingernails [Ali08a], embedded in collected human fingerprints [West09] and on US currency [Noonan05] (to name a few). Currency provides an especially challenging environment for Raman spectroscopy due to fluorescence from the ink. This fluorescent background must be removed from the observed spectra, either by photo bleaching (subjecting the bill to bright light) or by background subtraction (a computational approach). Noonan and coworkers developed procedures to accomplish this and demonstrated the viability of using Raman spectroscopy for these identification purposes [Noonan05].

The structures of cocaine (right) and benzocaine (left) are illustrated below:

ring structure of cocaine

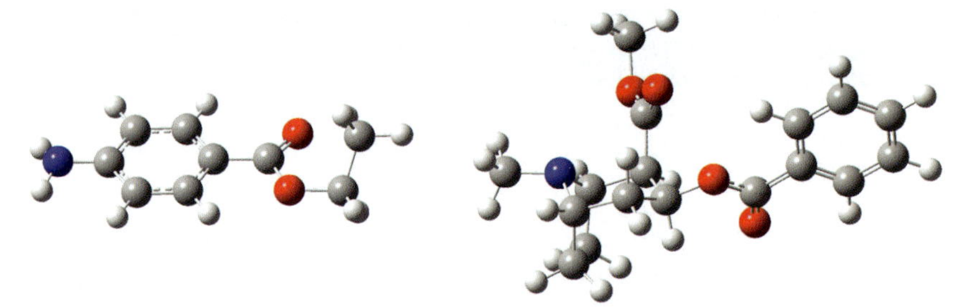

We optimized the geometries of these two compounds, and then performed **Freq=Raman** calculations. The predicted spectra are illustrated below. The lines at the top of the plot indicate the positions of the strongest peaks from the observed spectra ([Palafox89] for benzocaine and [Gamot85] for cocaine). The regions of most dramatic difference between the two spectra are indicated by green stars:

* IUPAC: cocaine: methyl (1R,2R,3S,5S)-3-(benzoyloxy)-8-methyl-8-azabicyclo[3.2.1]octane-2-carboxylate; benzocaine: ethyl 4-aminobenzoate.

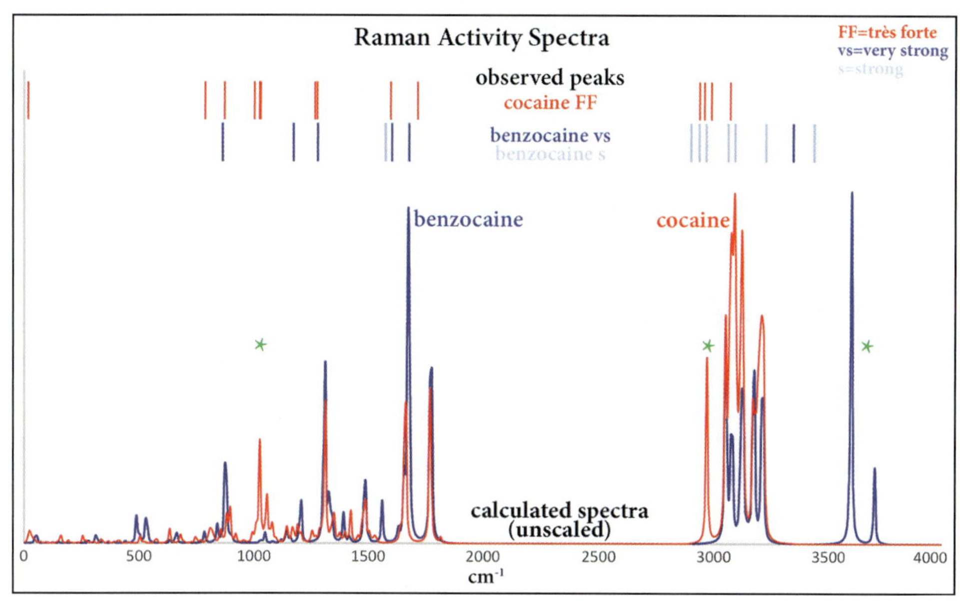

Although there is some downward shifting of peaks, the predicted spectra are generally in good agreement with experiment.

The following table summarizes these results. It includes the strongest peaks present in the experimental data—defined "très forte" (FF) in [Gamot85] (cocaine) and as "very strong" (vs) in [Palafox89] (benzocaine)—for which corresponding frequencies were identifiable in the corresponding predicted spectrum:

VIBRATIONAL MODE	MODE #	CALC. (scaled)	RAMAN ACTIVITY	EXP.
COCAINE				
benzene deformation	53	1012	41	1000
CH_3 rotation	54	1018	2.4	1021
benzene in-plane	56	1044	19	1026
CH_2 rock	77	1297	63	1266
benzene C–C stretch	100	1642	112	1596
C=O stretch	101	1751	97	1713
C–H stretch (piperidine)	105	3027	121	2935
C–H stretch (pyrrolidine)	107	3040	172	2956
CH_3 stretch (NH^+–CH_3)	103	2936	142	2986
BENZOCAINE				
C–H bend in plane (benzene)	36	1192	22	1171
O–C–C asymm. stretch	37	1295	110	1277
C–C stretch (benzene)	51	1656	228	1600
C=O stretch	52	1753	163	1677
C–H asymm. stretch (ethyl)	53	3017	190	2868
C–H symm. stretch (ethyl)	55	3083	108	2934
C–H stretch (benzene)	58	3141	90	3089
N-H symm. stretch	62	3559	289	3344
N-H asymm. stretch	63	3662	64	3434

As we noted previously, identifying substances on currency can pose some challenges due to interference from the bill itself. This effect is illustrated well in the work of Noonan and coworkers [Noonan05]. The following figure compares the Raman spectra for benzocaine for one sample on a microscope slide and for another sample on a one dollar bill:

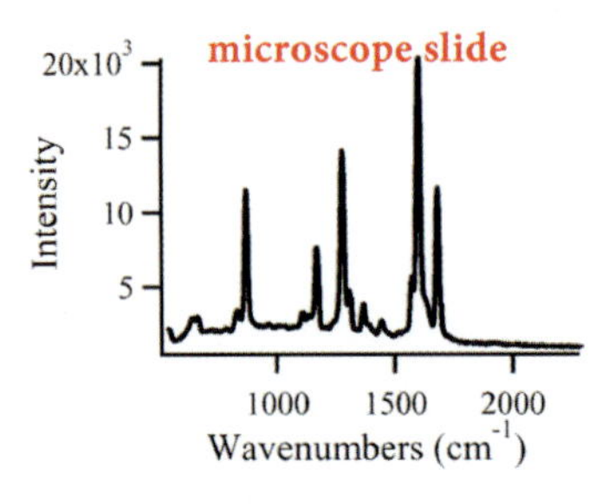

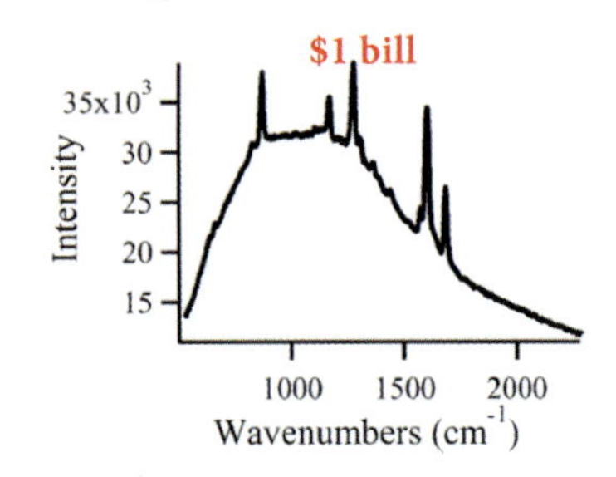

The spectrum for the sample on currency is distorted by a mass of signal from the bill itself, but the major peaks are still easily discernible, and their positions change only a small amount.

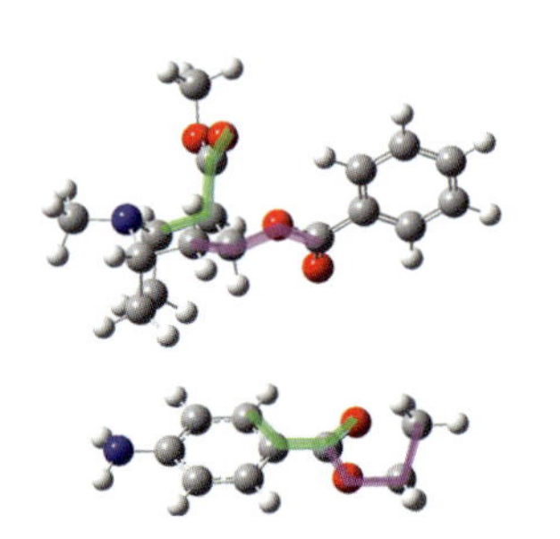

Conformations of Benzocaine and Cocaine

For this example, we have limited our consideration to a single conformation of each compound. In reality, both compounds have many possible conformations which differ in the values of the C-C-O-C and C-C-C-O dihedral angles, highlighted in purple and green (respectively) in the figure in the margin.

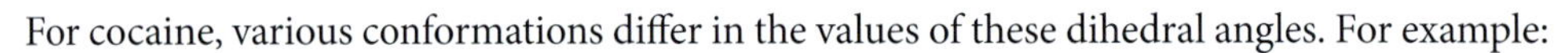

For cocaine, various conformations differ in the values of these dihedral angles. For example:

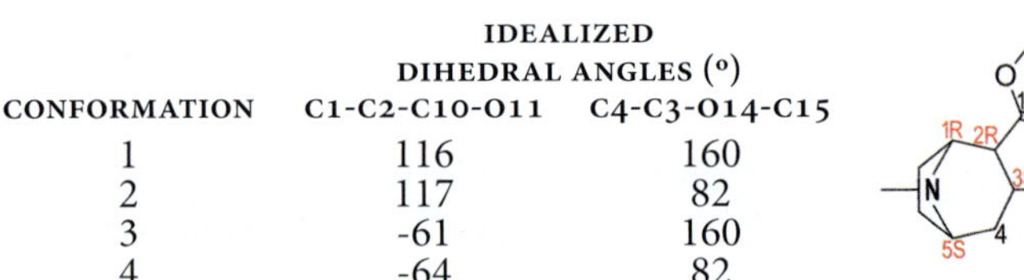

CONFORMATION	IDEALIZED DIHEDRAL ANGLES (°) C1-C2-C10-O11	C4-C3-O14-C15
1	116	160
2	117	82
3	-61	160
4	-64	82

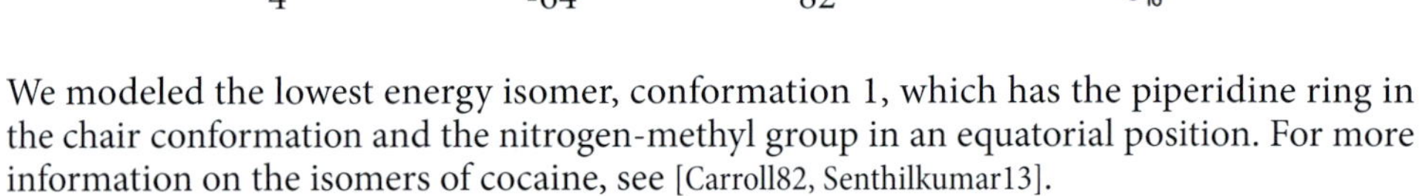

We modeled the lowest energy isomer, conformation 1, which has the piperidine ring in the chair conformation and the nitrogen-methyl group in an equatorial position. For more information on the isomers of cocaine, see [Carroll82, Senthilkumar13].

Similarly, benzocaine has three conformations: two gauche forms and a trans form:

- gauche 1: C-C-C-O≈0°, C-C-O-C≈47°
- gauche 2: C-C-C-O≈180°, C-C-O-C≈47°
- trans: C-C-C-O≈180°, C-C-O-C≈180°

Our calculations find that all three forms are very close in energy, with a thermal energy spread of ~0.2 millihartrees. The isomer we refer to as gauche 1 is the lowest in energy. However, all three isomers will be present in actual samples in roughly equal amounts. Fortunately, Raman spectra are not highly sensitive to conformation, and the predicted spectra are very similar for all three conformations, except for the very lowest modes, so this simplification is acceptable in this case. See [Balci08] for a detailed comparative study of the predicted Raman spectra for the conformations of benzocaine.

Actual research studies would need to take the various conformations of these substances fully into account. Example 4.8 (p. 164) performs such an analysis for the ^{13}C NMR spectrum of 2,2,4-trimethylpentane-1,3-diol.■

Substituting Isotopes

In general, Gaussian uses the most abundant isotope for each atom when computing vibrational frequencies. However, you can specify different isotopes if desired, and you can also recompute a frequency analysis with different isotopes very rapidly, provided that you have retained the job's checkpoint file.

You can specify the desired alternate isotope for any atom by using the **Isotope** keyword following the atom type. For example, the following molecule specification line indicates a tritium atom:

```
H(Iso=3) 0.00 0.00 0.00
```

Such specification can be added manually by editing a Gaussian input file. In GaussView, you can also specify isotopes using the **Atom List Editor**, available from the **Edit** menu (pictured below). The settings for **Isotopologue 0** will be used in any subsequently created input files.

The dialog allows you to specify multiple sets of isotopes for use in vibrational analysis. In the following example, we are defining a second isotope set—**Isotopologue 1**—in which the hydrogen atoms in formaldehyde are replaced with deuterium atoms:

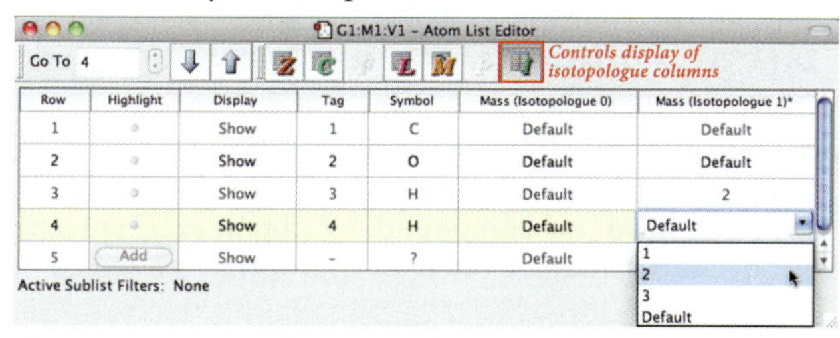

In order to view the IR spectrum and associated normal modes for an alternate set of isotopes, you must perform the following steps after opening the relevant checkpoint file and viewing the vibrational analysis results for the isotopes specified in the corresponding input file:

1. Click on the **Edit Isotopologues** button in the **Display Vibrations** dialog. This will open the **Atom List Editor** (as above).

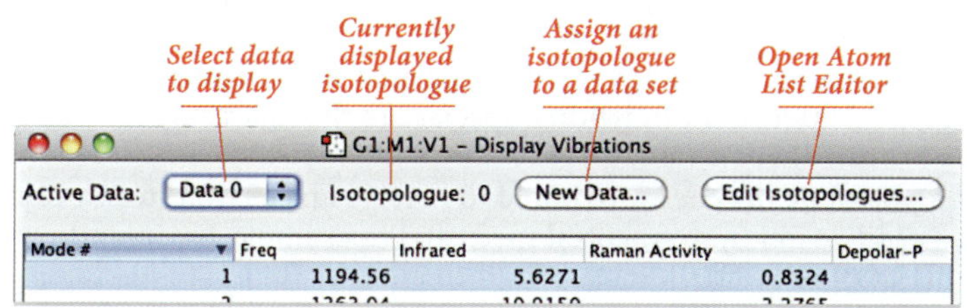

2. Specify the desired isotopes for **Isotopologue 1**, and then close the **Atom List Editor.**
3. Return to the **Display Vibrations** dialog and click **New Data**. This will display the **Display Vibrations: Run FreqChk** dialog:

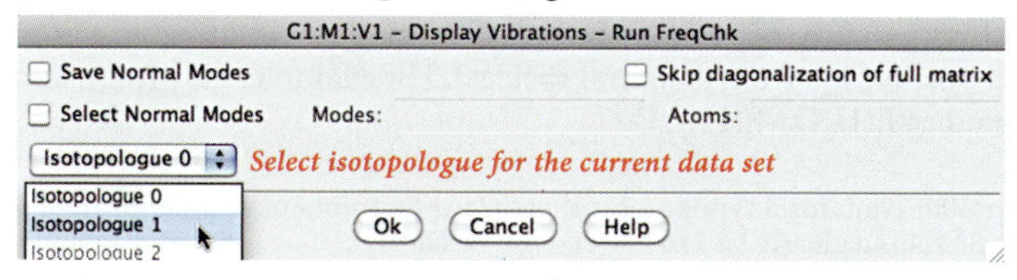

4. Select **Isotopologue 1** from the popup menu. This associates that set of isotopes with the vibrational analysis data set named **Data 1**. Close the dialog.

5 Select **Data 1** from the **Active Data** popup in the **Display Vibrations** dialog. The data in the table will now correspond to the alternate isotopes.

If you do not use GaussView, you can use the **FreqChk** utility to perform a vibrational analysis using different isotopes. Consult the *Gaussian Users Reference* for details.

EXAMPLE 4.4 SUBSTITUTING DEUTERIUM IN FORMALDEHYDE

We substituted deuterium for the hydrogen atoms in formaldehyde. The following table gives the predicted frequencies for the various normal modes (the observed values are from [Herzberg91 p. 300]*):

	OBSERVED		CALC. (scaled)		RATIO (D/H)	
MODE	CH_2O	CD_2O	CH_2O	CD_2O	OBS.	CALC.
1	1167.26	937.8	1185	949	.80	0.80
2	1249.09	986.7	1248	983	.79	0.79
3	1500.17	1101.5	1511	1098	.73	0.73
4	1746.01	1701.5	1822	1754	.97	0.96
5	2782.46	2057.2	2865	2098	.74	0.73
6	2843.33	2160.3	2924	2195	.80	0.75

Our computed values again deviate from experiment by about the same amounts as for the default isotopes. Notice, though, that when we compute the ratio of the frequencies for the deuterated and normal forms of the compound, we obtain excellent agreement with the observed data. Such derived values are often an appropriate way to compare theory and experiment.

For formaldehyde, the normal modes for the two isotopic variations happen to have the same ordering. Be aware that this is seldom the case for larger molecules, and the motions of the various modes must be examined to determine which ones correspond.

Modeling Thermochemistry

Thermochemistry studies the energy (heat) associated with chemical reactions. The major quantities of general interest for individual species are the following:

- Ionization potential: the energy required to remove an electron from a compound (e.g., forming a positive ion from a neutral species). For example, the ionization potential for water is $E(H_2O^+)$-$E(H_2O)$.
- Electron affinity: the energy released when an electron is added to a compound (as in forming a negative ion from a neutral species). For example, the electron affinity for CN is $E(CN)$-$E(CN^-)$, which yields a positive value in this case. Negative electron affinities are also possible, and they indicate that energy is required to attach the electron.
- Proton affinity: the energy released when a proton (H^+) is added to a compound, (e.g., forming a positive ion from a neutral species). For example, the proton affinity for water is computed as $E(H_2O)$-$E(H_3O^+)$.†

Other conventions define electron and proton affinities as the energy change for the relevant reaction and thus have the opposite sign.

* [Herzberg91 p. 300] contains a typographical error in the frequency value for the fourth mode for CD_2O: "1301.5" should clearly be 1701.5 (cf. e.g. [Wohar91]).

† Formally, the proton affinity is defined as the energy released for the reaction of, e.g., $H_2O + H^+ \rightarrow H_3O^+$; i.e., $E(H_3O^+)$-$E(H_2O)$-$E(H^+)$. However, the total energy of a proton is zero since it has no electrons.

- Atomization energy: the energy difference between a compound and its monoatomic components. For example, for water, this is the difference between its energy, E (H_2O), and the sum: E(O) + 2E(H).
- Heat of formation: the enthalpy change resulting from the formation of a compound in its standard state from its constituents in their standard states. For example, for carbon dioxide, this is the difference between its enthalpy and the combined enthalpies of graphite and oxygen gas (O_2). The heat of formation often cannot be computed solely from Gaussian results as it requires the use of experimental parameters for elements whise standard states are liquid or solid (see Example 4.6, p. 160). This value is sometimes called the enthalpy of formation.

Frequency calculations are required to compute these quantities because various thermochemical quantities are required as components of the computations. These include the zero point energy, the thermal correction and the enthalpy. Upcoming examples will describe the specific procedure. For a detailed discussion of computing thermochemical properties, as well as the theory upon which they are based, see [Ochterski00].

Why & How to Model Thermochemical Quantities

The various thermochemical properties listed above are relevant to different chemical processes. For example, the ionization potential, electron affinity and proton affinity are all important to study electrochemistry. They are experimental observables, and calculations can usually directly model the relevant reactions and predict the measured quantities under the same physical conditions.

When studying reactions, typically we need to predict quantities such as reaction energies, bond energies, isomerization energies, and so on. Whenever possible, it is best to model these reactions directly in order to predict the quantities of interest. However, heats of formation are used in computing, e.g., ΔH for reactions including components that cannot be calculated, as when oxygen removes carbon from a graphite sheet to form carbon dioxide.

We recommend the following best practices with respect to modeling thermochemical properties of compounds and reactions:

- Model what is observed, not what is easiest to calculate.
- Compute what is measured rather than an arbitrary thermochemical quantity.
- Model the reaction under the same environmental conditions as the experiment rather than "correcting" observed results to different conditions.

EXAMPLE 4.5 THERMOCHEMISTRY CALCULATIONS ON SMALL MOLECULES

In order to illustrate these concepts, we will predict the following thermodynamic quantities, using our standard model chemistry:

- Ionization potential of CN and Cl_2. The energy of the positive ion (e.g., CN^+) minus the energy of the neutral molecule: e.g., $E(CN^+)–E(CN)$.
- Electron affinity of CN and Cl_2. The energy of the negative ion minus the energy of the neutral molecule: e.g., $E(CN)–E(CN^-)$.
- Proton affinity of NH_3. The energy of the neutral molecule minus the energy of the compound with an added hydrogen: $E(NH_3)–E(NH_4^+)$.
- Atomization energy of benzene. The energy of the molecule subtracted from the sum of the energies of the component atoms: 6E(C)+6E(H)–E(benzene).

The experiments corresponding to the first three items are typically conducted at room temperature, so we use the predicted thermal-corrected energy to compute all of these quantities.

We will compute the atomization energy of benzene and compare it to the difference in the experimental heats of formation between the molecule and its component atoms. The experimental values are at STP, so we will use the predicted enthalpy value for benzene from an optimization+frequency calculation. The enthalpy of an atom is the total energy obtained from a single point energy calculation plus a constant correction:

$$\text{atomic thermal enthalpy correction} = {}^{3}/_{2}RT = 0.00236 \text{ hartrees}$$

Here are the predicted energies and resulting quantities:

MOLECULE	E (298.15 K) (hartrees)
NH_3	-56.49430
NH_4^+	-56.81882
CN	-92.64748
CN^+	-92.11662
CN^-	-92.79241
Cl_2	-920.17235
Cl_2^+	-919.75255
Cl_2^-	-920.27378

MOLECULE	H (298.15 K) (hartrees)
benzene	-232.00417

ATOM	TOTAL ENERGY (hartrees)
C	-37.81643
H	-0.50225

QUANTITY	CALCULATED (kcal/mol)	EXPERIMENT	$\lvert\Delta^{EXP}\rvert$ (kcal/mol)
CN IP	333.1	14.03±0.02 eV[a] = 323.6±0.5 kcal/mol	9.5
Cl_2 IP	263.4	11.50±0.01 eV[b] = 265.3±0.2 kcal/mol	-1.9
CN EA	90.9	89.1±0.1 kcal/mol[c]	1.8
Cl_2 EA	63.6	2.5±0.2 eV[d] = 55.1±0.2 kcal/mol	8.5
NH_3 PA	203.6	204.0[e]±0.08[f] kcal/mol	-0.4
benzene AE	1312.8	1320.5±0.7 kcal/mol[g]	-7.7

[a][Berkowitz69]. [b][VanLonkhuyzen84] and [Yencha95]. [c][Bradforth93]. [d][Bowen83]. [e][Hunter98].
[f]Error is that of the observed $\Delta_f H°$ of ammonia [Cox84] from which its PA was in part derived.
[g]$\Delta_f H°(298.15)$ values: benzene [Roux08], atoms [Lias88].

The traditional desired accuracy for such calculations is ±2.0* kcal/mol.† The ionization potential for Cl_2, the electron affinity of CN and the proton affinity of NH_3 achieve this level. However, the errors in the other predicted quantities exceed this goal, at times by more than an order of magnitude. The ionization potential of CN is a notoriously difficult case.

Deviations from experiment such as these motivated researchers to develop compound model chemistries designed to produce very accurate results at acceptable computational cost. We will consider some of them in the next section.■

* Some methods developers aim for ±1 kcal/mol, a level sometimes referred to as "chemical accuracy."

† Or the equivalent in the usual units. Note that in this and the following subsections, we will be reporting energy differences in kcal/mol, following the standard practice of compound model creators.

Evaluating Deviations from Experiment: What Does ±2 kcal/mol Actually Mean?

Let's take another look at the three largest deviations from experiment in the preceding example:

ITEM	EXP. VALUE (kcal/mol)	DEVIATION (kcal/mol)	% ERROR
CN IP	323.6	9.5	2.9
Cl_2 EA	55.1	8.5	15.5
C_6H_6 AE	1320.5	-7.7	-0.6

Although these three errors are about the same magnitude, the value for the electron affinity of Cl_2 actually deviates the most from experiment. In contrast, the ~8 kcal/mol difference from experiment for the atomization energy of benzene is an excellent result, nearly as good as that for the proton affinity of ammonia, despite the large difference in their raw magnitudes.

These results make it clear that the ±2 kcal/mol accuracy level cannot be taken literally. This may be a reasonable and achievable level for basic thermochemical properties of small molecules, but it makes little sense when the desired quantities are outside of that general numeric range, as is typically the case for actual experiments of research and/or commercial interest.

Compound Model Chemistries for High Accuracy Energies

Since very accurate energy predictions are important to modeling many processes, a considerable amount of research effort has been devoted to developing models for doing so. As we've seen earlier, extremely accurate theoretical methods do exist, but they are still prohibitively expensive for all but the smallest molecules. As a result, several families of compound models have been devised to approximate calculations using very accurate methods with large basis sets. We refer to such model chemistries as "compound" because they combine the results from a series of smaller, less expensive calculations in order to predict the total energy of a molecular system.

In order to predict the energy of a molecule, such methods generally function as follows:

- A geometry optimization is performed to obtain the structure to be used for a subsequent series of single point energy calculations. This calculation uses the least expensive model chemistry which will achieve the desired level of accuracy.
- The energy for this structure is predicted by a reasonably accurate model chemistry, generally more accurate than the one used to obtain the equilibrium geometry. We'll refer to this energy as $E^{initial}$.
- Single point energy calculations are run using various higher accuracy model chemistries. Their predicted energies are used to compute corrections to $E^{initial}$ designed to compensate for factors neglected by that first calculation. These corrections are computed as the difference between the various predicted energies and $E^{initial}$.
- Various additional corrections may also be applied to $E^{initial}$. These empirical corrections do not involve additional calculations but are rather computed from formulas involving general parameters, per-element parameters and the characteristics of the molecule under investigation. They may include:
    - A spin orbit correction computed from experimental data or high accuracy theoretical calculations. It applies only to calculations for single atoms.
    - An empirical correction based on the number of valence electrons in the system and one or more predefined parameters. For open shell systems, the numbers of alpha and beta electrons may be relevant.

- The zero-point energy is computed via a frequency job. The structure used for this calculation may be the same as the one used for the single point calculations or it may be one computed with a lower level model chemistry.

- The total energy predicted by the model is the sum of $E^{initial}$, the zero-point energy and all of the corrections defined by the model.

Referemces: Compound Models Used in this Book

CBS-QB3: [Montgomery99, Montgomery00]

G3: [Curtiss98]

W1BD: [Martin99, Parthiban01, Barnes09]

The first compound model in widespread use was Gaussian-1 theory (G1). It was devised by Pople and coworkers in the late 1980s [Pople89].* Today, there are several families of compound models, of which these are the most important for our purposes:

- Gaussian-*n* (G*n*) models: A series of models for high accuracy thermochemistry predictions. Successive models have made the calculation series more complex. Later versions, especially Gaussian-4 [Curtiss07] and its variations [Curtiss07a], also parametrize the method to the molecule set (a characteristic that makes G4 problematic for us).

- Complete Basis Set (CBS) models: Developed by George Petersson and coworkers, the family name reflects the fundamental observation underlying these methods: the largest errors in ab initio thermochemical calculations result from basis set truncation. These models also computed the total energy from the results of a series of calculations which are devised on the basis of the following principles and observations [Petersson88]:

  - The successive contributions to the total energy generally decrease with order of perturbation theory. For example, in order to compute the dissociation energy for O_2 to within 1 millihartree, the SCF energy must be correct to six figures, the first-order electron correlation MP2 contribution must be correct to three figures, and contributions from higher orders of correlation need only be correct to two figures. The CBS models accordingly use progressively smaller basis sets as the level of theory increases in order to minimize their computational requirements.

  - The CBS models take advantage of the known asymptotic convergence of pair natural orbital expansions. They extrapolate from ordinary, finite basis set calculations to the estimated complete basis set limit (see p. 496 for more details on this technique).

- Weizmann-*n* (W*n*) models: Developed by Jan Martin and coworkers [Martin99], these models are designed to achieve ultra high accuracy (~0.25 kcal/mol), almost irrespective of computational cost. Like the other families, the W*n* models approximate a coupled cluster calculation with a large basis set. However, they use the CCSD(T) method for all of the energy correction components rather than the less accurate Møller-Plesset perturbation theory. These methods also include relativistic corrections for elements beyond the first row and take core electron correlation into account (rather than always freezing the core electrons). W1 uses a single empirical parameter, and the other W*n* models use no empirical corrections. The most recent method is W4 [Karton06]. Gaussian 09 offers the W1 and W2 methods and the W1BD variation.

Calibration Sets and Their Pitfalls

As we noted previously, the goal of all of these compound models is predicting thermochemistry quantities with very high accuracy. In its initial presentation, Gaussian-1 theory was evaluated via a test set of 31 species used to compute 24 atomization energies [Pople89]. This initial collection was expanded to include species including second row elements as well as other thermochemical properties [Curtiss90].

Over the years, additional modifications were made: atomization energies were replaced by heats of formation, and many additional molecules were incorporated. The following diagrams illustrate the succession of test sets used to evaluate the various G*n* models:

* The first published compound model chemistry was CBS-APNO in 1988 [Petersson88].

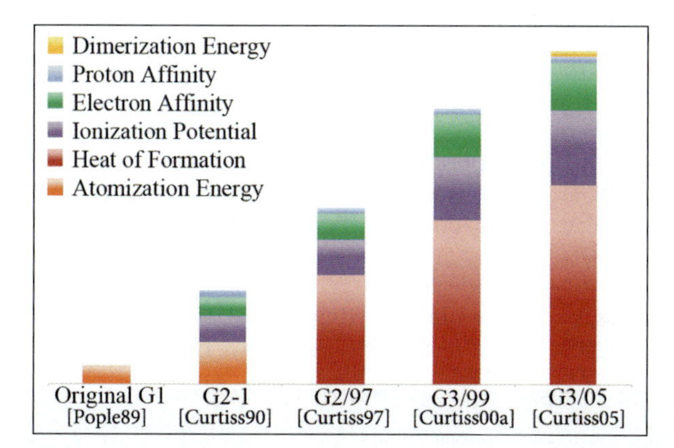

Although successive test sets expand the number, size and kinds of molecules, they become increasingly dominated by heats of formation over time, and hydrocarbons and substituted hydrocarbons are overrepresented in the latest sets. This lack of balanced diversity calls the usefulness of the composite error rates into question.

It should also be emphasized that heats of formation are a tool for understanding reactions: what happens when a bond breaks, whether a process is exothermic or endothermic and the like. Computing heats of formation ought not to become an end in itself.

Martin and Oliveira have also argued for the primacy of electron affinities in testing any high accuracy method:

> Because electron affinities involve a change in the number of electrons correlated in the system, they are very taxing tests for any electron correlation method; in addition, they involve a pronounced change in the spatial extent of the wavefunction, making them very demanding in terms of the basis set as well. [Martin99]

In addition, the experimental values with which the methods are compared are not always sufficiently accurate. In order to compare a method whose goal is a given level of accuracy, the uncertainty in the experimental standard must itself fall within that range. Parthiban and Martin make the following comment on this issue:

> The main problem with the G2-1 and G2-2 test sets* for heats of formation is the limited accuracy of the experimental data themselves. These were critically reviewed by Liebman and Johnson who concluded that less than half of the data [met] the 1 kcal/mol accuracy criterion. [Parthiban01 and ref. 26 therein]

Inaccurate and insufficiently accurate experimental values come in via several routes:

- Lingering values from older experiments for which later, more accurate studies exist.
- Values from the literature whose error bars turn out to be larger than the original researchers thought.
- Typographic and other errors propagated from the literature.

The G*n* test sets also modify some experimental values before comparing to them. For example, experimental electron affinities are "corrected" to 0K using empirical parameters. This violates the best practice of always comparing calculated values to what is actually observed.

* The combination of G2-1 and G2-2 comprises the G2/97 test set.

Finally, when the models themselves are formulated by fitting to the molecule set, the latter becomes an empirical training set, and any use as an evaluative test set is compromised.

EXAMPLE 4.6 USING & EVALUATING HIGH ACCURACY MODEL CHEMISTRIES

We will compute the ionization potentials, electron affinities and proton affinity from the previous example using G3 theory and the CBS-QB3 model chemistry. In addition, we will predict the C–H bond dissociation energy in benzene: benzene → benzene radical + neutral hydrogen atom.

We run all of the necessary calculations, and compute the various quantities using the thermal corrected energies, E(298.15 K). These are the results we obtained:

		CALCULATED		$\|\Delta^{EXP}\|$		
QUANTITY	**EXPERIMENT**	**G3**	**CBS-QB3**	**STD.**	**G3**	**CBS-QB3**
CN IP	323.6±0.5	329.3	325.1	9.5	5.7	1.5
Cl_2 IP	265.3±0.2	265.8	266.0	-1.9	0.5	0.7
CN EA	89.1±0.1	90.6	90.0	1.8	1.5	0.9
Cl_2 EA	55.1±0.2	56.7	57.4	8.5	1.6	2.3
NH_3 PA	204.0±0.08	203.1	202.7	-0.4	0.9	1.3
benzene C–H bond dissoc.	112.9±0.5[a]	114.0	114.8		1.1	1.9

[a][Blanksby03].

With the exception of the ionization potential of CN, all of the G3 results are within the ±2 kcal/mol accuracy range. CBS-QB3 performs almost as well, and is more accurate than G3 for the CN electron affinity. Both compound models are unambiguously more accurate than our standard model. This accuracy comes at significant additional computational cost, however. The following table compares the total execution time for all of the calculations in Example 4.5 and Example 4.6:

MODEL	**RELATIVE CPU TIME**
Standard model	1.0
G3	4.7
CBS-QB3	2.8

Note that the calculations on benzene and benzene radical consume about 70% of the total time. The compound models' requirements also increase more rapidly with system size than our standard model. While their relative accuracies are similar, G3's CPU requirements are about 1.7 times that of CBS-QB3, making the latter practical for a much broader range of system sizes.*

CBS-QB3 also has two technical advantages over G3 (and G4):

- It is size consistent.
- It takes spin contamination into account (see p. 479).

For these reasons, we recommend CBS-QB3 for general use when a high accuracy energy is required.

* There are variants of G3 that have different speed-accuracy tradeoffs. See, e.g., [Baboul99].

We will continue our discussion of these results when we examine the ionization potential of CN in detail in Advanced Example 4.10 (p. 196). This molecule is particularly tricky to model properly, and doing so illustrates many important points.

GOING DEEPER: CONNECTING THERMOCHEMISTRY TO STATISTICAL MECHANICS

It is important to be aware of how thermochemical properties arise from the energetics of vibrational frequencies. This connection is based upon partitioning the total energy of a macroscopic system among the constituent molecules. Nash's *Elements of Statistical Thermodynamics* provides an excellent discussion of the mathematical details of this transformation [Nash68].

Predicting Spectra: NMR

Gaussian can predict a wide variety of spectra. IR/Raman and NMR are among the most frequently used NMR shielding tensors that can be predicted by including the **NMR** keyword in the route section for the job.

NMR References

Method Definition: [London37, McWeeny62, Ditchfield74, Wolinski90]

Implementation & Review: [Gauss92, Gauss95, Cheeseman96]

Recent Review: [Teale13]

Here is the predicted shielding value for the carbon atom in methane as it appears in the Gaussian output file:

```
SCF GIAO Magnetic shielding tensor (ppm):
1  C     Isotropic =    193.2143   Anisotropy =   0.0000
  XX=    193.2143    YX=      0.0000    ZX=      0.0000
  XY=      0.0000    YY=    193.2143    ZY=      0.0000
  XZ=      0.0000    YZ=      0.0000    ZZ=    193.2143
  Eigenvalues:    193.2143    193.2143    193.2143
```

The output gives the predicted value for each atom in the molecule in turn. Here we see that the predicted value for the carbon atom is about 193.2 parts-per-million.

Shielding constants reported in experimental studies are usually shifts relative to a standard compound, often tetramethylsilane (TMS). In order to compare predicted values to experimental results, we also need to compute the absolute shielding value for TMS, using exactly the same model chemistry. The predicted absolute shielding values using the APFD/6-311+G(2d,p) model chemistry are:

- C: 187.4350 ppm
- H: 31.8036 ppm

To obtain the predicted shift for the carbon atom in methane, we subtract its absolute value from that of the reference molecule, resulting in a predicted shift of -6.6 ppm, which is in good agreement with the experimental value of -7.0. Note the sign convention for shifts: a negative number indicates that there is more shielding in the specified molecule than in the reference molecule (and a positive number indicates the reverse).

The following exercise examines the chemical shifts with respect to TMS for the carbon atom(s) in several small organic molecules.

EXAMPLE 4.7 ^{13}C NMR EXTREMES: METHANE, BENZENE, METHYL CATION

We are interested in the ^{13}C chemical shifts for carbon atoms in a range of electronic states. We considered the following compounds:

- Methane: The carbon atom has sp^3 hybridization, and its p orbital is completely filled.
- Benzene: The carbon atoms have sp^2 hybridization. The ring's π system contains one unpaired electron per carbon atom.
- Methyl cation* (CH_3^+): In this system, the carbon atom's hybridization is also sp^2, and it has an empty p orbital.

We optimized the geometry of each of these compounds, and verified that the resulting structures were minima with frequency calculations (via an **Opt Freq** calculation). We then ran **NMR** calculations on the resulting structures via a second job whose input files look like the following:

In order to achieve full symmetry in the benzene NMR results, the job's route section also included the **CPHF=UltraFine** option, which increases the size of the integration grid for the CPHF part of the calculation. Both job steps used the **UltraFine** integration grid as usual (**Int=UltraFine**).

```
%Chk=name
# APFD/6-311+G(2d,p) NMR Geom=AllCheck Guess=Read
                                          final blank line
```

This route section says to retrieve the optimized structure from the checkpoint file (**Geom=AllCheck**) and also the initial guess (**Guess=Read**). We could have run both jobs within a single input file by separating the two sections—known as job steps—with the following line:

```
--Link1--
```

Note that both job steps must still end with a blank line.

The following table lists the predicted ^{13}C chemical shifts with respect to TMS:

MOLECULE	^{13}C SHIFT (ppm)
methane	-5.8
benzene	132.7
methyl cation	407.3

Methyl cation has the largest chemical shift, and its sign is positive, indicating that the nucleus is the least shielded of them with respect to TMS; in other words, it has the maximum deshielding among this molecule group. In contrast, methane has a small negative chemical shift, indicating that its carbon nucleus is more shielded from the applied magnetic field than those in TMS. Benzene also has a positive chemical shift, but of a significantly lower magnitude than methyl cation. These results correspond to the molecules' differing electronic configurations.

Gaussian's predicted isotropic shielding values are an average of the three dimensional tensor elements. The reported anisotropy value in the Gaussian output indicates the degree of nonuniformity in the absolute shielding, as in this example for a carbon atom in benzene:

```
SCF GIAO Magnetic shielding tensor (ppm):
1  C    Isotropic = 54.7456   Anisotropy = 186.8909
 XX=    42.2658   YX=     0.0000   ZX=      0.0000
 XY=     0.0000   YY=   -57.3687   ZY=      0.0000
 XZ=     0.0000   YZ=     0.0000   ZZ=    179.3395
```

* IUPAC: methylium.

The carbon atoms in benzene exhibit significant anisotropy. If we examine the full shielding tensor (the 3 lines following the isotropic shielding value in the output), we see that it has positive shielding values in two directions, one 4 times larger than the other, and a negative shielding value in the 3rd direction. The large positive value corresponds to the magnetic field being perpendicular to the plane of the molecule. In this configuration, the magnetic field induces a current in the electrons in the π system, resulting in greater shielding than for the other two directions. In general, maximum shielding corresponds to the induced current density in the molecule being opposed to the applied magnetic field, while maximum deshielding involves the current density being aligned with the applied magnetic field. ■

Viewing NMR Data in GaussView and WebMO

GaussView and WebMO can also display results from NMR calculations. GaussView displays the following dialog when **NMR** is selected from the **Results** menu:

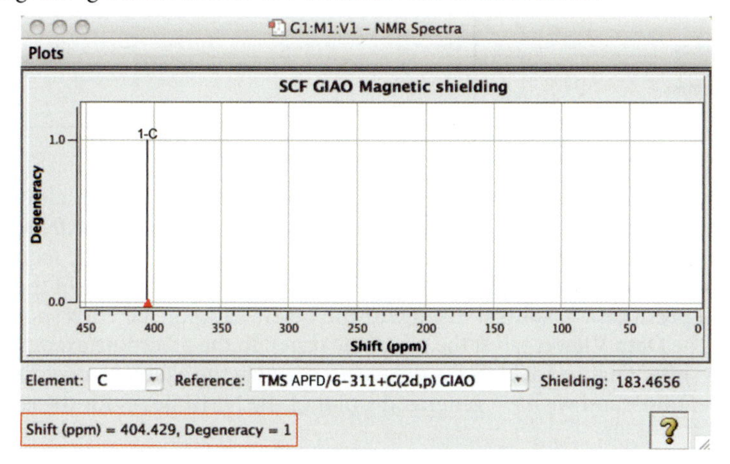

This display is for formaldehyde. We have limited the display to carbon atoms via the **Element popup** menu, and we have specified a reference calculation for use in computing chemical shifts (**Reference** popup menu).

The data for the selected peak appears in the bottom left of the windows (outlined in red above). In this case, the chemical shift is shown; if no reference were selected, then the absolute shielding would be displayed instead.

You can display each atom's shielding or chemical shift in a **View** window by right clicking in NMR plot area and then choosing **Show Shieldings in View** from the context menu:

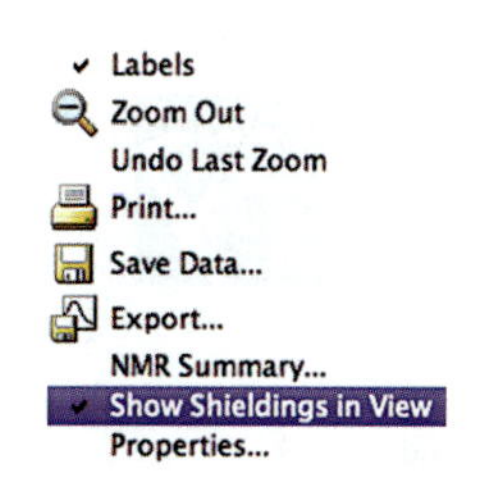

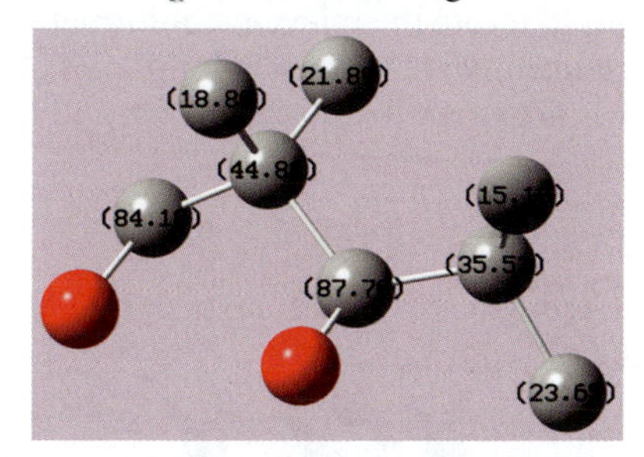

You can also create a text file containing the predicted NMR shieldings for each atom using the **NMR Summary** item on the same context menu. The contents of the resulting file reflects the current settings in the popup menus below the NMR plot.

WebMO includes the predicted chemical shifts in **Calculation Quantities** for NMR jobs, in the section labeled **Absolute Chemical Shifts**:

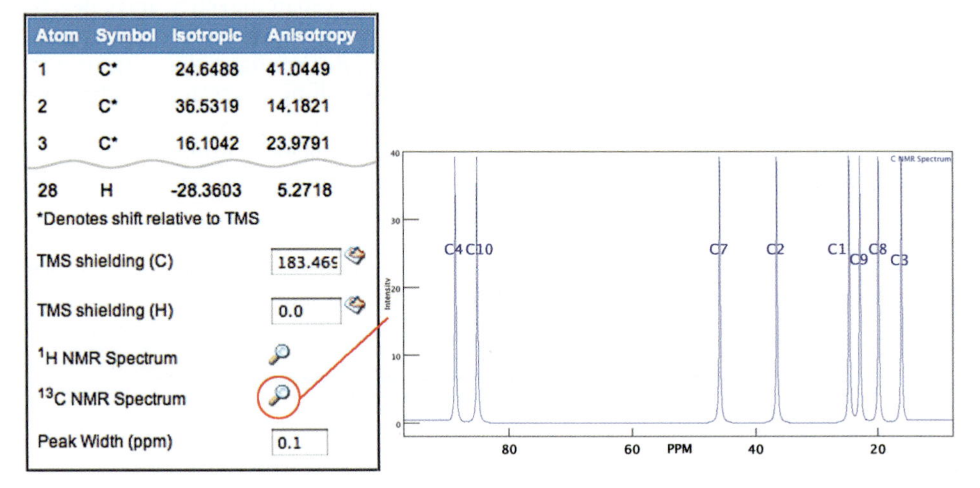

Atom	Symbol	Isotropic	Anisotropy
1	C*	24.6488	41.0449
2	C*	36.5319	14.1821
3	C*	16.1042	23.9791
28	H	-28.3603	5.2718

You can enter whatever reference values you want into the relevant **TMS shielding** fields, and chemical shifts will be calculated with respect to that standard. A value of **0.0** in the reference field will cause the absolute shielding values to be displayed. Click on the icon to the right of the field to recompute the displayed data. Finally, you can click on one of the magnifying glass icons to produce a graphical display of the NMR data for the corresponding atoms (viewable via the **Data Viewer** tab at the top of the page). In the preceding example, we display the chemical shifts relative to TMS for the carbon atoms, the absolute shielding values for the hydrogen atoms, and we have generated a plot of the NMR peaks for the carbon atoms.

Conformational Searching and Boltzmann Averaging

For many kinds of spectral data, including NMR, the observed spectrum comes from the mix of conformations present in the sample. In contrast, predicted spectra arise from the single electronic structure provided as input. Thus, in many cases, a single calculation will not adequately model the desired spectrum. Rather, the spectrum must be computed for all relevant conformations, and these separate results must be combined via a technique known as Boltzmann averaging; the final predicted spectrum is the weighted average of those from the various conformations, according to their relative abundance.

We use the Pcmodel program to locate the relevant conformations for the structure we are studying (available from Serena Software, *www.serenasoft.com*). Once we have located all of the relevant structures, we use Gaussian to optimize them and then compute their spectra.

EXAMPLE 4.8 TRIMETHYLPENTANEDIOL ^{13}C SPECTRUM

We will predict the NMR chemical shifts with respect to TMS for the carbon atoms in 2,2,4-trimethyl-1,3-pentanediol:

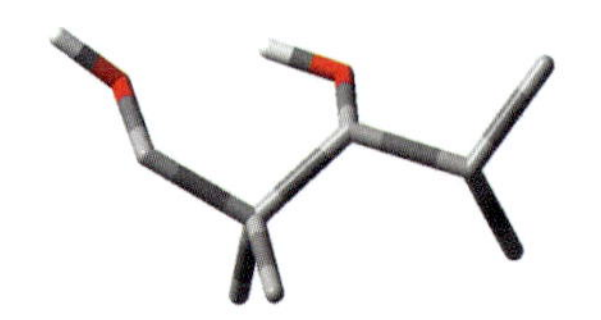

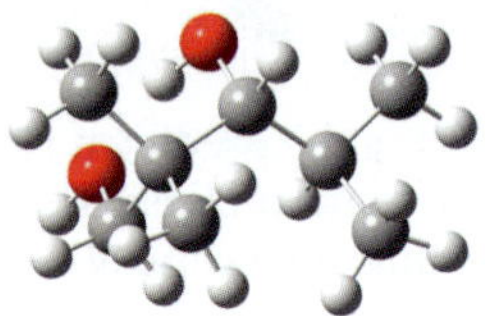

We used the B3LYP functional for this example (as it was completed prior to the release of APFD). The calculations also used the default integration grid.

◀1 We will begin by building this molecule in GaussView or WebMO, and we then import it into Pcmodel.

Using Pcmodel for Conformational Searching

We can open a Gaussian input file in Pcmodel using its **File→Open** menu item, specifying **Gaussian Input** in the **File Format** popup menu.

◀2 The next step is to specify the force field and display settings to Pcmodel. The program's main window is depicted in the illustration following, which shows several menu selections relevant to/helpful with conformational searching:

- The **Labels** item on the **View** menu opens the lower dialog, where we opt to display atom numbers.
- We select **Hydrogen Bonding** from the **Mark** menu. This causes hydrogen bonds to appear in yellow.
- We choose the **MMFF94** force field from the **Force Field** menu.
- The **H–A/D** button on the toolbar is used to display/hide hydrogen atoms (among other functions). It also has the side effect of regenerating atom types for the currently selected force field. You should press it twice whenever a new force field is specified.

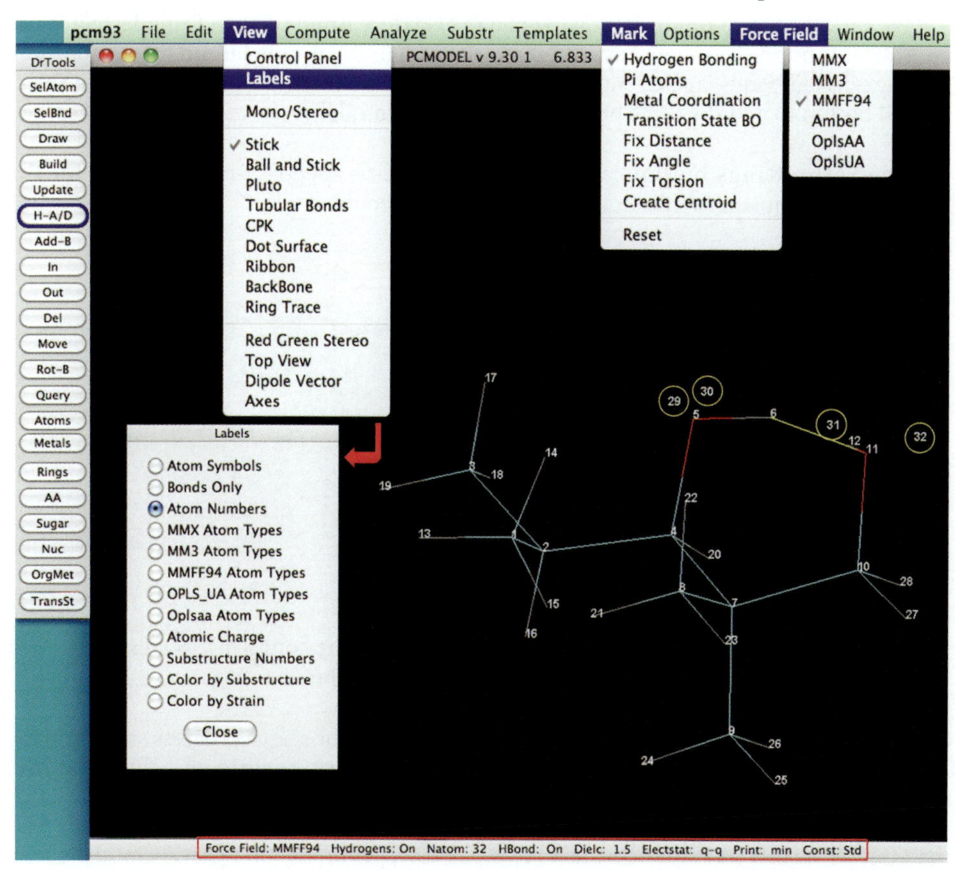

Note: Menu item names and other interface details may vary in your version of Pcmodel.

Note that we have also circled the lone pairs in this structure in the preceding illustration. The status area at the bottom of the window provides summary information about the molecule and our current selections.

3 We minimize the structure with molecular mechanics, via the **Compute→Minimize** menu item.

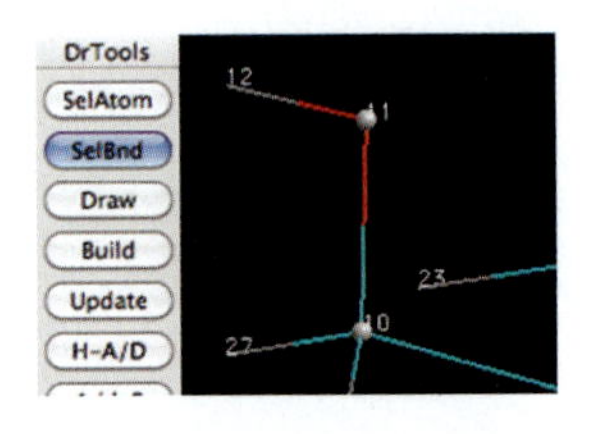

4 Next, we select the two C–O bonds, using the **SelBnd** button on the toolbar. The bonds will turn white when the mouse hovers over them. When selected, the two atoms involved turn into small grey spheres.

5 We are now ready to specify the parameters for the conformational search. We select the **Compute→GMMX** menu path, which opens the **GMMX Setup** dialog (in the center of the illustration on the next page). There are several items to specify:

- Click the **Comp Methods** button, and then select **Check Equivalent Atoms**. This will cause duplicates to be automatically removed from the final list of conformations.
- We click the **Options** button, and uncheck **Keep Conformational Enantiomers** in the resulting dialog. This will cause the search to locate only conformations with the same chirality (R or S) as the starting structure.
- We click the **Setup Bonds** button to open the **GMMX Bonds** dialog. Pcmodel will locate rotatable bonds automatically, and they are listed in the **Bonds Found** list. In addition, any selected bonds are pre-entered into the **Bonds To Search On** list. We click the **Select All** button to add the three bonds identified by Pcmodel to the two we selected earlier.

The **Setup Rings** button can be similarly used to specify ring selections for the conformation search. It is not relevant to our molecule.

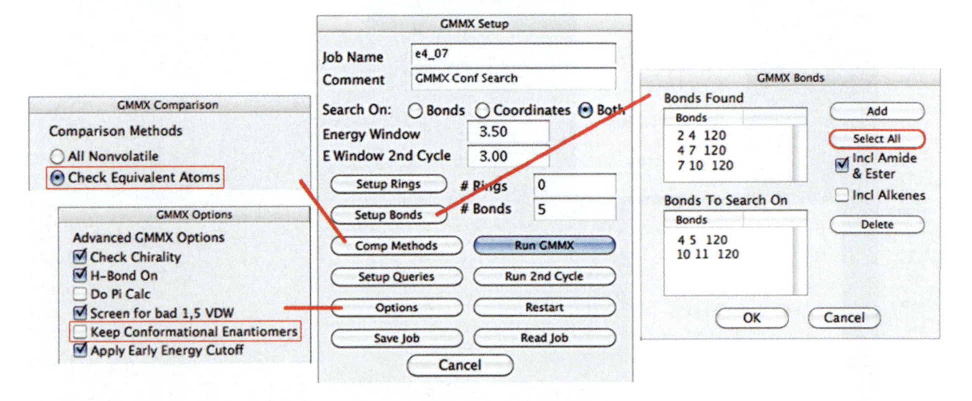

6 Once all parameters are set, we click **OK** to begin the conformational search. As it runs, the Pcmodel display appears similar to the figure following.

The **GMMX** window on the left displays information about the search progress so far, and the Output window on the right displays detailed data for the structure currently being examined. The buttons in the former can also be used to pause or terminate the search process.

The GMMX facility in Pcmodel runs in two passes. During the first pass, Pcmodel uses a fast algorithm to locate as many potential conformations as possible. The second pass then optimizes each structure in the list, paring it down to unique conformations.

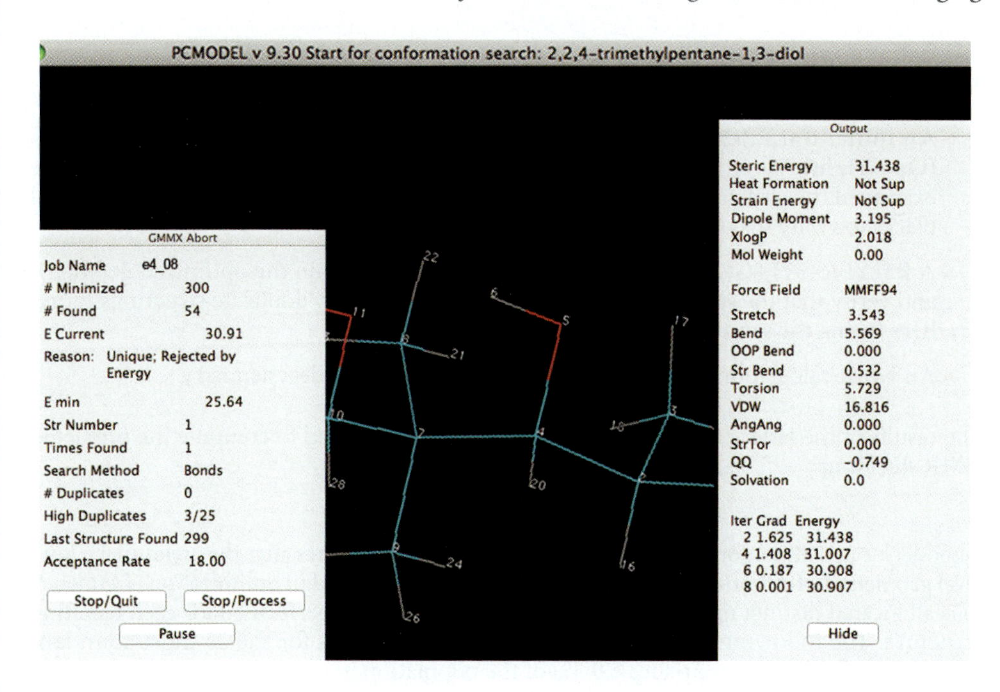

7 Once the conformational search is completed, you can examine the results by opening the PC model file corresponding to the second GMMX, which has a filename of the form *basename2.pcm*.

Opening a file of this type opens the **Substructure List** dialog (illustrated in the margin). It lists all located conformations along with their energies. Clicking on any item in the list will open that structure.

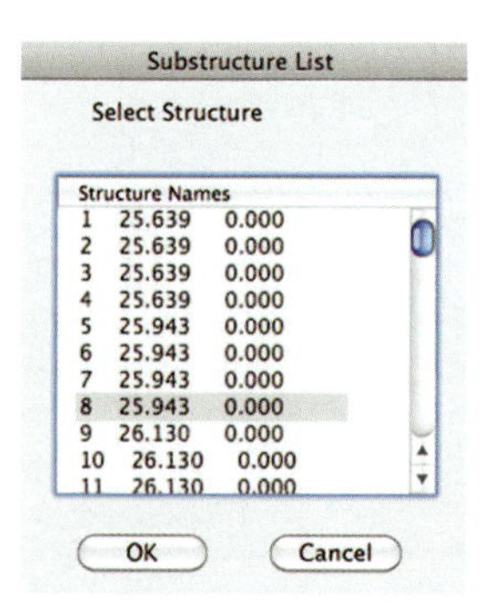

8 The final step in Pcmodel is to create a Gaussian input file for each substructure with a unique energy. We do this via the normal **File→Save** command, selecting **Gaussian Input (*.gjf)** from the **Save as type** popup menu, resulting in the following dialog:

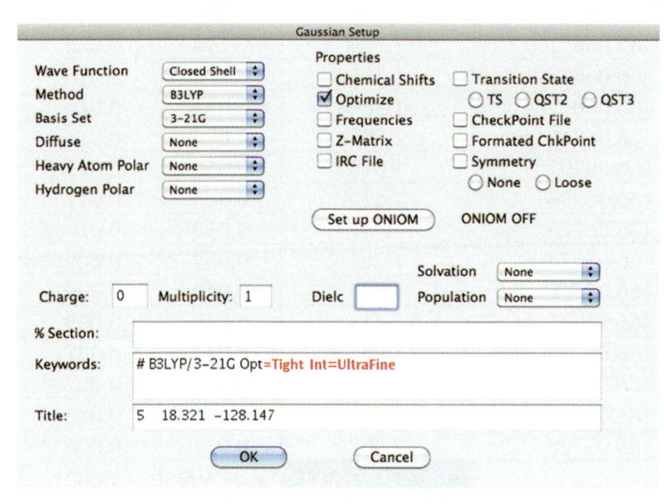

We select the desired job type and model chemistry from the various menus, and the corresponding keywords are automatically entered into the ***Keywords*** *field, which we can also edit as necessary. Clicking* ***OK*** *will generate the Gaussian input file.*

Selections in this dialog persist across uses, so we only need to specify the desired items the first time.

We repeat this process for each unique conformation.

Gaussian Calculations on Each Conformation

We will run the following three series of Gaussian jobs on each conformer:

- An initial B3LYP/3-21G geometry optimization with tight optimization cutoffs (**Opt=Tight**). We will use these optimizations to eliminate any duplicate structures: we examined the structures of ones having predicted energies identical through 6 decimal places to verify that they were in fact the same geometry.
- A B3LYP/6-311+G(2d,p) optimization+frequency, to obtain the optimized geometry and verify that the structure is a minimum. If there are any duplicate structures found here (using the same criterion), they are again eliminated.
- An NMR calculation using the B3LYP/6-311+G(2d,p) model chemistry.

The results of the latter calculation set will be analyzed and used to compute the predicted NMR shieldings.

Results

Pcmodel located 38 conformations. There were 6 duplicate structures after the preliminary, low-level geometry optimizations completed. The B3LYP/6-311+G92d,p) optimization+frequency jobs all located distinct minima, and an NMR calculation was performed for each resulting structure. The following table summarizes the energy results for the 25 most abundant conformations (those comprising ≥ 0.5% of the population, renormalized):

CONFORMER	GIBBS FREE ENER. (au)	REL. G (kcal/mol)	q_i	FRACTION
1	-466.07055	0.000	1.000	0.179
6	-466.07045	0.059	0.906	0.162
10	-466.06995	0.373	0.535	0.096
2	-466.06989	0.412	0.501	0.089
4	-466.06978	0.479	0.448	0.080
15	-466.06932	0.770	0.275	0.049
9	-466.06931	0.773	0.273	0.049
8	-466.06912	0.895	0.223	0.040
29	-466.06905	0.937	0.208	0.037
5	-466.06901	0.962	0.199	0.036
3	-466.06876	1.121	0.153	0.027
25	-466.06871	1.155	0.144	0.026
13	-466.06838	1.361	0.102	0.018
26	-466.06819	1.480	0.084	0.015
28	-466.06810	1.537	0.076	0.014
30	-466.06797	1.618	0.066	0.012
12	-466.06796	1.621	0.066	0.012
22	-466.06776	1.749	0.053	0.010
14	-466.06774	1.761	0.052	0.009
19	-466.06766	1.809	0.048	0.009
20	-466.06765	1.813	0.048	0.009
32	-466.06742	1.962	0.037	0.007
16	-466.06740	1.972	0.037	0.007
31	-466.06733	2.019	0.034	0.006
35	-466.06728	2.052	0.032	0.006
		TOTALS	q=5.599	1.000

The various coefficients, q_i, for the Boltzmann averaging operation, are computed as:

$$q_i = e^{(\Delta E_i/kT)}$$

where ΔE_i is the energy difference between conformer *i* and the lowest energy conformer. kT is equal to 0.596152 kcal/mol at STP. The values in the Fraction column in the preceding table are computed as q_i/q, and they represent the relative abundance of that conformer. q is the sum of all of the individual q_is.

The predicted chemical shift for an atom is computed as the weighted sum of the shifts for the various conformations:

$$\sum_i f_i \times \delta_i = \sum_i f_i \times (\nu_{TMS} - \nu_i)$$

where f_i is the fraction of the total population represented by conformer *i*. The carbon shielding for TMS using the B3LYP/6-311+G(2d,p) model chemistry is 183.4717.

The following table presents the computed chemical shifts for each conformer, along with the final Boltzmann averaged prediction:

		CHEMICAL SHIFT (ppm)							
CONFORMER	% POP	C1	C2	C3	C4	C5	C6	C7	C8
1	17.9%	85.19	45.86	88.71	36.53	24.65	16.11	22.89	19.87
6	16.2%	86.45	45.93	89.56	36.73	24.89	15.88	22.89	18.47
10	9.6%	82.39	46.37	92.96	34.16	24.33	15.81	24.60	18.53
2	8.9%	76.33	46.25	90.64	34.29	25.27	17.73	26.83	25.70
4	8.0%	85.08	46.30	93.20	36.60	23.42	15.49	23.99	19.26
15	4.9%	76.70	46.32	90.76	34.00	25.32	17.45	25.52	24.80
9	4.9%	76.70	46.33	90.78	34.00	25.32	17.45	25.53	24.80
8	4.0%	74.46	46.10	93.44	33.71	25.65	17.05	25.12	24.48
29	3.7%	75.41	46.07	84.77	33.93	25.15	18.28	24.67	20.16
5	3.6%	74.84	45.98	93.47	33.84	25.16	17.96	26.90	25.25
3	2.7%	75.26	46.25	93.37	36.53	23.47	15.42	29.09	23.82
25	2.6%	75.24	46.81	79.88	36.28	23.78	16.19	22.94	19.92
13	1.8%	76.28	47.25	83.03	33.98	25.35	16.83	20.19	25.61
26	1.5%	83.75	46.77	89.35	35.64	25.91	19.65	23.30	15.87
28	1.4%	82.13	46.94	92.96	35.00	25.29	18.79	24.43	16.59
30	1.2%	74.99	48.57	88.16	33.29	25.06	16.51	22.77	17.03
12	1.2%	82.84	46.80	89.41	35.84	25.90	19.86	24.04	17.24
22	1.0%	80.45	45.80	94.50	38.81	21.97	22.23	26.13	17.27
14	0.9%	75.26	46.62	84.50	35.20	23.73	16.27	19.81	27.34
19	0.9%	76.20	47.32	85.37	34.87	23.53	15.95	19.15	26.59
20	0.9%	76.32	48.09	87.38	36.13	23.66	16.00	21.50	16.58
32	0.7%	71.61	49.02	90.69	34.11	26.08	17.93	23.16	23.42
16	0.7%	74.21	46.21	92.99	34.67	25.38	19.48	26.84	25.04
31	0.6%	77.34	48.18	86.19	36.68	23.29	16.25	21.33	16.33
35	0.6%	78.67	46.46	85.19	37.28	22.79	23.71	25.46	18.50
BOLTZMANN AVG.		**80.90**	**46.27**	**90.05**	**35.43**	**24.71**	**16.72**	**24.20**	**21.00**
OBSERVED		**69.74**	**39.68**	**78.26**	**28.53**	**23.55**	**16.94**	**21.95**	**20.46**

The experimental spectrum was obtained in a solution of DMSO-d6 (200 mg/ml) [Pisarenko03].

The predicted chemical shifts are generally higher than the observed values. The error in the predicted shifts for C1, C2 and C3 is on the order of +15%; if the computed values are corrected for this error level, then the predicted shifts are all within 3 ppm of experiment.

Note that the predicted chemical shift for C1 derived from the complete set of conformations is more accurate than that for the most abundant (lowest energy) conformer, while the computed shifts for the other carbon atoms are comparable. This illustrates the importance of considering more than a single conformation for accurate prediction of NMR and other spectral properties.

Exercises

EXERCISE 4.1 FREQUENCIES OF STRAINED HYDROCARBONS

Perform frequency calculations for each of these strained hydrocarbon compounds:*

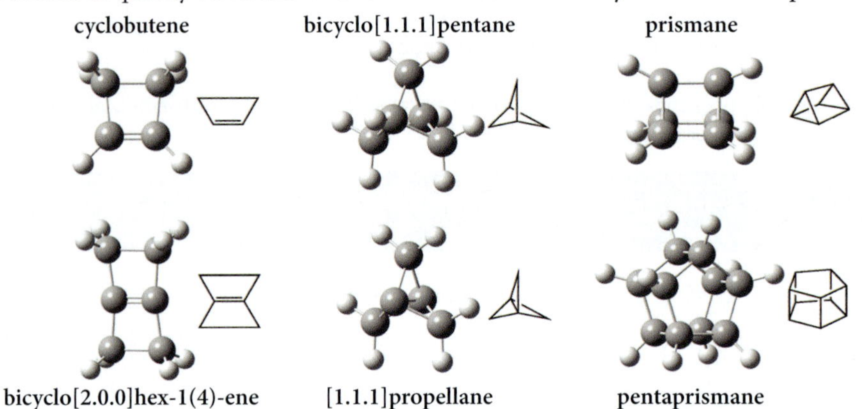

For the four smaller systems, determine how well the predicted frequencies compare to the experimental IR spectral data given below. Identify the symmetry type for the normal mode associated with each assigned peak. For prismane and pentaprismane, predict the dominant IR peaks.

MOLECULE	EXPERIMENT: PROMINENT IR PEAKS (cm^{-1})
cyclobutene	3058, 2955, 2916, 2933, 635 [Wiberg92b]
bicyclo(2.2.0)hex-1(4)-ene	2933, 2966, 1226 [Wiberg86]
bicyclo[1.1.1]pentane	2976, 2973, 2878 [Wiberg92c]
[1.1.1]propellane	611, 3020, 3079 [Wiberg85]

These compounds were among the first of their type whose frequencies were predicted with electronic structure calculations. In addition, in the case of propellane, theoretical predictions of its energy and structure preceded its synthesis.

Prismane and pentaprismane can provide some optimization challenges using the default settings. If your optimization has trouble, consider increasing the size of the integration grid.

SOLUTION

We ran optimization+frequency calculations on these six molecules. Our results are summarized in the following table:

* IUPAC: [1.1.1]propellane: tricyclo[1.1.1.$0^{1,3}$]pentane; prismane: tetracyclo[2.2.0.$0^{2,6}$.$0^{3,5}$]hexane; pentaprismane: hexacyclo[4.4.0.$0^{2,5}$.$0^{3,9}$.$0^{4,8}$.$0^{7,10}$]decane.

MOLECULE	FREQUENCY (cm⁻¹) CALC. (scaled)	EXP.	SYMMETRY & MODE	INTENSITY
cyclobutene	645	635	B1, all H rock	51.5
	3020	2933	A1, all H symm. stretch	44.7
	3015 3069	2955	B1, all H asymm. stretch	41.0 41.3
	3145	2916	B2, all H asymm. stretch	14.2
	3176	3058	A1, 2 H symm. stretch	19.9
bicyclo(2.2.0)hex-1(4)-ene	3017	2933	B3U, all H symm. stretch	167.0
	3072	2966	B1U, all H asymm. stretch	88.6
	3016	2933	B2U, all H asymm. stretch	93.9
	1209	1226	B2U, all H rock	16.7
bicyclo[1.1.1]pentane	3076	2976	A2", term. H asymm. str.	125.9
	3016	2878	E', mid. H asymm. stretch	162.3[a]
	3083	2973	E', all H stretch	122.5[a]
[1.1.1]propellane	618	611	A2", central bond rock	135.4
	3102	3020	E1, lower H asymm. str.	41.5[a]
	3190	3076	E', all H asymm. stretch	17.3[a]

[a]Doubly degenerate mode; reported intensity is the sum of the two.

These results are in good agreement with the experimental values.

Here are the strongest predicted peaks for prismane and pentaprismane:

MOLECULE	FREQUENCY (scaled, cm⁻¹)	INTENSITY
prismane	813	108.6[a]
	3185	40.7
	3179	44.4[a]
	1257	26.4
pentaprismane	3090	415.6[a]
	1256	12.6

[a]Triply degenerate mode.

Remember that intensities can be compared qualitatively, but should not be taken too literally. When we examine the other frequencies for these compounds, we find that most of the normal modes are not IR active (the intensity is 0).■

GOING DEEPER: FURTHER FREQUENCY DISCUSSION

We've only scratched the surface of the IR frequency data for these strained hydrocarbon compounds. More detailed treatments might include:

- A more detailed comparison with the entire experimental spectra.
- Identifying the modes associated with motion of the carbon atoms.
- Isotopic substitution and its effect on the frequencies. For example, substituting deuterium for hydrogen in propellane produces different IR peaks.

EXERCISE 4.2 CARBONYL STRETCH BY SUBSTITUENT

This exercise will investigate various compounds containing a carbonyl group. Examine the frequencies for the systems* pictured in the following illustration, and determine the frequencies associated with carbonyl stretch in each case. (We looked at this mode in formaldehyde in Example 4.1, p. 145) In addition, locate the characteristic peak produced by the single hydrogen attached to the carbonyl for the applicable systems.

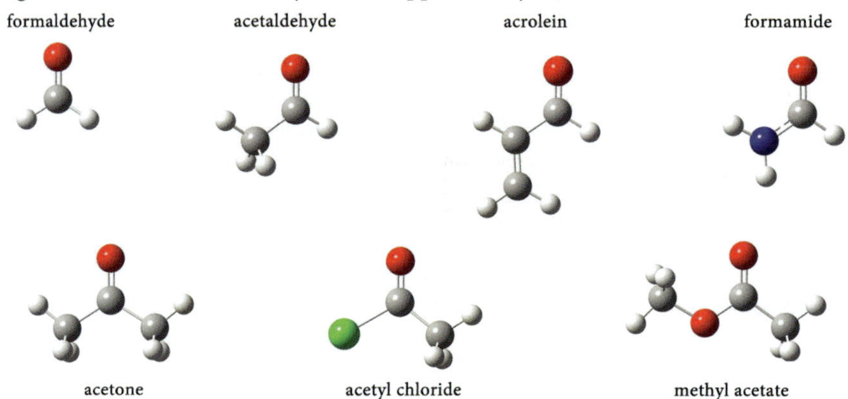

Optimization Challenges with Acetone

You may find acetone a bit challenging to optimize to a minimum (as opposed to a transition structure). Here are a couple of hints to guide you:

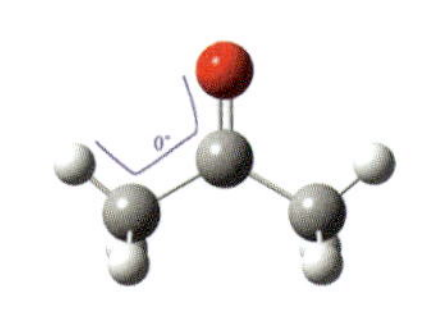

- Be sure that your starting structure has C_{2v} symmetry, with H-C-C-O dihedral angles of 0°, 120° and 120°. The clean features of molecule building software often create a structure with C_2 symmetry by setting one of the H-C-C-O dihedral angles to 90°.
- Including the **Int=UltraFine** option in the route section of the Gaussian input file may also be of use. It specifies the use of a more accurate integration grid during the calculation.

SOLUTION

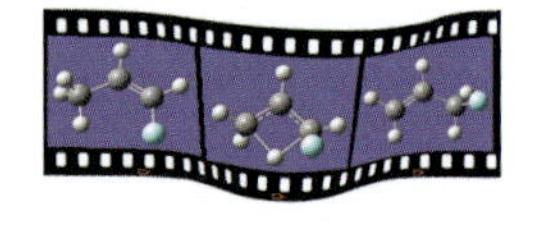

We ran optimization+frequency calculations for each compound and then examined the normal mode results. The following table summarizes the predicted locations of the peaks associated with carbonyl stretch and compares them with experiment as reported in [Wong91a], ordered by increasing observed frequency:

	C=O STRETCH FREQ. (cm^{-1})				
		CALCULATED		EXPERIMENT	
MOLECULE	RAW	SCALED	ΔH_2CO	PEAK	ΔH_2CO
acrolein	1797	1778	44	1723	23
acetone	1809	1789	33	1737	9
formamide	1808	1789	33	1740	6
formaldehyde	1842	1822	0	1746	0
acetaldehyde	1827	1807	15	1746	0
methyl acetate	1811	1791	30	1761	-15
acetyl chloride	1897	1876	-55	1822	-76

The largest shifts in the carbonyl stretch with respect to formaldehyde occur for acrolein and acetyl chloride (where the hydrogen atoms are replaced by a chlorine atom and a methyl group). For methyl acetate, theory predicts a shift in the opposite direction from what is observed.

* IUPAC: acetone: propanone; acetaldehyde: ethanal; acrolein: 2-propenal; formamide: methanamide.

Most of these values are in reasonable agreement with experiment, and scaling aligns the peaks more closely with the observation locations although generally the scaled peaks are still about 50 wavenumbers away from the corresponding observed ones. The outliers are formaldehyde and acetaldehyde. One possible explanation arises from the fact that frequency calculations compute only the harmonic vibrational frequencies, in the interest of reducing computational complexity and cost. For high frequency modes, the difference between the harmonic approximation and anharmonic analysis (including higher order terms) is about 5%.

George and coworkers [George00] have fit the original experimental data ([Wiberg95c, Kwiatkowski92, Bouwens96]) to a power series in order to partition it into harmonic and anharmonic terms, with the following results (all results are in cm^{-1}):

MOLECULE	CALC. (scaled)	OBSERVED	ESTIMATED HARMONIC
acetone	1759	1737	1760
formaldehyde	1792	1746	1778
acetaldehyde	1783	1746	1774

With this approach, these results are in reasonable agreement with experiment.

We will perform an anharmonic frequency analysis for these systems in Advanced Exercise 4.13 (p. 189).

Lone Hydrogen Modes

Here are the results for the peaks associated with a lone hydrogen within a carbonyl group:

MOLECULE	FREQ. (cm^{-1}, scaled)
acrolein	2864
formamide	2927
acetaldehyde	2856

The normal modes associated with these frequencies are characterized by motion limited to the hydrogen atom in question. The values of the frequencies are in reasonable agreement with observations which place this peak in the range 2745-2710 cm^{-1}. They are also similar to the symmetric and asymmetric stretching modes for the hydrogen atoms in formaldehyde (2865 and 2924 cm^{-1}, respectively).■

EXERCISE 4.3 ISOTOPE SUBSTITUTION EFFECTS ON BENZENE'S RAMAN SPECTRUM

Compare the IR spectrum for deuterated benzene with that of the normal molecule. Be sure to take care in identifying corresponding peaks between the two.

SOLUTION

We ran an optimization+frequency calculation for benzene, specifying the deuterium isotope for the hydrogen atoms.

The following table compares the experimental [Herzberg91 p. 364] and predicted Raman spectra for ordinary and deuterated benzene (all frequencies are in cm^{-1}; calculated values for doubly degenerate modes are averaged):

MOTION	OBSERVED PEAKS C_6H_6	C_6D_6	CALC. PEAKS (scaled) (corr. normal modes) C_6H_6	C_6D_6	RATIO (D/H) OBS.	CALC.
C-C opposite stretch	608.13	580.20	617 (3,4)	585 (4-5)	.95	.95
C-H wagging	845.00	659.00	852 (7,8)	657 (7-8)	.78	.77
ring breathing	993.06	945.58	1014 (13)	960 (17)	.95	.95
C-H scissoring	1177.78	867.00	1193 (17,18)	867 (15-16)	.74	.73
C-C adjacent stretch	1591.33	1558.30	1632 (23,24)	1583 (23-24)	.98	.97
asymmetric C-H stretch	3056.70	2272.50	3142 (26,27)	2331 (26-27)	.73	.74
symmetric C-H stretch	3073.94	2303.44	3168 (30)	2362 (30)	.75	.75

Note that it is vital to visualize and compare the normal modes in order to properly align the spectra. Modes are not always numbered the same, and at times they are not even in the same order.

The peak locations are in reasonable agreement with experiment in all cases. In addition, when we consider the ratio between the peaks corresponding to the deuterated and normal forms of benzene, we find excellent agreement with experiment.■

EXERCISE 4.4 NMR PROPERTIES OF ALKANES, ALKENES & ALKYNES

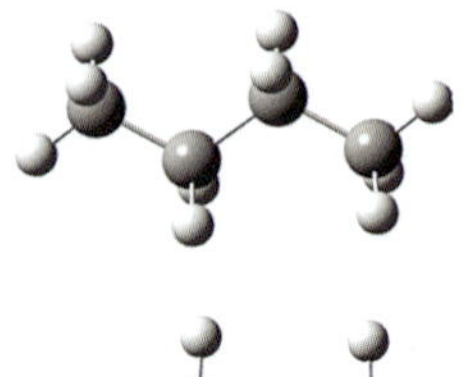

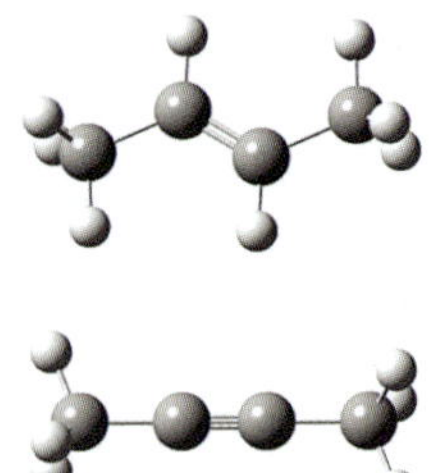

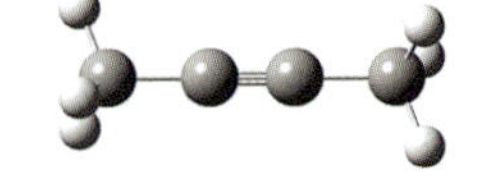

The NMR magnetic shielding for atoms like carbon is affected greatly by what it is bonded to and the type of bond to its neighbor. Use the inner carbon atoms of normal butane as the reference atom and calculate the shift in ^{13}C isotropic shielding for 2-butene* and 2-butyne* (pictured in the margin). Use our standard model chemistry.

Can you explain these shifts as a function of the changing molecular environments?

SOLUTION

The predicted absolute shielding value for the central carbons in butane is 158.11 ppm, which is what we will use as the reference value, subtracting the computed shielding values for the outer carbons in butane and for each type of carbon in the other two compounds from it.

Here are the predicted shifts with respect to the central carbon in butane:

MOLECULE	RELATIVE SHIFT (ppm) CENTRAL CARBON CALC.	EXP.	OUTER CARBON CALC.	EXP.
butane	0.0	0.0	-13.2	-11.8
2-butene	103.7	100.8	-8.9	-7.6
2-butyne	50.0	48.4	-25.4	N/A

The agreement with experiment [Silverstein91] is excellent for these cases.

* IUPAC: but-2-ene and but-2-yne (respectively).

For the central carbons, the shielding decreases greatly as we move from the alkane to the alkene. This is due to the fact that the sp^3 orbitals have a greater ability to oppose the applied magnetic field. The shift is much smaller when moving to the alkyne, which has been explained by the fact that the π bonding present in an sp environment creates a cylinder of electric charge that acts to oppose the applied magnetic field.■

GOING DEEPER: MAGNETIC PROPERTIES

For another dramatic illustration of chemical shifts, calculate the magnetic shielding of nitrogen in pyridine and compare it to its saturated cyclohexane analogue.

EXERCISE 4.5 ^{13}C SHIFTS IN NITROANILINES: A SURPRISE DEVIATION FROM ADDITIVITY

Assigning peaks in proton spectra is usually fairly straightforward for typical organic molecules. It can be accomplished by reference to the splitting of the signal due to nearby hydrogen atoms as well as the magnitude of the shift. In contrast, it may be much less obvious how to accomplish the same task for proton decoupled ^{13}C spectra. Traditionally, this was done via empirical rules derived from the spectra of similar compounds. Unfortunately, this approach sometimes led to incorrect assignments in some cases: e.g., where the signals are close to each other, where there is possible interaction between substituents, and the like. We will consider these issues with respect to the NMR spectrum of nitroaniline [Foresman13].

The following figure plots the ^{13}C NMR spectrum for 2-nitroaniline (chemical shifts with respect to TMS) [SDBS2336]:

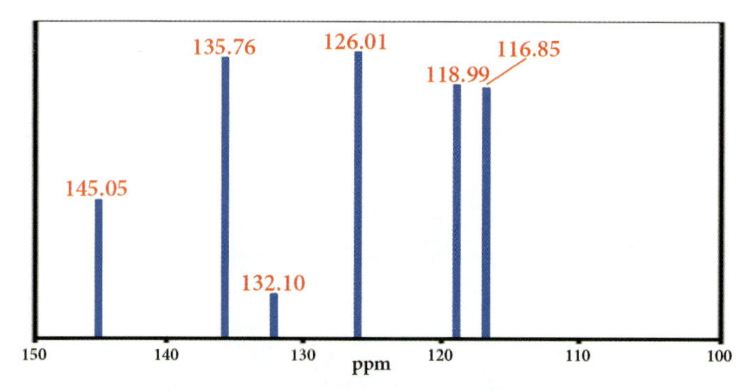

This is a spectrum that one can easily reproduce in the laboratory (see the "To the Teacher" box on page 178). Once the spectrum is obtained, the observed peaks need to be assigned to the six unique carbon atoms in this compound. There are several ways of accomplishing this:

- Additivity parameters can be used to estimate the substituent effects with respect to benzene at each carbon atom. Thus, these values must be added to the benzene base value of 128.5 ppm, its chemical shift with respect to TMS, yielding the predicted shift for each atom. The following table lists the additivity parameters for the two substituent groups in 2-nitroaniline [Cooper80]:

R	IPSO	ORTHO	META	PARA
NH_2	20.2	-14.1	0.6	-9.6
NO_2	20.6	-4.3	1.3	6.2

- The NMR spectrum can be predicted via a Gaussian NMR calculation. Chemical shifts for each carbon atom can be computed from these results.
- Two-dimensional NMR spectroscopy can be used to determine peak assignments via a hydrogen-carbon correlation spectrum.

Assign the peaks in the observed spectrum using the first two methods. How do the assignments compare? How do you account for any differences?

SOLUTION

We alluded to incorrect peak assignments arising from the traditional empirical rules for proton spectra in the opening paragraph of this exercise. The case for 2-nitroaniline provides an example of this.

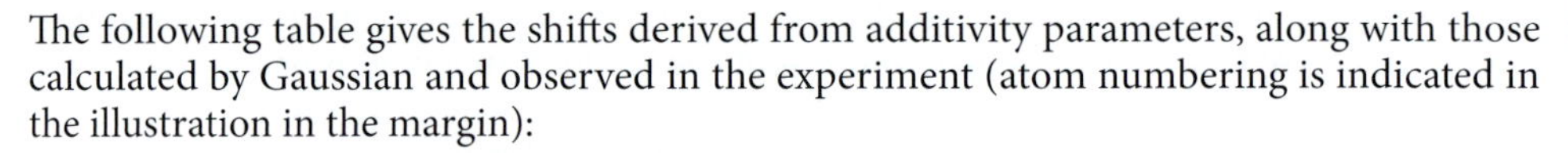

The following table gives the shifts derived from additivity parameters, along with those calculated by Gaussian and observed in the experiment (atom numbering is indicated in the illustration in the margin):

METHOD	C1	C2	C3	C4	C5	C6
Additivity	144.4	135.0	124.8	120.2	135.3	115.7
NMR calculation	148.9	135.1	132.9	118.5	140.2	119.5
Experiment	145.1	132.1	126.0	116.9	135.8	119.0

Using additivity parameters: In order to obtain the empirical value for C1, for example, we add the ipso value for NH_2 and the ortho value for NO_2 from the table to the benzene value: 20.2 - 4.3 +128.5. Similarly, the shift for C4 is computed as -9.6 (para NH_2) + 1.3 (meta NO_2) + 128.5.

Viewing the NMR chemical shifts in GaussView indicates the peak assignments directly:

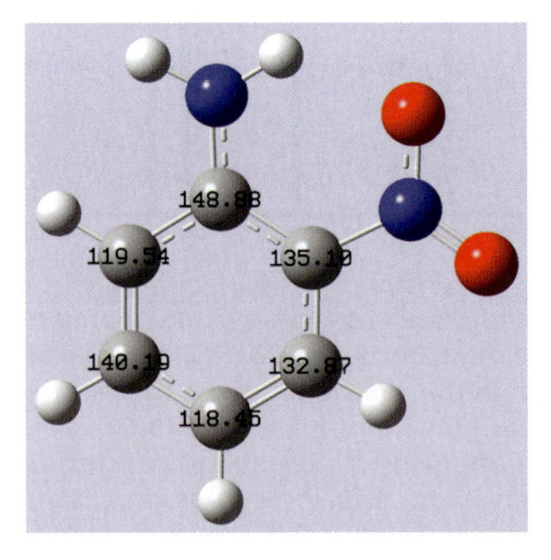

Additivity parameters and the Gaussian NMR calculation yield opposite assignments for peaks C4 and C6.

The Gaussian results can be confirmed via two-dimensional NMR spectroscopy, which yields the 2D spectrum shown in the illustration following. The ^{13}C spectrum appears at the top (containing only the peaks corresponding to carbons with attached hydrogens), and the proton spectrum appears on the left. The fact that the values for the C1 and C2 carbons are absent here verifies that the C1 and C2 carbons were correctly assigned as carbons containing substituents.

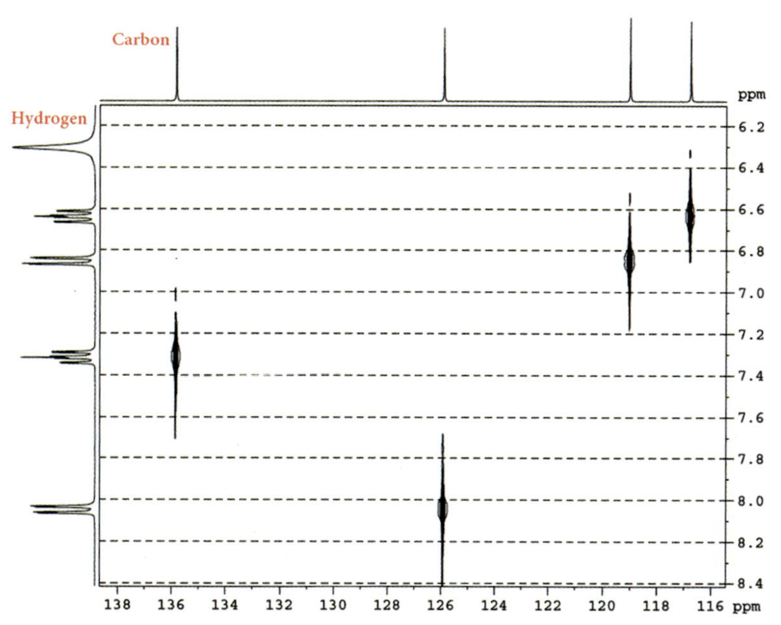

We will work our way through the ^{13}C peaks from right to left. The first and fourth carbon peaks each correlate with a triplet in the proton spectrum (indicated by the number of peaks in the matching proton spectrum). Therefore, they must correspond to C4 and C5. Since the first one is the less shielded of these two, it must correspond to C5; this means that the final peak is assigned to C4. The second and third peaks each correlate with a doublet: C3 and C6. Again, since the second peak is the less shielded, it must be C3, and the third matches with C6.

The predictions based on additivity parameters lead to incorrect assignments for C4 and C6 even though they are very close in value to the experimental chemical shifts. The fact that the ordering of C4 and C6 is reversed from that in the ^{13}C spectrum of aniline is surprising. The interaction between the 2 substituents alters the molecule's response to a magnetic field from that of the species containing only one of them.

What can account for this difference between the two carbon atoms? One suggestion, hydrogen bonding between the aniline hydrogen and the nitro oxygen, cannot be responsible for this effect since the reversal is also seen in (N,N)-dimethyl-2-nitroaniline. Resonance theory could be used to explain the buildup of charge at the ortho and para positions of 2-nitroaniline as well as greater delocalization at C6 than C4, resulting in greater lost charge at the former. However, such an approach would predict that N-methyl-2-nitroaniline should behave the same way, but it does not; in the methyl substituted compound, C6 is once again closer to TMS than C4, as in aniline.

Comparing the eigenvalues of the shielding tensors for aniline and 2-nitroaniline gives one possible answer to this puzzle. These are reported in the Gaussian output file.

	C4	C6
aniline	-30.9 +61.8 +179.9	-14.9 +68.8 +161.2
2-nitroaniline	-27.3 +58.6 +175.7	-19.2 +65.3 +157.6

Notice that the negative eigenvalue becomes more negative for C6 upon the addition of the nitro group whereas for C4 it becomes less negative. The negative eigenvalue is a measure of the paramagnetic contribution to the chemical shift: these additional electrons are adding

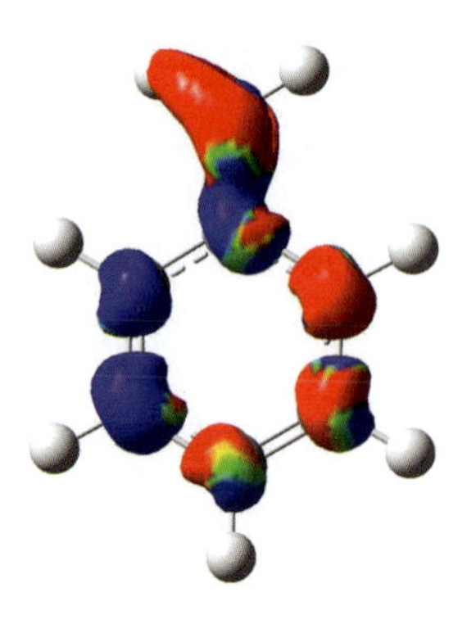

to the magnetic field at C6 instead of shielding the applied field, resulting in a greater shift (away from TMS).

This effect can also be seen graphically. The figure in the margin shows a 0.0005 isosurface of the current density (induced by the magnetic field applied in a direction colinear with the C-NO_2 bond) for aniline, painted with the difference in this density between 2-nitroaniline and aniline. The colors span the values of -1E-5 (red) to +1E-5 (blue). We see that C6 (and C5) has increased sigma electron density relative to C4 in the presence of the nitro group, increasing the paramagnetic contribution to the chemical shift for these atoms. This is consistent with the notion that the ^{13}C signal from C4 stays relatively constant while the signal from C6 is shifted away from TMS in 2-nitroaniline as a result of deshielding.

For information on how to create visualizations like this one, see the book's website.■

TO THE TEACHER: COMBINING LABORATORY EXPERIMENTS & CALCULATIONS

This exercise provides an excellent opportunity for combining experimental laboratory work with calculations. Students measure the ^{13}C NMR spectra of 2-nitroaniline and other substituted benzenes and then must identify the observed peaks. The results obtained using additivity parameters can be compared with those from quantum calculations, and the latter can be verified via a standard two-dimensional HCCOSW NMR experiment. In this way, students learn to connect substituent effects with changes in electronic structure, and they gain practice with electronic structure calculations in the context of solving chemical problems. Additional details about this exercise, including a set of recommended compounds, are available on the Exploring Chemistry website.

Similarly, Exercise 4.8 provides the opportunity for students to compare small differences in heat of combustion for two hydrocarbon isomers via direct measurement with a semimicro bomb calorimeter and via a high accuracy electronic structure calculation. See [Salter98] for details.

EXERCISE 4.6 THE ^{13}C NMR SPECTRUM OF PROPELLANE

Predict the ^{13}C chemical shifts with respect to TMS for [1.1.1]propellane (see Exercise 4.1, p. 170). You will need to start with a structure that is of the proper symmetry in order to produce the correct carbon atom equivalencies. Reoptimize the structure you obtained earlier with the **Opt=Tight** option,* and then perform the NMR calculation.

SOLUTION

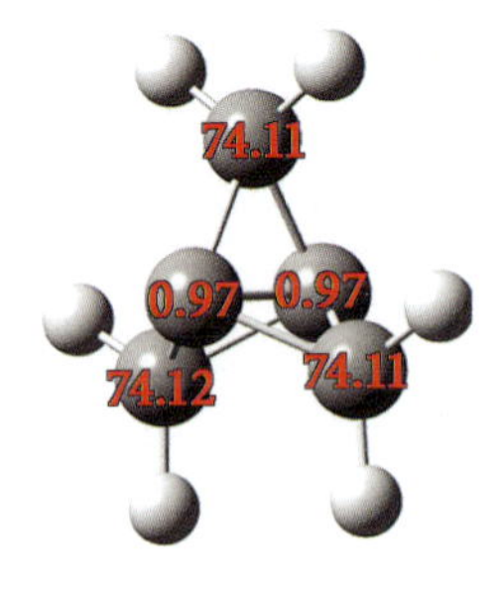

Our calculation computed a shift of 74.1 ppm for the carbon atoms within methylene groups and a shift of 1.0 ppm for those along the molecule's backbone. These results are in good agreement with experimental values of 79.3 ppm and 4.3 ppm (respectively) [Orendt85].■

* Also include **Int=UltraFine** if you do not do so as a matter of course.

EXERCISE 4.7 AZULENE/NAPHTHALENE HEAT OF ISOMERIZATION WITH CBS-QB3

Since they are isomers, azulene and naphthalene* have the same combustion reaction:

$$C_{10}H_8 + 12O_2 \rightarrow 10CO_2 + 4H_2O$$

The difference in heats of combustion directly reflects the difference in the energies of the two molecules, and it can be thought of as the enthalpy change for the isomerization of solid azulene to solid naphthalene. This can be computed as the difference in predicted enthalpies for the two compounds since the other terms in the shared combustion reaction cancel.

Predict the isomerization enthalpy for azulene and naphthalene using the CBS-QB3 high accuracy model chemistry. Which isomer is the more stable?

SOLUTION

The following table gives the results of our CBS-QB3 calculations, as well as ones with two other model chemistries for comparison:

MODEL	AZULENE (hartrees)	NAPHTHALENE (hartrees)	ΔH (kcal/mol)
B3LYP/6-311++G(d,p)			-33.15[a]
APFD/6-311+G(2d,p)	-385.49832	-385.54887	-31.7
CBS-QB3	-385.08781	-385.14306	-34.7
Experiment			-35.3±2.2[b]

[a][Salter98]; [b][Salter98] and references therein.

Azulene has the larger (less negative) heat of combustion and is therefore the higher-energy (less stable) isomer. All of the models produce very good agreement with the observed enthalpy difference, and CBS-QB3 results are within 1 kcal/mol of experiment. However, the significance of the latter should not be overstated, given that the experiment itself has an uncertainty of ±2.2 kcal/mol.■

EXERCISE 4.8 COST & ACCURACY OF CBS-QB3 VS. G3/G4: BENZENE HEAT OF COMBUSTION

The enthalpy of combustion for benzene in the gas state is 757.53±0.1 kcal/mol.† How well do the CBS-QB3 and G3/G4 methods predict this quantity (Select G3 and/or G4 based on your available computing resources). How do they differ in their computational requirements?

SOLUTION

We will need to predict the enthalpies of benzene, water, carbon dioxide and oxygen gas in order to calculate the enthalpy of combustion:

$$C_6H_6 + 7.5O_2 \rightarrow 6CO_2 + 3H_2O$$

The table following presents the results of our calculations (all energy quantities in the top section of the table are in hartrees, and those in the bottom section are in kcal/mol).

* IUPAC: azulene: bicyclo[5.3.0]deca-1,3,4,6,8-pentaene; naphthalene: bicyclo[4.4.0]deca-1,3,5,7,9-pentaene.

† Computed from the liquid phase heat of combustion of -780.98±0.1 kcal/mol [Good69] and the values of $\Delta H^{liquid \rightarrow gas}$: water=10.517±0.01kcal/mol [Chase98], benzene=8.1±0.02 kcal/mol [Roux08].

	CBS-QB3	G3	G4
benzene	-231.78431	-232.04674	-232.08857
water	-76.33371	-76.37827	-76.39347
CO_2	-188.36852	-188.49672	-188.53171
O_2	-150.16129	-150.24491	-150.27535
$\Delta H^{combustion}$	-764.45	-772.87	-763.67
$\Delta^{experiment}$	6.98 0.9%	15.37 2.0%	6.17 0.8%
relative CPU time	1.0	1.9	5.9

All of these methods overestimate the heat of combustion slightly. CBS-QB3 and G4 predict values within ~7 kcal/mol of experiment, and the G3 value differs from experiment by about 15 kcal/mol. Despite the "large" magnitudes of these deviations—significantly more than 1-2 kcal/mol—they are all in fact excellent results, within 1-2% of the value derived from observables. In general, it is necessary to take into account the magnitude of the results—here, hundreds of kcal/mol—when interpreting computational accuracy, rather than relying on a fixed, arbitrary, maximum desired error. For this problem, all three methods produce essentially equivalent results.

CBS-QB3 has by far the lowest computational requirements. It is almost twice as fast as G3 and almost six times as fast as G4 for this problem.■

EXERCISE 4.9 $C_{60}O$ **ISOMERS REVISITED**

Here is the experimental IR spectrum for $C_{60}O$ at STP [Cardini94]:

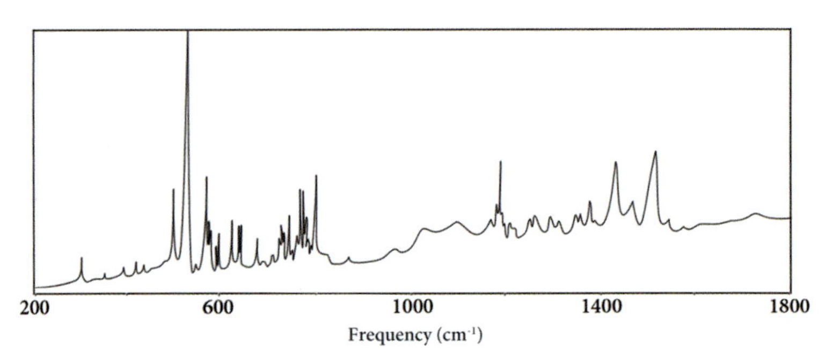

Predict the frequencies for the two isomers of this compound and determine which one(s) contribute(s) to the observed spectrum. We optimized these structures in Exercise 3.2 (p. 112). Run your frequency calculations with the APFD/6-311G(2d,p) model chemistry; diffuse functions are likely unnecessary for this system, and they add considerably to the computational cost.

SOLUTION

The following illustration combines the observed IR spectrum for $C_{60}O$ with the two we calculated for the open and epoxide isomers (the frequencies are unscaled):

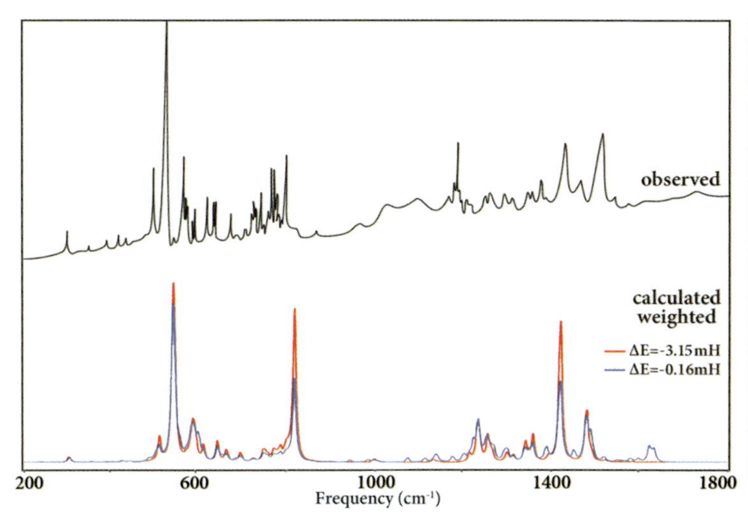

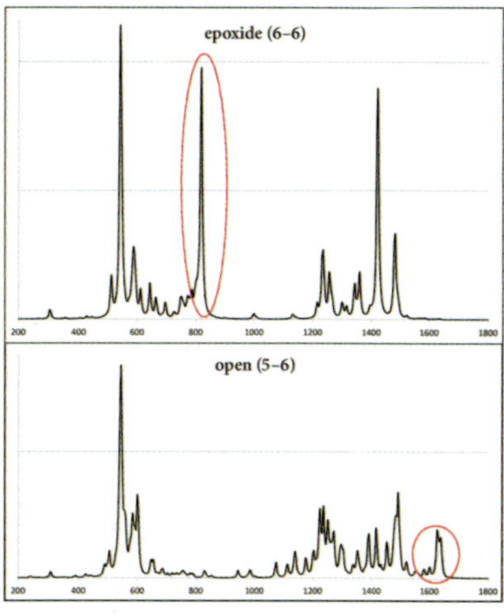

The individual spectra for the two isomers are shown at the right. Each contains features not present in the other; the most significant are circled in red. The larger plot on the left shows the observed (upper) and calculated (lower) spectra, with the latter obtained by Boltzmann weighting of the two isomers. Two different isomeric mixtures are plotted: the red line represents the energy difference between the epoxide and open form predicted by our standard theoretical method, and the blue line corresponds to the energy difference found by [Sohn10].

The observed spectrum contains the peak at around 800 cm^{-1} predicted for the epoxide isomer and seems to lack the peak at around 1600 cm^{-1} predicted for the open isomer. These features tend to reinforce the view that the epoxide isomer is the lower energy one, dominating the observed spectrum. However, there are some other aspects of the data which are less definitive:

- The peaks in the 1100-1300 cm^{-1} range in the observed spectrum would seem to come from the open isomer, but there is the possibility of baseline drift in the experimental results.
- The relative heights of the peaks at ~550 cm^{-1} and ~800 cm^{-1} are very different in the observed spectrum and the calculated ones. In particular, the spectrum predicted by the APFD/6-311G(2d,p) model chemistry overestimates the intensity of the latter peak compared to the former. The best fit to the intensity ratio that can be interpolated from the observed spectrum corresponds to an energy difference of about -0.21 millihartrees.

  Intensities predicted at this level of theory are only semi-quantitative, so one must avoid overinterpreting their specific values. In addition, anharmonic vibrational features may be significant for this system, and an anharmonic vibrational analysis would diffuse some of the intensity in the one strong mode to several less intense ones as happens in the observed spectrum.

The epoxide form seems very likely to be the lower energy one but this is still not completely confirmed by theoretical calculations.■

EXERCISE 4.10 **PROTON NMR OF CHLOROCYCLOHEXANE CONFORMATIONS**

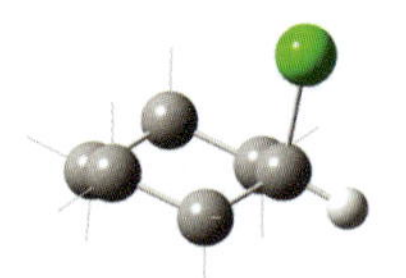

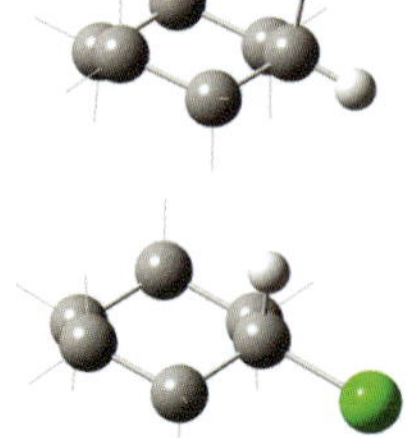

The axial and equatorial isomers of chlorocyclohexane are shown in the margin. The equatorial form is the lower energy isomer. However, at STP, even the equatorial form is highly unstable, having a half-life of only a fraction of a second. However, at lower temperatures, both forms can be observed (e.g., the half-life of the equatorial isomer increases to 23 min. at -120°).

The NMR proton spectrum for equilibrated axial and equatorial chlorocyclohexane at -115° exhibits two peaks, corresponding to shifts of 4.5 and 3.8 ppm with respect to TMS [Jensen69].

Predict the proton shifts for these two species, and then go on to use the results along with their relative energies to compute the expected peak location at higher temperatures, via Boltzmann averaging. The proton shielding value for TMS was reported earlier (see page 161).

SOLUTION

We ran optimization+frequency calculations to obtain the two structures and verify that they are minima, followed by NMR calculations. The following table summarizes our results:

ISOMER	G (hartrees)	PROTON SHIELDING	SHIFT REL. TMS (ppm)
axial	-695.13397	27.25	4.6
equatorial	-695.13462	27.94	3.8

The proton shielding results are in very good agreement with experiment, and the difference between the peaks is the same 0.7 ppm.

The predicted difference in free energy between the two forms is very small, only 0.65 millihartrees (0.65 kcal/mol, 2.7 kJ/mol). The following table summarizes the conformational averaging results:

ISOMER	ΔG (hartrees)	q_i	FRACTION	RAW SHIFT (ppm)	COMPONENT (ppm)
axial	-0.000651	1.98	0.66	4.6	3.1
equatorial	0.0	1.00	0.34	3.8	1.3

Adding the components for the two conformers yields a predicted proton shift of ~4.4 ppm at STP.■

Advanced Topics & Exercises

More About Computing Vibrational Properties

The peaks in IR and Raman spectra reflect the energy differences between vibrational states of the molecule being observed. These transitions are typically in the range of 100-5000 cm^{-1} (equivalent to ~0.3-15 kcal/mol or ~1-65 kJ/mol), although this whole range is not always scanned. A photon with energy in this range is in the infrared part of the electromagnetic spectrum.

IR spectroscopy probes the vibrational energy levels by inducing the absorption of a photon, which produces the vibrational transition. Spectral peaks result when the energy of the incoming photon matches the transition energy of one of the molecule's vibrational modes.

These energy differences are much, much lower than those between different electronic states of the molecule (e.g., compared to the energy required to excite it from its electronic ground state to the lowest excited state).

Raman spectroscopy involves the inelastic scattering of a photon, meaning that it leaves the molecule with less energy that it had when it entered. This energy difference again corresponds to the transition energy of some vibrational mode.

IR: $E(\gamma^{in}) = \Delta E^{vib}$
Raman: $E(\gamma^{out}) = E(\gamma^{in}) - \Delta E^{vib}$

For the case of IR, the photon energy is small because the vibrational transition energies are small. In contrast, for Raman interactions, $E(\gamma^{out})$ and $E(\gamma^{in})$ are typically much larger than ΔE^{vib}, usually in the visible or UV part of the electromagnetic spectrum.

Modeling these phenomena involves mathematically describing both the vibrations of the nuclei and the molecule's interactions with the electromagnetic field produced by the incoming photons. For low energy photons, this interaction is limited. However, for the energy range of photons involved in Raman scattering, the photon energy has a non-trivial effect on the predicted interaction with the molecule.

What is done by default in frequency calculations is called the *double harmonic approximation*. It is the simplest description of both the vibrations and the interactions with the electromagnetic field (i.e., the photons). It has two distinct parts:

- First, vibrations are assumed to be *harmonic*. That is, the vibrational modes are treated as simple harmonic motion about the minimum energy structure. Mathematically, this approach depends only on the second derivatives of the energy with respect to nuclear coordinates. The normal modes of the resulting force constant matrix yield the vibrational modes, and the corresponding eigenvalues give the transition energies. In this approximation, normal modes don't couple.

  For example, if E_i is the vibrational energy required for the vibrational transition to normal mode *i*, then the energy corresponding to the second vibrational state above the ground state—i.e., the second excitation of that mode—is simply $2E_i$. Similarly, the energy to go from the vibrational ground state to the lowest vibrational state having both mode *i* and mode $_j$ active is E_i+E_j.

  In reality, the energy difference between states *n* and *n*+1 of a normal mode decreases as *n* increases, and the energy difference between the ground state and the vibrational state corresponding to a combination band is less than the sum of the component modes' energies. These effects are not included in the harmonic approximation, which would affect its treatment of overtones and combination bands. However, these bands are not predicted at all due to the second part of the approximation.

- The second part of the double harmonic approximation involves the treatment of the molecule's interaction with the electromagnetic field of the photons. For IR absorption, the interaction is described by the dipole moment operator, which is assumed to change at a constant rate throughout the motion, which means that the transitions between successive excitations of a given normal mode are described by the first derivative of the dipole moment with respect to motion in the direction of that mode. For Raman scattering, the same approximation relates the scattering amplitude to the derivative of the molecular polarizability with respect to motion in the direction of the particular normal mode.

- In the double harmonic approximation, the only allowed vibrational transitions are ones starting from the ground state and resulting from the interaction with a single photon (i.e., from state 0 to state 1 of a single mode):

  all modes in state 0 → one mode in state 1, all other modes in state 0

  Under this approximation, the intensities of overtones (the absorption/scattering of a photon resulting in a transition from the ground state to the second vibrational state for a given normal mode) and combination bands (the excitation of two vibrational modes simultaneously from interaction with a single photon) are automatically 0. Thus, frequency analysis is limited to predicting intensities of fundamental bands only.

For absorption (IR), the dipole moment involved does not depend significantly on the photon energy, because the photons have very low energy. Accordingly, modeling it at the zero frequency limit—also known as the static limit—is usually sufficient. For Raman scattering, the photons have much higher energy, and there is significant dependence of the polarizability and its normal mode derivative on the photon energy (frequency), which must be included in the calculation even within the double harmonic approximation. This results in the somewhat increased computational requirements when a frequency calculation predicts Raman modes as well as IR.

Anharmonic Frequency Analysis

Anharmonic Frequency Analysis Referemces

Definition: [Califano76, Miller80, Papousek82]

Algorithms/Implementation: [Clabo88, Page88, Barone04, Barone05, Bloino12]

An anharmonic frequency analysis computes additional terms—higher derivatives—which relax both parts of the double harmonic approximation, with these results:

- The locations of the fundamental (0→1) frequencies change.
- The 0→1 and 1→2 transitions within a mode no longer have identical energy differences.
- The analysis predicts non-zero intensities for overtones (0→2 within one mode) and combination bands (e.g. 0→1 in mode 1 and 0→1 in mode 2 at the same time). For example, in an IR spectrum, the first overtone in the first mode corresponds to absorbing one photon (just as for the fundamental), but the photon will have energy:

  $$E(\gamma^{in}) = E_{i_0 \to_2} = E_{i_2} - E_{i_0} < 2E_i^{harmonic}$$

  Similarly, the same overtone mode in a Raman spectrum would correspond to scattering one photon with incoming and outgoing energies $E(\gamma^{out}) = E(\gamma^{in}) - (E_{i_2} - E_{i_0})$.

In practical terms, anharmonicity typically changes stretching modes by ~5%; the effect can be larger on low-frequency high amplitude modes such as a functional group wagging. You request an anharmonic frequency analysis with the **Freq=Anharmonic** option. We'll use this feature in several subsequent examples and exercises.

Frequency-Dependent Raman Intensities

Raman spectra often have different modes showing stronger peaks than IR because the interaction of the molecule with the photons' electromagnetic field is different. This is easiest to see for a symmetric molecule: the polarizability tensor has different symmetry properties than the dipole moment and so some modes may be forbidden by symmetry for one process but not the other. Even in non-symmetric molecules, the different properties of the different interactions will favor different modes.

Incident light at different frequencies will result in different intensities for the peaks in the Raman spectrum (the peak locations themselves do not change). As we've seen, computing the Raman spectrum depends on the polarizability, which is the second derivative of the energy with respect to an electromagnetic field.

Raman intensities depend on the polarizability derivative with respect to the nuclear coordinates, a third derivative of the molecular energy. Gaussian can predict Raman intensities at specific wavelengths of incident light via a dynamic frequency calculation (also known as a pre-resonance Raman calculation). This is accomplished via the **RdFreq** option to the **CPHF** keyword (for "read frequency"), which requires a list of desired frequencies in the input, separated with spaces or commas and terminated with a blank line. The following example illustrates these requirements.

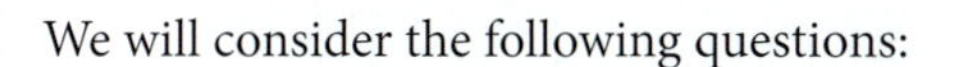

RAMAN SPECTRA OF SMALL WATER CLUSTERS 1

The properties and characteristics of liquid water are substantially different from those of an isolated water molecule. Thus, in order to model this essential substance, one must consider multiple water molecules. Calculations on small clusters of water molecules have yielded interesting and important results [Cybulski07].

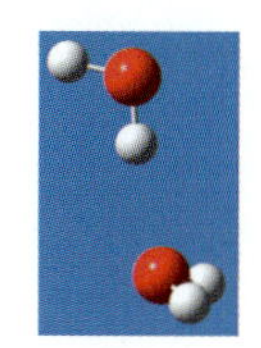

The most stable conformations of small water clusters are cyclic, built around the basic structural unit of a single proton donor-single proton acceptor water molecule. We are interested in studying the Raman of several such clusters with increasing numbers of water molecules: 2 (illustrated in the margin), 4 and 6.

We will focus on the region of intramolecular OH stretching vibrations between 3130 and 3930 cm^{-1}, specifically the lowest frequency OH stretching mode of the hydrogens involved in hydrogen bonding. We will predict Raman intensities at 3 different wavelengths of incident light commonly used for experimental spectra: 351.1, 435.8, and 514.5 nanometers.

We will consider the following questions:

- How does the intensity vary with the frequency of the incident light?
- How does the intensity vary as the cluster size increases?

We begin with water dimer. We optimized the structure, and then ran a frequency calculation, specifying three frequencies for the incident light. The input file for the calculation was structured as follows:

```
%Chk=name
# APFD/6-311+G(2d,p) Opt Freq=Raman CPHF=RdFreq

Water dimer frequency-dependent Raman spectrum

molecule specification

351.1nm, 435.8nm, 514.5nm                    incident light wavelengths
                                             final blank line
```

We specify the wavelengths in nanometers by including the **nm** suffix (the default units are hartrees).*

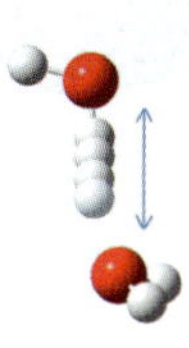

The mode with the largest Raman intensity is the OH stretch along the hydrogen bond (illustrated in the margin); the corresponding peak is located at 3686 cm^{-1} (scaled). The following table lists the predicted Raman intensities at the static limit and with the various wavelengths of incident light that we are considering:

* Versions of Gaussian 09 prior to D.01 will need to set up this calculation as a two-step job. See the provided alternate input file for details.

INCIDENT LIGHT	RAMAN INTENSITY (a^4/amu)
static limit	152.4
514.5 nm	170.5
435.8 nm	178.6
351.1 nm	196.1

As expected, the Raman intensity increases as the wavelength of the incident light decreases. We will compare these results with the corresponding ones for cyclic water tetramer and hexamer in the following exercise. See [Cybulski07] for a detailed computational investigation of this topic.■

ADVANCED EXERCISE 4.11 RAMAN SPECTRA OF SMALL WATER CLUSTERS 2

Predict the Raman intensities for lowest OH stretching mode for cyclic clusters of 4 and 6 water molecules. Do the results vary according to the number of water molecules included in the cluster?

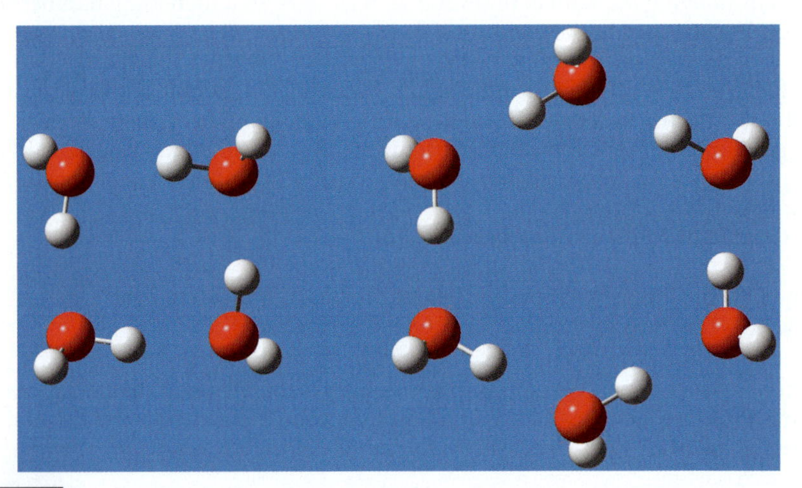

SOLUTION

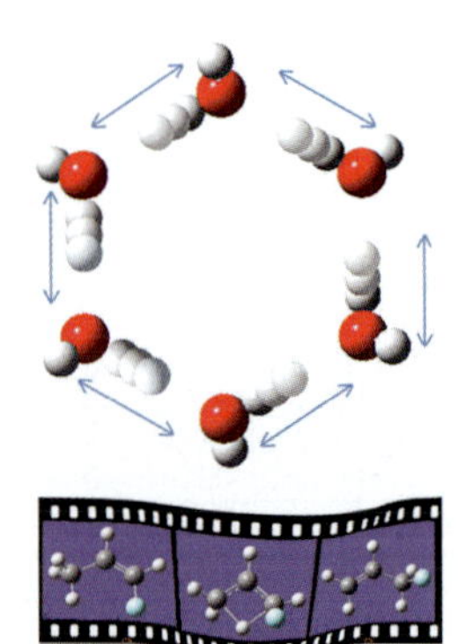

The OH stretching mode along the hydrogen bonds was again the one with the highest Raman intensity for both clusters. The following table summarizes the results of all three sets of calculations:

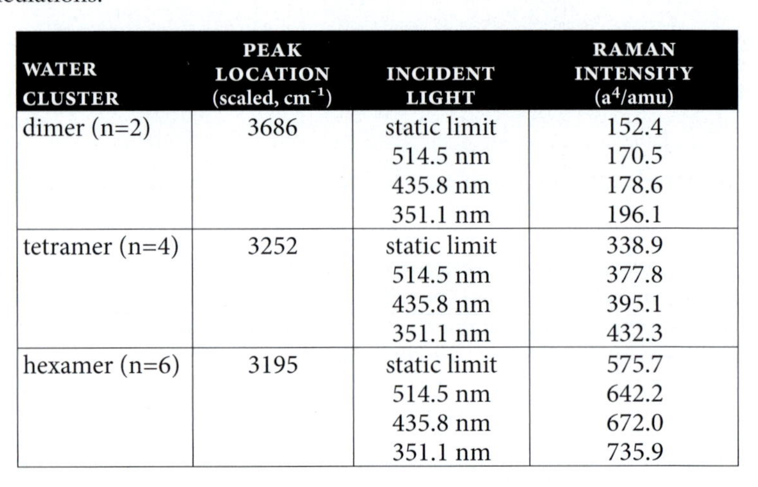

WATER CLUSTER	PEAK LOCATION (scaled, cm^{-1})	INCIDENT LIGHT	RAMAN INTENSITY (a^4/amu)
dimer (n=2)	3686	static limit	152.4
		514.5 nm	170.5
		435.8 nm	178.6
		351.1 nm	196.1
tetramer (n=4)	3252	static limit	338.9
		514.5 nm	377.8
		435.8 nm	395.1
		351.1 nm	432.3
hexamer (n=6)	3195	static limit	575.7
		514.5 nm	642.2
		435.8 nm	672.0
		351.1 nm	735.9

Although the number of data points is small, there is a near-linear relationship between the number of water molecules in the cluster and the Raman intensities. The relationship exists for all three wavelengths of incident light and for the static limit, as illustrated in the following graph:

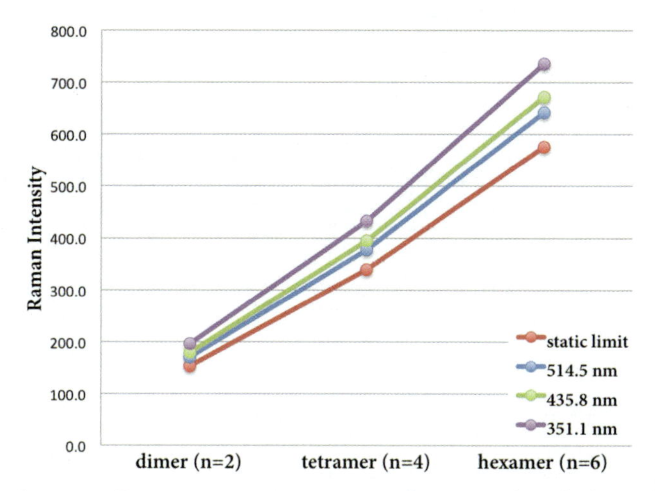

In general, the dynamic Raman intensities arising from incident light with the shortest wavelength are a little less than 30% larger than those computed at the static limit.■

ADVANCED EXERCISE 4.12 FORMALDEHYDE ANHARMONIC FREQUENCY ANALYSIS

Herzberg includes an overtone band in his formaldehyde frequency data [Herzberg91], at 2995.5 cm^{-1}. Perform an anharmonic frequency calculation on formaldehyde. Is this band predicted? How do the peak locations change with respect to those predicted by the default frequency analysis (i.e., the harmonic approximation)?

SOLUTION

The key section of the anharmonic frequency analysis output follows (see the next page). Its three sections list the fundamentals, overtones (identified by the mode that is multiply excited) and combination bands (identified by their component modes).

The following table summarizes the results for the fundamentals (all values are in cm^{-1}):

OBSERVED	HARMONIC (scaled)	ANHARMONIC (unscaled)
1167.26	1180	1180
1249.09	1247	1242
1500.17	1512	1495
1746.01	1791	1814
2782.46	2857	2751
2843.33	2915	2830

With the exception of the observed 1746 cm^{-1} mode, the anharmonic frequencies are lower than the harmonic (even with scaling), and they are in excellent agreement with experiment.

The anharmonic frequency analysis does predict the observed overtone band. It is identified as a second excitation of the fifth fundamental mode (via the notation **5(2)** in the Mode column).

```
==================================================
      Anharmonic Infrared Spectroscopy
==================================================

Units: Energies (E) in cm^-1
       Integrated intensity (I) in km.mol^-1

Fundamental Bands
-----------------
 Mode(Quanta)   E(harm)  E(anharm)   I(harm)   I(anharm)
     1(1)       2961.04   2830.11    113.86     133.10
     2(1)       1261.72   1242.29     10.87      11.44
     3(1)       2899.63   2751.30     69.66      68.98
     4(1        1839.85   1814.32    118.47     123.05
     5(1)       1527.26   1495.44     10.04       9.38
     6(1)       1197.29   1180.22      5.91       5.95

Overtones
---------
 Mode(Quanta)   E(harm)  E(anharm)             I(anharm)
     1(2)       5922.07   5496.40                 1.36
     2(2)       2523.45   2481.15                 0.40
     3(2)       5799.25   5441.48                 0.03
     4(2)       3679.60   3609.94                 4.22
     5(2)       3054.53   2994.28                 4.54
     6(2)       2394.58   2354.50                 0.11

Combination Bands
-----------------
 Mode(Quanta)   E(harm)  E(anharm)             I(anharm)
     2(1) 1(1)  4222.76   3996.33                 0.03
     3(1) 1(1)  5860.66   5407.28                 0.53
     3(1) 2(1)  4161.35   3986.63                 0.34
     4(1) 1(1)  4800.89   4600.88                 4.99
     ...
```

There are also other overtones and combination bands predicted by the calculation which are not reported in the reference. Such additional bands may be present for several reasons:

- They are outside of the experimental range. For example, the third overtone listed in the preceding output—a second excitation of mode 3—is located at 5441 cm^{-1}, which is outside of the range scanned in the experiment (e.g., ~1000-4000 cm^{-1}).
- They are forbidden modes, indicated by an intensity of exactly or near 0. The first combination band listed in the preceding output is an example (it is a combination of fundamental modes 3 and 1).
- They lie very close to another, more intense mode into which they are subsumed.
- Their intensity is overestimated by the calculation (relevant when the predicted intensity is small).

Anharmonic frequency analysis is much more computationally expensive than the harmonic approximation. Since the latter usually produces reasonable results, anharmonic analysis is not performed by default.■

ADVANCED EXERCISE 4.13 ANHARMONIC ANALYSIS OF CARBONYL STRETCH

In Exercise 4.3 (p. 173), we studied the carbonyl stretch vibrational mode in a series of compounds. Most of the results were in very good agreement with experiment. However, the peak locations for two of the compounds deviated quite a bit. Perform anharmonic frequency analyses for the same series of compounds. Can we conclude that the error was due to the neglect of anharmonic effects?

SOLUTION

The following table lists the results of our harmonic and anharmonic frequency calculations as well as some additional calculations that we ran with different theoretical methods (all calculations used our standard basis set). We've included the unscaled harmonic frequencies predicted by our standard model for consistency of comparison (all results are in cm^{-1}):

	APFD		B2PLYPD		
MOLECULE	**HARM.**	**ANHARM.**	**HARM.**	**ANHARM.**	**EXP.**
acrolein	1797	1754	1753	1707	1723
acetone	1809	1728	1728	1707	1737
formamide	1808	1789	1789	1739	1740
formaldehyde	1840	1814	1814	1749	1746
acetaldehyde	1827	1843	1843	1757	1746
methyl acetate	1811	1788	1788	1744	1761
acetyl chloride	1897	1894	1894	1817	1822
mean absolute deviation	4.2%	2.9%	2.9%	0.7%	

The most problematic cases with the default frequency analysis are formaldehyde and acetaldehyde. An anharmonic frequency calculation using our standard model improves all of the predicted frequencies except for acetaldehyde. Acrolein, acetone and methyl acetate are modeled well using this method. However, formaldehyde, acetyl chloride and acetaldehyde still differ substantially from experiment. Thus, the anharmonic correction ameliorates some but not all of the problems. Note that the peak ordering of both the harmonic and anharmonic treatments differ from the observed one.

We ran the same calculations with other DFT functionals in order to determine how pervasive such problems are. Results from some functionals showed no significant improvement over our standard model. B3LYP was an exception, with results almost as good as B2PLYPD (see below): MADs: 2.8% harmonic, 1.1% anharmonic.

We next considered whether this problem is a difficult one for traditional density functional theory in general. Accordingly, we performed optimization+frequency calculations for the series of compounds using the double hybrid DFT method keyword **B2PLYPD**, which includes both dispersion and an MP2-like correction to DFT. The **Freq=Anharmonic** calculation using this method produced very good agreement with experiment. However, even at this high level of theory with anharmonic corrections included, methyl acetate appears out of order with the other carbonyl-containing compounds. These results illustrate the challenges in obtaining high accuracy values for the C=O stretching frequency.■

References: Frequency-Dependent Polarizabilities & Hyperpolarizabilities

Definition: [Buckingham67, Buckingham67a]

Algorithms/Implementation: [Olsen85, Sekino86, Rice90, Rice91, Rice92]

Materials Properties: Polarizability, Hyperpolarizability & Gamma

Higher derivatives of the energy beyond those used in frequency analyses can be computed for many theoretical methods in Gaussian, enabling the prediction of polarizabilities (α), hyperpolarizabilities (β) and higher polarizability derivatives such as γ (the second hyperpolarizability), in order to model nonlinear optical properties. These quantities may be computed either at the static limit or with a specified frequency of incident light. They are requested with various options to the **Polar** keyword (possibly in combination with **CPHF=RdFreq**). Static polarizabilities and hyperpolarizabilities are computed by default during frequency calculations using methods for which the relevant analytic derivatives are available. Such methods include all hybrid DFT functionals. Gamma is computed also when the **Polar=Gamma** option is specified.* See the *Gaussian User's Reference* for full details.

ADVANCED EXERCISE 4.14 PREDICTING NONLINEAR OPTICAL PROPERTIES

We will consider three small molecules for our first foray into nonlinear optical properties. Compute the first and second hyperpolarizability for the molecules listed below using our standard model chemistry. How well do the predicted values compare with experiment?

- CH_3CN: $\beta_{||}$=26.3, $\gamma_{||}$=4629 measured via ESHG at 514.5nm [Stahelin93]
- CH_3F: $\beta_{||}$=−57.0, $\gamma_{||}$=2617 measured via ESHG at 514.5nm [Shelton82]
- CH_3Cl: $\beta_{||}$=13.3, $\gamma_{||}$=6860 measured via ESHG at 694.3nm [Miller77]

All of the preceding β and γ values are in atomic units.†

When preparing input for and examining the Gaussian output from frequency-dependent **Polar** calculations, care must be taken to ensure that what is calculated is the same quantity that was measured by the experiment. There are several considerations to keep in mind:

- The molecule specification must place the molecule in the same orientation with respect to the Cartesian axes as in the experiment. For molecules with nonzero dipole moments, these quantities are measured parallel to the dipole moment. By default, Gaussian reports alpha, beta and gamma in both the input orientation and the dipole orientation for such molecules. For molecules without dipole moments, it may be necessary to align the structure manually so that it has the same orientation as that of the experiment. We will see an example of this in Advanced Exercise 4.15 (p. 191).
- The appropriate quantity must be located within the output file, which typically provides several values corresponding to the static limit and to multiple experimental scenarios. The following list identifies the variations present for beta and gamma:
  - Beta(0;0,0): The static limit.
  - Beta(-w;w,0): dc Pockel's effect experiment at the specified frequency of incident light.
  - Beta(-2w;w,w): Second harmonic generation experiment at the specified frequency of incident light.
  - Gamma(0;0,0,0): The static limit.
  - Gamma(-w;w,0,0): dc Kerr effect experiment at the specified frequency of incident light.
  - Gamma(-2w;w,w,0): Second harmonic generation experiment at the specified frequency of incident light. Such experiments are variously known as ESHG, dcSHG and EFISH.

---

* Equivalent to **Polar=(dcSHG,Cubic)**, which must be used in early revisions of Gaussian 09.

† As rescaled by and reported in [Shelton94].

All three sets of values are presented in both the input orientation and the dipole orientation when relevant. [Shelton94] provides a helpful list of the main nonlinear optical processes used in gas phase hyperpolarizability experiments.

- Once the proper output section is identified, the correct quantity must be used for comparison with experiment. The output lists the parallel and perpendicular values of beta and gamma as well as all of their individual components. The parallel value is generally what is relevant, but in some cases other quantities are appropriate. Thus, understanding the details of the experimental measurement is vital.

In this case, since all of the molecules have a nonzero dipole moment, the molecular orientation should not be an issue in our calculations. In comparing with experiment, we will use the quantities computed in the dipole orientation, specifically "Beta(-2w;w,w) ||" and "Gamma(-2w;w,w,0) ||."

SOLUTION

The following table summarizes the results of our calculations and the experiments (all values are in au):

	$\beta_{\|\|}$		$\gamma_{\|\|}$	
MOLECULE	CALC.	EXP.	CALC.	EXP.
CH_3CN	28.1	26.3	4567	4619
CH_3F	43.0	-57.0	1766	2617
CH_3Cl	13.2	13.3	3404	6860

The calculated values are in rough, qualitative agreement with experiment. All of the predictive quantities are of the same order of magnitude as the observed values. In general, the trends among the three molecules are reproduced. The exception is the gamma value for CH_3Cl, which is far too small and results in incorrect ordering among the 3 compounds. While some of the computed values are remarkably similar to the measured ones, this is fortuitous. Given what is known about the method and basis set requirements for accurate prediction of nonlinear optical properties, these values must be seen as Pauling points reflecting a lucky cancellation of error. The next exercise will explore these model chemistry considerations in more detail. ■

ADVANCED EXERCISE 4.15 PREDICTING GAMMA FOR POLYACETYLENES

Introduction

The search for materials suitable for nonlinear optical applications focuses on ones having a very large hyperpolarizability, organic crystals and polymers among them. Such compounds have received a great deal of experimental and theoretical attention. In this exercise, we will consider a series of all-trans polyacetylenes (PA) (IUPAC: polyethynes).

These systems exhibit a great deal of electron delocalization, a phenomenon used to explain many properties in π conjugated systems, including the polarizability and its derivatives. The compounds with n=1, 2 and 5 are shown in the illustration following, with the unit cell bracketed in red in the latter:

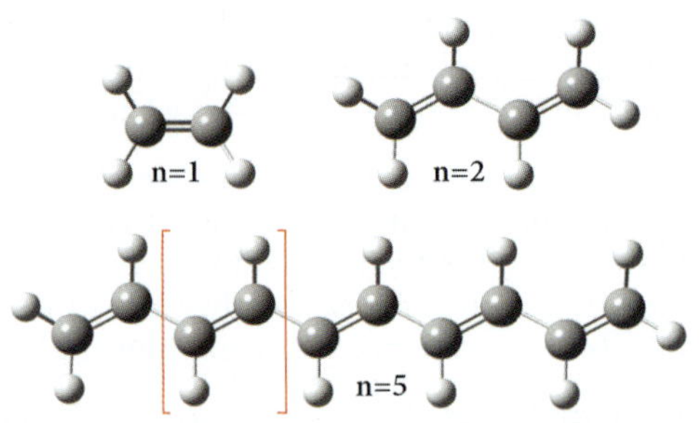

We will study the second hyperpolarizability of a series of PA compounds in two ways:

- By comparing the predicted second hyperpolarizability with experiment.
- By analyzing the rate of increase of the longitudinal component of γ—the component aligned with the polymer's direction of growth—as the chain length increases. This quantity is assumed to follow a power law with respect to the number of unit cells in the polymer chain [Shuai91]: $\gamma \propto n^a$. NB: This proportionality holds only for this component of γ.

As we have seen, computing these properties is computationally intensive. Conventional hybrid DFT functionals perform adequately for compounds with moderate electron delocalization, for example, polyphosphazene chains (units of PH_2=N) [Jacquemin07]. However, they are known to overestimate these nonlinear optical-related properties for systems with significant electron delocalization, increasingly so as the polymer chain length increases [Champagne98].

The problem with conventional functionals is that the non-Coulomb part of exchange functionals typically dies off too rapidly, becoming very inaccurate at large distances, making them unsuitable for modeling processes such as electron excitations to high orbitals and nonlinear optical processes. Both pure and hybrid functionals fail to correctly describe the asymptotic behavior of an electron as its distance from the nucleus becomes large.

We see these limitations with our standard model chemistry and similar ones with PA compounds. For example, the following table gives the predicted second hyperpolarizability for the PA compounds with n=1–3 predicted by two hybrid functionals,* along with the experimental values [Ward78] and published CCSD calculations [Limacher09]:

	$\gamma_{\|\|}$ [esu]		
MODEL§	ETHENE	BUTADIENE	HEXATRIENE
B3LYP/6-311++(3df,2p)	0.67	2.6	13.4
APF/6-311++(3df,2p)	0.60	2.3	11.6
CCSD/POL	0.83	2.5	8.6
Experiment	0.758	2.3	7.53

§ Geometries were optimized with B3LYP/6-311+G(2d,p) for the hybrid functionals and with CAM-B3LYP/cc-pVDZ for CCSD. POL is the basis set of Sadlej, designed specifically for computing polarizabilities and related properties [Sadlej88].

The following illustration plots the longitudinal component of gamma, γ_{xxxx}, for these model chemistries:

* When comparing with observed values and some theoretical studies, it is necessary to be aware of the different conventions regarding the factor of $1/n!$ in reporting γ. Here, we multiply the predicted values from Gaussian by a factor of 1/6. Similar considerations can also apply to the hyperpolarizability β, where the factor is ½.

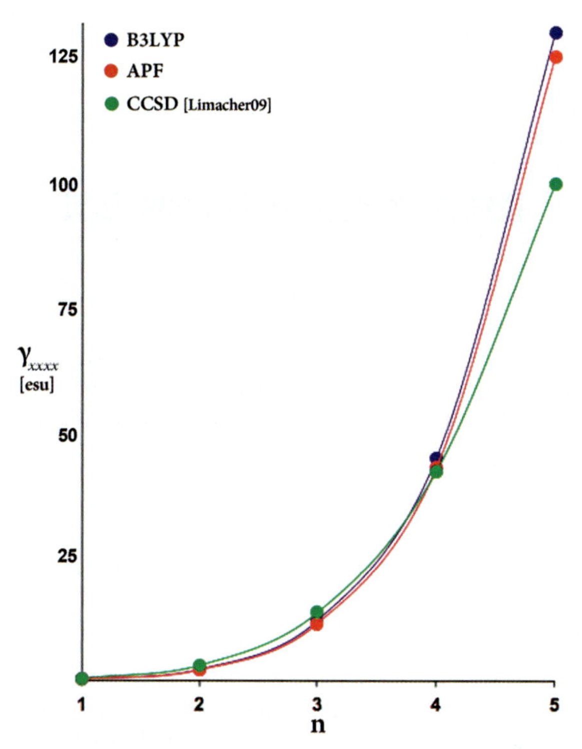

A strong curvature can be seen for both hybrid functionals, indicating a systematic deviation from a power law resulting in an excessively large hyperpolarizability, growing more so as the polymer lengthens. In contrast, the CCSD results from [Limacher09] conform well to a power law. CCSD is a very accurate method that includes electron correlation and also provides correct asymptotic behavior as distance increases.

Of course, CCSD is prohibitively expensive for most real-world compounds. Fortunately, various schemes have been devised to improve the handling of such cases within the DFT framework, and we will consider two of them in this exercise.

The Problem

Compute γ for the PA compounds with n=1–5, with an incident light wavelength of 694 nm. Use the 6-311++(3df,2p) basis set with the following functionals, which provide two different approaches for long range corrections to DFT:

- LC-BLYP: The long-range corrected version of the BLYP pure functional [Iikura01]. The LC technique adds exact exchange to a pure functional (here, BLYP) at long range only. These methods exhibit correct asymptotic behavior with increasing distance but retain all of the weakness of pure functionals at normal ranges.
- CAM-B3LYP: The Coulomb-attenuating model employing the B3LYP hybrid functional [Yanai04]. This functional uses different coefficients for the exchange part of the hybrid function at different distances. This results in partially correct asymptotic behavior combined with the accuracy advantages of hybrid functionals over pure functionals.

First, compare the predicted results to the experimental values. How well do these models correct the deficiency of standard hybrid functionals? Second, determine whether the longitudinal component of gamma conforms to a power series with increasing polymer chain length.

If you do not do so by default, include the **Int=UltraFine** option in your input files, specifying high accuracy integration grids. These are essential to accurate modeling of nonlinear optical properties.

You will need to optimize the structures and verify that they are minima as usual, and our standard model chemistry is fine for this task, as evidenced by comparing the predicted structure of butadiene with experiment [Craig06]:

PARAMETER	CALCULATED	OBSERVED
C=C (Å)	1.34	1.33508
C-C (Å)	1.45	1.4539
C-C-C (°)	124.3	123.6

The optimized structure is in excellent agreement with experiment, and the frequency calculation confirms that it is a minimum. Adamo and coworkers found equally good agreement for the bond length alteration for both PA and polymethineimine (PMI) chains [Jacquemin07].

Molecule Orientation

As we mentioned in the preceding exercise, it is necessary for molecules to be aligned appropriately with respect to the Cartesian axes in order for properties to be computed correctly. For these compounds, the x-axis needs to be aligned with the longitudinal axis of the polymer. In many cases, this means that you will need to align the structure manually within your molecule building program prior to creating the Gaussian input file.

GaussView

We have displayed the Cartesian axes along with the structure for butadiene in GaussView in the following illustration. In the molecule on the left, the axes are not aligned as required. The result of selecting **Reorient** from the **Edit** menu is illustrated on the right.

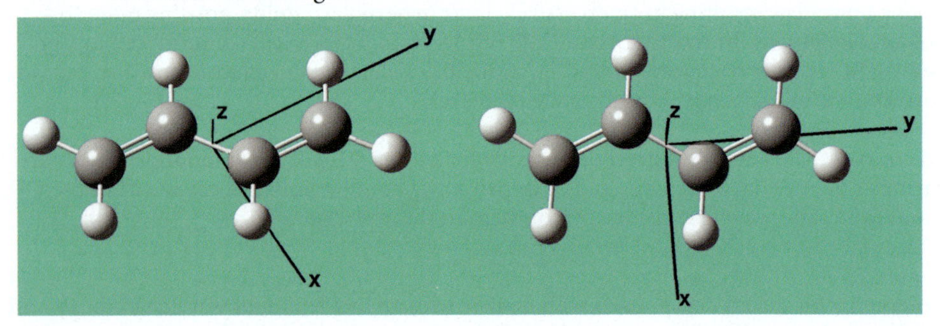

Note that the molecule is now aligned with the y-axis. You can change this to the x-axis by creating the Gaussian input file as usual and then swapping the x and y coordinates with the molecule specification with any text editor.

SOLUTION

The following table presents the predicted gamma values for the two model chemistries we are considering, along with the CCSD results of [Limacher09] and the observed values:

	$\gamma_{\|\|}$ [esu]				
MODEL	ETHENE	BUTADIENE	HEXATRIENE	OCTATETRAENE	DECAPENTAENE
CAM-B3LYP/ 6-311++(3df,2p)	0.56	2.1	9.8	55.1	460.1
LC-BLYP/ 6-311++(3df,2p)	0.43	1.6	7.1	32.5	150.6
CCSD/POL	0.83	2.5	8.6		
Experiment	0.758	2.3	7.5		

For the small polymers for which we have experimental data, both DFT methods designed to handle long-range interactions perform well, significantly better than the conventional hybrid functionals we considered previously. For the small systems, LC-BLYP consistently underestimates γ, although its results are good qualitative agreement with experiment. CAM-B3LYP is closer to experiment for those systems whose γ value has been measured. However, it is possible that CAM-B3LYP is overestimating γ somewhat for the larger systems, probably because it still contains a finite amount of DFT exchange. However, given that conventional hybrid functionals grossly overestimate gamma for larger systems (e.g., the APF value for decapentaene is >20,000), it is clear that both CAM-B3LYP and LC-BLYP ameliorate a great many of the shortcomings of traditional pure and hybrid functionals in treating systems with significant electron delocalization.

Both CAM-B3LYP and LC-BLYP perform well when predicting the longitudinal component of the second hyperpolarizability. The following table summarizes these results as provides the CCSD/POL (n=1–6) and CAM-B3LYP/POL//CAM-B3LYP/cc-pVDZ (n=6) results from [Limacher09] for comparison:

	γ_{xxxx} (esu)					
MODEL	**1**	**2**	**3**	**4**	**5**	**6**
CCSD/POL	0.48	3.1	13.5	41.5	98.5	214.2
LC-BLYP/6-311++(3df,2p)	0.43	1.6	7.1	34.6	93.5	
CAM-B3LYP/6-311++(3df,2p)	0.26	2.1	11.2	40.4	114.1	
CAM-B3LYP/POL						264.8

The following graph plots the preceding data, as well as the remainder of the CAM-B3LYP/POL values (through n=12) provided in [Limacher09]:

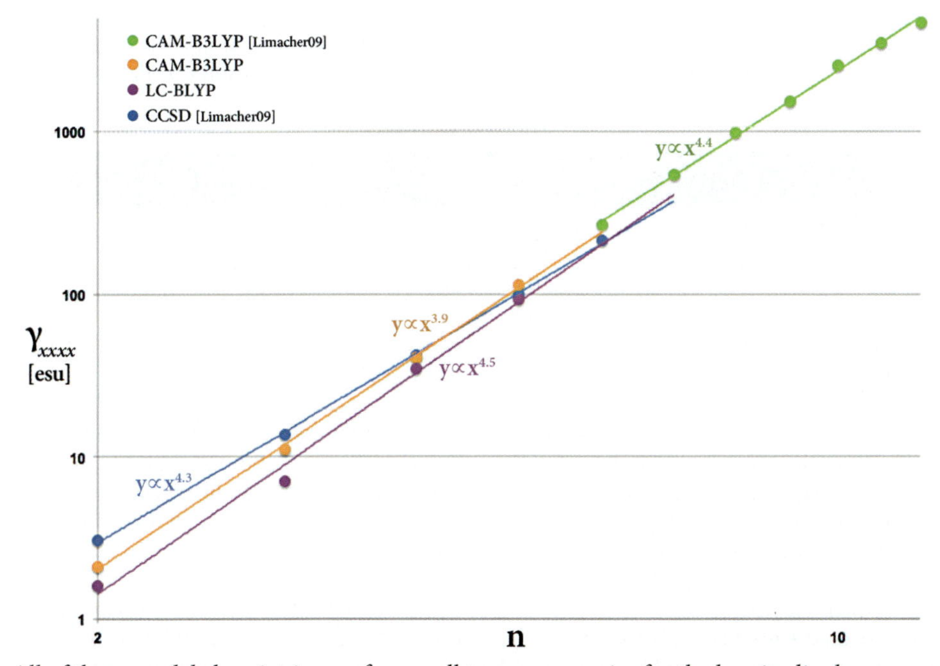

All of these model chemistries conform well to a power series for the longitudinal component of γ. The curves appear as straight lines on this log-log plot. The computed best fit exponents appear on the graph next to the corresponding line.

Similar considerations apply to predicting the polarizability α: see [Limacher09] for CAM-B3LYP results and [Sekino07] for LC-BLYP results. The latter paper also computed γ_{xxxx} for PA compounds from n=2-10, but they were presented graphically only, preventing inclusion or comparison here. Adamo and coworkers found similar results for polarizabilities for long polyphosphazene and PMI oligomers [Jacquemin07]. In all of these studies, both of the special-purpose functionals correct the major part of B3LYP faults.

We shall consider these functionals and the related issues again when we discuss modeling excited states (see Chapter 8).■

Highly Accurate Thermochemistry & Wavefunction Stability

In our final topic in this chapter, we reconsider the ionization potential of CN and the effect that the molecular wavefunction has on the subsequent predictions.

ADVANCED EXAMPLE 4.10 THE WAVEFUNCTION FOR THE CN CATION

A *biradical* is a compound having an even number of electrons with two independent free radical centers.

The ionization potential of CN* is a particularly difficult property to model with high accuracy. This is due to the biradical nature of the CN cation. Although this molecule is a singlet, it must be modeled with an appropriate, open shell wavefunction (see pp. 444ff for more on open shell systems). Running a **Stable=Opt** calculation on the closed shell wavefunction reveals an RHF to UHF instability.

When we presented the predicted ionization potentials for the various model chemistries in Example 4.5 and Example 4.6, we reported values computed by starting with a proper wavefunction rather than the one generated by default for each method. We chose to do so in order to focus on the thermochemistry and the methods rather than a single problematic case.

The table below compares the differing values for this ionization potential with and without specifying the initial wavefunction for the various models (IP values are in kcal/mol and energy differences are in hartrees):

MODEL	CLOSED SHELL IP (ERROR)	OPEN SHELL IP (ERROR)	$\Delta E^{CLOSED-OPEN}$
APFD/6-311+G(2d,p)	351.2 (27.6)	333.1 (9.5)	0.02882
CBS-QB3	317.4 (-6.2)	325.1 (1.5)	-0.01223
G3	319.6 (-4.0)	329.3 (5.7)	-0.01540
BD/6-31G(d)// APFD/6-311+G(2d,p)	320.8 (-2.8)	319.2 (-4.4)	0.00261
Experiment	323.6±0.5		

Using the default, closed shell wavefunction for this singlet system results in significantly poorer values for the ionization potential for our standard model chemistry and CBS-QB3. For G3, the magnitude of the error is similar, although the sign changes. However, single point energy calculations using the very accurate Brueckner Doubles (BD) method produced similar results for both closed shell and open shell cases.

The predicted electronic state for the optimized structure with the open shell wavefunction is the expected sigma state. This can be checked in several ways. The simplest is to locate the electronic state predicted by Gaussian. Search the output file for the string "State=" to determine this. In this case, we find: **State=1-SG**. (Unfortunately, the program is not always able to determine the electronic state.)

* IUPAC: carbon mononitride.

Another method is to examine the computed spin density. If you specify the **Population=Full** option for the calculation, then the population analysis within the output file will contain a great deal of information about the spin density. We have excerpted the key portions in the following listing. It confirms that the spin density is primarily located in the s and p_z atomic orbital components and that there is an unpaired electron on each nucleus as expected.

```
   Gross orbital populations:
                          Total       Alpha       Beta        Spin
 1 1   C   1S           1.10484     0.55132     0.55353    -0.00221
 2         2S           0.88422     0.44391     0.44032     0.00359
 3         2PX          0.18483     0.06403     0.12080    -0.05677
 4         2PY          0.18483     0.06403     0.12080    -0.05677
 5         2PZ          0.17660     0.05679     0.11980    -0.06301
 6         3S           0.85566     0.29158     0.56407    -0.27249
 7         3PX          0.41903     0.14812     0.27090    -0.12278
 8         3PY          0.41903     0.14812     0.27090    -0.12278
 9         3PZ          0.34766     0.09299     0.25467    -0.16169
10         4S           0.40908     0.07979     0.32929    -0.24950
11         4PX          0.15861     0.05515     0.10346    -0.04831
12         4PY          0.15861     0.05515     0.10346    -0.04831
13         4PZ          0.03912    -0.00450     0.04362    -0.04812
...                    The majority of the spin density is in the σ-comprising AOs.

The spin density for the N atom has analogous characteristics for spin up.

 Mulliken charges and spin densities:
                      2
     1  C  ...   -1.215637   One unpaired electron on each atom.
     2  N  ...    1.215637
```

The spin density can also be visualized in GaussView as a surface. For example, in the following illustration, the plot on the left shows the spin density for this molecule and it clearly illustrates its biradical nature:

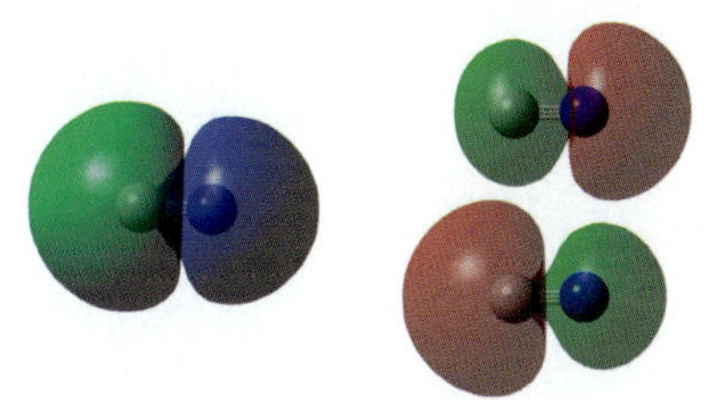

The molecular orbitals will also include α and β singly occupied σ orbitals. Those for the CN cation appear on the right above. Note, however, they may not appear as the highest occupied molecular orbitals, but rather be placed below one or more doubly occupied π orbitals. This is not uncommon for biradical systems.■

5

Modeling Chemistry in Solution

In This Chapter:

Continuum Models of Solvation

Structural Changes in Solution

Predicting Properties of Solvated Systems

Advanced Topics:

Predicting Free Energies in Solution

Including Explicit Solvent Molecules

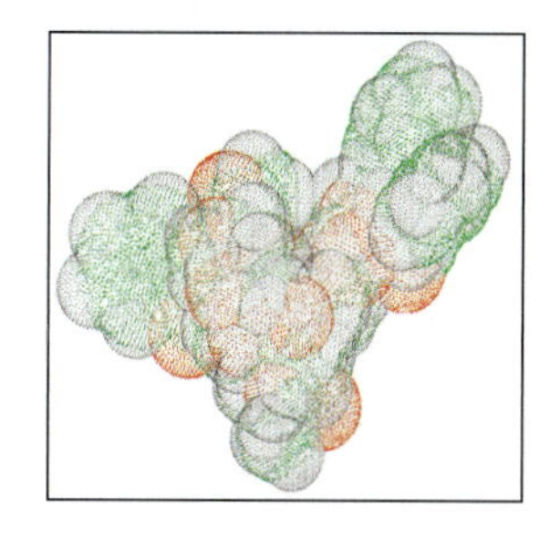

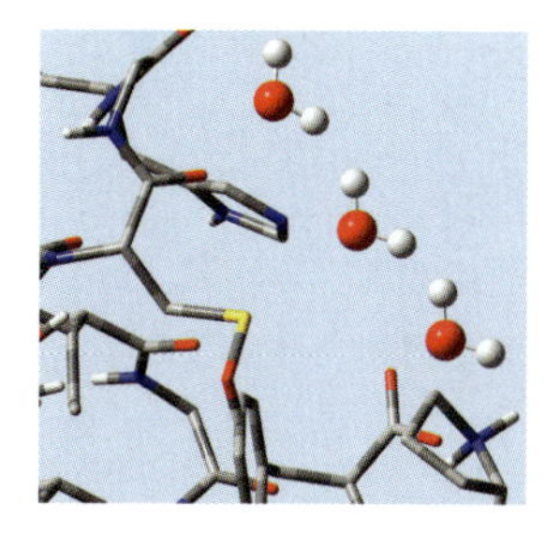

5 ◆ Modeling Chemistry in Solution

Continuum Models of Solvation

In the preceding chapters, we have performed calculations exclusively in the gas phase. While gas phase predictions are appropriate for many purposes, there are also a variety of systems and environments where they fail to reproduce the chemistry adequately. A substantial part of interesting chemical processes occur in solution. In this chapter, we focus on modeling systems in the presence of a solvent.

References: Self-Consistent Reaction Field Solvation Models

Prehistory: [Kirkwood34, Onsager36]

Initial Definitions: [Huron72, Rinaldi73, Tapia75, Miertus81]

IEFPCM: [Cances97]

Review: [Tomasi05]

Implementation: [Cossi96, Tomasi05, Scalmani10]

Of the many possible approaches to treating solvation, the models used in Gaussian belong to the family of Self-Consistent Reaction Field (SCRF) methods ([Tomasi05] is an excellent review article). These models are distinguished by several characteristics:

- The solvent is treated as a continuous, uniform dielectric medium characterized by its dielectric constant ε.

- The solute is typically modeled as a single molecule/complex, corresponding to a very dilute solution. It is treated using normal quantum mechanical approaches. In some circumstances, the solute may be supplemented by some explicit solvent molecules.

- The solute is placed inside an empty cavity within the solvent dielectric medium. The interaction between the solute and the solvent consists primarily of electrostatic interactions: the mutual polarization of the solute and the solvent.* The charge distribution of the solute inside the cavity polarizes the dielectric continuum, which in turn polarizes the solute charge distribution. For this reason, SCRF models like these are referred to as Polarizable Continuum Models (abbreviated PCM in the singular). Because of the mutual polarization of the solute and solvent, they are members of the family of self-consistent Reaction Field methods.

  The purpose of the cavity is to exclude the solvent from the solute, allowing the solute-solvent electrostatic interaction to be reformulated mathematically in terms of apparent charges at the solute-solvent interface (which exert a reaction field on the solute). This serves to significantly simplifying the computation. The overall electrostatic interaction is solved self-consistently using iterative numerical integration techniques.

The first SCRF method was proposed by Onsager in the mid 1930s [Kirkwood34, Onsager36]. In this model, the cavity is a sphere, and the solute molecule is treated as a dipole. The solute dipole induces a dipole in the medium, and the electric field applied by the solvent dipole will interact in turn with the molecular dipole. Iterative treatment of this interaction ultimately achieves self-consistency (net stabilization).

PCM models differ in the way that the cavity is defined and in the specific formulation of the electrostatic equations. We'll consider these two aspects individually in the following subsections. Over the years, several models have been implemented in Gaussian: e.g., [Miertus81] and [Cossi96]. The default SCRF method in Gaussian is the PCM modeling using the integral equation formalism, and it is known as IEFPCM [Cances97].

* In some models, other interactions which are not electrostatic in nature may also be included through parametrization.

Cavity Shapes

As we noted above, the first SCRF model used a spherical cavity. While this cavity shape worked well for some molecules, it was less accurate for the vast majority of molecules, whose shapes are not close to spherical. Devising a cavity always involves a tradeoff between two competing factors:

- Cavities composed of/from simple shapes like spheres and ellipsoids simplify the computations, enabling them to be performed more rapidly.
- Inaccuracies in the cavity shape produce deformations in the solvent reaction field and result in significant inaccuracies in the final predicted molecular properties.

Various SCRF models have used a variety of cavity shapes:

- The single sphere of Onsager.
- A surface of constant electron density, also known as an isodensity surface (illustrated in the upper image in the figure following).
- The solvent accessible surface, defined by the center of a spherical solvent molecule rolling over the topological boundary of the solute molecule.
- The superposition of interlocked spheres centered on the atoms in the molecule. The radii of the individual spheres are close to the van der Waals values (as defined by the UFF force field [Rappe92]). See the lower image in the following illustration.

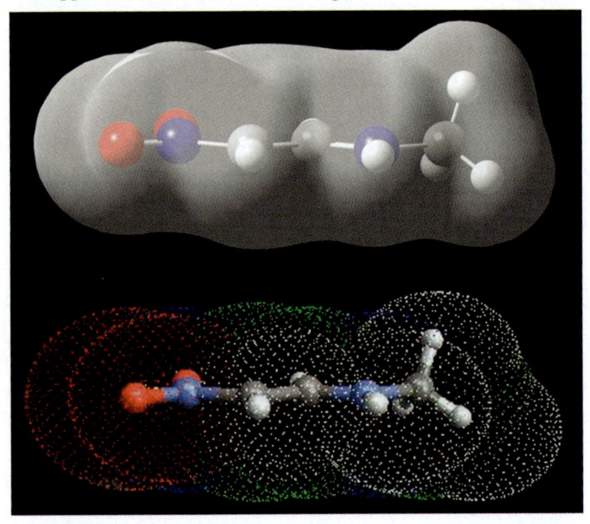

The PCM models in Gaussian use the latter formulation for the cavity as it provides the best tradeoff between computational simplicity and accuracy.

Escaped Charge and the Electrostatic Equations

An earlier version of PCM, now referred to as DPCM, uses equations for the electrostatic effect of the solvent dielectric which assume that all of the solute charge density is contained within the cavity. This is not strictly true for quantum mechanical wavefunctions, which decay slowly away from the molecule but reach zero only at very long distances.

In Gaussian, the IEFPCM model [Cances97]—the default SCRF method—is implemented to account for the so-called escaped charge density outside the cavity [Tomasi05], giving the correct solvent response within the cavity. This implementation uses a continuous surface charge formalism that ensures continuity, smoothness and robustness of the reaction field, which also has continuous derivatives with respect to atomic positions and external perturbing fields. This is achieved by expanding the apparent surface charge that builds up at the solute-solvent interface in terms of spherical Gaussian functions located at each surface element in which the cavity surface is discretized [Scalmani10]. The innovations in this latest implementation draw on the formalism of Karplus and York [York99]. Discontinuities in the surface derivatives are removed by effectively smoothing the regions where the spheres intersect.

Setting Up Solvation Calculations with Gaussian

In general, setting up Gaussian jobs for solvation calculations requires only the addition of a keyword to the corresponding gas phase input file. Both GaussView and WebMO make doing so simple by including menu items for this purpose.

In GaussView, the **Solvation** panel in the **Gaussian Calculation Setup** dialog contains a popup menu for specifying the desired solvent:

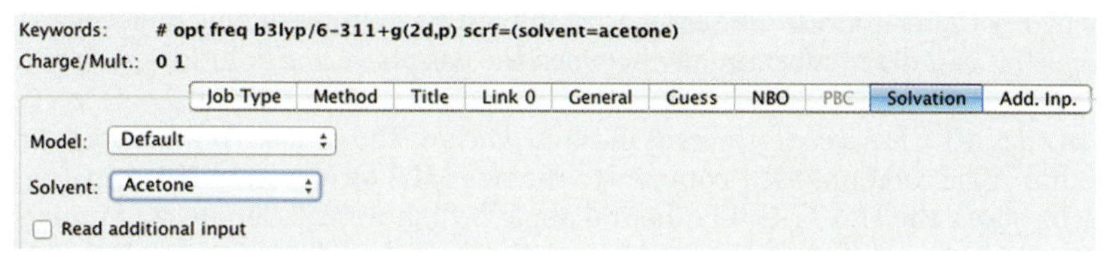

In WebMO, the **Solvent** field is located on the **Advanced** tab of the **Configure Gaussian Job Options** page:

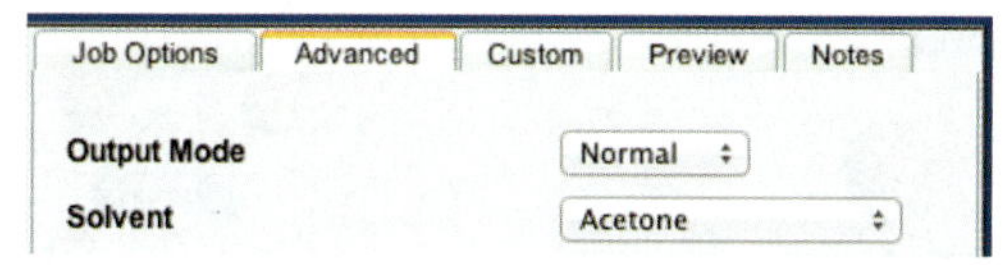

If the solvent you want to use is not in the **Solvent** list in your visualization program, you can specify it via the program's additional keywords feature, using the **Solvent** option to the **SCRF** keyword (described below). Alternatively, additional solvents can be added to WebMO's **Solvent** menu by editing the *gaussian.html* file in the *webmo* subdirectory of the web server's top-level content directory on the system where WebMO is installed. Search for the first line below, and then add additional entries like the third OPTION line (in red) for the desired solvent(s). Here we have added **Toluene** to the **Solvent** menu:

```
<SELECT NAME="solvent">
<OPTION VALUE="">None
<OPTION VALUE="SCRF=(PCM,Solvent=Water)">Water
...
<OPTION VALUE="SCRF=(PCM,Solvent=Toluene)">Toluene
</SELECT>
```

Note that many of the menus on WebMO's **Configure Gaussian Job Options** page are defined in this file, and you can customize any and all of them to your liking.

Within a Gaussian input file, the **SCRF** keyword is used to specify a calculation in solution, using the IEFPCM model by default. The desired solvent can be specified with its **Solvent** option.* For example, the following route section specifies an optimization+frequency calculation using our standard model chemistry in solution with acetone:

```
# APFD/6-311+G(2d,p) Opt Freq SCRF(Solvent=Acetone) ...
```

* See the *Gaussian User's Reference* for a complete list of defined solvents. If you cannot find the solvent you want, be sure to check the list for alternative names.

Next, we consider a specific example where we compare the predicted structures for the conformations of methyl lactate in the gas phase and in solution with methanol.

EXAMPLE 5.1 METHYL LACTATE CONFORMERS IN METHANOL

Concerns over the effects of chemicals used in industrial processes and everyday life have made environmental chemistry an important subdiscipline of the field. Since most traditional solvents have been shown to have serious detrimental effects on both the environment and on people who come into contact with them, searching for new solvents which are nontoxic and environmentally friendly is an important ongoing area of research.

The lactate ester family has great potential for providing these desirable, "green" solvents. These compounds function well as solvents and are also biodegradable and nontoxic. The simplest member of the lactate ester family is methyl lactate. This compound has been studied in detail by Aparicio [Aparicio07]. It serves as an interesting example of how molecular structure and properties can differ substantially between the gas phase and solution environments.

Methyl lactate (ML)* has 5 conformers as illustrated below. The various conformers interconvert via dihedral angle rotation. ML1 converts to the two ML2 forms, and ML3 converts to ML4 by rotating about the HO-C-C=O dihedral angle, which we will denote ϕ_1. Similarly, ML1 converts to ML3 by rotating about the H-O-C-C dihedral angle, ϕ_2 (as do ML2a and ML2b into ML4).

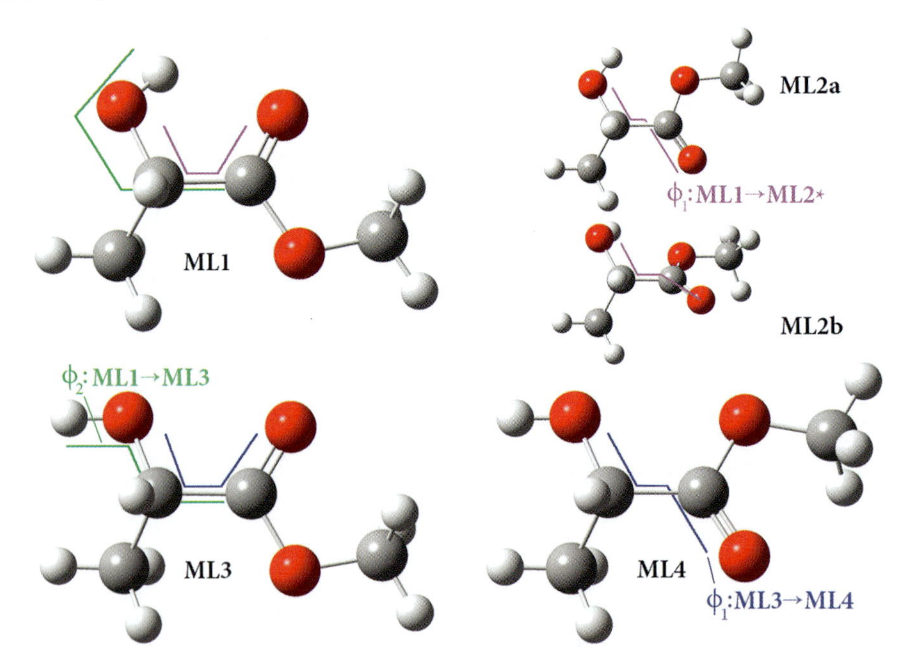

The two forms of ML2 exist because the intramolecular hydrogen bond leads to two non-equivalent geometries.

We optimized these conformers in the gas phase and in solution with methanol in order to determine how their structures and properties change in methanol solution. The following table presents these results. It lists the predicted free energy and the values of the key structural parameters for the various optimized structures in each environment.

* IUPAC: methyl (2S)-2-hydroxypropanoate.

	G (hartrees)		KEY DIHEDRAL ANGLES (°)			
			GAS PHASE		METHANOL SOLUTION	
FORM	GAS PHASE	METHANOL SOLUTION	ϕ_1	ϕ_2	ϕ_1	ϕ_2
ML1	-382.64971	-382.65728	-6.7	6.1	-6.4	6.6
ML2a	-382.64633	-382.65490	-162.8	-37.0	-171.5	-46.5
ML2b	-382.64632	-382.65509	152.8	48.7	154.3	59.2
ML3	-382.64166	-382.65342	-20.2	169.8	-15.3	177.8
ML4	-382.64307	-382.65359	162.7	170.3	164.1	176.6

The structure of the ML1 changes only slightly in solution with respect to the gas phase. However, for the other conformers, one or both angles change quite substantially.

For both environments, ML1 is the lowest energy conformer. However, the range of energies for the series of conformers is reduced from about 21 kJ/mol in the gas phase to about 10 kJ/mol in solution with methanol.

The relative free energies of the various conformers with respect to ML1 and the Boltzmann population percentages are given in the following table:

	$\Delta G^{\text{relative to ML1}}$ (kJ/mol (~kcal/mol))			BOLTZMANN %	
FORM	GAS PHASE	METHANOL SOLUTION	$\Delta G^{\text{gas-solution}\dagger}$ (kJ/mol (~kcal/mol))	GAS PHASE	METHANOL SOLUTION
ML1			19.9 (~4.8)	94.5	82.0
ML2a	8.87 (~2.12)	6.3 (~1.5)	22.5 (~5.3)	2.7	6.7
ML2b	8.90 (~2.13)	5.8 (~1.4)	23.0 (~5.5)	2.7	8.2
ML3	21.1 (~5.1)	10.1 (~2.4)	30.9 (~7.4)	<0.1	1.4
ML4	17.4 (~4.2)	9.7 (~2.3)	27.6 (~6.6)	<0.1	1.7

†Positive values indicate a lower energy in solution.

The free energy differences among the various conformers are small in the gas phase, and they become even smaller in solution. The range of energies for the series of conformers is reduced in solution.

Molecular properties predicted in the gas phase versus solution will also differ, since what is observed in the laboratory is the Boltzmann-weighted average of the properties of the individual conformers. In this case, ML1 almost completely dominates in the gas phase. However, in the methanol solution environment, other conformations make non-trivial contributions to observed properties.

The following table reports the Boltzmann-averaged values for several important properties of this compound:

PROPERTY	GAS PHASE	SOLUTION
Dipole moment (D)	3.0	3.7
C=O stretch: frequency (cm^{-1})	1794	1761
C=O stretch: intensity (KM/mol)	245	457

The dipole moment is significantly larger in solution. The data for the carbonyl stretch mode illustrates how IR frequencies shift in solution and also have different intensities. These effects are observed in all spectroscopic data. ■

The next subsection provides additional examples of the changes in molecular properties that typically accompany structural changes in solution.

Properties in Solution

The structural changes that occur in solution also affect the system's molecular properties. Even small changes in geometry can change the chemistry of a system or reaction. In this section, we will compare the predicted properties in solution to those in the gas phase for some molecules we studied previously.

EXAMPLE 5.2 FORMALDEHYDE IR SPECTRUM IN ACETONITRILE

We optimized the structure of formaldehyde in acetonitrile and then computed the vibrational frequencies. The following table lists the results (frequencies are scaled):

		VIBRATIONAL FREQ.		SHIFT
MODE	MOTION (SYMMETRY)	GAS PHASE	SOLUTION	(solvent effect)
1	CH_2 wag (B1)	1180	1198	+18
2	CH_2 rock (B2)	1247	1247	0
3	CH_2 scissors (A1)	1512	1495	-2
4	C=O stretch (A1)	1791	1761	−30
5	CH asymm. stretch (A1)	2857	2886	+29
6	CH symm. stretch (B2)	2915	2957	+42

The peaks fall in the same order as in the gas phase. As the values in the table indicate, the peak locations for peaks 1 and 4-6 in the IR spectrum in solution are shifted with respect to the gas phase, in some cases significantly. ■

EXAMPLE 5.3 VITAMIN E OXIDATION MODEL IN SOLUTION

We will now reconsider the gas phase models of Vitamin E oxidation from Chapter 2 in a solvent environment (see Example 2.9, p. 62 and Exercise 2.7, p. 69) in order to determine whether the favored conformations and the energy gaps between them are affected by the solvent environment. We will optimize the structures of α-tocopherol OQM and modification I in solution with acetonitrile. We will use the ONIOM facility in combination with an SCRF calculation, using the same partitioning as before. The low ONIOM layer is displayed in tube format in the following illustration of the up form of α-tocopherol ortho quinone:

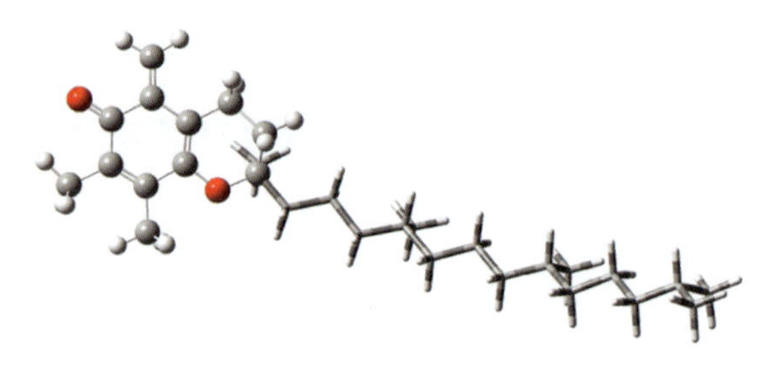

The optimizations for both structures locate minima. The results of our calculations are presented in the following table (we again compare the free energies):

	$\Delta G^{up-down}$	
MOLECULE	**GAS PHASE**	**ACETONITRILE SOLUTION**
α-TOH ortho quinone	-2.0 kcal/mol ≈ -8.4 kJ/mol	-2.9 kcal/mol ≈ -12.1 kJ/mol
modification I	+2.4 kcal/mol ≈ +9.9 kJ/mol	+3.3 kcal/mol ≈ +13.8 kJ/mol

The up form is again the lower in energy for the vitamin E oxidation product, while the down form is lower in energy in modification I. In fact, the energy difference between the two forms increases in solution for both compounds.■

Predicting Thermodynamic Quantities

In a broad chemical sense, any measurable thermodynamic quantity must be thought of as an energy difference between two stationary points on the molecular potential energy surface. Therefore, to calculate such a quantity always requires a geometry optimization. To know whether or not the quantity refers to an equilibrium process or an activation process further requires verifying the type of stationary point. This can only be done using a frequency calculation starting from the optimized geometry. The importance of this does not change when studying systems in solution. Therefore, our standard model chemistry whenever a thermodynamic quantity in solution is required will be:

```
# APFD/6-311+G(2d,p) Opt Freq SCRF
```

We may need further options depending on the particular quantity or system. For example, a transition structure would require suboptions to the **Opt** keyword, we will typically require **SCRF** options specifying the solvent environment and sometimes the solvation model, and so on.

Predicting Free Energies in Solution: The SMD Method

Reference: SMD Method
Definition: [Marenich09]

Thermochemistry is as relevant to solution chemistry as it is for molecules and reactions in the gas phase. An often desired quantity is the free energy, which can be used to compute the *solvation energy* of a molecule: the energy change going from the gas phase to solution. The solvation energy can be computed for the same compound with several solvents in order to understand its relative solubility in different environments. The predicted free energy can also be used to predict reaction energies in solution. In this section, we present the recommended technique for computing ΔG in solution as well as the rationale underlying it.

For such calculations, we use the SMD method, a parametrized SCRF-based solvation model developed by Truhlar and coworkers [Marenich09]. It was developed specifically to predict free energies of solvation. It uses different values for the radii and non-electrostatic terms than the default SCRF model, which were chosen specifically to reproduce free energies of solvation.We have also found the SMD method to perform well for predicting reaction energies in solution.

For a given solute, the SMD solvation energy is calculated as a free energy of transfer from an ideal gas to an ideal solution, at temperature of 298.15 K in both cases, and using the same concentration in the gas phase and in solution. Specifically, the SMD model was parametrized using experimental solvation free energies corrected to 1 mol/L (1 M) for both the gas phase* and the solution phase.

* 1 M concentration is equivalent to 1 mole of an ideal gas at a pressure of 24.5 atm.

In this work, when we refer to free energy changes, we will consistently mean energy changes for processes for which the initial and final states use the same standard reference (see the box "Going Deeper" on p. 108). When a solute transitions from the gas phase to solution, *what is important is that the reference be the same for both phases*, which is the case for the free energies predicted by the SMD model. This means that they can be compared to other measurements and predictions of this quantity for which the gas phase and solution reference states are consistent with one another.

A small wrinkle arises with experimental ΔG^{solv} values reported in the literature in this respect. For reported values, the solution phase G is typically referenced to a standard state of 1 M. However, the gas phase G may be referenced to an ideal gas at a pressure other than 24.5 atm, i.e., at a concentration other than 1 M. Commonly-used gas phase references include pressures of 1 atm and 1 bar. In such cases, you will need to apply a concentration correction to the experimental numbers before comparing them to the Gaussian SMD results. The correction required for a 1 atm gas phase pressure is -1.89 kcal/mol (-7.93 kJ/mol); for a pressure of 1 bar, it is -1.90 kcal/mol (-7.96 kJ/mol).*

In contrast, when computing reaction energies in solution, the concentration corrections typically cancel and may be omitted provided the number of moles remains constant (Δn). The same holds true for many energy comparisons in solution (e.g., between isomers). Comparison with experiment may be performed with the reported literature values.

EXAMPLE 5.4 SOLUBILITY OF ACETIC ACID IN CHLOROFORM AND IN WATER

We want to compare the solubility of acetic acid in several solvents. The free energy of solvation can be used for this purpose. Here, we will consider the solvents chloroform and water.

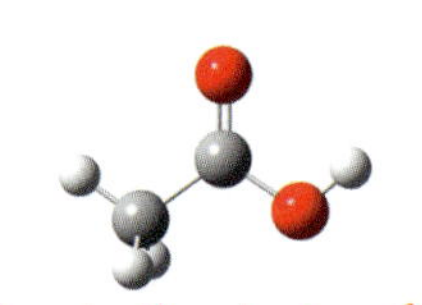

Your starting structure for acetic acid should have C_s symmetry with dihedral angles H-O-C=O and H-C-C=O equal to 0°.

In order to predict the free energy of solvation in each environment, we will optimize the geometry of acetic acid in the gas phase and in each solvent (with the second optimization starting from the gas phase-optimized structure). We will compute **ΔG** of solvation as the difference of the predicted Gibbs free energy values in the gas phase and in solution, taken from the two frequency calculations.

The following table presents the results of our calculations and compares them with the experimental values for the free energies of solvation [Marenich09 Supp. 1]:

		ΔG^{solv} (kcal/mol)	
ENVIRONMENT	**G (hartrees)**	**CALC.**	**OBS.†**
gas phase	-228.94720		
chloroform	-228.95275	-3.48	-4.74
water	-228.95497	-4.88	-6.70

†Reported values include the concentration correction to 1 M.

The calculated values are in good agreement with experiment. They also correctly predict that acetic acid will be more soluble in water than in chloroform.

In an effort to improve these values, we will use the SMD method. This method is requested with the **SMD** option to the **SCRF** keyword. We ran optimization+frequency calculations

* These corrections were applied to the experimental solvation free energies used as the training set for the SMD model [Marenich09].

in the two solvent environments, and obtained the following results:

ΔG^{solv} (kcal/mol)			
ENVIRONMENT	IEFPCM	SMD	OBS.
chloroform	-3.48	-5.03	-4.74
water	-4.88	-6.68	-6.70

The computed values for ΔG^{solv} are now in excellent agreement with experiment. This level of accuracy is not surprising since acetic acid in these solvent environments is present in the training set for the SMD method. That notwithstanding, this SCRF model incorporating an SMD-based correction provides an excellent approach for computing free energies of solvation, and we will continue to use it for the relevant examples and exercises in this book.

We will expand this study to include additional solvents in Exercise 5.5 (p. 215).■

SMD Transition Sructure Optimizations

Searching for transition states with the SMD method can be even more challenging than usual since SMD introduces additional noise into the numerical integration. An effective approach is to perform such calculations as a two-step job: an optimization+frequency job with IEFPCM solvation first (i.e., **SCRF** without the **SMD** option), followed by a second one with **SCRF=SMD** that reads in the force constants from the previous job via **Opt=RCFC**.

Alternate Solvation Models Based on SMD

Other models incorporating the SMD method can also be constructed. For example, the simplest way to calculate the SMD free energy of solvation is to subtract the solute's gas phase total energy from the SMD total energy, both computed at the gas phase optimized geometry:

$$\Delta G^{solv} = E^{SMD} - E^{gas\ phase}$$

Such an approach was used for the original parametrization of the SMD model. Note that this procedure does not involve an optimization+frequency calculation in solution, and may accordingly be useful for very large problems and/or in computing environments where resources are limited.

Another possible model relies on the fact that SMD incorporates only a few thermodynamic terms within its formulation. One could use the SMD prediction for the *total energy*—without any thermochemical corrections—in place of the value predicted by the default IEFPCM method. Doing so effectively replaces the IEFPCM total energy component of the computed G value with the improved SMD energy, while retaining the other thermochemical components determined by the IEFPCM frequency calculation:

$$\begin{aligned} G^{ESMD} &= G^{IEFPCM} - E^{IEFPCM} + E^{SMD} \\ &= G^{IEFPCM} + (E^{SMD} - E^{IEFPCM}) \end{aligned}$$

where E^{IEFPCM} and E^{SMD} are the total energies computed by the IEFPCM and SMD methods, and G^{IEFPCM} is the free energy computed by the frequency calculation in solution.

The following table summarizes the results of this model for the preceding example:

					ΔG^{solv} (kcal/mol)	
ENV.	G^{IEFPCM}	E^{IEFPCM}	E^{SMD}	G^{ESMD}	CALC.	OBS.
gas phase	-228.94720					
chloroform	-228.95275	-228.98735	-228.98974	-228.95515	-4.99	-4.74
water	-228.95497	-228.98942	-228.99177	-228.95731	-6.20	-6.70

This model also produces excellent agreement with experiment for this problem. It may be useful for cases where the SMD method has difficulty optimizing the molecular geometry.

EXAMPLE 5.5 REACTION FREE ENERGY: HYDROLYSIS OF METHYL ACETATE

Ester hydrolysis reactions are iconic reactions in organic chemistry. In a neutral aqueous environment, a water molecule combines easily with the carboxylate ester, breaking the COO-R bond and thereby creating an acid and an alcohol as products. The reaction is catalyzed in basic solution; under these conditions, it is the conjugate base of the acid which dominates as the product along with an alcohol:

R'C(=O)OR + HOH ⟶ R'C(=O)OH + ROH *neutral solution*

R'C(=O)OR + OH^- ⟶ R'C(=O)O^- + ROH *basic solution*

In this example, we will study the base catalyzed reaction where R=R'=CH_3: methyl acetate and hydroxide ion going to acetate ion and methanol [Takano05]:

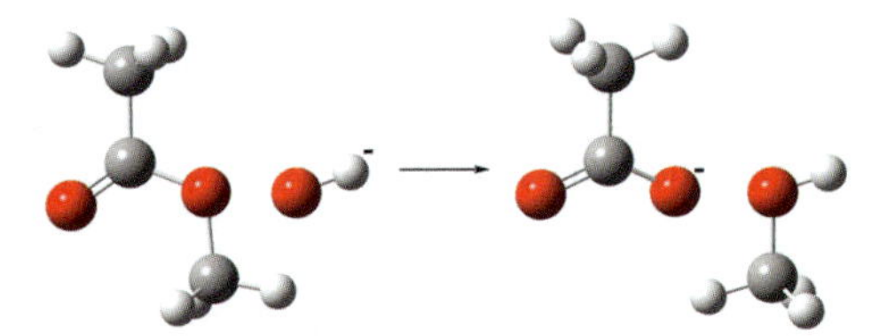

We want to predict the Gibbs free energy of the reaction. In order to do so, we need the free energies of the four compounds in question, which we will compute in the gas phase and from SMD optimization+frequency calculations.

The following table lists the computed values of ΔG for the reaction, along with the value estimated by the original experiment [Guthrie73]:

	GAS PHASE	SMD	EXP.
ΔG (kcal/mol)	-43.6	-19.9	-14.4

The equilibrium favors the products over the reactants in both the gas phase and in solution. However, in water, the reaction is predicted to have only half the free energy change as in the gas phase. The calculations in solution recover ~80% of the observed solvent effect. The results deviate from the experiment by ~5.5 kcal/mol (~23 kJ/mol). The remaining difference is most likely due to effects not included in the SCRF model: solute-solvent interactions treated too simply by the solvation model, the non-static nature of the solvation shell, and the like.

We will consider more advanced methods for modeling this reaction in Advanced Example 5.6 (p. 217) and Advanced Exercise 5.6 (p. 219).

Exercises

EXERCISE 5.1 FORMALDEHYDE FREQUENCIES IN CYCLOHEXANE

Predict the structure and vibrational frequencies for formaldehyde in solution with cyclohexane. Do you expect the solvent effect to be larger or smaller than it was for acetonitrile?

SOLUTION

The following table summarizes the solvent effect on the IR spectrum of formaldehyde for the two solvents:

	FREQUENCY SHIFT	
MODE	ACETONITRILE	CYCLOHEXANE
1	+18	+8
2	0	+1
3	−2	+1
4	−30	+11
5	+29	+8
6	+42	+13

The positions of peaks 2 and 3 are quite similar for both solvents and are quite close to the corresponding gas phase peak locations. However, for the other four modes, the peak locations in acetonitrile are shifted more than they are for cyclohexane. Acetonitrile is much more polar than cyclohexane; the two solvents have ε values of 35.69 and 2.02, respectively. The strong interaction between the polar solvent and a molecule with a large dipole moment like formaldehyde is not surprising.■

EXERCISE 5.2 FURFURALDEHYDE CONFORMERS IN VARIOUS SOLVENTS

Furfuraldehyde* can exist in either a syn or anti conformation. The two conformations are illustrated in the margin (the syn form is at the top). Which form is favored will depend on the solvent environment. Using the Gibbs free energy values computed by the IEFPCM method, predict the more favored species in each of the following solvents:

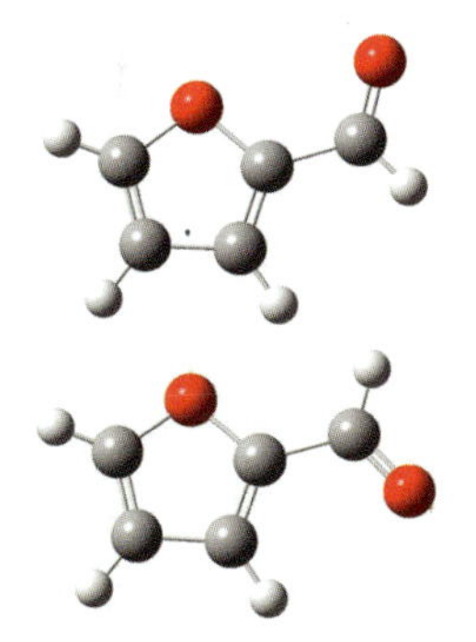

dielectric constant given in parentheses for reference

- methylcyclohexane (ε=2.02)
- dibutylether (ε=3.05)
- diethylether (ε=4.24)
- tetrahydrofuran (ε=7.4)
- acetone (ε=20.5)

SOLUTION

The following table summarizes the results of our calculations (energy differences are in kcal/mol). We have also included gas phase results for comparison purposes:

	CALC. (25 °C)		OBSERVED (-85 °C)	
SOLVENT	ε	$\Delta G^{syn\text{-}anti}$	ε	$\Delta G^{syn\text{-}anti}$
gas phase	1.0	0.70	1.0	0.82[a]
methylcyclohexane	2.02	0.25	2.19	0.24[b]
dibutylether	3.05	0.01	4.6	-0.10[b]

* IUPAC: furan-2-carbaldehyde.

SOLVENT	CALC. (25 °C) ε	$\Delta G^{syn\text{-}anti}$	OBSERVED (-85 °C) ε	$\Delta G^{syn\text{-}anti}$
diethylether	4.24	–0.15	7.1	–0.28[b]
tetrahydrofuran	7.4	–0.36	11.4	–0.53[b]
acetone	20.5	–0.57	32.5	–0.73[b]

[a][Little89]; [b][Foresman96]

These results are in excellent agreement with experiment. The anti conformer is lower in energy than the syn conformer in the gas phase and in solution with methylcyclohexane, the solvent with the lowest ε value. The reverse is true in the other solutions with ε>4. The calculations also reproduce the trend of increasing energy difference as the value of ε increases.

Exactly where the crossover point at which the syn conformation becomes favored over the anti form depends on the temperature used in the lower dielectric solvents. The experimental numbers in the preceding table are obtained from integrating low temperature (-85 °C=188K) NMR spectra. This was necessary in order to see the aldehyde protons as separate signals. The low temperature will slow down the syn-anti conversion, but should not alter the concentration ratios significantly. However, the dielectric constants of the various solvents will be different at this temperature (see corrected values given in the table).

Although Gaussian can perform frequency analysis at any temperature, the IEFPCM model does not account for temperature differences in the ε values and other parameters defining the solvent (everything has been parameterized for 298K). Therefore, in order to assess the theory in this case it is useful to view the trend via an appropriate mathematical function. The theoretical data fits well to a quadratic function of the Onsager (O) function:

$$O(\varepsilon)=\frac{\varepsilon-1}{2\varepsilon+1}$$

In the figure below, we use this function to overlay the experimental (points) and theoretical (line) results on a graph:

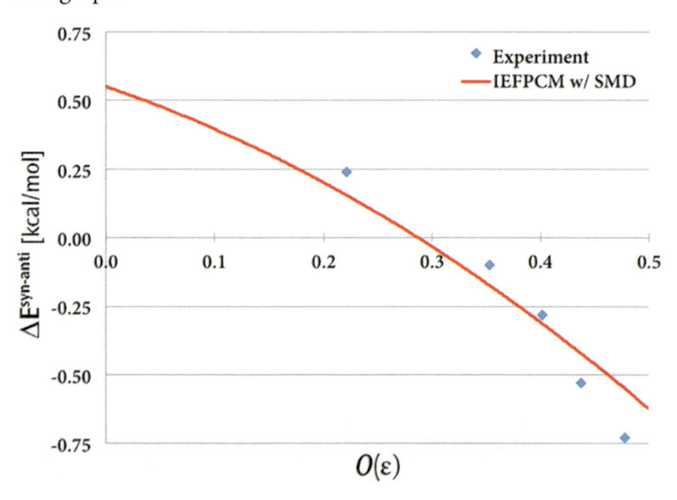

From the graph we can see that the SCRF model nicely predicts the observed trend. The syn conformer becomes more stable at a hypothetical $O(\varepsilon)$ value of about 0.29, corresponding to a dielectric constant of $\varepsilon \approx 3.8$.■

EXERCISE 5.3 METHYL LACTATE IN WATER

Repeat the study of methyl lactate from Example 5.1, p. 204, using water as the solvent. How do the results change in a different solvent environment?

SOLUTION

The following tables summarize the results of our calculations, along with the previous ones in the gas phase and in solution with methanol. As the following table indicates, the various conformers do not differ very much between the two solvent environments. In fact, the structure of ML1 optimized in methanol is also a minimum in water.

	KEY DIHEDRAL ANGLES (°)						$\Delta G^{\text{relative to ML1}}$ (kJ/mol)		
	GAS PHASE		METHANOL SOLUTION		WATER SOLUTION				
FORM	ϕ_1	ϕ_2	ϕ_1	ϕ_2	ϕ_1	ϕ_2	GAS PHASE	METHANOL	WATER
ML1	-6.7	6.1	-6.4	6.6	-6.5	6.6			
ML2a	-162.8	-37.0	-171.5	-46.5	-172.2	-47.2	8.87	6.3	6.4
ML2b	152.8	48.7	154.3	59.2	154.5	59.4	8.90	5.8	5.6
ML3	-20.2	169.8	-15.3	177.8	-15.3	-177.8	21.1	10.1	9.6
ML4	162.7	170.3	164.1	176.6	164.1	176.6	17.4	9.7	9.5

When we examine the predicted free energy values, we note that the energy ordering of the conformers is the same in water as it is in methanol. As in methanol, there is smaller range of energies for the series of conformers compared to the gas phase.

The next table reports the predicted energy differences among the conformers in the three environments, along with the Boltzmann analysis results:

	$\Delta G^{\text{gas-solution}}$ (kJ/mol (~kcal/mol))		BOLTZMANN %		
FORM	METHANOL SOLUTION	WATER SOLUTION	GAS PHASE	METHANOL SOLUTION	WATER SOLUTION
ML1	19.9 (~4.8)	20.5 (~4.9)	94.5	82.0	81.2
ML2a	22.5 (~5.3)	23.0 (~5.5)	2.7	6.7	6.3
ML2b	23.0 (~5.5)	23.8 (~5.7)	2.7	8.2	8.6
ML3	30.9 (~7.4)	32.0 (~7.7)	<0.1	1.4	1.7
ML4	27.6 (~6.6)	28.4 (~6.8)	<0.1	1.7	1.8

The percentages of the various conformations are essentially the same for the two solvent environments. The following table reports the Boltzmann-averaged values for the properties we considered earlier:

PROPERTY	GAS PHASE	METHANOL	WATER
Dipole moment (D)	3.0	3.7	3.8
C=O stretch: frequency (cm^{-1})	1794	1761	1760
C=O stretch: intensity (KM/mol)	245	457	466

The dipole moment in water is also essentially the same as in methanol, indicating that these compounds are polarized to the same degree in both solvent environments. The carbonyl stretch frequency shift and intensity increase are also similar in methanol and in water.■

EXERCISE 5.4 A MENSHUTKIN REACTION

Menshutkin reactions consist of the alkylation of tertiary amines with alkyl halides:

$$NR_1R_2R_3 + R_4{-}X \longleftrightarrow N^+R_1R_2R_3R_4 + X^-$$

The effects of solvation on such reactions were first described in 1890 by Menshutkin [Menshutkin1890, Menshutkin1890a], for whom the reaction class is named. Consider the reaction:

$$NH_3 + CH_3Cl \rightarrow NH_3CH_3^+ + Cl^-$$

(R_1=R_2=R_3=H, R_4=CH_3, and X=Cl). This reaction has been frequently studied [Gao91, Truong97, Amovilli98, Castejon99]. Predict the free energy difference between the reactants and products: the free energy change for the reaction, in both the gas phase and in aqueous solution. In addition, compute the predicted activation energy for the reaction in solution by locating the transition structure between the reactants and the products. The reaction proceeds through an S_N2 transition structure in solution. In the gas phase, the reaction is most likely a two-step process.

Use the SMD model to predict the free energy values for these computations. Compare your results to the observed values:

- $\Delta G^{reaction}$=110±5 kcal/mol (gas phase) [Gao91]
- $\Delta G^{reaction}$=34±10 kcal/mol (in water) [Gao91]
- $\Delta G^{activation}$=23.5 kcal/mol (in water) [Okamoto67] (performed with CH_3I)

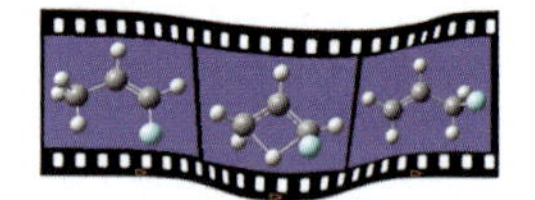

Building the TS Initial Structure

Here is one method for building the starting structure* for the transition state optimization (the illustrations use GaussView, but the same steps can be carried out in WebMo):

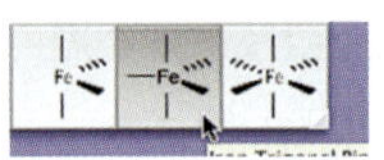

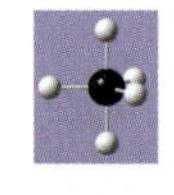

1. Select the **Fe Trigonal Bipyramid** atom.
2. Click within a new molecule window to place the atom.

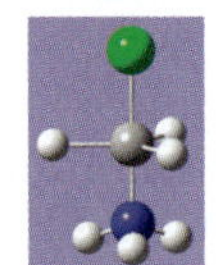

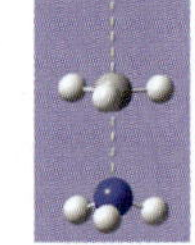

3. Change the axial hydrogens to nitrogen and chlorine and the iron to carbon. Add three hydrogens to the nitrogen atom.
4. Change both heavy atom bonds to the carbon to half bonds. Clean the structure.

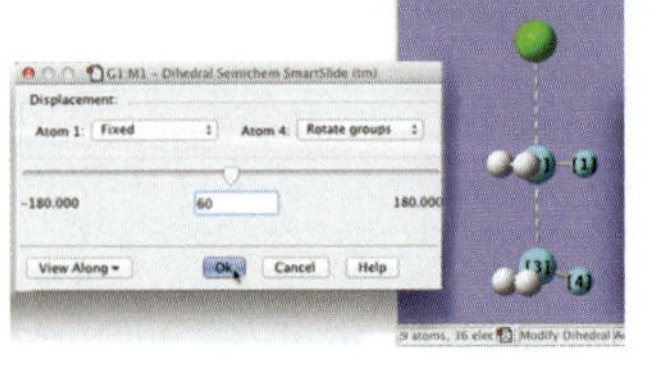

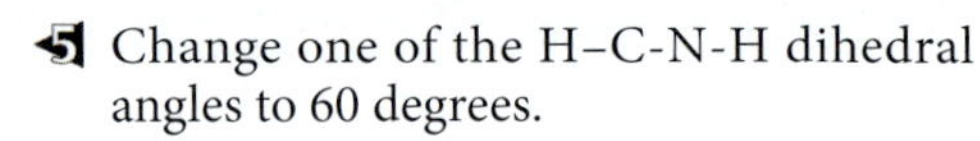

5. Change one of the H–C-N-H dihedral angles to 60 degrees.
6. Use the point group symmetry feature to impose C_{3v} symmetry on the structure. The completed molecule is at right.

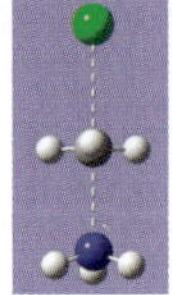

* To be used as the initial structure for an **Opt=(TS,CalcFC)** calculation or as the third structure for an **Opt=QST3** calculation.

SOLUTION

Our transition state optimization located the structure depicted in the margin. The Cl–C–N angle is linear, and the hydrogen atoms in the methyl group point slightly toward the nitrogen atom.

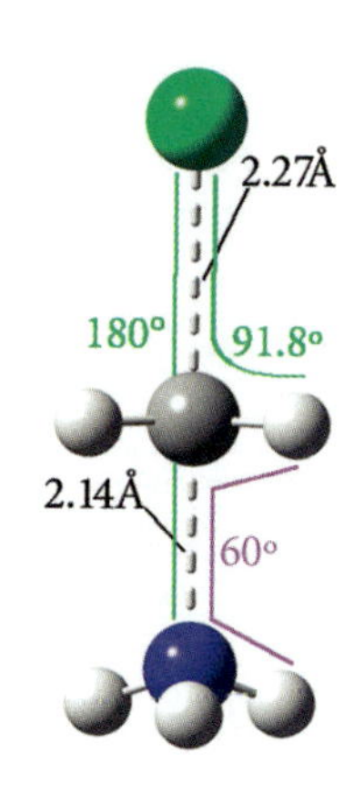

The following table lists the predicted values for the reaction energy ΔG and the forward and reverse activation barriers.

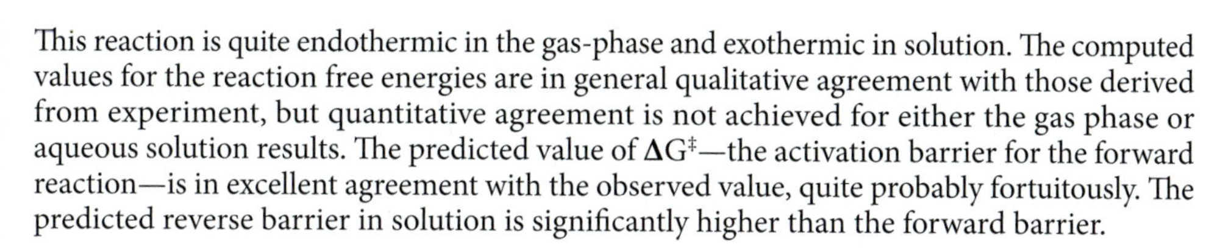

QUANTITY (all values in kcal/mol)	GAS PHASE	SMD	EXP.
ΔG	122.5		110±5
		-7.3	-34±10
Forward barrier		25.4	23.5
Reverse barrier		32.7	

This reaction is quite endothermic in the gas-phase and exothermic in solution. The computed values for the reaction free energies are in general qualitative agreement with those derived from experiment, but quantitative agreement is not achieved for either the gas phase or aqueous solution results. The predicted value of $\Delta G^{\ddagger}$—the activation barrier for the forward reaction—is in excellent agreement with the observed value, quite probably fortuitously. The predicted reverse barrier in solution is significantly higher than the forward barrier.

Such results are not unusual for reactions in aqueous solutions which often require both a continuum solvation method and the inclusion of one or more explicit water molecules to achieve high accuracy results. We will consider the use of explicit solvent molecules in the Advanced Topics section later in this chapter (p. 216).■

EXERCISE 5.5 COMPARING FREE ENERGIES OF SOLVATION

In this exercise, we will expand on the work in Example 5.4 (p. 208) by comparing the free energy of solvation for acetic acid in several solvents and also comparing the free energy of solvation for some other compounds in water.

Compute the free energy of solvation for the following compounds in the indicated solvents, following the procedure in Example 5.4. Compare your results to the experimental values* given in parentheses [Marenich09 Supp. 1; Marenich09a]:

- acetic acid in cyclohexane (-1.73 kcal/mol), in tributylphosphate (-7.11 kcal/mol) and in nitrobenzene (-4.78 kcal/mol);
- benzamide (-10.9 kcal/mol), benzene (-0.87 kcal/mol), hexafluoroethane (+3.94 kcal/mol), urea (-13.8 kcal/mol) and propene (+1.27 kcal/mol) in water.

Can you identify any trends in the results?

SOLUTION

We performed optimization+frequency calculations for acetic acid in the two solvents and for the other compounds in the gas phase and in water.

The following table lists the results of this study:

* The values reported in the references include the concentration correction to 1 M.

		$\Delta G^{solvation}$ (kcal/mol)	
MOLECULE (μ^{gas} (debye))	**SOLVENT (ε)**	**SMD**	**EXP.**
acetic acid	cyclohexane (2.0)	-3.10	-1.73
acetic acid	chloroform (4.7)	-5.03	-4.74
acetic acid	tributylphosphate (8.2)	-7.20	-7.11
acetic acid	nitrobenzene (34.8)	-6.08	-4.78
acetic acid	water (78.4)	-6.68	-6.70
urea (3.8)	water (78.4)	-14.44	-13.80
benzamide (3.7)		-10.56	-10.90
acetic acid (2.2)		-6.68	-6.70
benzene (0.0)		-1.32	-0.87
propene (0.4)		0.74	1.27
hexafluoroethane (0.0)		4.01	3.94

For acetic acid, $\Delta G^{solvation}$ is generally higher in the more polar solvents, although nitrobenzene is an exception. Compounds with larger gas phase dipole moments have more negative free energies of solvation in water than ones with small or no dipole moment.

The SMD results are in very good agreement with experiment. The results for acetic acid in the various solvents differ from experiment by ~1-2 kcal/mol, and their ordering by solvent is correct. Note that the medium to high ε-value solvent environments yield essentially identical results for acetic acid. The ordering of the relative solubilities for the series of compounds in water is also reproduced by the solvation model, and the predicted values for ΔG are in close agreement with experiment.■

Advanced Topics & Exercises

Including Explicit Solvent Molecules

In this chapter, we have explored a model chemistry which is useful for predicting energies and properties of neutral and charged molecules that are in solution. The Gaussian program makes it very easy to add a solvent to a calculation via the **SCRF** keyword, making the process very similar to performing a gas phase computation on the same solute. However, this simplicity involves several assumptions that you may need to investigate further depending on the complexity of your research problem.

The first such assumption involves the interaction between the solute and the solvent:

> ASSUMPTION 1A: *The use of a cavity model assumes that the solute's electron density is not dramatically altered by specific interactions* with individual solvent molecules.*

It is possible that the environment you are modeling contains intermolecular interactions, either between the solvent and the solute or among two or more solute molecules. While some of these effects (e.g.: hydrogen bonding in water) are included via the empirical parameters present in the SCRF models, in general a continuum model alone is insufficient for treating them and therefore may not produce quantitatively correct results for your system.

A related assumption involves reaction participation by individual solvent molecules:

> ASSUMPTION 1B: *The mechanism/reaction being modeled in solution does not itself involve specific solvent molecules.*

* Technically referred to as "first solvation shell effects" or "non-bulk electrostatic effects."

For instance, in the case of modeling a chemical reaction, it may be true that solvent molecules play a part in the actual mechanism, in which case the continuum model alone will not yield correct information about the process.

The limitations resulting from these assumptions can be addressed by including explicit solvent molecules and/or multiple solute molecules inside the cavity. We will demonstrate how to do this in the advanced exercises which follow.

The next assumption relates to how the solvent molecules behave over the course of a reaction:

> ASSUMPTION 2: *The property or process being modeled in solution requires only the static information from an equilibrium representation of the solvent molecules.*

In this chapter, we have seen considerable evidence that many reactions in solution can be studied successfully with SCRF-based solvation models. Nevertheless, there exist chemical phenomena that depend on how the solvent reorganizes during the time period that it is being observed in the experiment; temperature may also play a critical role in solvent organization. In other words, the solvent environment is in fact dynamic, changing on a time scale that is comparable or shorter than that of the experiment. In these cases, a continuum model will not correctly represent the property or process you are studying because it assumes a single, specific solvent configuration for a given molecular configuration of the solute under specific physical conditions. In other words, while a continuum method can model, say, the various steps along a reaction path, it treats each point as a specific molecular and solvent configuration. It cannot treat the myriad movements of solvent molecules on the scale of nano- or picoseconds.

Such problems will require techniques such as molecular dynamics in order to properly model the solvent effects. Molecular dynamics is a topic beyond the scope of this book. However, we will examine one type of solvent reorganization as part of modeling excited states in chapter 8 (see p. 371).

The final assumption concerns the method used for predicting the Gibbs free energy of a molecular system in solution:

> ASSUMPTION 3: *The thermal contributions to the Gibbs free energy of the solvated system can be approximated using the same statistical formulas used for an ideal gas rotating as a rigid rotor and vibrating as a harmonic oscillator.*

The value of G, as reported in the output file from a frequency calculation, uses a formula which assumes that the molecule is in the gas phase. However, translational and rotational motion in the solution may be very different from the same solute acting as an ideal gas. When looking at differences in free energies, the translational and rotational contributions largely cancel, and so this assumption is benign.* We will discuss this assumption in more detail in the closing section of this chapter.

ADVANCED EXAMPLE 5.6 METHYL ACETATE HYDROLYSIS WITH EXPLICIT WATERS

We will add explicit water molecules to the constituents of our previous study of the base catalyzed hydrolysis of methyl acetate (see Example 5.5, p. 210) in an effort to improve the results. The natural question that follows this choice is where exactly the water molecules should be located.

* One might also argue that these effects are accounted for to some extent in the parameterization of the SCRF solvents and/or the SMD model.

The initial placement of explicit water molecules to form a "supermolecule" inside of a continuum model cavity can be guided by chemical intuition (based on their role within the specific reaction), knowledge from experiment, the results of molecular dynamics simulations or some combination of these. Preliminary electronic structure calculations can also aid in making this decision.

In the case of the methyl acetate hydrolysis reaction, it might be reasonable to assume one or more water molecules will hydrogen bond to the acetate component. We will add two water molecules per oxygen atom. We can place them in any reasonable orientation, at a distance of ~1.4 angstroms away, and then optimize their positions as we would any other geometric parameter.

In GaussView, you can make a half bond connection between the hydrogen and oxygen atoms; when the input is created, information will be placed in the connectivity matrix which helps the optimizer identify that there is a potential hydrogen bond present.

The following figure shows the initial structures for the reactants and products with two water molecules per oxygen atom.

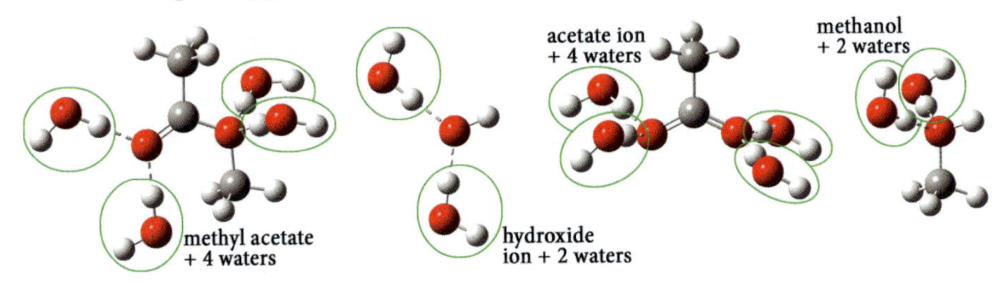

We attempted to locate minima for each of these complexes, beginning in the gas phase. However, all four of our geometry optimizations exceeded the maximum number of steps before locating a minimum*. When we examined the optimization results in detail, we found that one of the water molecules had changed position drastically in two of the complexes:

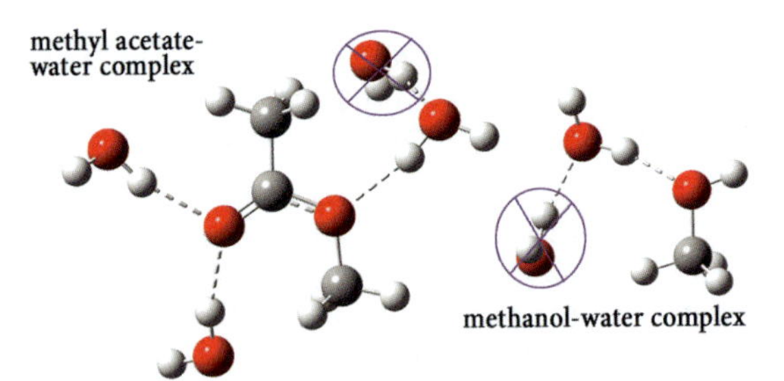

The circled water molecules are hydrogen bonded to the other water molecule rather than the carbonyl oxygen atom in the methyl acetate and methanol complexes. We eliminated these water molecules and then restarted the optimizations for both complexes adding the **Opt=CalcFC** option (computing the force constants at the first point to aid the optimizer). We also resumed the optimizations for the other two complexes from their final points, adding the **Opt=CalcAll** option, which directs the optimizer to compute the force constants at every point during the optimization (a strategy which is computationally expensive but necessary in this case).

* Of course, better practice is to monitor optimizations and stop them if it becomes apparent that they are not making progress.

The second series of optimizations all completed successfully. We then optimized these gas phase minima in solution with water via **SCRF=SMD**, again including **Opt=CalcFC**. We also imposed more stringent optimization criteria with **Opt=Tight**. The latter calculations also completed successfully.

Here are the results of using five explicit water molecules in the computation of the free energy of reaction for the hydrolysis of methyl acetate, along with those from Example 5.5 and experiment [Guthrie73] for comparison. The reactant energy used is the sum of the energies of methyl acetate complexed with three water molecules and the hydroxide-two water complex. Similarly, the product energy is the sum of the energies of the acetate ion-four water complex and the methanol-water complex.

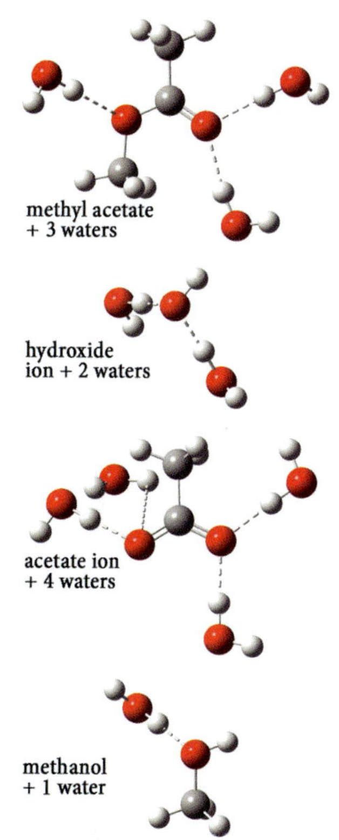

	ΔG (kJ/mol)	
MOLECULES	**GAS PHASE**	**SMD**
isolated species	-182.6	-83.1
+5 water molecules	-145.1	-61.2
experiment	-60.2	

Simply adding explicit water molecules to the gas phase calculation significantly lowers the predicted energy difference, but the calculated ΔG value is still over twice as exothermic as it should be. Using SMD alone, without explicit water molecules, recovers 81% of the solvent effect. Adding five explicit water molecules within the SMD cavity model produces the most accurate prediction: within 2% of the experiment in this case.

Examination of the optimized structures shows that the acetate complex in SMD is somewhat different than the gas phase complex (see the margin for the solution phase minima). One of the four water molecules has moved and is binding to the oxygen of the other water molecule. To test the significance of the placement of these water molecules, we used the gas phase optimized complexes and performed single point SMD calculations, taking the thermal free energy contributions from the gas phase. That model gives a result of -57.5 kJ/mol for the reaction free energy change. Therefore, we conclude that the absolute position of the water molecules is not all that significant. They are free to move within the cavity. However, their presence in the cavity leads to a better result.

We will continue to study this hydrolysis process in the following exercise.■

ADVANCED EXERCISE 5.6 HYDROLYSIS REACTANT AND PRODUCT COMPLEXES

In Chapter 6 we will discuss how to study the mechanism of reactions, including the hydrolysis of methyl acetate (see p. 265). In preparation for this, create reactant and product complexes for this reaction which include five explicit water molecules. You can use the optimized structures located in Example 5.6 and combine the two reactant complexes into one "super" molecule, and then do the same for the products. After combining the structures, you may need to reposition one or more water molecules. Alternatively, you can place the two optimized reactant isolated species in proximity and then add five water molecules to the combined complex.

Optimize the reactant and product complexes in the gas phase and in aqueous solution, using the SMD model for the latter. Predict the reaction free energy in each case.

SOLUTION

Here are the initial structures we used for the gas phase optimizations (the reactant complex is on the left). Note that we have reduced the size of the water molecules for clarity.

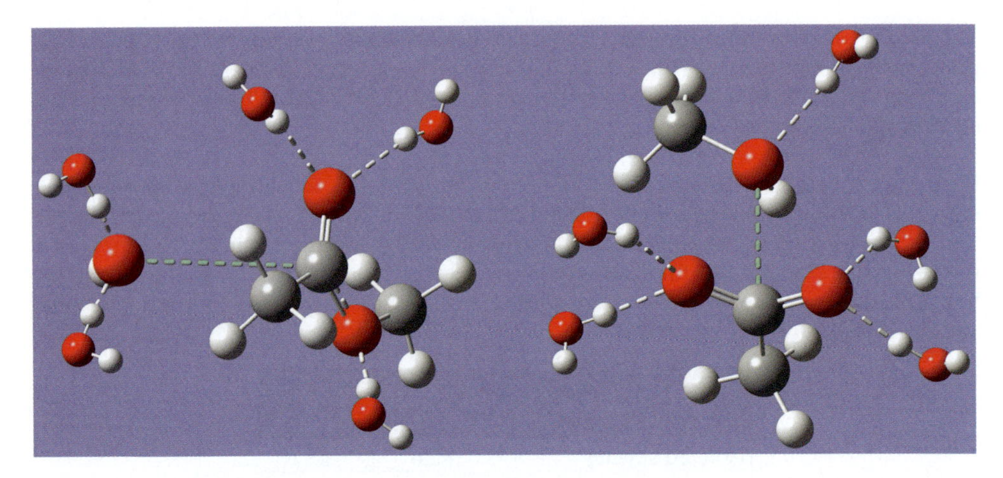

For both complexes, we placed the ion's oxygen atom opposite the central carbon atom in the larger compound. The dotted green lines in the figure highlight these placements. We added two water molecules to each oxygen atom bonded to only a single other atom, and we added one water molecule to the remaining oxygen atom.

For the reactant complex, we specified the **Opt=(CalcFC,Tight)** options, which compute the force constants at the first optimization point and specify stricter convergence criteria (respectively). The optimization completed successfully in 115 steps. For the product complex, we first optimized the structure with normal convergence criteria and then ran a second optimization with tight criteria (and **Opt=CalcFC**). Both optimizations converged, in 107 and 14 steps (respectively).

These are the final minimized structures in the gas phase:

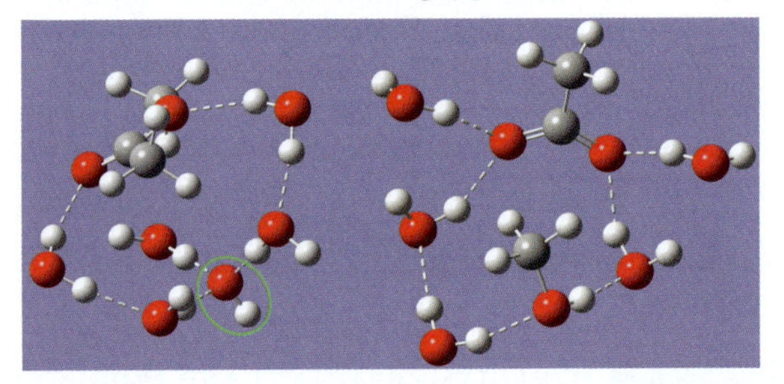

In the reactant complex, most of the water molecules are situated near the hydroxide ion (circled), and only two of them remain hydrogen bonded to the methyl acetate molecule. This is an expected result given the nature of hydroxide. In the product complex, two of the water molecules hydrogen bonded to the acetate ion are similarly linked to methanol or another water molecule hydrogen bonded to methanol. Thus, in both cases, we have located a plausible complex for the collection of molecules.

We then optimized the gas phase minima using the **SCRF=SMD** method, specifying aqueous solution, tight optimization criteria and computing the force constants at the initial point. These optimizations in solution both completed successfully, requiring 125 steps for the reactant complex and 100 steps for the product complex.

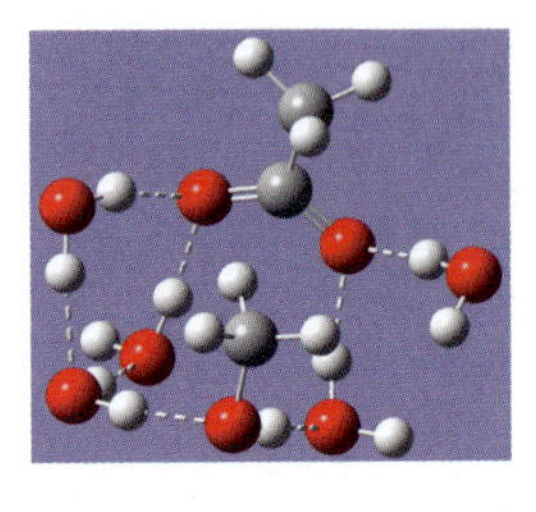

The optimized reactant complex structure is similar to the gas phase structure. In contrast, in the product complex in solution, the methanol molecule has a different orientation with respect to acetate ion than in the gas phase structure. As the structure in the margin illustrates, the O–C bond in methanol is more-or-less aligned with the C–C bond in acetate ion. The hydrogen atoms in the methyl group in the acetate ion have also rotated.

The following table summarizes the energetic results for these calculations as well as those from our previous considerations of these systems:

	ΔG (kJ/mol)	
MOLECULES	GAS PHASE	SMD
isolated species	-182.6	-83.1
+5 water molecules	-145.1	-61.2
reactant/product complexes	-70.3	-53.7
experiment	-60.2	

Both treatments of the complexed reactants and products lead to a fair prediction of the reaction free energy change. The predicted value of ΔG is lowered in both cases, dramatically so in the gas phase. For the SMD model, this results in a poorer prediction that the previous effort.

We will continue our consideration of this reaction in Advanced Exercise 6.7 (p. 265).■

Handling "Bend failed" Errors

In a preliminary calculation we ran while developing the preceding exercise, we encountered an optimization where the reactant complex stopped after 18 steps because of the presence of a linear angle within the structure (illustrated in the margin). This is not an uncommon situation, so we describe how to handle such cases briefly here.

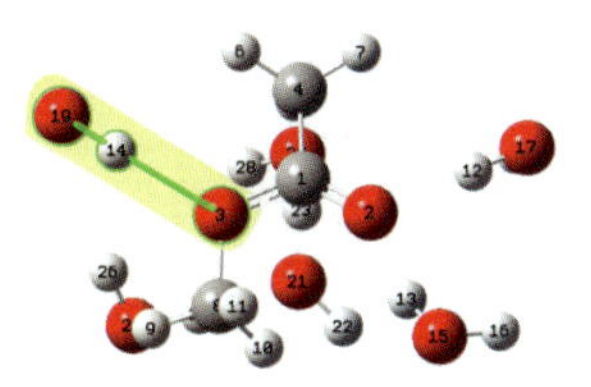

When such an error occurs, the error messages from Gaussian are similar to the following:

```
Bend failed for angle      3 -     14 -     19
Tors failed for dihedral      1 -      3 -     14 -     19
Tors failed for dihedral      8 -      3 -     14 -     19
Tors failed for dihedral      3 -     14 -     19 -     20
FormBX had a problem.
Error termination via Lnk1e ...
```

The first message identifies the atoms comprising the linear angle. The additional messages beginning with "Tors" refer to dihedral angles involving the same three atoms.

At this point, we have two choices. We can restart the optimization from the beginning using the **Opt=ModRedundant** option to remove the angle that will eventually become linear from the set of redundant internal coordinates (the associated dihedral coordinates will be removed authmatically). Alternatively, we can simply restart the optimization from the final structure of the failed job. When Gaussian begins the new optimization, it will regenerate the redundant internal coordinates and handle the linear angle that is present automatically.

Explicit Solvent Molecules in Other Solvent Environments

Water is not the only solvent which may be modeled with the SCRF method in combination with explicit solvent molecules. See Exercise 7.11 (p. 313) for an example in which a VCD spectrum is modeled in the presence of explicit chloroform molecules.

For a detailed discussion of calculating thermochemical quantities in Gaussian, see J.W. Ochterski, "Thermochemistry in Gaussian," gaussian.com/g_whitepap/thermo.htm.

More About Free Energies in Solution

As we have seen, the Gibbs free energy is one of the thermodynamic quantities computed by a frequency calculation. This quantity is defined by the following equation:

$$G = H - TS$$

where H is the enthalpy, T is the temperature and S is the entropy.

Both H and S have components arising from the molecule's electronic, translational, rotational and vibrational motion:

$$H = E_{total} + (E_{elec} + E_{trans} + E_{rot} + E_{vib}) + KT$$
$$S = S_{elec} + S_{trans} + S_{rot} + S_{vib}$$

E_{total} is the uncorrected total energy reported prior to the thermochemistry data within the Gaussian output, and K is Boltzmann's constant. The sum $E_{elec}+E_{trans}+E_{rot}+E_{vib}$ is the thermal correction to the total energy we discussed early in this book.

Thus, the reported value for G includes contributions from all of the molecule's vibrational, translational and rotational modes, in both the energy and entropy components. Each component is reported separately within the Gaussian output. Here is part of the thermochemistry section of the output for a frequency calculation on methanol:

Corrections to E, H and G arising from the molecule's electronic, translational, rotational and vibrational motion:

```
Thermal correction to Energy=                 0.054356
Thermal correction to Enthalpy=               0.055300
Thermal correction to Gibbs Free Energy=      0.028178
```

Preceding corrections added to the total energy (appears earlier in the output), H and G:

```
Sum of electronic and zero-point Energies=    -115.71655
Sum of electronic and thermal Energies=       -115.71318
Sum of electronic and thermal Enthalpies=     -115.71224
Sum of electronic and thermal Free Energies=  -115.73934
```

Breakdown of the individual components to the thermal energy, heat capacity & entropy:

```
                  E (Thermal)         CV              S
                   KCal/Mol    Cal/Mol-Kelvin  Cal/Mol-Kelvin
Total               34.160          8.865          57.051
Electronic           0.000          0.000           0.000
Translational        0.889          2.981          36.324
Rotational           0.889          2.981          18.994
Vibrational         32.382          2.904           1.733
```

Each vibrational component is then further decomposed into contributions from the individual vibrational modes for vibration temperature < 900 K.

Note that the energy corrections and energy sums are reported in atomic units (hartrees), while the units for the components of E and S are kcal/mol and cal/mol-K (respectively).

These computations rely on some significant assumptions. In the context of gas phase calculations, Ochterski describes them as follows:

> One of the most important approximations to be aware of throughout this analysis is that all the equations assume non-interacting particles and therefore apply only to an ideal gas. This limitation will introduce some error, depending on the extent that any system being studied is non-ideal. Further, for the electronic contributions, it is assumed that the first and higher excited states are entirely inaccessible. This approximation is generally not troublesome, but can introduce some error for systems with low lying electronic excited states. [Ochterski00]

Most relevant to our discussion here is the fact that the harmonic approximation and other approximations used in frequency calculations treat the translational and rotational modes in ways that are only appropriate for an ideal gas at STP. The approximation is cruder for these modes for molecules in solution compared to the gas phase due to the fact that molecular motion is much more constrained in solution.

In principle, it might be possible to remove some of the subcomponents of the thermal energy correction and/or the entropy on the grounds that these motions are more-or-less inhibited in solution.* However, it is far from clear that doing so is desirable or even necessary. Consider Jorgensen's comments on computing the entropy for bimolecular reactions in solution:

> There is fundamental uncertainty in comparing activation entropies for bimolecular reactions in the gas phase and solution. For a 1 M solution, there is restricted translational motion owing to the presence of the solvent molecules compared to the 1 M gas. One way of estimating the effect is to consider that the "free volume" in a liquid is about 1% of the molar volume. This leads to a roughly R ln 100 = 9 cal/mol·K more negative activation entropy for a bimolecular reaction in the gas phase than in solution owing to the -1 change in the number of molecules between the reactants and transition state.... However, this difference is not clearly apparent in comparing the limited experimental data for reactions that have been studied both in the gas phase and in solution. A definite decision on the magnitude of the corrections cannot be made in the absence of uniformly high- accuracy determinations of activation entropies for several ... reactions in the gas phase and different solvents. [Jorgensen93]

Jorgensen underscores the points that both the method and the necessity for solution-phase corrections to thermochemical quantities are poorly understood. This remains true as of this writing.

To be precise, retaining and removing the various rotational and translational components from the computation of thermochemical quantities both involve an assumption. Removing a component assumes that the corresponding motion is completely negligible in the solution phase. Retaining a component assumes that the difference in the corresponding motion between the gas phase and solution is negligible. Both assumptions can easily be invalid for any particular molecular system.

The Truhlar group's SMD method takes these issues into account via its parametrization from corrected experimental values.† Ribeiro, Truhlar and coworkers also state that errors arising from the use of gas phase thermochemical calculations with a solution model "are expected to be smaller than the mean error intrinsic to the continuum solvation model" [Ribeiro11], meaning that they are a secondary factor considered in the context of the inherent accuracy level of the SCRF model.

In light of all of these factors, at the present moment the best approach seems to be to use the unmodified thermochemistry data reported by Gaussian from frequency calculations at the geometry optimized in solution using whichever SCRF method is appropriate to your problem. When this procedure cannot deliver the required accuracy, then other approaches such as molecular dynamics may be in order.

* One might also consider a correction to the value of G reflecting change in reference from 1 atm (gas phase) to 1 mol/L (solution phase). The latter would amount to about -1.89 kcal/mol and is representative of the relatively small magnitudes of such potential modifications.

† The SMD method accounts for the change of reference noted above.

ADVANCED EXAMPLE 5.7 **THE COMPONENTS OF FREE ENERGIES IN SOLUTION**

In this final example, we will examine the individual contributions to the free energy in detail. We will do so by considering a preliminary question about the mechanism for the methyl acetate hydrolysis reaction we considered earlier.

The base-catalyzed hydrolysis of methyl acetate going to acetate ion and methanol can proceed either via an S_N2 mechanism involving a complex of the reactants or via a $B_{AC}2$ mechanism with a tetrahedral intermediate. We will consider this question in detail in chapter 6. For now, we will compare the components of the free energies of two potential transition structures corresponding to these mechanisms:

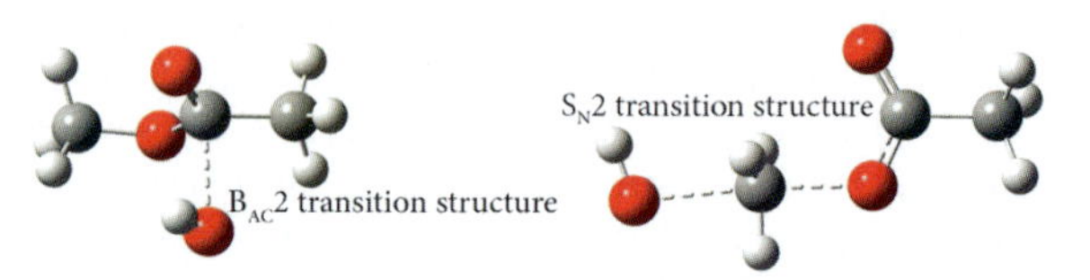

These transition structures were located in solution without the addition of any explicit water molecules. The $B_{AC}2$ mechanism involves the attack of the hydroxide on the alpha carbon atom, while the S_N2 mechanism involves the attack of the hydroxide to the terminal methoxy group. For the S_N2 TS, there are three positions from which the OH could approach: HCOH = -60°, +60° and 180°. These are all essentially equivalent, so we selected the latter to optimize.

The following table presents the various contributions to the free energy (at STP) for the two transition structures, along with the differences in value between the two structures:

CONTRIBUTION TO $G^{solvation}$	$B_{AC}2$ TS (hartrees)	S_N2 TS (hartrees)	DIFFERENCE (kJ/mol)
SCRF energy: E^{SCRF}	-344.43636	-344.41814	-47.84
E_{trans}: translational energy (1.5RT)	0.00142	0.00142	0.00
E_{rot}: rotational energy (1.5RT)	0.00142	0.00142	0.00
E_{vib}: vibrational energy	0.10476	0.10394	2.15
translational entropy term: $-TS_{trans}$	-0.01874	-0.01874	0.00
rotational entropy term: $-TS_{rot}$	-0.01270	-0.01304	0.89
vibrational entropy term: $-TS_{vib}$	-0.00861	-0.01262	10.53
degeneracy contribution	-0.00065	-0.00104	1.02
correction to constant pressure: RT	0.00094	0.00094	0.00
G^{SCRF}	-344.36852	-344.35586	-33.24

Most of the data presented in this table is taken or derived from the values in the Gaussian output file. The entropy terms multiply the output values by -T (-298.15) and convert the units to hartrees; the energy terms are simply converted to atomic units. The degeneracy contribution relates to the degrees of freedom (ω*) of the molecule and is computed as: ln(ω)RT. Finally, the correction to constant pressure is simply RT.

These results suggest that the S_N2 mechanism is not significant compared to the $B_{AC}2$ mechanism since its transition structure lies well above the other in energy. Examination of

* ω is 2 for the $B_{AC}2$ transition structure and 3 for the S_N2 transition structure. Note that the value of the free energy reported in the Gaussian output does not include this term.

the various components of the free energy reveals that it is the total energy from the SMD model that is most important in determining this preference.

The other term contributing significantly to the magnitude of the free energy difference is the vibrational entropy change. A few other components differ slightly between the two transition structures. For example, the degeneracy contribution of the S_N2 transition structure is slightly more negative (-ln(3)RT) since there are now three independent routes from reactants to products. The translational entropy term is also much smaller for the S_N2 transition structure.

It is not always the case that the total energy term dominates the free energy change. The following table lists the components of the free energy with respect to the reactants:

CONTRIBUTION TO $G^{solvation}$ (kJ/mol)	$B_{AC}2$ TS	S_N2 TS
SCRF energy change	39.20	87.04
translational energy change	-3.72	-3.72
rotational energy change	-2.47	-2.47
vibrational energy change	9.75	7.60
translational entropy change	42.17	42.17
rotational entropy change	6.61	5.72
vibrational entropy change	-4.23	-14.75
degeneracy change	-1.72	-2.74
RT term change	0.00	0.00
ΔG relative to reactants	75.23	108.47

Viewed in this manner, it is the loss of translational entropy arising from two molecules going to a single-molecule transition structure that contributes the largest component of $\Delta G^{reactants}$. The transformation from two molecules to a single molecule results in a positive change in the free energy of about 42 kJ/mol for the transition states with respect to the reactants. Such a change in free energy is present in any bimolecular reaction that forms a single molecular product. Neglecting it would clearly lead to significant errors in the results.■

6

Studying Reaction Mechanisms

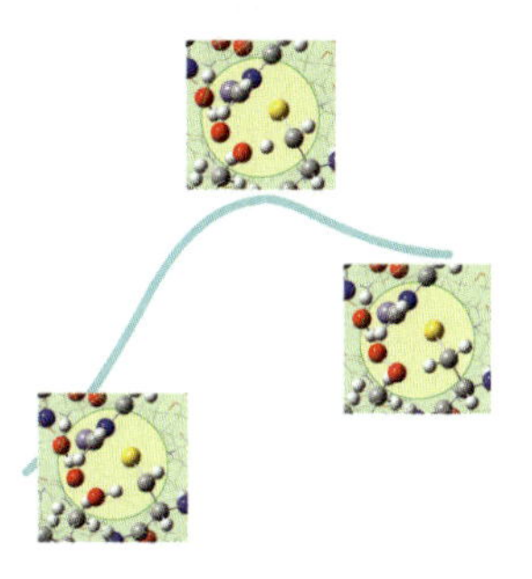

In This Chapter:

Molecular Orbitals Revisited

Electrostatic Potentials

Potential Energy Surface Scans

Following Reaction Paths

Advanced Topics:

Two Dimensional Scans

S_N2 Reactions

6 ◆ Studying Reaction Mechanisms

In this chapter, we consider techniques for studying chemical reactions. The calculation types are the same as those for predicting the properties of specific compounds. However, there are some additional considerations when modeling full reactions.

We considered enthalpies of reaction and activation barriers in the latter part of the previous chapter on modeling chemistry and solution. We will see similar problems in this chapter. We begin by considering visualization techniques which can give quick, general insight into reactivity. We then go on to consider a pair of new calculation types: potential energy surface scans and intrinsic reaction coordinate (IRC) calculations. The latter can be used to model the energetics and relationships between reactants, products and the transition structures that connect them.

Visualizing Surfaces to Understand Reactivity

Examining graphical representations of molecular properties can be a very useful first step in studying the properties of compounds. Relevant properties are displayed as three-dimensional surfaces by visualization software, allowing numerical data to be examined in ways that are more straightforward and intuitive. Such properties include molecular orbitals, the electron density and the electrostatic potential. In this section, we consider examples where visualizing such surfaces is of great help in understanding the chemistry involved in some process or reaction, as well as one case where visualization is inadequate by itself and additional, more in-depth analysis is required.

EXAMPLE 6.1 DIELS-ALDER REGIOSELECTIVITY

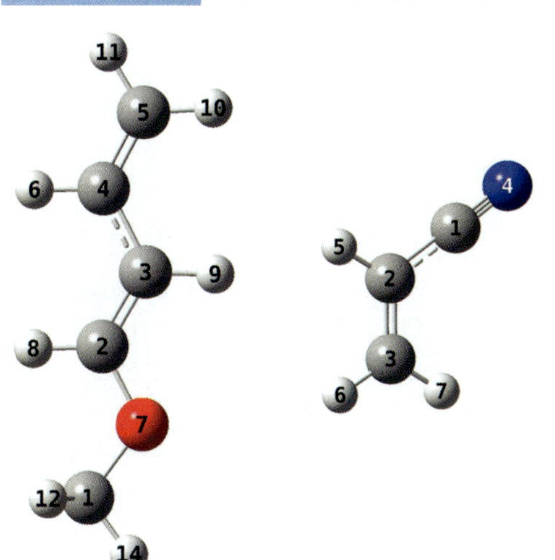

Diels-Alder reactions [Diels28] consist of a diene and a dienophile combining to form a six-membered ring. In the process, three π bonds in the reactants break, and the product structure contains two C-C σ bonds and one C-C π bond, with these changes occurring in a concerted process.

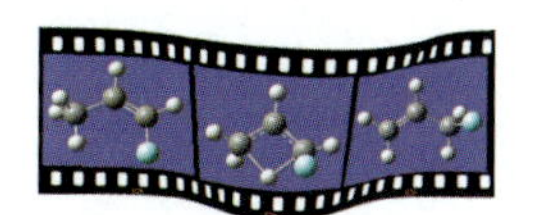

Predicting the products for these reactions is a common task for students learning organic chemistry, and examining molecular orbitals and electron densities can be of use.

Consider the reaction between 1-methoxy-1,3-butadiene and acrylonitrile.* We include the atom labels in the illustration because we will need to know the numbering for the carbon atoms later.

This reaction and related Diels-Alder reactions are well studied in the literature. For an experimental study, see [Sustmann96].

We ran geometry optimization+frequency calculations for both substances to locate and verify the minima; we included the **Population=Full** option on the job so that the individual contributions from each atom to the molecular orbitals—the molecular orbital coefficient—will be included in the output.

* IUPAC: acrylonitrile: 2-propenenitrile.

Orbital energies can be viewed in GaussView using the **MO Editor** and in WebMO on the **MO Viewer** panel of the **View Job** page.

We then examine the molecular orbitals for both reactants and estimate the energy difference between the two possible HOMO-LUMO combinations in order to determine which molecule is the nucleophile and which is the electrophile from the reported orbital energies:

QUANTITY	VALUE	MOS
$HOMO_{diene}$–$LUMO_{dienophile}$	-0.15	23, 15
$HOMO_{dienophile}$–$LUMO_{diene}$	-0.29	14, 24

Because the HOMO of the diene and the LUMO of the dienophile are closer in energy than the other combination, we can conclude that these orbitals will interact to form the product. These reacting orbitals are illustrated in the margin.

In order to determine the orientations of the two molecules, we next examine the coefficients of the relevant MO perpendicular to the plane of the molecule (the Z direction in our case) for the various carbon atoms that might potentially interact: atoms 2 and 5 in 1-methoxy-1,3-butadiene and atoms 2 and 3 in acrylonitrile. Reactions occur such that the largest orbital components react with one another, providing the best overlap during the reaction and lowering the energy of the transition state.

We can locate them by examining the output file section headed Molecular Orbital Coefficients and locating the desired MO and atoms. For example, the part of the output corresponding to carbon atoms 2 and 3 in the LUMO of acrylonitrile (MO 15) is listed below, with only the coefficients with significant (absolute) magnitude are listed beyond the second shell for carbon atom 2:

```
Molecular Orbital Coefficients:
...
                          11         12         13         14         15
                           O          O          O          O          V
     Eigenvalues --   -0.40038   -0.38236   -0.35606   -0.30902   -0.06797
 28 2   C   1S         0.00000   -0.00556    0.00041    0.00000    0.00000
 29         2S         0.00000   -0.00903    0.00051    0.00000    0.00000
 30         2PX        0.00000    0.05454   -0.01269    0.00001    0.00000
 31         2PY        0.00000   -0.02919   -0.05609    0.00000    0.00000
 32         2PZ        0.12430    0.00000    0.00000    0.13959   -0.14384
 36         3PZ        0.20015    0.00001    0.00000    0.21808   -0.22064
 40         4PZ        0.13657    0.00000    0.00000    0.17337   -0.32637
 44         5PZ       -0.00856    0.00000    0.00000    0.03647   -0.31035
...
 55 3   C   1S         0.00000    0.00190   -0.01109    0.00000    0.00000
 59         2PZ        0.07700    0.00000    0.00000    0.15656    0.17890
 63         3PZ        0.12402   -0.00001    0.00001    0.24728    0.27760
 67         4PZ        0.09021    0.00000    0.00000    0.21353    0.41277
 71         5PZ        0.00518    0.00000    0.00000    0.02107    0.35165
```

Atom 3 has the larger contribution to the LUMO.

Here is the relevant corresponding output for 1-methoxy-1,3-butadiene:

```
 Molecular Orbital Coefficients:
                          21         22         23         24         25
                          O          O          O          V          V
     Eigenvalues --   -0.35533   -0.31155   -0.21514   -0.01851    0.01281
  28 2   C  1S        -0.00915    0.00000    0.00000    0.00000    0.01117
  32        2PZ        0.00008    0.00579   -0.12371    0.16473    0.00001
  36        3PZ        0.00013    0.01030   -0.19438    0.25562    0.00001
  40        4PZ        0.00009    0.01904   -0.19782    0.37951    0.00006
 109 5   C  1S        -0.00431    0.00000    0.00000    0.00000    0.00183
 113        2PZ        0.00004    0.11496    0.14176    0.14096    0.00001
 117        3PZ        0.00006    0.18286    0.21970    0.21831    0.00002
 121        4PZ        0.00004    0.14951    0.21982    0.31904    0.00004
```

Carbon atom 5 makes the larger contribution to the HOMO (MO 23). Thus, we can conclude that the two carbon atoms we have identified will interact in forming the cycloaddition product, and the orientation of the product becomes clear.

The electrostatic potential provides another way of considering these effects. This quantity is a measure of the potential energy of a proton (or other positive charge) near a molecule, and its value differs for different regions within the molecule. A negative electrostatic potential corresponds to high electron density and the ability to attract a proton. Similarly, a positive electrostatic potential corresponds to low electron density and shielding of the nuclear charge, repulsing a proton.

The electrostatic potential is typically visualized as a mapped surface. An electron density isosurface is colored according to the value of the electrostatic potential at each point on it. Regions of large positive and negative electrostatic potential conventionally appear red and blue (respectively) within the resultant display.

In GaussView, you can generate the electrostatic potential mapped surface by creating an electron density cube: use the **Results→Surfaces** menu path and then select **New Cube** from the **Cube Actions** popup in the resultant dialog, specifying the **Type** as **Total Density**. Once the cube has been generated, select it from the list and then select **New Mapped Surface** from the **Surface Actions** popup. Finally, set the **Type** to **ESP**, and select **Generate values only at surface points**.

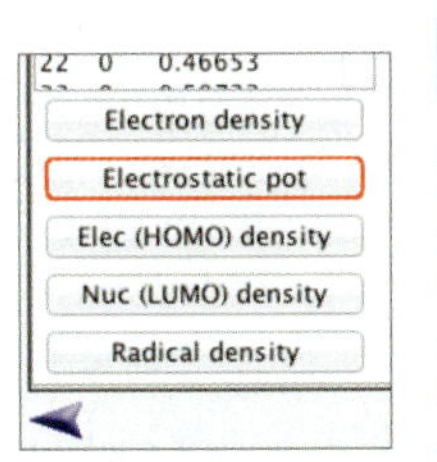

In WebMO, you must first run a job to generate the molecular orbitals: set **Calculation** to **Molecular Orbitals** in the **Job Options** panel of the **Configure Gaussian Job Options** page. Once the job has completed, go to its **View Job** page and click on the magnifying glass icon next to any molecular orbitals in the **Calculated Quantities** section, which will populate the **MO Viewer** panel. The mapped electrostatic potential can then be viewed using the **Electrostatic pot** button.

The following illustration shows the mapped electrostatic potential for the two reactants we are considering:

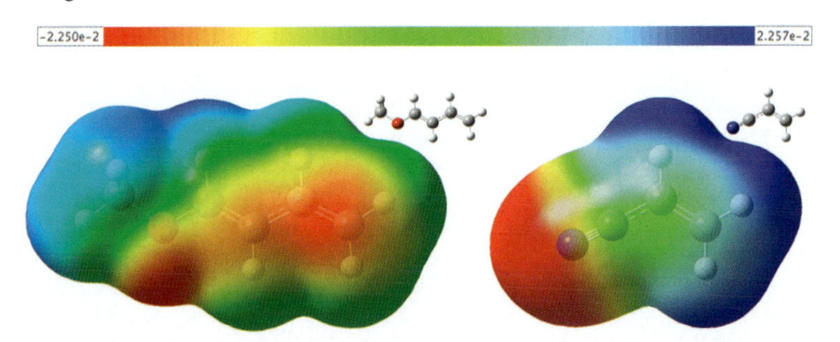

The electrostatic potential has a large negative value in the region of carbon atom 3 in acrylonitrile, and the region around carbon atom 5 in 1-methoxy-1,3-butadiene has a positive electrostatic potential. This visualization allows us to easily identify the bonding orientation for the two molecules.

GOING DEEPER: MOLECULAR ORBITALS

Although we ran these calculations using our standard model, much lower accuracy calculations are often adequate for visualizing orbitals and the electrostatic potential and examining MO coefficients. For example, we also ran PM6 semi-empirical optimizations for these species, obtaining values of -0.31 for $HOMO_{diene}$–$LUMO_{dienophile}$ and -0.43 for $HOMO_{dienophile}$–$LUMO_{diene}$. The orbital coefficients are greatly simplified using this model chemistry, consisting of a single set of s and p orbitals for each atom.

Molecular orbitals can sometimes be useful in understanding why substances react as they do. In the next example, we consider the Al_5O_4 anion reacting with potassium.

EXAMPLE 6.2 REACTIVITY OF $Al_5O_4^-$

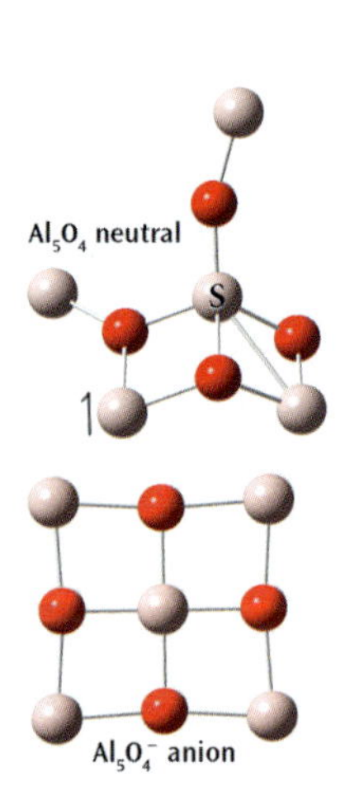

The Al_5O_4 (pentaaluminum tetroxide) molecule is pictured in the margin (upper structure), along with its anion (tetraoxy pentaaluminate, lower). Das and Krishnan Raghavachari have studied this system in detail in the context of cluster building blocks for new solid state materials [Das08]. In our discussion, we will focus on this cluster's reactivity and bonding with potassium.

Adding an electron to the neutral Al_5O_4 structure produces an anion with higher symmetry (D_{4h} vs. C_1) and stability. In fact, the latter is a "magic cluster": one with "exceptionally high stability compared to its immediate neighbor [and characterized by] high symmetry in the structure, saturated electronic shell, chemical inertness, and a large energy separation between the highest occupied and the lowest unoccupied molecular orbitals" (p. 2011).

The neutral form is a chiral molecule. We modeled the S enantiomer, but it does not matter for this example which one is used. The neutral species is a doublet. We first tested the stability of the wavefunction with a stability calculation. It indicated we had a stable SCF solution. We also computed other spin states to determine if they were lower in energy. This analysis confirmed that the doublet is the lowest energy spin state. It also revealed that the unpaired electron is mostly located on the aluminum atom indicated in the figure in the margin.

We optimized both structures* and computed the thermal energy difference between the neutral molecule and the anion as 71.3 kcal/mol (3.09 eV).

In attempting to design periodic systems based on this anion cluster, it is necessary to determine the orientations of the potassium counterions. There are two possibilities: bonding to the outer aluminum atoms and bonding to the oxygen atoms:

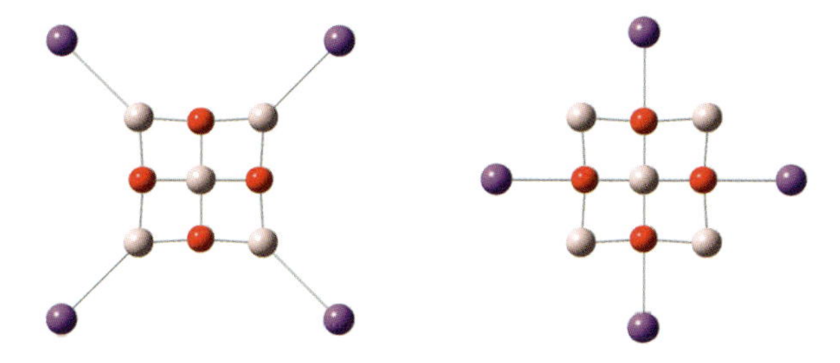

In Al_5O_4K, the oxygen binding position is 7.4 kcal/mol lower in energy than that for the terminal aluminum atom. We can examine the molecular orbitals for $Al_5O_4^-$ to determine whether the same will be true for $Al_5O_4K_4^{3+}$:

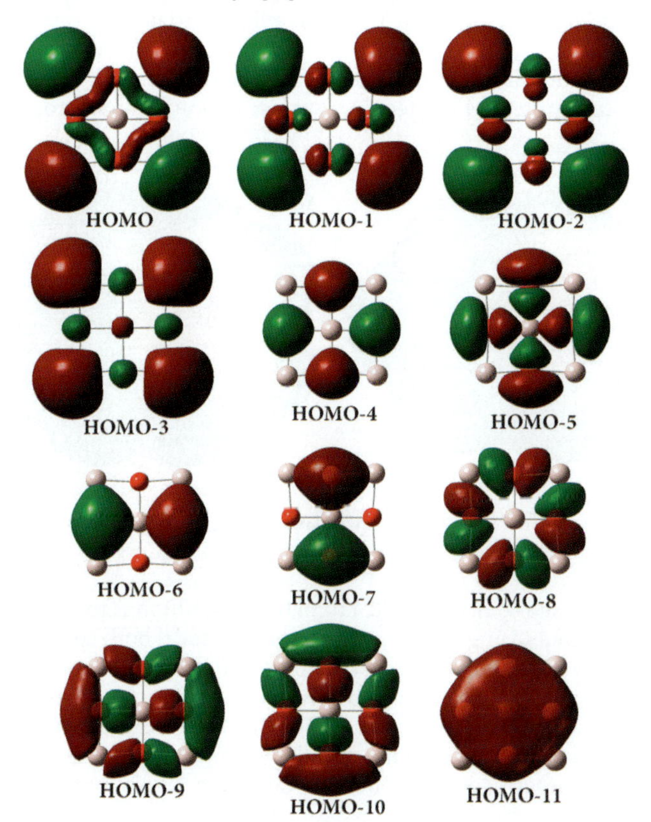

* We obtained a structure for the anion with D_{2d} symmetry; the planar D_{4h} structure is a saddle point, and the imaginary frequency corresponds to motion of the oxygen atoms perpendicular to the plane of the aluminum atoms. A similar result was obtained by the original researchers using a different basis set. However, based on several factors, they "conclude that the vibrationally averaged structure of $Al_5O_4^-$ is planar" [Das08, p. 2012]. They computed the energy difference between the planar and slightly out-of-plane structures to be less than 1 kcal/mol.

The four highest occupied molecular orbitals correspond to lone pair electrons on the terminal Al atoms. In contrast, the orbitals associated with the electron pairs on the oxygen atoms are much lower in energy (HOMO-6, HOMO-7 and HOMO-11). As a result, strong electrophilic interactions between K^+ ions and the lone pairs on the terminal Al atoms suggests that, in contrast to Al_5O_4K, the corner positions will be preferred by the potassium ions when forming $Al_5O_4K_4$. This is in fact the case, and the energy difference between the two forms was computed by these researchers to be about 36 kcal/mol. ■

We've seen how visualizing various surfaces can be helpful to understanding the chemistry of substances and reactions. In the next example, we provide a cautionary tale in which this is not the case.

EXAMPLE 6.3 INDANE AND TETRALIN

Indane* and tetralin† both consist of two fused rings: one six- and one five-membered ring for indane and two six-membered ones for tetralin (indane is the upper molecule in the margin). The substances are similar to part of the vitamin E structure we are considering periodically in this book. We are interested in determining which of the available carbon atoms in the aromatic ring are most reactive (available to electrophilic substitution): the α carbon closest to the second ring or the β carbon adjacent to it.

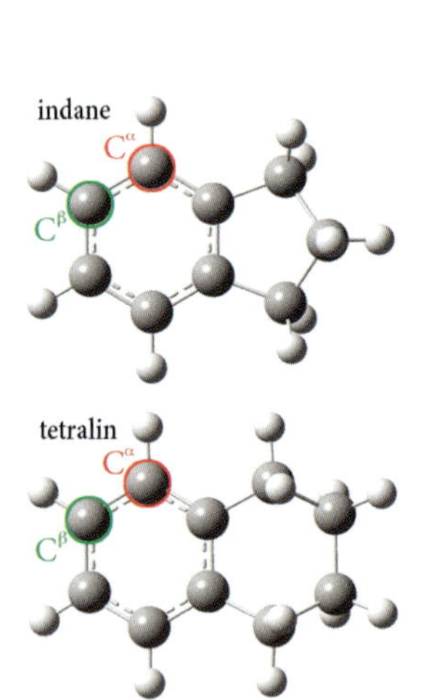

We optimized both structures to minima using our standard model. The illustration below shows the electrostatic potential plotted on a surface of constant electron density for both molecules:

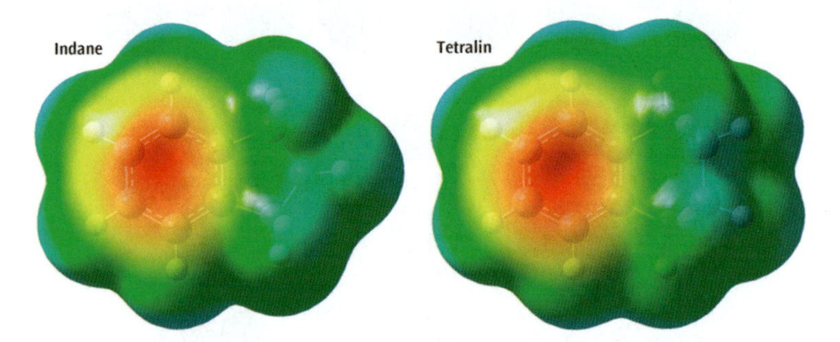

There is essentially no difference between the two surfaces. We will have to look elsewhere for the answer to our question.

One way to predict the site of electrophilic aromatic substitution is to study the compound's reaction with a deuterated strong acid. Therefore, we will model both compounds undergoing deuteration at the two locations in question by predicting the energies of each cation intermediate having a hydrogen and a deuterium on the relevant carbon atom. For both indane and tetralin, we will add a deuterium atom to each location in turn, optimize the structures and verify that they are minima, and then compute the energy difference between them. The carbocation intermediates used in this study are shown in the margin on the next page (only one resonance structure is shown for simplicity).

The hydrogen atom can be added in the usual way in the molecule builder of your choice.

---

* This compound was formerly known as indan.

† IUPAC: tetralin: 1,2,3,4-tetrahydronaphthalene.

Specifying alternate isotopes in GaussView is accomplished via the **Atom List Editor** (select **Atom List** from the **Edit** menu):

G1:M1:V1 – Atom List Editor

Go To 20

Row	Highlight	Display	Tag	Symbol	Mass (Isotopologue 0)
18		Show	18	H	Default
19		Show	19	H	Default
20		Show	20	H	2
21	Add	Show	-	?	Default

Active Sublist Filters: None

Be sure that the icon labeled with a green I is selected in the tool bar, which causes the **Mass (Isotopologue 0)** column to be visible. You can then locate the relevant hydrogen atom and change it to deuterium by setting its mass to 2. This action has the effect of replacing the simple **H** symbol for this atom within the molecule specification to **H(Iso=2)**. Of course, this change may also be made by manually editing the generated Gaussian input file.

The following table presents the total energy results of our four optimization+frequency calculations:

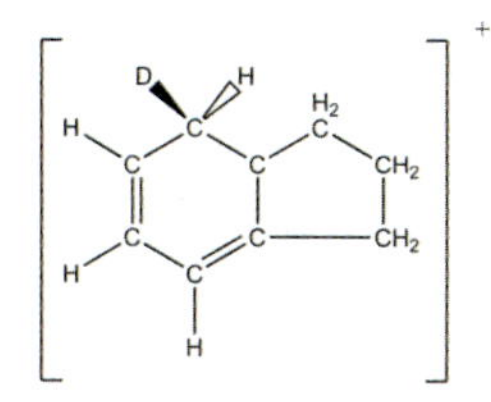
indane cation intermediate at alpha position

COMPOUND	G (hartrees) c^{α}	c^{β}	$\Delta G^{\alpha\text{-}\beta}$ (kJ/mol)
indane carbocation	-348.96724	-348.97103	10.0
tetralin carbocation	-388.24031	-388.24215	4.8

For indane, the intermediate involving the β carbon has a significantly lower energy than the one involving the α carbon. For tetralin, the energy difference is much smaller. Observed reactivity is in line with these results: in indane, deuteration at the β carbon is about four times faster than that at the α carbon while in tetralin it is only 7% faster [Selander71].

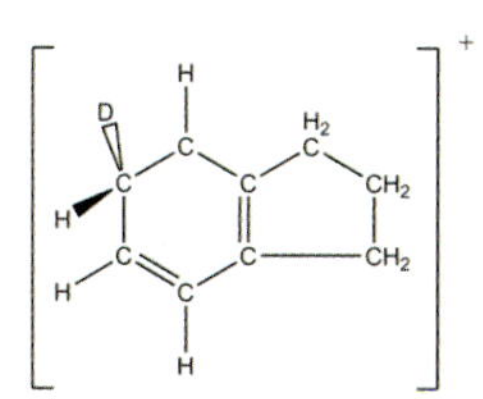
indane cation intermediate at beta position

The electrophile can approach the molecule from either above or below the ring, and these positions are not equivalent because of the aliphatic ring. We modeled the approach from above, taking care to be consistent when comparing the two deuterium positions. However, the results between the two approaches are very similar when using deuterium; the energy differences are 9.9 and 4.6 kj/mol for the indane and tetralin carbocations (respectively) when the electrophile approaches from below the ring.

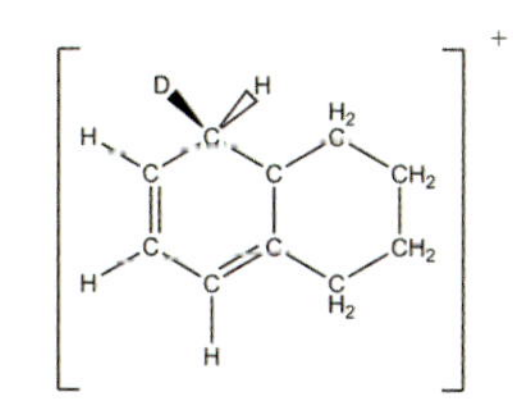
tetralin cation intermediate at alpha position

Notice that the carbocation intermediates are chiral, which would add –RT·ln(2) to the free energy of each intermediate. We ignored that in the data table since we are focusing on the energy difference, and this term will cancel.

tetralin cation intermediate at beta position

It has been suggested that the site of electrophilic substitution in aromatic compounds correlates with carbon-13 NMR shifts [Schnatter14], with the lower chemical shift (great shielding) being more reactive. Here are the results for the isotropic shielding (ppm) using our standard model. We can compare these results to the calculated shielding value of 54.75 ppm for a carbon atom in benzene (see Example 4.7, p. 162):

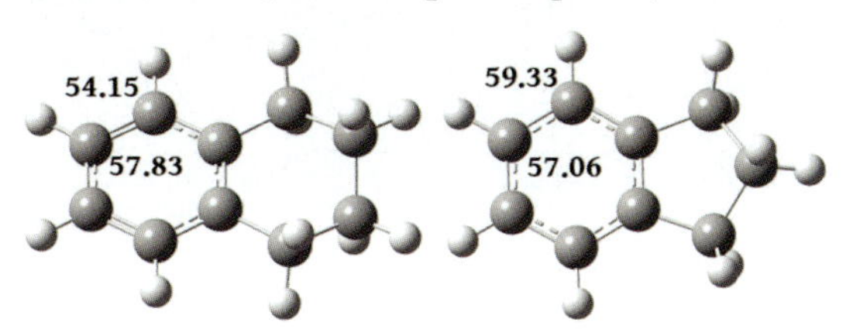

These results are surprising for a number of reasons. The tetralin shieldings are very close to one another and would be in line with those of o-xylene where the aliphatic ring has the same effect as single methyl groups. In indane, however, the shieldings are quite different, indicating that there is substantial change in the environment of the aromatic carbons due to the strain of the five-membered ring. The carbon-13 assignments agree with experiment [Adcock74]; however, they do not correctly predict the favored site of deuteration.

The anomalous behavior of indane was first discussed in 1930 by Mills and Nixon [Mills30]. Such preferential reactivity is accordingly known as the Mills-Nixon effect.

Studying Potential Energy Surfaces

References: Potential Energy Surfaces
Definition: [Marcelin1915, Eyring31, Eyring35]
Review: [Shaik92, Lewars11]

Thus far in our discussion, we have treated the study of reactivity by focusing on molecular geometries such as reactants, products, intermediates, and transition structures. The relative energies of these are all we need to predict thermochemical properties such as the enthalpies of reaction or activation free energies. However, these are only stationary points on a much larger potential energy surface (PES). The actual landscape of this surface can also be explored to see how the various stationary points connect. Details about these pathways are important in validating mechanisms especially when more than one pathway between reactants and products can exist.

In general, the PES for a reaction is a $3N$-dimensional surface where N is the number of atoms. It is impossible to calculate or visualize the entire PES for any but the simplest of reactions. A judicious choice of internal coordinates to explore is the first step in any investigation. We begin by illustrating a pitfall encountered in trying to calculate the activation energy for a chemical process and how a careful look at the PES can offer an explanation.

Theoretical predictions of potential energy surfaces and reaction paths can sometimes yield quite surprising results. Consider rotational isomerism in allyl cation. We are interested in knowing how difficult it would be to twist this molecule given that it is held together by a double bond. One suggested path between the two forms is via a perpendicular transition structure having C_s symmetry. A plausible way to begin an investigation of this reaction is to attempt to locate a saddle point on the potential energy surface corresponding to this hypothesized transition structure.

EXAMPLE 6.4 ROTATIONAL ISOMERIZATION IN ALLYL CATION

Consider a chemical process where the reactants and products are both allyl cation,* and one set of external hydrogen atoms are exchanged, as in the left two structures in the following illustration:

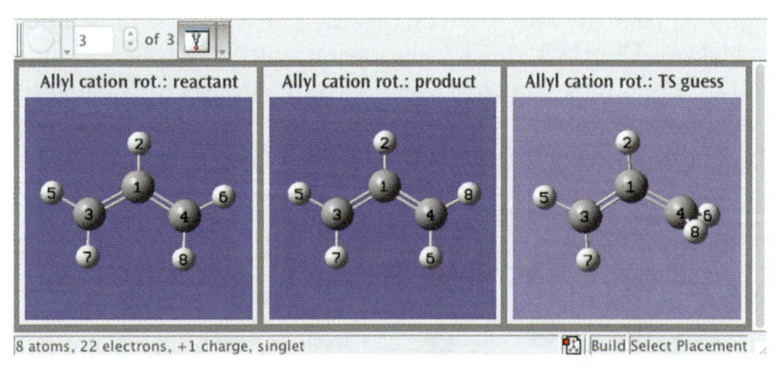

* IUPAC: 1-propen-3-ylium.

The three structures shown above comprise the required input for a QST3 transition structure search for this process. We run this calculation using the small 6-31G(d) basis set, calculating the force constants at each step. We do the latter because this is a tricky optimization which failed to converge with several other less costly approaches:

```
# APFD/6-31G(d) Opt(QST3,CalcAll)
```

In a subsequent job step, we read the force constants from the smaller basis set (**Opt=ReadFC**) and perform a tight, high-accuracy transition structure optimization with our standard model:

```
# APFD/6-311+G(2d,p) Opt(TS,ReadFC,Tight) Freq
```

When we animate the imaginary frequency from this calculation, we are disappointed to see that the reaction coordinate is not what we sought, but instead involves the hydrogen atom of the middle carbon migrating to the terminal carbon, suggesting that the exchange of the two terminal hydrogen atoms does not occur by the mechanism we thought. Instead, it could involve a stepwise process:

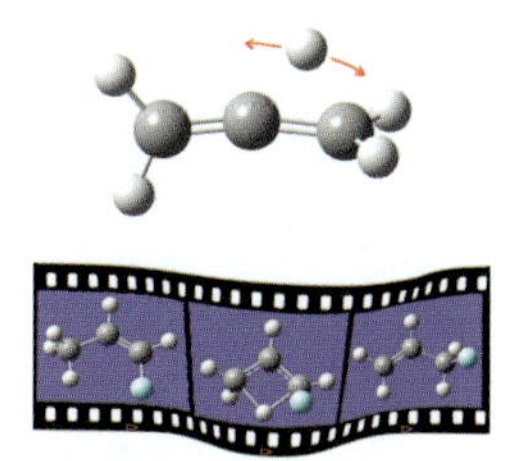

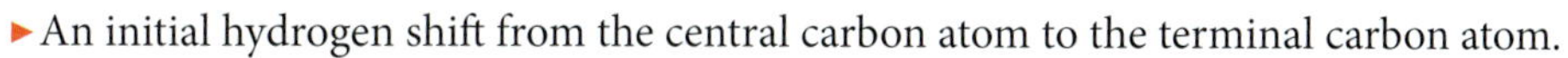

- An initial hydrogen shift from the central carbon atom to the terminal carbon atom.
- Methyl rotation about the C-CH_3 single bond.
- A second hydrogen shift from the terminal carbon atom back to the center.

Each of these steps would involve its own transition structures with associated activation barriers. For steps 1 and 3, the TS is the same because of symmetry. For step 2, the methyl rotation passes through two barriers, but these are presumably much smaller than the migration barrier.

The following table presents the energy results for the various structures:

STRUCTURE	E (hartrees)	BARRIER (kJ/mol)
$[CH_2\text{-}CH\text{-}CH_2]^+$	-116.89039	115.1
H atom transfer TS	-116.84656	
$[CH_2\text{-}C\text{-}CH_3]^+$	-116.87774	81.9

The barrier to hydrogen transfer from the central carbon atom to the terminal carbon atom is higher than the barrier for the return process by ~23 kJ/mol. The barrier to concerted twist must be even higher in energy and can be estimated by a single point calculation using our standard model and the first step in the QST3 calculation (-116.83034 hartrees) to be ~158 kJ/mol.

In this case, the failed transition structure search was due to our poor guess at how this process actually occurs. This may not be a general result for C=C bond rotation since environmental effects (e.g., gas phase versus solution) and/or changes in the chemical structure (e.g., methyl groups instead of hydrogen atoms) may change the landscape of the PES.

Potential Energy Surface Scans

Geometry optimization algorithms allow you to find points of interest on the PES (minima and maxima corresponding to equilibrium and transition structures) through sophisticated use of energy derivatives. However, it is also possible to simply sample the PES in a region that corresponds to the process in which you are interested. This type of calculation is called a *scan*.

There are two types of scan procedures:

- A *rigid* scan takes a geometric structure and freezes all the coordinates in place except for the particular coordinate being scanned. A single point energy calculation is performed for each generated structure.
- A *relaxed* scan does a partial optimization at each point of the scan, freezing the scan coordinate and optimizing all others. In other words, each optimization locates the minimum energy geometry with the scanned parameters set to specific values.

For instance, in the previous example, we wanted to find the TS for twisting the C=C in allyl cation. We could have forced the twist angle to take on a set of values in the range from 0° to 180°, evaluating the energy or partially optimizing the geometry at each point. The following figure shows the results of our two scans, which each incremented the scan variable—the H-C-C-H dihedral angle—by 10° at each step.

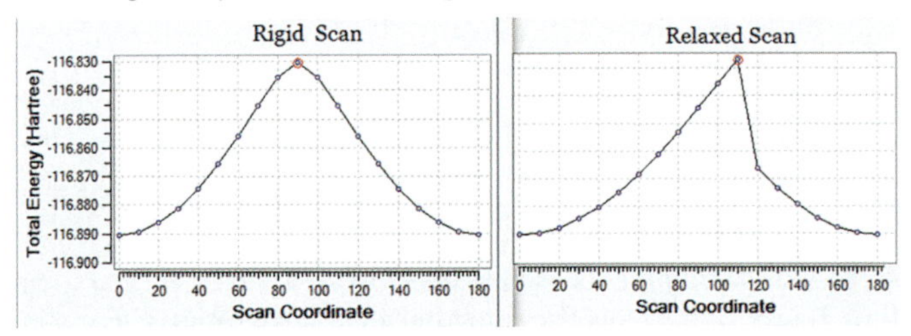

The two scans are quite different. A smooth PES is suggested by the rigid scan, but that is misleading because we are only examining the effect of one degree of freedom. Real molecules travel on relaxed potential energy surfaces with many degrees of freedom. The relaxed scan shows a discontinuity for our hypothetical reaction process, suggesting that a more complicated set of atomic motions is involved in the actual reaction coordinate.

While scan calculations provide considerable insight into the structure of the PES, they do not define the lowest energy path between two structures. In order to plot the actual lowest energy pathway, we need to follow our transition structure downhill to the reactants and to the products rather than simply stepping across the PES. An intrinsic reaction coordinate (IRC) calculation does precisely this; we will discuss this calculation type later in the chapter.

Relaxed potential energy surface scans are requested by the **Opt=ModRedundant** option and one or more lines specifying the structural parameters to be scanned in an additional input section following the molecule specification.*

Setting up Relaxed Scan Calculations with Graphics Programs

GaussView includes features for setting up relaxed potential energy surface scan calculations. To do so, you build or import the desired molecule in the usual manner, and then you use the **Redundant Coordinate Editor** (accessed via the

* Rigid scan calculations use the **Scan** keyword. See the *Gaussian User's Reference* for details on its use.

Edit→Redundant Coordinates menu path) to specify what structural parameters should be scanned. This dialog is illustrated below:

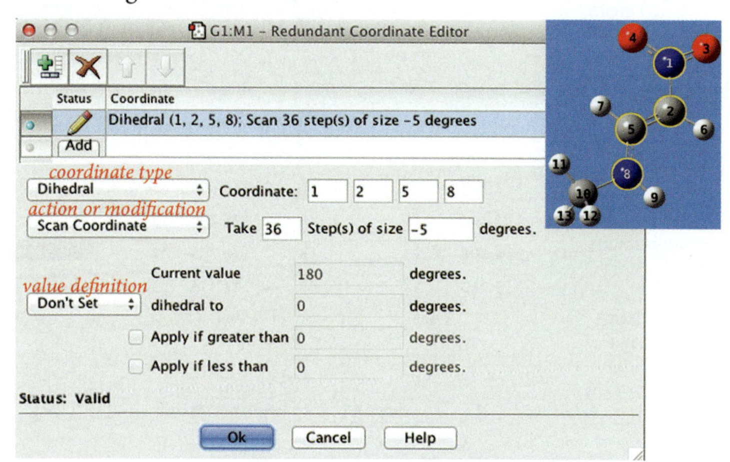

Initially, the coordinate list is empty. Click on the **Add** button to place an item into the list. The details about that coordinate—structural parameter—appear in the fields below the coordinate list. The first popup defines the type of coordinate (here we are defining a dihedral angle), and the fields to its right, labeled **Coordinate**, contain the atom numbers corresponding to it. You can enter atom labels directly into these fields, or you can click on the relevant atoms in the **View** window (we did the latter).

To complete the specification for this scan parameter, we select **Scan Coordinate** from the second popup menu (indicating the specific action/modification to be performed for this coordinate). We indicate the number of steps and step size in the fields that appear to its right. If appropriate, you can add additional scan coordinates in the same way. Note that the value definition popup can no longer be used to change the coordinate's value and so should remain at **Don't Set**.

When you set up the Gaussian calculation, select the **Scan** job type and **Relaxed (Redundant Coordinate)** subtype in the **Gaussian Calculation Setup** dialog. These are the default when you have defined scan variables with the **Redundant Coordinate Editor**. GaussView will add the appropriate **Opt** keyword and option to the job's route section.

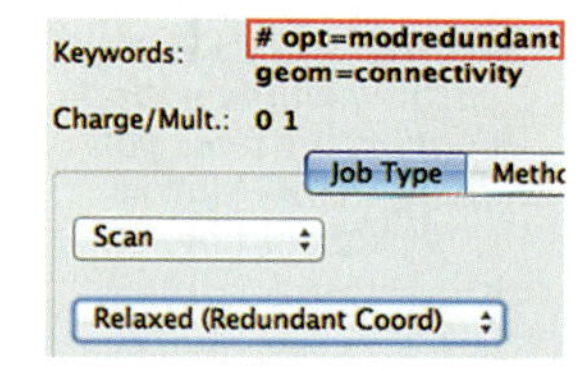

WebMO In WebMO, you set up relaxed potential energy scans in two steps. First, you use the Z-matrix editor to specify the scan parameters. Second, you specify **Coordinate scan** as the **Calculation** type on the **Configure Gaussian Job Options** page.

You access the Z-matrix editor in the **Build Molecule** page via the **Tools→Edit Z-Matrix** menu path. It is illustrated in the following figure:

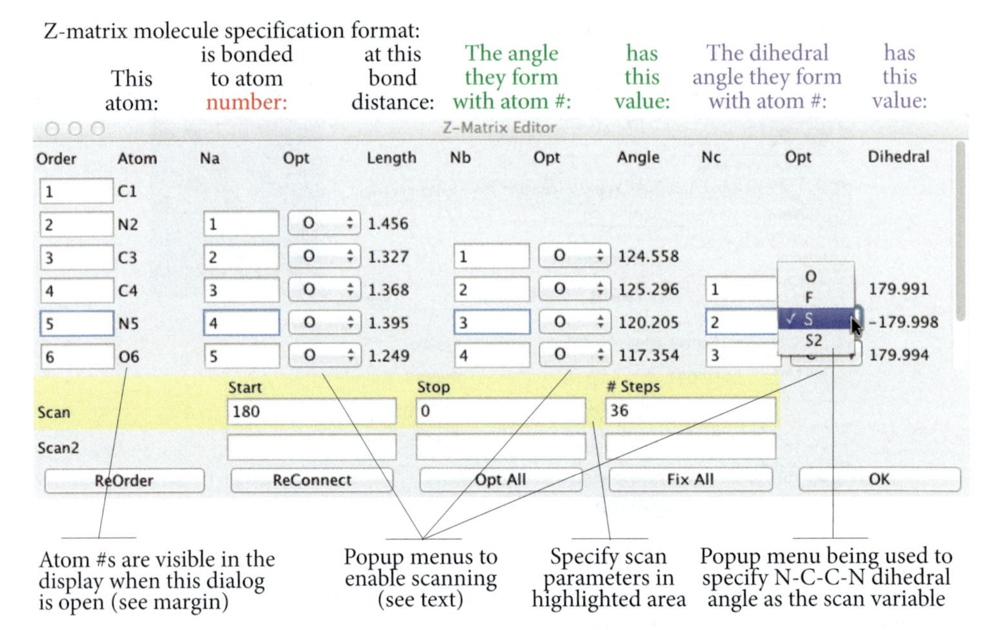

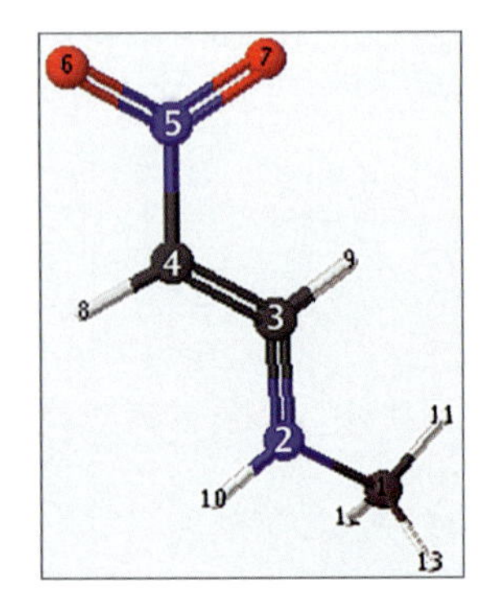

Z-matrix input is explained in the callouts at the top of the illustration. This format describes a molecular structure by specifying the bond distance, bond angle and dihedral angle for each atom with respect to one, two or three previously defined atoms (respectively).*

WebMO indicates the scan variable with highlighting once you leave the Z-Matrix Editor dialog.

You do not have to understand all of the subtleties of Z-matrices in order to set up a scan calculation. All that is necessary is that you identify the coordinate(s) that you want to vary. In the preceding example, the scan will vary the N-C-C-N dihedral angle. This coordinate is specified as the bond distance in the line for the oxygen atom (line 2). The popup menu between the **Na** and **Length** columns is used to specify how each coordinate should be treated during the scan calculation. The **S** and **S2** selections on this menu are used to indicate the first and second variables to be scanned, and the corresponding value limits are entered into the fields labeled **Scan** and **Scan2** near the bottom of the dialog. Here, we have set the dihedral angle variable to **S** and entered the starting value, ending value and number of steps into the fields in the highlighted area. For this calculation, we will not use the second scan variable. Note that Gaussian allows you to scan more than two variables, but you can only set up two of them via WebMO's interface.

The default value for the scan variable popup menus (their columns are labeled **Opt**) is **O**, which says to optimize that coordinate at each scan point. These values should not be modified for coordinates other than the scan variable when performing a relaxed potential energy surface scan.

When the Coordinate You Want to Scan Isn't in the Z-matrix

WebMO There are many possible Z-matrix representations for all but the tiniest molecular structures. As a result, sometimes the coordinate that you want to scan is not present in the Z-matrix generated automatically by WebMO. However, it is very simple to modify the Z-matrix in such cases.

* The first three lines of a Z-matrix have fewer items specified as there are not yet enough other atoms defined to include all three parameters.

In the Z-matrix editor modify one of the existing entries to include the items you require. As an example, consider a scan on formaldehyde in which we want to vary the value of the C-H-H bond angle. However, the Z-matrix generated by WebMO defines the position of the hydrogen atoms via the O-C-H bond angle:

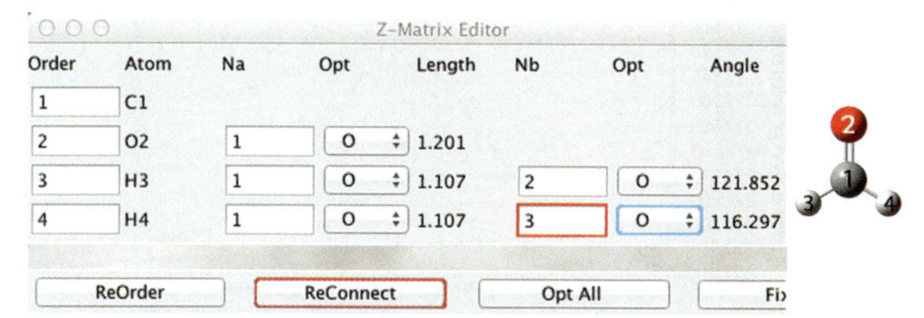

In order to specify the scan variable we want, we modify the atom specifying the bond angle in line 4—which defines the second hydrogen atom—changing the value to correspond to the first hydrogen atom (i.e., **Nb** is set to **3**). When you click the **ReConnect** button, the value in the **Angle** column will be updated.

You can then continue to set up the scan as normal, using the various popup menus (the one for the scan variable is highlighted in blue) and the dialog's scan parameter fields (the latter are omitted in the illustration above).

EXAMPLE 6.5 SCAN CALCULATIONS: ROTATIONAL ISOMERIZATION

In general, the difference in energy between the E and Z forms of n-methyl-(2-nitrovinyl) amine* is quite small. The E form is on the left in the following illustration:

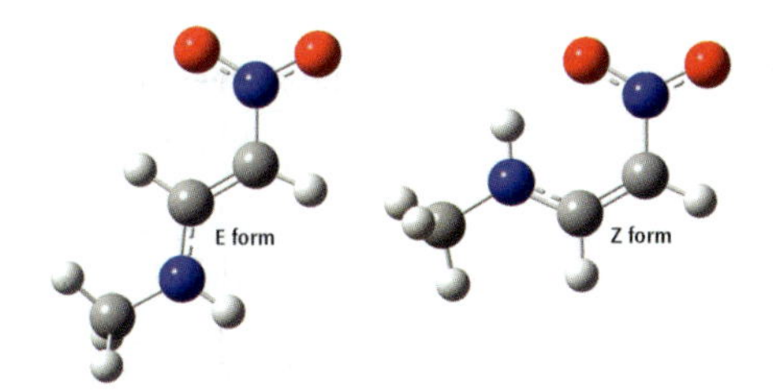

We are interested in investigating the rotational barrier between these two isomers: rotation around the N–C–C–N dihedral angle. We will model the process in solution with ortho-dichlorobenzene and N,N-dimethylformamide. One way to do so is to perform a relaxed potential energy surface scan that varies this dihedral angle. We ran such a scan from 180° to 0°, in -5° steps.

* IUPAC: 1-methoxy-N-methyl-2-nitroethenamine.

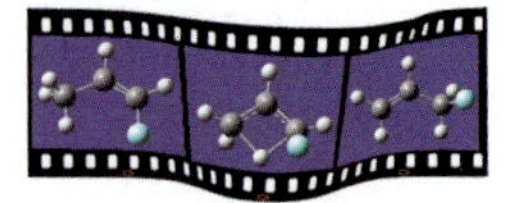

The following graph displays the energy results from the relaxed scan in the solvent ortho-dichlorobenzene using the SMD solvation method introduced in chapter 5:

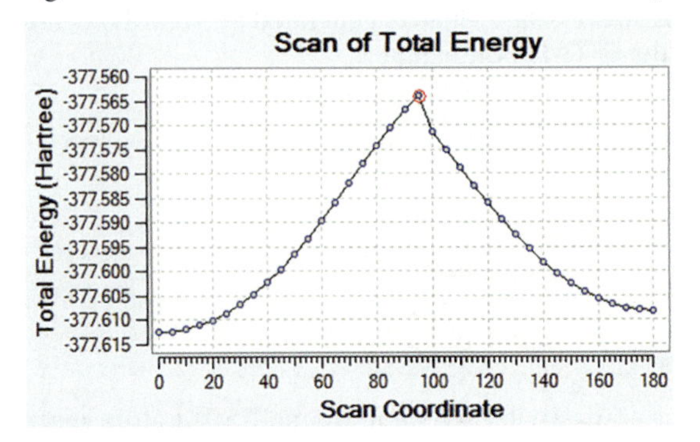

The discontinuity at 95 degrees is due to the sudden inversion of pyramidalization at carbon 2. We can observe this by plotting the dihedral angle N1-H6-C2-C5 as a function of the scan coordinate with the **Plot Molecular Property** item from the **Plots** menu in GaussView:

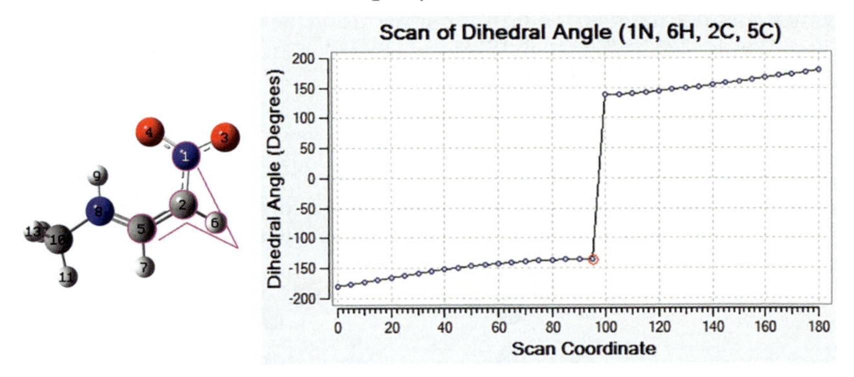

This angle needs to be positive at the 95 degree scan coordinate in order for a smooth PES to be achieved (that geometry would be reached if the scan were computed in reverse).

We proceeded as follows in both environments. We started with the 90° structure as the initial guess for the transition state, setting the previously indicated dihedral angle to 180°. We then set up a QST TS optimization. We created the reactant and product structures needed for QST3 by modifying the N1-C2-C5-N8 dihedral by ±15° (respectively). We then performed an optimization+frequency calculation using our standard model chemistry and the SMD solvation model. We also optimized the E and Z forms themselves, again with our standard model chemistry and the SMD solvation model.

The results of our calculations appear in the following table:

	ortho-dichlorobenzene			N,N-dimethylformamide		
		BARRIER (kcal/mol)			BARRIER (kcal/mol)	
MOLECULE	G (hartrees)	CALC.	OBS.[a]	G (hartrees)	CALC.	OBS.[a]
E form	-377.53928	28.1		-377.53949	26.7	
TS	-377.49454			-377.49700		
Z form	-377.54324	30.6	21.1	-377.54368	29.3	19.2

[a][Pappalardo93, Krowczynski83, Chiara92]

Both the experimental observations and the calculated results indicate that there is little difference in the barriers to rotation in the two solvents, predicting that the barrier will be about 1.5 (calculated) to 2.0 (observed) kcal/mol lower in N,N-dimethylformamide solution than in ortho-dichlorobenzene solution. This system is also an example of the lowering of rotational barriers in solution. The predicted barrier in the gas phase is over 40 kcal/mol.■

EXAMPLE 6.6 BOND DISSOCIATION IN METHANE

In this example (derived from [Duchovic82]), we will explore the potential energy surface associated with a single C-H bond rupture in methane. In so doing, we will compare several levels of theory to discover their similarities and stark differences in the bond breaking region. What we are studying is the reaction of methane going to methyl radical and a hydrogen atom. We begin with a methane molecule with tetrahedral symmetry and a bond distance of 0.75 Angstroms. We will then perform a relaxed scan over one of the C-H bonds and increasing it in steps of 0.1 Angstroms for 26 steps.

We will use an unrestricted* model—indicated by a **U** prefixed to the method keyword—as well as the **Guess(Mix,Always)** keyword. The former removes the restriction that all electrons remain paired (since the products include a radical species), while the latter is necessary to create an appropriate wavefunction throughout the bond distance range covered in the scan. At distances beyond 1.5 Angstroms, the two electrons that had occupied the sigma bonding orbital are now in separate, singly occupied orbitals each with opposite spin. We describe this as an *open-shell singlet*, since the total spin is still zero, but not all orbitals are doubly occupied.

The figure following shows the results of this scan calculation, along with ones from several other theoretical methods:

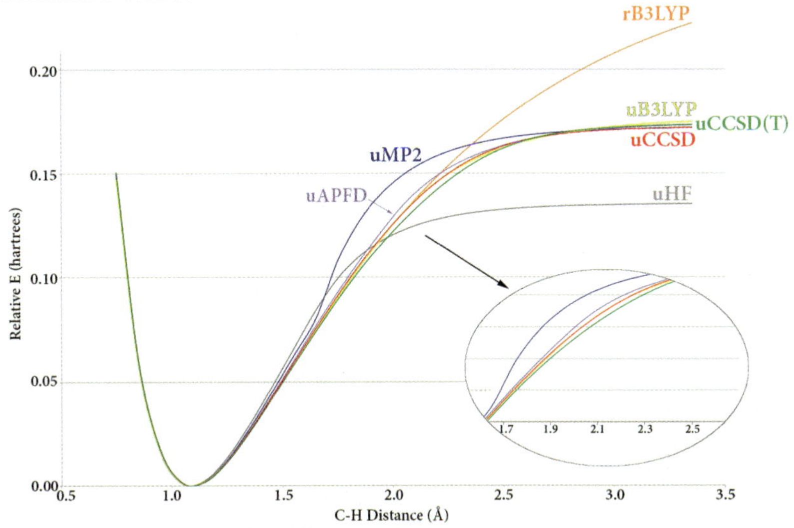

The illustration plots the energy relative to the energy of methane computed at the equilibrium geometry at the particular level of theory. The following features of this graph are noteworthy:

- At distances of ±50% of the equilibrium bond distance, the shape of the PES is essentially identical for all methods.
- Beyond 1.5 times the equilibrium bond length, substantial deviation occurs among the PESs computed at different theories.

* See chapter 9 for a detailed discussion of unrestricted calculations (p. 444ff).

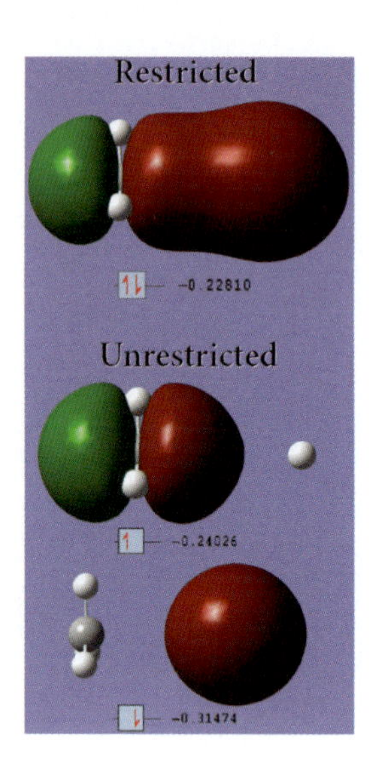

- The PESs computed with a restricted wavefunction—all orbitals doubly occupied—will not go to the correct products. The energy will continue to slowly rise even at very long distances due to the inability of a restricted wavefunction to properly describe this region. The figures in the margin show what the HOMO looks like for the case of stretched methane in both the restricted and unrestricted cases.
- All of the models using unrestricted electrons go to the correct products, but the UHF wavefunction underestimates the dissociation energy by 20%.
- The DFT methods and CCSD(T) are very consistent in terms of their predicted dissociation energies and the overall shape of the PES.
- UMP2 theory predicts a similar dissociation energy to CCSD(T). However, the shape of the PES in the bond-breaking region is quite different than the other methods. ■

Reaction Paths

As we've noted previously, successfully completing a transition structure optimization does not guarantee that you have found the right transition structure: the one that connects the reactants and products of interest. One way to determine the minimum to which a transition structure connects is by examining the normal mode corresponding to the imaginary frequency and determining whether or not the motion tends to deform the transition structure as expected. However, it is often difficult to tell for certain. In this section, we will discuss a more precise method for determining what points on a potential energy surface are connected by a given transition structure.

References: Intrinsic Reaction Path Methods
Definition: [Fukui81]
Review: [Hratchian05a]
Implementation: [Hratchian04, Hratchian05b]

An intrinsic reaction coordinate (IRC) calculation examines the reaction path leading down from a transition structure on a potential energy surface. Such a calculation starts at the saddle point and follows the path in both directions from the transition state, optimizing the geometry of the molecular system at each point along the path. In this way, an IRC calculation definitively connects two minima on the potential energy surface by a path which passes through the transition state between them.

Note that two minima on a potential energy surface may have more than one reaction path connecting them, corresponding to different transition structures through which the reaction passes. From this point on, we will use the term *reaction path* to designate the intrinsic reaction path predicted by the IRC procedure, which can be qualitatively thought of as the lowest energy path, in mass-weighted coordinates, which passes through a given saddle point.

Reaction path computations allow you to verify that a given transition structure actually connects the starting and ending structures that you think it does. Once this fact is confirmed, you can go on to compute an activation energy for the reaction by comparing the appropriate energy values of the reactants and the transition state.

Running IRC Calculations

In Gaussian, a reaction path calculation is requested with the **IRC** keyword in the route section. Before you can run one, however, certain requirements must be met. An IRC calculation begins at a transition structure and steps along the reaction path a fixed number of times (the default is 10) in each direction, toward the two minima that it connects. However, in most cases, it will not step all the way to the minimum on either side of the path.

Here is the procedure for running an IRC calculation:

- Optimize the starting transition structure.

- Run a frequency calculation on the optimized transition structure. This is done for several reasons:
  - To verify that the first job did in fact find a transition structure.
  - To determine the zero-point energy for the transition structure.
  - To generate force constant data. IRC calculations require force constants to proceed. These can be read from the results of a previous frequency calculation (using the **IRC=RCFC** option) or generated prior to beginning the IRC (with the **IRC=CalcFC** option).
- Perform the IRC calculation. This job will help you to verify that you have the correct transition state for the reaction when you examine the structures that are downhill from the saddle point.

To accurately predict the barrier for the reaction, you need to perform some additional computations in order to collect all required data. Optimization+frequency calculations for the reactants and the products will predict the thermally-corrected energy, enthalpy or free energy (depending on your specific requirements as well as any experimental data with which you plan to compare your results).

The entire process can be repeated for a different reaction path, starting from a different saddle point, in order to explore other possible ways to move from the reactants to the products. In this way, you can perform a comprehensive exploration of a potential energy surface.

In Example 6.5, we saw an odd result when we scanned the value of a dihedral angle (see p. 242). The plot on the left in the following figure is the IRC computed* for the transition structure we located in solution with o-dichlorobenzene:

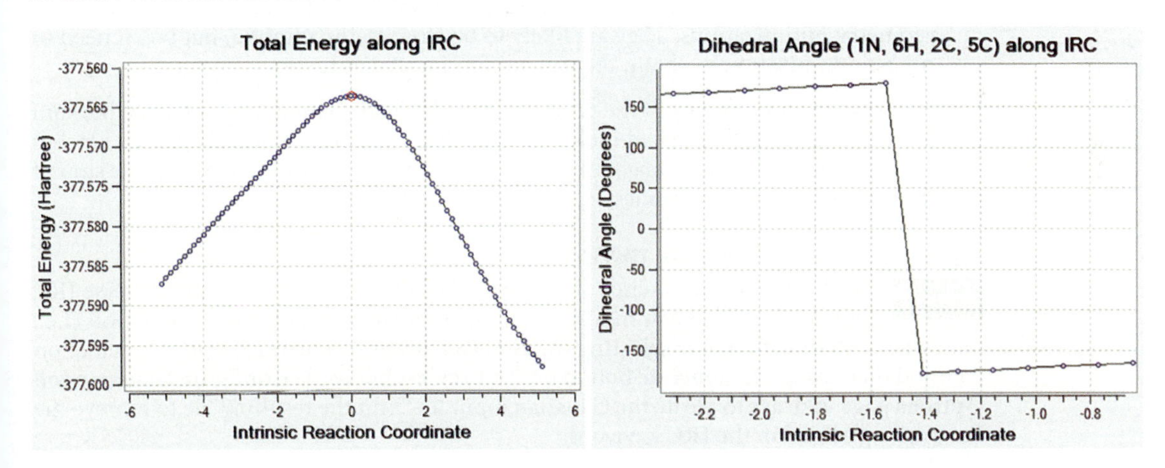

In the figure on the previous page, the plot on the right indicates that the value of the dihedral angle changes smoothly from positive to negative during the twist (going through 180°).

Tips for Running IRC Calculations

- Occasionally, you may need to increase the number of steps taken in the IRC in order to get closer to the minimum; the **MaxPoints** option specifies the number of steps to take in each direction as its argument (the default is 10). You can also resume a completed IRC calculation from its checkpoint file by using the **IRC=(Restart,MaxPoints=*n*)** keyword, setting *n* to some appropriate value.

* We used the **IRC=(RCFC,MaxPoints=40,Recorrect=Never)** options and our standard model chemistry to calculate the IRC. These options are discussed in a bit.

- If you want to follow the IRC all the way to the reactants and products, you can include the **Recorrect=Never** option in addition to increasing the value of **MaxPoints**; the former option suppresses a corrective action which is taken when the computed value of the relevant IRC parameter exceeds a threshold. Using this option results in a significantly less accurate reaction path, and so using it is appropriate only when your goal for the IRC calculation is verification of the endpoints rather than the reaction path itself. The **IRC=Step=***n* option can also be used to increase the step size between reaction path points in order to reach the endpoints more quickly (the default value is 10). However, it will also increase the uncertainty in the reaction path.

- By default, IRC calculations follow the reaction path in both directions from the transition structure. You can limit the direction to the forward or reverse direction with the **IRC=Forward** and **IRC=Reverse** options (respectively).

In GaussView, you can combine the results of separate forward and reverse IRC calculations on the same structure by setting the **Target** popup menu in the **Open** dialog to **Single new molecule group for all files** when you open the relevant output/checkpoint files.

- You can use the **IRC=Phase** option to specify explicitly which direction is "forward"; this direction is computed first when the IRC follows both the forward and reverse directions. The argument specified to the **Phase** option specifies a structural parameter within the molecule whose increasing value defines the forward direction. For example, **Phase=(1,2,3)** says that the forward direction in the IRC follows increase in the angle formed by atoms 1, 2 and 3.

- The endpoints of the IRC generally do not correspond to the minima as obtained via geometry optimizations. They are likely to be close to the minima, but you'll need to run an optimization to obtain the true minimum structure.

- Note that the final energy of the products in an IRC calculation may not equal the sum of the energies of the isolated molecules. An IRC terminates when the energy reaches a minimum for the molecular complex, a level which may be different than the sum of the isolated product molecules.

Setting Up IRC Calculations in WebMO

WebMO's default menus allow you to select either a forward or a reverse IRC calculation. You can modify a job to perform a default IRC calculation (i.e., following both directions) by selecting the **IRC Calculation (Forward)** option calculation type and then using the **Generate** button on the **Preview** tab of the **Configure Gaussian Job Options** page in order to create the Gaussian input file. Edit the resulting file to remove the **Forward** option from the **IRC** keyword.

You can modify this job type permanently by editing the *gaussian.tmpl* file in the *cgi-bin/webmo/interfaces* subdirectory of the web server's top-level content directory on the system where WebMO is installed. Copy the entire section, and then delete the text indicated in red below within the copied text:

```
=========================================
IRC Calculation (Forward)          You can label the menu item as you choose.
[%IF nodes > 1 && gaussianVersion != "g09" %]%NProcLinda=...
[%IF ppn > 1 %]%NProcShared=$ppn\n[%END-%]
...
```

```
#$outputMode $theory[%IF theory ... "AM1" %]
$basisSet[%END%] $basisType IRC=(CalcFC,Forward) $solvent
$additionalKeywords ... Geom=Connectivity[%END%] $noSymmetry
...
[%IF !checkpointFile %]$geometry\n[%END-%]
```

The new menu item will set up a default IRC calculation, which follows the reaction path in both directions.

Unfortunately, WebMO's visualization facilities are set up to handle only unidirectional IRC job output, so visualizing the reaction path from a normal bidirectional IRC calculation in WebMO produces somewhat erratic animations.

EXAMPLE 6.7 STUDYING THE H_2CO POTENTIAL ENERGY SURFACE

We will use Gaussian's reaction path following facility to explore the H_2CO potential energy surface. There are several minima on this surface—formaldehyde, hydroxycarbene (HCOH: cis and trans), and hydrogen molecule plus carbon monoxide—each corresponding to different reactant/product combinations. We will consider these two reactions in this example:

- Molecular dissociation of formaldehyde: $H_2CO \longleftrightarrow CO + H_2$
- 1,2 hydrogen shift reaction: $H_2CO \longleftrightarrow$ trans HCOH

We will use the results of our previous formaldehyde calculation (see Example 2.6, p. 58) and will run optimization+frequency calculations for the other molecules in these reactions. We will then locate the transition states using the **Opt=QST2** procedure and run IRC calculations beginning from the resulting structures.

We successfully completed all of the optimizations, and all of the stationary points were verified to be of the correct type (minimum or transition state). The IRC calculations also completed successfully.

Molecular Dissociation Reaction

We ran the IRC calculation for molecular dissociation reaction in two ways: using the default options and using the **IRC=(MaxPoints=40,Step=20,Recorrect=Never,Phase=(1,4))** options. In the latter case, we increased the maximum number of points to be computed in each direction, increased the step size, suppressed the correction recomputation and defined the forward direction as increasing in the C1-H4 bond length, which corresponds to formaldehyde as the reactant (the marginal illustration identifies the involved atoms).

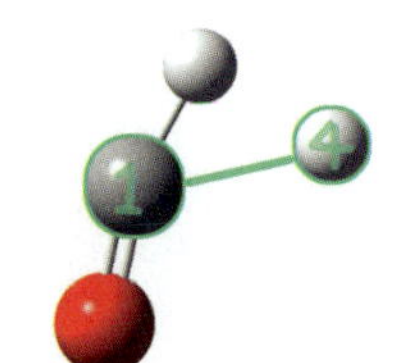

This is the transition structure and the two farthest points located by a default IRC calculation:

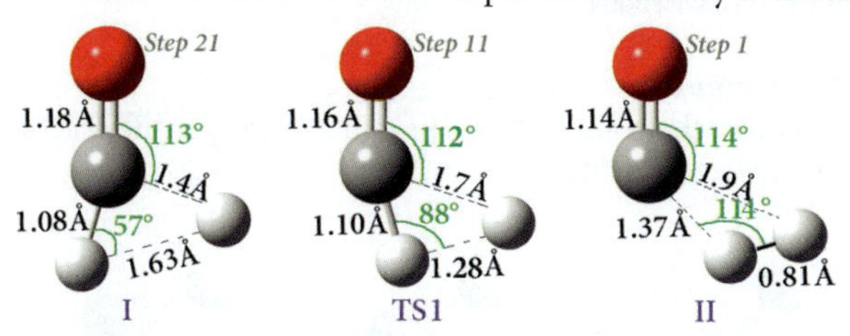

In structure I (which is the final point in the computed IRC), we find a formaldehyde-like structure with the O-C-H bond angles are distorted from the equilibrium geometry. However, we can identify the minimum along this side of the path as formaldehyde.

In structure II, the C-H bond has lengthened with respect to the transition structure (1.37 Å vs. 1.10 Å), while the C-O bond length has contracted slightly. Both changes are what would be expected as formaldehyde dissociates to form carbon monoxide and hydrogen molecule. In fact, many chemical visualization programs display this structure as two separate molecules.

The TS and two final points from the second IRC calculation, which followed the reaction path much further in order to approach the reactants and products as closely as possible, are shown below:

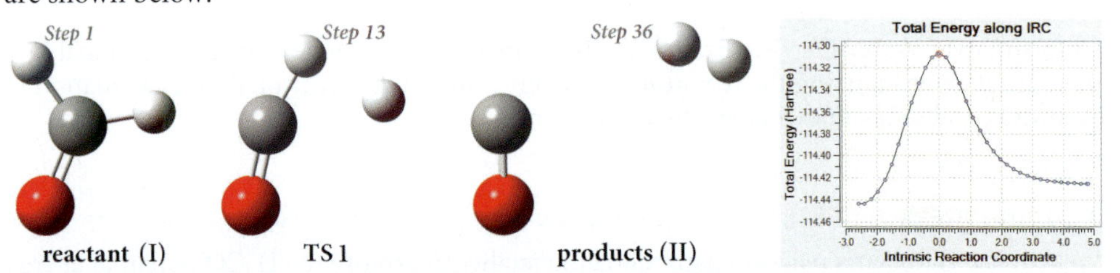

Formaldehyde and carbon monoxide plus hydrogen molecule are easily recognizable as the reactant and products (respectively). Note that not all 40 steps were required to reach the endpoints in this IRC calculation.

We can conclude that this is the transition structure which connects these reactants and products. Accordingly, we can now compute the activation energies for the reaction:

SYSTEM	ENTHALPY (hartrees)	RELATIVE TO REACTANT (kcal/mol)	RELATIVE TO PRODUCTS (kcal/mol)
reactant: formaldehyde (I)	-114.41357	0.0	-6.3
dissociation transition structure (TS1)	-114.28465	80.9	74.6
products: H_2+CO (II)	-114.40346†	6.3	0.0

†H_2: -1.15908, CO: -113.24438

These values indicate that the reaction is nearly thermoneutral and thus the barriers are similar in both directions. They are in excellent agreement with the observed value for the forward reaction of 80.6±0.3 kcal/mol [Horowitz78].

While formaldehyde dissociation was studied extensively in the late 1970s and early 1980s (see, e.g., [Gelbart80] and [Ho82]) and the corresponding potential energy surface was among the first to be modeled theoretically [Harding80], it is far from completely understood even today. For example, in an exciting article in *Science*, Townsend, Harding, Suits, Bowman and coworkers discuss alternate dissociation pathways which test the limits of traditional transition-state theory [Townsend04].

1,2 Hydrogen Shift Reaction

The following figure shows the results of the IRC calculation for the reaction of formaldehyde deforming into trans hydroxycarbene:

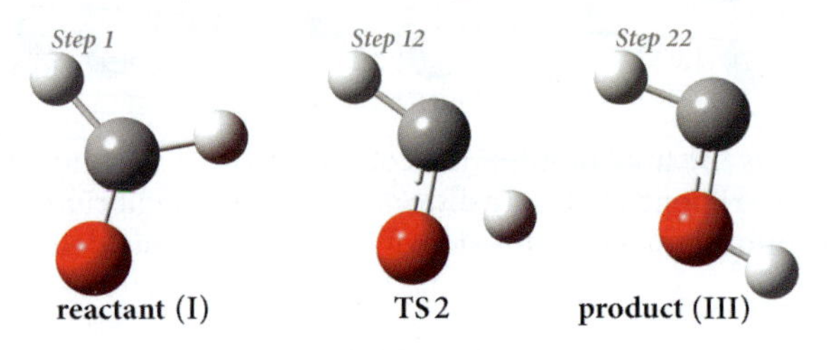

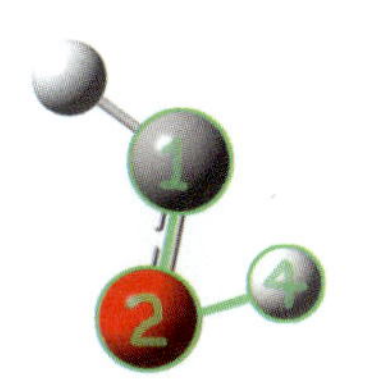

We included the **MaxPoints=40**, **Step=20**, **Recorrect=Never** and **Phase=(1,2,4)** options to the **IRC** keyword; the latter specifies the increasing C1–O2–H4 bond as the forward direction, again corresponding to formaldehyde as the reactant.

Since we have verified that the transition structure connects the reactant and product, we can now compute the activation energy for the reaction. The results we obtained are listed in the following table:

SYSTEM	ENTHALPY (hartrees)	RELATIVE TO REACTANT (kcal/mol)	RELATIVE TO PRODUCTS (kcal/mol)
reactant: formaldehyde (I)	-114.41357	0.0	-53.2
transition structure (TS2)	-114.28192	82.6	29.4
product: trans HCOH (III)	-114.32874	53.2	0.0

The forward barrier for the 1,2 hydrogen shift is very similar to that for the molecular dissociation (82.6 vs. 80.9 kcal/mol), but the endothermicity of the shift is much greater for the former: 53.2 kcal/mol compared to 6.3 kcal/mol.

The following diagram illustrates the results we've computed so far for the H_2CO potential energy surface:

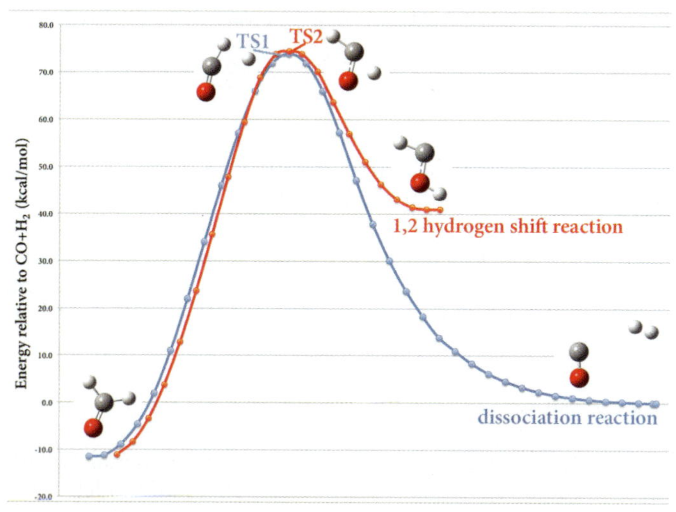

We'll continue our study of this potential energy surface in Exercise 6.3 (p. 256). Subsequent exercises will explore more complex potential energy surfaces.

A Final Note on IRC Calculations

We'll close this section with the following reminder from Shaik, Schlegel and Wolfe [Shaik92], describing both the usefulness and the limitations of the IRC method:

> Although intrinsic reaction coordinates like minima, maxima, and saddle points comprise geometrical or mathematical features of energy surfaces, considerable care must be exercised not to attribute chemical or physical significance to them. Real molecules have more than infinitesimal kinetic energy, and will not follow the intrinsic reaction path. Nevertheless, the intrinsic reaction coordinate provides a convenient description of the progress of a reaction, and also plays a central role in the calculation of reaction rates by variational transition state theory and reaction path Hamiltonians.
>
> *Theoretical Aspects of Physical Organic Chemistry*, pp. 50-51

Isodesmic Reactions

An isodesmic reaction is one in which the total number of each type of bond is identical in the reactants and products. Here is a simple example:

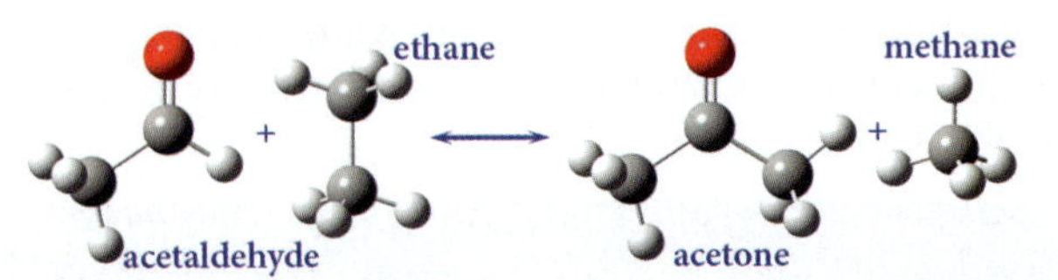

In this reaction, there are two C–C bonds, ten C–H bonds and one (C-O) double bond in both the reactants and products. Because of the conservation of the total number and types of bonds, even modest sized calculations yield very good results due to the cancellation of errors on the two sides of the reaction. In other words, comparing very similar systems enables us to take maximum advantage of cancellation of error.

Isodesmic reactions may be studied in the same manner as any other reaction: energy differences between the reactants and products may be used to predict ΔH, transition structures can be located and their energies used to predict activation barriers, and so on. In addition, isodesmic reactions may be used to predict the heats of formation for compounds of interest by predicting ΔH for the reaction and then computing the desired heat of formation by subtracting the experimental heats of formation for the other compounds from this quantity, a technique which relies on Hess's Law. We will look at an example next.

EXAMPLE 6.8 CO_2 ENTHALPY OF FORMATION

In this example, we will use an isodesmic reaction to predict the enthalpy of formation of carbon dioxide. It will illustrate how an enthalpy of formation can be predicted with a combination of theoretical and experimental quantities.

The enthalpy of formation of carbon dioxide is the heat associated with forming the substance from its elements in their standard states: $O_2^{gas} + C^{solid} \rightarrow CO_2^{gas}$. Calculating this enthalpy difference entirely from theory would involve estimating the enthalpy of solid carbon in its graphite form.

As an alternative, one can instead view this reaction as a sum of three other reactions:

▸I: $2H_2CO^{gas} \rightarrow CO_2^{gas} + CH_4^{gas}$	ΔH_I can be calculated using our standard model
▸II: $2H_2^{gas} + 2C^{solid} + O_2^{gas} \rightarrow 2H_2CO^{gas}$	$\Delta H_{II} = 2\times\Delta_f H^{expt}(H_2CO)$
▸III: $CH_4^{gas} \rightarrow C^{solid} + 2H_2^{gas}$	$\Delta H_{III} = -\Delta_f H^{expt}(CH_4)$

Adding these three equations together yields the desired reaction. Therefore, according to Hess's Law, the sum of the three enthalpy changes will yield the heat of formation of carbon dioxide.

Equation I is the isodesmic reaction whose enthalpy change we will calculate entirely from theory, taking H of products minus H of reactants. Equation II is twice the formation reaction for formaldehyde, so we use twice the experimental enthalpy of formation for ΔH_{II}. Equation III is the reverse of the formation reaction for methane, so we use -1 times the enthalpy of formation of methane for ΔH_{III}.

Combining I, II and III algebraically into one equation, the enthalpy of formation of carbon dioxide is thus:

$$\Delta_f H(CO2) = H(CO_2) + H(CH_4) - 2\times H(H_2CO) + 2\times\Delta_f H^{expt}(H_2CO) - \Delta_f H^{expt}(CH_4)$$

The experimental heats of formation for methane and formaldehyde are -74.5±0.4 kJ/mol (-0.02840 hartree) [Pittam72] and -115.9 kJ/mol (-0.04414 hartree) [Chase98], respectively.

Optimization+frequency calculations include a calculation of H. We can use the results for formaldehyde from the preceding example, as well as the earlier optimization+frequency computation for methane (Example 4.7). After running the calculation for carbon dioxide, we have all of the required data:

SYSTEM	H (hartrees)
formaldehyde	-114.41357
methane	-40.44222
carbon dioxide	-188.48363

Our calculations lead to a predicted value of -416.5 kJ/mol for the heat of formation of CO_2, compared to the observed value of -393.51±0.13 kJ/mol [Cox84].* ■

Limitations of Isodesmic Reactions & Hess's Law

Isodesmic reactions can be very useful for modeling systems and reactions. However, this approach is not without its limitations as well, which include the following:

- Good experimental values must be available for all but one reaction component.
- The predicted heat of formation is no more accurate than the least accurate of the experimental values used to compute it.
- This technique cannot be applied to activation barriers.
- This technique cannot be applied to reactions which do not happen to be isodesmic (for example, the destruction of ozone by atomic chlorine).
- Different reactions often predict different values for the same heat of formation. Since this technique does not produce a uniquely defined value for the heat of formation, it is not a model chemistry and cannot be systematically evaluated quantitatively.

The final point is illustrated in our next example.

EXAMPLE 6.9 TESTING HESS'S LAW

We will compute the heat of formation for ethane in two different ways, using the following sets of reactions:†

- propane + H_2 → ethane + methane
  3 C + 4 H_2 → propane
- 2 methane → ethane + H_2
  2 C + 4 H_2 → 2 methane

* Interestingly, B3LYP performs better than APFD on this problem, predicting a heat of formation of -404.1 kJ/mol.

† Not isodesmic.

In each case, the sum of the reactions gives the formation reaction for ethane. Therefore, according to Hess's Law, we can add the component ΔH values to obtain the heat of formation of ethane.

We will use the H_2 results from Example 6.7 (p. 247), the methane results from Example 6.7 (p. 247), and run optimization+frequency calculations for propane and ethane. The following table summarizes our results:

SYSTEM	ENTHALPY (hartrees)	$\Delta_f H$ (kJ/mol)[a]
H_2	-1.15908	0.0
propane	-118.96342	-104.7±0.5
methane	-40.44222	-74.5±0.4
ethane	-79.70110	?

[a][Pittam72]

The following table presents the computed values of ΔH for the reaction and $\Delta_f H$ for ethane:

REACTION	ΔH (kJ/mol)	$\Delta_f H$ (kJ/mol)
propane + H_2 → ethane + methane	-54.7	
3 C + 4 H_2 → propane	-104.7	
CH_4 → 2 H_2 + C	74.5	
ethane		-84.9
2 methane → ethane + H_2	63.7	
2 C + 4 H_2 → 2 methane	-149.0	
ethane		-85.3

Both ways of computing the heat of formation of ethane agree well with the observed value of -83.8±0.3 kJ/mol [Pittam72]. Be aware that this is not always the case. Therefore, when using Hess's law, we recommend trying different reference reactions to verify that there is agreement.■

Exercises

EXERCISE 6.1 ELECTRON DENSITIES OF SUBSTITUTED BENZENES

The ways in which functional groups affect electrophilic aromatic substitution is a well-studied topic in organic chemistry:

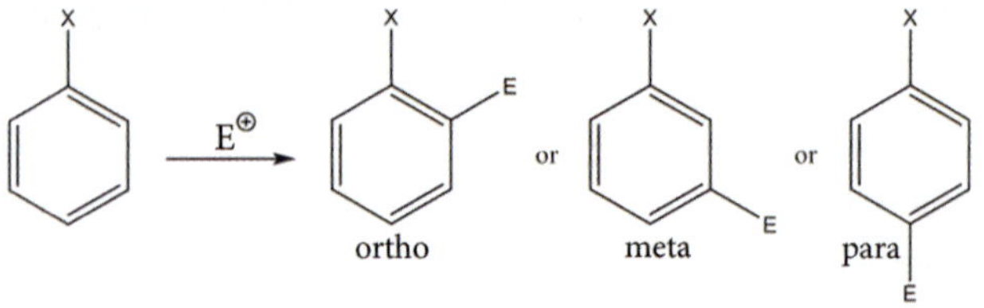

$E^{\oplus}$
ortho or meta or para

In this exercise, you will use a visualization technique known as a difference density as an aid to understanding this phenomenon. A difference density plots the difference between two electron density distributions as a surface with positive and negative lobes. It enables you to visualize the shift in electron density between two compounds or two environments.

The nitration of nitrobenzene and of chlorobenzene are known to occur via the same mechanism: the ring is initially attacked by NO_2^+, yielding a cation intermediate for each isomer. When the nitration process is fully complete, the distribution of the various isomers of the final product varies greatly for the two compounds (all percentages below should be considered approximate):

SYSTEM	ORTHO	META	PARA
$C_6H_4NO_2Cl$	29%	1%	70%
$C_6H_4N_2O_4$	6%	92%	2%

Experimental population results vary with reaction conditions. For example, [Reusch13] reports a range for nitrobenzene:
ortho 5%-8%,
meta 90%-95%,
para 0%-5%.

We will explore this phenomenon by examining the electron densities of the two species, as well as their difference density. The required procedure is described in detail below using GaussView. While WebMO has the ability to visualize the electron density and to save cube files, there is currently no way to load in an external cube file, so this technique cannot be performed with this package.

The first task is to create an electron density cube for benzene to be used as the reference orientation for the two substituted derivatives.

GaussView

1. Create a benzene molecule, and clean the structure. Make sure that the structure has D_{2h} symmetry (use the **Edit=Point Group** facility). Disable symmetry during the calculation to ensure that the program does not reorient your structure (see instructions in the margin); the corresponding Gaussian keyword is **NoSymm**. Then run a single point calculation using our standard model (without optimizing the molecule). Save the resulting checkpoint file.

Disabling Symmetry for a Gaussian Calculation

Next, run single point energy calculations for the two compounds we are studying.

2. Starting from the benzene structure just created, replace the hydrogen atom at carbon 1 with a chlorine atom. Your building program will place the atom at an appropriate distance, so do not clean further. Launch a single point calculation using our standard model, again with **NoSymm**.

   Repeat the preceding step using a nitro group instead of a chlorine atom.

We will create a reference cube file for the total density of benzene. We will use it to impose the identical orientation on the density data for the two derivatives.

3. In GaussView, open the checkpoint file from the benzene calculation. Create a new total density cube using the SCF density.* Save the cube to an external file using the **Save Cube** option on the **Cube Actions** menu in the **Surfaces and Contours** dialog.

4. Use the **cubegen** utility included with Gaussian to generate a cube for each compound having the same dimensions as the one for benzene. Note that you will need to create a formatted version of the checkpoint file prior to running **cubegen**.

See the *Gaussian User's Reference* for details about **cubegen**.

* The corresponding command for the **cubegen** utility is (using arbitrary example file names):

```
cubegen 0 density=scf benz.fchk benz.cub -2 h
```

Here are the required commands for chlorobenzene:

```
formchk clb.chk
cubegen 0 density=scf clb.fchk clb.cub -1 h benz.cub
```

The **cubegen** command says to produce a file containing volumetric ("cube") data for the SCF electron density from the file *clb.fchk*, saving the generated data into the file *clb.cub*. The dimensions for the generated cube are taken from the cube file specified as the final parameter to the command: here, *benz.cub*, the cube file saved from GaussView for benzene.

Perform the same operation for nitrobenzene.

These two cube files can now be loaded into GaussView for visualization.

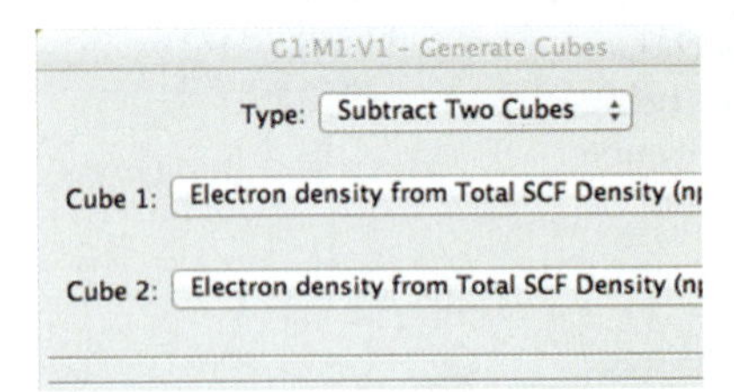

5 Open the cube file for chlorobenzene with GaussView. Load the generated cube file for nitrobenzene using the **Load Cube** option on the **Cube Actions** menu in the **Surfaces and Contours** dialog. The new cube will appear in the **Cubes Available** list.

6 Create a new cube of type **Subtract Two Cubes**. Select the cube corresponding to chlorobenzene as **Cube 1** and the one corresponding to the nitrobenzene as **Cube 2**. Note that the two cubes will typically have identical names; you will have to select the correct one by position in the list. Finally, create and display the surface for the difference density cube.

Examine and compare the density visualizations for the individual molecules as well as the difference density. Do they help illuminate which products are favored in each case?

SOLUTION

The resulting surface is displayed in the margin. The clipping at the top occurs because the underlying grid was generated for benzene. The positive nodes are colored blue, and the negative nodes are colored yellow.* The former indicate areas within the molecule where chlorobenzene has greater electron density than nitrobenzene, while the latter indicate the opposite. The difference density indicates the preference for the ortho and para positions in chlorobenzene.

The following figure illustrates the difference densities between benzene and each of the two derivatives as well as repeating the difference density plot for the two compounds themselves. We have removed the portions of the surfaces around the substituents in each case in order to focus on the density in the ring:

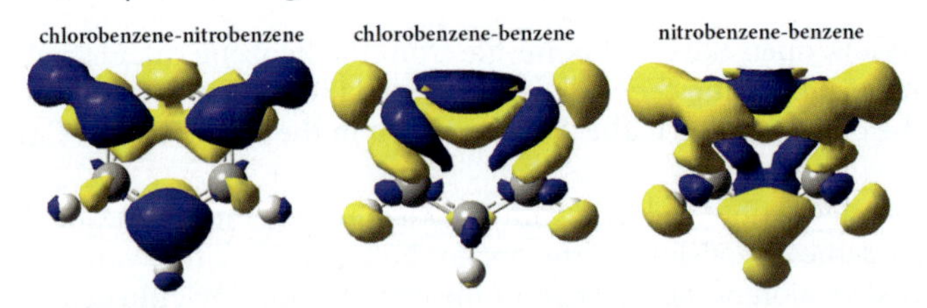

Once again, blue is positive, and yellow is negative. Blue indicates where electrons have gone as a result of the substituent, and yellow indicates where they had been in benzene. The blue area in the meta position for chlorobenzene-benzene is not perpendicular to the ring (but rather lies in the same plane). Thus, there is no enhancement in the π electron density at

* Colors for surfaces can be customized in GaussView using the **Preferences**

the meta positions in chlorobenzene compared to benzene. The increases in π density in chlorobenzene occur at the ortho and para positions, which are the preferred positions for electrophilic substitution.

The nitrobenzene-benzene plot shows a large decrease π electron density at the ortho and para positions in nitrobenzene with respect to benzene, while there is essentially no change in π electron density at the meta positions, which explains the preference for electrophilic substitution for meta over ortho and para for this compound.■

EXERCISE 6.2 RELAXED SCANS: ROTATIONAL BARRIERS IN ACETOPHENONES

This exercise will consider the rotational barriers for a series of para-substituted acetophenones (1-phenylethanones):

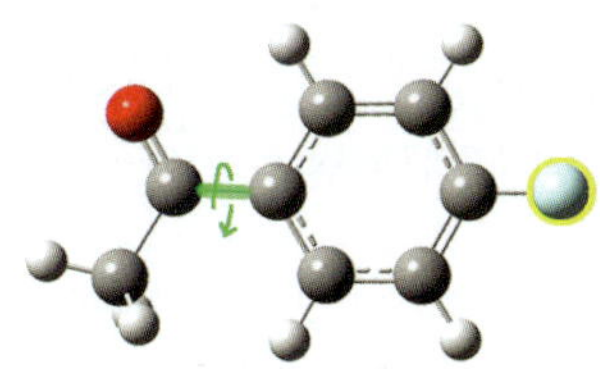

These compounds vary in the substituent on the right (indicated in yellow): fluorine in this case. We will model the barrier to rotation about the C–C bond between the acetyl group and the ring (highlighted in green above), in a chloroform solvent environment, for the substituents: H, F, CF_3 and $N(CH_3)_2$:

- Perform a relaxed PES scan for each compound, varying the O=C(CH_3)–C–C dihedral angle from 0° to 180°, in 10° increments. Since the curve will be symmetric about 90°, we could end the scan at that value, but we chose to compute both parts of the curve for easier plotting.
- Calculate the rotational barrier by predicting the Gibbs free energy for the reactant/product and transition structure for each substituent. Compute the frequencies at the experimental temperature of 140 K. Note that you can specify the temperature for a frequency calculation using the **Temp** keyword in the job's route section.

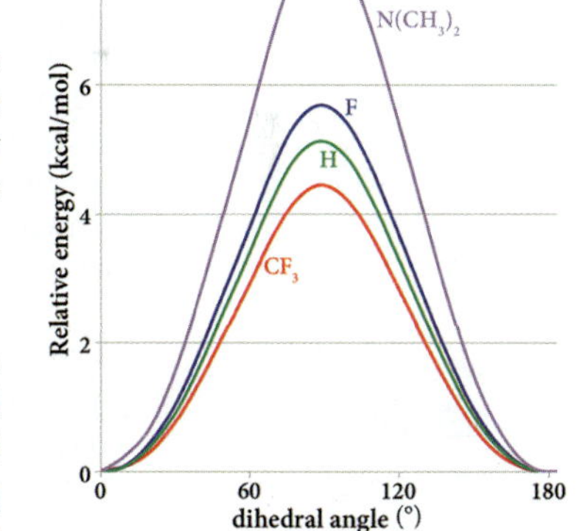

SOLUTION

The figure at the right plots the results of our relaxed scan calculations. The maximum energy structure occurs at the dihedral angle value of about 90°. We constructed initial structures for the various transition states using this value. We located the TS for the H, F and $N(CH_3)_2$ substituents by first running an **Opt=(TS,CalcFC) Freq SCRF** calculation with the 6-31G(d) basis set, followed by an **Opt=(TS,RCFC) Freq SCRF=SMD** job with our standard basis set.

The TS for the CF_3 substituent was more challenging. We reached it in three phases:

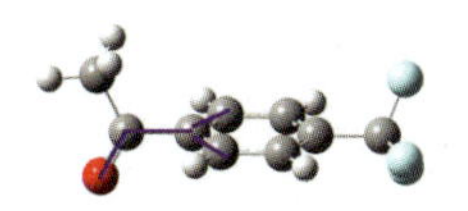

- An initial constrained optimization with the 6-31G(d) basis set and IEFPCM solvation. The O-C-C-C two dihedral angles into the ring were frozen (see figure in the margin).
- A second constrained optimization with our standard basis set, again using the default IEFPCM solvation model.
- A full transition structure optimization with our standard model chemistry and SMD solvation.

Frequency calculations were performed for each stationary point, and **Opt=RCFC** was used to read in the force constants for the second and third optimizations.

We also ran **Opt Freq** calculations for each minimum. We specified **Temp=140** for all frequency calculations.

The following table lists the results of our calculations for the rotational barriers and compares them to experiment:

	ΔG (kcal/mol)	
R	**SMD$^{140\,K}$**	**EXP.**[a]
CF_3	4.1	4.7
H	4.7	5.4
F	5.2	5.4
$N(CH_3)_2$	7.9	8.3

[a][Haloui10, Drakenberg76]

The substituent trend seen in the experiment is reproduced in our calculations.■

GOING DEEPER: ADDITIONAL ROTATIONAL BARRIERS

The work from which the preceding exercise is drawn [Haloui10] considers additional substituents (Cl, Br, I, CH_3, NO_2, CN, OCH_3 and NH_2), as well as a second solvent environment: dimethyl sulfoxide (DMSO). Thus, this exercise can be easily extended.

EXERCISE 6.3 THE H_2CO POTENTIAL ENERGY SURFACE

In this exercise, we will continue our investigation of the H_2CO potential energy surface begun in Example 6.7 (p. 247). Previously, we considered two reactions; in doing so, we studied five stationary points: three minima—formaldehyde, trans hydroxycarbene, and carbon monoxide plus hydrogen molecule—and the two transition structures connecting formaldehyde with the two sets of products. One obvious remaining step is to find a path between the two sets of products.

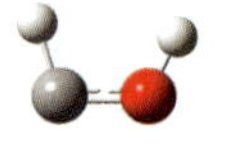

Determine the reaction path connecting trans hydroxycarbene and H_2 + CO, and predict the activation energy. This reaction occurs via a two-step process, with trans HCOH first converting to the cis form and then dissociating into carbon monoxide and hydrogen molecule:

$$\text{trans HCOH} \leftrightarrow \text{cis HCOH} \leftrightarrow H_2 + CO$$

SOLUTION

This study will require these steps for each of the two reactions:

- Finding the transition structure.
- Verifying that the stationary point is a transition structure, and computing its enthalpy.
- Determining which minima the transition structure connects.

We began with the trans-cis transition structure (TS3). We ran an **Opt=QST2** calculation and located a transition structure in 12 steps; the subsequent frequency job found one imaginary frequency. The IRC calculation (submitted with the **MaxPoints=40**, **Step=20**, **Recorrect=Never** and **Phase** options) located the following structures at the ends of the reaction path:

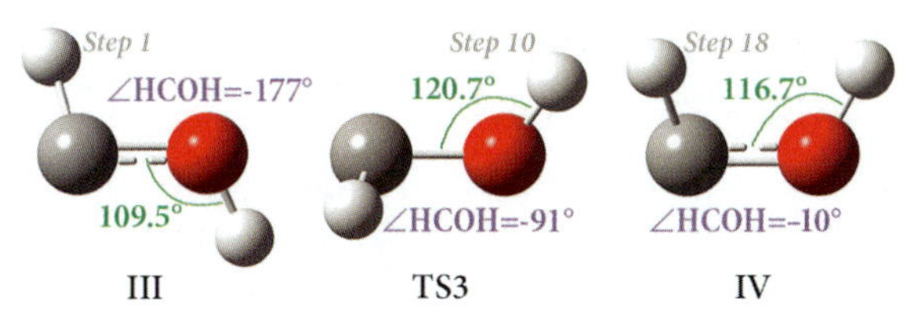

Structures III and IV clearly resemble the trans and cis forms of hydroxycarbene (respectively), and so we can conclude that this is the transition structure we were seeking.

One prominent feature distinguishing the various structures is the dihedral angle, and the transition structure's value is midway between that of the cis (IV) and trans (III) forms. The bond lengths remain more or less constant across the three structures. In contrast, notice how the bond angle changes as the conversion takes place. The following table summarizes the results for this process:

SYSTEM	ENTHALPY (hartrees)	RELATIVE TO REACTANT (kcal/mol)	RELATIVE TO PRODUCTS (kcal/mol)
reactant: trans HCOH (III)	-114.32874	0.0	-4.7
transition structure (TS3)	-114.28443	27.8	23.1
product: cis HCOH (IV)	-114.32121	4.7	0.0

These values suggest that the two hydroxycarbene isomers convert into one another relatively easily at temperatures not far above room temperature, with the trans isomer being the most stable.

Next, we consider the dissociation of the cis isomer into the products carbon monoxide and molecular hydrogen. We again ran an **Opt=QST2 Freq** calculation and located a transition structure in 11 steps. The IRC calculation (using the same options as the previous one) yielded the following results:

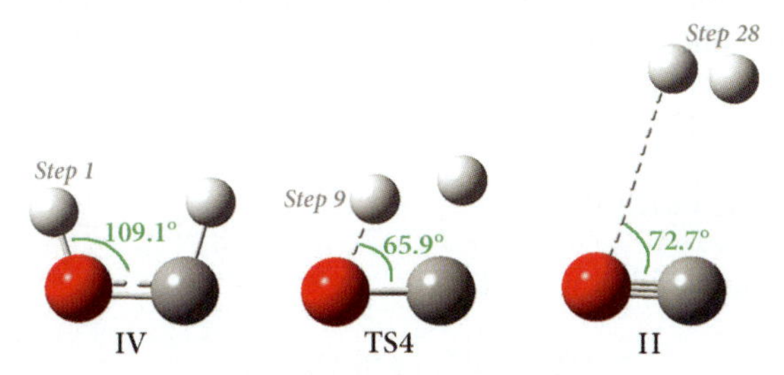

The following table summarizes the enthalpy results for the decomposition of cis-hydroxycarbene:

SYSTEM	ENTHALPY (hartrees)	RELATIVE TO REACTANT (kcal/mol)	RELATIVE TO PRODUCTS (kcal/mol)
reactant: cis HCOH (IV)	-114.32121	4.7	0.0
transition structure (TS4)	-114.24792	46.0	97.6
product: CO + H_2 (II)	-114.40346	-51.6	0.0

The barrier to molecular dissociation of the cis form is formidably high.

The following plot shows the relationships between the structures that we have studied in this exercise:

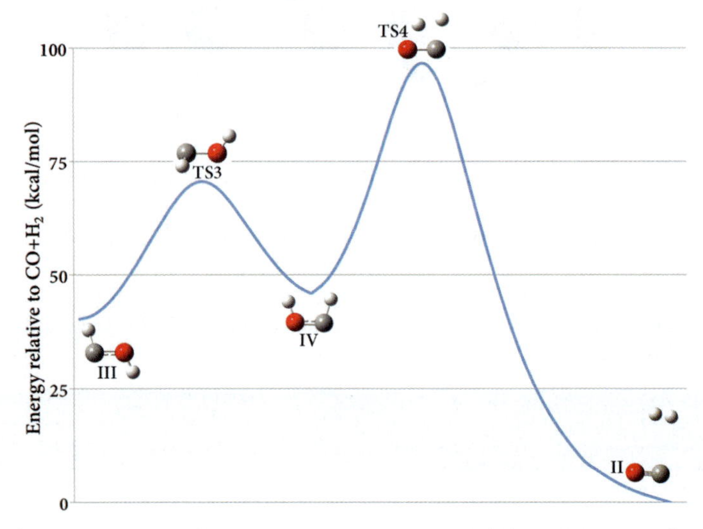

Combining the previous PES for H_2CO isomerization and dissociation with current study reveals that dissociation would occur from the H_2CO structure; it is more favorable for HCOH to revert to H_2CO and then dissociate than to dissociate directly.

While H_2CO PES is a quite simple potential energy surface, studying it familiarizes you with the tools that can be used for even the most complex ones.■

EXERCISE 6.4 THE SILICON CATION + SILANE POTENTIAL ENERGY SURFACE

Silicon cluster reactions have been an important area of research for several decades, and they are very amenable to study by electronic structure methods. This exercise will examine the potential surface for silicon cation reacting with silane (SiH_4). Such reactions are central to the growth of large silicon clusters, which occurs by sequential additions of $-SiH_2$:

$$Si^+ + SiH_4 \rightarrow Si_2H_2^+ \rightarrow Si_3H_4^+ \rightarrow Si_4H_6^+ \rightarrow Si_5H_{10}^+ \rightarrow ...$$

with H_2 also produced in each step.

We will examine the first addition reaction:

$$Si^+ + SiH_4 \rightarrow Si_2H_2^+ + H_2$$

The original researcher, Krishnan Raghavachari, found the following minima (in addition to the reactants) when he investigated the $Si_2H_4^+$ potential energy surface [Raghavachari88]:

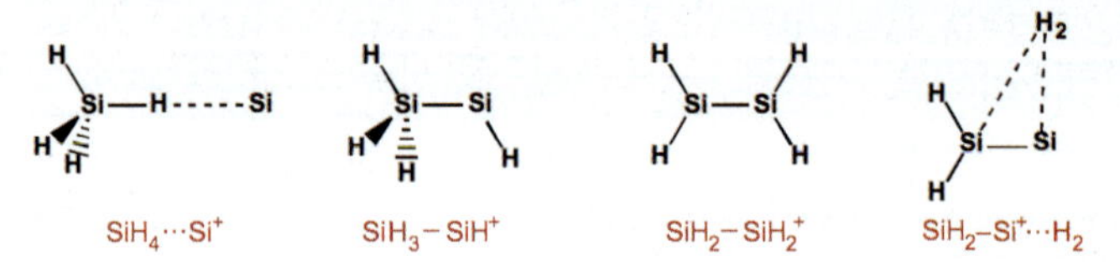

The rightmost structure is a weak complex of the products (having a binding energy of about 1 kcal/mol), and for our purposes it may be construed as the reaction end point.

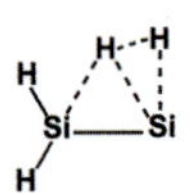

Raghavachari also found the transition structure in the margin.

Determine which of the minima are connected by this transition structure and predict the activation enthalpy for the reactions at STP. When you optimize the transition state in preparation for an IRC calculation, perform an **Opt=TS** calculation starting from this structure, and include the **CalcFC** option to compute the force constants at the first point. Also include the **MaxPoints**, **Step**, **Recorrect** and **Phase** options (as before).

When you have finished with this first transition structure, go on to locate the transition structures connecting the other minima that were located on this potential energy surface.

Make a plot of the relative enthalpies of the various systems,* indicating the known paths between them.

Note: The predicted enthalpy for H_2 using our standard model chemistry is -1.15908 hartrees (from Example 6.7).

SOLUTION

The structures/complexes on this potential energy surface have a charge of +1 and one unpaired electron (doublet spin multiplicity). For the reactants, the charge and radical are isolated on the Si^+ atom; for the products, this shifts to the $Si_2H_2^+$ species. The exceptions are SiH_4 and H_2, which are neutral singlets.

Our geometry optimization of the transition structure above completed in four steps, and the optimized structure had one imaginary frequency. This optimization and the subsequent IRC calculation produced the following results:

H_2 ELIMINATION REACTION		
OPTIMIZED TS	IRC RESULTS	
	reactant	products
IRC Step 15	$SiH_2 \cdots H \cdots SiH^+$ Step 1	$Si_2H_2^+ \cdots H_2$ Step 20

The structure on the left is the optimized transition structure; we've also noted which step in the follow-on IRC corresponds to the TS. The two structures on the right (green backgrounds) correspond to the endpoints of the IRC calculation, which are identified with their stoichiometries and IRC step number.

We'll consider the products first. It is the weakly bound SiH_2–$Si^+ \cdots H_2$ complex. When we examine the entire reaction path computed by the IRC calculation, it seems that the hydrogen molecule in the products consists of one hydrogen atom from the silicon atom on the right in the illustrations and another hydrogen atom initially shared between the two silicon atoms (view the animation for details). The movement from the reactant through the transition state to the products sees the bridged hydrogen atom moving away from the SiH_2 group as the Si-H bond distance in SiH^+ increases and the Si-Si-H bond angle decreases. While the reactant produced by the IRC calculation is suggestive of SiH_3–SiH^+ (silylsilene ion), it is not this molecule; rather, it is an intermediate compound not present in our initial list of candidates. The transition structure connects the two minima in a H_2 elimination reaction.

* Obviously, you will also need to optimize the minima and run frequency calculations for them. For Si^+, only a frequency calculation is needed—there is nothing to optimize—or you can predict only the total energy and manually add the translational contribution to the enthalpy of 2.5RT=0.00236 hartrees (the sole relevant thermal contribution for an atom).

We also optimized the two minima in order to compute the activation enthalpies (discussed a bit later).

We went on to consider other reactions on this potential energy surface.

Hydrogen Migration Reactions

An obvious follow-up to the preceding results is to ask where the $SiH_2\cdots H\cdots SiH^+$ bridged intermediate fits into the PES we are studying. It could conceivably connect to either $SiH_3\cdots SiH^+$ or to $SiH_2\cdots SiH_2^+$, depending on the ultimate destination of the bridged hydrogen atom. In addition, SiH_3–SiH^+ and SiH_2–SiH_2^+ differ only in the position/bonding of one of the hydrogen atoms. Thus, it is reasonable to assume that one or more of them may participate in H migration reactions.

We began exploring this possibility by setting up an **Opt=QST2 Freq** calculation with the latter two compounds as the reactant and product. This calculation located a saddle point in 18 steps, and the frequency calculation produced a single imaginary frequency. We then ran an IRC calculation beginning from the TS. The results of these calculations are summarized in the following table:

HYDROGEN MIGRATION REACTION 1				
QST2 CALCULATION		OPTIMIZED TS	IRC RESULTS	
reactant	product		reactant	product
$SiH_3\cdots SiH^+$	SiH_2–SiH_2^+	IRC Step 10	$SiH_2\cdots H\cdots SiH^+$ Step 1	SiH_2–SiH_2^+ Step 29

The IRC ends with a structure that resembles the SiH_2—SiH_2^+ minimum, but the final point in the other direction is the bridged intermediate complex rather than the reactant with which we started. The transition state is a structure that is very close to the intermediate in form and energy.

Next, we ran a **QST3** optimization using the bridged intermediate and SiH_2—SiH_2^+ as reactant and product, along with a starting structure for the transition state. The results of that optimization+frequency job and the subsequent IRC calculation are given in the following table:

HYDROGEN MIGRATION REACTION 2					
QST3 CALCULATION			OPTIMIZED TS	IRC RESULTS	
reactant	product	TS guess		reactant	product
$SiH_2\cdots H\cdots SiH^+$	SiH_2–SiH_2^+		IRC Step 10	SiH_2–SiH_2^+ Step 1	SiH_2–SiH_2^+ Step 29

The optimized transition structure and IRC endpoints are the same as for our first TS optimization, confirming that this transition state connects these minima. The optimized structure is close to our starting guess; it is presented in a different orientation than the optimized TS in order to illustrate its structure more fully.

Finally, we ran a **QST2** optimization+frequency calculation using the bridged intermediate and SiH_3–SiH^+ as reactant and product. The optimization successfully located another transition structure in 11 steps (with one imaginary frequency), and the follow-on IRC calculation confirmed that this TS connects these reactant and product:

HYDROGEN MIGRATION REACTION 3				
QST2 CALCULATION		OPTIMIZED TS	IRC RESULTS	
reactant	product		reactant	product
SiH_2···H···SiH^+	SiH_3–SiH^+	IRC Step 6	SiH_2···H···SiH^+ Step 1	SiH_3–SiH^+ Step 16

Silicon Insertion Reaction

The SiH_4···Si^+ and SiH_3–SiH^+ minima differ in the atom bonded to the silicon atom in silane, and they thus correspond to a silicon insertion reaction. After optimizing the minima, we set up an **Opt=QST3 Freq** calculation to locate the transition structure connecting them, using the structure in the table following as our initial guess for the transition state. We distorted the SiH_3–SiH^+ structure so that the hydrogen atom in the SiH^+ group bent toward the other group; this was a sufficient modification due to the fact that the lone silicon atom in the optimized SiH_4···Si^+ complex lies midway between two hydrogen atoms in the silane group rather than forming a Si–H–Si linear bond (see the illustration in the table following). We also set the Si–Si bond distance to be halfway between that in SiH_3–SiH^+ and the Si-H bond length in SiH_4···Si^+. This calculation located a transition structure in 25 steps (with a single imaginary frequency).

The following illustration shows the input for the optimization job as well as results of the IRC calculation beginning from the optimized transition structure:

SILICON INSERTION REACTION					
QST3 CALCULATION			OPTIMIZED TS	IRC RESULTS	
reactant	product	TS guess		reactant	product
SiH_4···Si^+	SiH_3···SiH^+		IRC Step 16	SiH_4···Si^+ Step 1	SiH_3···SiH^+ Step 30

The hydrogen atom clearly shifts from one silicon atom to the other in the course of the IRC, along with a few other, less dramatic changes in the H–Si–H bond angle and the silicon-silicon bond length.

We can now make a plot of the silicon cation + silane potential energy surface:

MINIMA	REL H (kcal/mol)
$SiH\cdots H\cdots SiH_2^+$	0.0
$Si_2H_2^+\cdots H_2$	17.4
$SiH_3\cdots SiH^+$	-0.9
$SiH_2—SiH_2^+$	-14.9
$SiH_4\cdots Si^+$	41.7
TRANSITION STATES	
H_2 elimination	20.7
Si insertion	13.4
H migration I	-0.1
H migration II†	3.0

†Migration reaction 3 in the text.

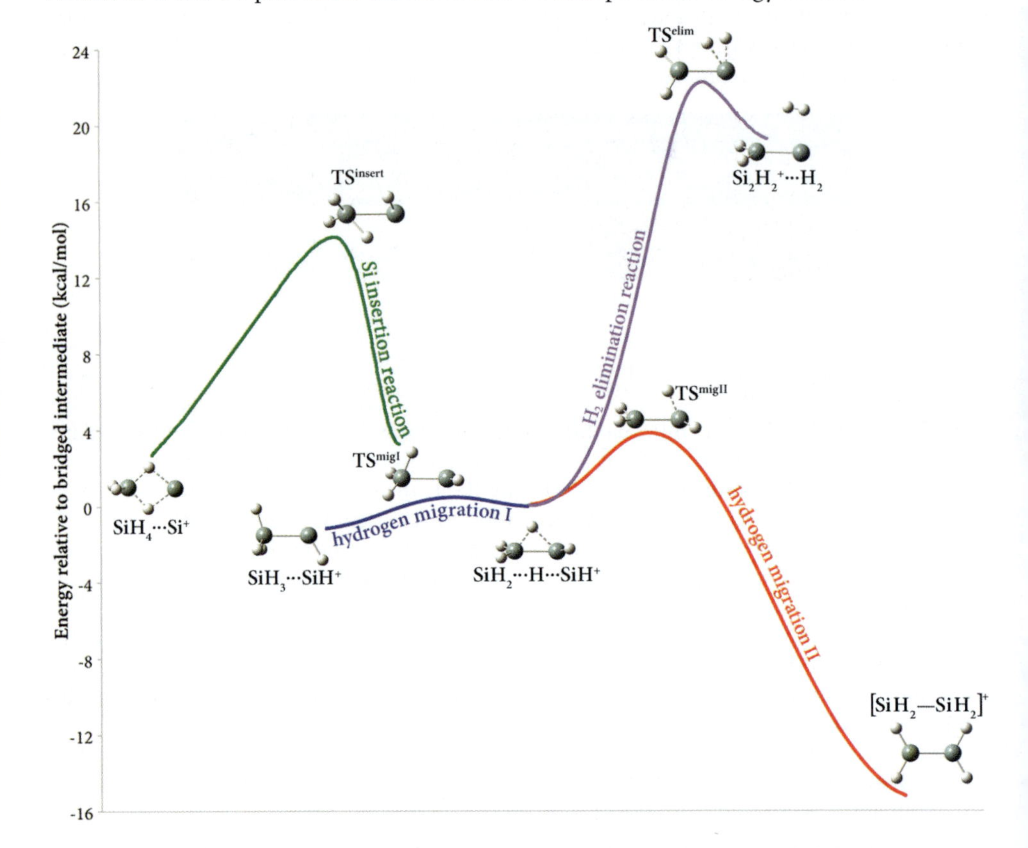

The numerical values of the enthalpies relative to the bridged intermediate for the various structures are found in the table in the margin.

We can now answer the question posed in the problem statement for this exercise: the indicated transition structure connects the bridged intermediate and the H_2 elimination reaction products. This intermediate is itself involved in another hydrogen migration process from the product of the Si insertion reaction.■

EXERCISE 6.5 ISODESMIC REACTIONS

In this exercise, we will consider a series of isodesmic reactions of the form:

▸ $CH_3CXO + CH_3\text{–}CH_3 \leftrightarrow (CH_3)_2CO + CH_3\text{–}X$

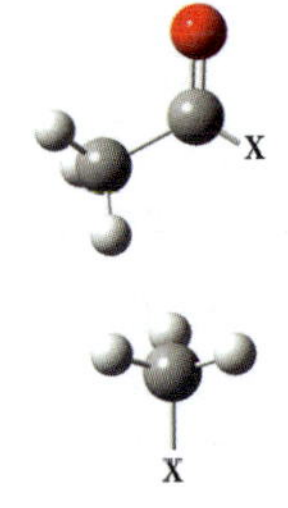

where the substituent X varies according to the following table:

X	Reactants	Products
H	acetaldehyde + ethane	acetone + methane
F	acetyl fluoride + ethane	acetone + methyl fluoride
Cl	acetyl chloride + ethane	acetone + methyl chloride

Compute ΔH^{298} for each reaction using our standard model chemistry. Which of these reactions are exothermic and which are endothermic?

Note: We have studied some of these molecules before, so you can draw on past results in these cases if you ran the corresponding calculations: acetaldehyde, acetone and acetyl chloride: Exercise 4.2; methane: Example 4.7; ethane: Example 6.9.

SOLUTION

We ran optimization+frequency calculations for all of the relevant compounds. The following table summarizes the results, as well as the predicted [Wiberg92d] and observed values for ΔH^{298}:

	H^{298} (hartrees)			
ethane	-79.70110			
acetone	-192.96538			
			ΔH (kcal/mol)	
X	CH_3CXO	CH_3X	CALC.	OBS.
H	-153.69127	-40.44222	-9.6	-9.9±0.3
F	-252.93127	-139.63804	18.2	17.9±1.3
Cl	-613.22951	-499.95288	7.8	6.6±0.3

The predicted values are all in very good agreement with experiment. The theoretical calculations correctly predict the direction of each reaction: only the first one is exothermic.■

GOING DEEPER: ADDITIONAL ISODESMIC REACTIONS

You can expand the consideration of isodesmic reactions by considering the additional reactions: X= NH_2, SiH_3, PH_2, CN, SH, CF_3.

EXERCISE 6.6 HEAT OF FORMATION FOR TETRAFLUOROSILANE

Compute the heat of formation for SiF_4 using the following experimental heats of formation and Hess's Law:

COMPOUND	$\Delta_f H^{exp}$ (kJ/mol)
F_2	0.0
HF	-273.3±0.7[b]
H_2	0.0
SiF_2H_2	-790.78[a]
$SiFH_3$	-376.56[a]
SiF_3H	-1200.81[a]
SiH_4	+34.31[a]

[a][Chase98] [b][Cox84]

How do the predictions from each Hess's Law computation compare?

Note: We will use the results for hydrogen molecule from Example 6.7 and for silane from Exercise 6.4.

SOLUTION

We addressed this problem by studying the following reactions:

- $SiH_4 + 2F_2 \rightarrow 2H_2 + SiF_4$
- $SiF_2H_2 + 2HF \rightarrow 2H_2 + SiF_4$
- $SiF_3H + HF \rightarrow SiF_4 + H_2$
- $SiFH_3 + 3HF \rightarrow 3H_2 + SiF_4$
- $SiH_4 + 4HF \rightarrow 4H_2 + SiF_4$
- $SiH_4 + 4F_2 \rightarrow SiF_4 + 4HF$

The following table summarizes the results of our calculations and the reaction enthalpies (at 298.15 K) for the preceding reactions:

COMPOUND	H (hartrees)
F_2	-199.42739
HF	-1.15908
H_2	-100.39650
SiF_2H_2	-490.32243
SiF_3H	-589.60954
SiF_4	-688.89085
$SiFH_3$	-391.03866
SiH_4	-291.76533

REACTION	$\Delta H^{reaction}$ (kJ/mol)
ΔH_1: $SiH_4 + 2F_2 \rightarrow 2H_2 + SiF_4$	-1546.14
ΔH_2: $SiF_2H_2 + 2HF \rightarrow 2H_2 + SiF_4$	-245.70
ΔH_3: $SiF_3H + HF \rightarrow SiF_4 + H_2$	-115.24
ΔH_4: $SiFH_3 + 3HF \rightarrow 3H_2 + SiF_4$	-367.39
ΔH_5: $SiH_4 + 4HF \rightarrow 4H_2 + SiF_4$	-461.68
ΔH_6: $SiH_4 + 4F_2 \rightarrow SiF_4 + 4HF$	-2630.60

The following table summarizes the computation of the heat of formation for tetrafluorosilane corresponding to each reaction. These expressions combine the computed reaction enthalpy with the appropriate experimental heats of formation to estimate $\Delta_f H$ for our target molecule. The accepted value for the heat of formation for SiF_4 is -1615.0±0.8 kJ/mol [Cox84]; the final column in the table gives the error for each predicted value.

EXPRESSION	$\Delta_f H$ (kJ/mol)	% ERROR
$\Delta H_1+\Delta H_f(SiH_4)$	-1511.8	-6.4%
$\Delta H_2+\Delta H_f(SiF_2H_2)+2\Delta H_f(HF)$	-1583.1	-2.0%
$\Delta H_3+\Delta H_f(SiF_3H)+\Delta H_f(HF)$	-1589.3	-1.6%
$\Delta H_4+\Delta H_f(SiFH_3)+3\Delta H_f(HF)$	-1563.8	-3.2%
$\Delta H_5+\Delta H_f(SiH_4)+4\Delta H_f(HF)$	-1520.6	-5.8%
$\Delta H_6+\Delta H_f(SiH_4)-4\Delta H_f(HF)$	-1503.1	-6.9%

Several of these silicon compounds used as reference have large uncertainties in their experimental heats of formation.* As a result, the theoretical estimates for the heat of formation of tetrafluorosilane vary quite a bit in accuracy. ■

* Although the NIST online database no longer always includes the uncertainty in its reported values, the original experimental work does.

Advanced Topics and Exercises

ADVANCED EXERCISE 6.7 METHYL ACETATE HYDROLYSIS REVISITED

In Chapter 5 (pp. 210ff and 217ff), we computed the free energy of reaction for the base-catalyzed hydrolysis of methyl acetate going to acetate ion and methanol. We now will explore the mechanism for this reaction.

This reaction can theoretically proceed either via a concerted base-catalyzed alkyl bimolecular—$B_{AL}2$—mechanism involving a complex of the reactants and an S_N2 transition structure or via a two-step base-catalyzed acyl bimolecular—$B_{AC}2$—mechanism involving a tetrahedral intermediate with acyl cleavage that connects to the reactants and products via separate transition structures [Haeffner99, Takashima78, Pranata94]. These two mechanisms differ as to which carbon-oxygen bond is broken (alkyl vs. acyl, respectively).

The following figure illustrates the $B_{AC}2$ mechanism; the separate reactant and product molecules are included for reference. The intermediate and transition states include one explicit water molecule, which are displayed at reduced size to make the structures easier to decipher.

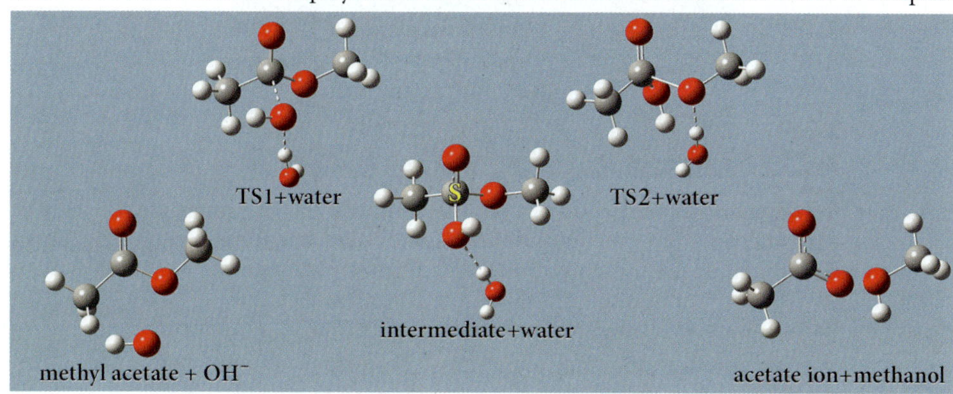

Notice that the intermediate is chiral. The attack of the hydroxide can equivalently occur from either side of the molecule. We include one water molecule in the mechanism because it plays an explicit role in the second step of this mechanism [Haeffner99]. As the Acyl C-O bond is broken, a proton from a nearby water molecule will attach to the methoxy group to form methanol. At the same time, acetate ion and a new water molecule can form from the remaining pieces.

The following figure illustrates the $B_{AL}2$ mechanism. The transition structure includes one explicit water molecule as above.

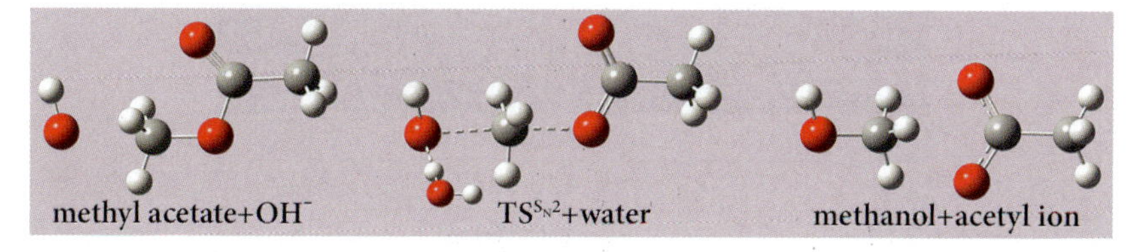

Evaluate each of these pathways and predict which one is dominant in aqueous solution. Use the SMD solvent model to optimize the geometries in water for TS1, the intermediate, TS2 and TS^{S_N2}. Use the relative Gibbs free energies of the various molecules to answer the question.

Hint: In constructing the input for the intermediate species, first build the molecule without a water molecule. Clean this structure (or optimize it using a small basis set). Then add a

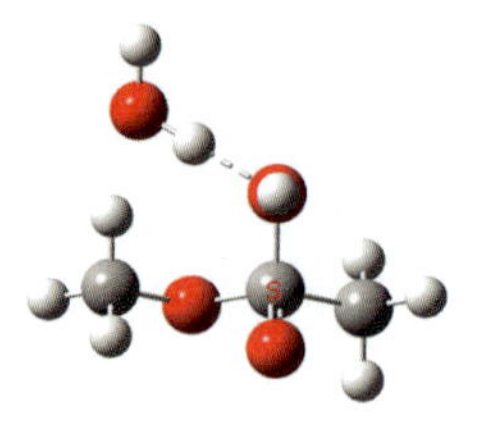

valence at the hydroxide oxygen. Move this new atom away to a distance of 1.4 Å, and replace the bond with a half-bond. Then add a valence to the hydrogen you just added, and replace it with a tetravalent oxygen oriented as shown in the figure in the margin.

For the transition structure searches, begin with **QST3** optimizations using the smaller 6-31+G(d) basis set, followed by frequency calculations. These force constants can then be used in a subsequent optimization with the large standard basis set. These two optimization+frequency calculations should be performed using the default IEFPCM solvation model. Finally, optimize the transition structure with the SMD solvation method.

SOLUTION

We optimized the intermediate and verified that it was a minimum with the frequency calculation. We used the following input for our QST3 optimizations for the $B_{AC}2$ mechanism:

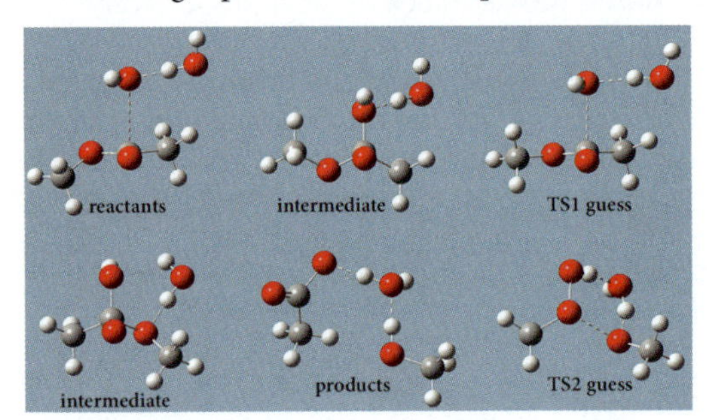

Note that the water molecule is positioned differently in the intermediate for the two calculations. The transition structure optimizations succeeded, and the subsequent frequency calculations both verified the presence of a single imaginary frequency.

We use the following input for the QST3 calculation for the $B_{AL}2$ mechanism:

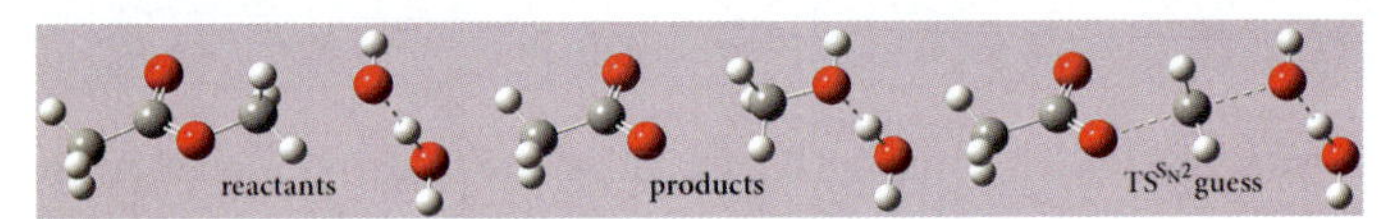

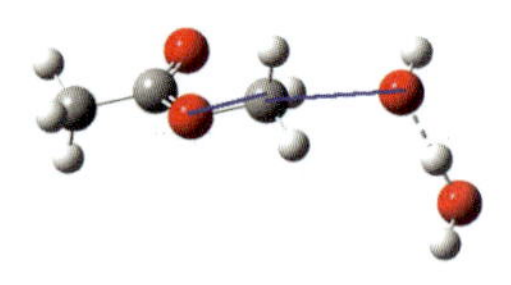

This optimization was more challenging than the others. It was necessary to perform a constrained optimization with SMD following the IEFPCM optimization with our standard model chemistry. The two C-O bonds indicated in the margin were frozen. Once this optimization succeeded, we followed it by an unconstrained optimization with SMD solvation.

The following table lists the predicted free energies* for the various species:

SPECIES	G (hartrees)	REL. TO INTERMED. (kJ/mol)
intermediate	-420.50820	0.0
TS1	-420.50142	17.4
TS2	-420.50187	16.6
TS^{S_N2}	-420.47762	80.3

* The actual free energies of all compounds will have an additional -ln(2)RT term because of their chiral nature. However, since we are comparing only relative energies, we can safely ignore this.

Our results predict that the $B_{AC}2$ mechanism is favored in aqueous solution. This is in agreement with the findings of experiment [Isaacs87]. In fact, the intermediate has been observed [Bender67].■

Scanning Multiple Variables

Potential energy surface scans can be performed for more than one scan variable. The first point of a two-variable scan uses the initial values of both variables. In subsequent steps, the first variable is incremented by its specified value for the number of steps requested in the scan, all the while keeping the second variable at its initial value. When the scan point where the first variable has its maximum* value is complete, the second scan variable is incremented, and the first variable retains its current (final) value. The scan then proceeds backward over the value range for the first variable while the second variable remains constant, until the first variable attains its initial (usually minimum) value. At this point, the second variable receives its second increment, and the cycle begins again. This process repeats over all of the possible combinations of values for the two variables, ending with the final scan point corresponding to the maximum values of both variables. Potential energy surface scans with more than two variables proceed analogously.

The results from two-variable scans can be plotted as a surface, with the two scan variables as the X and Y axes and the energy as the Z axis. The following example will illustrate this.

ADVANCED EXAMPLE 6.10 **THE O_3 POTENTIAL ENERGY SURFACE**

We will perform a rigid scan of the O_3 potential energy surface, which includes ordinary and cyclic ozone (illustrated in the margin). The latter compound has not been observed in nature. However, one research group has observed evidence of minute quantities of cyclic ozone at the surface of magnesium oxide crystals annealed at high temperature in air [Plass98].

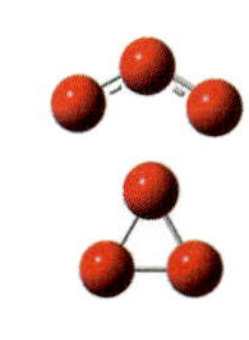

For this scan, we will vary both the O–O distance (from 1.0 to1.4 Å) and the O–O–O angle (from 55° to 125°).

The following illustration (produced by GaussView) plots the scan results as a surface:

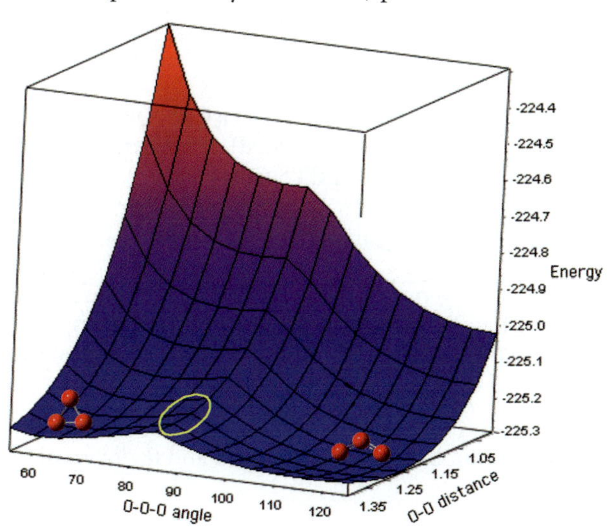

* More precisely, its terminal value since the specified increment value may be negative.

The minimum corresponding to ordinary ozone appears in the right side of the plot, and the minimum for cyclic ozone appears near the left front corner of the bounding box. It is easy to see that ordinary ozone is the lower energy structure.

The region highlighted in yellow contains the transition structure connecting the two minima. The saddle shape of this region is evident: it is a maximum in energy as one moves between the two transition structures, and it is a minimum in energy along the ridge corresponding to an O-O-O bond angle of about 85°. We could use a point from this region as the starting structure for a transition state optimization, either as the third structure in an **Opt=QST3** calculation or in an **Opt=TS** job.■

ADVANCED EXERCISE 6.8 STUDYING KETO-ENOL TAUTOMERISM

The keto-enol tautomerism of 2-pyridone (PY) and 2-hydroxypyridine* (HY) has been extensively studied, both experimentally and computationally.

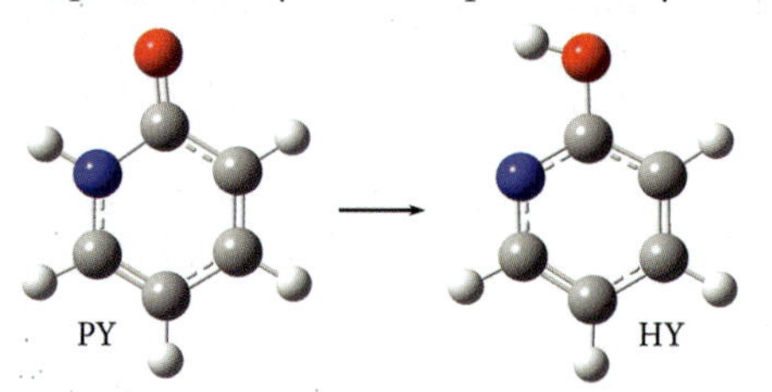

In an aqueous solution, the enthalpy of reaction has been measured as 14.2 kJ/mol at 298 K [Cook76]. Use our standard model chemistry and the CBS-QB3 high accuracy method to predict the enthalpy of reaction in both the gas phase and aqueous solution.† How do the results compare between the gas phase and aqueous phase? How well do they compare to experiment? How important is the inclusion of an explicit water molecule (see below)?

Here are some suggestions for the gas phase part of this exercise:

- Perform an **Opt Freq** calculation on PY and HY using our standard model chemistry.
- Perform a two-variable relaxed PES scan to help locate a starting structure for the transition state connecting these two forms. Then use our standard model in an **Opt(TS,CalcFC) Freq** calculation.
- Using the three stationary points found, perform a **CBS-QB3** calculation. In general, this is a standalone keyword (no basis set is needed). However, in the case of the transition structure, you need to also include **Opt(TS,CalcFC)**.

Here are some suggestions for the calculations in the aqueous phase:

- Begin from the gas-phase geometries found above for the three stationary points and perform an optimization+frequency calculation on each species, using the SMD solvent model for water. First use our standard model, and then repeat the process with the CBS-QB3 model.
- Next, use one explicit water molecule within the SMD model in an attempt to help shuttle the hydrogen atom between the nitrogen and the oxygen. This catalytic water

* IUPAC: 2(1H)-pyridinone.

† When using SCRF in conjunction with compound model chemistries such as CBS-QB3, the effects of the solvent are included only on the initial MOs arising from the SCF part of the calculation.

molecule serves to shuttle the hydrogen atom between the two heavy atoms, acting as a bridge connecting the donor and acceptor sites. One of its hydrogen atoms will be pointing at the nitrogen atom for the keto form and at the oxygen atom for the enol form [Barone95a]. Optimize all three structures with the explicit water included, first using the standard model and then the CBS-QB3 model. Finally, predict the frequencies for each optimized structure.

SOLUTION

We optimized the structure of the two minima in the gas phase. In order to locate the transition structure, we performed a relaxed PES scan, beginning from the HY structure and stepping through a range of distances for the key bonds:

- O-H: 15 increments of 0.1 Å starting from 1.0 Å.
- N-H: 15 decrements of 0.1 Å starting from 2.36 Å.

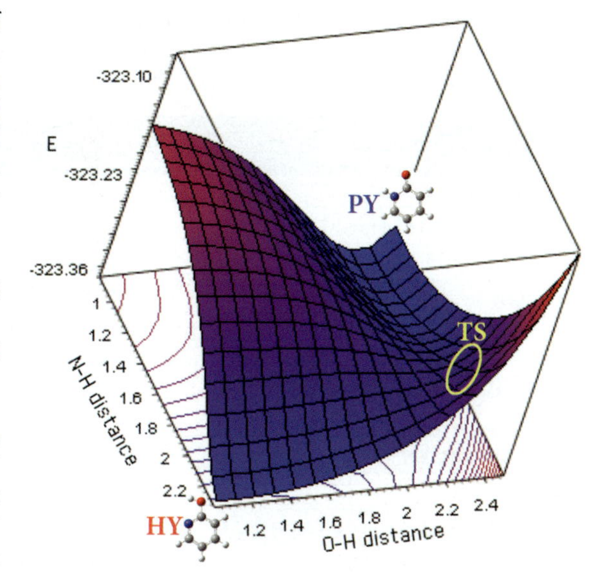

The plot on the right presents the results of the PES scan. The HY minimum is located in the front left corner of the plot while the PY minimum is located near the back right corner. We've highlighted the region containing the transition structure in the front right corner. The saddle shape of this region is evident from the plot.

We selected the structure at the lowest energy point in the region of the saddle point and optimized it, locating the transition structure in the margin in four steps. We verified that the structure is a transition state with a frequency calculation.

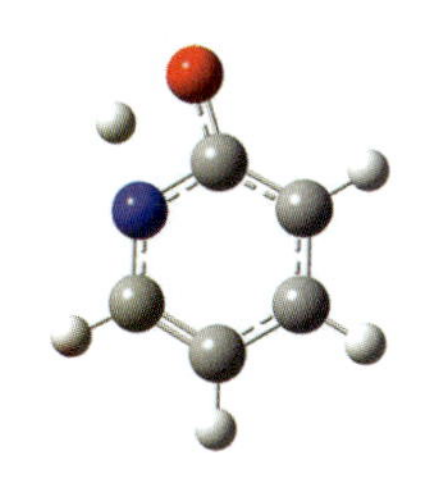

We reoptimized these structures with the CBS-QB3 method in the gas phase and in aqueous solution with SMD solvation, using both our standard model chemistry and CBS-QB3.

The final phase of the study involved adding an explicit water molecule. We placed the water molecule opposite the nitrogen atom with one of its hydrogen atoms positioned toward the receiving atom for the two minima. For the transition state's initial structure, we placed the water molecule in a similar location and lengthened one O-H bond so that there were hydrogen atoms about the same distance from both the oxygen and nitrogen atoms:

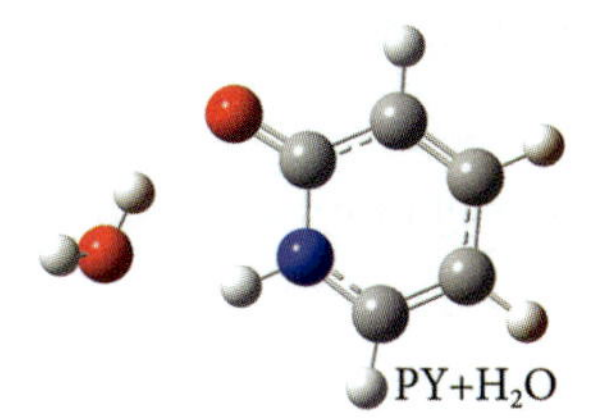

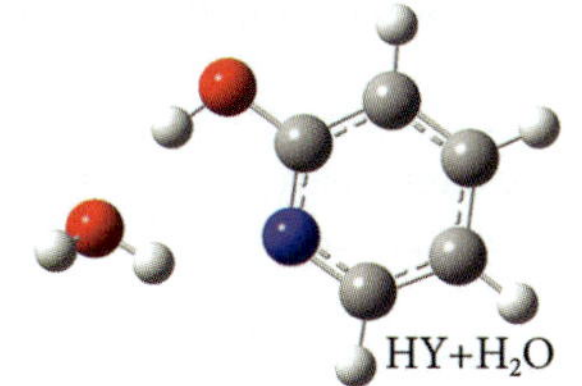

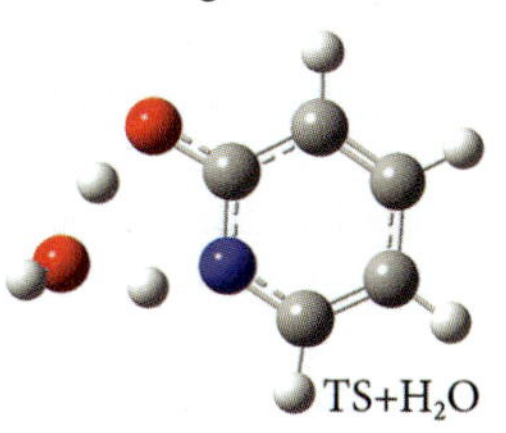

We optimized the resulting structures and verified the natures of the stationary points with frequency calculations using our standard model chemistry. The TS optimization was done in two phases: first with the default IEFPCM model and them with the SMD model.

Finally, we ran an IRC calculation starting from the resulting transition state. The animation resulting from the IRC shows the proton shuttling process very clearly. Finally, we ran CBS-QB3 calculations for these systems.

The following table presents the results of our calculations. ΔH is the reaction enthalpy change, and ΔH_a is the activation enthalpy: the enthalpy difference between the transition structure and PY. The results of the original computational study, which used the more expensive G3 method, are also included for comparison. Note that the latter reports results at 0 K.

MODEL CHEMISTRY	#WATERS	ΔH^{298} (kJ/mol)	ΔH_a^{298} (kJ/mol)
GAS PHASE			
APFD/6-311+G(2d,p)		1.2	135.8
CBS-QB3		-5.9	141.0
AQUEOUS SOLUTION			
APFD/6-311+G(2d,p) SCRF=SMD	0	21.5	153.4
	1	20.1	42.7
CBS-QB3 SCRF=SMD	0	12.9	156.5
	1	14.7	55.1
EXPERIMENT			
[Cook76]		14.2	
ORIGINAL COMPUTATIONAL STUDY†	**#WATERS**	ΔH^0	ΔH_a^0
G3 [Sonnenberg09a]	0	12.4	155.0
	1	10.8	62.2
	2	10.9	67.5
	3‡	13.2	*

† Enthalpies reported at 0 K (equivalent to the thermal energy reported by Gaussian).
‡ The original researchers' animation of the three-water case is available at: pubs.acs.org/doi/media/10.1021/ct800477y/ct800477y_weo_004.avi
* Calculation impractical at time of original research.

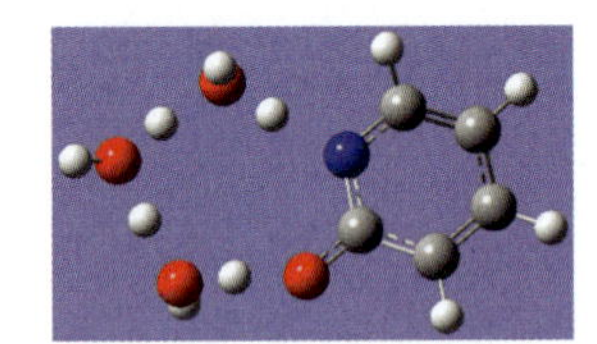

The gas-phase calculations predict a nearly isoenthalpic reaction with a high enthalpy barrier. In aqueous solution, the reaction becomes endothermic. Explicit waters reduce the activation enthalpy substantially. Using the highly accurate CBS-QB3 compound model chemistry leads to a result for ΔH that is very close to the experimental estimate. The CBS-QB3 results are also consistent with the published G3 calculations [Sonnenberg09a]. ■

Pitfalls of Multicoordinate Relaxed PES Scans

Relaxed potential energy surface scans involving multiple coordinates must be set up with some care, and their output must be checked to ensure that all of the component optimizations successfully located stationary points. Beware of variable combinations which result in extremely distorted or even unphysical starting molecular geometries as the corresponding constrained optimizations may fail.

Searching the output file from a relaxed scan for the string "Non-Optimized" will indicate whether any of the component optimizations failed to converge.

On UNIX/Linux systems, you can use the following command to quickly summarize the individual optimization results from a relaxed PES scan. The compound command below displays the final step number for each successive optimization performed in the course of the scan:*

```
$ grep Step\ num filename.log | tac | uniq -s 57 -w 5 |\
  sort -k+13 -n | more
```
Lists final step number for each optimization — *Indicates the particular point in the PES scan*
```
Step number 5 out of a max 62 on scan point 1 out of 256
Step number 4 out of a max 62 on scan point 2 out of 256
```

Points where the step number reaches the maximum should be scrutinized especially carefully to ensure that the optimization in fact converged.

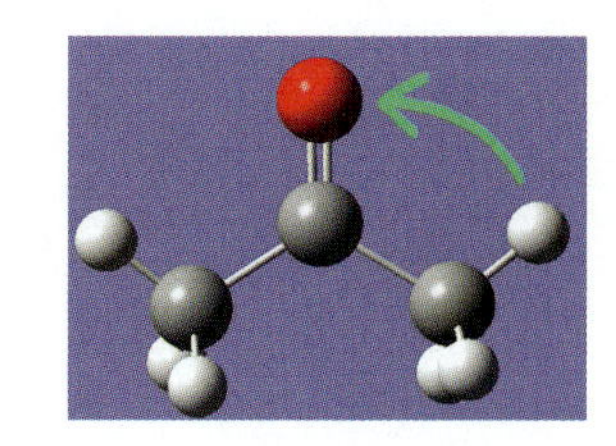

When defining variables for relaxed PES scans, be sure that the variables are truly independent. Consider the case of a proton transfer between the oxygen and carbon atoms for the species in the margin. You can define the two scan variables as the C-H and O-H distances, each ranging from about 1.0 to 2.5 Å. However, this scan will encounter problems, and some optimizations which fail to converge. In fact, these two variables both relate to the same structural parameter—the position of the shifting hydrogen atom—and some combinations of values generated by the proposed scan are problematic.

A better approach is to define the scan variables as the C-H bond distance and the C-C-H bond angle. These variables allow you to consider the same range of structures as before but specified via two less tightly coupled structural parameters. The resulting relaxed PES scan calculation completes successfully.

S_N2 Reactions

S_N2 reactions are characterized by an exchange of substituents between two species and have the general form: $N^- + RX \rightarrow RN + X^-$. These reactions proceed via the nucleophilic species attacking the opposite side of the molecule with respect to the substituent that it liberates; the latter is referred to as the "leaving group." Such a process proceeds through a transition structure in which the ion and neutral reactants are weakly bound.

References: S_N2 Reactions
Definition: [Ingold53]
Review: [Shaik92]

The potential energy surface for such a reaction has the following general shape in the gas phase:

Idealized S_N2 Reaction

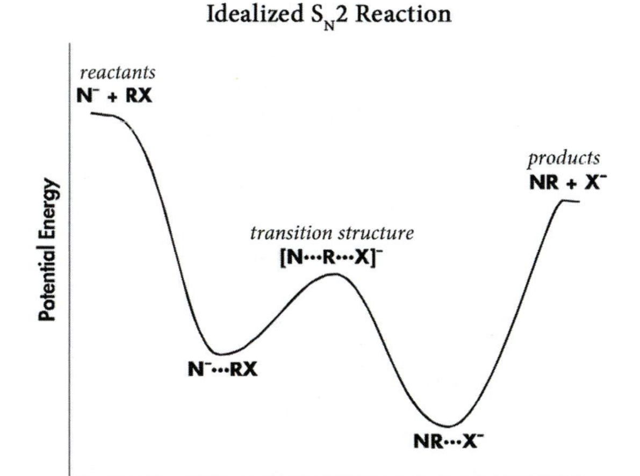

* On systems without the **tac** command (e.g., Mac OS X), substitute this command string:
... | **awk '{print NR,$0}' | sort -nr | sed 's/^[0-9]* //'** | ...
In a similar vein, omit the **-w** option from the **uniq** command if it is not supported.

The reactants and products are at the two ends of the curve. The transition structure for the reaction connects two minima. These minima are two ion-molecule complexes: intermediate species through which the reaction proceeds. Note that the relative positions of the peaks will vary in different S_N2 reactions.

We previously considered an S_N2 reaction in Advanced Exercise 6.7 (p. 265). We will consider several more in the following example and exercise.

ADVANCED EXAMPLE 6.11 A SIMPLE S_N2 REACTION

We want to study the S_N2 reaction: $F^- + CH_3Cl \rightarrow CH_3F + Cl^-$, in solution with acetonitrile.* Doing so will require predicting the structures of the transition state and the two intermediate ion-molecule complexes as well as computing the various energy barriers. We will run the following calculations:

- An optimization+frequency calculation to locate the transition structure.
- An IRC calculation starting from the transition structure (to verify that it connects the minima we think it does).
- Two geometry optimization+frequency calculations to find the intermediate minima. Build each complex by adding the Cl- or F- to the alkyl halide at a distance of 2.59 Å and forcing C_{3v} symmetry.

The following table gives the results we obtained in this study:

MOLECULAR SYSTEM	G (hartrees)	RELATIVE G (kcal/mol)
Reactants: $F^- + CH_3Cl$	-600.13649	18.6
Products: $CH_3F + Cl^-$	-600.16568	0.3
Transition structure	-600.12240	27.4
Reactant complex: $F^-\cdots CH_3Cl$	-600.13632	18.7
Product complex: $CH_3F\cdots Cl^-$	-600.16620	0.0

The following plot depicts the potential energy surface for this reaction:

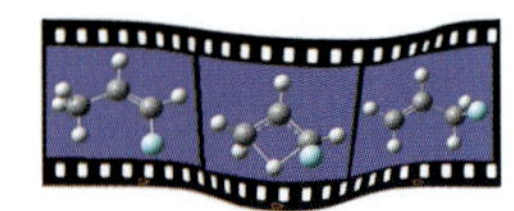

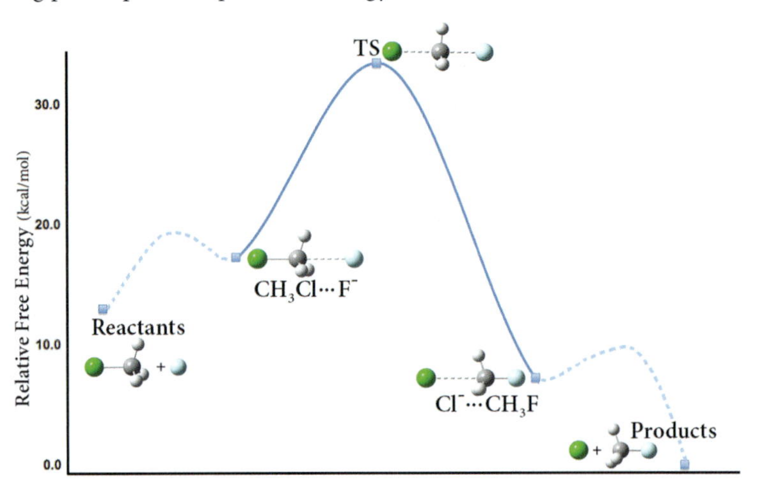

* The original gas phase study is [Shi89].

For both the reactants and the products, the complex is higher in energy than the separate molecules: 4.1 kcal/mol for the reactants, and 7.2 kcal/mol for the products (about 0.4 and 1.3 kJ/mol). The predicted value for the reaction energy is 13.0 kcal/mol (~54.5 kJ/mol), and the computed reaction barrier for the reactant complex is 15.9 kcal/mol (~66.6 kJ/mol).

The barriers between the separate reactant/product molecules and the associated complexes are approximated by the light blue dashed lines in the plot.

The arrangement of the heavy atoms in both complexes and in the transition structure is nearly linear. In both complexes, the substituent bonded to the carbon atom retains the bond length in the isolated methyl compound (1.40 Å for C-F and 1.80 Å for C-Cl), while the distance to the ion is significantly lengthened (3.62 Å in the case of the chloride ion and 3.48 Å for the fluoride ion), indicating a fairly weak linkage. In contrast, the transition structure bond lengths are about 25-40% elongated with respect to the isolated species: C-Cl=2.23 Å, C-F=1.95 Å.■

ADVANCED EXERCISE 6.9 **LEAVING GROUP EFFECTS IN ETHYL HALIDE S_N2 REACTIONS**

In this exercise, we consider the S_N2 reactions between acetate ion and various ethyl halide compounds, in DMSO solution. We will study the reactions involving chlorine and bromine as the substituent.

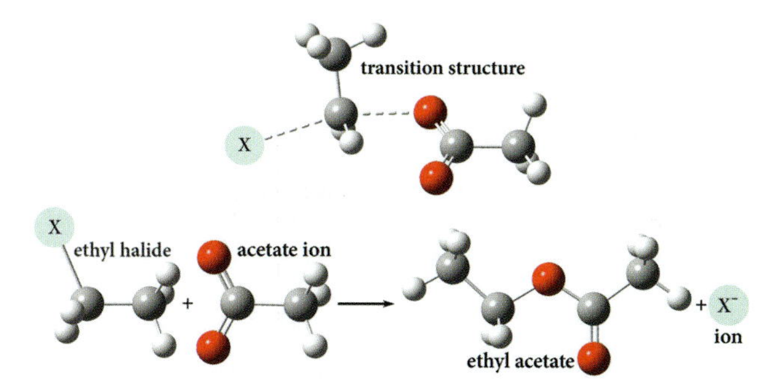

Predict the activation barrier and reaction energy for each of the reactions using our standard model chemistry. Use the predicted values of the Gibbs free energy to compute these quantities. Compare your results to the observed values for $\Delta G^{\ddagger}$: 22.3 kcal/mol for the chlorine reaction and ~20.0 kcal/mol for the bromine reaction [Tondo05].

It is a good idea to optimize these structures with the 6-31+G(d) basis set before moving on to our standard model chemistry. You can then use the force constants computed by the frequency calculations in the subsequent larger optimizations.

SOLUTION

Each reaction requires several optimization+frequency calculations—for the reactants: acetate ion and the ethyl halide compound; for the products: ethyl acetate and the ion; and for the transition structure. An IRC calculation should also be run to verify that each located saddle point is the desired transition state.

The optimizations of the minima were performed in two steps: a preliminary optimization with the 6-31+G(d) basis set followed by a second one with our standard model chemistry. The acetate ion optimization with the small basis set required the use of the **CalcFC** option to the **Opt** keyword. The transition states were located via QST3 calculations.

The following table reports the predicted activation barriers and reaction energies:

		(kcal/mol) Cl	Br
$\Delta G^{\ddagger}$	calc.	21.8	20.8
	exp.	22.3	~20.0
ΔG	calc.	19.5	15.7

These results are in very good agreement with the experiment values.* As expected, the barrier for ethyl bromide is somewhat lower than for ethyl chloride.■

* The value for the bromine reaction is an interpolation from data for related reactions.

7

Predicting Spectra

In This Chapter:

NMR Spin-Spin Coupling Constants

Vibrational Circular Dichroism

Raman Optical Activity

Optical Rotations

Advanced Topics:

Microwave Spectra &
Hyperfine Coupling Constants

7 ◈ Predicting Spectra

In general, spectroscopy is the study of the interaction of matter and electromagnetic radiation. Molecules are bombarded with radiation, and the resultant changes in their characteristics are detected and measured. These interactions can, for example, lead to molecular vibrations, nuclear spin reorientations, electronic excitations or other effects which can be observed by specialized instruments.

The following table summarizes the electromagnetic spectrum and the chemical spectroscopy associated with various types of radiation in common use:

RADIATION	λ (Å)	~ν (Hz)	EFFECTS
gamma ray	<0.1	>10^{19}	nuclear transformations
x-ray	0.1-10	10^{17}-10^{19}	inner electron shell transitions
ultraviolet visible	10-4000 4000-7000	10^{14}-10^{17} 10^{14}	outer electron shell transitions
infrared	7000-10^6	10^{12}-10^{14}	molecular vibrations
microwave	10^6-10^9	10^9-10^{12}	molecular rotations, electronic spin orientation transitions
radio†	>10^9	<10^9	nuclear spin orientation transitions

†Includes radio, broadcast television, etc.

The following are the main types of spectra that can be predicted with Gaussian:

- IR: Molecules are exposed to infrared light, inducing molecular vibrations. Photons are absorbed at specific frequencies based on molecular structure, producing a characteristic spectrum.
- Raman: Light in the visible range induces molecular vibrations via inelastic scattering of photons. The spectrum results from the photons' energies being shifted up or down in characteristic ways.
- NMR: Molecules are exposed to radio range radiation in the presence of an external magnetic field, causing nuclear spin flips. Absorption of these photons produces a characteristic spectrum.
- VCD (vibrational circular dichroism): IR spectroscopy carried out with circularly polarized light in the infrared range. The difference in photon absorption between left- and right-polarized light produces the VCD spectrum, allowing enantiomers to be distinguished in chiral molecules.
- ROA (Raman optical activity): ROA is Raman spectroscopy carried out with circularly polarized light. As with VCD, different enantiomers of chiral molecules yield different spectra.
- OR (optical rotation) and ORD (optical rotatory dispersion): Spectroscopy in which molecules are exposed to polarized visible light, and photons are elastically scattered. OR measures the specific angle—denoted [α]—that polarized visible light of a particular frequency is elastically scattered by a molecule containing a chiral center. Usually, the yellow sodium D line of 589 nm is used: $[\alpha]_D$. ORD observes the variation in the optical rotation of the incident light as a function of its frequency.

- Hyperfine (microwave): Molecules interact with microwave-range radiation. Photon absorption induces rotational transitions and produces characteristic transition data. This data is analyzed via standard quantities known as hyperfine coupling constants.
- UV/visible: Molecules undergo electronic transitions when photons in the ultraviolet and/or visible range are absorbed.
- ECD (electronic circular dichroism): UV/visible spectroscopy performed with polarized light.

We have already considered IR and Raman spectra as well as NMR chemical shifts. In this chapter, we will consider some advanced aspects of NMR as well as several other kinds of spectra. Spectra involving electronic transitions—molecules in their excited states—are the subject of the next chapter.

NMR Spectroscopy: Beyond Chemical Shifts

We considered the calculation of NMR chemical shifts earlier in this book (see p. 161). In this section, we will discuss two additional aspects of NMR.

NMR Shielding Tensor Components

In NMR experiments, a substance* is placed in a magnetic field and then exposed to radiation in the radio wave range of the electromagnetic spectrum (~60-1000 MHz). This radiation is absorbed and then re-emitted at specific resonance frequencies—corresponding to changes in nuclear spin state—that depend on the strength of the applied magnetic field and the magnetic properties of the component atoms (isotopes). The actual magnetic field experienced by the nucleus is the result of the interaction of the external field and the local electron density near the nucleus; the difference between the strength of the applied field and the field experienced by the nucleus is known as the *shielding*,† and it is very small (typically measured in parts per million). Overall shielding is used to compute *chemical shifts*: the difference in location for a given resonance frequency with respect to a reference compound such as TMS for ^{1}H and ^{13}C NMR. Shielding and the resultant chemical shifts for a given nucleus type vary according to many structural features within the molecule, enabling NMR spectroscopy to provide detailed information about bonding, substituents and the like.

The isotropic shielding value used to compute chemical shifts represents the average effect seen by the nucleus. Computationally, it is derived from a tensor describing the shielding in three dimensions. Examining the tensor's individual components is sometimes interesting.

EXAMPLE 7.1 NMR SHIELDING SUBSTITUENT EFFECTS IN SUBSTITUTED ACETYLENES

In this example and the follow-up exercise, we will examine the components of the NMR shielding tensor in a series of substituted acetylenes: H–C≡C–X [Wiberg04c]. We will begin by considering the compounds with H and F substituents. We ran optimization+frequency calculations for each compound to find the ground state structure and then ran an NMR calculation at that geometry.

Here is the output from an NMR calculation on fluoroacetylene‡ corresponding to the alpha carbon atom:

---

* Specifically, substances possessing a nuclear magnetic moment: nuclei with an odd atomic number or odd mass (e.g., ^{13}C but not ^{12}C).

† More precisely, *shielding* refers to a weaker resultant field than the applied field, and *deshielding* refers to a stronger resultant field.

‡ IUPAC: fluoroethyne.

```
   1  C   Isotropic =   92.3724   Anisotropy =   291.8176
 XX=      -4.9001    YX=      0.0000     ZX=       0.0000
 XY=       0.0000    YY=     -4.9001     ZY=       0.0000
 XZ=       0.0000    YZ=      0.0000     ZZ=     286.9175
 Eigenvalues:      -4.9001     -4.9001     286.9175
```

Negative values indicate deshielding and positive values indicate shielding, relative to the bare nucleus.

The full shielding tensor is listed below the isotropic shielding and anisotropy values in the standard orientation of the molecule. In this case, only the diagonal elements are nonzero, and the XX and YY values are identical (by symmetry). For this molecule, the ZZ component is aligned with the carbon-carbon triple bond, and we will use this designation to refer to the parallel component in our discussions of the whole set of results. In some cases, however, the actual orientation of the molecule will result in a different component being aligned with the molecular axis (e.g., XX).

The following table lists the computed values for the overall shielding and the parallel and perpendicular components for acetylene and fluoroacetylene, as well as the experimental values [Duncan97]:

		CALCULATED			EXPERIMENT		
SUBSTITUENT	C	SHIELDING	XX,YY	ZZ	SHIELDING	XX,YY	ZZ
H		113	30	280	116	36	276
F	α	92	-5	287	102		
	β	174	120	282	177		

In both cases, the shielding component along the C=C axis is much larger than in the perpendicular directions, especially for the α carbon in fluoroacetylene. The computed values are in very good agreement with experiment.

We will continue this study in Exercise 7.1 (p. 299).■

NMR Spin-spin Coupling Constants

In general, a spinning charged particle—such as a proton in an atomic nucleus—generates a magnetic field. In the presence of an external magnetic field, as happens in an NMR experiment, half of the protons will align with the external field and half will align against it. A proton adjacent to another that is aligned with the external field will experience a slightly stronger field than it would in the second proton's absence. Similarly, a proton adjacent to another aligned against the field will experience a slightly weaker field. These effects result in additional peaks in the NMR spectrum, and the splitting of the resonance into two peaks is called *spin-spin coupling*.* The distance between them is known as the *coupling constant*. Spin-spin coupling falls off very quickly with distance and depends primarily on the bonding relationship between the two nuclei, allowing a great deal of structural information to be obtained. Spin-spin coupling constants of carbon nuclei are widely used to study the nature of chemical bonds, the electronic effects of substituents and related phenomena.

References: NMR Spin-Spin Coupling
Review: [Gunther96]
Implementation: [Helgaker99, Sychrovsky00, Barone02, Peralta03, Deng06]
Best Practices: [Deng06]

The accuracy of predicted spin-spin coupling constants is closely related to the basis set used to compute them. As Deng and coworkers note [Deng06, p.1028]:

> The accuracy of a spin-spin coupling constant calculation is highly dependent on the Gaussian basis set employed. The basis sets of quantum chemistry are well-developed for the valence electrons. However, NMR experiments probe the electron density closer to the nuclei, where many standard basis sets of ab initio theory will give erroneous results.

* Note that "spin" in this context refers to nuclear spin and not to electronic spin angular momentum.

These researchers created basis sets designed specifically for modeling NMR spin-spin coupling by adding functions that address this weakness in standard basis sets while minimizing additional computational cost.*

The **NMR=Mixed** option is useful for performing these types of calculations. It performs two functions. First, it requests that spin-spin coupling constants be predicted as part of the NMR calculation. Second, it causes the basis set specified for the job to be modified appropriately for computing spin-spin coupling constants.† We will specify it as the second job step after our usual optimization+frequency calculation, and we will use the **aug-cc-pVTZ** basis set with it. Thus, an input file for predicting NMR spin-spin coupling constants will have this general form:

```
%Chk=file
# APFD/6-311+G(2d,p) Opt Freq ...

title

molecule specification

%Chk=file
# APFD/aug-cc-pVTZ NMR=Mixed Geom=AllCheck Guess=Read ...
                                                    final blank line
```

As usual, the second job step can be added within GaussView or WebMO or directly into the input file with any text editor.

Gaussian output reports the total spin-spin coupling constant values as well as their breakdown into components. The output reports two matrices for each component, labeled J and K. We are interested in the former, which takes the specific isotopes in the job into account (whether specified explicitly or by accepting the default values); the latter gives the isotope-independent values.

Here is an example of the listing of the total isotope-dependent spin-spin coupling term:

```
Total nuclear spin-spin coupling J (Hz):
             1             2             3             4
  1  0.00000D+00
  2  0.10833D+02  0.00000D+00
  3  0.10835D+02  0.10833D+02  0.00000D+00
  4  0.15606D+03 -0.24185D+01 -0.24185D+01  0.00000D+00
  ...
```

The results are listed as a lower triangular matrix with atom numbers going from left to right and top to bottom. Thus, if we are interested in the value for the coupling between carbon nuclei 2 and 3, we find the corresponding entry in the matrix: in this case, the value is 10.8 Hz.

* Specifically, by "uncontracting a standard basis set, such as correlation-consistent aug-cc-pVTZ, and extending it by systematically adding tight s and d function" for the core, which "allows the high sensitivity of this property to the basis set to be handled in a manner which remains computationally feasible" [Deng06, p.1028].

† It also generates and runs an additional job step. See the *Gaussian User's Reference* for full details on the computational procedure.

EXAMPLE 7.2 SPIN-SPIN COUPLING CONSTANTS

NMR spin-spin coupling constants reveal significant information about molecular structure, including the stereoelectronic nature of carbon-carbon bonds and the electronic influence of substituents. In this example, we consider two 3-membered ring compounds and observe how their differing results reflect the varying properties of their substituents.

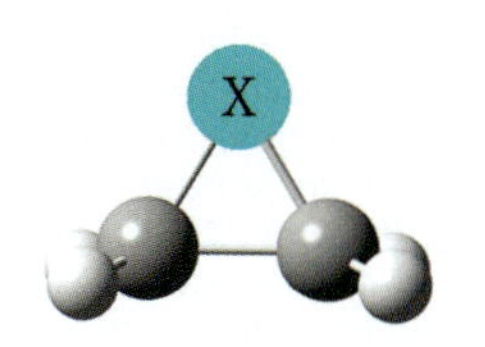

We optimized cyclopropane and ethylene oxide (oxirane)—X is CH_2 and O in the illustration in the margin—and then verified that the resulting structures were minima. We then predicted the spin-spin coupling constants for each compound.

The following table lists the results for the coupling between two adjacent ^{13}C nuclei from these calculations, as well as experimental values [Rusakov13, Krivdin03] for the total spin-spin coupling (all values are in Hz):

	COMPONENTS OF J				TOTAL J	
X	J-DSO	J-PSO	J-SD	J-FC	CALC.	EXP.
CH_2	0.15	-0.82	-0.19	11.6	10.8	12.4
O	0.17	-0.73	-0.53	25.6	24.5	28.0

The components of the total spin-spin coupling are:

- J-DSO: diamagnetic spin orbit term
- J-PSO: paramagnetic spin orbit term
- J-SD: spin-dipolar term
- J-FC: Fermi contact term

For these two compounds, the calculated results are in good agreement with experiment. The differences between them are due primarily to the Fermi contact term along with a small difference in the spin-dipolar term; the DSO and PSO terms are essentially identical.

We will continue this study in Exercise 7.3 (p. 300). It was drawn from [Rusakov13], which provides an excellent review of this general topic.■

Vibrational Circular Dichroism

Various spectroscopic techniques have been developed to study molecules with one or more chiral centers. Chiral molecules are affected differently by left and right circularly polarized light. Vibrational circular dichroism (VCD) spectroscopy exposes substances to light in the IR range and measures the dipole strength resulting from the differential absorption of left vs. right circularly polarized light.

References: VCD

Phenomenon: [Holzwarth72, Schellman73]

Method Definition: [Nafie83, Nafie83a, Stephens85, Galwas83, Buckingham87]

Review: [Nafie11, Stephens12]

Implementation: [Devlin96, Cheeseman96a]

Solvation Treatment: [Mennucci11]

As in ordinary IR spectroscopy, VCD spectra are sensitive to the specific groups present in the molecule and their orientations, and this technique is accordingly capable of providing structural information about the molecule. Because R and S chiral centers produce VCD spectra with opposite signs,* comparing observed and computed spectra allows quick identification of the absolute configurations of enantiomers. For a general review of using VCD calculations to determine absolute configuration, see [He11].

* For molecules with only a single chiral center, the VCD spectra corresponding to the R and S enantiomers will be exact mirror images of one another. The same is true for the (R,R) and (S,S) enantiomers of a molecule with two chiral centers, as it is for the (R,S) and (S,R) enantiomers, although the relationship between, say, the spectra corresponding to the (R,R) and (R,S) diastereomers is more complex.

Absolute Configurations

Compounds are labeled as (+) or (–) based on the sign of their measured optical rotation. The absolute configuration—the R vs. S "handedness"—of an enantiomer must be determined from X-ray diffraction data or from VCD or ROA spectroscopic data in combination with theoretical calculations.

As is true for all types of spectroscopy, observed VCD spectra can arise from a mixture of conformations. Boltzmann averaging is often necessary to reproduce the details of an experimental spectrum.

A VCD calculation is requested with the **Freq=VCD** option (which can be combined with **Opt**). Such jobs are easy to set up in both GaussView and WebMO. In GaussView, a checkbox labeled **Compute VCD** is included in the **Job Type** panel for all frequency calculations (illustrated on the left below). Checking this box will cause the job to predict the VCD spectrum along with the IR spectrum (and the Raman spectrum if computed by default or request).

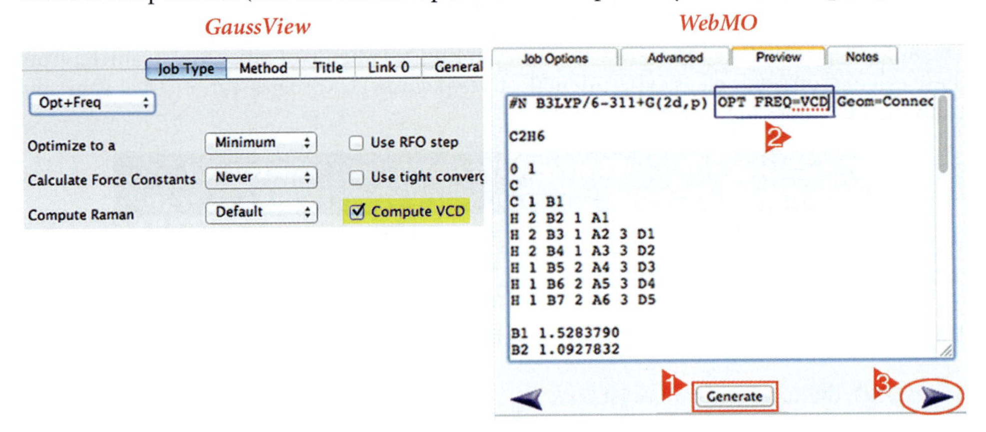

In WebMO, after selecting a job type which includes a frequency calculation on the **Job Options** tab, activate the **Preview** tab (illustrated on the right above). Click on the **Generate** button. Once the generated Gaussian input file appears in the panel, edit the **Freq** keyword, adding the **VCD** option. Finally, click the right arrow button in the bottom right corner of the panel to submit the job. There will be a message confirming that your edited version of the input file is what will be submitted.

GaussView will label the chiral centers within a molecule when you select the **Stereochemistry** option from the **View** menu. GaussView can display the predicted spectra from VCD calculations, via the **Vibrations** item on the **Results** menu; click the **Spectrum** button in the resulting window to open the spectrum.

Two techniques for working with plots in GaussView are useful for VCD spectra:

- *Reversing the direction of the X-axis*: Right click in the plot area, and then select **Properties** from the context menu. Click the relevant **Invert Axis** box to modify the spectrum display:

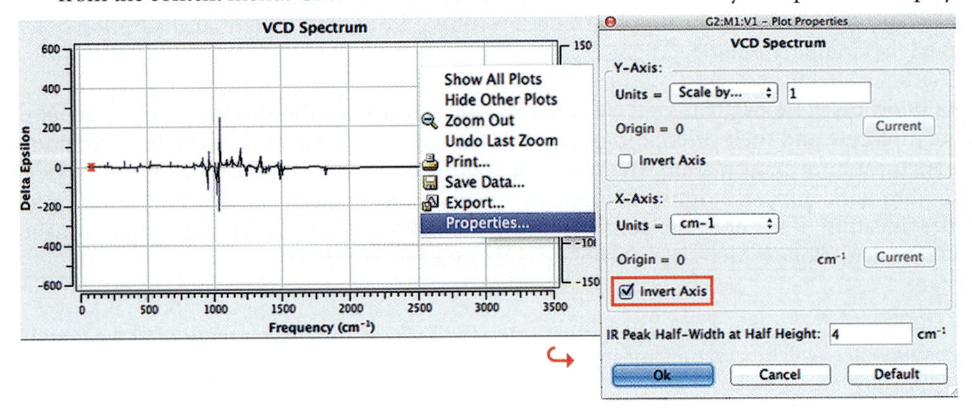

- *Enlarging a portion of the plot*: You can zoom in on a specific area within any plot by dragging a selection rectangle around the desired portion of the plot area. The selected region will be expanded to fill the entire available plot area. The example below illustrates focusing on the plot area between about 1500 and 1000 cm^{-1}:

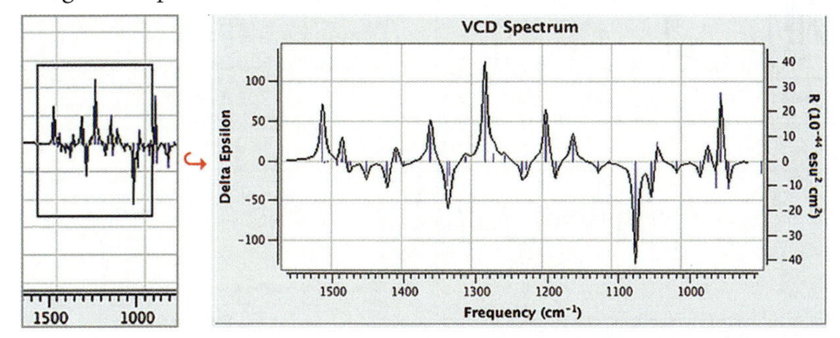

One or both of these operations often makes it easier to compare the predicted spectra with experiment.

IR and VCD spectra are plotted in terms of ε and Δε (respectively), where ε is the molar absorptivity in units of $M^{-1}cm^{-1}$. ε is related to the absorbance, A, through Beer's law:

$$A = \varepsilon \times \textit{concentration} \times \textit{pathlength}$$

For VCD, Δε is defined as $\varepsilon^{left}-\varepsilon^{right}$, where left and right refer to the circularly polarized light. Note that the vertical Δε axis is actually $\Delta\varepsilon\times10^4$. Plotting in terms of ε and Δε allows for a direct comparison between the calculated and experimental spectra.

GaussView follows the standard practice of using a Lorentzian band shape to plot the calculated spectrum. The vertical blue sticks represent the magnitudes of the dipole strenghts (for IR absorption) and rotational strengths (for VCD) for each normal mode frequency (as labeled on the right side of the plot). The dipole strengths are in units of $10^{-40}esu^2cm^2$, and the rotational strengths are in units of $10^{-44}esu^2cm^2$.

EXAMPLE 7.3 ABSOLUTE CONFIGURATION OF CAMPHOR

Camphor* is a natural product found in several species of large evergreen trees in the laurel family that are native to Asia (most notably, *cinnamomum camphora*). It is a significant ingredient in the spice rosemary. It is a chiral molecule with two chiral centers (the full molecule for the (1R,4R) enantiomer and the backbones for both enantiomers are illustrated in the margin). We optimized the (1R,4R) and (1S,4S) forms of this molecule and then predicted their VCD spectrum.

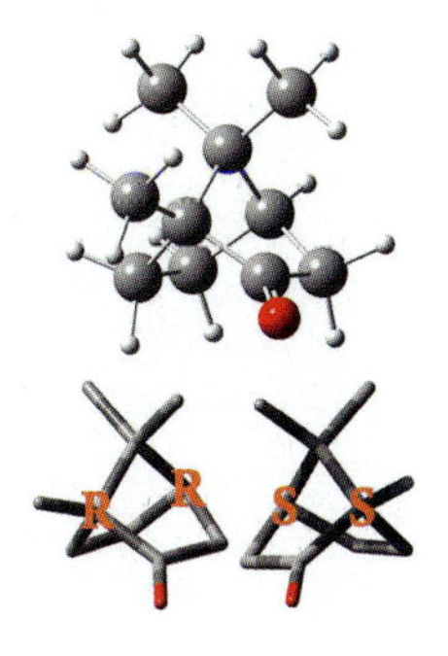

The following figure shows the experimental IR and VCD spectra for (+)-camphor on the left [Devlin97] and the predicted spectra for the (1R,4R) enantiomer on the right:

* IUPAC: 1,7,7-trimethylbicyclo[2.2.1]heptan-2-one.

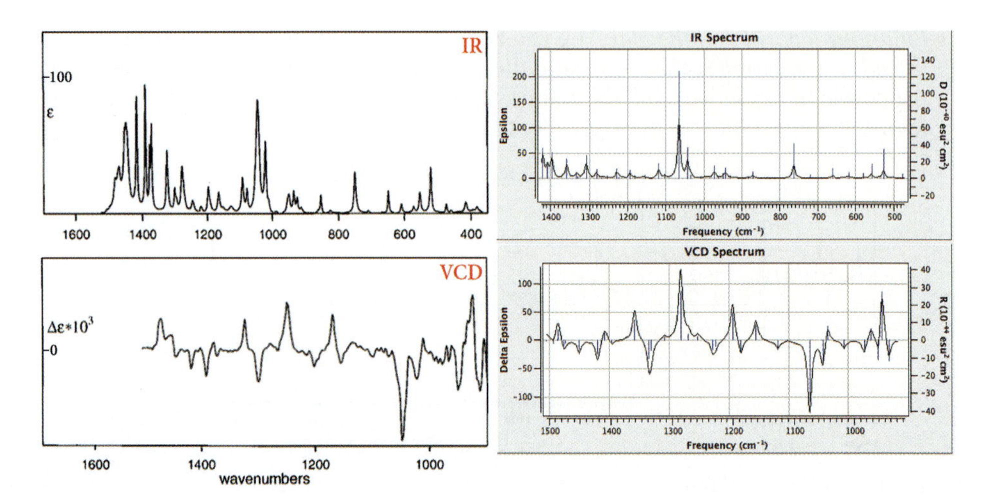

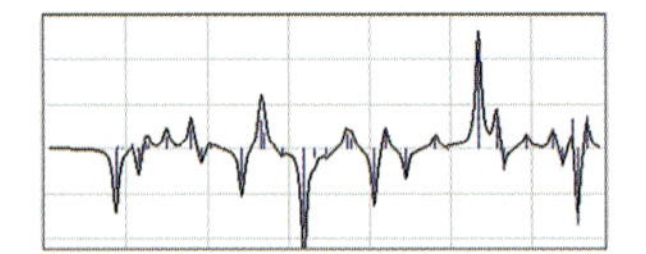

Remember that the IR spectra is the same for opposite enantiomers while the VCD spectra are mirrored. The computed VCD spectrum matches the one observed. Contrast this to the predicted spectrum for the (1S,4S) enantiomer in the margin, which is its mirror image. Therefore, we only need do the calculations for one of the enantiomers. The experimental spectrum would overlay either with the computed spectrum (if it is the same enantiomer) or with the mirror image of the computed spectrum (if it is the opposite enantiomer).

The following figure overlays the predicted spectrum for the (1R,4R) enantiomer (yellow) on the observed spectrum (black):

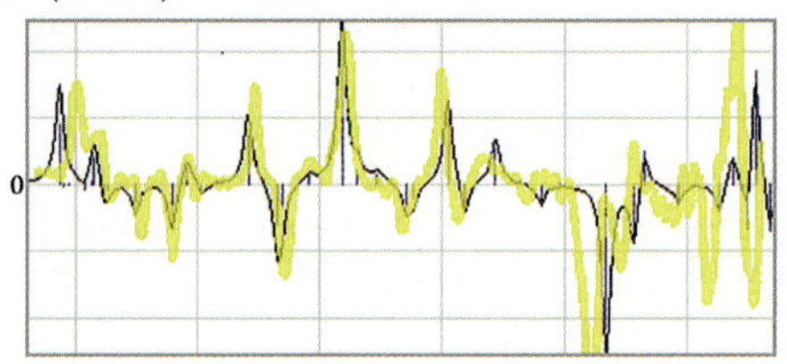

The agreement between experiment and theory is very good.■

Conformational Averaging

As is the case for NMR spectra, experimental VCD spectra can also be comprised of features from the individual spectra of various isomers of the compound in question. For such molecules, the conformational averaging technique is necessary in order to reproduce the observed spectrum. As we saw earlier in the case of NMR (p. 164), this consists of locating and verifying the minimum for the various conformations, predicting the VCD spectrum for each one, and finally combining all of them into a single composite spectrum by weighting the individual spectra according to their relative energies. The next example will illustrate this technique in the context of VCD.

EXAMPLE 7.4 VCD SPECTRUM OF DESFLURANE

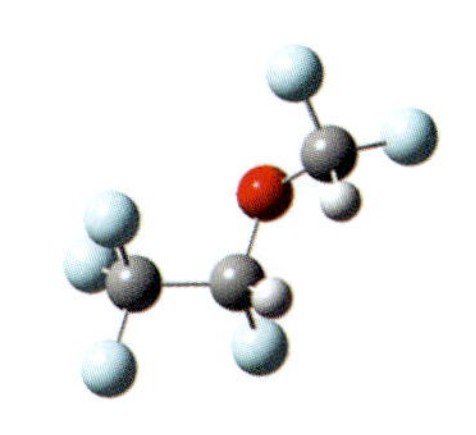

Desflurane* is a fluoronated ether compound which is used as a medical anesthetic. It has one chiral center, and its absolute configuration is the (R)-(-) form. Desflurane is typically administered as a racemic mixture, but, as with many chiral pharmaceutical species, single enantiomer versions are also being explored. This compound has been examined experimentally and computationally by several groups. Here, we draw on [Biedermann99, Polavarapu99].

This compound has multiple isomers that must be modeled in order to predict its VCD spectrum (and compare to the observed one). In general, there are several methods for locating the various isomers for a compound:

- Sometimes a compound's isomers can be formulated in a systematic way based on its structure.
- A conformational searching package such as Pcmodel may be used to locate starting structures for potential conformations.
- A potential energy surface scan in Gaussian can be used to locate minima by varying one or more structural variables across a range of values.

All three of these approaches are feasible for desflurane. The illustration following shows the six conformations that are possible for this compound; they are constructed by varying two dihedral angles within the molecule [Polavarapu99]. The configuration numbering follows the reference.

We built these six structures in GaussView and then ran optimization+frequency calculations for each one. Before considering the results, we will briefly consider the other two approaches to conformational searching for this molecule.

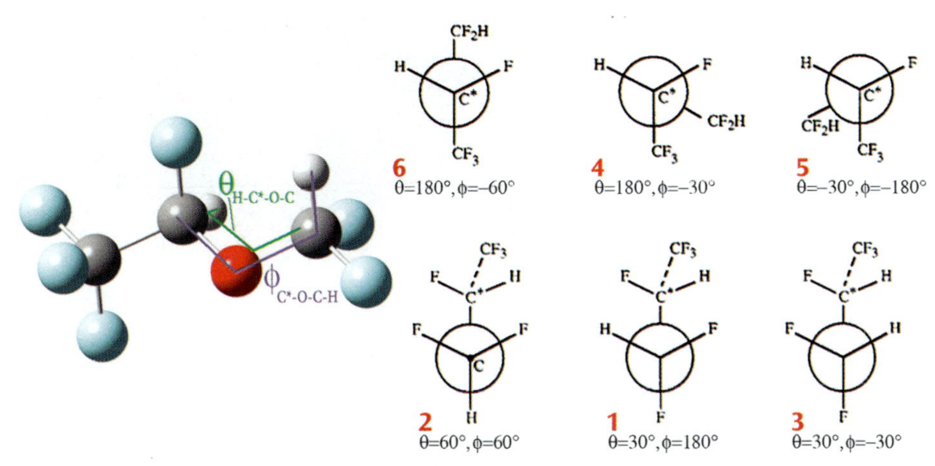

We ran a conformational search in Pcmodel, which located four conformational candidates. We've included our starting structure among the job files for this example.

We also performed a potential energy surface scan over two variables: the same two dihedral angles as mentioned above. The following illustration indicates the parameters for the scan. This scan searches the same conformational space as the analysis based on structure above.

* IUPAC: 2-(difluoromethoxy)-1,1,1,2-tetrafluoroethane.

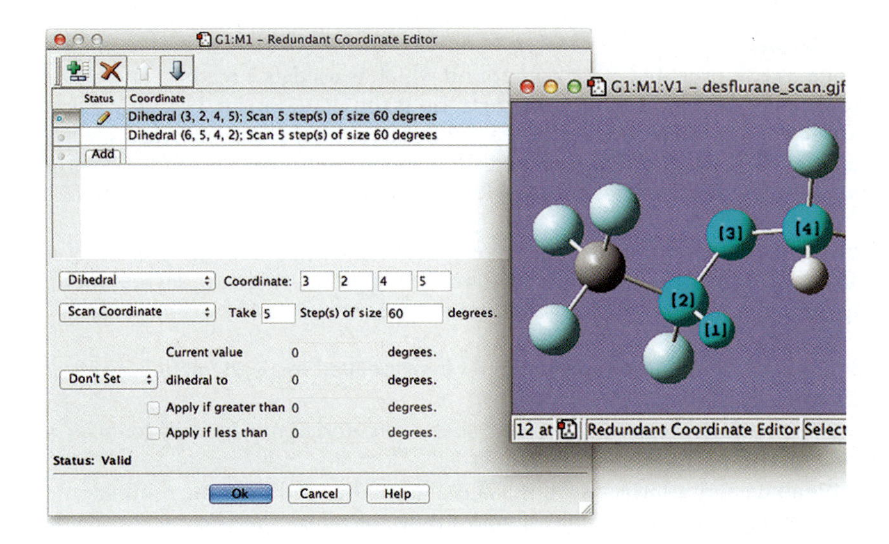

The potential conformers are the minima located by the scan. In order to identify and access them, it is helpful to rotate GaussView's two-dimensional scan display by 180° (i.e., turn it upside down), interchanging the valleys and peaks:

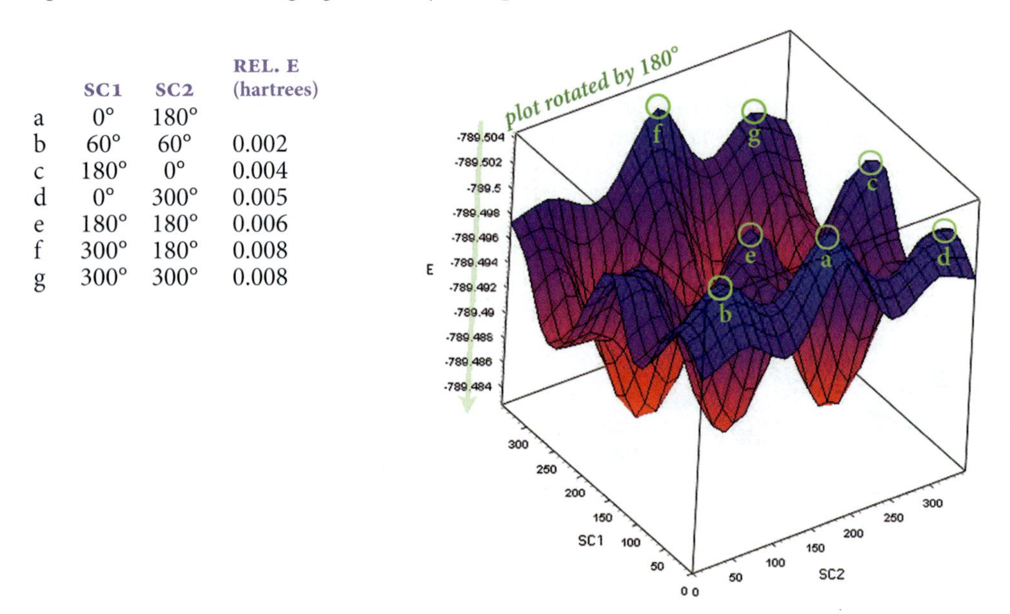

	SC1	SC2	REL. E (hartrees)
a	0°	180°	
b	60°	60°	0.002
c	180°	0°	0.004
d	0°	300°	0.005
e	180°	180°	0.006
f	300°	180°	0.008
g	300°	300°	0.008

In this way, the relevant points are easier to click on in order to examine each minimum structure and set up a follow-up calculation for that structure. As you can see in the illustration above, our scan located seven possible minima, of which four turn out to be important: a-d (see the next page). We've included the scan input file among this example's job files.

As we noted earlier, we chose to run optimizations for the six isomers constructed via the first approach. When these calculations were complete, we found that there were only four unique conformations: structures 2 and 5 optimized to the same minimun, as did structures 4 and 6.

The following table summarizes the minimized structures and relative energies of the conformations of desflurane:

	DIHEDRAL ANGLES (°)				
STRUCTURE	H-C*-O-C	H-C-O-C*	G (hartrees)	q_i	%
2	25.2	176.5	-789.47801	1.00	85.3%
1	48.2	57	-789.47610	0.13	11.4%
3	38.7	-41.4	-789.47458	0.03	2.3%
4	176.9	-23.9	-789.47376	0.01	1.0%

These structures correspond to a-d from the scan results. Thus, optimizing the minima found by the scan produces the same results as the approach we took.

The population is dominated by conformation 2 with some contribution from conformation 1:

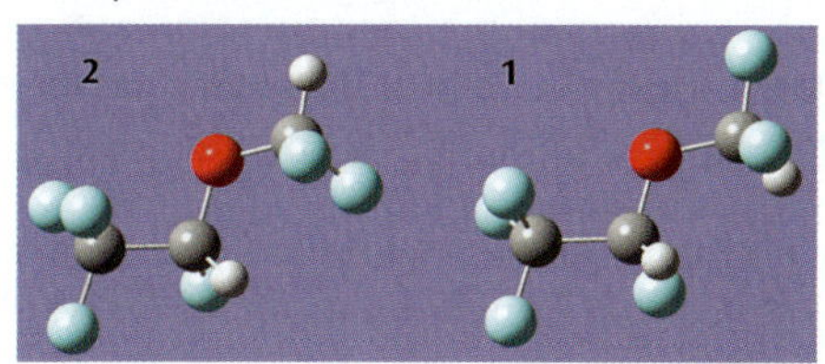

We now need to generate the weighted VCD spectrum. GaussView makes this process easy, via the following technique:

1. Open each output (or checkpoint) file in GaussView. Display the VCD spectrum (select **Vibrations** from the **Results** menu. Click on the **Spectrum** button, and then scroll to the desired spectrum).

2. Right click in the plot area, and then select **Save Data** from the context menu:

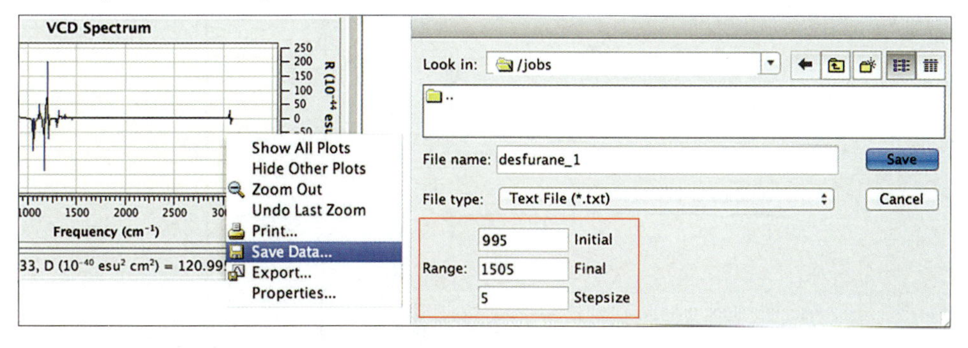

3. Enter a file name into the result **Save** dialog, specifying a file type of **Text File**. Use the fields in the **Range** area in the bottom left of the dialog to specify the range of wavenumbers for which to include data as well as the separation in wavenumbers between consecutive data points. As an example, the specified values will save the data for the range 1000-1500 cm^{-1}, with values recorded every 5 cm^{-1}.

4. Repeat this process for each conformation. When all of the files are prepared, import each one into a separate column of a spreadsheet. Place the factor computed from the Boltzmann analysis in convenient proximity to each column (see the values in red in the next figure). Finally, add a new column which computes the weighted average of the spectrum components for the set of conformations. For example, in this spreadsheet,

column F contains the weighted sum of the values for the four conformations at each point:

*If the row containing the weighting factors is row 2, then the formula in cell F4 is B4*B2 + C4*C2 + D4*D2 + E4*E2.*

A	B	C	D	E	F
	Weights				
	0.11	0.85	0.02	0.01	
Frequency	desfl1	desfl2	desfl3	desfl4	Weighted
1000	-2.70	0.61	-0.25	9.94	0.30
1005	-3.25	0.99	-0.61	15.83	0.62
1010	-4.48	1.62	-0.68	29.19	1.14
1015	-5.52	2.71	-0.78	68.99	2.33
1020	-6.85	4.86	-0.91	222.33	5.51
1025	-8.98	9.96	-0.97	212.92	9.53

5 The weighted spectrum—as well as any of the individual spectra—can now be plotted using the spreadsheet application's graphing capabilities.

The following figure presents the VCD spectra for the four conformations of desflurane, the composite spectrum generated from the weighted contributions of the conformations and the experimental spectrum:

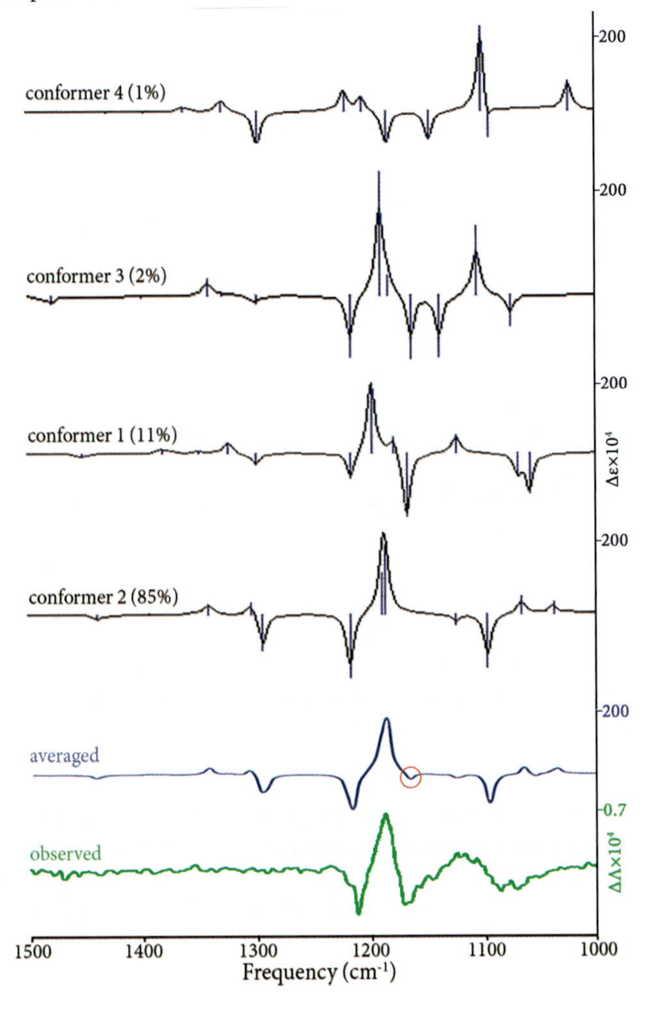

While the spectrum corresponding to conformation 2 resembles the experimental spectrum most closely, there are features in the latter which come from the spectra of other conformations. For example, the (negative) peak circled in red in the averaged spectrum is not found in the spectrum of conformation 2, but is present in those for conformations 1, 3 and 4 and in the observed spectrum. Thus, the data from at least the two most abundant conformations is needed in order to accurately reproduce the experimental data. The averaged spectrum is in fact in excellent agreement with experiment.■

Raman Optical Activity

Raman optical activity (ROA) is another type of spectroscopy for studying chiral molecules. ROA is related to VCD in the same way as Raman relates to IR spectroscopy. Like ordinary Raman spectroscopy, ROA observes the inelastic scattering of light as it interacts with the molecule's vibrational modes. In this case, the light source is circularly polarized light—as with VCD—and the instrument measures the differential scattering for left vs. right circularized polarized light. The resulting spectrum reveals information about the chiral centers in the molecule as well as other structural features.

References: ROA
Phenomenon: [Barron73]
Method Definition: [Nafie92, Nafie94]
Review: [Nafie11]
Implementation: [Helgaker94, Ruud02a, Barron04, Thorvaldsen08, Cheeseman11a]
Solvation Treatment: [Mennucci11]

ROA is defined as the difference between the left- and right-polarized scattered intensity. GaussView plots peak heights as I^{right} and I^{left} for Raman spectra and as $(I^{right}\text{-}I^{left})\times10^4$ for ROA, both mutiplied by $(\nu^{incident}-\nu^{i})^4$, in arbitrary units. This means that you typically must scale the calculated spectrum in order to compare with experiment.

VCD and ROA provide complimentary information about molecular structure. One important difference between them is that ROA experiments can be conducted for samples in aqueous solution while VCD typically requires an organic solvent. This practical consideration makes ROA more convenient for many biological systems. In addition, for some systems, one of these spectroscopic techniques will be more revealing than the other. We will consider an example of this in the final portion of the section.

Accurate computation of ROA spectra requires a basis set with diffuse functions, specifically more diffuse functions than are typically needed for geometry optimizations and ordinary frequency calculations. For this reason, ROA calculations are typically run as a second job step following the optimization+frequency, using a basis set with additional diffuse functions. The aug-cc-pV*Z basis sets are obvious candidates for this role since they contain s, p and d diffuse functions on all atoms. The member of this family corresponding to our standard basis set is the very large aug-cc-pVTZ. [Cheeseman11] explores this issue in detail, arriving at two important findings. First, the smaller aug-cc-pVDZ basis set performs nearly equivalently to the larger triple zeta basis set. Second, removing the d diffuse functions from the basis set has only a very small effect on the accuracy of the predicted ROA spectra, but brings a substantial savings in computational cost. Thus, the recommended basis set for predicting Raman optical activity is a modified version of aug-cc-pVDZ known as spaug-cc-pVDZ.

The general format for computing an ROA spectrum starting from the results of an optimization+frequency calculation is as follows:

```
%Chk=chkpt_file                                          opt+freq checkpoint file
# method/spAug-cc-pVDZ Polar=ROA Guess=Read Geom=Check

ROA                                                      title line

0 1                                                      charge & spin multiplicity

532nm                                                    frequency of incident light
                                                         final blank line
```

Here, the ROA calculation is requested with the **ROA** option to the **Polar** keyword.* ROA calculations require that you specify the frequency of the incident light in a separate input section following the molecule specification. ROA experiments use incident light with a wavelength of 532 nanometers. The example above illustrates the format and placement of this parameter within the input.

Note that using the **Compute ROA** controls on the **Job Type** panel in the **Gaussian Calculation Setup** dialog will not set up a job in this recommended manner in GaussView version 5 or earlier. Instead, it must be done manually by adding a second job step using the **Add. Inp.** panel.

By default, several ROA spectra are displayed in the **Vibrational Spectra** window in GaussView. Generally, the one of interest (i.e., comparable to experiment) is labeled **ROA ICP$_u$/SCP$_u$(180) Spectrum (incident freq. 18797 cm^{-1})**.

Depending on the exact calculation chosen, it may also be necessary to select the frequency of the incident light from the **Properties calculated at** popup menu in the **Display Vibrations** dialog:

EXAMPLE 7.5 OBSERVING α-PINENE EPOXIDATION WITH ROA

α-pinene† is a cycloalkene found in the oils of many conifers, most notably pine trees (as hinted at in its name). It is a molecule with two chiral centers. In nature, its different enantiomers can be found in different species of pine and other trees, and racemic mixtures are also present in other substances.

The absolute configuration for (+)–α-pinene was identified as the (1R,5R) enantiomer by VCD spectroscopy [Devlin97a]. More recently, researchers [Qiu10] have studied the ROA spectra of α-pinene and the products of its epoxidation and subsequent hydrolysis: α-pinene oxide and 2,3-pinanediol:‡

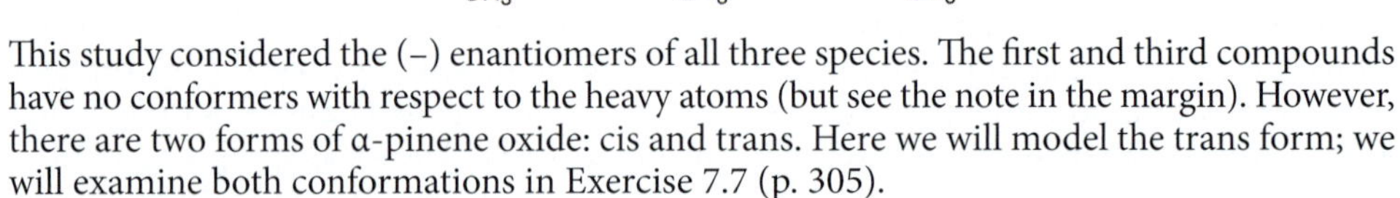

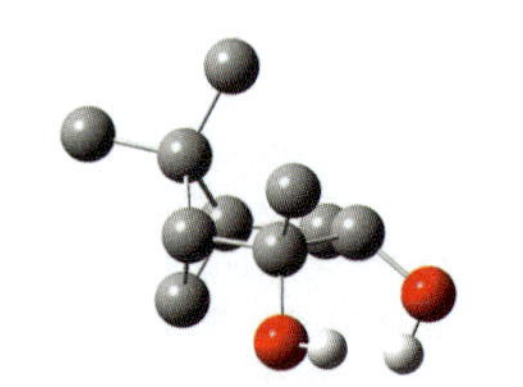

We are ignoring any hydroxyl rotation in 2,3-pinanediol (the structure we used is above with non-hydroxyl H atoms hidden). In fact, the differing structures resulting from hydroxyl rotation yield different ROA spectra. We will examine this effect in Exercise 7.8 (p. 305).

This study considered the (–) enantiomers of all three species. The first and third compounds have no conformers with respect to the heavy atoms (but see the note in the margin). However, there are two forms of α-pinene oxide: cis and trans. Here we will model the trans form; we will examine both conformations in Exercise 7.7 (p. 305).

We computed the ROA spectrum for each of the three compounds by optimizing the molecule using our standard model chemistry, verifying that the stationary point was a minimum and then computing the spectrum in a separate job step as outlined previously.

* ROA can be computed as part of frequency calculations using the **Freq=ROA** option (for methods with analytic second derivatives).

† IUPAC: 2,6,6-trimethylbicyclo[3.1.1]hept-2-ene.

‡ IUPAC: 2,7,7-trimethyl-3-oxatricyclo[4.1.1.0(2,4)]octane and 2,6,6-trimethylbicyclo[3.1.1]heptane-1,2-diol (respectively).

The following figure shows predicted (blue lines) and observed ROA spectra:

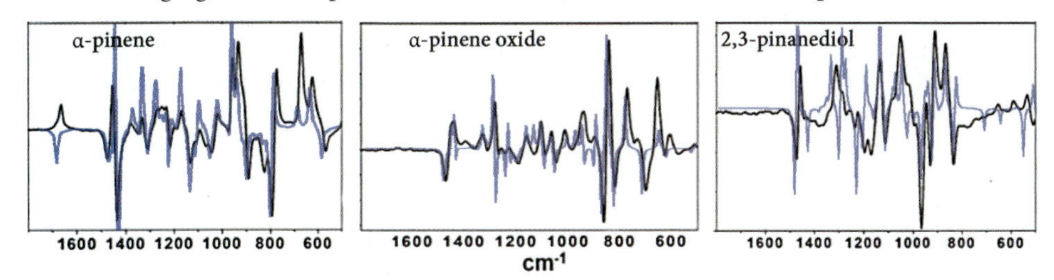

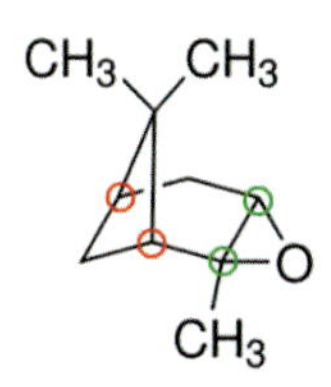

The predicted spectra are in good agreement with experiment (although there are offsets and other deviations in some areas for 2,3-pinanediol). Interestingly, the addition of the oxygen to α-pinene results in the formation of two additional chiral centers at the carbon atoms bonded to the oxygen atom(s) in both subsequent species (the former location of the C=C bond). The figure in the margin shows the existing chiral centers circled in red and the new centers circled in green in α-pinene oxide (they are the same for 2,3-pinanediol).

Moving beyond considerations of absolute configuration, the original researchers go on to discuss the three ROA spectra in comparison with one another, focusing on how changes in the molecular structure affect specific features in the different spectra. In doing so, they demonstrate how ROA can be used in combination with ordinary Raman to interpret spectroscopic results:

> Due to the additional stereo-specific information contained in ROA, in some cases it resolves overlaps in the parent Raman spectra into the bands of different signs or different relative intensities, and so helps to assign the vibrational modes more clearly [Qiu10, p. 3007].

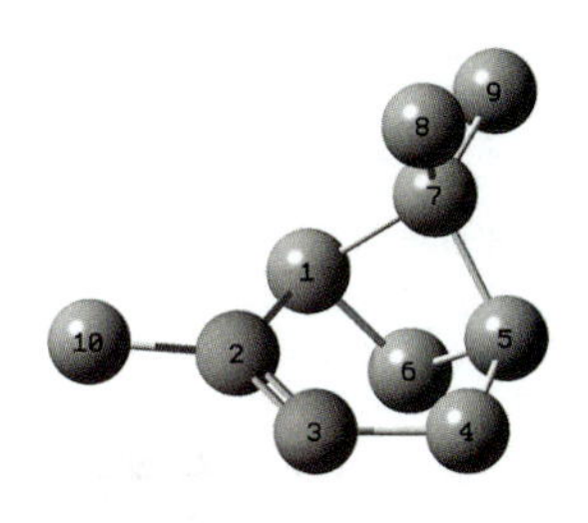

There are many differences between the three spectra, some obvious and others more subtle. One clear change from α-pinene is the absence of the peak at 1731 cm^{-1} in the predicted spectrum for the two products, corresponding to the stretching of the C=C bond (carbon atoms 2 and 3 following the numbering in the figure in the margin); this peak appears in the observed spectrum at 1666 cm^{-1}. Similarly, the band associated with C=C twisting is present only in the spectrum for α-pinene (located at 580 cm^{-1} in the predicted spectrum and at 566 cm^{-1} in the observed spectrum).

Two regions of the spectra are of particular interest. The first is from about 1550 cm^{-1} to 1400 cm^{-1} (illustrated at right). This region is dominated by CH_2 scissoring and CH_3 asymmetric deformation vibrational modes, and its major feature in all three spectra is a negative-positive couplet—negative peak(s) adjacent to positive peak(s), moving to the right, from higher to lower wavenumbers—although its position varies somewhat for the three compounds (they are highlighted in the figure).

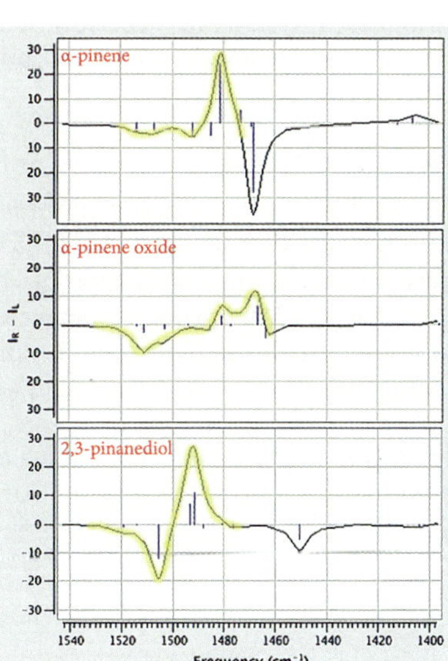

The modes corresponding to these peaks are very similar in the bands for α-pinene (at 1473, 1481, 1485, 1492 cm^{-1}) and 2,3-pinanediol (at 1488, 1491, 1493, 1505, 1519 cm^{-1}): they all involve H scissoring motion at carbons 4 and/or 6, along with asymmetric bond angle changes for the methyl groups including carbons 8 and 9. The peaks at 1481 and 1485 cm^{-1} in α-pinene also include asymmetric deformation in the methyl group at carbon 10, and the peak at 1505 cm^{-1} in 2,3-pinanediol also includes motion of the hydroxyl group at carbon 3.

In contrast, the modes corresponding to the peaks in this region in α-pinene oxide (1464, 1467, 1481, 1503, 1511 cm^{-1}) involve motion of hydrogen atoms throughout the compound, especially those at carbons 4, 8, 9 and 10. The peak at 1464 cm^{-1} also includes breathing motion of the C–C–O ring. The methyl group containing carbon 10 is much more active in this compound than in the other two.

There is a solitary peak at the lower end of the region for both α-pinene and 2,3-pinanediol. In the former case, it corresponds to symmetric deformation of the bond angles in the methyl group containing carbon 10 (at 1407 cm^{-1}). The peak at 1451 cm^{-1} in 2,3-pinanediol is unrelated, and its primary motion corresponds to H wagging parallel to the C–O bond at carbon 3.

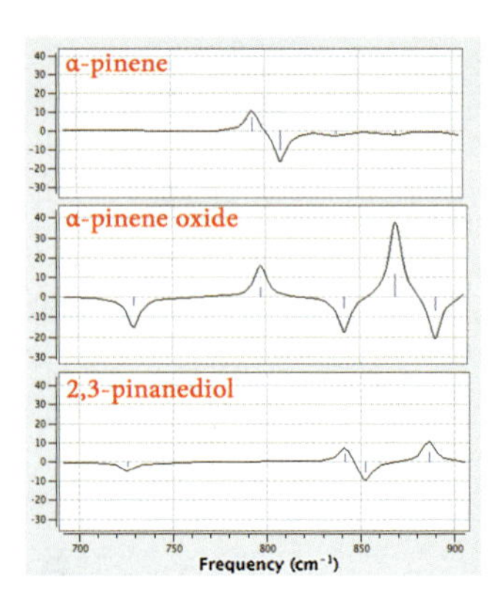

The second region of particular interest is 900-700 cm^{-1}. This region is dominated by ring breathing and ring deformation vibrational modes. The peaks in the α-pinene spectrum (870, 838, 808, 793 cm^{-1}) also include some hydrogen wagging motion, especially in the ones near the upper end of the region. The same is true for those in the 2,3-pinanediol spectrum (886, 852, 842, 726 cm^{-1}). Its peak at 886 cm^{-1} also includes wagging in the hydroxyl group at carbon 3. Additional modes involving hydroxyl group wag lie above this region between 1100 and 1050 cm^{-1}.

In α-pinene oxide, the ring deformation motion is frequently accompanied by C–O stretching involving one of the carbon atoms in the three-membered ring. The mode involving deformation of the six-membered ring at the lower end of this region that we see in spectra of the two product compounds is absent in the spectrum for α-pinene (792 and 726 cm^{-1} for α-pinene oxide and 2,3-pinanediol).

As this comparison makes clear, the ROA spectra can provide considerable elucidation of the vibrational modes in addition to identifying absolute configuration. In this way, it serves as an important complement to traditional Raman spectroscopy for chiral species.■

Solvent Effects on VCD and ROA Spectra

References: Chiroptical Properties in Solution
Review: [Mennucci11]

In the context of predicting spectra, the solution environment can have significant effect both on the predicted geometry for compound as well as on the spectrum itself. In addition, the distribution of conformations can differ between the gas phase and solution. Since VCD and ROA spectra are typically measured in solution, a solvation model is often used when modeling them. In this section, we discuss predicting ROA spectra in solution, but the considerations raised here apply equally to other kinds of spectra (including VCD).

EXAMPLE 7.6 EPICHLOROHYDRIN ROA SPECTRUM: GAS PHASE VS. CYCLOHEXANE

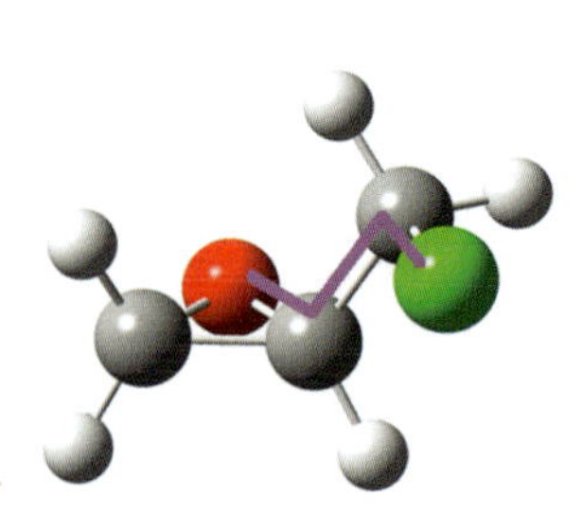

We will begin with a small chiral molecule: epichlorohydrin* (the R form is pictured in the margin). This molecule has three conformations (per enantiomer), corresponding to 0° (pictured), 120° and 240° values of the Cl–C–C–O dihedral angle (indicated in purple in the illustration).† We optimized each conformation both in the gas phase and in cyclohexane, computed the Boltzmann weights and then predicted the ROA spectrum at each optimized geometry.

The following table lists the values of the key dihedral angle for each conformation in the two environments, as well as the corresponding relative energies with respect to the lowest energy conformation and Boltzmann factors (i.e., percentage expressed as a decimal). Energy

* IUPAC: chloromethyloxirane.

† These conformations are also referred to as *gauche* I (120°), *gauche* II (0°) and *cis* (240°).

differences in this example are based on the zero-point energy, following what was done in the original paper [Pecul06]:

	GAS PHASE			CYCLOHEXANE		
FORM	OPT. θ (°)	ΔE_0 (kJ/mol)	Bf	OPT. θ (°)	ΔE_0 (kJ/mol)	Bf
0°	160.3	0.00	0.563	160.3	0.00	0.452
120°	-82.5	1.17	0.352	-81.1	0.01	0.450
240°	49.4	4.70	0.086	48.6	3.82	0.098

The differences among the optimized geometries, both within the conformations in a given environment and comparing the same conformations across environments, are quite modest in general. In both the gas phase and in solution with cyclohexane, the 0° conformer has the lowest energy, followed closely by the 120° conformer (in fact, the two lowest energy conformers in cyclohexane have essentially the same energy). The 240° conformer lies ~4-5 kJ/mol higher in energy than the 0° conformer (with the larger difference in the gas phase), thereby providing only a minute contribution to the ROA spectrum.

The following figure shows the conformationally-averaged ROA spectrum as predicted in the two environments. The vertical scale has been compressed by 50% above 3000 cm^{-1} for display purposes:

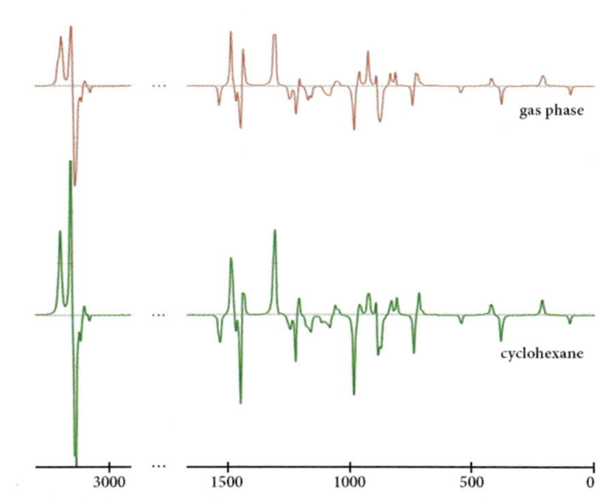

The two spectra are very similar, exhibiting essentially the same shape across the entire range of wave numbers. The differences are subtle, consisting primarily of increased activity in the solution environment (although the attenuation of the peak at 924 cm^{-1} in the gas phase is an exception).

This compound illustrates the way that solvent effects on a spectrum can be threefold:

- The solvent can induce changes in the molecular geometry.
- The solvent can influence the values of the optical tensors leading to the spectrum.
- The solvent can alter the relative populations of the various conformations.

We will continue this study in Exercise 7.9 (p. 309), where these three effects of the solvent environment are again in evidence.

As we saw previously in chapter 5, modeling compounds and chemical processes in aqueous solution can present challenges. This is again true when modeling chiroptical properties in solution with water. In fact, in some cases, electronic structure theory alone is not capable of treating the problem in fine detail and must be combined with molecular dynamics simulations in order to reproduce extremely faithful experimental intensity profiles and other features. In the following example, we will explore the capabilities and limitations of modeling ROA spectra in aqueous solution. You will be introduced to such problems yourself in Exercise 7.10 (p. 310), which includes consideration of the use of explicit water molecules.

EXAMPLE 7.7 MODELING ROA SPECTRA IN WATER

In this example, we model the ROA spectrum of methyl-β-D-glucose.* The experiment is performed in an aqueous environment, and we will predict the molecule's ROA spectrum in the gas phase and in water.

This compound has three conformations, depending on the dihedral angle formed by the ring oxygen atom, the adjacent carbon atom, and the carbon and oxygen atoms in the hydroxymethyl group. Three staggered rotamers exist: gauche-gauche (gg), gauche-trans (gt), and trans-gauche (tg), corresponding to dihedral angles of 60°, –60° and 180° (respectively). X-ray and NMR studies have determined that the rotamers exist in ratios of 60% gg and 40% gt (with negligible amounts of tg). The gg and gt forms are illustrated below (the dihedral angle is identified in purple):

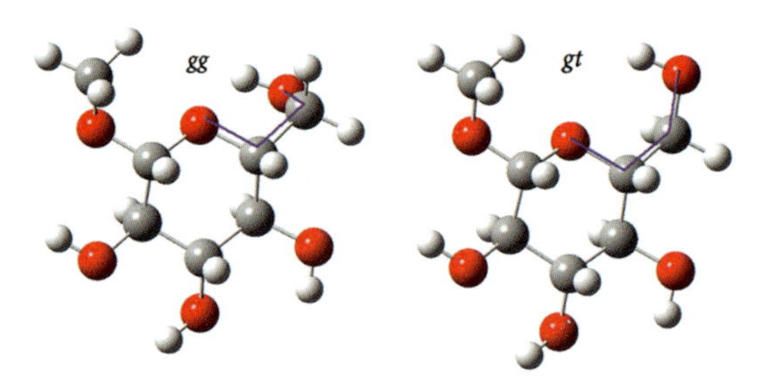

We optimized both structures in both environments and computed the ROA spectra for the resulting structures. We produced each composite spectrum using the experimental ratio of 60%-40%. The resulting spectra are shown below, along with the experimental results [Cheeseman11]:

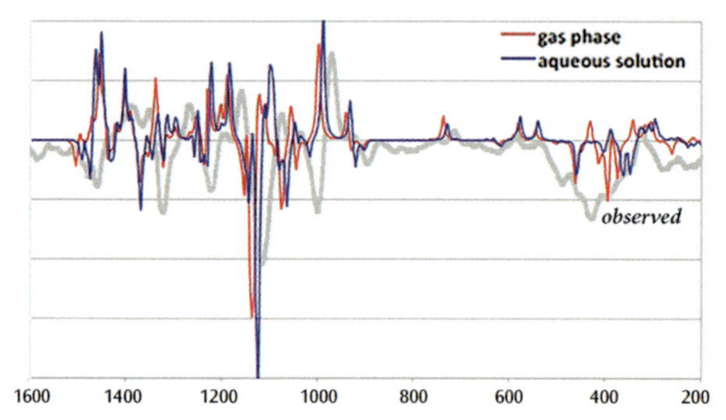

The most striking differences between the spectra in the gas phase and in aqueous solution are found in the range of 300-500 cm^{-1}. The spectrum predicted including solvent effects is a marginally better fit to the observed data, but both spectra reproduce the general shape of the experimental spectrum while deviating significantly in many details.

The original researchers went on to perform molecular dynamics (MD) simulations of this problem in an effort to improve on the electronic structure results. As indicated in the following figure, they achieved considerable success by doing so:

* IUPAC: 2-O-methyl-D-glucose.

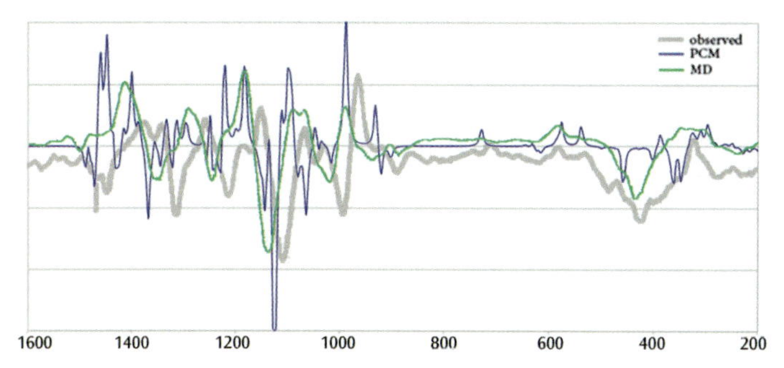

While there are some deviations, in general, the MD results are in very good agreement with experiment, both in terms of overall shape and many fine details. As illustrated in this example, sometimes electronic structure theory alone is incapable of simulating an aqueous environment. Molecular dynamics must be used to achieve very high accuracy for some properties. ■

Optical Rotations

Optical rotation is another sort of optical activity that can be observed and predicted. When polarized light travels through some materials, it is rotated with respect to the direction of motion. Such materials include crystals, spin-polarized molecules in the gas phase and chiral molecules. Here, we are concerned with the latter, which can yield important information about molecular structure. As with other observed effects of chirality, optical rotation differs for left and right circularly polarized light.

> **References: ORD**
> Theoretical Definition: [Rosenfeld28, Condon37]
> Implementation: [Karna91, Pedersen95, Kondru98, Stephens01, Ruud02, Stephens03]
> Solvation Treatment: [Mennucci02]

Observed optical rotations vary for the same substance according to the wavelength of the incident light. This variation is known as optical rotatory dispersion (ORD). ORD results are presented as plots rotation at various incident light wavelengths.

This phenomenon can be modeled in Gaussian using the **Polar=OptRot CPHF=RdFreq** keyword combination; **Polar=OptRot** directs Gaussian to predict optical rotation, and **CPHF=RdFreq** tells the program to read the desired incident light frequencies in a separate input section following the molecule specification. The optical rotation calculation should be performed as a second job step following an optimization+frequency calculation. As was true for many spectral properties, modeling optical rotation requires a larger basis than is needed for geometry optimizations and frequencies, specifically one including extra polarization and diffuse functions [Stephens01]. We will use the **6-311++G(2d,2p)** basis set; the **aug-cc-pVTZ** basis set is also an appropriate choice.

The following sample input file illustrates the method for performing an ORD calculation. Its final line provides the input required by the **CPHF=RdFreq** option: a list of one or more frequencies for the incident light. Here, we specify them in nanometers, indicated with the **nm** suffix.

```
%Chk=ord
# APFD/6-311+G(2d,p) Opt Freq

Optimization + frequency
```

Molecule specification

```
--Link1--
%Chk=ord
# APFD/6-311++g(2d,2p) Polar=OptRot CPHF=RdFreq
  Geom=Check Guess=Read
```

```
Predict optical rotation at several frequencies
```

Charge & spin multiplicity

```
633nm 589nm 578nm 546nm 436nm 365nm 355nm
```

final blank line

Note that neither GaussView nor WebMO provide a full interface for this calculation type. You will need to specify the required keywords in your tool's additional keywords area and add the list of desired frequencies manually.

The predicted optical rotation appears in the Gaussian output as follows:

```
                                              frequency in au
Dipole-magnetic dipole polarizability for W= 0.071980:
                1                2                3
      1    0.890037D+00   0.379592D+01 -0.493224D+01
      2   -0.586890D+01   0.698045D+00 -0.883638D+01
      3    0.133444D+01   0.948776D+01 -0.163500D+01
w= 0.071980 a.u., Optical Rotation Beta= 0.0156 au.
Molar Mass= 62.0435 g/mol, [Alpha] (6330.0 A)= 8.45 deg.
                              frequency in Å    [α]25°C 633nm
                                                 predicted
                                                 specific rotation
```

Note that the actual units of [α] are degrees/(dm·g/cm^{-3}).

There will be a section like this one for each frequency requested in the input.* Searching the output file for the string "Molar" is a quick way to list the ORD results, as in this example from a UNIX-based system:

```
$ grep Molar e7_08_f.log
 Molar Mass= 62.0435 g/mol, [Alpha] (6330.0 A)= 8.45 deg.
 Molar Mass= 62.0435 g/mol, [Alpha] (5895.0 A)= 10.11 deg.
 ...
```

The predicted specific rotations at the various frequencies can be plotted using any available spreadsheet or plotting software.

EXAMPLE 7.8 OPTICAL ROTATIONS: SUBSTITUTED OXIRANES

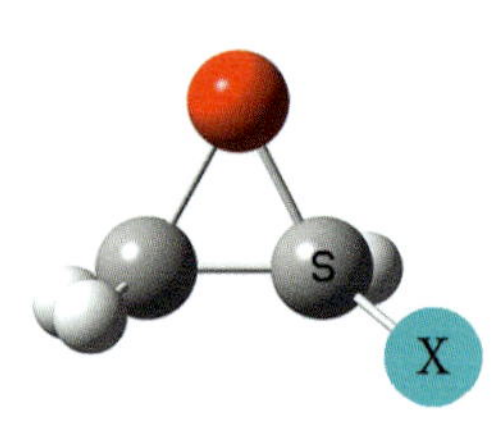

In this example, we will predict the optical rotation for two oxirane compounds: methyloxirane and fluorooxirane† (X=CH_3 and F, respectively) [Wilson05, Wiberg07]. We will run optimization+frequency calculations on both molecules and then run **Polar=OptRot** jobs at the following frequencies: 633nm, 589nm, 578nm, 546nm, 436nm, 365nm and 355nm. Both molecules have only a single conformation and are studied via their S enantiomer. All calculations were run in the gas phase.

The following table lists the key structural parameters, dipole moment and predicted optical rotations for each molecule, along with the available gas phase experimental results [Wilson05]:

* If you are interested in the full optical rotation tensors and not just the values of the specific rotations ([α]), then include **IOp(10/46=7)** in the route section. See the discussion of the **Polar=OptRot** option in the *Gaussian User's Reference* for more information.

† IUPAC: 2-methyloxirane and 2-fluorooxirane (respectively).

X	C-X (Å)	C-O-C-X (°)	μ (D)	OPTICAL ROTATIONS 355nm	365nm	436nm	546nm	578nm	589nm	633nm
F calc.	1.36	-110.5	2.30	45.0	40.9	23.3	12.4	10.6	10.1	8.5
CH_3: calc.	1.50	-112.8	2.02	3.2	-1.3	-13.8	-14.0	-13.2	-12.9	-11.8
exp.				7.49 ±0.3						-8.39 ±0.2

The predicted values are in reasonable agreement with the experimental results for the methyl substituent.

The following figure plots the curves corresponding to the predicted optical rotations and indicates the experimental observations for X=CH_3 as black squares:

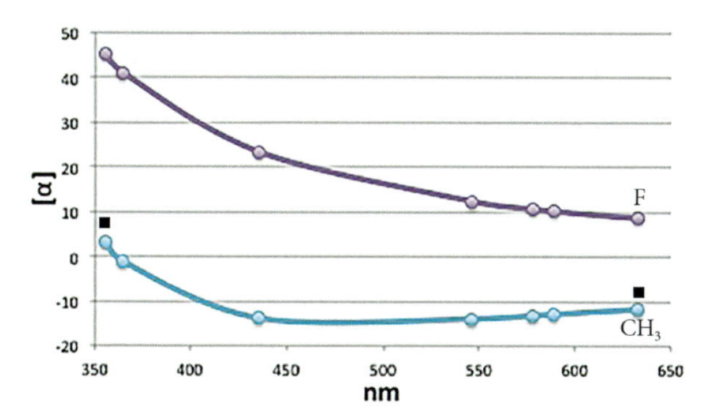

The graph illustrates the change in sign in the optical rotation for methyloxirane observed in the experiment. The shape of the curve for the fluorine substituent is quite similar to that of methyloxirane (although the predicted values are higher). Both substituents show an increase in specific rotations with decreasing wavelength.

We will continue this study in Exercise 7.12 (p. 316).

Modeling ORD in Solution

The solvent environment can also affect optical rotation results. The following plot shows the variation in optical rotatory dispersion for methyloxirane in a variety of solvents [Wilson05]:

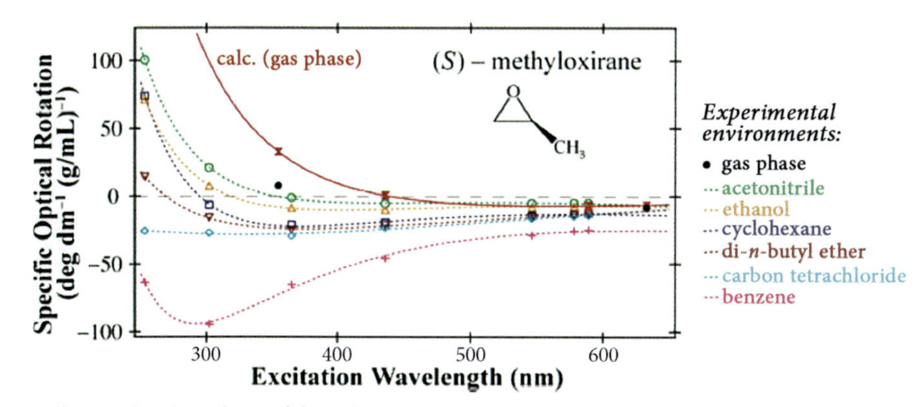

We will consider the effects of the solvent environment on this property for a related compound in Exercise 7.13 (p. 317).

A Final Note on Boltzmann Averaging

The VCD, ROA and ORD examples, along with the accompanying exercises, provide many examples where Boltzmann averaging is necessary in order to accurately model the desired spectra. In general, for many molecular properties, it is necessary to explore the conformational space in order to identify the set of conformers which contribute significantly to the observed spectrum. Many times, this is a straightforward process. However, some cases require a bit more consideration, especially when the energy range separating the conformers is quite small.

Consider the following sets of five conformations, displayed in table form and graphically below:

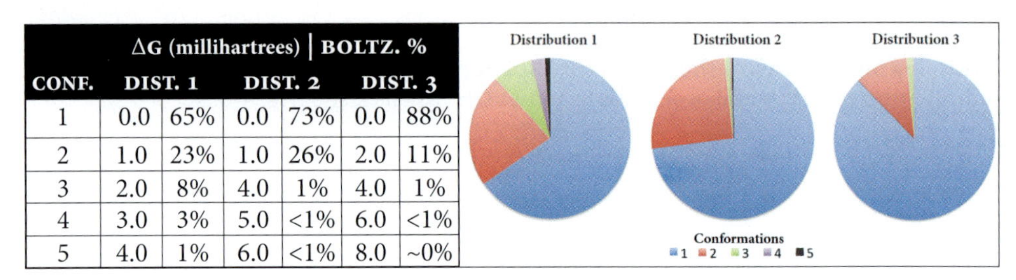

	ΔG (millihartrees) | BOLTZ. %					
CONF.	**DIST. 1**		**DIST. 2**		**DIST. 3**	
1	0.0	65%	0.0	73%	0.0	88%
2	1.0	23%	1.0	26%	2.0	11%
3	2.0	8%	4.0	1%	4.0	1%
4	3.0	3%	5.0	<1%	6.0	<1%
5	4.0	1%	6.0	<1%	8.0	~0%

In distribution 1, their energies are uniformly spaced at 1 millihartree (about 0.5 kcal/mol). In this case, the three lowest energy conformations comprise the bulk of the Boltzmann population, with the most abundant form making up 65%. Contrast this with distribution 3 in which the energies are uniformly spaced at two millihartrees. Now only the two most abundant conformations contribute significantly to the observed properties, and the most abundant conformation now dominates it, at 88%.

Distribution 2 modifies distribution 1, placing the three higher energy conformations farther from the top two (3 vs. 1 millihartree). This results in significant changes in the Boltzmann population: only the two lowest energy conformations contribute significantly, and their relative contributions change as well.

The important point to notice from these hypothetical scenarios is that small changes in relative energies can produce significant modifications to the Boltzmann population—and hence to the predicted spectrum. Keep in mind that the energy differences noted here are much smaller than the uncertainty in the energy predictions themselves. The Boltzmann averaging technique assumes that any uncertainty in energy predictions is systematic across a set of conformations, enabling the differences in energy between them to be considered highly quantitatively reliable, much more so than the uncertainty in commonly used model chemistries would justify for the predicted total energies. This assumption is generally reasonable for problems like those we have considered, but you should be aware of it. Nevertheless, in the vast majority of cases, it is better to perform conformational averaging—even if it is approximate—than to just use the lowest energy conformation alone.

Finally, the relative abundance of conformations need not be determined from theoretical calculations. It is common to use experimentally-determined conformational distributions as the weighting factors for theoretical predictions of spectra or other molecular properties.

Exercises

EXERCISE 7.1 NMR SHIELDING TENSORS: SUBSTITUENT EFFECTS

Continue the study examining the NMR shielding tensor components that we began in Example 7.1 (p. 278) for the following substituents: Cl, CN, CH_3, SiH_3, $SiMe_3$, NO_2, $CH{=}CH_2$.* You will need to run an optimization+frequency job for each compound, followed by an NMR calculation. What trends do you observe in the results? How do the various substituents change the results with respect to acetylene?

SOLUTION

The following table presents the results of our calculations along with the observed overall isotropic shielding values [Wiberg04c, Duncan97] (the latter have an error of ±5 ppm):

		CALCULATED			OBSERVED
X	C	SHIELDING	XX,YY	ZZ	SHIELDING
H		113	30	280	116
F	α	92	-5	287	102
	β	174	120	282	177
Cl	α	117	30	291	
	β	127	50	282	
CN	α	125	43	287	129
	β	114	31	280	112
CH_3	α	104	10	293	106
	β	116	44	261	117
SiH_3	α	106	15	289	
	β	84	1	252	
$Si(CH_3)_3$	α	95	-3	292	96
	β	90	11	247	93
NO_2	α	103	5,15	287	
	β	129	91,69	227	
$CH{=}CH_2$	α	102	3,22	280	103
	β	102	45,64	195	106

All of the computed values are in very good agreement with experiment. Differences in shieldings at the carbon atoms generally vary according to the electronegativity of the substituent.

The following table lists the overall chemical shifts and changes in the axial components for each carbon atom relative to acetylene, along with the observed values for the former:

	CALC.: α,β		OBS.†: α,β
X	ISO.	ZZ	ISO.
F	21,-61	7,2	14,-61
Cl	-4,-14	11,2	
CN	-12,-1	7,0	-13,4
CH_3	9,-3	13,-19	10,-1
SiH_3	7,29	9,-28	
$Si(CH_3)_3$	18,23	12,-33	20,23
NO_2	10,-16	7,-53	
$CH{=}CH_2$	11,11	0,-85	13,10

†±5 ppm

* IUPAC: chloroacetylene: chloroethyne; cyanoacetylene: prop-2-ynenitrile; methylacetylene: 1-propyne; SiH3-acetylene: ethynylsilane; trimethylsilane-acetylene: ethynly(trimethyl)silane; nitroacetylene: nitroethyne; ethenyl-acetylene: but-1-en-3-yne.

The largest differences are generally seen in the β (outer) carbon. The largest shift in isotropic shielding comes in fluoroacetylene even though the axial component remains essentially unchanged. Chloroacetylene exhibits only a small decrease in the shielding and increase in the ZZ component. The three axially-symmetric, non-planar compounds have small to moderate changes in the chemical shift for the β carbon and corresponding decrease in the corresponding axial component.

Large changes in the axial tensor component do not necessarily lead to corresponding ones in the chemical shift. For example, the β carbon atom in the vinyl substituent sees its ZZ component decreased substantially—approaching 100 ppm—but the chemical shift is only slightly increased. The same effect is present for the NO_2 substituent, albeit less dramatically.■

EXERCISE 7.2 SPIN-SPIN COUPLING CONSTANTS: THREE-MEMBERED RING SYSTEMS

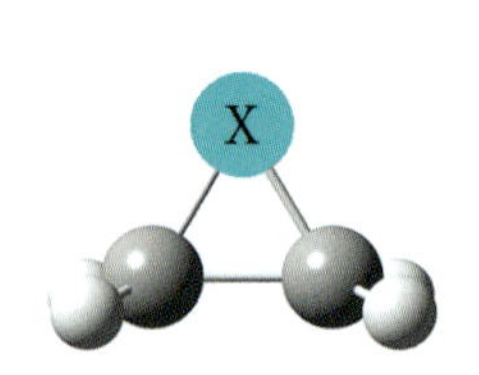

Compute the spin-spin coupling constants for the C–C bond in the three-membered ring systems where X is NH, S and SiH_2. How do the results vary with the substituent for the various three-membered rings we have considered?

SOLUTION

The following table lists the predicted spin-spin coupling terms and total ^{13}C-^{13}C constants for each compound, along with the experimental values when available [Rusakov13, Krivdin03] (all values are in Hz):

	COMPONENTS OF J				TOTAL J	
X	J-DSO	J-PSO	J-SD	J-FC	CALC.	EXP.
CH_2	0.15	-0.82	-0.19	11.6	10.8	12.4
O	0.17	-0.73	-0.53	25.6	24.5	28.0
NH	0.16	-0.73	-0.36	19.6	18.6	21.0
S	0.20	-2.10	-0.05	24.8	22.8	
SiH_2	0.18	0.09	0.28	7.1	7.0	

The value for the NH substituent is again in good agreement with experiment, and the calculations correctly reproduce the trends among the three compounds for which there is data. The lowest value occurs for the SiH_2 substituent, due to the electropositive silicon atom. The highest values are in oxirane and thiirane (O and S respectively).

The diamagnetic spin orbit term is not sensitive to the heteroatom in the substituent. There is some variation in values in the paramagnetic spin orbit and spin-dipolar terms, but it is the Fermi contact term whose contribution once again dominates the total term.■

EXERCISE 7.3 SPIN-SPIN COUPLING CONSTANTS: HIGHLY STRAINED SYSTEMS

Spin-spin coupling constants are interpreted as indicating the degree of steric strain for C-C bonds, thereby revealing the nature of their hybridization. Compute the spin-spin coupling constants for the bridgehead carbon atoms in bicyclobutane* and propellane. What can you conclude about the natures of the two bonds from the results?

* IUPAC: bicyclo[1.1.0]butane.

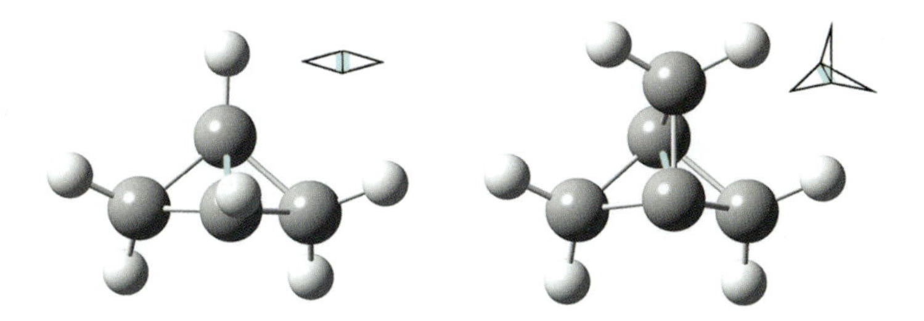

SOLUTION

The following table lists the results for these two compounds (values are in Hz):

	COMPONENTS OF J				TOTAL J	
	J-DSO	J-PSO	J-SD	J-FC	CALC.	EXP.†
propellane	0.17	-0.70	2.60	26.0	28.1	*
bicyclobutane	0.18	-1.00	-0.90	-12.4	-14.2	-17.5

†[Rusakov13, Finkelmeier78].

*This value cannot be measured experimentally for symmetry reasons.

The individual natures of the bonds in these compounds are reflected in their spin-spin coupling constants. In the case of bicyclobutane, the coupling constant for the central bond is negative; this compound and its derivatives provide the only known such examples between covalently bound carbon atoms. This result also indicates a very high level of involvement of the p orbitals in the corresponding carbon atoms in forming the bond.

In contrast, the central bond in propellane has a large positive spin-spin coupling constant. This value indicates that what is sometimes termed a "phantom bond" does physically exist in this molecule and its character is more or less that of an ordinary aliphatic carbon-carbon bond.■

EXERCISE 7.4 ABSOLUTE CONFIGURATION OF FENCHONE

Fenchone* is a compound similar in structure to camphor. Fenchone is a colorless liquid found in fennel while camphor is a white or transparent solid. Like camphor, fenchone has two chiral centers. Its (1S,4R) configuration is illustrated in the margin.

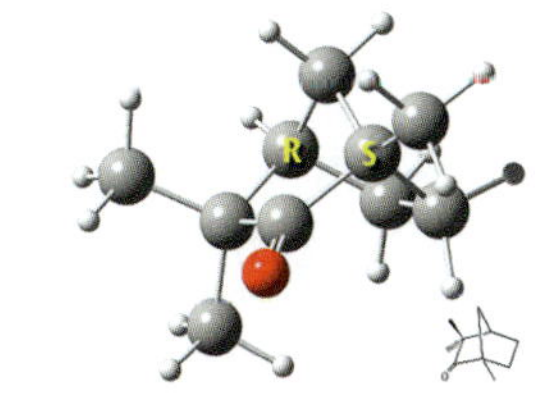

Experimental IR and VCD spectra for (+)-fenchone are reproduced below [Devlin97]. Determine the absolute configuration for fenchone—(1S,4R) vs. (1R,4S)—by predicting the VCD spectrum for the relevant molecule(s).

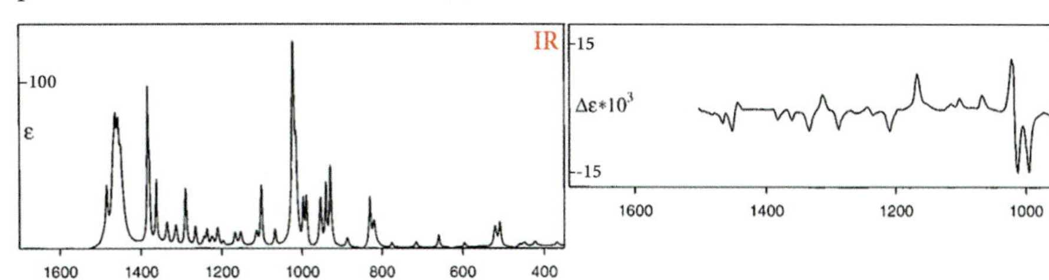

* IUPAC: 1,3,3-trimethylbicyclo[2.2.1]heptan-2-one.

SOLUTION

We ran an optimization+frequency calculation including prediction of the VCD spectrum for the (1S,4R) enantiomer. The predicted IR and VCD spectra are displayed below, with the experiment spectra included as insets:

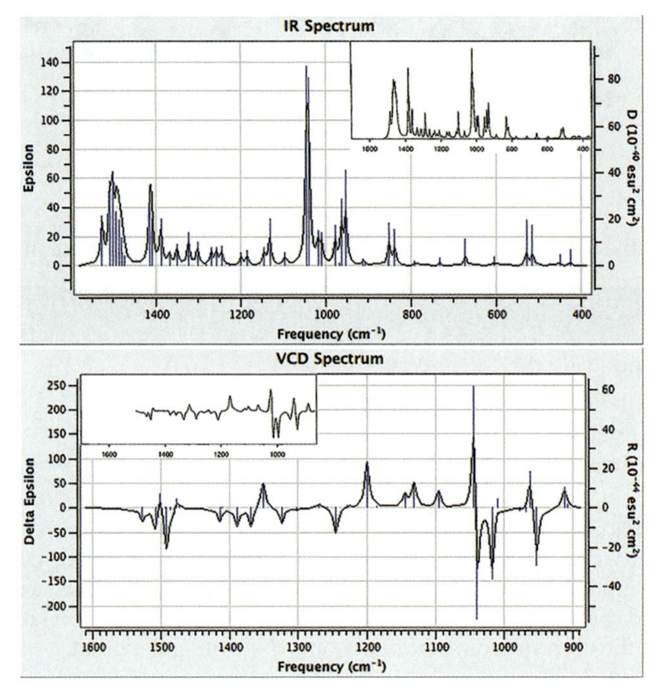

Both spectra are in excellent agreement with experiment, with the VCD results indicating that the experimental spectra correspond to the (1S,4R) enantiomer. Thus, the absolute configuration of fenchone is (1S, 4R) (+) and (1R,4S)-(–).■

EXERCISE 7.5 DISTINGUISHING PRODUCTS WITH VCD

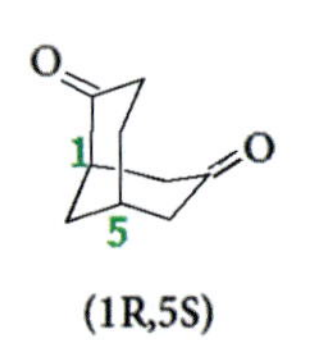

In this example, we will use VCD spectra to distinguish between structural isomers. Consider the Baeyer-Villiger oxidation of (+)-(1R,5S)-bicyclo[3.3.1]nonane-2,7-dione (pictured in the margin).

There are four possible keto-lactone products:*

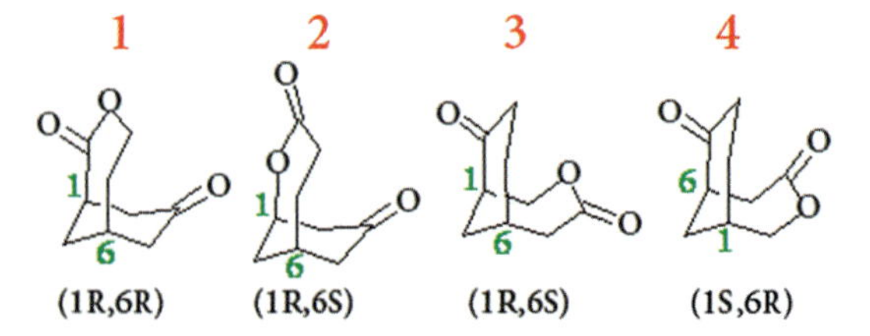

The standard mechanism of the Baeyer-Villiger reaction would predict that structure 2 will be the product, based on the greater migratory aptitude of tertiary alkyl over secondary alkyl groups [Stephens05a].

* **1**: (1R,6R)-3-oxabicyclo[4.3.1]decane-2,8-dione; **2**: (1R,6S)-2-oxabicyclo-[4.3.1]decane-3,8-dione; **3**: (1R,6S)-3-oxabicyclo[4.3.1]decane-4,9-dione; **4**: (1S,6R)-3-oxabicyclo[4.3.1]decane-4,7-dione.

Perform VCD calculations to determine which isomer is in fact the product by comparing the predicted spectra to the experimental spectrum reproduced below:

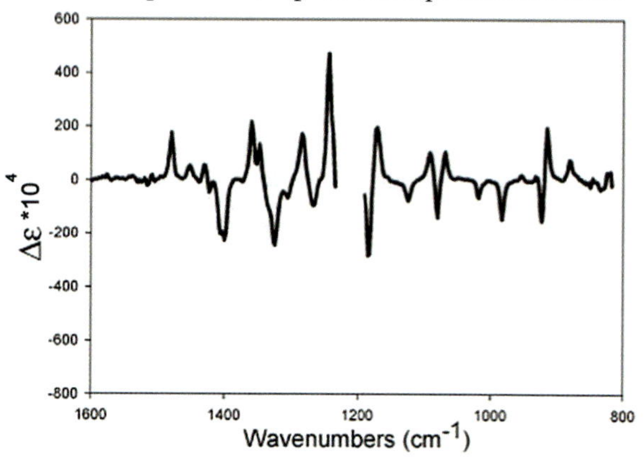

SOLUTION

After optimizing the four isomers and verifying that they are minima, we can examine their predicted VCD spectra:

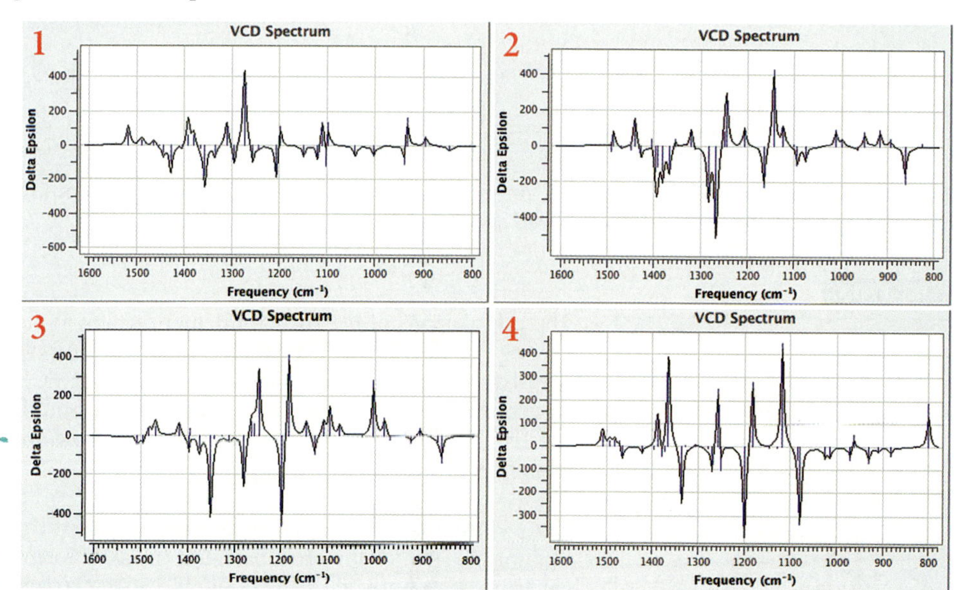

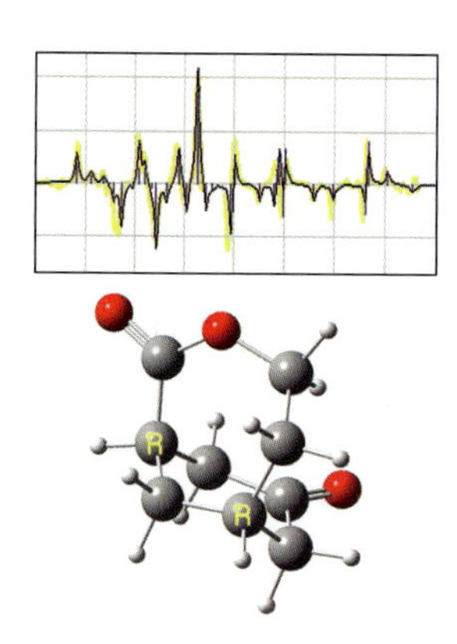

These results indicate that structure 1 is the observed product. The experimental and predicted spectra are overlayed in the illustration in the margin (experimental spectrum in yellow). It illustrates the very good agreement between theory and experiment. The molecular structure of isomer 1 is illustrated in the margin next to the predicted spectra.■

EXERCISE 7.6 (R)-3-METHYLCYCLOHEXANONE VCD SPECTRUM

This exercise will predict the VCD spectrum for (R)-3-methylcyclohexanone,* a compound whose enantiomers have axial and equatorial forms. The two conformations for the indicated enantiomer are shown in the following illustration:

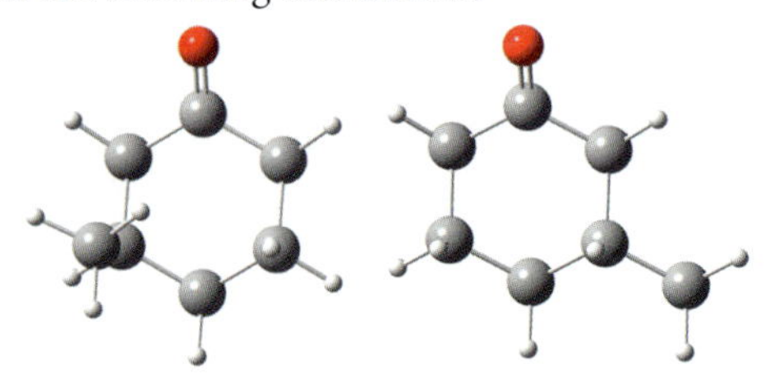

If you find your structure has the wrong stereochemistry, you can invert it via GaussView's **Mirror Invert** feature (on the **Edit** menu).

When constructing these molecules in a graphical interface, the methyl group must move to the opposite side of the molecule (as shown here) when transforming one form into the other (e.g., axial into equatorial) in order for the two forms to retain the same stereochemistry (as illustrated here).

Here is the experimental spectrum [Devlin99]:

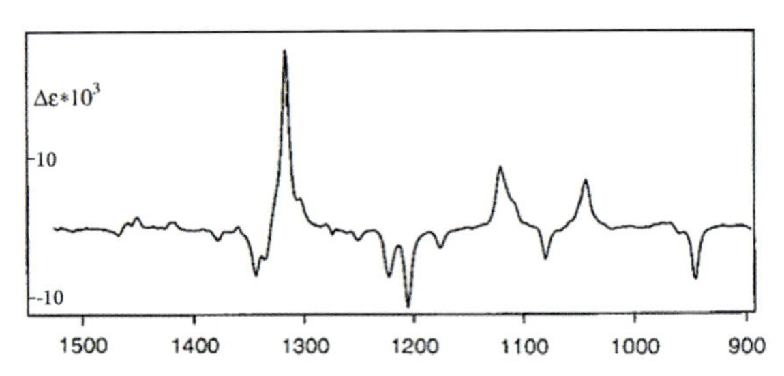

Locate the minima for the two structures, compute their individual spectra and then combine them appropriately to predict the observed spectrum.

SOLUTION

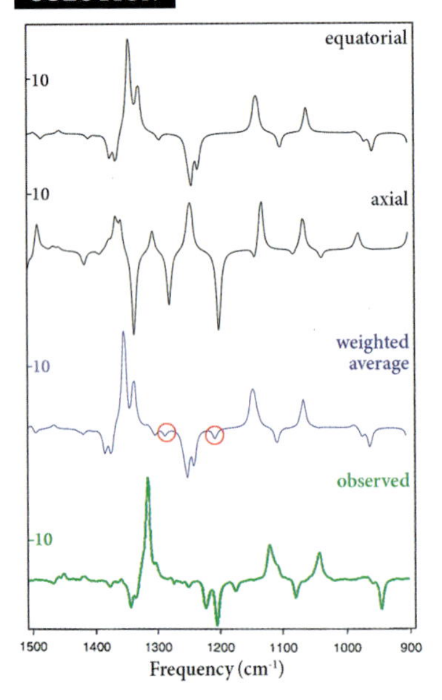

The populations of the two conformations are about 89% equatorial and 11% axial at our standard model chemistry. The illustration shows the individual VCD spectra for the two forms, the predicted (weighted) spectrum and repeats the experimental spectrum for reference:

The weighted average spectrum follows that of the equatorial form for the most part, but the axial form's spectrum contributes in small but significant ways (most obviously at the points circled in red). There is excellent agreement between the composite predicted spectrum and experiment.■

* IUPAC: (3R)-3-methylcyclohexanone.

EXERCISE 7.7 α-PINENE OXIDE DIASTEREOMERS

α-pinene oxide has two diastereomers: cis and trans forms (illustrated without hydrogen atoms in the margin). The experimental ROA spectrum of alpha-pinere oxide obtained from epoxidation of (1S, 5S)-(-)-alpha-pinere [Qiu10] is reproduced below:

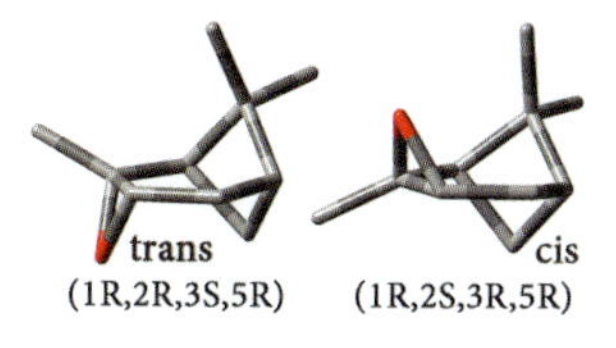

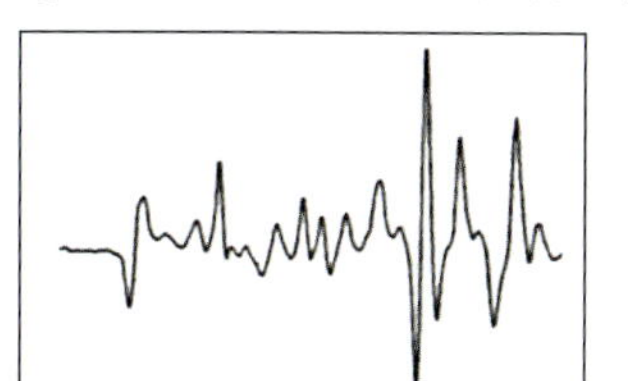

Determine which diastereoisoner—trans or cis—corresponds to the experimental spectrum.

SOLUTION

Comparison of the predicted ROA spectra for the two diastereomers indicates that the experimental spectrum corresponds to the trans isomer (see the figure on the left below). The predicted energies for the two diastereomers are given in the following table:

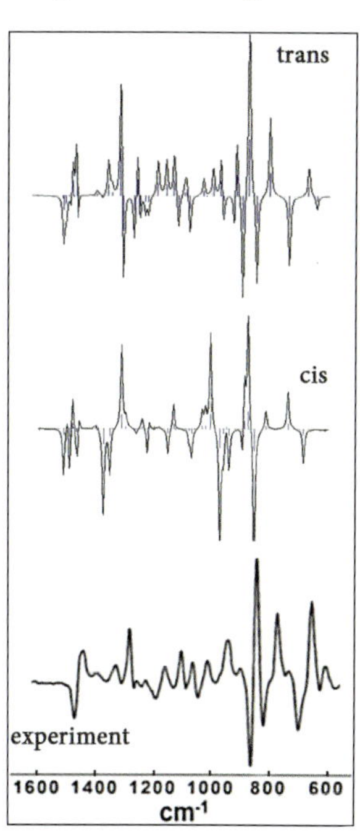

CONFORMATION	G (hartrees)	%
cis	-465.41831	0.2
trans	-465.42412	99.8

As these results indicate, the cis diastereomer contributes essentially nothing to the observed spectrum, and so the predicted spectrum and other properties for the trans conformation can be used directly for this compound. No conformational averaging is needed.■

EXERCISE 7.8 CONFORMATIONAL ELUCIDATION OF A CHIRAL DRUG

The biological process by which new blood vessels are formed from pre-existing vessels is known as *angiogenesis*. This process is a normal part of growth and wound healing in animals, including human beings. It is also an important step by which tumors transform from benign to malignant. Accordingly, antiangiogenic agents are an important part of cancer treatment and a focus of ongoing research.

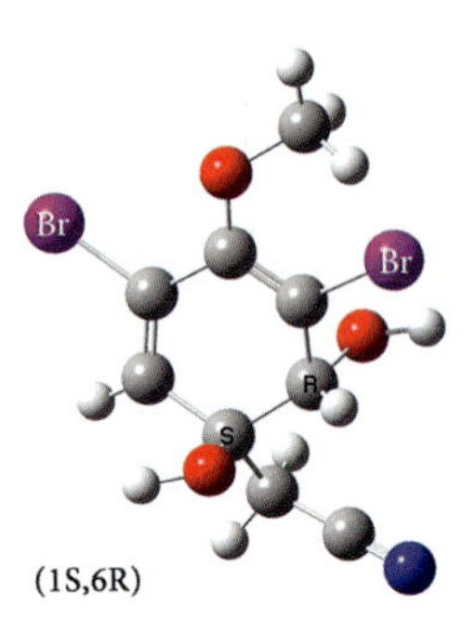

(1S,6R)

Aeroplysinin-1* is a brominated compound that is of interest as an antiangiogenic drug. This compound occurs naturally in the marine sponge *Aplysina cavernicola* found in the Mediterranean Sea. Aeroplysinin-1 is a chiral molecule with two chiral centers: the two asymmetric carbon atoms indicated in the figure in the margin. It is common that only one of the stereoisomers of a chiral drug is primarily responsible for its therapeutic effects. In such cases, it is important to identify the appropriate stereoisomer. In this exercise, you will investigate the various conformations of aeroplysinin-1 in order to determine which ones contribute to the observed ROA spectrum, which is reproduced below [NietoOrtega11]:

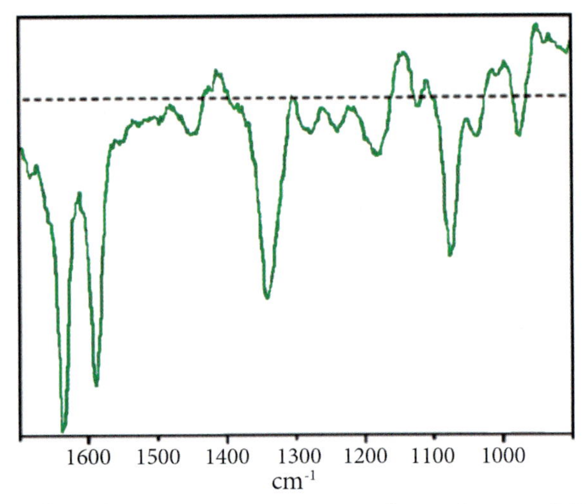

This compound has four diastereomers, two pairs of enantiomers: (1R,6S), (1S,6R) and (1R,6R), (1S,6S). Predict the ROA spectrum for a member of each pair of enantiomers and determine the absolute configuration for the molecule. The experimental spectrum was obtained in an aqueous environment.

SOLUTION

We chose to model the (1S,6R) and (1S,6S) diastereomers. The conformational space for this compound is large and complex, and exploring it required several steps to ensure that all of the relevant structures were examined. The key structural parameters for this compound are the two dihedral angles highlighted in the illustration in the margin and the orientations of the hydroxy groups.

We used the following procedure:

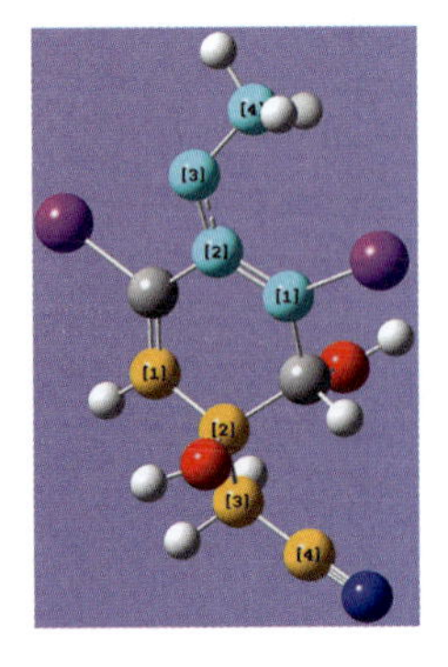

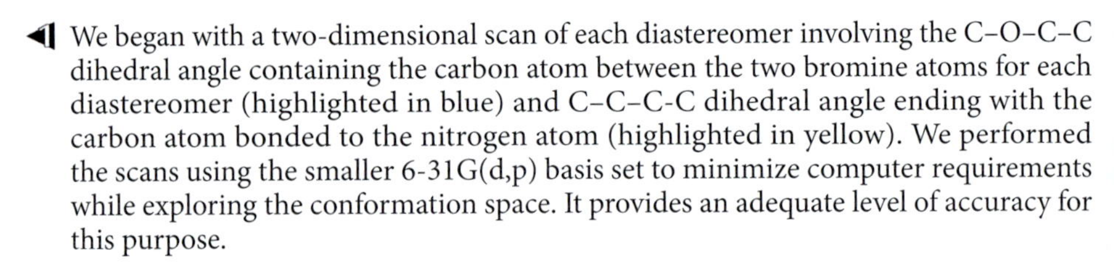

◀ We began with a two-dimensional scan of each diastereomer involving the C–O–C–C dihedral angle containing the carbon atom between the two bromine atoms for each diastereomer (highlighted in blue) and C–C–C–C dihedral angle ending with the carbon atom bonded to the nitrogen atom (highlighted in yellow). We performed the scans using the smaller 6-31G(d,p) basis set to minimize computer requirements while exploring the conformation space. It provides an adequate level of accuracy for this purpose.

The scan results yielded six minima for each diastereomer that we located using GaussView's plot of the scan results. We have inverted the plot as we did previously in order to make identifying the minima straightforward.

* IUPAC: (3,5-dibromo-1,6-dihydroxy-4-methoxy-2,4-cyclohexadien-1-yl)acetonitrile.

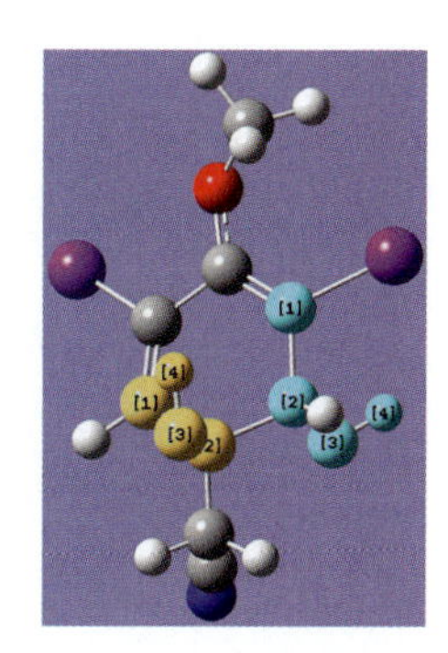

2. We optimized the structure of each located minimum using the same smaller basis set as in the previous step.

3. For each of the 12 optimized structures, we performed a second two-dimensional scan over the two C–C–O-H dihedral angles (see the illustration in the margin). The positions of these groups can have a significant influence on the ROA spectrum. The scans again used the smaller 6-31G(d,p) basis set.

4. We compared the energies of all of the minima located by the six scans for each diastereomer. We eliminated duplicate structures and discarded ones constituting less than about 1% of the Boltzmann distribution. This yielded 19 structures for the (1S,6R) diastereomer and six structures for the (1S,6S) diastereomer. These structures were then optimized using our standard model chemistry.

5. We ran ROA calculations for those structures from the previous step that constituted more than 5% of the Boltzmann distribution. We saved the checkpoint files from the frequency calculations performed in the preceding step.

   Here is the input file for the ROA calculation:

```
%OldChk=chkpt_from_freq_calc
%Chk=chkpt_for_this_calc
# APFD spAug-cc-pVTZ Polar=ROA Guess=Read Geom=Check
  SCRF(Solvent=Water)

ROA

0 1

532nm
```
final blank line

   We use the **%OldChk** directive to specify the checkpoint file from the frequency calculation from which information will be retrieved only. A different checkpoint file, specified as usual with **%Chk**, will be used for the current calculation. This allows the optimized structure and frequency data to be retained unmodified for future use.

6. The predicted ROA spectra for the set of conformations for each diastereomer were combined using Boltzmann averaging.

All calculations were performed in solution with water via Gaussian IEFPCM facility.

The following illustrations present the results of our calculations for the (1S,6R) diastereomer. The eight structures whose data we incorporated into the predicted spectrum are shown on the left, along with their relative contributions. Structures which differ only by rotation of one or both of the hydroxy groups are given the same number followed by a unique letter.

The experimental and predicted spectra are shown on the right. There is good agreement between them, and this diastereomer can be identified as the absolute configuration for aeroplysinin-1.

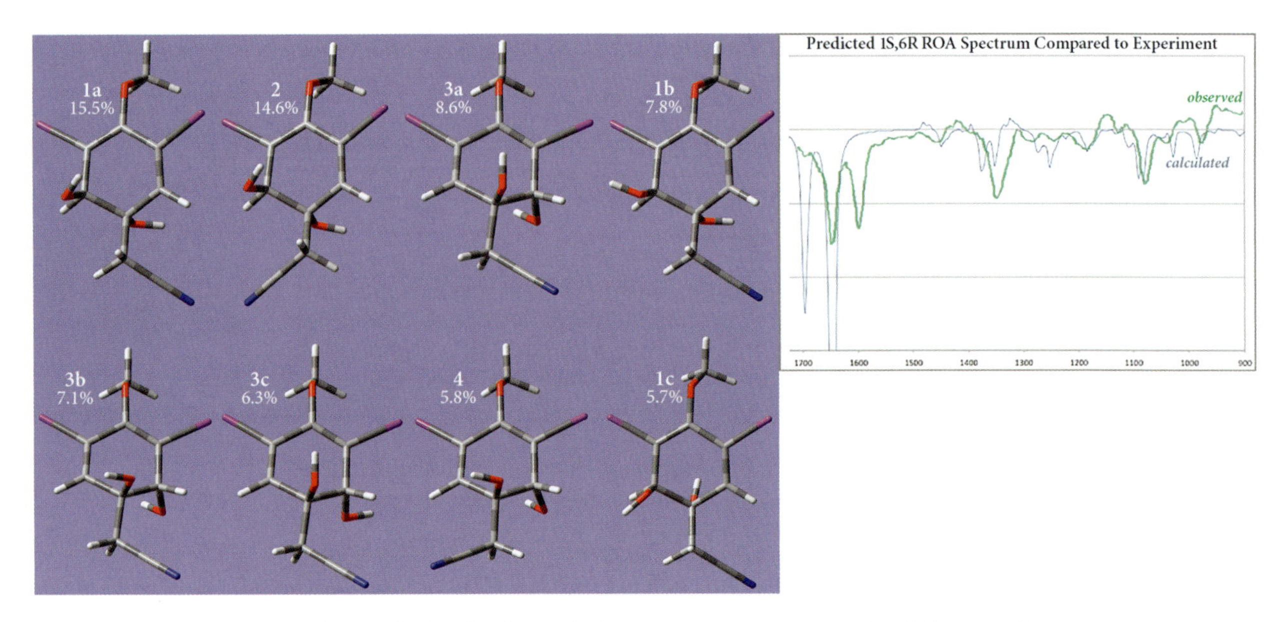

The results for the (1R,6R) diastereomer are given in the following illustrations:

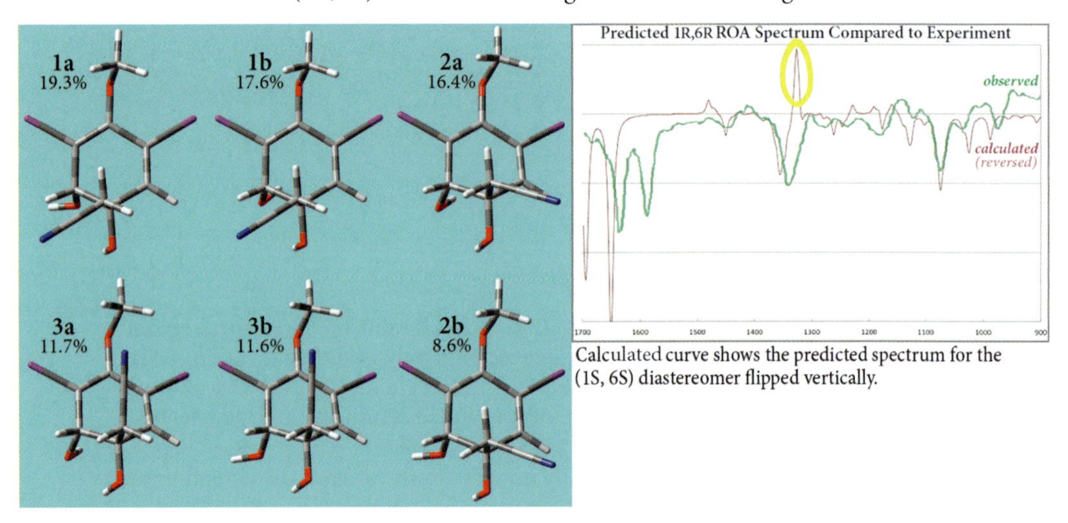

Calculated curve shows the predicted spectrum for the (1S, 6S) diastereomer flipped vertically.

As before, the structures whose spectra contributed to the predicted spectrum are shown on the left, and the predicted and experimental spectra are shown on the right. The comparison shows that the spectrum for this enantiomer does not match the observed one; the former includes several significant reversals of peak direction with respect to experiment, most notably at ~1325 cm^{-1}.

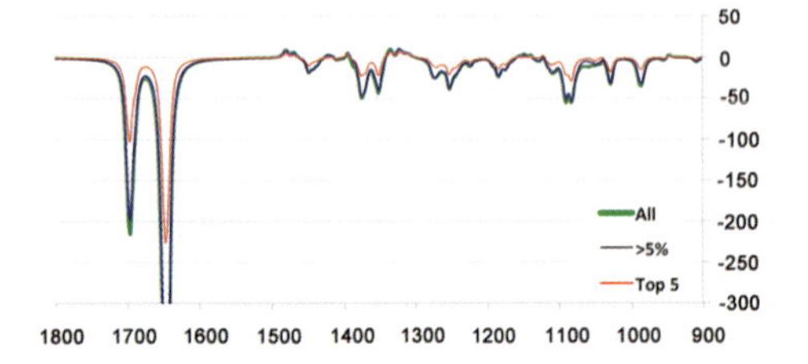

We included all of the conformations that comprised more than about 5% of the Boltzmann distribution within each composite weighted spectrum. You may wonder what the effects of different cutoffs might be. The figure in the margin illustrates three versions of the (1S,6R) composite spectrum. The green line indicates the spectrum produced by including all of the conformations for which the ROA calculation was performed. The blue line indicates the effect of eliminating data for conformations contributing <~5% to the distribution, and the orange line indicates the spectrum resulting from considering only the five most abundant conformations.

There is little difference between the green and blue spectra. There are some differences in detail between the blue and orange spectra although the general shape is retained even in the latter.■

EXERCISE 7.9 EPICHLOROHYDRIN ROA SPECTRUM IN ACETONITRILE

In this exercise, you will continue to study the ROA spectrum of (R)-epichlorohydrin [Pecul06]. Predict the ROA spectrum in solution with acetonitrile, and compare your results with those in the gas phase and in cyclohexane given earlier (see p. 292ff).

SOLUTION

The following table presents the structural and energetic data for (R)-epichlorohydrin in the three environments:

	GAS PHASE			CYCLOHEXANE			ACETONITRILE		
FORM	θ (°)	ΔE_0 (kJ/mol)	Bf	θ (°)	ΔE_0 (kJ/mol)	Bf	θ (°)	ΔE_0 (kJ/mol)	Bf
0°	160.3	0.00	0.563	160.3	0.00	0.452	160.1	2.98	0.204
120°	-82.5	1.17	0.352	-81.1	0.01	0.450	-78.5	0.00	0.675
240°	49.4	4.70	0.086	48.6	3.82	0.098	46.8	4.28	0.121

The molecular structures optimized in acetonitrile are similar to the ones in the other environments. Interestingly, the energetic order of conformations changes in acetonitrile, with the 120° conformer now the lowest in energy. The 240° conformer again is significantly higher in energy than the other two and contributes in only a minor way to the composite spectrum.

Here are the predicted spectra in the three environments:

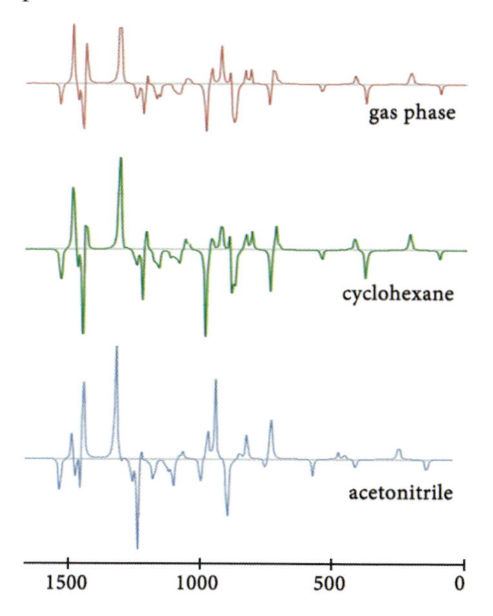

The gas phase spectrum is altered much more in acetonitrile than in cyclohexane. For example, while the general shape is similar in the range 1000-1500 cm^{-1}, both the magnitude and the sign of the intensity varies greatly. In general, this solvent has a more significant effect on both the conformer populations and the optical tensors than did cyclohexane. Although the molecular structural differences are minor, acetonitrile does again produce a threefold solvent effect on the ROA results.■

EXERCISE 7.10 LACTAMIDE ROA SPECTRUM IN WATER

This exercise studies the ROA spectrum of lactamide* in an aqueous environment. In general, polar molecules can be challenging to model in solution due to the complex interactions between the solute and the solvent. Predict the ROA spectrum for lactamide in the gas phase and in solution with water, taking into account the conformational flexibility of the molecule. Does adding explicit water molecules have an effect on the predicted spectrum? The experimental ROA spectrum for lactamide appears below [Hopmann11]:

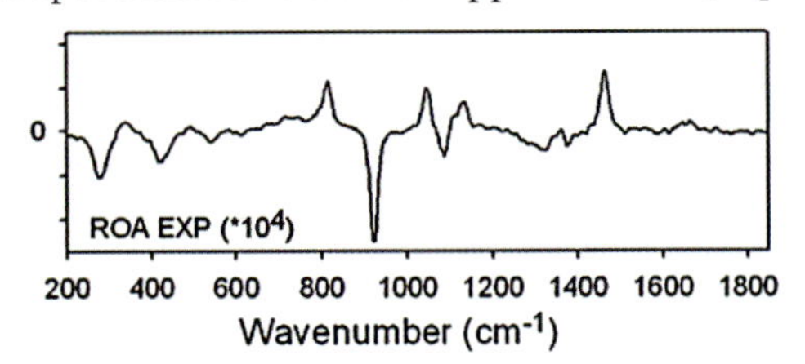

SOLUTION

We ran a two-dimensional scan of the R form of lactamide in the gas phase with our standard model chemistry, varying the O=C-C-H and C-C-O-H dihedral angles; the scan identified seven minima. Optimizing these structures yielded four unique conformations, the same results as those found by the original researchers using a different model chemistry [Hopmann11]. The resulting conformations appear below [Hopmann11]:

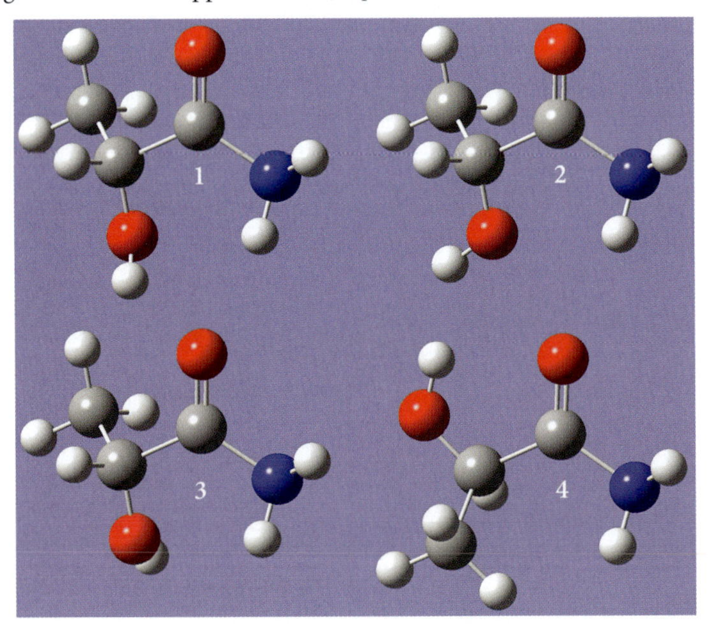

Conformations 1 through 3 are of the anti form. These structures are very similar except for the orientation of the hydroxyl proton. Conformation 4 is a syn form. The anti conformations contain a hydrogen bond between the amine and hydroxyl groups, while the syn conformation exhibits one between the double-bonded oxygen atom and the hydroxyl group.

* IUPAC: 2-hydroxypropanamide.

We repeated the optimizations for the four conformations in solution with water. The following table lists the various conformations' key parameters—the dihedral angles corresponding to the carbon-nitrogen chain and to the orientation of the hydroxyl proton—as well as their relative energies and Boltzmann populations:

	GAS PHASE				WATER			
CONF.	N-C-C-C (°)	C-C-O-H (°)	ΔG (kJ/mol)	%	N-C-C-C (°)	C-C-O-H (°)	ΔG (kJ/mol)	%
1	-106.5	79.9	3.9	10.7	-106.7	80.0	1.8	17.2
2	-115.5	-161.0	1.8	24.7	-110.6	-159.0	0.0	35.6
3	-131.9	-93.3	3.3	13.8	-126.7	-93.5	1.6	18.6
4	65.8	-6.9	0.0	50.9	68.6	-10.4	0.6	28.5

While the structural changes between the gas phase and the aqueous environment are small, the relative energies of the conformations are quite different. In the gas phase, conformer 4 has the lowest energy, corresponding to approximately half of the Boltzmann population (the latter is in excellent agreement with the estimate from a free jet experiment [Maris02]). The remaining conformers are in the descending order 2, 3, 1, with the latter two forms extremely close in energy. Indeed, all four conformations lie in a range of less than 4 kJ/mol in the gas phase, a span which is compressed to less than 2 kJ/mol in solution. In the aqueous environment, conformer 2 is the lowest in energy, followed closely by conformers 4, 3 and 1. In both environments, all four conformations contribute non-trivially to the Boltzmann population.

The composite ROA spectra for the gas phase and solution environments are shown below (the experimental spectrum is colored yellow):

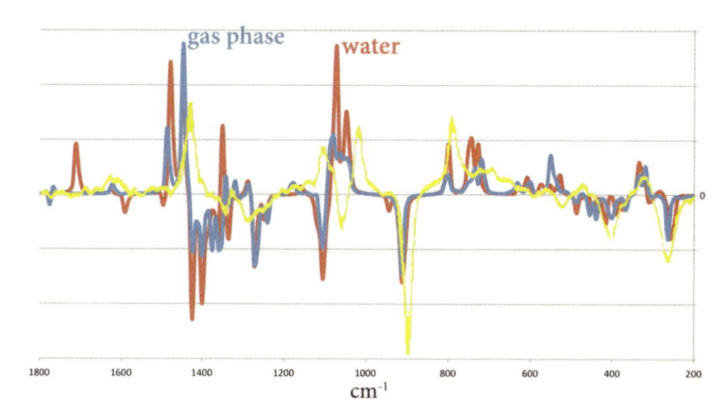

For this compound, including the solvent environment has only minor effects on the predicted spectrum, yielding sometimes slightly better and sometimes slightly worse agreement with experiment. Nevertheless, both computed spectra reproduce the general shape of the experimental spectrum reasonably well—certainly sufficiently well to identify the R form as the absolute configuration for this molecule.

We attempted to improve on these results by adding four explicit water molecules to the structure for each conformation. The positions of the water molecules in each case are indicated in the following figure:

CONF.	ΔG (kJ/mol)	%
$1{+}4H_2O$	5.1	7.9
$2{+}4H_2O$	3.4	15.2
$3{+}4H_2O$	0.0	60.2
$4{+}4H_2O$	3.2	16.8

Interestingly, the energetic range of the conformations increases somewhat when explicit water molecules are added to the aqueous solution model, and the energy ordering also differs.

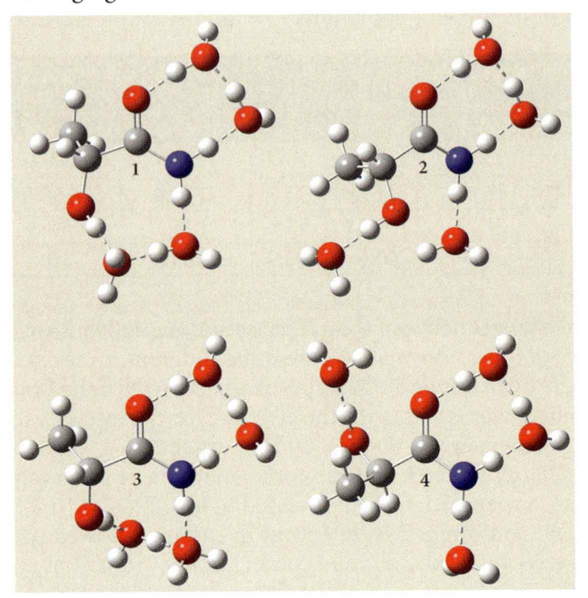

The resulting complexes were optimized, and their ROA spectra were predicted, both in aqueous solution. The relative energies of the conformations are given in the table in the margin, and the predicted spectrum is displayed below, along with the previously computed spectrum in solution and the observed spectrum:

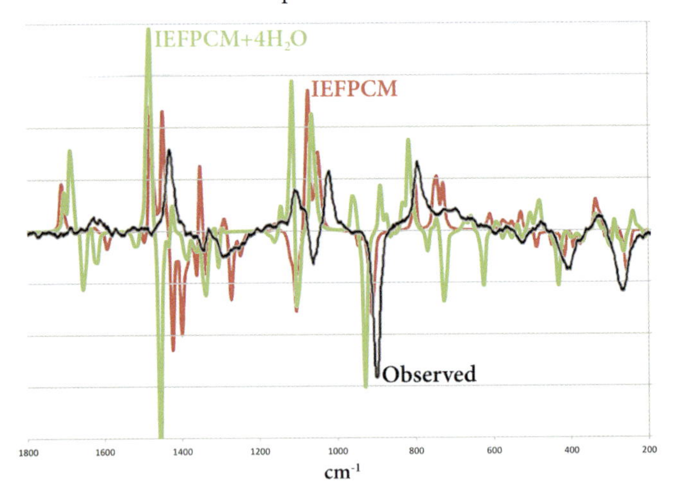

The addition of the four water molecules serves mainly to amplify the range of the one modeled with IEFPCM alone. As we saw when moving from the gas phase prediction to that in solution, there are areas of closer coincidence with experiment, but the overall results are not an unequivocal improvement. Moreover, the original researchers found that "using a different number of water molecules (e.g., three or five) or a different hydrogen-bonding pattern, however, leads to different spectra" (p. 4132). Thus, for this problem, adding water molecules does not lead to more accurate computational results.

The original researchers undertook this study with the goal of obtaining qualitatively accurate agreement with experiment. To that end, they continued this work by performing molecular dynamics simulations. Their results obtained with Car-Parrinello molecular dynamics appear below. Although the predicted spectrum generally follows both the shape and the intensity of the observed spectrum, there are several areas of significant deviation as well:

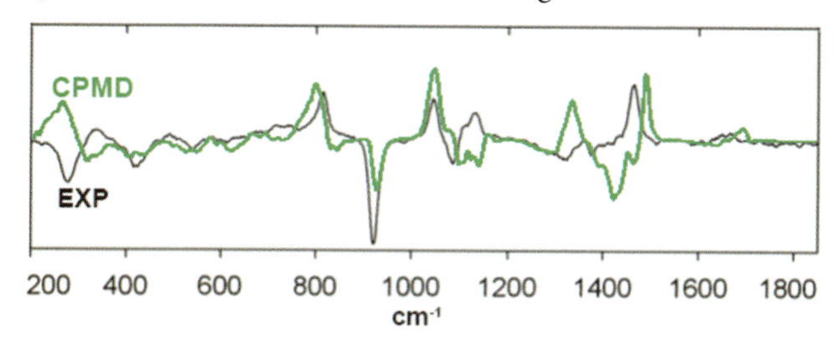

Modeling chiral species in aqueous solution remains an area of active ongoing theoretical research. Nevertheless, currently available methods and techniques achieve quite reasonable qualitative accuracy and are of proven benefit in interpreting experimental spectra and identifying absolute configurations.■

EXERCISE 7.11 INDUCED CHIRALITY: CAMPHOR VCD SPECTRUM IN CHLOROFORM

In this exercise, we study the VCD spectrum of camphor dissolved in deuterated chloroform* (see the structure in the margin). It will illustrate how a chiral compound can induce chirality in solvent molecules.

The experimental IR spectrum of deuterated chloroform contains a peak at 2254 cm^{-1}; the associated normal mode corresponds to C–D stretching. Since chloroform is not a chiral molecule, this mode is not VCD active. However, this mode is present in both the IR and VCD spectra of camphor in chloroform.

Predict the IR and VCD spectra of camphor in solution with chloroform. Identify the C–D stretching mode, and compare its characteristics with experiment. Is this mode active in either spectrum? Do the results surprise you?

Repeat the calculations for a complex of camphor and a single chloroform molecule, with the latter hydrogen bonded to the oxygen atom (see the illustration below, where we've hidden the hydrogens on the camphor molecule). How do the results change?

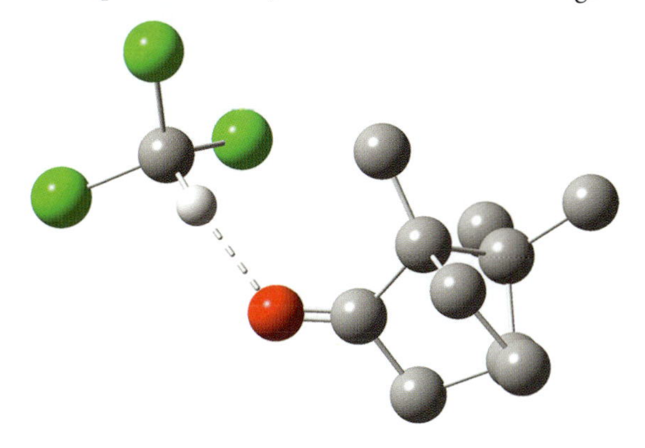

* IUPAC: chloroform: trichloromethane.

Be sure to specify the proper isotope for the deuterium atom in the solvent molecule. This may be done with the **Iso** keyword within the molecule specification as in this example:

```
H(Iso=2)        -4.38159840    -0.94375367     0.26428064
```

In GaussView, you can also use the **Atom List Editor** to specify the desired isotope (via the **Atom List** item on the **Edit** menu). Click on the **I** icon at the top of the window, and then select the desired isotope value from the popup menu in the **Mass (Isotopologue 0)** column. Note that isotope values are specified as integers, but Gaussian will use the corresponding actual exact isotopic mass (e.g., deuterium is specified as **2**, and Gaussian will use the value of 2.014101787).

G3:M1:V1 - Atom List Editor

Go To 2

Row	Highlight	Tag	Symbol	Mass (Isotopologue 0)*	X
1		1	C	Default	-0.0000
2		2	H	Default	0.00018
3		3	Cl	1	-1.1602
4		4	Cl	2 / 3	1.64517
5		5	Cl	Default	-0.4849

SOLUTION

We optimized the structure of camphor in solution with chloroform using IEFPCM. After optimizing an isolated chloroform molecule, we constructed and optimized the camphor-chloroform complex, running the calculations both in the gas phase and in the chloroform solvent environment. All of our optimizations completed successfully, and frequency calculations confirmed that the structures are minima. VCD spectra were predicted for all species/environments.

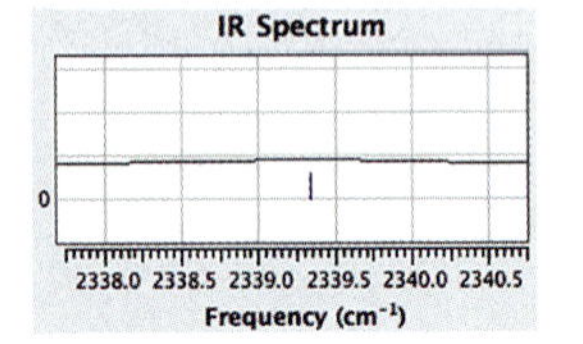

The predicted IR spectrum of chloroform contains a peak at ~2339 cm^{-1} with a tiny dipole strength of $0.46\times10^{-40}esu^2cm^2$ (reported in the Gaussian output as 0.27 KM/mol). The mode is not VCD active. The predicted IR and VCD spectra of camphor in solution with chloroform contain no peak in the range 2000-3000 cm^{-1}.

The key structural parameters for the camphor-chloroform complex in the two environments are listed in the following table:

PARAMETER	GAS PHASE	IEFPCM
D–C (Å)	2.0	2.0
C=O–D (°)	127.6	127.9
C=O–D–C (°)	67.8	67.7

The two structures are quite similar.

The following table summarizes the C–D stretching mode IR and VCD results for the various systems under investigation:

SYSTEM/ENVIRONMENT	WAVENUMBER (cm^{-1})	INTENSITY (KM/mol)	
		IR INTENSITY (KM/mol)	VCD ROT. STRENGTH ($10^{-40}esu^2cm^2$)
chloroform: gas phase	2339	0.3	—
camphor: IEFPCM	—	—	—
camphor+chloroform: gas phase	2297	80.9	16.0
camphor+chloroform: IEFPCM	2287	112.2	20.1

The left side of the following figure shows the experimental IR and VCD spectra for (S,S)-camphor dissolved in various mixtures of deuterated chloroform and deuterated dichloromethane (where the red lines correspond to 100% chloroform) [Debie08]; the right side of the figure shows the calculated spectra for the camphor-chloroform complex in solution:

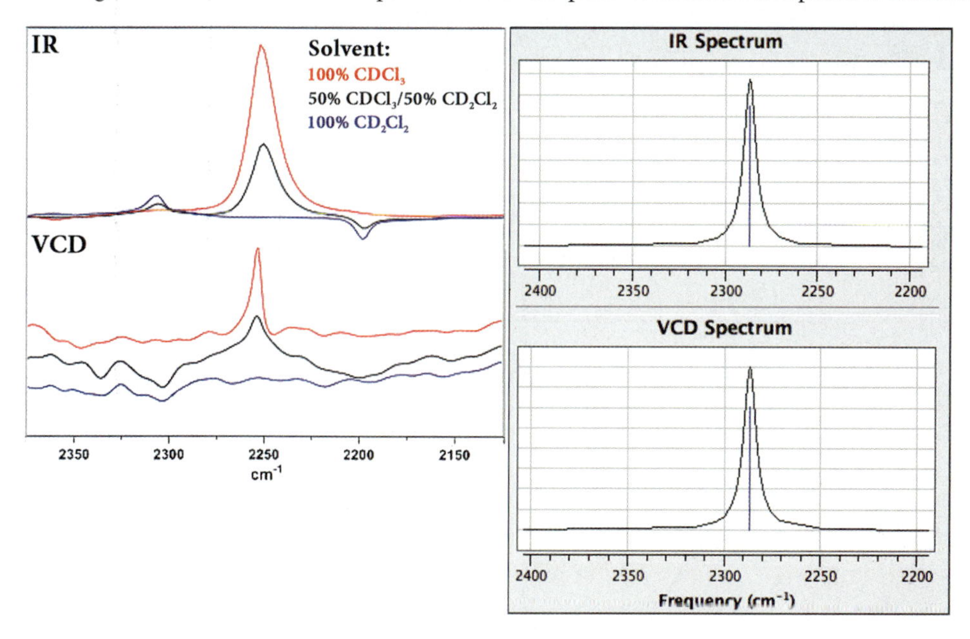

The experimental results reveal an increased intensity in the peaks as the concentration of chloroform increases. As the tabulated data and the illustration show, the predicted spectra are in excellent agreement with experiment.

These results illustrate how a chiral compound in solution can induce chirality in the molecules of the solvent. The D···O hydrogen bond causes the C-D stretching mode to be VCD active. It also greatly increases the intensity of the corresponding IR mode, as is usual with hydrogen bonds.

A note about conformations. The camphor molecule has only a single form, so we did not need to worry about conformational searching/averaging when modeling its optical properties. However, the same is not true for the camphor-chloroform complex. The best solution to this exercise would include a search for the lowest energy conformations leading to a spectrum derived from their weighted-average. We focused on a single conformation for simplicity in illustrating the chemical phenomenon of induced chirality.■

EXERCISE 7.12 OPTICAL ROTATIONS: SUBSTITUTED OXIRANES

In this exercise, we continue the study begun in Example 7.8 (p. 296). Predict the optical rotations for the substituted oxiranes where X=CCH,* CN and Cl, at the same frequencies used previously: 633nm, 589nm, 578nm, 546nm, 436nm, 365nm and 355nm. Model the S enantiomers in all cases.

How do the results for these compounds compare to the previous ones?

SOLUTION

The following table lists the results of the geometry optimization and specific rotation calculations for all studied compounds, along with the experimental optical rotations for X=CN observed in the liquid phase (i.e., neat solution) [Wiberg07]:

				OPTICAL ROTATIONS						
X	**C-X (Å)**	**C-O-C-X (°)**	**μ (D)**	**355nm**	**365nm**	**436nm**	**546nm**	**578nm**	**589nm**	**633nm**
F	1.36	-110.5	2.30	45.0	40.9	23.3	12.4	10.6	10.1	8.5
CH_3	1.50	-112.8	2.02	3.2	-1.3	-13.8	-14.0	-13.2	-12.9	-11.8
CCH	1.44	-111.4	1.83	532.0	482.6	274.1	147.6	127.7	121.8	102.0
Cl	1.77	-111.4	2.26	-260.9	-244.0	-161.4	-98.2	-86.9	-83.4	-71.6
CN: calc.	1.45	-110.4	3.85	358.0	327.1	192.0	105.8	91.9	87.7	73.8
exp.					353±13	214±8	119±4	104±4	99±4	

The predicted and observed results for cyanooxirane are in reasonable agreement.

The following illustration plots the predicted optical rotation results:

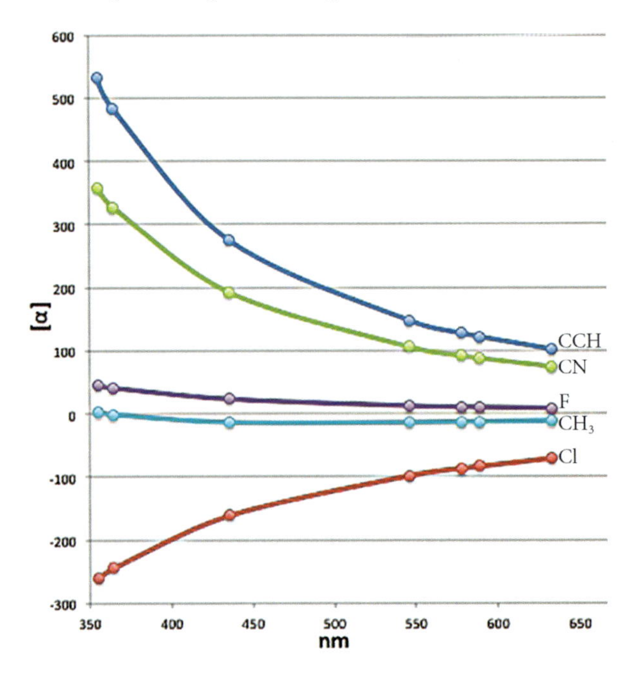

* IUPAC: 2-ethynyloxirane.

The curves for the compounds including the CCH and CN substituents exhibit the same behavior as the two we considered previously: increased specific rotations with decreasing incident light frequency. The chlorine compound is an exception, and it exhibits an opposite change.

The original researchers attribute this deviation to the interaction between the oxygen and the chlorine in chlorooxirane:

> Such interactions are well-known and are referred to as anomeric effects where an electron-withdrawing substituent antiperiplanar to an atom with a lone pair leads to energetic stabilization and the bond to the substituent is found to have an increased length. This results from the donation of electron density from the electron-rich lone pair to the σ^* orbital of the C–X bond. (p. 6209)

These researchers ruled out the possibility of this change arising from the fact that the chlorine is attached to a three-membered ring by examining a series of related compounds containing four-membered rings: trans-1-methy-2-X-cyclopropanes. In the latter case, the pattern of change in specific rotation with decreasing wavelength is the same for all of the compounds.■

EXERCISE 7.13 SOLVENT EFFECTS ON ORD: EPICHLOROHYDRIN

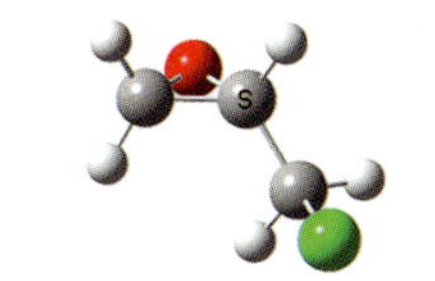

In this exercise, you will study the optical rotatory dispersion for the (S) enantiomer of epichlorohydrin, a compound we examined previously in Example 7.6 (p. 292) and Exercise 7.9 (p. 309). This compound has three conformations, corresponding to 0°, 120° and 240° values of the Cl–C–C–O dihedral angle.

Predict the optical rotation for the following wavelengths of incident light: 633nm, 589.5nm, 436nm and 355nm. Perform the calculations in both the gas phase and in solution with acetonitrile. How do the results change in the two environments? Compare your observations with the results for methyloxirane presented previously (p. 297).

SOLUTION

The following table presents the results of our geometry optimization calculations for the three conformations of (S)-epichlorohydrin in the two environments:

	GAS PHASE				ACETONITRILE		
CONF.†	μ (D)	O-C-C-Cl (°)	ΔG (kcal/mol)	Bf	O-C-C-Cl (°)	ΔG (kcal/mol)	Bf
0°	0.53	-160.3	0.00	58.6%	-160.1	0.00	22.3%
120°	3.25	82.5	0.32	34.4%	78.6	-0.66	67.7%
240°	2.68	-49.4	0.95	6.9%	-46.7	0.48	10.0%

† The original researchers [Wilson05] refer to the conformations as *gauche* I (120°), *gauche* II (0°) and *cis* (240°).

Although all of the conformations are quite close in energy in both environments, their energy ordering changes when moving from the gas phase to acetonitrile solution.

The following figure shows the conformationally-averaged ORD results, along with the corresponding experimental values [Wilson05]:

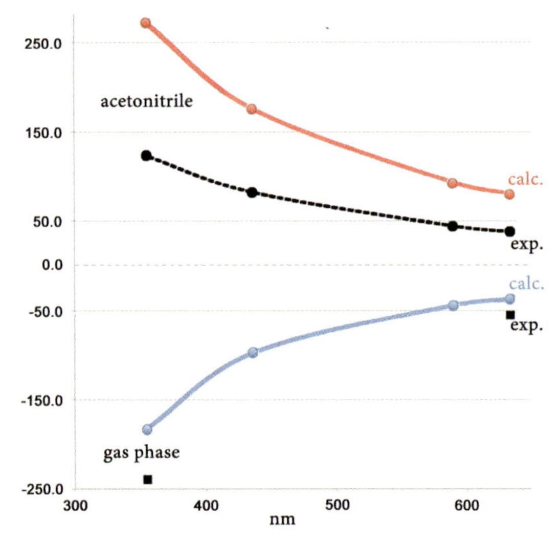

The calculated results reproduce the overall shape of the experimental ORD curves, as well as the signs and variation with increasing wavelengths of the specific rotations. For a flexible molecule like epichlorohydrin, the gas phase and different solution environments give rise to varying stabilities for the individual conformers, which in turn leads to quite different total optical activity.

These results can be contrasted with rigid species such as the substituted oxiranes we considered previously. In that case, the results for the gas phase and solution in acetonitrile were similar although the magnitude of the latter was significantly higher for the lower wavelengths (see the graph on p. 297). For epichlorohydrin, the same two environments produce specific rotations of opposite sign and opposite trends with increasing wavelength of the incident light.■

TO THE TEACHER: ADDITIONAL SOLVENTS

The original paper corresponding to the preceding exercise [Wilson05] studied epichlorohydrin in several additional solvent environments: cyclohexane, carbon tetrachloride, di-*n*-butyl ether, ethanol and benzene. For larger classes, different students/group of students could perform the calculations for each solvent. The results can be analyzed to determine how the specific rotation varies with the dielectric constant of the solvent.

Advanced Topics and Exercises

Microwave Spectroscopy and Hyperfine Coupling Constants

Microwave spectroscopy observes rotational transitions in molecules via the absorption of photons in the microwave range: 10^7-10^{10} Å=1-1000 mm; this energy range is comparable to that of the rotational states. Samples are studied in the gas phase. Microwave spectrometers are highly sophisticated apparatuses incorporating very sensitive electronic circuits. Light in the GHz range is required to detect transitions with energies of MHz or kHz, which necessitates extremely accurate control over the frequency of the incident light.

References: Hyperfine Coupling Constants

Theoretical Definition: [Frosch52]

Review: [Hirota92]

Implementation: [Miller80, Clabo88, Page88, Miller90, Page90]

The basic rotational spectrum is determined by the molecule's moments of inertia. However, these rotational transitions are split by a variety of interactions between the overall rotational motion of the molecule and both the electron distribution and spins of the nuclei, yielding a *fine structure* which can be measured with very great accuracy.

For molecules with unpaired electrons, the fine structure can be further split as a result of additional factors including the magnetic interactions between the magnetic moment and induced field of the electron and the magnetic moment and induced field of the nucleus. The effect is known as *hyperfine coupling*, and the data is called the *hyperfine structure*. Experimentally, hyperfine structure is studied with microwave spectroscopy. It can also be modeled by Gaussian.

Raw data from microwave spectroscopy experiments are analyzed using a numerical fitting procedure. One widely used technique is due to Pickett [Pickett91]; the suite of computer programs which implements it, called SPFIT, can also employ data computed by electronic structure calculations and/or fixed constants in the fitting process. Gaussian can provide input in the format required by this package.

Analysis of microware spectroscopic data ultimately yields values for a set of parameters known generically as *coupling constants*; those related to hyperfine structure are thus the hyperfine coupling constants. Specific constants have individual names denoting their individual characteristics (e.g., "isotropic Fermi contact couplings"). The values of the various constants are determined by fitting the observed data to predicted and/or preset values for the same items; the final result is the set of predicted values for the coupling constants. Thus, these experimental studies typically include a theoretical component.

Open shell energy calculations report the most common hyperfine coupling constants by default. In addition, all available hyperfine coupling constants may be requested with the **Freq=(VCD,VibRot)** options; the keyword **Output=Pickett** is also included as it provides a helpful summary section of the hyperfine results at the end of the output file. As was true for NMR spin-spin coupling, certain of these constants—specifically, the Fermi contact terms and the spin-spin (dipole-dipole) tensor—benefit from a larger basis set with additional basis functions in the area of the nucleus [Deng06]. For this reason, a single point energy calculation is typically run as a second job step after the freqency calculation, using a larger basis set and the **Int=FCMod1** option.*

* In versions of Gaussian prior to Gaussian 09 Rev E.01, use **IOp(3/10=1100000)** instead of this keyword.

Here is the general structure of an input file for predicting hyperfine coupling constants:

```
%Chk=hyperfine
# APFD/6-311+G(2d,p) Opt Freq=(VCD,OptRot) Output=Pickett

Predict hyperfine coupling constants
```

molecule specification

```
--Link1--
%Chk=hyperfine
# APDF/aug-cc-pVTZ Int=FCMod1 Output=Pickett
  Geom=Check Guess=Read

Use augmented and uncontracted basis set to predict
Fermi contact terms and electron-nuclear spin-spin tensor
```

charge & spin multiplicity

final blank line

The following table lists the principal fine and hyperfine spectroscopic constants, along with their labels in the Gaussian **Output=Pickett** output section. The job step column indicates which job step to use for the predicted results. Items marked for the second job step will also appear in the output of the first job step, but the second set of results is the one to use.

Constants	*Output label*	*Pickett input table labels*	*Per-atom?*	*Job step*
A_0, B_0, C_0	Rotational Constants	A, B, C	no	2
$\mathbf{\Delta}_N$, $\mathbf{\Delta}_K$, $\mathbf{\Delta}_{NK}$, $\boldsymbol{\delta}_N$, $\boldsymbol{\delta}_K$†	Quartic Centrifugal Distortion Constants	-DELTA N, -DELTA K, -DELTA NK, -delta N, -delta K	no	1
T (tensor)	electron spin-nuclear spin (dipole-dipole) [T]	3/2*T_aa, 1/4*(T_bb - T_cc), T_ab	yes	2
a_F	Isotropic Fermi Contact Couplings [a_f]	b_f‡	yes	2
ε (tensor)	electronic spin-molecular rotation tensor [epsilon]	epsilon_aa, epsilon_bb, epsilon_cc	no	1
C (tensor)	nuclear spin-molecular rotation tensor [C]	C_aa, C_bb, C_cc, (C_ab+C_ba)/2, (C_ab-C_ba)/2	yes	1

† Typically reported multiplied by 10^3.
‡ Labeled "a_f" everywhere else in the Gaussian output.

Items marked as tensors are typically reported/used via individual tensor elements (e.g., T_{aa}) or expressions thereof (e.g., T_{bb}-T_{cc}).

The following excerpt illustrates the structure of the **Output=Pickett** section of the Gaussian output:

```
Rotational constants (MHZ):
 11170.2604507   3927.4267780   2905.7682376          A0, B0, C0
electron spin - nuclear spin (dipole-dipole) [T] (MHz):
 1  C12
  aa=      0.0000   ba=      0.0000   ca=      0.0000
  ab=      0.0000   bb=      0.0000   cb=      0.0000
  ac=      0.0000   bc=      0.0000   cc=      0.0000
 4  H1
  aa=     20.5775   ba=      0.0000   ca=      0.0000
  ab=      0.0000   bb=    -20.9905   cb=      0.0000
  ac=      0.0000   bc=      0.0000   cc=      0.4130
 5  F19
  aa=   -157.8510   ba=    -18.9414   ca=      0.0000
  ab=    -18.9414   bb=   -172.8777   cb=      0.0000
  ac=      0.0000   bc=      0.0000   cc=    330.7287

Isotropic Fermi Contact Couplings [a_f](MHz):
 1  C12          0.0000
 4  H1         -47.7853
 5  F19        101.1906
Nuclear quadrupole coupling constants [Chi] (MHz):
Dipole moment (Debye):
    -1.5276093 0.0000000 0.0000000  Tot= 1.5276093

--------------------------------------------------------
1,1-difluoroprop-2-ynyl radical hyperfine coupling job title
  99  1000  50  0.1760E-11  0.1000E+07  0.1000E+01 ...
 ...   0.11170260E+05  0.0  /          A                 /
       0.39274268E+04  0.0  /          B                 /
       0.29057682E+04  0.0  /          C                 /
       0.30866251E+02  0.0  / (4 H-1)  3/2*T_aa          /
      -0.53508708E+01  0.0  / (4 H-1)  1/4*(T_bb - T_cc) /
      -0.23677652E+03  0.0  / (5 F-19) 3/2*T_aa          /
      -0.12590160E+03  0.0  / (5 F-19) 1/4*(T_bb - T_cc) /
      -0.18941372E+02  0.0  / (5 F-19) T_ab              /
      -0.23677652E+03  0.0  / (6 F-19) 3/2*T_aa          /
      -0.12590160E+03  0.0  / (6 F-19) 1/4*(T_bb - T_cc) /
       0.18941372E+02  0.0  / (6 F-19) T_ab              /
      -0.47785264E+02  0.0  / (4 H-1)  b_f  Fermi contact term /
       0.10119063E+03  0.0  / (5 F-19) b_f               /
       0.10119063E+03  0.0  / (6 F-19) b_f               /
--------------------------------------------------------
    value       atom info: input #, symbol, atomic #    constant/expression
```

The output consists of several sections listing the individual values of the tensor elements and individual constants, followed by a final section set off by dashed lines. The latter is the input for Pickett's SPFIT suite of programs [Pickett91], in their required format. The lines in the SPFIT input above are shortened in that the initial identifier field is omitted and the zero value is truncated. Here is a sample line showing all the fields:

```
120010000  0.30866251E+02  0.00E+00 / (4 H-1) 3/2*T_aa /
```

The various items in the SPFIT input section often contain desired tensor components and component expressions, but may also include a factor. For example, if you want the value of T_{aa}, the first diagonal component of the electron spin magnetic dipole-nuclear spin magnetic dipole tensor, you can retrieve the value from the sixth line of the preceding output—approximately 30.87—and multiply it by ⅔ to obtain the desired quantity. Be aware that the quartic centrifugal distortion constants reported as SPFIT input are multiplied by a factor of -1 (indicated in the label: e.g., "-DELTA N"), and that the Fermi contact terms are labeled "b_f" rather than "a_f."

ADVANCED EXAMPLE 7.9 1,1-DIFLUOROPROP-2-YNYL RADICAL

In this example, we predict various hyperfine coupling constants for the 1,1-difluoroprop-2-ynyl* radical. After optimizing the molecule (resulting in a planar structure), we run the **Freq=(VCD,VibRot)** calculation, followed by the single point calculation, both with **Output=Pickett**.

The following table lists the calculated values for a variety of quantities. It also includes the fitted values from the original paper [Kang06], which are derived from microwave spectroscopy experimental data. All values are in MHz.

CONSTANT	CALC.	FITTED
A_0	11170.26	11126.26
B_0	3927.427	3927.522
C_0	2905.768	2904.245
$\Delta_N \times 10^3$	0.43	0.57
$\Delta_{NK} \times 10^3$	18.0	19.1
$\Delta_K \times 10^3$	30	30†
$\delta_N \times 10^3$	0.13	0.19
$\delta_K \times 10^3$	10.2	14.6
ε_{aa}	-42.712	-43.674
ε_{bb}	-18.884	-17.589
ε_{cc}	2.173	0.359
a_F (H)	-47.785	-32.542
T_{aa} (H)	20.5775	16.1589
T_{bb}-T_{cc} (H)	-19.02	-16.47
a_F (F)	101.19	145.73
T_{aa} (F)	-157.851	-153.502
T_{bb}-T_{cc} (F)	-495.60	-486.82
T_{ab} (F)	18.9	13.4

† Held constant in the fit.

In general, the predicted values do not deviate very much from the final fitted values.

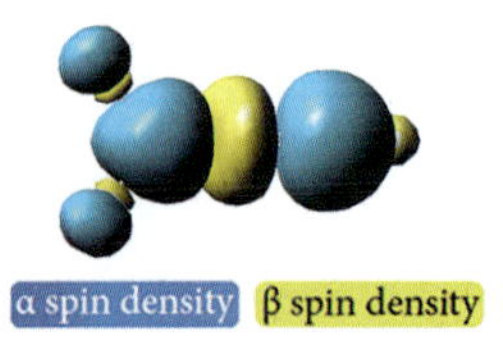

It is often useful to examine plots of the spin density in conjunction with studies of hyperfine structure (we used a isosurface of 0.002). The spin density for this molecule appears in the margin (the molecule is in an orientation similar to that of the preceding illustration although tipped slightly less out-of-plane). Note the abundance of α spin density in both potential locations for the unpaired electron: the two end carbon atoms.

* IUPAC: 3,3-difluoro-1-propyne.

The fitted data in the preceding table did not take into account any coupling between the two resonance configurations of this molecule:

$$F_2\text{–}\dot{C}\text{–}C{\equiv}CH \text{ and } F_2\text{–}C{=}C{=}\dot{C}H$$

In their paper, the original researchers fit the experimental data in two ways, the second of which takes the two resonance configurations into account. There was no difference in the results between the two schemes.■

ADVANCED EXERCISE 7.14 PROP-2-YNYL RADICAL HYPERFINE COUPLING

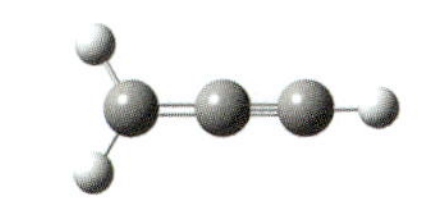

In this exercise, you will study prop-2-ynyl* radical, which is the base molecule before fluoride substitution for the compound in the previous example. Predict the values for the Fermi contact terms (a_F) and the electron spin magnetic dipole-nuclear spin magnetic dipole tensor (T). Compare the values of the following quantities to the corresponding ones for 1,1-difluoroprop-2-ynyl:

- Fermi contact terms for each type of hydrogen atom.
- T_{aa} and T_{bb}-T_{cc} for each type of hydrogen atom.

What variations do you observe, and what is their significance?

SOLUTION

The following table lists the results of the calculations for prop-2-ynyl, the corresponding data from the calculations on 1,1-difluoroprop-2-ynyl, as well as the fitted values from the experimental observations (all values are in MHz):

	F_2-C-C≡C-H		H_2-C-C≡C-H	
CONSTANT	CALC.	FITTED EXP. [Kang06]	CALC.	FITTED EXP. [Tanaka97]
a_F (H)	-47.785	-32.542	46.93	-36.32
T_{aa} (H)	20.5775	16.1589	21.09	17.40
T_{bb}-T_{cc} (H)	-19.02	-16.47	-20.29	-17.04
a_F (F,H)	101.19	145.73	-58.93	-54.21
T_{aa} (F,H)	-157.851	-153.502	-14.79	-14.12
T_{bb}-T_{cc} (F,H)	-495.60	-486.82	11.92	11.64

The calculated values correspond quite well to the results of the fitting procedure. They also reproduce the trends in the experimental values in moving from prop-2-ynyl to the fluoridated compound.

The values of the various constants for the lone hydrogen atom are similar in both compounds. If we examine the spin densities for the two molecules (pictured in the margin), it is clear that the spin density in this region is similar in both radicals.

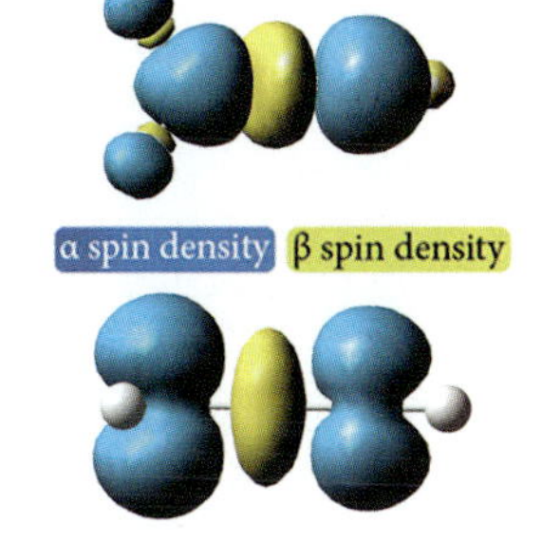

The hyperfine constants for the fluorine atoms in 1,1-fluoroprop-2-ynyl differ quite substantially from those of the corresponding hydrogen atoms in prop-2-ynyl. The Fermi contact terms have much greater magnitude (a factor of about 2.7) and opposite sign for the fluorine atom than for the hydrogen atom. The spin density surface in the region of the former shows π spin density on the p electrons, a feature which is absent in the spin density of prop-2-ynyl.

* Also known as propargyl.

These researchers note that such a spin density configuration leads to an effect known as "core polarization," resulting in a Fermi contact term with a positive value for the corresponding atom (see p. 054314 in [Kang06] for details).

The components of the T tensor are also much greater for the fluorine substituent than for hydrogen. In addition, the tensor exhibits significantly more anisotropy in the former case than the latter.■

ADVANCED EXERCISE 7.15 **CF^+ IN INTERSTELLAR SPACE**

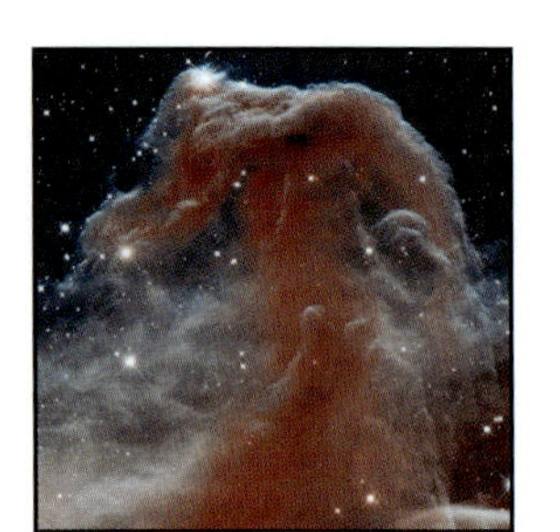

Horsehead nebula
Photo credit: NASA/ESA/Hubble

Microwave spectroscopy plays an important role in the field of astrochemistry. Radio astronomy detects a variety of molecules in interstellar space; microwave spectroscopy can study the same species in the laboratory in order to determine their rotational constants and other hyperfine properties with high accuracy. Once these parameters have been determined, it is possible to predict line positions for the relevant molecules in order to aid, verify and/or interpret radio astronomical observations.

In 2012, Guzmán and coworkers [Guzman12a] detected CF^+ in the Horsehead mane area of the Horsehead nebula. One of the lines in the spectrum was double peaked, a result which was both unexpected and difficult to reconcile with the identification of the observed compound as CF^+.*

In order to test the hypothesis that this phenomenon was due to discrete hyperfine components arising from the fluorine nucleus, they computed the value for the relevant spin rotation constant. Using this result, their fit of the observed data verified the presence of two hyperfine components in the relevant electromagnetic region, thereby verifying the identification of the observed compound as CF^+ [Guzman12].

Predict the value for the nuclear spin-molecular rotation coupling constant for the fluorine atom in this compound.

SOLUTION

After completing the preliminary geometry optimization, the frequency calculation predicted a value of 244 kHz for this parameter. This is in reasonable agreement with the value of 229.2 kHz computed by the original researchers using a different model chemistry incorporating a coupled cluster method.

The previous estimated value for the constant had been 165 kHz [Guzman12a], and the results here are in general agreement with its modification [Guzman12].■

ADVANCED EXERCISE 7.16 **HYPERFINE COUPLING CONSTANTS: ARSENIC COMPOUNDS**

In this exercise, you will predict the electron spin electronic dipole-nuclear spin nuclear dipole tensor (T) for a set of small arsenic compounds: AsH_2, $AsOH_2$ and AsO_2.† Perform the two-step calculation sequence we have used for computing hyperfine coupling constants with the following modifications:

- For elements with heavy nuclei, such as arsenic, there are significant relativistic effects on the spin density in the core region of the nucleus (see, e.g., [Stojanovic12], a study which suggested this exercise). For hyperfine interactions, these effects can be modeled using a scalar-relativistic Hamiltonian. Gaussian provides the Douglas-Kroll-Hess Hamiltonian

* IUPAC: carbon monofluoride cation, fluoromethylidynium.

† IUPAC: arsenic dihydride AKA arsanylium, arsinite, (oxoarsino)oxidanyl (respectively).

for this purpose. Its use is requested with the **Integral=DKH** keyword. Include this keyword in the second job step (the single point calculation).

- A larger integration grid is often necessary for calculations with an uncontracted basis set and heavy nucleus as well as when using a scalar-relativistic Hamiltonian. For these reasons, include the **Integral=(Grid=399590)** keyword in the root section for the second job step (the single point calculation). Also include the **Int=FCMod1** option.

Here are the experimental values of the tensor components:

- AsH_2: T_{aa}=-287.694, T_{bb}=-322.150, T_{cc}=609.855 [Hughes00]
- AsO_2: T_x=-130, T_y=171, T_z=-42 [Knight95]

What is the difference between T_{aa} and T_x? There are two sets of axes in which the components of this tensor may be expressed. The output from the **Output=Pickett** keyword always uses the moment-of-inertia axis system since that is what is required by the SPFIT package. In many cases, this representation will be the same as that in the principal axis system. In such cases, the T tensor reported in this output will be a diagonal matrix, and T_{aa} and T_x coincide.

On the other hand, if there are off-diagonal elements present in the listing (e.g., T_{ac}), then you will need to find the required tensor components earlier in the Gaussian output file, where they are presented in the other axis system. They are listed in a section introduced with the heading "Anisotropic Spin Dipole Couplings in Principal Axis System." In this part of the Gaussian output, the tensor elements are labeled B rather than T, and they are presented in several different units. The axes themselves may also be permuted, and so the value as T_x in the literature may correspond to, say, the value B_{bb} in the output. You will need to compare the orientations in the article and in the program to identify the various components.

SOLUTION

The following table presents the predicted hyperfine coupling constants along with the observed values (all values are in MHz):

	CALC.	OBS.
AsH_2:		
T_{aa}	-292.070	-287.694
T_{bb}	-308.201	-322.150
T_{cc}	600.271	609.855
$AsOH_2$:		
T_x	-107.292	
T_y	-92.341	
T_z	199.634	
AsO_2:		
T_x	-119.473	-130
T_y	183.711	171
T_z	-64.238	-42

These values are in very good agreement with the available experimental data.

The following table compares the results for AsH_2 using several different model chemistries. All calculations used the APFD method, with the indicated basis set and additional keywords:

	6-311+G(2d,p)	aug-cc-pVTZ Int=FCMod1	6-311+G(2d,p) Int=DKH	aug-cc-pVTZ Int=(DKH, FCMod1)	Obs.
T_{aa}	-267.147	-269.021	-280.725	-292.070	-287.694
T_{bb}	-283.418	-284.978	-296.785	-308.201	-322.150
T_{cc}	550.566	553.999	577.510	600.271	609.855

While both the use of the larger, uncontracted basis set and the scalar-relativistic Hamiltonian improve on the results obtained with our standard model chemistry, the latter has a much larger effect; comparing results columns 1 and 3 and columns 2 and 4 shows that the Douglas-Kroll-Hess Hamiltonian improves the results for both basis sets significantly more than the larger, uncontracted basis set does on its own (cf. results columns 1 and 2).■

Deducing Structure from Microwave Spectra

Studies of hyperfine splitting can also be used to probe information about molecular structure. This is done in a manner generally analogous to traditional physical chemistry laboratory exercises in which students calculate bond lengths and nuclear-nuclear distances from observation data of vibrational and rotational transitions. For an example of such research, see [Lin09] for a study of equitorial vs. axial cyclobutanol.

8

Modeling Excited States

In This Chapter:

Modeling UV/Vis Spectroscopy

Electric Circular Dichroism

Studying Fluorescence

Advanced Topics:

Franck-Condon Analysis

State-Specific Solvation

Multireference Models

Conical Intersections

8 ◆ Modeling Excited States

Chemistry and Light: Excited States

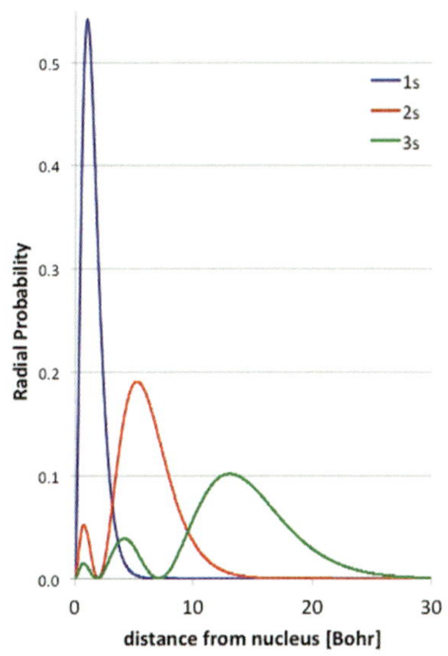

Up to this point, we have considered only molecules in the ground state: the lowest energy electronic configuration. An *electronic state* of a molecule is defined as a particular probability distribution describing the electrons (which we also refer to as the *electron density*). According to the laws of physics, only certain probability distributions—electronic states—can exist. Distinct electronic states have different energies, and the ground state is the one with the lowest energy. Other electronic states are known as *excited states*, and molecules change to an excited state as the result of having gained energy, typically by absorbing a photon. Systems generally remain in an excited state for only a short period of time, after which they can relax back to the ground state in several ways.

These concepts are easy to visualize for a simple quantum mechanical system, for which there is an exact solution to the Schrödinger Equation: the hydrogen atom. The figure above plots the allowed normalized radial probability distributions of one electron travelling around a positive charge (i.e., the hydrogen atom):

Note that the curves are normalized: each of them will integrate to 1. In other words, the area under each curve is the same and corresponds to 100% probability of observing an electron. The 1s distribution is the ground state for hydrogen and resembles the Bohr atom in that the probability of observing the electron is large at the Bohr radius and small elsewhere. The 2s electronic state has two regions of probability with an intervening node (point of zero probability). The 3s electronic state is quite diffuse by comparison and has three separate regions of probability. The 2s and 3s distributions constitute the excited states of the hydrogen atom.

Excited states are relevant to the study of a wide variety of phenomenon, including:

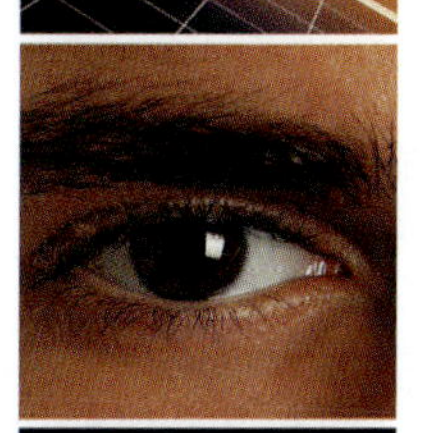

- *Photosynthesis*: Many phases of the photosynthesis process are photochemical in nature. For example, in the beginning stages of this process, a photon from sunlight is absorbed by a chlorophyll molecule, causing it to enter an excited state. This molecule soon undergoes nonradiative decay whereby the energy of the photon is transferred to an adjacent chlorophyll molecule, sending it into its excited state. This process is known as *resonance energy transfer*. It continues until it reaches a chlorophyll molecule that is adjacent to a reaction center: a slightly modified chlorophyll having an excited state at a slightly lower energy than normal. The reaction center can absorb the energy from the ordinary chlorophyll molecule via an electron transfer process, which creates charged species and fuels the further reactions in the photosynthesis process.

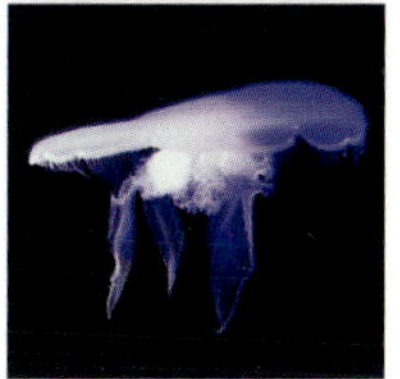

- *Photodecomposition*: Exposure to light can cause some materials to be broken down in part or entirely. For example, when fabric dyes absorb photons, they become susceptible to oxidation, and eventually they break down, and the textiles fade. This happens because the dye molecule undergoes structural changes when it moves to its excited state. Some harmful substances such as certain pesticides also eventually break down as a result of prolonged exposure to light.

- *Photochemical generation of electricity*: Dye-sensitized solar cells (DSSCs) rely on an organic dye that enters an excited state upon exposure to light. The excited electron is then rapidly transferred to the electrolyte in which the dye is immersed, after which the solar cell functions analogously to a conventional battery, employing an anode and a cathode on either side of the electrolyte.

- *Perception of light*: Rhodopsin is a pigment that is present in the photoreceptor cells of the retina in the eyes of many animals. It is composed of two components: scotopsin and cis-retinol. When exposed to light, the retinol component enters an excited state where it undergoes cis-trans isomerization. The formation of trans-retinol causes the rhodopsin molecule to decompose, forming metarhodopsin II (also known as activated rhodopsin). This latter substance transmits an electrical impulse to the optic nerve as it returns to its ground state.

- *Bioluminescence*: In a few animals, chemical reactions involving excited states lead to the emission of photons. For example, in the abdomens of fireflies, an oxidation reaction involving the luciferin pigment (catalyzed by the enzyme luciferase) causes the pigment to enter an excited state. When it returns to the ground state, a photon is emitted, and the firefly glows.

We will examine several photochemical applications in detail in the course of this chapter.

The spectroscopic technique unique to excited states is UV/Visible spectroscopy, which studies molecular structure by bombarding substances with light in the ultraviolet and visible range of the electromagnetic spectrum, thereby inducing electronic transitions at specific wavelengths.

Model Chemistries for Modeling Excited States

Gaussian provides several different methods for studying excited state systems:

- ZIndo: A semi-empirical method useful for basic calculations on very large systems. The **ZINDO** keyword requests this method. No basis set keyword is needed for this method as the basis set is part of the method. Note that ZIndo is parametrized only for a limited set of elements, and there is no mechanism for inputting additional parameters. See [Zerner91] for a review.

- CI-Singles [Foresman92]: A zeroeth order method for studying excited states. Although mostly supplanted by TD (see below), CI-Singles can still be useful for the preliminary stages of excited state geometry optimizations for difficult cases. The **CIS** keyword specifies this method, and it is given in combination with a basis set keyword in the job's route section. CI-Singles is sometimes referred to as TDA in the literature.

- TD: Time-dependent methods, which can be applied to Hartree-Fock- and DFT-based model chemistries (referred to as TD-HF* and TD-DFT, respectively). TD-DFT is the usual model to be used for excited state calculations. The **TD** keyword is specified in addition to the DFT method and basis set keywords in the route section of the job.

- EOM-CCSD: A method based on CCSD for high accuracy modeling of excited states via the equation of motion coupled cluster method. The keyword for this method is **EOMCCSD** (abbreviable to **EOM**), and it is combined with a basis set keyword in the job's route section. References for EOM-CCSD are given on p. 342.

References: TD-DFT

Method Definition: [Runge84, Bauernschmitt96a, Casida98]

Review: [Adamo13]

Implementation: [VanCaillie99, VanCaillie00, Furche02, Scalmani06]

Solvation Treatment: [Scalmani06]

* TD-HF is also known as RPA.

Modeling Excited States in Solution

In general, the solvent responds in two different ways to changes in the state of the solute: it polarizes its electron distribution, which is a very rapid process, and the solvent molecules reorient, a much slower process on the same timescale as molecular rotations and vibrations.

The IEFPCM facility in Gaussian can treat both kinds of responses, via equilibrium and non-equilibrium solvation. An *equilibrium* solvation calculation models a situation where the solvent had time to fully respond to the solute in both ways: e.g., a geometry optimization (a process that takes place on the same time scale as molecular motion in the solvent). A *non-equilibrium* calculation is appropriate for processes which are too rapid for the solvent to have time to fully respond. In non-equilibrium solvation, only the rapid, electronic changes are modeled, while in equilibrium solvation, the slower, orientational changes are also taken into account.

The electronic excitations we are considering in this chapter fall into the second class of phenomenon as they are very, very rapid. Accordingly, the solvent molecules can be considered as fixed with respect to excited state transitions. This means that a molecule that has transitioned to an excited state will see the same solvent reaction field as was present for the originating state. Modeling such scenarios is referred to as *non-equilibrium* solvation.

By default, excited state calculations in solution, modeled with the CIS and TD methods, are handled as follows:

- Single point energy calculations to predict vertical transition energies, UV/visible spectra and other properties are modeled with non-equilibrium solvation.
- Geometry optimizations are modeled with equilibrium solvation.

These defaults mean that no special actions are required for most excited state calculations in solution. Techniques for handling special cases where non-default treatments are required as introduced beginning on p. 371.

Predicting Vertical Transition Energies & UV/Visible Spectra

Setting up excited state calculations is straightforward and quite similar to ones involving the ground state. Both GaussView and WebMO include features for setting up excited state calculations.

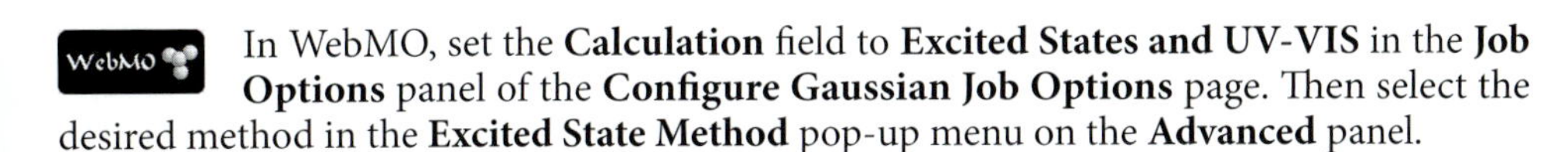

In WebMO, set the **Calculation** field to **Excited States and UV-VIS** in the **Job Options** panel of the **Configure Gaussian Job Options** page. Then select the desired method in the **Excited State Method** pop-up menu on the **Advanced** panel.

In GaussView, select the desired method in the first pop-up menu in the **Method** line of the Method panel in the **Gaussian Calculation Setup** dialog. The selection **TD-SCF** should be used for TD-DFT calculations. If appropriate, choose the desired model chemistry from the remaining pop-up menus on that line.

The following options, which apply to CIS, TD and EOM, are often useful:

NStates=*n* Solve for the specified number of excited states. The default is 3.

Root=*n* Specify the state to use for predicting properties: population analysis, electron density, etc. The default is 1, which refers to the lowest energy excited state.

Triplets Solve for triplet excited states from a closed shell singlet ground state.

By default, Gaussian solves for excited states of the same type as the ground state; i.e. closed or open shell, with the same spin multiplicity. The **Triplets** option applies to closed shell singlet ground states and allows you to solve for its triplet excited states. For example, for a closed shell singlet ground state, **TD(NStates=8,Root=2,Triplets)** will solve for eight triplet excited states, treating the second lowest energy excited state as the state of interest.

Gaussian also provides an option named **50-50** for the various excited state method keywords. However, we do not recommend using it because doing so makes unwarranted assumptions about how many spin states of each type are present within the lowest *n* excited states. Instead, we recommend solving for singlet and triplet states in separate calculations.

GaussView provides support for these options in the **Method** panel of the **Gaussian Calculation Setup** dialog:

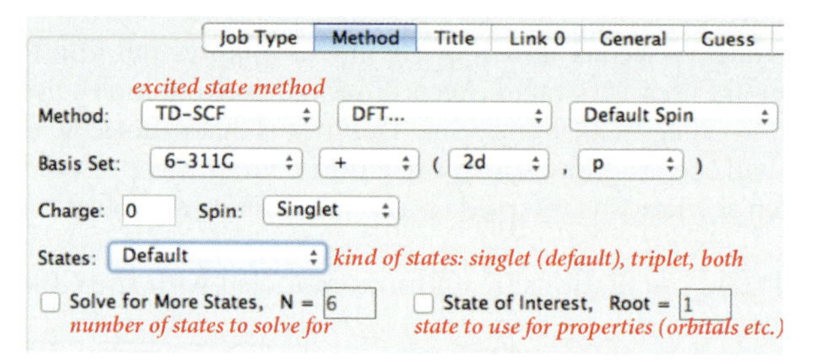

Another useful keyword is **Density=Current**. It specifies that the population analysis should use the electron density matrix for the excited state of interest rather than the ground state. It will also cause the excited state electron density to be stored in the checkpoint file for use in visualization programs.

The output from excited state calculations is of three types:

- Information about the various excited states, including their symmetries, excitation energies and oscillator strengths, is found in the Gaussian output file.
- GaussView produces plots of the UV/Visible spectrum and any other computed spectra (e.g., ECD). Excited state spectra can be viewed using the **Results→UV-VIS** menu path.
- Molecular orbitals, electron difference density surfaces and other volumetric data may be visualized in GaussView and WebMO.

We will examine the relevant Gaussian log file sections in the course of the first example in this chapter: modeling the excited states of benzene.

EXAMPLE 8.1 BENZENE EXCITATION ENERGIES

Benzene is the classic excited state problem for organic chemists. We would like to predict the six lowest excited states of benzene and compare them with experiment.We will use the TD method with our standard model chemistry.*

Once the job completes, we will need to examine the output file in order to retrieve the information about the predicted states. The following excerpt shows the header line, which

* Our input file requests additional excited states with the **TD(NStates=20)** option. Because of the D_{6h} symmetry of this molecule, there are many possible reducible representations for the electronic excited states. The program will only find those that are present in the initial guess. For molecules with high symmetry like benzene, you need to request more than the default number of states to ensure that you have all the symmetry types present in the solution.

begins the section of the output, as well as the data corresponding to the excited state with the lowest predicted energy. We have labeled the key data within the output.

spin multiplicity · *symmetry* · *excitation energy* · *oscillator strength* · S^2

```
Excitation energies and oscillator strengths:

Excited State 1:  Singlet-?Sym   5.4863 eV  225.99 nm  f=0.0  <S**2>=0.0
      20 -> 22          0.49950
      21 -> 23          0.49950
This state for optimization and/or 2nd-order correction.
Total Energy, E(TD-HF/TD-KS) =  -231.908393253
Copying the excited state density for this state as the
  1-particle RhoCI density.
```

excitation: from → to · *wavefunction coefficients (normalized)* · *indicates this is the state of interest*

The initial line identifies each state, listing its spin multiplicity and symmetry, the excitation energy in units of eV and the wavelength of the transition in nanometers. These are followed by the oscillator strength *f* (dimensionless) and the expectation value of the spin squared operator (S^2) in atomic units. Subsequent lines indicate the most important orbital transitions that participate in this excitation along with their respective coefficients; the entries with the largest coefficients contribute most significantly to the excitation. The final lines in the entry above indicate that this state is the "state of interest" for subsequent property analysis or geometry optimization.

Although Gaussian is usually able to identify the symmetry of each excited state, it cannot do so in this case. We describe how to determine the symmetry of an excited state immediately following this example.

Since the above example involves a restricted, closed-shell ground state, only the transitions between orbitals with alpha spin are listed. Combined with the equivalent set of beta-to-beta transitions (which are not reported), the complete set of coefficients would be *normalized*: the sum of their squares would equal 1.

The following table presents the results of our calculation and compares them to the experimental values [Lassettre68, Doering69, Nakashima80, Nakashima80a]:

We identify the transition for each state by examining the associated MOs (see the next section).

	STATE	TRANSITION	EXCITATION ENERGY (eV)	*f*	EXP. (eV)
1	$^1B_{2u}$	π-π^*	5.49	0.0	4.90
2	$^1B_{1u}$	π-π^*	6.16	0.0	6.20
3-4†	$^1E_{1g}$	π-R‡	6.62	0.0	6.33
5-6†	$^1E_{1u}$	π-π^*	7.06	0.598	6.94
7	$^1A_{2u}$	π-R	7.16	0.063	6.93
8-9†	$^1E_{1u}$	π-R	7.21	0.0	*forbidden*
10	$^1A_{2u}$	π-R	7.30	0.0	6.95

†Doubly degenerate state (appears twice in the Gaussian output). ‡R≡Rydberg.

The predicted excitation energies have inconsistent errors (between 1 and 13%). State 8-9 illustrates how theory is able to locate excited states that would not be observed in an optical experiment.

Many of the predicted states possess zero oscillator strengths but nonzero experimental intensities as a result of vibrational coupling. State 5-6 has a relatively large oscillator strength (due to a transition moment in the plane of the molecule) and corresponds to the very intense peak in the optical spectrum. ■

Determining the Symmetry of an Excited State

Whenever possible, Gaussian reports the symmetry for each excited state. In cases where Gaussian cannot assign a point group to an excited state, it prints **?Sym** in the output (as we saw with benzene). In these cases, you will have to determine the state's symmetry yourself. How to do so is the subject of this subsection.

Running the excited state calculation with the **Population=Full** and **IOp(9/40=3)** keywords makes this process easier. These keywords request that all molecular orbitals (occupied and virtual) be included in the population analysis and that all function coefficients greater than 0.001 be included in the excited state output (the default cutoff is 0.1), respectively.

The molecular symmetry of excited states is related to how the orbitals transform with respect to the ground state. From group theory, we know that the overall symmetry is a function of the symmetry products for the orbitals and that only singly occupied orbitals are significant in determining the symmetry of the excited state (since the fully occupied sets of symmetry-related orbitals are totally symmetric). It follows from this that an orbital with the same symmetry as totally symmetric representation of the ground state will transform into one with the symmetry of the excited state.

The symmetry of the ground state is listed in the output in a line like the following:

```
The electronic state is 1-A1G.
```

Therefore, we can determine the symmetry of the first excited state of benzene if we can find an excitation from an orbital with A_{1g} symmetry within the transition list. The symmetry of the virtual orbital into which it is excited will give us the symmetry for that excited state.

Both visualization programs report the symmetry for MOs. In GaussView, it is displayed in the orbital list in the **MOs** dialog. In WebMO, see the appropriate column of the MO list in the **Calculated Quantities** section of the job results.

Gaussian reports orbital symmetries as part of the molecular orbitals listing within the population analysis when **Population=Full** is specified for the job.* Here is a section of that output for orbitals 6-10 for benzene:

```
     6          7          8          9          10
 (B1U)--O   (A1G)--O   (E1U)--O   (E1U)--O   (E2G)--O
```

Orbital 7 has the desired symmetry, and the first entry in the transition list for the first excited state involves this orbital:

```
Excited State 1:   Singlet-?Sym     5.4863 eV ...
      7 -> 34          0.00143
      7 -> 61         -0.00203
      8 -> 52          0.00287
      ...
```

If we examine orbital 34, we find that its symmetry is B_{2u}, and we can assign the symmetry to the excited state. Note that the coefficient for this transition is very small, illustrating that we needed the **IOp** keyword requesting that larger range of coefficients be printed.

Another method for determining the symmetry of an excited state involves examining only the transition with the largest coefficient. In the case of the second excited state of benzene, this is 20→22 (or equivalently 21→23, the other halves of two pairs of doubly degenerate orbitals). These two orbitals are of symmetry types E_{1g} and E_{2u}, respectively, and their cross

* You can determine the type of orbital transformation for this transition using the orbital listing labeled **Orbital Symmetries** if you did not request a full population analysis, but you will have to manually count through the list to determine the number for an orbital with the desired symmetry.

product will also yield the symmetry for the excited state. You can determine this by consulting the character table for the appropriate totally symmetric point group: D_{6h} in the case of benzene. Character tables are widely available online (e.g., *www.webqc.org/symmetry.php*).

If you want to determine the specific type of orbital transformation for this transition, it is necessary to examine the molecular orbitals for the largest components of the transition (i.e., those with the largest wavefunction coefficients). In this case, we can examine orbitals 20 and 22 (see the illustration in the margin). It is easy to see that these are π orbitals and that this excited state corresponds to the π→π* transition. Such transitions are seen as orbitals involving the p_z atomic orbitals of the carbon atoms with density above and below the plane of the ring. π* orbitals have additional nodes compared to π orbitals, but look generally the same. R refers to a Rydberg orbital: a diffuse orbital that looks a lot like an excited atomic orbital (s, p or d).

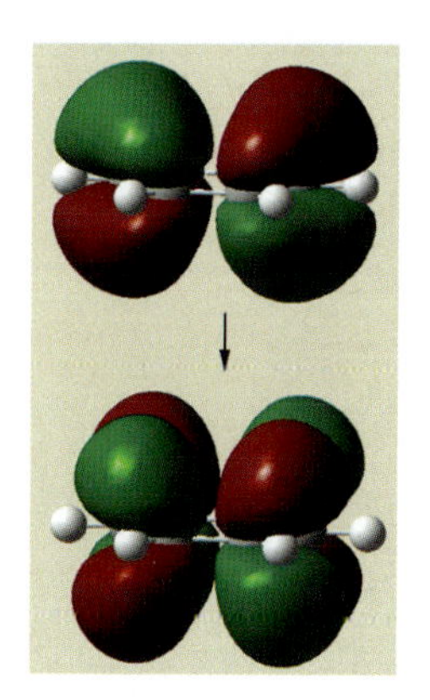

Oscillator Strengths

The oscillator strength (f) is a unitless quantity proportional to the strength of an observed electronic transition (i.e., proportional to the molar absorptivity as measured by a spectrometer). A zero value indicates that the transition is optically forbidden. Using atomic units, we can evaluate this quantity two ways:

Velocity form: *Dipole form:*

$$f^{\nabla} = \frac{2}{3\omega}\left|\left\langle \Psi_{ground} \middle| \hat{\nabla} \middle| \Psi_{excited} \right\rangle\right|^2 \qquad f^{r} = \frac{2\omega}{3}\left|\left\langle \Psi_{ground} \middle| \hat{\mu}_e \middle| \Psi_{excited} \right\rangle\right|^2$$

ω is the excitation energy for the transition, and the two bracketed quantities are the velocity and electric dipole transition moments (respectively) operating on the ground and excited wavefunctions. These two equations give slightly different results only because we are using approximate quantum mechanical models. In the limit of an infinite basis set and full configuration interaction, the two forms would yield exactly the same values.

The values of the two different oscillator strengths along with their X, Y, Z components are listed for each state in the output file:

```
Grnd to excited state transition electric dipole moments:
  state      X         Y         Z       Dip. S.     Osc.
    1    -4.3254   -0.0157   -0.0414   18.7112     1.2035
    2     0.1414   -0.4826    0.3372    0.3666     0.0347
    3    -1.1681   -0.3086   -0.1294    1.4765     0.1431
Grnd to excited state transition velocity dipole moments:
  state      X         Y         Z       Dip. S.     Osc.
    1     0.4174    0.0015    0.0040    0.1742     1.2037
    2    -0.0199    0.0685   -0.0485    0.0074     0.0350
    3     0.1700    0.0448    0.0188    0.0313     0.1434
```

The values of f that are listed next to the excitation energies in the output are derived using the dipole form of the operator:

```
Excited State 1: Singlet-A  2.6254 eV ... f=1.2035 ...
```

Oscillator strengths are used by graphical interface programs to simulate the UV-Vis spectrum of a molecule by overlapping Gaussian line shapes for each transition, forcing the relative heights of each peak to have the same proportional values as f. Note that this technique is very approximate. An improvement in modeling an electronic spectrum can be achieved considering vibrational energy levels of both ground and excited state. This is known as Franck-Condon analysis, and it will be discussed later in this chapter.

EXAMPLE 8.2 MODELING DYES FOR SOLAR CELLS

In this example, we will examine a dye used in a research environment dye-sensitized solar cell (DSSC). DSSCs are solar energy devices that exploit the photochemical properties of an organic dye in order to generate electricity. Like traditional electrochemical devices—batteries—DSSCs contain an anode, a cathode and an electrolytic medium, with the former consisting of a thin film of photoactive dye attached to a titanium-based nanostructure. The dyes used in these devices are metal-free organic compounds. The first of these devices was invented by Grätzel and O'Regan in the late 1980s [ORegan91].

The functioning of DSSC devices is based on exciting the dye from the ground state to the first excited state by exposing it to light. Its lowest energy excited state is a charge transfer state, and it results in an electron being injected into the conduction band of the titanium-based electrode, which then continues on to the cathode. An external load can be attached to this circuit and draw power. In the course of this process, the dye is oxidized; it is ultimately reduced/returned to the ground state by accepting an electron from the anode [Xu08].

The specific dye molecules consist of three regions with distinct purposes: an *anchor* area bound to the substrate, a *linker* section and a *donor* area at the free (unbound) end. These three features of the dyes can be varied (different donors, different length chains and different anchors), and they control the speed at which electrons can be transferred and other aspects that are important in the design of a solar cell. When excited by light, electrons transfer from the free end of the molecule to the bound end. For this reason, the anchor area is also referred to as the *acceptor*, indicating its role as the destination for intramolecular charge transfer.

The specific dye we are considering here has a diphenylamine (DPA) moiety at the free end (the donor), then a short styrene chain as the linker, and then a cyanoacrylic acid (CAA) end as the anchor to the titanium substrate [Xu08, Peng10]:

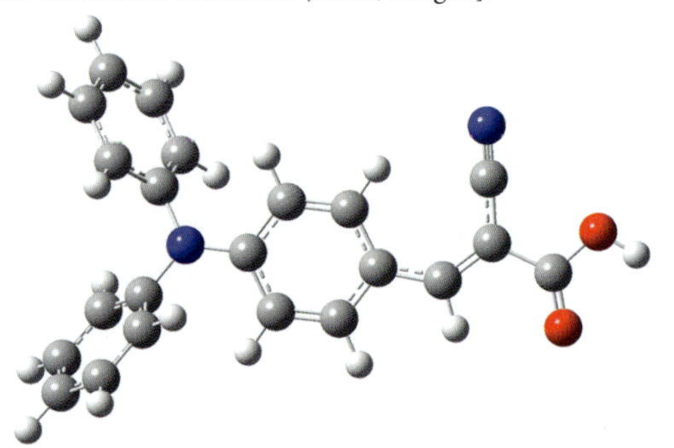

Note that there are two conformations possible by rotation of the carboxylic C-C bond. The one depicted above has a slightly lower energy.

We will begin by modeling this molecule in isolation and studying its excited states. We will perform our calculations in methanol as a model for the conditions inside the DSSC.

We optimized the geometry of the molecule in solution, using our standard model chemistry and procedure. We use this geometry in a subsequent TD calculation to examine the vertical excitation energies (again using our standard model chemistry). We included **Density=Current** in the TD-DFT job's route section so that we would be able to visualize the excited state electron density.

Note that solvation is handled differently in the two phases of this investigation. During the geometry optimization, equilibrium solvation is used, and the solvent field is allowed to react to the presence of the solute as the optimization proceed since this effect occurs within the same timeframe as motion of the nuclei. During the TD calculation, however, non-equilibrium solvation is used since the solvent field will not change within the time frame of the electronic excitation. The SCRF facility applies the methods for the appropriate type of solvation by default for these calculations.

The following table lists the three lowest energy excited states predicted by our calculation along with the observed values for the excitation energies [Xu08]:

	EXCITATION ENERGY (nm)	*f*	TRANSITION	STATE	OBSERVED EX. ENERGY (nm)
1	440.23	0.9874	HOMO-LUMO	donor-acceptor CT	~400
2	316.96	0.0222	HOMO-LUMO+1	π-π^*	~300
3	299.74	0.1549	HOMO-1-LUMO	π-π^*	

The strongest peak—the largest value of *f*—is the first excited state. This state is largely described as a HOMO to LUMO transition (molecular orbitals 89 to 90). Contributions from other orbitals are very small: in this case, below the default threshold for inclusion in the output.

```
Excited State 1: Singlet-A   2.816 eV   440.23 nm   f=0.987
     89 -> 90               0.70409
```

The value of 0.70409 reported in the output above is the normalized coefficient of that particular transition. For a closed shell singlet ground state, the percentage that a given transition contributes is $2x^2$ where x is the normalized coefficient. Thus, if the excited state were made up 100% of a single transition, its coefficient would be $\frac{1}{2}\sqrt{2} \approx 0.7071$.* Therefore, ~99.1% of the transition for the first excited state is due to electron density moving from orbital 89 to orbital 90.

The following figure shows the HOMO and LUMO for this molecule:

Electron density is clearly moving from the triphenylamine/donor side of the molecule to the cyanoacrylic acid/acceptor end of the molecule. Note that the electron density in the linker section is essentially an extended π system.

* A pure transition has two spin components: α-α and β-β. The coefficient values reported for a closed shell calculation include only the α-α spin components. After normalization, the coefficient of each one is $\frac{1}{2}\sqrt{2}$.

Technique: Plotting an Electron Difference Density

Difference densities between ground and excited state may also be used to describe this excitation. In GaussView, we can create such a difference density as follows:

1. Create a cube corresponding to the ground state total electron density. Open the checkpoint file from the **TD Density=Current** job. Create a **Total Density** cube, leaving the value of **Density Matrix** as **SCF**.
2. Create a second **Total Density** cube, this time specifying the **CI** density in the **Density Matrix** pop-up menu. This will cause the excited state electron density to be used.

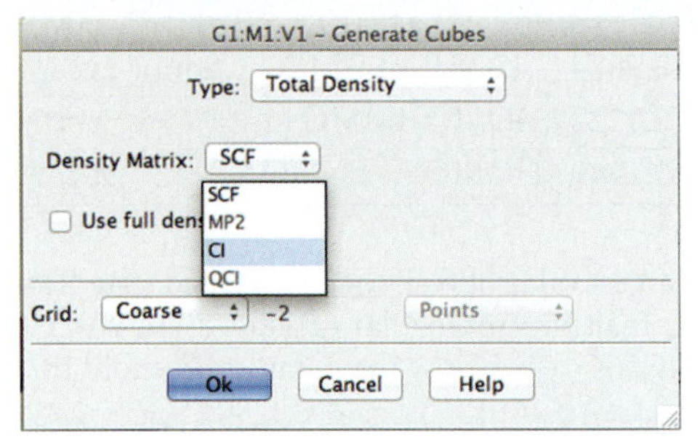

3. Create a third difference cube by specifying the **Type** as **Subtract Two Cubes**. Subtract the **SCF** (ground state) density from the **CI** (excited state) density using the pop-up menus in the resulting dialog. Finally, create a surface using the difference density cube.

The following figure shows two representations of the difference density between the excited state minus the ground state. The figure on the left plots the difference density as an isosurface having a specific value of the density, here ±0.0001. The yellow area plots the surface where the value of the difference density is -0.0001, and the blue area corresponds to where it is +0.0001. The visualization also uses GaussView's transparent surfaces option. Electron density moves from the yellow region to the blue region when moving from the ground state to the first excited state. Thus, for the most part, the transition can be described as a charge transfer originating from two of the rings and arriving at the opposite end of the molecule.

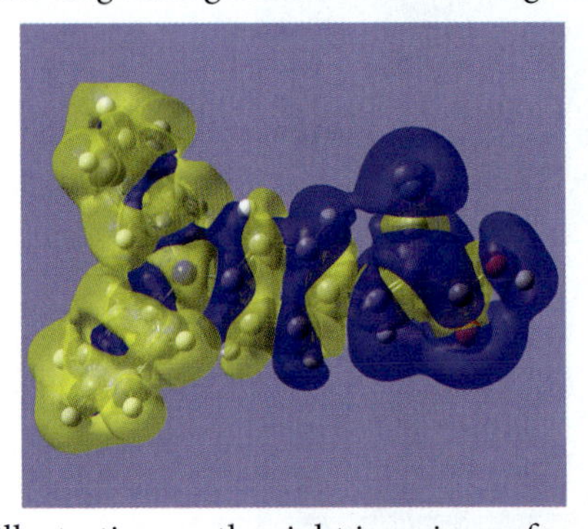

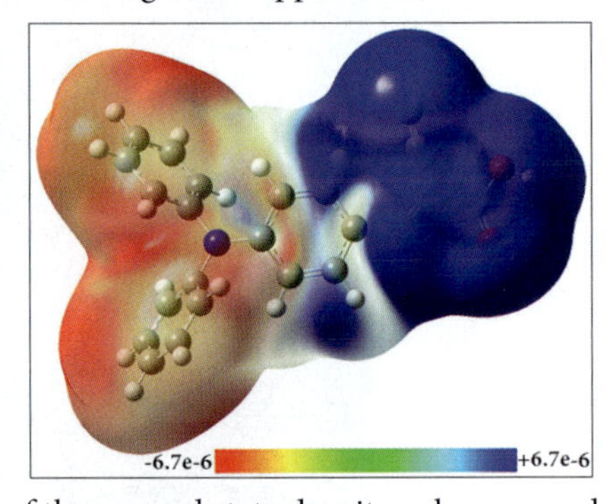

Creating a color mapped surface is described in detail on p. 231.

The illustration on the right is an isosurface of the ground state density color mapped using the value of the difference density (with the end values selected to produce the maximum contrast). An isosurface of 0.0001 was used for this surface as well. In this case, the blue region again indicates where the difference density is positive, meaning that the electron density in the excited state is larger than it is in the ground state. This figure conveys the same information as the difference density via a simpler surface. Which kind of figure to use is a matter of personal preference.

Exercise 8.1 (p. 350) will continue this study by modeling this compound bound to a titanium oxide complex in order to simulate how it is situated in an actual solar cell.■

EXAMPLE 8.3 EXCITED STATES OF $V(H_2O)_6^{2+}$

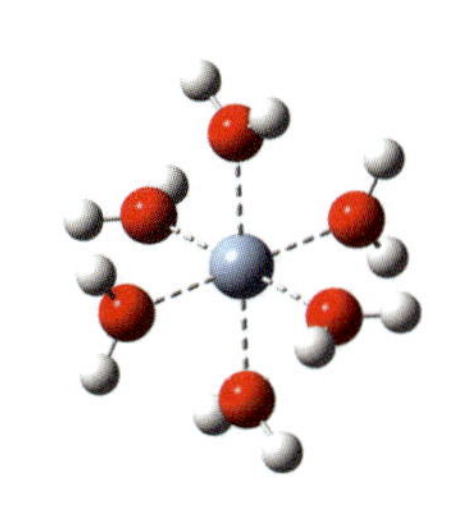

Ligand field theory has been a central theme of inorganic chemistry for many years. Interpretation of d-d transitions in transition metal complexes is discussed in many advanced inorganic chemistry textbooks, with the emphasis frequently placed on extraction of values of Δ and the Racah B parameter (describing how a set of degenerate d orbitals of a metal are split in a particular molecule) from experimental spectra. In this example, we illustrate how visualizing the results of excited state calculations can be useful in interpreting them [Anderson01u]. In this example and the accompanying exercise (p. 353), we will use TD-DFT to model a classic inorganic chemistry experiment involving the synthesis of $V(H_2O)_6^{2+}$ and $V(H_2O)_6^{3+}$ and the investigation of their UV/Visible spectra [Ophardt84].

We will begin by modeling the hexaaquavanadium (II) cation $V(H_2O)_6^{2+}$. At the time of this writing, a theoretical study with a more sophisticated approach than we are using here was just released [Yang14].* The compound has also been studied experimentally [Jorgensen62].

As a dication, vanadium will have three unpaired electrons, and so the multiplicity of the complex should be set to 4 (a quartet). We ran an optimization calculation for this molecule, followed by a frequency calculation as a separate job step. We included the **Population=NPA Density=Current** keywords in the route section of the frequency job.† They request a natural population analysis (NPA) procedure to compute the atomic charges for the ground state. NPA is part of the natural bond order (NBO) analysis technique [Foster80, Reed85, Weinhold88].

The metal-oxygen bond distances were 2.1925 Å in the optimized structure, and the molecule belongs to the T_h point group (the ground state is ^4A_g).

Next we created a multi-structure file which performed a series of single point excited state calculations, starting from the ground state checkpoint file:

```
%OldChk=v6aqua_gs                                   ground state checkpoint file
%Chk=v6aqua_xs                                      first excited state checkpoint file
# TD(NStates=20,Root=1) APFD 6-311+G(2d,p) Pop=NPA
  Density=Current Geom=AllCheck Guess=Read

--Link1--
%OldChk=v6aqua_xs                                   first excited state checkpoint file
%Chk=v6aqua_xs_state2                               second excited state checkpoint file
# TD(NStates=20,Read,Root=2) APFD 6-311+G(2d,p) Pop=NPA
  Density=Current Geom=AllCheck Guess=Read

--Link1--
%OldChk=v6aqua_xs                                   first excited state checkpoint file
%Chk=v6aqua_xs_state3                               third excited state checkpoint file
# TD(NStates=20,Read,Root=3) APFD 6-311+G(2d,p) Pop=NPA
  Density=Current Geom=AllCheck Guess=Read
file continues for all 20 states ...
```

The first job performs the energy calculation for 20 excited states. It also performs the natural population analysis for the first excited state (specified with the **Root=1** option). The subsequent job steps read the excited state results from the first job's checkpoint file—via the **Read** option to **TD**—and obtain the electron density of the specified excited state, using it to perform the NPA. The input file continues in the same manner until all states are processed.

* See especially table 12. Note that calculations must constrain the structure to D_{2h} symmetry in order to compare to the state labels in this paper.

† You could also add these keywords to an **Opt Freq** job, but it would result in a lot of extraneous output.

This technique will create a checkpoint file for each state that can be used in GaussView to perform further interpretation and visualization.

Here is a summary of our TD-DFT results with comparison made to experiment [Jorgensen62] (performed in solution) and the multireference configuration interaction (MRCI) study of [Yang14]:

			EXCITATION ENERGY (eV)				
		IDEALIZED	CALC.			NPA CHARGE	
STATES	SYMMETRY	DESCRIPTION	TD-DFT	MRCI	EXP.	V	O
	^4A_g	ground state				1.1	-0.9
1-3	^4T_g	d-d	1.58	1.28	1.53	1.1	-0.9
4 -6	^4T_g	d-d	2.30	2.07	2.29	1.1	-0.9
	^4T_g	d^2-d^2		3.41	3.46		
7-9	^4T_g	d-R_{4s}	4.61			1.0	-0.9
10-18	^4T_g	d-R_{4p}	5.9-6.0			1.1	-0.9

The first column gives the number listed in the output file for the particular state. These states are triply degenerate in energy because of the three equivalent occupied d orbitals. States 10-18 involve these three occupied electrons going to three equivalent 4p orbitals (so there are nine states in total).

Gaussian will not identify the symmetries of all of these states because of degeneracies in irreducible representations of the T_h point group. However, assignments can be made by examining the orbitals of the transition as explained previously (p. 334). Notice that the charges on vanadium and oxygen in the various excited states are not significantly different from the ground state, indicating no charge transfer type states.

We can qualitatively describe each of these states as one of the following:

- d-d: electrons are transferred from occupied to unoccupied d orbitals of the V atom.
- d^2-d^2: double excitation involving the d orbitals (not accessible by our calculation).
- d-R: diffuse Rydberg type states that resemble the excited states of an atom (both s and p type atomic orbitals are found).

We can get further insight into these transitions by visualizing the total electron density difference between ground and excited states. Here we show the ±0.002 isosurfaces, where once again electrons move from yellow regions to blue regions in the transition:

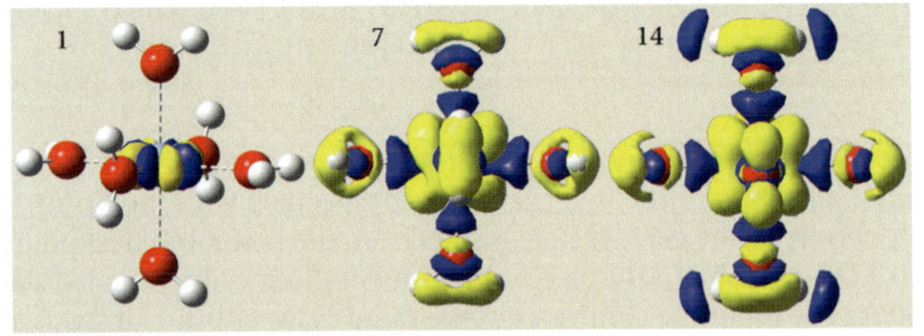

In excited state 1, the plot clearly identifies a d-d transition, with electrons moving from one vanadium orbital to another, but staying on the metal atom. The plots for excited states 7 and 14 are more complicated to interpret, but clearly involve electrons moving on the ligands as well as the metal. As we will see in the next section, we can get a clearer picture for these transitions by looking at the natural transition orbitals.■

Natural Transition Orbitals

Difference densities are often a quite useful method of analyzing the transitions corresponding to excited states. In some cases, however, they are not clear or definitive. Another technique which may be helpful is to examine the *natural transition orbitals* (NTOs). They can be computed rapidly using the transition densities produced by TD energy calculations.

NTOs are a transformed version of the canonical orbitals that attempt to isolate a specific excited state transition to one or two pairs of orbitals [Martin03]. Consider excited state 6 in the preceding example. Here is the output listing the transition coefficients for the largest contributions to this state:

```
Excited State 6: 4.002-?Sym  2.3031 eV …
     40A -> 45A          0.73835
     40A -> 50A          0.12967
     42A -> 44A         -0.56982
     42A -> 45A          0.30529
```

No one transition clearly dominates, and examining the relevant canonical orbitals is not particularly helpful in this case.

We can calculate the NTOs using a job like the following, enabling them to be visualized in GaussView. We use the checkpoint file from the TD calculation for this compound as the source of the excited state data, and we save the orbitals to a different checkpoint file, giving it a name which clearly indicates its contents. Here is a sample input file for state 6, which uses the **ChkBasis** keyword to retrieve the basis set from the checkpoint file:

```
%OldChk=v6aqua_xs
%chk=v6aqua_nto6
# APFD ChkBasis Guess=(Only,Read) Pop=SaveNTO
  Geom=AllCheck Density=(Check,Transition=6)
```

When we examine the molecular orbitals using GaussView's **Edit→MO** dialog, the resulting list of orbitals will have the transition density for that excited state concentrated in a single pair of orbitals (or sometimes two pairs), as illustrated in the margin.

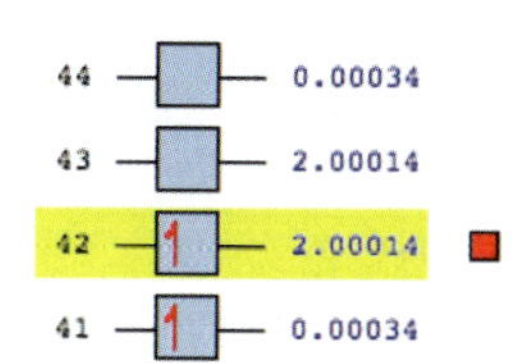

The NTOs for states of interest from the example we're currently working on—states 1, 7 and 14—are illustrated in the figure following. The lower row of the illustration shows the highest occupied transition orbitals (HOTOs), and the upper row shows the lowest unoccupied transition orbitals (LUTOs) (also referred to as the *hole* and the *particle*, respectively). The isovalue used for these surfaces is ±0.02. We can clearly see the nature of the transitions, confirming the assignments of d-d and d-R type states. Notice that none of these originating orbitals is purely a d orbital of vanadium, but they have some admixture of electron density from the ligands.

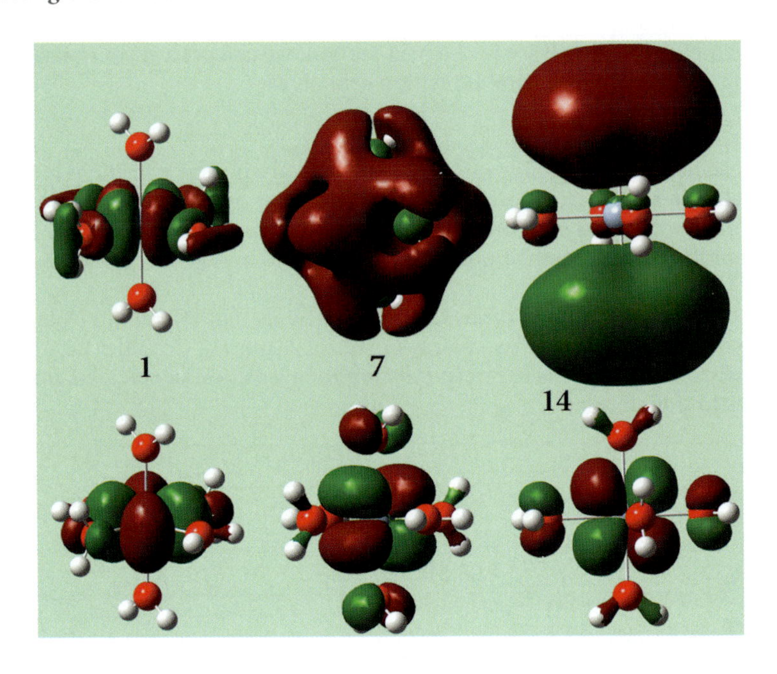

High Accuracy Excited States

References: EOM-CCSD
Method Definition:
[Stanton93, Bartlett07]
Implementation:
[Caricato12b, Caricato13, Caricato13b, Caricato14]

The EOM-CCSD method in Gaussian can be used to perform high accuracy calculations of excited states. It provides CCSD-level accuracy and requires comparable computational cost (scaling as N^6 with problem size just like CCSD*). The following example and its follow-up exercise will illustrate the capabilities of this method and the issues that arise with such calculations.

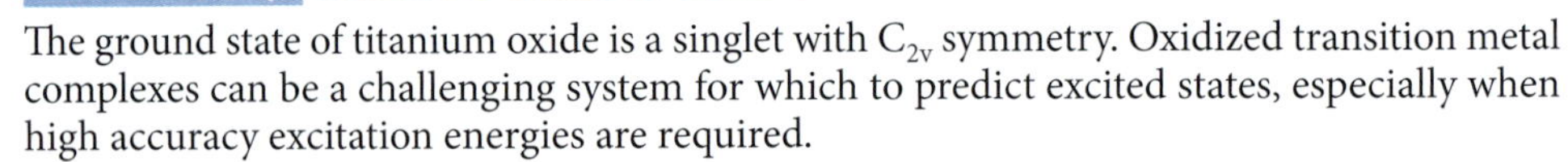

EXAMPLE 8.4 TITANIUM OXIDE EXCITED STATES

The ground state of titanium oxide is a singlet with C_{2v} symmetry. Oxidized transition metal complexes can be a challenging system for which to predict excited states, especially when high accuracy excitation energies are required.

We optimized this molecule using our standard model chemistry and then computed the excited states using both the TD-DFT method (with our standard model chemistry) and the EOM-CCSD method. We used several different basis sets for various calculations:

- Our standard basis set: 6-311+G(2d,p). For this molecule, it uses a total of 119 basis functions, with 10s,7p,4d,2f for titanium (of which 1s,2p,1d are diffuse).
- cc-pVTZ: The triple zeta member of the family of "correlation consistent" basis sets of Dunning and coworkers, designed for use with electron correlation methods (128 total basis functions; 7s,6p,4d,2f,1g for titanium).
- aug-cc-pVTZ: The cc-pVTZ basis set with added diffuse functions (185 total basis functions; 1s,1p,1d,1f,1g diffuse functions added to cc-pVTZ, yielding 8s,7p,5d,3f,2g for titanium).

As these descriptions indicate, there is substantial variation in the descriptions of the titanium atom with the different basis sets which are not entirely reducible to a question of size.

* See p. 483 for a brief theoretical description of this method.

We ran TD-DFT calculations with our standard basis set and aug-cc-pVTZ and EOM-CCSD calculation with all three basis sets. The following table reports the results of these calculations:

	EXCITATION ENERGY (eV)					
	TD-DFT		EOM-CCSD →			
STATE	6-311+G(2d,p)	aug-cc-pVTZ	6-311+G(2d,p)	cc-pVTZ	aug-cc-pVTZ	[Taylor10]
1: $^3B_2$	2.88	2.89	2.46	2.48	2.53	2.33
2: $^1B_2$	3.01	3.02	2.51	2.53	2.58	2.39
3: $^3A_2$	3.38	3.38	3.22	3.24	3.27	3.05
4: $^1A_2$	3.52†	3.52†	3.22	3.25	3.28	3.02
5: $^3A_1$	3.49	3.47	3.30	3.29	3.31	3.08
6: $^3B_2$	3.43†	3.43†	3.32	3.34	3.38	
7: $^1B_2$	3.65	3.65	3.37	3.39	3.43	3.21

†The symmetries of the 4th and 6th excited states are reversed for TD-DFT compared to EOM-CCSD.

The published results in the rightmost column were computed with EOM-CCSD and a basis which is approximately comparable to aug-cc-pVTZ.

Examining the preceding table leads to several observations about the results:

- All of the states lie within an energy range of less than 1 eV. States 3 through 6 are very close in energy.
- The ordering of states 4 and 6 are reversed for TD-DFT compared to EOM-CCSD.
- Increasing the basis set size has essentially no effect on the TD-DFT results. In contrast, increasing the basis set size and adding diffuse functions both raise the predicted excitation energies for EOM-CCSD.
- There is qualitative agreement between the excitation energies predicted by EOM-CCSD/aug-cc-pVTZ and the published results, but the former are generally ~0.2 eV higher than the latter.

It is difficult to know what to make of the final point in that there is little experimental data for this problem. The observed energy value for the first excited state is 2.3 eV [McIntyre71]. However, this value corresponds to the adiabatic (0–0) transition rather than the vertical excitation energy.* One might expect the latter to be some few tenths of an eV higher than the former, but there is considerable uncertainty.

In Exercise 8.3 (p. 355), we will explore other sources of divergence between the published results and our initial calculations. ■

* Modeling the adiabatic transition for the S_1 state would involve optimizing the geometry at the excited state and predicting its energy. Excited state optimizations are discussed later in this chapter.

Electronic Circular Dichroism

> **References: ECD**
> Method Definition: [Helgaker91]
> Implementation: [Bak95, Olsen95, Hansen99, Autschbach02]

Electronic Circular Dichroism is the effect observed when you measure the electronic spectrum of a chiral sample using circularly polarized light. The spectrum that results shows both positive and negative features as it is the difference between the absorption obtained from left- and right-hand polarized light. Enantiomers exhibit mirror image spectra.

The information needed for modeling an ECD spectrum is the rotary strength of each electronic state, which is reported by default in the output of any TD-DFT calculation (no additional option is required). However, you need to be careful to request a sufficient number of excited states so that the results cover the entire energy range in the experimental spectra with which you want to compare. If you have only a few states, the generated spectrum will not be precise enough to reliably identify the enantiomer that generated the data.* For these reasons, the default number of states is never sufficient. We recommend solving for at least 20 states or even more for molecules with many chromophores.

EXAMPLE 8.5 PLUMERICIN ECD

Plumericin† is an iridoid natural product with anti-fungal and anti-bacterial properties. It was first isolated and studied in the 1960s. The molecule has five chiral centers.

We will consider (+)-plumericin, which has the two enantiomers:

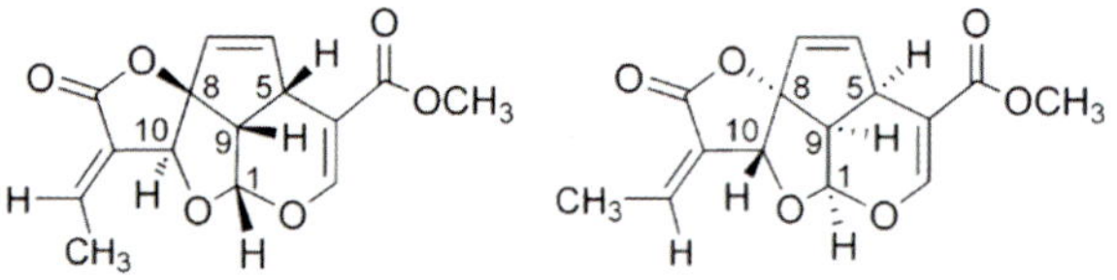

(1R,5S,8S,9S,10S) (1S,5R,8R,9R,10R)

We want to determine the molecule's absolute configuration. We will run our calculations on the (1R,5S,8S,9S,10S) enantiomer. We will perform all of them in solution with chloroform.

The experimental ECD spectrum for this molecule incudes the following peaks [Stephens07]:

WAVELENGTH (nm)	ROTATORY STRENGTH (10^{-40} erg-esu-cm/Gauss)
192	+15
215	-5
237	+13
268	-0.13

There are four geometric conformations for each enantiomer for this molecule: s-cis or s-trans orientations of the methyl ester group with respect to the adjacent C=C. The five-membered ring containing an oxygen atom can be deformed in one of two ways. The four structures have the following key dihedral angles:

* Indeed, as [Stephens07] points out, given that the values of the band half-width at half-height used in plotting are arbitrary, it is possible to fit a too-small set of data to the spectrum to either enantiomer!

† IUPAC: Methyl (3E,3aS,4aR,7aS,9aS,9bS)-3-ethylidene-2-oxo-3,3a,7a,9b-tetrahydro-2H,4aH-1,4,5-trioxadicyclopenta[a,hi]indene-7-carboxylate. Note that we use a different numbering scheme in this example.

Conformation	Idealized Dihedral Angle (°) C_5-C_4-C_{12}-O_{14}	C_8-C_9-C_1-O_{11}
a	180	-20
b	0	-20
c	180	+20
d	0	+20

Here are the four conformations for the (1R,5S,8S,9S,10S) enantiomer:

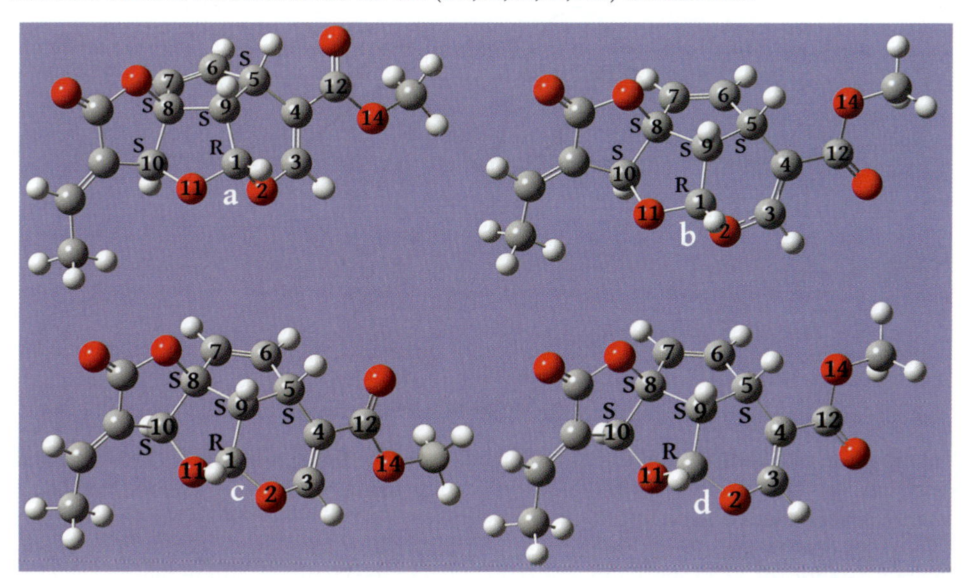

A proper simulation of the experimental spectrum will require Boltzmann averaging of all four conformations.

After optimizing and verifying the geometries of the four conformations, we ran a TD-DFT calculation for each one, solving for 20 states. Predicting the ECD spectrum is done automatically as part of the excited states calculation, with one rotatory strength produced for each computed excited state.

Here is the ECD spectrum for the most abundant conformer, conformation a:

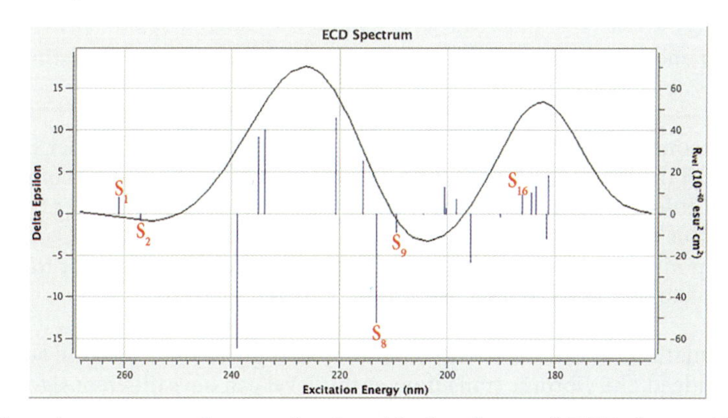

The labeled peaks correspond most closely with the observed ECD data, with the labels indicating the corresponding excited state. We used the GaussView default of 0.333 eV for the **UV-Vis Peak Half-Width at Half Height** value (you can see this parameter in the **Plot Properties** dialog, accessed via the **Properties** item on the plot's context (right click) menu.

Here is the Boltzmann-averaged ECD spectrum (in black), along with those of the individual conformations:

Conformation Populations

Conf.	ΔG (kcal/mol)	%
a	0.00	55.1
b	0.30	33.3
c	1.22	7.1
d	1.49	4.5

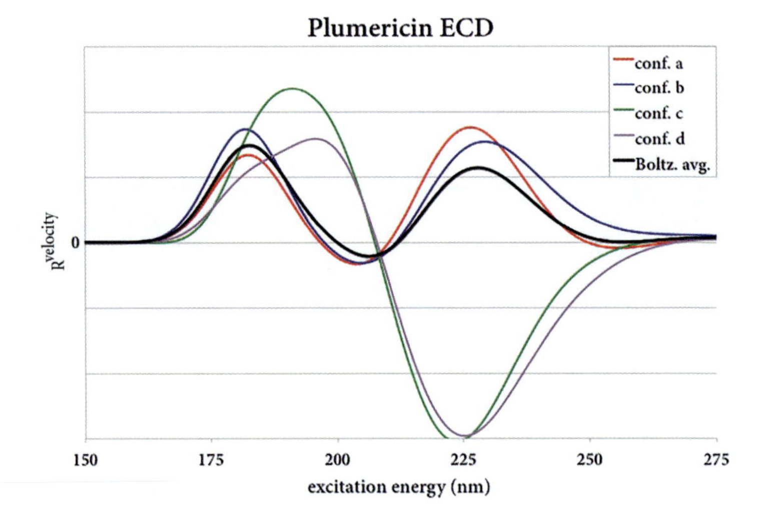

Note that the x-axis direction in this plot is reversed from the earlier one. The theoretical spectrum matches the experimental progression of peaks as wavelength increases: up, slight down, up, very slight down.

The raw data for the ECD spectrum is also included in the Gaussian output file, just prior to the listing of the excited states. The relevant value is identified by state number in the table and appears in the column labeled **R(velocity)**:

```
<0|del|b> * <b|rxdel|0> + <0|del|b> * <b|delr+rdel|0>
Rotatory Strengths (R) in cgs (10**-40 erg-esu-cm/Gauss)
state        XX        YY        ZZ    R(velocity)  E-M Angle
  1      4.9730   -8.8284   -5.6082      -3.1545     109.61
  2     12.8808    0.8708    3.5571       5.7696      67.26
 ...
 20     11.1325   -1.7273  -10.6692      -0.4213      91.41
```

The excitation energy corresponding to the excited state is given in the listing of the states (as usual).

We have a reasonable match with experiment, suggesting that this isomer is the absolute configuration of the molecule. However, this determination was based on comparing at best four pieces of information: four peaks from the ECD spectrum. In order to have greater confidence in this result, we should also compare other data: e.g., ORD, VCD or ROA spectra. If all of these point in the same direction, we would have a strong case for the assignment. You will complete this study in Exercise 8.4 (p. 358).■

Vibronic calculations at the Herzberg-Teller level can greatly improve the accuracy of the band shape. Indeed, the vibronic transitions at this level can have different signs. Even at low definition, the resulting band shape can be quite different.

Predicting Fluorescence: Optimizing Excited State Geometries

Excited state calculations using the ground state geometry will predict the vertical excitation energies of the molecule, which are necessary to understand the process of electronic absorption. However, if we want to study an emission process like fluorescence, the potential energy surface of the excited state must be explored. The lowest energy predicted for the optimized excited state structure corresponds to the energy emitted when the molecule fluoresces as it returns back to its ground state.

EXAMPLE 8.6 DMABN EXCITED STATE GEOMETRY

One of the most heavily studied excited molecules is N,N-(dimethylamino)-benzonitrile (DMABN):

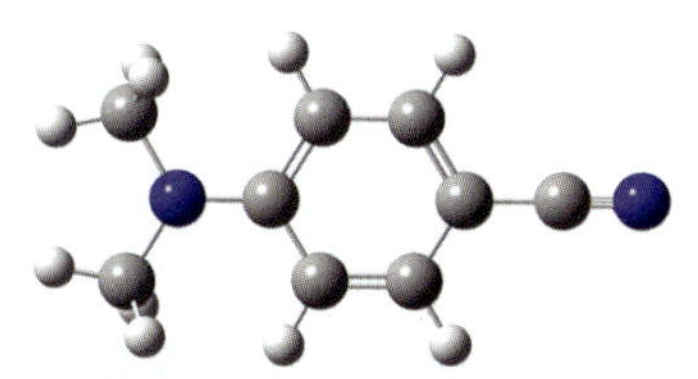

This small molecule is a simple push-pull chromophore whose fluorescence spectrum contains emission from both a local excited state (LE) typical of a benzene ring and a second peak whose origins are hotly debated. Two recent experimental investigations [Coto11, Rhinehart12] have confirmed that the process leading to the anomalous peak most certainly involves competing geometric distortions that are strongly influenced by solvent. Establishing an authoritative mechanism which explains this process is extremely difficult and would require knowledge of the time scale of various events that occur while the molecule is in its excited state. The authors of [Rhinehart12] make the point in this way:

> As is typical with most studies of DMABN over the past 50 years, this data invites further experimental and theoretical studies into the microscopic mechanism of the CT process.

Of course, we will not be able to solve these issues definitively in this example. What we can do is demonstrate how to find stationary points on the excited state surface of this molecule using TD-DFT. There are certainly other important features of this surface that we will not have time to explore.

The observed UV-Vis spectrum of DMABN has a feature at ~291 nm and can also exhibit shoulder peaks on either side depending on the polarity of the solvent; these features correspond to the absorption of a photon. Fluorescence occurs at ~350 nm; again depending on the solvent, the spectrum may include a second fluorescence peak at ~475 nm [Lipinski80]. (In eV, these values are ~4.26, ~3.54 and ~2.61, respectively.)

This figure from [Rettig99] shows the spectra in two different solvents:

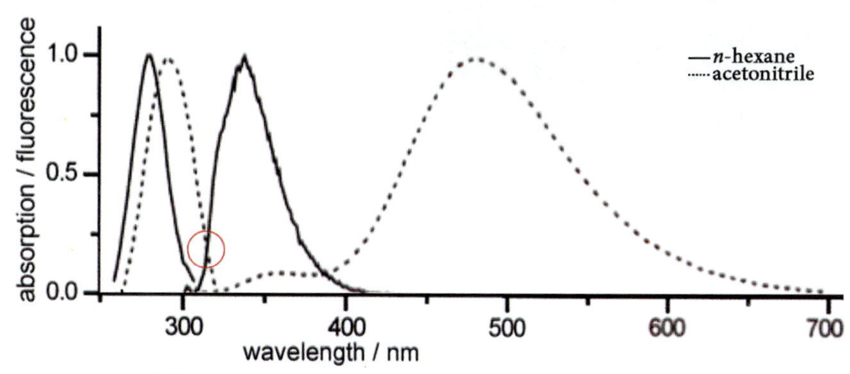

In the non-polar solvent *n*-hexane, there is an absorption peak and a second fluorescence peak. In contrast, in the polar solvent acetonitrile, the first peak is slightly red-shifted. It also contains a small shoulder in the area outlined in red (it is obscured by the line for *n*-hexane). There are also two fluorescence peaks in this environment: one with small intensity at ~350 nm and the more obvious second peak at the longer wavelength ~475-485 nm.

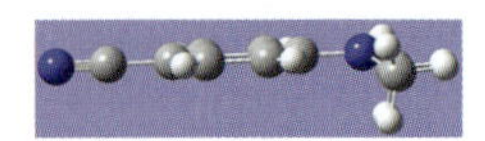

Our first step was to optimize the ground state of this molecule. Notice the position of the methyl hydrogens in the illustration in the margin. Molecular builders—including GaussView and WebMO—may position the hydrogen atoms differently, and the resulting optimization may locate a transition structure with an imaginary frequency corresponding to methyl rotation. We were careful to begin the optimization from a structure with the C-N-C-H dihedral angles equal to 180°, -60° and +60° on both methyl groups. We also placed the nitrogen atom in a slightly pyramidal position (not planar) and made sure that the point group was C_s. Finally, we included **Opt=CalcFC** in this optimization because the force constant associated with the out-of-plane wagging of the nitrogen atom is not well estimated.

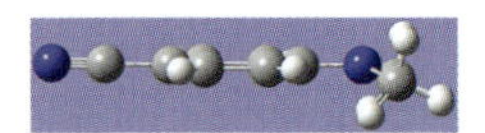

Our optimization was successul, and it found a structure with C_s symmetry that has a very slightly pyramidal nitrogen atom (the angle is ~4 degrees), so the predicted geometry is nearly planar except for two sets of methyl hydrogen atoms.* The lengths of the middle C-C bonds within the ring are increased by ~0.4 Å in the optimized structure, and the C-C bond between the ring and the amide group is longer by ~0.2 Å.

Our next step was to run a TD-DFT calculation on the optimized ground state geometry, which predicted the following two lowest singlet excited states:

	EXCITATION ENERGY (eV)			
STATE	CALC.	EXP.	TYPE	EXCITATION
S_1	4.45	4.25	LE	HOMO→LUMO+1
S_2	4.71	4.56	CT	HOMO→LUMO

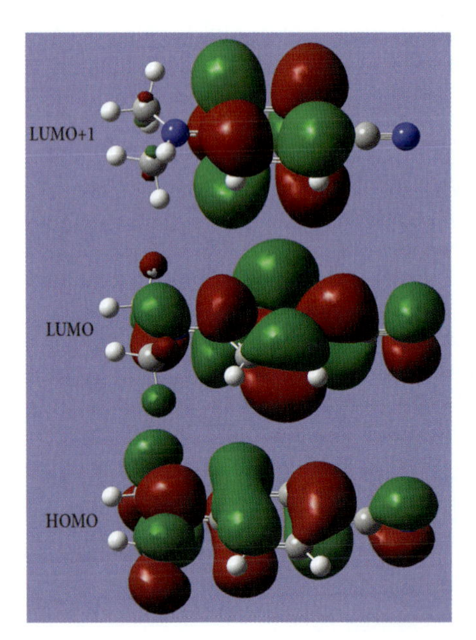

These calculated values are in reasonable agreement with gas phase experiments [Bulliard99].

Examining the molecular orbitals allows us to identify the primary characters of these excited states as follows:

- The lowest energy state is a local excitation (LE) involving the electrons in the π orbital of the benzene ring being excited to a π* orbital.
- The second excited state is primarily an intramolecular charge transfer state from the donor amino group nitrogen atom to the cyano group (acceptor).

We continue this study by examining the excited state potential energy surface. We will begin by optimizing the geometry of the excited state, which will lead us to an approximate value for the emission wavelength coming from the local excited state. Later on, we will locate a second stationary point on the excited state PES.

In order to optimize the structure of the first excited state, we begin with the optimized ground state structure. We impose C_{2v} symmetry on the molecule,

* Experiments on the ground state observed a small pyramidalization of the amino nitrogen atom, by an angle of ~12-15° [Kajimoto91, Heine94]. We optimized the geometry of several pyramidal structures for this molecule using our standard model chemistry; all of them resulted in a nearly planar structure for the minimum. Increasing the basis set size and the size of the integration grid had no effect on the results. There is one report in the literature of locating a pyramidal ground state structure [Parusel98]; that calculation used a basis set comparable in size to our standard basis set and a smaller integration grid. We did not succeed in locating such a structure. In any case, [Parusel98] demonstrated that whether or not the amino nitrogen atom was pyramidal had virtually no effect on the predicted properties.

and then perform an optimization plus frequency calculation using the **TD(Root=1)** keyword and our standard model chemistry. The optimization locates a C_{2v} minimum.

Next, we will locate a stationary point on the excited state surface that is known to be important in understanding the dual fluorescence phenomenon in this compound. The so-called *twisted intermolecular charge transfer* (TICT) state is one in which electrons of the amino group have been donated to the ring, allowing the hybridization at the nitrogen atom to be more purely sp^2 and allowing rotation about the N-C bond such that the methyl groups are now above and below the plane of the ring:

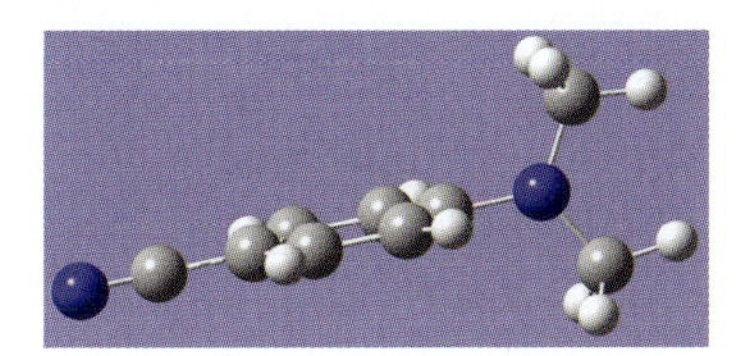

We created the guess structure by taking the optimized ground state geometry and twisting the CCNC dihedral angles to be 90°. The resulting structure can be fixed to C_s symmetry before running the optimization.

The following table summarizes the values of the key parameters in the structures on the excited state PES:

	DIHEDRAL ANGLE (°)		C-C BOND LENGTHS (Å)	
STRUCTURE	C-C-N-C (red)	C-C-C-N (green)	RING (from NC)	RING-AMIDE
ground state	-4.7	179.9	1.40, 1.39, 1.41	1.37
LE minimum	0.0	180.0	1.40, 1.43, 1.41	1.39
TICT minimum	90.0	180.0	1.43, 1.37, 1.42	1.43

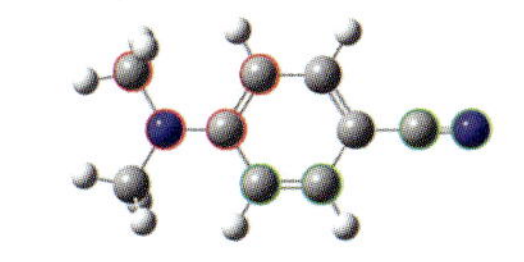

The C-C bond changes with respect to the ground state are even more pronounced in the TICT minimum than they were in the LE minimum. In the ring, the shortest bonds are the middle ones; this is also the case for the ground state, although the length difference between these bonds and the others is larger in the excited state. The C-N bond distance lengthening between the ring and the amide group is also significant in the TICT minimum.

The transition wavelength reported in the output file from the TICT optimization, 436 nm (2.61 eV), is actually a vertical emission energy since we are now at a stationary point on the excited state surface. This approximates the peak position for the anomalous peak in the fluorescence spectrum but by no means completely explains it, as we have not investigated how the molecule would reach this state after excitation.

However, we can tell that it is a charge-transfer type state, clearly indicated by the following difference density plot:

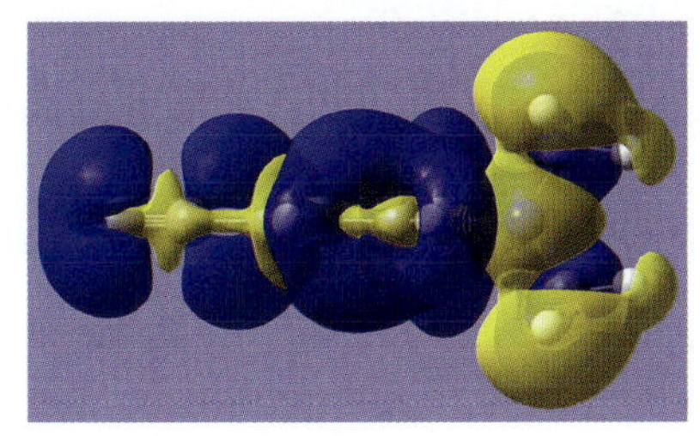

Electron density moves from the yellow to the blue areas as the molecule transitions from the ground state to the first excited state.

These calculations have all been performed in the gas phase. In Exercise 8.5 (p. 359), you will continue this study by modeling this molecule in solution with a polar solvent—the environment in which the dual fluorescence is observed—which will provide an interesting comparison. ■

Fluorescence can be modeled in a more sophisticated way using Franck-Condon/Herzberg-Teller analysis, which will be discussed in the advanced examples and exercises section of this chapter, beginning on page 364.

Exercises

EXERCISE 8.1 EVALUATING DYES FOR DSSC DEVICES

In this exercise, we continue our study of a prototypical dye molecule for DSSC devices from Example 8.2. In an actual device, the CAA end of the dye molecule is bound to a titanium oxide nanostructure. The original researchers validated that the metal surface in this case can be adequately represented as a $Ti(OH)_3H_2O$ moiety: the titanium atom is an octahedral complex with two positions taken up by the oxygen atoms of the acrylic acid anchor, three hydroxyl groups and one water molecule.

For purposes of binding, the acidic hydrogen is relocated from the dye to the titanium complex itself. It is for this reason that one position is occupied by a water molecule rather than a hydroxide group, keeping the system neutral with an even number of electrons and allowing the ground state to be modelled as a closed shell singlet.

This molecular structure is illustrated below:

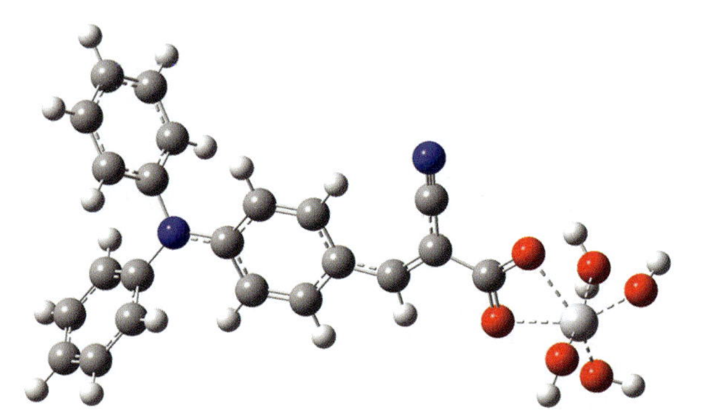

Optimize the geometry of this complex, following our standard procedure. Then predict its first three excited states. Analyze the numerical and visualization data from the excited state calculation in light of the following questions:

- Does the nature of the excited state change as a result of being bound to the titanium complex?
- Are any of the molecule's properties modified in the new environment?

SOLUTION

The following table reports the excitation energies, oscillator strengths and assignments for the predicted excited states:

	EXCITATION ENERGY (nm)	BOUND-FREE SHIFT (nm)	*f*	TRANSITION	STATE
1	447.28	7.05	1.1143	HOMO-LUMO	donor-acceptor/Ti CT
2	333.47	16.51	0.0009	HOMO-LUMO+1	π-π*
3	326.95	27.21	0.0032	HOMO-LUMO+2	π-π*

The three states are characterized in much the same way as for the unbound molecule. The bound complex experiences a modest red shift to higher wavelengths for all excited states, which increases at lower wavelengths.

When we examine the molecular orbitals corresponding to the first excited state, we learn more details about it:

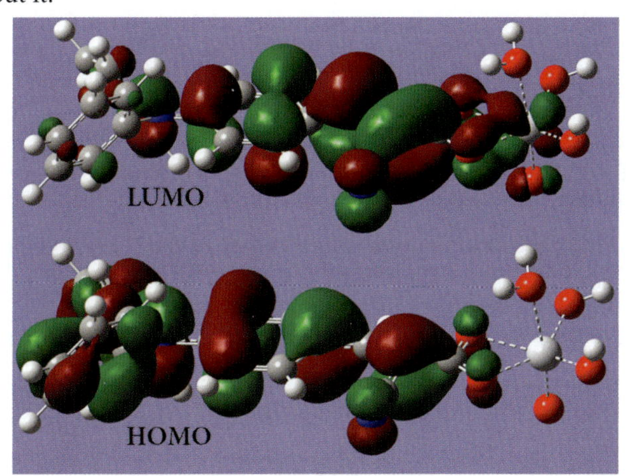

The fact that it is a charge transfer state is again evident. Electrons are moving from the DPA end to the CAA end of the molecule. Interestingly, the titanium complex is also involved in the latter region, becoming part of the acceptor. The 3d orbital from the titanium atom is visible in the LUMO but not in the HOMO, indicating that electrons in the Ti complex do not interfere with the electronic transition. They receive electrons, but do not donate them. The original researchers explain (referring to the dye molecule by the designation TC1):

> The similarity of excited transitions in free and bound states is probably due to the fact that the energies of the Ti 3d orbitals lie beyond the range of the free TC1 frontier orbitals. [Peng10, p. 034312]

In addition, the carbonyl group in the CAA which attaches to the titanium atom also plays an important role in the charge transfer by contributing to both the HOMO and the LUMO.

The following figure shows the difference density plotted in the two ways. The ±0.0001 isodensity surfaces for the difference density are plotted on the left, and the electron density surface is colored by the difference density on the right:

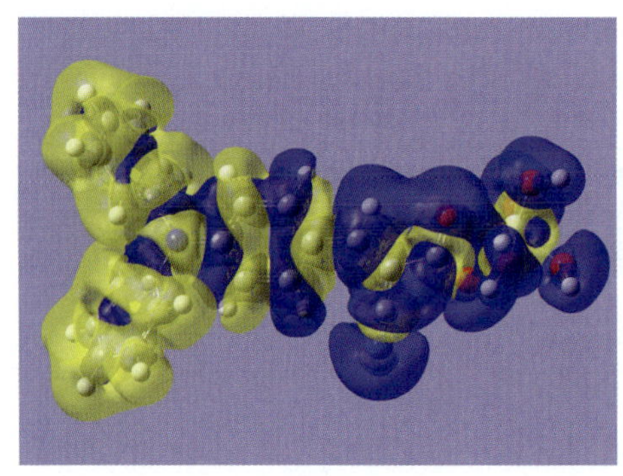

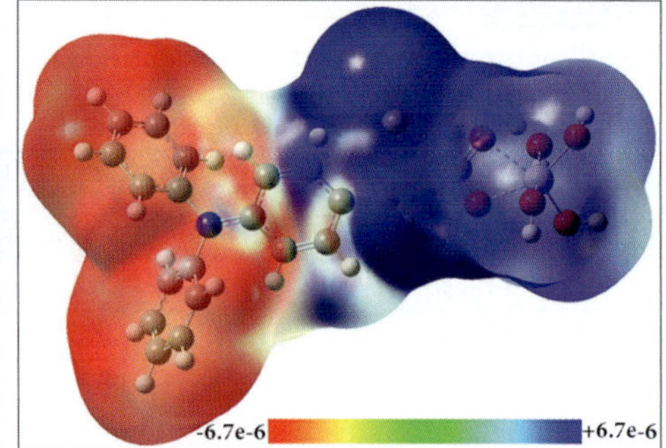

Both visualizations confirm the charge transfer character of the first excited state.

The dipole moment of the molecule also reveals its interesting character. The predicted value of the component of the dipole moment along the axis of the molecule—and perpendicular to the titanium substrate—is -8.9 debye in the unbound case and -11.5 in the bound form.

In a related paper [Xu08], these researchers also studied several related species to TC1, among them a compound which they designate TC4. It was designed to improve the photovoltaic performance of TC1. Two additions were made to the structure:

- A vinyl group was added to one of the terminal phenyl rings in an effort to enhance the molecule's ability to function as an electron donor.
- Another vinyl unit was added to the linker to extend the π-conjugated bridge area.

The following figure plots the excited state spectrum for TC1 (black), TC4 (red) and two other variations; the structure of the latter appears in the upper right corner:

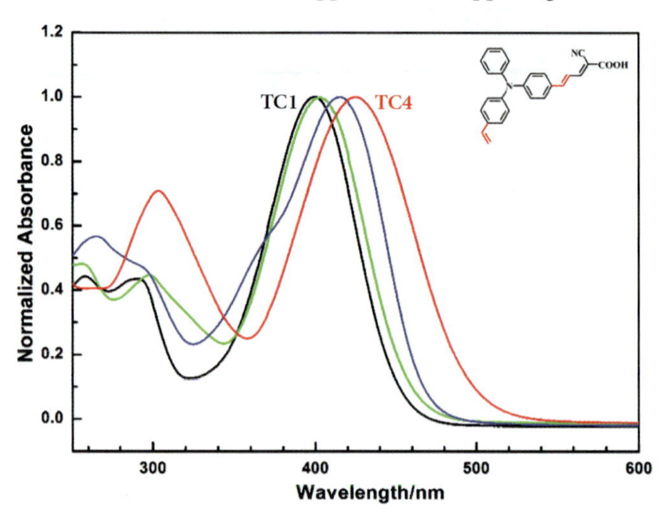

The TC4 spectrum is red shifted somewhat with respect to TC1. It is also stronger and broader around 300 nm, indicating a "bathochromic shift in the absorption spectra of the dyes, which are desirable for harvesting the solar spectrum" [Xu08, p. 878]. Overall, TC4 exhibits better energy conversion efficiency. Such compounds are promising candidates for use in DSSCs and are the subject of ongoing research.■

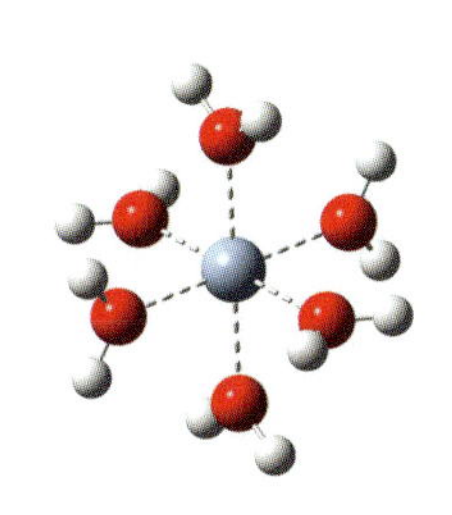

EXERCISE 8.2 EXCITED STATES OF VANADIUM-WATER COMPLEXES

In this exercise, we will continue our study of vanadium-based water complexes by studying hexaaquavanadium (III) cation. $V(H_2O)_6^{3+}$ has one less electron than the species we considered in Example 8.3. How would you expect this to change the excited state results? What transitions would you expect to see for this molecule?

Model this compound in the ground state and the excited state using our standard model chemistry and the TD-DFT method to predict the excited states. Solve for at least 10 excited states, and compute NPA charges for both ground and excited states for comparison purposes. Use difference densities and/or natural transition orbitals to help describe the states found. The +3 charge for this complex will create a d^2 electron configuration and a triplet multiplicity resulting from the two unpaired electrons.

Tip: Start your ground state optimization from a structure with S_6 symmetry.

SOLUTION

The following table summarizes the key structural parameters for the two optimized molecules:

MOLECULE	V-O (Å)	O-V-O (°)	POINT GROUP
$V(H_2O)_6^{2+}$	2.19	90.0	T_h
$V(H_2O)_6^{3+}$	2.03	89.3, 90.7	S_6

Although the V-O bond distances in the V(II)-based compound are somewhat longer, the two hexaaquavanadium complexes are quite similar in structure.

The following table summarizes the lowest predicted excited states for the V(III)-based compound:

			EXCITATION ENERGY (eV)				
		IDEALIZED	CALC.			NPA CHARGE	
STATES	SYMMETRY	DESCRIPTION	TD-DFT	MRCI[a]	EXP.[b]	V	O
	^3A_g	ground state				1.3	-0.8
1-2	^3E_u	d-d	0.51			1.3	-0.8
3	^3A_g	d-d	2.25	1.50	2.0-2.2	1.4	-0.9
4-5	^3A_g	d-d	2.55			1.3	-0.9
6	^3A_g	d-d	2.95	3.06	3.0-3.2	1.3	-0.8
	^3A_g	d^2-d^2		3.17			
7-8	^3E_u	charge transfer	4.81		4.6	0.9	-0.7

[a][Yang14] [b][Jorgensen62, Ophardt84]

We included other theoretical and experimental results in the preceding table. Comparison to literature needs to be done with care:

- States 1-2 involve transitions between orbitals that were degenerate in energy for the higher symmetry hexaaquavanadium (II) complex, which had a d^3 rather than a d^2 electron configuration. Unrestricted DFT is based on a single electron configuration for the ground state, and this will not allow two electrons to be distributed equivalently among three orbitals. The result is that Jahn-Teller type distortions are exaggerated and the ground state geometry optimizes to a structure with S_6 symmetry.* The reported MRCI calculations for a structure constrained to D_{2h} symmetry preserve the degeneracy of these states through the use of a multi-configurational ground state wavefunction.

- Our states 3-6 represent the single electron d-d type transitions that would be seen experimentally in a typical UV-Vis spectrum. There would be only two such transitions from the higher symmetry MRCI calculation.

- The d^2-d^2 transition is a double excitation and can be calculated by MRCI but cannot be found with TD-DFT. Experimentally, it is difficult to observe this state as well, since it would be masked by the single electron excitation in the region 3.0-3.2 eV.

- The degenerate excited states 7-8 computed by our standard model chemistry are characterized as charge-transfer states (ligand-to-metal), as indicated by the greatly reduced NPA charge on vanadium. MRCI calculations do not find this state, since they are restricted to finding the d-d states. Experimentally, this state is seen in the aqueous UV spectrum. Previously, [Ophardt84] assigned it to the double excitation, but with the recent MRCI calculation showing that state to be much lower in energy, we can propose this to be a CT state.

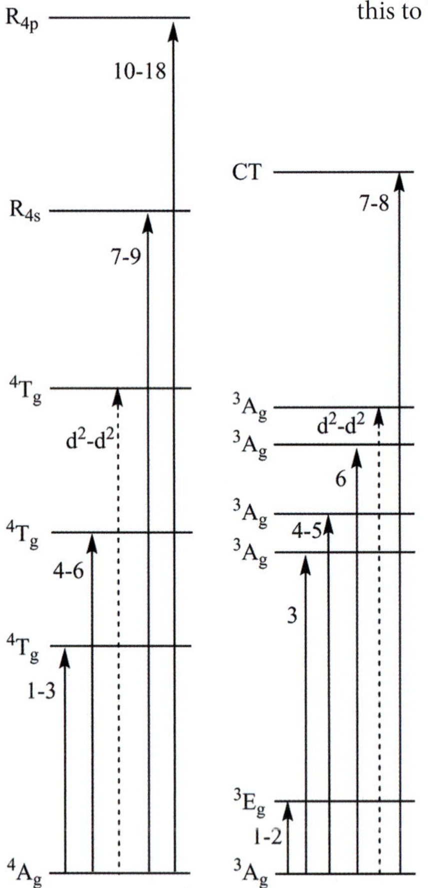

The comparison between the vanadium (III) and vanadium (II) water complexes can be summarized in an energy level diagram with our TD-DFT states labeled as in the output file.

The predicted NBO charges for the charge transfer state reflect the shift in electron density, in this case from oxygen atoms to the vanadium atom.

Here are the difference densities corresponding to these excited states as well as the HOTO and LUTO for state 7:

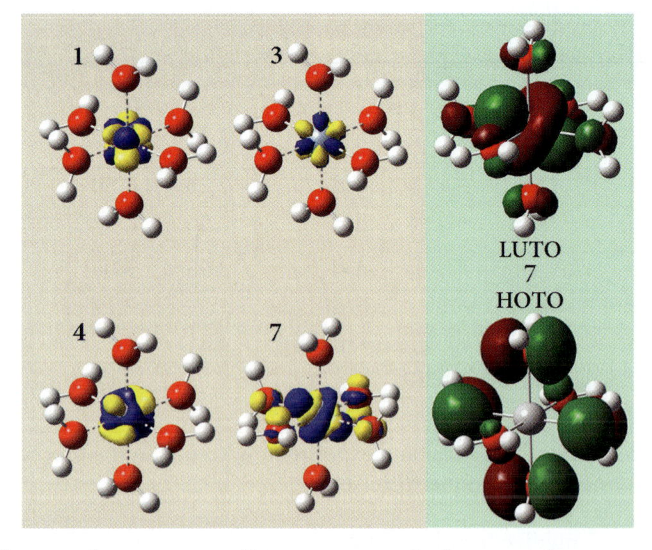

The difference densities are all consistent with the preceding assignments. Interestingly, the plot for the charge transfer state illustrates how electron density

* For a detailed discussion of the geometric distortions possible for transition metal complexes, see [Kallies01].

shifts from the four oxygen atoms in plane perpendicular to the V-O bond to the metal atom, a more complex configuration than was present in the other complex.

The HOTO and LUTO of state 7 further clarify the nature of the charge transfer state (we used an isovalue of ±0.03 to simplify the plots). This charge-transfer state is created by ligand electrons moving to the unoccupied d orbital of the metal. Contrast this with the Rydberg states found for the Vanadium (II) complex.■

EXERCISE 8.3 HIGH ACCURACY EXCITED STATES: TITANIUM OXIDE

This exercise continues our study of TiO_2 from Example 8.4, where we modeled the excited states of this system using a variety of TD-DFT- and EOM-CCSD-based model chemistries. Here, you continue the calibration study we began earlier. Keep in mind that completing a process like this one is a vital early step for every research problem you undertake.

The accuracy of the previous EOM-CCSD calculations could potentially be increased in a variety of means, including the following:

- By using a more accurate geometry in the calculation.
- By using a larger basis set.
- By including some of the core electrons within the correlation configurations (see below).

You may be able to think of additional ways.

Perform additional calculations on titanium oxide in order to produce more accurate predictions of the excitation energies of the molecule's six lowest excited states.

The Frozen Core Approximation

By default, Gaussian uses the *frozen core approximation* in electron correlation calculations, which assumes that only the valence electrons make significant contributions to the molecular properties. The core orbitals, closest to the nucleus, are constrained to be doubly occupied. For example, for silicon, the core consists of the $1s^2 2s^2 2p^6$ orbitals, and the valence region is the $3s^2 3p^2$ orbitals (containing 10 and four electrons respectively); this configuration is conventionally denoted $[1s^2 2s^2 2p^6]3s^2 3p^2$. The frozen core approximation reduces both the number of configurations considered and the computational complexity of evaluating those that remain, resulting in significant time savings with minimal effect on accuracy.

For some calculation types, however, this approximation is not appropriate. Modeling excited states is one such case. Gaussian provides several options for specifying which electrons are active (they are described in the *Gaussian User's Reference*). The one which is relevant here is the *read window* option, which allows you to specify which orbitals are treated as part of the valence region (retained) in electron correlation calculations. The keyword for this feature is **Window=(*n*,*m*)**, where *n* and *m* specify the range of orbitals to be retained: treated as part of the valence region with the calculation. A value of **0** for *n* or *m* indicates the first/last orbital (respectively).

In the case of titanium, the valence region normally consists of the highest two orbitals (four electrons): i.e., $[1s^2 2s^2 2p^6 3s^2 3p^6]4s^2 3d^2$. The **Window=(8,0)** keyword retains four orbitals within the electron correlation calculation—orbitals 8 through 11. This has the effect of including 3s and three 3p orbitals from Ti in the correlation treatment. The 1s, 2s and three 2p orbitals from Ti and the two 1s orbitals from O remain frozen, a total of seven orbitals (which is why the window starts at orbital 8).

SOLUTION

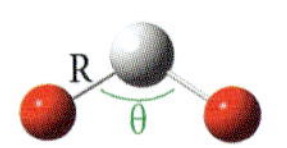

We began by reoptimizing the geometry with the CCSD/aug-cc-pVTZ model chemistry. The following table compares the geometry we obtained with the one from our standard model chemistry used for the previous calculation on this molecule:

PARAMETER	TD-DFT	CCSD
$R^{Ti\text{-}O}$ (Å)	1.630	1.640
$\theta^{O\text{-}Ti\text{-}O}$ (°)	111.6	113.5

The bond length is slightly longer with CCSD, and the bond angle is somewhat smaller. The following table lists the predicted excitation energies obtained at the EOM-CCSD/aug-cc-pVTZ level for the two geometries, along with those predicted by the reference for this study:

	EXCITATION ENERGY (eV)		
STATE	TD-DFT GEOM.	EOM-CCSD GEOM.	[Taylor10]
1: $^3B_2$	2.53	2.50	2.33
2: $^1B_2$	2.58	2.56	2.39
3: $^3A_2$	3.27	3.19	3.05
4: $^1A_2$	3.28	3.20	3.02
5: $^3A_1$	3.31	3.27	3.08
6: $^3B_2$	3.38	3.29	
7: $^1B_2$	3.43	3.34	3.21

The excitation energies of the first, second and the fifth excited states decrease very slightly with the geometry predicted by the higher accuracy model chemistry, while the differences for the higher energy states are generally in the range of ~0.8-0.9 eV. In all cases, the predicted values remain substantially higher than those obtained in the original study.

We also investigated the effect of using a larger basis set. The following table compares the results of the aug-cc-pVQZ basis set compared to EOM-CCSD/aug-cc-pVTZ:

	EXCITATION ENERGY (eV)		
STATE	TZ	QZ	[Taylor10]
1: $^3B_2$	2.50	2.57	2.33
2: $^1B_2$	2.56	2.62	2.39
3: $^3A_2$	3.19	3.24	3.05
4: $^1A_2$	3.20	3.25	3.02
5: $^3A_1$	3.27	3.32	3.08
6: $^3B_2$	3.29	3.34	
7: $^1B_2$	3.34	3.39	3.21

The larger basis set—300 basis functions, 9s,8p,6d,4f,3g,2h on titanium—predicts slightly higher excitation energies than the triple zeta version: ~0.05-0.07 eV. You can see the same effect by comparing the results of EOM-CCSD using the cc-pVTZ and cc-pVQZ basis sets. Interestingly, for TD-DFT, the predicted energies are also higher with the larger basis set, but the difference with respect to aug-cc-pVTZ is more than an order of magnitude smaller: ~0.002-0.004 eV.

The most dramatic change in the excitation energies for TiO_2 came when we included some of the core electrons within the electron correlation configurations. These are the results for an **EOM-CCSD/aug-cc-pVTZ Window=(8,0)** calculation:

	EXCITATION ENERGY (eV)		
STATE	FC	Window=(8,0)	[Taylor10]
1: $^3B_2$	2.50	2.37	2.33
2: $^1B_2$	2.56	2.42	2.39
3: $^3A_2$	3.19	3.03	3.05
4: $^1A_2$	3.20	3.05	3.02
5: $^3A_1$	3.27	3.11	3.08
6: $^3B_2$	3.29	3.14	
7: $^1B_2$	3.34	3.21	3.21

The predicted values from this calculation are now essentially the same as those in the original reference; those researchers included all of the core electrons within the electron correlation configurations (the equivalent Gaussian keyword would be **Window=(0,0)** which can also be specified as **Window=Full**).

The overall results for the principal methods and model chemistries are plotted in the following figure:

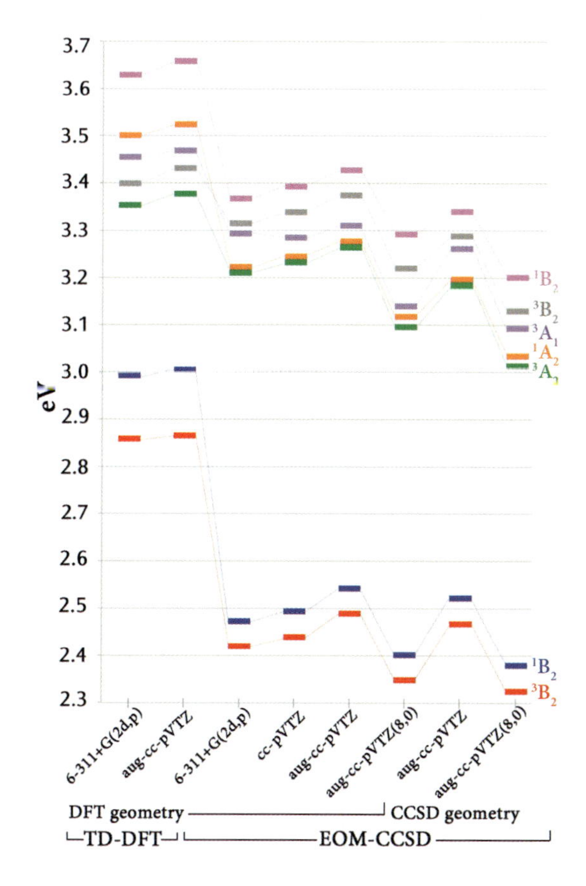

TD-DFT generally predicts much larger excitation energies for this molecule for all of the excited states we are considering. Larger basis sets increase the values slightly. This method also predicts a different ordering of the fourth and sixth excited states than EOM-CCSD. Finally, the energy gaps between states 1 and 2 and between states 6 and 7 are larger for TD-DFT than EOM-CCSD, while the gap between states 2 and 3 is much smaller for TD-DFT.

The plot illustrates the three factors we considered for the EOM-CCSD method, in order of increasing magnitude of the effect:

- Using the CCSD geometry lowers the predicted excitation energies a modest amount, regardless of basis set.
- Adding diffuse functions increases the predicted excitation energies for EOM calculations at the DFT geometry.
- Increasing the basis set size increases the predicted excitation energies somewhat for the calculations using the TD-DFT geometry. This effect is also seen for the CCSD geometry when comparing aug-cc-pVTZ vs. aug-cc-pVQZ, although it is much less pronounced (the latter is not included in the preceding diagram).
- Including some of the core electrons within the electron correlation configurations has the largest effect on the predicted excitation energies, lowering them by ~0.12-0.16 eV (the magnitude varies with the type of transition).

When studying excited states with EOM-CCSD, all of these factors must be taken into consideration as you plan and perform your calculations.■

EXERCISE 8.4 ECD RESULTS ANALYZED IN CONJUNCTION WITH VCD & OR

In Example 8.5, we demonstrated how to use ECD spectra to make an assignment of absolute configuration. Because there is limited data being compared in an ECD experiment, it is necessary to follow that with other comparisons to increase the reliability of the assignment. Compute the optical rotation (measured using the sodium D line: 589.3 nm) and VCD spectra of (1R,5S,8S,9S,10S)-plumericin and compare them to the experimental results given below. To do this properly requires that all four conformations be modeled and Boltzmann weighted. Perform your calculations in solution with chloroform.

The $[\alpha]_D$ specific rotation for this compound has been measured in a variety of solvents, and its valued ranged from +170 to +201 degrees [Stephens07].

Here is the VCD spectrum of (+)-plumericin reported by [Stephens07]:

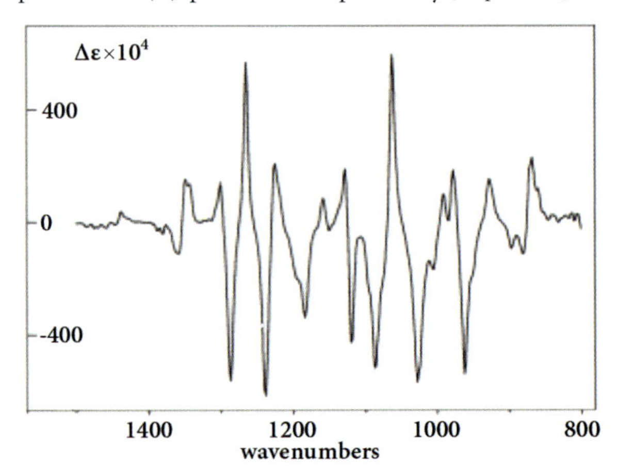

SOLUTION

The following illustration displays the experimental and calculated VCD spectra for plumericin (the latter is the Boltzmann-averaged results of all four conformations). The solid green line shows the calculated spectrum shifted by ~3% toward lower wavenumbers for easier comparison with experiment (the shifted data is plotted as a dotted line).

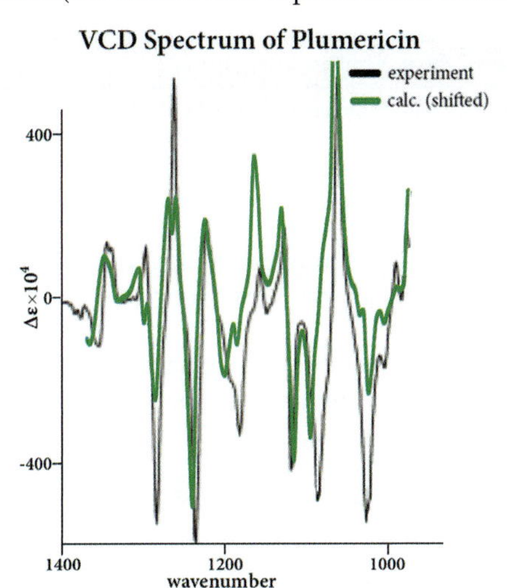

The computed spectrum agrees well with the observed one.

The following table presents the OR results for the sodium D line:

CONF.	$[\alpha]_D$
a	258.0
b	237.4
c	-29.8
d	-85.2
Average	215.3

The predicted value for $[\alpha]_D$ is in good agreement with experiment. Therefore, the OR data confirms the assignment suggested by the ECD data.

With all three techniques—ECD, OR and VCD—pointing at the same stereoisomer, we have confidence that we have assigned the absolute configuration of plumericin correctly: (1R, 5S, 8S, 9S, 10S)-(+).■

EXERCISE 8.5 DMABN EXCITED STATE GEOMETRY IN SOLUTION

Repeat the study of DMABN, this time using acetonitrile as the solvent. Predict the vertical excitation energies and optimized geometries of the LE and TICT states with our standard model and the SCRF(IEFPCM) model. How does the solvent environment affect the results as compared to the gas phase?

A note about TD-DFT frequencies. The version of Gaussian that is current as of this writing does not offer analytic TD-DFT second derivatives. Numerical TD-DFT frequency calculations can be quite lengthy for all but the smallest molecular systems. Nevertheless, for original research destined for publication, running the TD-DFT frequency calculation is not optional.

SOLUTION

We started with the optimized gas phase geometry and re-optimized in the presence of acetonitrile. This structure was then used to compute the vertical excitation energies with the TD-DFT model (again in solution). The following table summarizes our results compared to gas-phase:

STATE	EXCITATION ENERGY (eV) GAS PHASE	ACETONITRILE
LE (state 1)	4.45	4.37
CT (state 2)	4.72	4.43

The CT state is now much closer in energy to the LE state, suggesting that both states are important in understanding the observed fluorescence.

The following figure illustrates the stationary points that we located in solution (the ground state and LE excited state are not distinguishable visually):

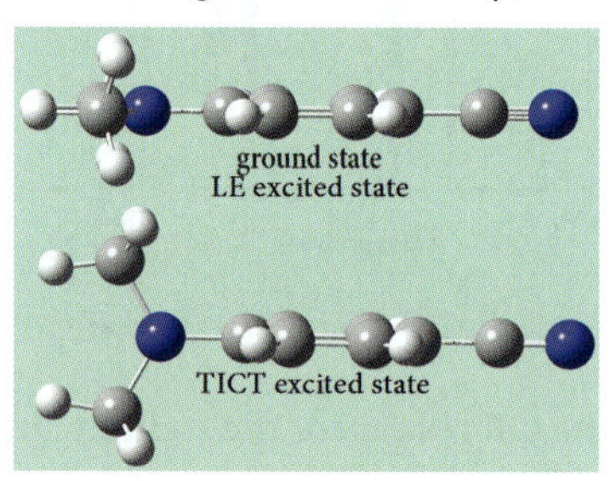

The following table gives the key structural parameters for these structures as well as their gas phase counterparts:

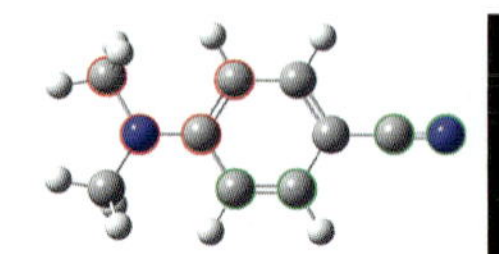

	DIHEDRAL ANGLES (°) C-C-N-C gas phase	C-C-N-C acetonitrile	C-C-C-N gas phase	C-C-C-N acetonitrile	BOND LENGTHS (Å) (ring from N, C-amide) gas phase	acetonitrile
ground state	-4.7	0.0	179.9	180.0	1.40,1.38,1.42,1.36	1.40,1.38,1.42,1.36
LE minimum	0.0	0.0	180.0	180.0	1.40,1.43,1.41,1.39	1.41,1.43,1.40,1.38
TICT minimum	90.0	90.0	180.0	180.0	1.43,1.37,1.42,1.43	1.42,1.37,1.44,1.41

The geometrical parameters indicate that the gas phase and solution structures vary only slightly in all three instances. Thus, the same trends in terms of bond lengthening in the excited states are present in solution.

The following figure shows the molecular orbitals involved in these transitions as well as the difference density plot for the TICT TD-DFT excited state. In solution, the molecular orbitals show the character of the charge transfer state even more distinctly than in the gas phase. The difference density confirms this characterization.

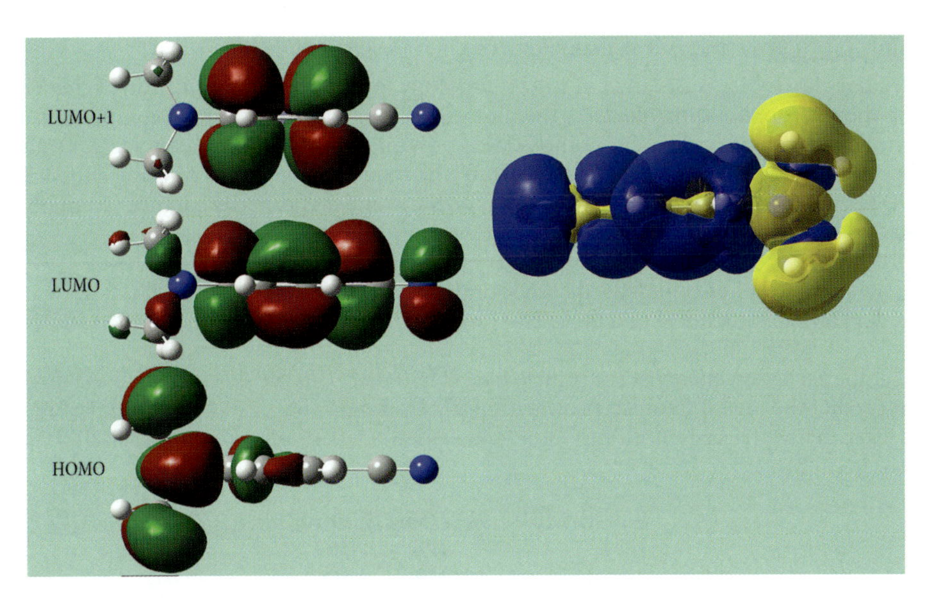

The following figure plots the energies of the predicted excited states as a function of the amino group twisting angle θ (blue and green lines). It also shows the difference in energy for each structure with respect to the ground state minimun (red line).*

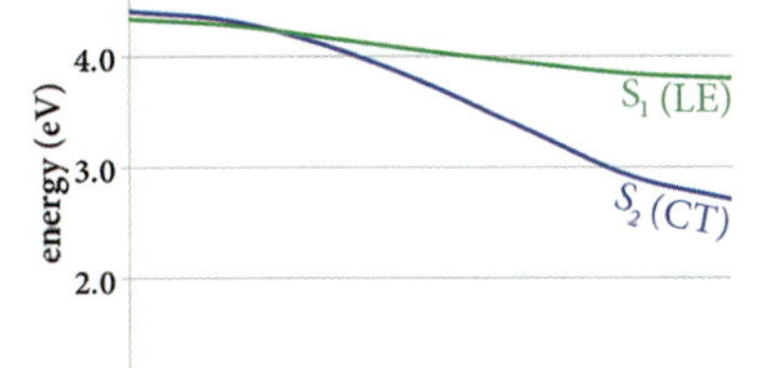

In contrast to the plot for the gas phase, there is a distinct separation between the two lowest excited states throughout the range of angles considered; the energy difference between them again increases with the twisting angle. See [Mennucci00] for more details on this result.■

GOING DEEPER: FOLLOW-ON STUDIES TO DMABN

The DMABN study discussed here can be expanded in two ways:

- The same molecule can be modeled in cyclohexane. The experimental results for this solvent find the peaks shifted to lower wavelengths with respect to acetonitrile: fluoresence is observed at ~388 nm and ~344 nm, respectively.
- The related molecule 4-aminobenzonitrile (ABN) can also be modeled. It differs from DMABN in that the methyl groups are replaced by hydrogen atoms. ABN shows a different pattern of state ordering using the gas phase and excited state geometries than DMABN. Experimentally, ABN does not exhibit dual fluorescence.

* The 0° structures are close to—but are not exactly—the minima, so their ΔE values are >0.

EXERCISE 8.6 MODELING FLUORESCENCE OF NANOFIBERS

Molecular electronics is a branch of nanotechnology focused on the fabrication of electronic components from building blocks at the molecular level. The fundamental components can be single molecules or clusters of molecules/polymers at the micrometer scale. Organic molecules having a rod-like shape have been one focus of research, including substituted quaterphenylenes. In this exercise, you will model the optical properties of three compounds from this family.

In general, quaterphenylene-based compounds have the general form X–(Ph)4–Y. We will consider the substituents H and NH_2: H–$(Ph)_4$–H, NH_2–$(Ph)_4$–H and NH_2–$(Ph)_4$–NH_2.

Predict the excitation energy of the principal peak in the UV/Visible spectrum. For that excited state, predict the lowest fluorescence wavelength. Compare your results to the experimental data listed below ([Finnerty10] citing [Vogtle91, Keegstra96, Schiek07, Schiek07a, Wallmann08]). Energy values are in nm:

Substituents	UV/Vis.	Fluoro.
H,H	298	367
NH_2,H	311	386
NH_2,NH_2	324	419

SOLUTION

We optimized the ground state structures of all three compounds, followed by a TD-DFT calculation during which we request six states. We then optimized the state with the largest oscillator strength—which was the first excited state for all three molecules—with TD-DFT in order to predict the fluorescence spectrum.

The following figure shows the optimized structures of the ground state and first excited state for NH_2–$(Ph)_4$–H:

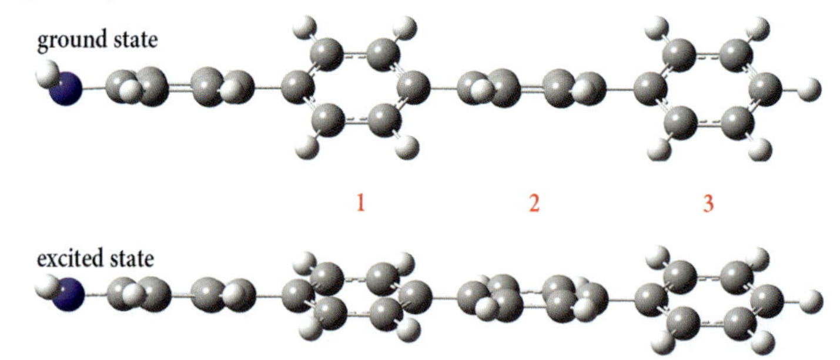

In the ground state, the structure is characterized by a twisting of every other phenyl ring: rings 1 and 3 in the figure. In the excited state, all three labeled rings are twisted out of the plane of the first (unlabeled) phenyl ring.

The following table lists the twisting angles for the three phenyl rings in question for both states of the compounds we are studying (all values are in degrees):

	GROUND STATE			EXCITED STATE		
SUBSTITUENTS	θ_1	θ_2	θ_3	θ_1	θ_2	θ_3
H,H	40.2	1.2	41.5	18.4	9.9	28.4
NH_2,H	38.2	-0.6	39.6	19.1	7.2	26.6
NH_2,NH_2	38.3	-0.3	37.8	17.6	8.6	26.2

The structures for all three compounds are similar in both the ground state and the lowest excited state. In the latter geometries, the two farthest distant phenyl rings are offset the most with respect to one another, with the other two rings taking intermediate positions.

The following table lists the predicted and observed excitation energies (in nm) for the lowest energy transitions in the UV/Visible spectrum and fluorescence spectrum for each compound:

	OBSERVED		CALCULATED	
SUBSTITUENTS	UV/VIS.	FLUORO.	UV/VIS.	FLUORO.
H,H	298	367	301	388
NH_2,H	311	386	324	404
NH_2,NH_2	324	419	326	410

These values for the excitation energies are in reasonable agreement with experiment, with the one for the NH_2,H substituents overestimated somewhat. For the fluorescence wavelengths, our standard model chemistry overestimates the values for the H,H and NH_2,H substituents and slightly underestimates the value for the third compound.

The following figure plots these predicted results along with the experimental values reported in the paper from which this exercise is drawn [Finnerty10]):

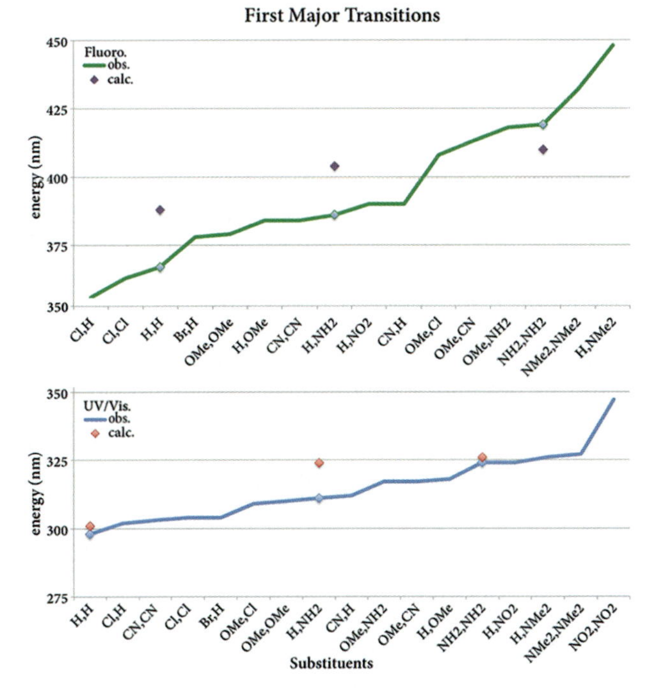

In both cases, the predicted values are in reasonable agreement with the experimental trends.■

Advanced Topics & Exercises

Franck-Condon Analysis

References: Franck-Condon
Method Definition: [Franck26, Condon26]
Implementation: [Barone09]

So far, we have focused our discussion of excited state on excitation energies. As we've seen, the quantity that is computed in a standard TD-DFT calculation is the energy change for a molecule going from some point on the ground state (GS) potential energy surface—ideally the minimum—to the point on the excited-state (XS) potential energy surface that is at the same geometry as the ground state. This is known as a *vertical transition*. The transition labeled "absorption" in the following figure provides an example, with the length of the arrow representing the excitation energy:

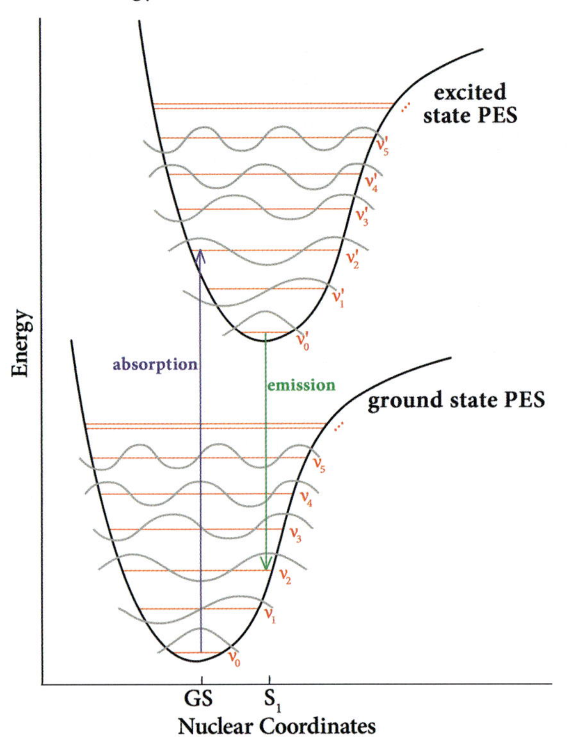

In this example, the geometry of the excited state immediately after absorption of a photon is not the minimum on its potential energy surface. Moreover, the vibrational state is not the lowest one either. Because the time scale in which electronic transitions occur is so small compared to nuclear motions, the nuclei are nearly unaffected, and the electric dipole can be considered constant during the transition. For this reason, the geometry of the excited state is initially unchanged from the ground state, and the same is true of its vibrational states. When the molecule transitions to the excited state, it moves to a vibrational state that is instantaneously compatible with its current vibrational state.

These effects are the basis of the Franck-Condon principle [Franck26, Condon26]. It states that the molecule may undergo a transition to a different, higher vibrational level in an excitation to an excited state or a relaxation back to the ground state. The probability of a vibrational transition to a particular vibrational state is proportional to the overlap integral between the initial and final vibrational states. In qualitative terms, this means that the final vibrational state is the one which most closely resembles the initial vibrational state. In the absorption example in the figure, the molecule transitions from the lowest vibrational state at the ground state geometry to the third vibrational state as it absorbs a photon and enters the first excited state.

Shortly thereafter, the geometry will relax to the minimum on the excited state potential energy surface (labeled "S_1" in the diagram). The molecule may then relax back to the ground state, emitting a photon in the process. This process will once again happen essentially instantaneously with respect to nuclear motion, and it may also involve a vibronic transition. The arrow labeled "emission" in the preceding figure provides an example. Note that the Franck-Condon principle applies to both absorption and emission.

This description, while more nuanced than our initial descriptions of excitation, is still a simplification. In reality, a sample present in a spectrometer at room temperature will have molecules with energies distributed among the various vibrational levels as dictated by the Boltzmann distribution formula. Therefore, the actual energy transition observed in the experiment is actually a series of many energy transitions from the various vibrational levels of the ground state to the various vibrational levels of the excited state. The strength of each individual transition will depend on the population of that state (which is temperature dependent) and the overlap between the ground and excited vibrational wavefunctions for that transition. The following figure captures some of this complexity:

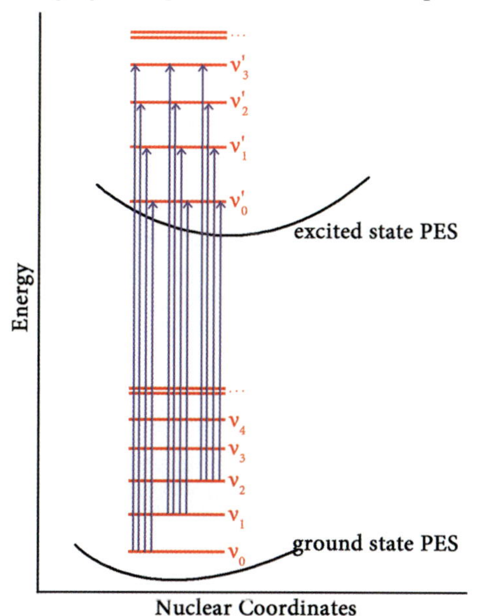

This figure shows the potential transitions for the four lowest vibrational states, considering only the same vibrational states for the excited state. The effect on the observed spectrum is to give the peak associated with an electronic transition additional structure in the form of sub-peaks: many lines instead of only a single absorption line. The Franck-Condon analysis features in Gaussian can be used to model this phenomenon.* The keyword corresponding to this method is **Freq=FC**; it is used in conjunction with the **TD** method. The following example will acquaint you with this feature and its use, and the subsequent exercises will allow you to explore it for yourself.

* Gaussian also provides the related Herzberg-Teller and Franck-Condon-Herzberg-Teller analyses, (beyond the scope of this book).

ADVANCED EXAMPLE 8.7 FRANCK-CONDON ANALYSIS: A UV ABSORPTION SPECTRUM

Here is the observed UV/Visible spectrum for trans,trans-1,4-diphenyl-1,3-butadiene (DPB), collected in cyclohexane:

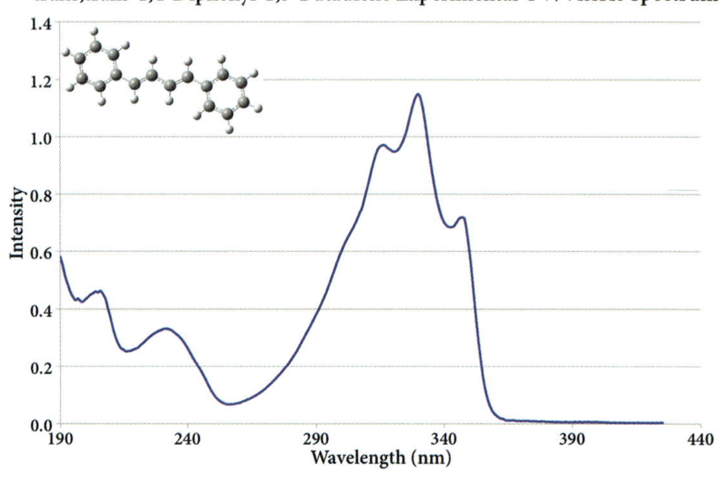

We will begin this study by optimizing the ground state structure and then examining the vertical excited states to identify the state of interest for the Franck-Condon calculation. Using this analysis technique, we can only simulate one region of the spectrum at a time.* We will focus on the 260 to 360 nm region of the spectrum and assume that only one excited state contributes significantly to the absorption recorded. A TD-DFT calculation will give us a preliminary sense of the excited states predicted, their strengths and their characters.

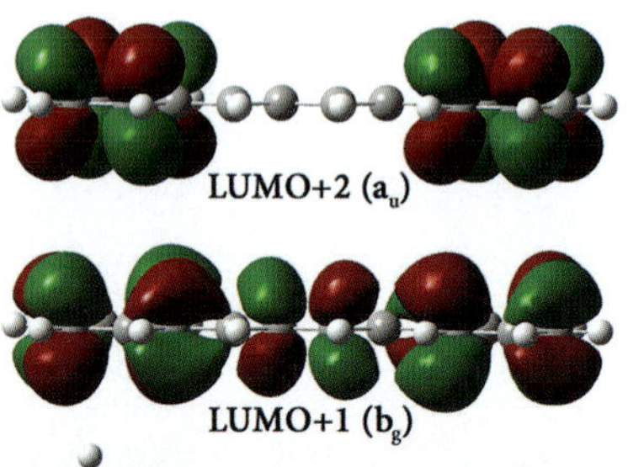

HOMO (b_g)

When we use the CAM-B3LYP/TZVP model chemistry and compute the TD-DFT excited states, we find two excited states with non-zero oscillator strengths:

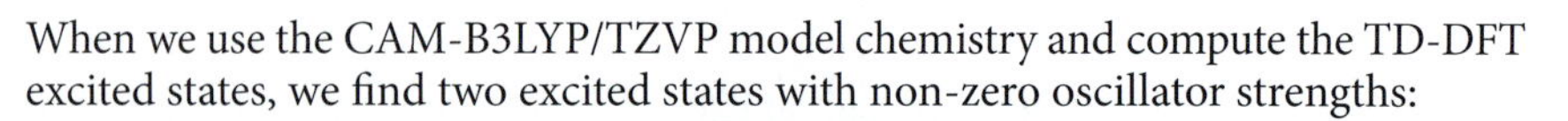

STATE	SYMM.	EX. ENERGY	f	CHARACTER	PRINCIPAL TRANSITION
1	^1B_u	340 nm	1.7	strong	HOMO→LUMO
2	^1A_g	256 nm	0.0	forbidden	HOMO→LUMO+1
3	^1B_u	254 nm	0.007	weak	HOMO→LUMO+2

All involve transitions among the π orbitals of the molecule.

The weaker, lower wavelength peak corresponds to the absorption seen in the experiment at 230 nm. The stronger, higher wavelength peak corresponds to the 260 nm to 360 nm region in the spectrum.

The additional structure seen in the larger peak is due to the individual transitions involving multiple vibrational levels of both the ground and excited states. The ordinary TD calculation cannot simulate the actual absorption spectrum. To do so, we will employ a Franck-Condon analysis.

Performing a Franck-Condon analysis involves multiple calculations:

* For molecules with excited states having overlapping spectral regions, the separate analyses for the various states/regions can be added to simulate the entire spectrum.

- We began with an optimization plus frequency job for the ground state. We will use the B3LYP functional with the TZVP basis set. We choose this functional since we plan to use the CAM-B3LYP functional for the excited state calculations (CAM-B3LYP [Yanai04] includes some long range corrections which are helpful for excited state calculations). The TZVP basis set was chosen as the best tradeoff of accuracy and cost.
- A TD-DFT calculation at the ground state geometry was performed next (as a second job step) in order to identify the excited state of interest. In this case, it is the first excited state. This calculation used the CAM-B3LYP/TZVP model chemistry.
- An optimization plus frequency job for the excited state was the next calculation. It included the **Freq=SaveNormalModes** option, which saves information that the Franck-Condon analysis will require in the checkpoint file. This calculation used the CAM-B3LYP/TZVP model chemistry, and it specified a different checkpoint file from the first job.
- The final calculation performs the Franck-Condon analysis, using the results of the two frequency calculations. It was performed using the B3LYP/TZVP model chemistry.

Here is the input file for the Franck-Condon calculation:

```
%Chk=DPB_gs.chk                                          Ground state checkpoint file
# B3LYP/TZVP Freq(ReadFC,FC,ReadFCHT) NoSymm                          Route section*
  SCRF(Solvent=Cyclohexane) Geom=Checkpoint Guess=Read

FC simulation of UV absorption spectrum for                            Title section
trans,trans-1,4-diphenyl-1,3-butadiene

0 1                                                        Charge & spin multiplicity

SpecHwHm=400 SpecRes=20 InpDEner=0.133        Specifications for the FC analysis

DPB_ex.chk                                              Excited state checkpoint file
                                                                     Final blank line
```

As indicated in the input file, we ran all calculations in solution with cyclohexane.

The **Freq** keyword has three options. Despite its name, the **Freq=ReadFC** option is not related to Franck-Condon analysis. It says to retrieve the **F**orce **C**onstants computed by a previous frequency job from the checkpoint file for use in the current calculation. We have seen this option before in the context of geometry optimizations.

The **Freq=FC** option says to perform the Franck-Condon analysis, and **ReadFCHT** says to read an additional input section of keywords which specify the parameters of the analysis; it appears following the charge and spin multiplicity line and is terminated by a blank line. The **FC** option requires the names of the checkpoint files for the ground state and the excited state. One is specified via the **%Chk** directive (or **%OldChk**), and the second is identified in an additional input section. When both **FC** and **ReadFCHT** are specified, the input section for the latter precedes the name of the second checkpoint file.

In this input file, the **ReadFCHT** parameters have the following meanings:

SpecHwHm The half-width at half-maximum value of the Gaussian function used for the spectra band of each vibration-vibration transition (in units of cm^{-1}).

* Technically, the method (**B3LYP**), basis set (**TZVP**), solvent (**SCRF**) and **Guess** keywords are all ignored in a **Freq=ReadFC** job because it proceeds directly from link 202 directly to link 716 (i.e., no energy or energy derivative calculation is done). However, we include them here for clarity.

SpecRes=*n* Compute a value in the spectrum every n cm^{-1}.

1 hartree = 27.21 eV
= 45.56 nm
= 219474.6 cm^{-1}

InpDEner Specify the value for the energy difference (in atomic units) between the initial and final states: here, between the ground state and the excited state. This is the adiabatic excitation energy: the difference between the energies of the ground and excited states at their respective equilibrium geometries.

We chose the **SpecHwHm** value of 600 cm^{-1} because it is a good match based on the resolution of our spectrometer. We picked the **InpDEner** value of 0.133 after examining the data and finding that it would produce a better overlap with the experimental spectrum. If you omit this parameter, the theoretical spectrum would be translated somewhat with respect to experiment.

The output of the Franck-Condon calculation includes details about the various vibration-vibration transitions as well as the data for the spectrum. The latter appears as follows in the output:*

```
 ==================================================
                    Final Spectrum
 ==================================================

 The broadening of spectral lines is done by Gaussian
 functions with a Half-Width at Half-Max of 600.0 cm^-1.
 Axis X = Energy (in cm^-1)
 Axis Y = Molar absorption coeff (in dm^3.mol^-1.cm^-1)
          Energy (cm-1)              Intensity (T=0K)
 -------------------------------------------------------
       27589.6643          0.622302D+04   Convert D → E for Excel etc.
       27609.6643          0.676358D+04
       27629.6643          0.734054D+04
       27649.6643          0.795510D+04
       ...
```

Convert D to E

Many mass market software packages, including Excel, will not accept the "D±nn" double precision scientific notation. You will need to replace D with E before importing the data.

Neither of the graphic programs we are considering currently includes features for plotting Franck-Condon predicted spectra, so this must be done manually in a spreadsheet or graphing package. When we extracted and plotted the calculated Franck-Condon spectrum data, we obtained the following results:

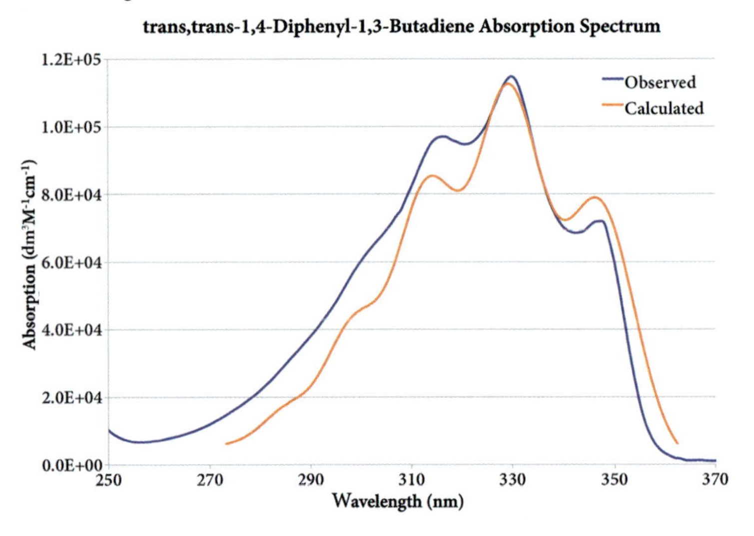

*The output format varies slightly in different Gaussian revisions.

We should not have to scale the absorbance values of our experiment to match the units used by Gaussian to report the spectrum. The scale factor was chosen such that the largest peak is the same intensity as in the theoretical spectrum. We also converted the X-axis of the calculated data to the units used by the instrument. As the plot indicates, the Franck-Condon analysis produces excellent agreement with the observed spectrum.■

Ionization Spectra and Franck-Condon Analysis

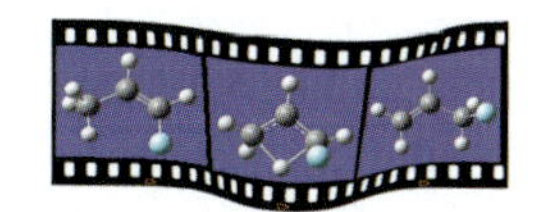

The Franck-Condon analysis applies equally well to studying ionization spectra. Here the "ground state" is a neutral species, and the "excited state" is the cation formed by removing an electron.

ADVANCED EXERCISE 8.7 FRANCK-CONDON ANALYSIS: ACROLEIN

The peak corresponding to the first excited state in the predicted UV/Visible spectrum for acrolein is plotted in the margin. It has a very small oscillator strength and appears as only a tiny blip in the plot.

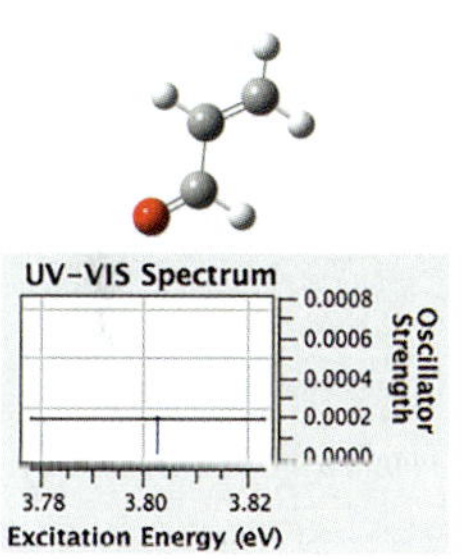

Perform a Franck-Condon analysis for this system to predict the UV/Visible absorption spectrum. Use the B3LYP DFT method for the ground state calculations and the CAM-B3LYP DFT method for the excited state calculations, in conjunction with the TZVP basis set in both cases.

The original study from which this exercise is drawn [Barone09] specified an fwhm value of 1000 cm^{-1} (equivalent to a half-width at half-maximum value of 500 cm^{-1}). You must specify the energy difference between the two states because two different computational methods are used for the ground and excited states, which means that the direct energy difference is not meaningful. Doing so causes the resulting plot to be properly placed in the energy region corresponding to this peak.

SOLUTION

After optimizing the ground state, the TD-DFT calculation verified that the first excited state was the one of interest. It is an n→π* transition, HOMO to LUMO (the orbitals are pictured in the margin).

We next ran an optimization plus frequency job for the first excited state. The predicted excitation energy at the optimized geometry is 3.2 eV, with an oscillator strength of 0.0002. Using this information, we ran a Franck-Condon analysis with the following parameters: **SpecHwHm=500 InpDEner=0.118**. Here is predicted spectrum:

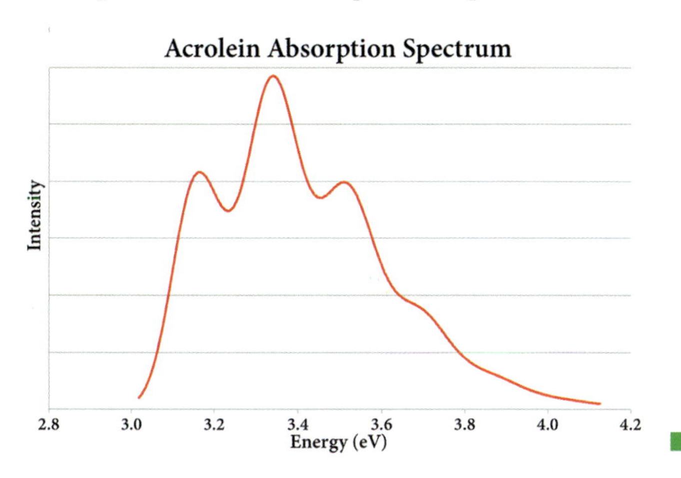

■

GOING DEEPER: ACROLEIN ABSORPTION SPECTRUM IN WATER

The original researchers [Barone09a] noted that a blue shift had been observed for this transition in going from the gas phase to aqueous solution [Brancato06]. This study can be repeated in solution with water and the result compared to the preceding gas phase results.

ADVANCED EXERCISE 8.8 ABSORPTION SPECTRUM OF ANOTHER DIPHENYL COMPOUND

Predict the UV/Visible absorption spectrum for all trans-1,8-diphenyl-1,3,5,7-octatetraene (DPO). Here is its experimental spectrum, collected in cyclohexane, shown in green below. The original 1,4-diphenyl-1,3-butadiene is shown in blue for comparison.

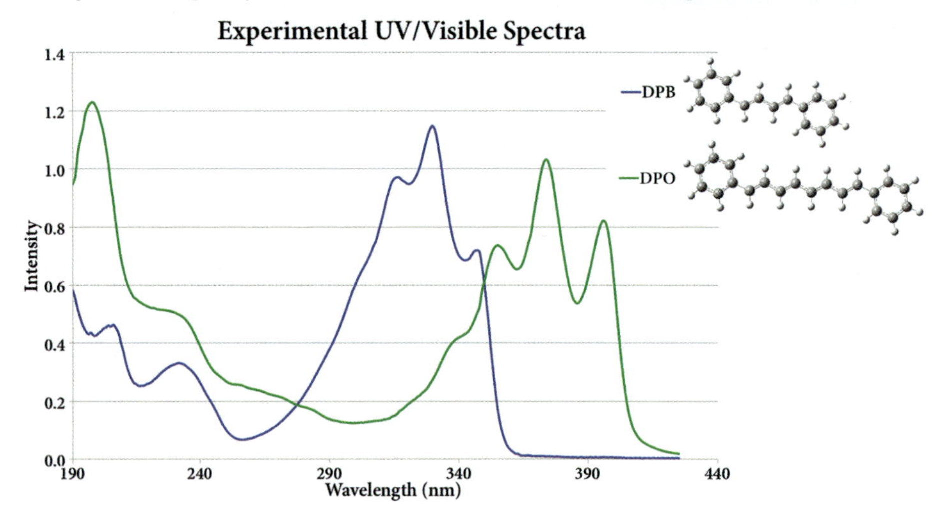

Your study will follow the same pattern as in Advanced Example 8.7. Use the same model chemistries as in the example: B3LYP/TZVP for the ground state and Franck-Condon calculations, and CAM-B3LYP/TZVP for the TD-DFT excited state calculations. The two spectra were obtained on the same instrument, so the values for the **SpecHwHm** and **SpecRes** parameters should be the same for this molecule as for DPB. You will need to determine an appropriate value for **InpDEner**.

SOLUTION

To account for the shift to longer wavelength, we used a value of 0.113 for the **InpDEner** parameter in the Franck-Condon calculation.

The following plot shows the predicted absorption spectrum for this compound in comparison with experiment:

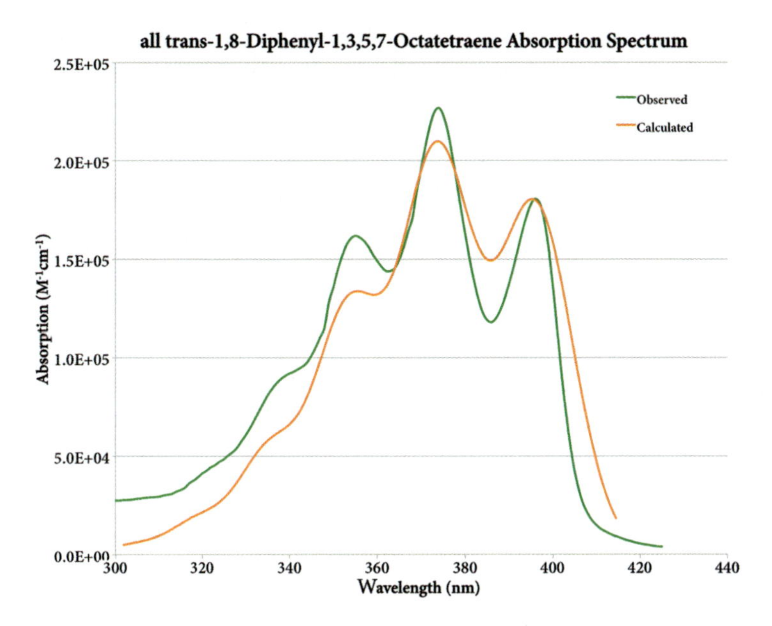

Notice the difference in the fine-structure between DPO and DPB. There are still three dominant peaks, with the central one being the strongest. However, the first peak—the one with the longest wavelength—is the second strongest for DPO while it is the least strong for DPB. This is in perfect agreement with the experiment.■

Modeling Emission with State-Specific Solvation

Thus far, we have modeled excited states in solution by the simple inclusion of the standard SCRF keyword. This defaults to non-equilibrium solvation unless the excited state geometry is being optimized in which case equilibrium solvation is implied. In both cases, however, the reaction field is defined using the ground state electron density. In this section, we will explain these terms more fully and show how to properly include the effects of the excited state's electron density on the reaction field.

SCRF continuum models like IEFPCM are referred to as *equilibrium* solvation models, a name which foregrounds the fact that they model the solvent in equilibrium with the solute. Therefore, processes which occur on longer timescales than the time required for the solvent to equilibrate to the introduction of the solute are treated as equilibrium processes with respect to solvation. In contrast, for phenomena which occur on a faster time scale than the solvent-solute equilibration—such as the electronic excitations we are considering in this chapter—the solvent molecules can be considered fixed (as with the nuclear positions).

This means that a solute molecule that has transitioned to an excited state will see a solvent reaction that has only partially responded to the change in the solute's electron density, while part of the reaction field is still in equilibrium with the originating state. Modeling such scenarios is referred to as *non-equilibrium* solvation. In order to properly model both equilibrium and non-equilibrium solvation, the implementation of the solvation model is separated into discrete parts which interact with the solute appropriately with respect to the differing timescales [Cossi00].* In short, the terms "equilibrium solvation" and "non-equilibrium solvation" refer to the differing time dependence of solvent polarization in

* This description has benefited greatly from [Cramer04]. For a detailed discussion of equilibrium vs. non-equilibrium solvation and related topics, see section 11.4.6 (pp. 421-22) and the follow-on text in chapters 14 and 15.

response to changes in the solute. In non-equilibrium solvation, only the solvent reaction field's rapid, electronic changes respond to the change in the solute's electron density, while in equilibrium solvation, the slower, orientational changes are also allowed to respond to changes in the solute's electron density.

Another term which occurs frequently in discussions of modeling excited state properties is *state-specific* (SS); it refers to any calculation which focuses on modeling the properties of a specific excited state. Such calculations involve the relaxation of the ground state density with respect to the specific excitation under investigation; as a result, the predicted excitation energy will reflect the updated density. The molecular orbitals are also altered from the ground state reference as a side effect.

When a molecule in an excited state emits a photon and thereby relaxes back to its ground state, the process is again instantaneous compared to nuclear motion, including within the solvent molecules. Therefore, the ground state after emission will both retain the geometry of the excited state and see a solvent reaction field that only partially responds to the change in solute's electron density. We need to use the Gaussian SCRF facility's *external iteration* (EI) procedure [Improta06, Improta07] to run a non-equilibrium solvation calculation on the ground state to model this phenomenon.

A non-state-specific non-equilibrium solvation calculation—such as a TD calculation in solution computing the excitation energies for several excited states—is also known as a *linear response* (LR) calculation, reflecting the fact that the default IEFPCM method uses linear response theory.

To summarize, for the absorption process, there are two approaches:

- LR: appropriate for looking at solvent effects on bright absorptions (excitations with relatively large oscillator strengths). This approach consists of running a TD energy calculation in solution on a ground state reference (with the latter having previously been optimized in solution).
- EI: appropriate for modeling solvation effects on dark absorptions (excitations with relatively small oscillator strengths). For these cases, the LR approach would not recover the solvation effects on the absorption energies. Instead, it is necessary to perform a more complex, two-part solvation calculation using the EI facility (described later).

For the emission process, only the external iteration approach is appropriate. We perform a non-equilibrium solvation calculation on the ground state, employing a solvent reaction field that keeps the slow component in equilibrium with the excited state density but allows the fast component to respond to the change back to the ground state.

The whole point of the availability of two different approaches to absorptions is that some transitions would be better modeled by LR and some other EI. A typical example is a charge transfer transition characterized by a small intensity. Since the LR approach depends on the transition dipole (to which the intensity is proportional), the LR solvent effect would grosssly underestimate the change in solute-solvent interaction resulting from the change in the dipole moment from ground state to excited state. In contrast, this effect would be taken into account by an EI treatment.

The following diagram illustrates the cycle of excitation and emission for the first excited state of coumarin 153 and identifies the principal Gaussian calculation required to model each component in green. We will study this cycle in the upcoming example, where full details of the required calculations will be discussed.

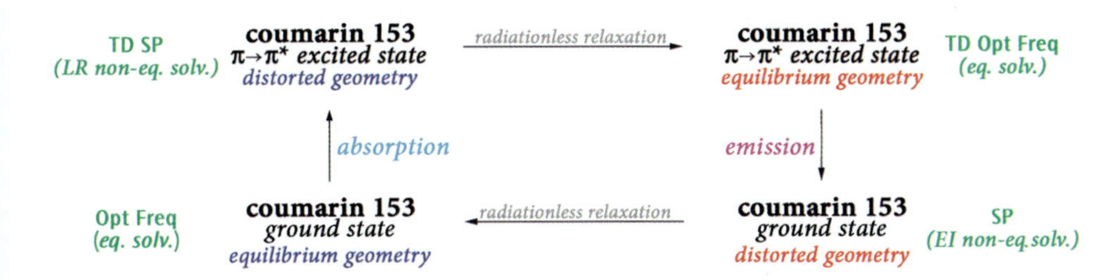

The following Gaussian keywords will be needed for studying emission in solution:

- The **SCRF=External Iteration** option computes the energy in solution by making the solute's electrostatic potential self-consistent with the solvent reaction field. When used for a non-equilibrium solvation process, self-consistency is achieved with only the fast component of the solvent polarization, a solvent reaction field calculated previously must be provided as input for the calculation (see the next bullet). When used for an equilibrium solvation process, self-consistency is achieved with both fast and slow components of solvent polarization.

- **SCRF=Read** says to read SCRF-specific parameters from the input file. They are specified in a separate input section following the molecule specification (terminated by a blank line).

  For studies incorporating state-specific solvation, the **NonEq** SCRF additional input keyword will be used to save/retrieve a computed reaction field to/from the checkpoint file. **NonEq=Write** says to save the reaction field computed during an SCRF calculation to the checkpoint file at the completion of the job. In a subsequent job, **NonEq=Read** will be similarly specified in order to read data for the previously computed reaction field for use in the current calculation.

ADVANCED EXAMPLE 8.8 STUDYING FLUORESCENCE IN COUMARIN 153

Coumarin 153 (C153)* is a dye that exhibits strong fluorescent emission. It is a well-studied molecule, especially for studying the influence of environment of fluorescence properties. Its structure and HOMO and LUMO are given in the figure following:

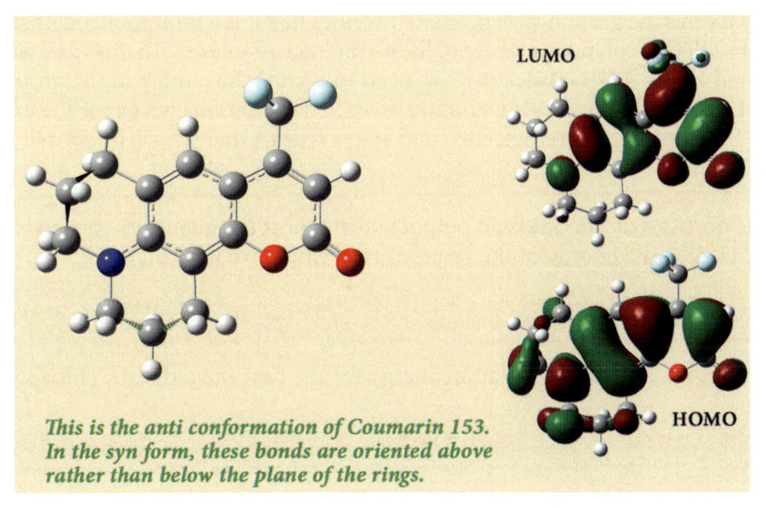

This is the anti conformation of Coumarin 153. In the syn form, these bonds are oriented above rather than below the plane of the rings.

* IUPAC: 9-(trifluoromethyl)-2,3,6,7-tetrahydro-1H,5H,11H-pyrano[2,3-f]pyrido[3,2,1-ij]quinolin-11-one.

The *syn* conformation is slightly energetically favored in the gas phase; in most solution environments, the energy gap between the conformers decreases to near zero. Accordingly, we will consider the two conformers to exist in equal proportions.

The first excited state of this compound is a $\pi \rightarrow \pi^*$ transition involving the HOMO and LUMO. [Improta06] describes the nature of the transition in this way:

> While the HOMO is delocalized on the whole molecule, with significant contribution by the π orbitals of the "central" benzenic ring and of the nitrogen atom, the LUMO is mainly localized on the "quinonelike" terminal ring with significant contribution of the π^* orbital of the carbonyl group. As a consequence the $S_0 \rightarrow S_1$ transition has a partial intramolecular charge transfer character (from the nitrogen atom to the carbonyl group), and S_1 has a partial zwitterionic character (with the nitrogen atom and the oxygen of the carbonyl group bearing formal positive and negative charges, respectively).

We will model the entire cycle of excitation/absorption and emission for this compound in solution with cyclohexane. We will run the following sequence of calculations:

JOB	DESCRIPTION	OPT. GEOMETRY	ENERGY RESULT	LOCATION IN OUTPUT
1	Step 1: ground state **Opt Freq SCRF**	yields ground state	ground state energy (E_{GS_0})	SCF Done
	Step 2: **TD(NStates=*k*) SCRF**	ground state	non-eq. solv. vertical excitation energy	Excited State *n*:
2	*State-specific solvation of absorption:*			
	Step 1: **SCRF(Read)** *with* **NonEq=Write**	ground state		
	Step 2: **TD(Root=*n*) SCRF(Read,ExternalIter)** *with* **NonEq=Read**	ground state	excited state energy (E_{XS_0})	After PCM corrections
3	**TD(Root=*n*) Opt Freq SCRF**	yields excited state		
4	*State-specific solvation of emission:*			
	Step 1: **TD(Root=*n*) SCRF(Read,ExternalIter)** *with* **NonEq=Write**	excited state	excited state energy (E_{XS_1})	After PCM corrections
	Step 2: **TD(Root=*n*) SCRF(Read)** *with* **NonEq=Read**	excited state	ground state energy (E_{GS_1})	SCF Done

There are seven calculations organized as four Gaussian jobs (counting an optimization plus frequency as a single calculation). We will perform the sequence for both conformations.

The first job locates the ground state geometry, verifies that it is a minimum and then performs an initial TD-DFT calculation in order to locate the state of interest. In this case, we will solve for six excited states: **TD(NStates=6)**. We need to record the energy of the ground state at the completion of the optimization and the predicted excitation energy for the excited state of interest. Examining the predicted excited states reveals that we are interested in the first excited state.

Here are the portions of the Gaussian output which report the quantities we need for the *anti* conformer. Here is the energy of the ground state, which we will call E_{GS_0}:

```
SCF Done: E(RAPFD) =  -1122.3799184 A.U. after 2 cycles
```

Here is the predicted vertical excitation energy for the first excited state: (LR approach)

```
Excited State   1:  Singlet-A 3.1768 eV … f=0.4740 …
```

The second job performs a non-equilibrium solvation analysis for the absorption energy, attempting to improve upon the predicted vertical excitation energy from the preceding TD-

DFT calculation. We include this calculation here for completeness, but we would only need to do it to recover the solvation effects on absorption energies of dark transitions.

The input file for the calculation looks like this:

```
%OldChk=e8_08_anti_chex_1                  Contains the optimized GS geometry
%Chk=e8_08_anti_chex_2
# APFD/6-311+G(2d,p) SCRF=(Solvent=Cyclohexane,Read)
  Geom=Check Guess=Read

Coumarin/anti/chex: prepare for EI non-eq solvation by
saving the solvent reaction field in equilibrium with the
ground state density

0 1

NonEq=write      Input for SCRF=Read: save the GS reaction field to the checkpoint file

--link1--
%Chk=e8_08_anti_chex_2
# TD(NStates=6,Root=1) APFD/6-311+G(2d,p) Geom=Check
  SCRF=(Solvent=Cyclohexane,Read,ExternalIter) Guess=Read

Coumarin/anti/chex: read solvent rx field from ground
state equilibrium solv. calculation and compute energy
of the first excited state with the EI method and non-
equilibrium solvation

0 1

NonEq=read       Input for SCRF=Read: retrieve & use the reaction field in the checkpoint file
                                                                  final blank line
```

Both jobs include SCRF parameters in a separate input section following the molecule specification (which here consists of just the charge and spin multiplicity). In the first job step, the SCRF input says to save a portion of the computed reaction field to the checkpoint file; in the second job step, that data is retrieved from the checkpoint file and used in the current calculation. Note that the option to the **SCRF** keyword is **Read** in both cases, which simply says to read the Gaussian input file for SCRF additional input; the actual operations are specified by the separate input section's contents.

At the conclusion of the second job step, the predicted energy computed by the IEFPCM method including the read-in non-equilibrium reaction field is reported in the output as follows:

```
After PCM corrections, the energy is -1122.26793841 a.u.
```

This output section appears at the very end of the log file.

The difference between this energy—we'll call it E_{XS_0}—and E_{GS_0} gives the absorption energy for the first excited state for this conformation including the non-equilibrium external iteration treatment of solvation. We'll look at the computed value below.

The third job performs an optimization plus frequency calculation for the excited state in order to locate its minimum energy structure.

The fourth and final job performs the non-equilibrium solvation modeling of the emission, by saving a portion of the reaction field computed in equilibrium with the excited state and applying it in the ground electronic state calculation. Here is the input file:

```
%oldchk=e8_08_anti_chex_3              Contains the optimized excited state geometry
%chk=e8_08_anti_chex_4
# TD(NStates=6,Root=1,Read) APFD/6-311+G(2d,p)
  SCRF(Solvent=Cyclohexane,ExternalIter,Read)
  Geom=Check Guess=Read

Coumarin/anti/chex: EI calc. on first excited state; save
the solvent reaction field for use in non-eq. solvation
calc. on the ground state

0 1

NonEq=write      Input for SCRF=Read: save excited state reaction field to checkpoint file

--link1--
%chk=e8_08_anti_chex_4
# APFD/6-311+G(2d,p) SCRF(Solvent=Cyclohexane,Read)
  Geom=Check Guess=Read

Coumarin/anti/chex: read reaction field from excited state
and compute energy of the ground state with it

0 1

NonEq=read       Input for SCRF=Read: retrieve & use the reaction field in the checkpoint file
                                                                             final blank line
```

The **SCRF=Read** keyword and its parameters function in the same way as in the previous EI solvation job for the absorption.

The predicted energy from the first job step—E_{XS_1}—appears as follows:

```
After PCM corrections, the energy is -1122.27556274 a.u.
```

and the predicted energy from the second job step—E_{GS_1}—is:

```
SCF Done: E(RAPFD) =  -1122.37211556 A.U. after 2 cycles
```

The following table summarizes the results of these calculations and compares them with experiment (unless otherwise labeled, energies are in hartrees):

	ABSORPTION		EMISSION	
ENERGY	ANTI	SYN	ANTI	SYN
ground state	-1122.26794	-1122.26799	-1122.27556	-1122.27552
excited state	-1122.37991	-1122.38007	-1122.37212	-1122.37231
difference	0.11197 3.047 eV	0.11208 3.050 eV	0.09655 2.627 eV	0.09679 2.634 eV
experiment	3.22 eV[a]		2.26 eV[b]	

[a][Horng95]; [b][Kim12] measured in ethanol

Let's take a moment to review which calculations the various raw energy values came from.

- All calculations summarized here used the **SCRF=Read** option.
- Values indicated with blue shading were computed with the **NonEq=Write** SCRF additional input: the relevant portion of the reaction field from the job was saved to the checkpoint file.
- Values indicated with red shading were computed with the **NonEq=Read** SCRF parameter: the reaction field data previously written to the checkpoint file was retrieved and applied to the indicated calculation.
- All excited state energy values were computed with the **SCRF=ExternalIter** option.

We can compare the predicted EI absorption energy to the corresponding one from the LR approach (i.e., as computed by the ordinary TD-DFT calculation): 3.18 eV for absorption. This is a bright transition, which means that the LR approach is adequate, and the predicted energy is closer to the experimenal value than the external iteration absorption energy. The external iteration emission energy is in reasonably good agreement with experiment (although the experimental value is in a different solvent). Note that the excitation energy reported at the end of the excited state geometry optimization (2.78 eV) is an energy difference between ground excited state. However, it does not adequately model emission, because the ground state energy is not computed in a non-equilibrium solvation process. The external iteration emission energy is a clear improvement over such energy.

For this problem, conformational averaging has only a small effect on the values of the predicted quantities the two conformations. In the ground state, the conformer populations are 55% *syn* and 45% *anti*. In the excited state, these become 37% *syn* and 63% *anti*.

EI solvation analysis can play a key role in modeling the excited state characteristics in varying solvent environments. This topic will be explored in detail for coumarin 153 in Advanced Exercise 8.10.■

ADVANCED EXERCISE 8.9 ACETALDEHYDE ABSORPTION & EMISSION

Predict the absorption and emission energies for the first excited state of acetaldehyde in ethanol. This state is known to be an $n \rightarrow \pi^*$ transition.

Compare your results with experiment:

- absorption energy: 287.5 nm [NIST08] citing [Fillet56]
- emission energy: 338.2-500 nm with a broad maximum at 405-420 nm [Parmenter63]

SOLUTION

We ran the following series of jobs on acetaldehyde:

Job 1: Ground state geometry optimization; initial TD-DFT calculation

```
%chk=x8_09_gs+td
# APFD/6-311+G(2d,p) Opt Freq SCRF(Solvent=Ethanol)

Acetaldehyde ground state opt freq

0 1
```

molecule specification

final blank line for this job step

```
--link1--
%Chk=x8_09_gs+td
# APFD/6-311+G(2d,p) TD(NStates=6) SCRF(Solvent=Ethanol)
  Geom=Check Guess=Read

Acetaldehyde: Equilibrium solv. vertical excited states

0 1
```
final blank line

Job 2: Absorption energy in solution

```
%OldChk=x8_09_gs+td                                   Use optimized GS geometry
%Chk=x8_09_ss_abs
# APFD/6-311+G(2d,p) SCRF(Solvent=Ethanol,Read)
  Geom=Check Guess=Read

Acetaldehyde: Prepare for non-eq solvation by saving the
solvent reaction field from the ground state

0 1

NonEq=write                  Input for SCRF=Read: save GS reaction field
                                     final blank line for this job step
--link1--
%chk=x8_09_ss_abs
# APFD/6-311+G(2d,p) TD(NStates=6,Root=1) Geom=Check
  Guess=Read SCRF(Solvent=Ethanol,ExternalIter,Read)

Acetaldehyde: Read non-eq solvation from ground state and
compute energy of the first excited state w/ EI solvation

0 1

NonEq=read      Input for SCRF=Read: retrieve & use saved reaction field
```
final blank line

Job 3: Excited state optimization

```
%chk=x8_09_tdopt
# APFD/6-311+G(2d,p) TD=(Read,NStates=6,Root=1) Opt Freq
  SCRF(Solvent=Ethanol)

Acetaldehyde: Excited state optimization
Modify GS geometry to break Cs symmetry

0 1
```
modified molecule specification

final blank line

Job 4: Emission energy in solution

```
%OldChk=x8_09_tdopt                       Use optimized excited state geometry
%Chk=x8_09_ss_emiss
# APFD/6-311+G(2d,p) TD=(Read,NStates=6,Root=1) Geom=Check
  Guess=Read SCRF(Solvent=Ethanol,ExternalIter,Read)

Acetaldehyde emission in solution at first excited state
optimized geometry

0 1
```

```
NonEq=write
```
Input for SCRF=Read: save EI reaction field
final blank line for this job step

```
--link1--
%chk=x8_09_ss_emiss
# APFD/6-311+G(2d,p) SCRF=(Solvent=Ethanol,Read)
Geom=Check Guess=Read

Acetaldehyde: GS non-equilibrium at excited state geom

0 1

NonEq=read
```
Input for SCRF=Read: retrieve & use EI reaction field
final blank line

In the third job, it is necessary to break C_s symmetry in the starting structure for the excited state optimization since its symmetry is A″. This can be accomplished by modifying the two dihedral angles indicated in green in the figure in the margin:

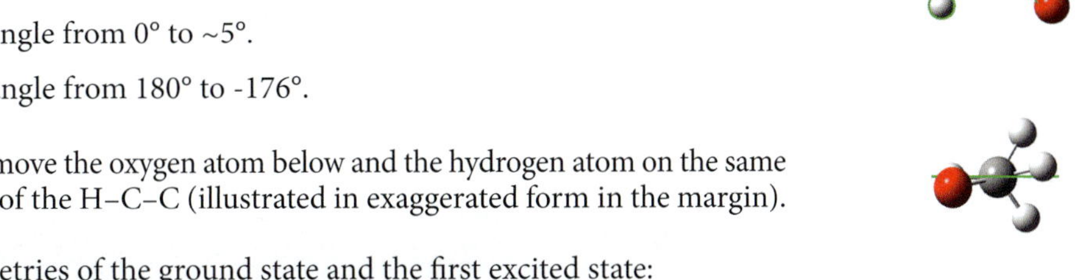

- Change the H–C–C–O angle from 0° to ~5°.
- Change the H–C–C–H angle from 180° to -176°.

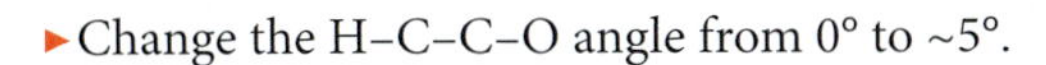

Note that these modifications move the oxygen atom below and the hydrogen atom on the same carbon atom above the plane of the H–C–C (illustrated in exaggerated form in the margin).

Here are the optimized geometries of the ground state and the first excited state:

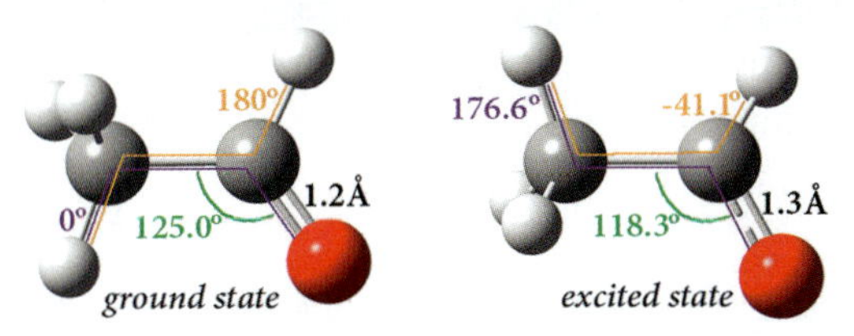

The following table summarizes the energy results from our acetaldehyde calculations:

	JOB	DESCRIPTION	E (hartrees)	EX. ENER. (nm)
ABSORP.	1	optimized ground state energy eq. solvation absorption energy	-153.75748	280.1
	2	excited state energy incl. SS solvation	-153.59169	274.8
EMISSION	3	eq. solvation emission energy		374.4
	4	excited state energy (non-eq. solvation) ground state energy incl. SS solvation	-153.61030 -153.72882	384.4

The computed values for the absorption and emission energies are in very good agreement with experiment. The differences in their predicted emission energies again demonstrates the importance of non-equilibrium solvation for modeling this phenomenon.■

ADVANCED EXERCISE 8.10 COUMARIN 153 EMISSION IN DMSO

Predict the absorption and emission energies for the $\pi \rightarrow \pi^*$ transition in coumarin 153 in DMSO, following the procedure in Advanced Example 8.8. How well do your predicted values agree with experiment? What trends do you observe across different solvent environments?

Here is the experimental data:

- Absorption energy (eV): cyclohexane: 3.22, DMSO: 2.94 [Horng95]; ethanol: 2.92 [Kim12]
- Emission energy (eV): ethanol: 2.26 [Kim12]

SOLUTION

The following table lists the energy results of our calculations in DMSO (energies are in hartrees unless otherwise indicated):

JOB	ENERGY	ANTI	SYN
1	energy of optimized ground state eq. solvation absorption energy	-1122.38713 3.039 eV	-1122.38724 3.045 eV
2	excited state energy incl. EI solvation	-1122.28420	-1122.28386
3	eq. solvation emission energy	2.554 eV	2.562 eV
4	excited state (non-eq. solvation) ground state energy incl. EI solvation	-1122.29490 -1122.37890	-1122.29463 -1122.38400

The following table presents the results of all of the calculations on coumarin 153 and compares them to the experimental values. We assumed 50%-50% populations of the two forms for conformational averaging purposes. All energy values are in eV:

	CYCLOHEXANE (ε=2.02)	DMSO (ε=46.83)		ETHANOL (ε=24.85)
QUANTITY	CALC.	CALC.	OBS.	OBS.
absorption solvent shift w.r. chex	0.0	-0.24	-0.28	-0.30
emission solvent shift w.r. chex	0.0	-0.27		
emission–absorption	0.42	0.45		0.66

The predicted absorption energy results in DMSO are similar to those in cyclohexane in that state-specific solvation lowers the calculation value slightly. The same is true of the corresponding predicted values for the emission energy, which would seem to bring this quantity in better agreement with experiment (assuming that the near agreement between the observed absorption energies in DMSO and ethanol also holds for the emission energies). The shift in energies in the two different solvent environments is in very good agreement with the available experiment values. The energy gap between the absorption and emission lines is underestimated somewhat with respect to that observed in ethanol.

For a detailed study of this problem and an extended comparison of the various solvation treatments, see [Improta06].■

Studying Ground States & Excited States with CASSCF

CASSCF References
Definition: [Roos80]
Algorithms/Implementation: [Siegbahn84, Robb90, Frisch92, Yamamoto96, Klene00]

In this section, we will introduce the *complete active space self consistent field* method (CASSCF), which is also known as *multi-configuration SCF* (MC-SCF).* A CASSCF calculation is a combination of an SCF computation with a full configuration interaction (full CI) calculation involving a subset of the orbitals (see chapter 10 for details on the full CI method). The orbitals involved in the CI are known as the *active space*. The CASSCF method optimizes

* From an expert's point of view, there are subtle differences between CASSCF and MC-SCF in that the former selects orbitals while the latter selects configurations, leading to quite different implementations. For MC-SCF, see [Almlof78] and later references which cite it. "Configurations" in this sense refers to configuration state functions of the CI rather than to electronic states.

the orbitals appropriately for the electronic state of interest, in contrast to the Hartree Fock orbitals, which are always optimized for the ground state.

The CASSCF method is useful for modeling processes where electron correlation is important, especially ones involving non-dynamic or static (near-degeneracy) electron correlation. It can be used to study both ground state and excited state systems. It is also able to study reactions involving both the ground state and excited state potential energy surfaces, and it can locate *conical intersections*: points where the excited state and ground state PESs coincide [Teller37].

Analytic second derivatives for the CASSCF method are available in Gaussian, so it can be used to perform single point energy calculations, geometry optimizations and frequency calculations. For an introduction to the CASSCF method, see chapter 10 (p. 499).

The Active Space

The active space for a CASSCF calculation consists of a contiguous subset of the molecular orbitals, and it includes both occupied and virtual orbitals. For example, the illustration in the margin depicts a three-orbital active space for formaldehyde that contains four electrons (the orbitals in the green box). The active space can be thought of as a window into the orbital space that encompasses a specific number of orbitals and electrons (see below).

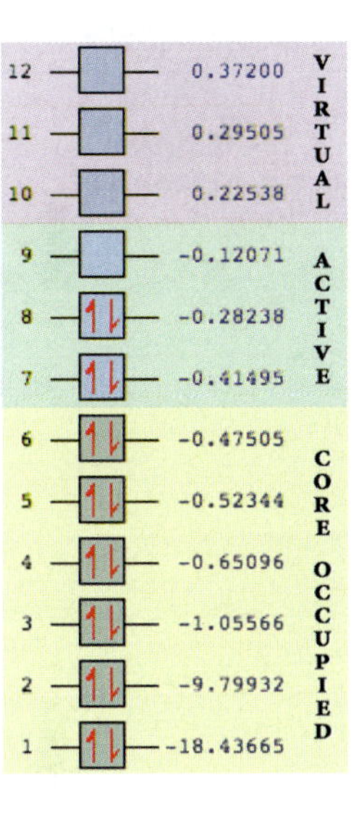

The CASSCF method proceeds by optimizing a linear combination of configuration state functions (CSFs) that comprise all possible occupations of the orbitals in the active space. In the converged wavefunction, the optimized active space orbitals will have fractional occupations, and all other orbitals remain empty or full (as appropriate), but they too will have been optimized.

CASSCF calculations are requested with the **CASSCF** keyword (abbreviable to **CAS**), which takes two required numerical options: **CAS(*n*,*m*)**. *n* is the number of electrons in the active space, and *m* is the number of orbitals in the active space. In the window metaphor, *m* is the height of the window: the number of orbitals that will be encompassed within it. The window is positioned within the orbital space so that there are *n* electrons inside it. For example, a calculation specified as **CAS(4,6)** would use an active space of six orbitals, two occupied (holding a total of four electrons) along with four virtual orbitals.

Selecting the active space is the most important part of designing a CASSCF computational study. The active space orbitals must be appropriate for the specific chemistry under investigation. It must include the orbitals that are directly involved in the chemical process, as well as any others that may interact strongly with the reacting orbitals.

Orbitals are typically examined and selected based on a Hartree Fock calculation with a small basis, often the STO-3G minimum basis set. The orbitals resulting from larger basis sets are generally too diffuse for this purpose, making it difficult to identify orbitals that are appropriate for the active space. In contrast, active spaces are readily constructed from minimum basis set orbitals selected to have the proper shape and other characteristics whose occupancy changes during the course of the excitation/reaction.

Alternatively, the orbitals for an active space can be selected based on orbital occupation rather than shape. Natural bond orbitals (NBOs) work well for this purpose, and we illustrate their use later in this section.

Orbitals must frequently be reordered in order to move the desired ones inside the window of the active space, using Gaussian's **Guess=Alter** facility (see below).

Consider CASSCF calculations on formaldehyde employing four electrons. Different active spaces will be appropriate for this molecule, depending on exactly what we want to model. If we want to study its lowest energy vertical excitations, we will need orbitals that can describe the associated $\pi \rightarrow \pi^*$ and $n \rightarrow \pi^*$ transitions.

If we examine the orbitals from an HF/STO-3G calculation (illustrated in the margin), we recognize that the occupied orbital immediately below the HOMO in energy is a π orbital encompassing both heavy atoms, the HOMO describes the oxygen lone pair, and the LUMO is an anti-bonding π^* orbital. Thus, these orbitals form an appropriate active space for modeling formaldehyde's first two vertical excitations. The calculation will specify the **CAS(4,3)** keyword.

Another process that might be of interest is the dissociation of formaldehyde to hydrogen gas and C=O. In this case, it is appropriate to select an active space containing the σ-σ^* pairs on each C-H bond. In addition, we will want to include the n orbital on the oxygen atom since it will mix with the C-H σ^* orbitals as these bonds break, ultimately forming a C-O π bond.

The following illustration depicts the range of orbitals starting at orbital 4, through the HOMO and LUMO (orbitals 8 and 9) and ending at orbital 11. We see that desired orbitals are the third and fourth orbitals energetically below the HOMO, the HOMO and the two orbitals energetically above the LUMO. The default active space for a **CAS(6,5)** calculation is HOMO–2, HOMO-1, HOMO, LUMO, LUMO+1: orbitals 6-10. However, we need orbitals 4-5, 8 (HOMO) and 10-11 in the active space (enclosed by the magenta line in the figure).

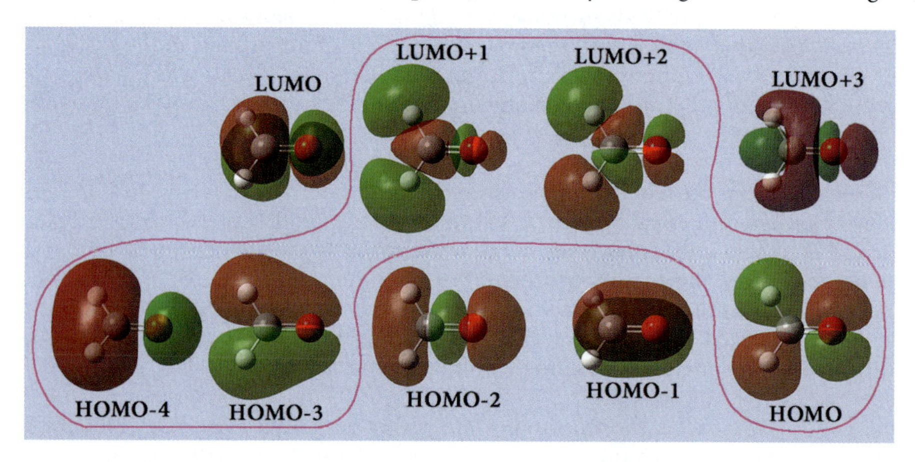

We can reorder the orbitals using the **Guess=Alter** option when we set up the calculation. This option reads a list of orbitals to swap from a separate input section following the molecule specification:

```
%Chk=form_orbs                                   checkpoint file from HF/STO-3G calculation
# CAS(6,5) Guess=(Read,Alter) Geom=Check ...

CAS calculation: dissociation of formaldehyde

0 1
                                                  blank line
4,6                                               list of orbitals to swap
5,7
9,11
                                                  blank line ends orbital alterations
```

As we've seen, different problems can require not only different orbitals but also differently sized active spaces. It is also vital to consider not only the reactants but also the products when designing the active space. The reactant orbitals need to be able to convert smoothly into the product orbitals.

Consider the dissociation of formaldehyde to H_2C and oxygen atom. In this case, we want the three 2p orbitals in the dissociated oxygen atom to be treated equally, so we again perform a **CAS(6,5)** calculation, this time using the unmodified orbital ordering. However, if we want to study the isomerization of formaldehyde to HCOH, we need an active space that describes all of the bond breaking and molecule rearrangement involved. In this case, a **CAS(8,8)** calculation would be required, and the active space will encompass orbitals 5-12: HOMO−3 through LUMO+3 in the preceding figure.

ADVANCED EXERCISE 8.11 ACTIVE SPACE FOR BENZENE

Run a Hartree-Fock calculation on benzene with a minimum basis set, and examine the orbitals. What orbitals should be included in a six-electron active space for modeling the molecule's $\pi \rightarrow \pi^*$ excited states? Set up a job to predict these excited states.

SOLUTION

The 10 orbitals centered on the HOMO and LUMO are illustrated below:

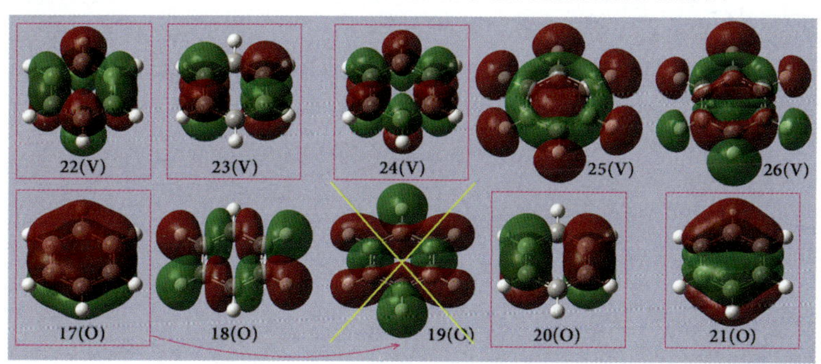

Orbitals 20 through 24 are appropriate for modeling electronic excitations, but orbital 19 is not. However, it can be swapped with orbital 17 to create a **CAS(6,6)** active space.

Here is the basic structure of a CASSCF calculation using this active space:

```
%Chk=benz_orbs                                  checkpoint file from HF/STO-3G calculation
# CAS(6,6) Guess=(Read,Alter) Geom=AllCheck ...

17,19                                           orbital alterations section
                                                blank line
```

We will run several CASSCF computations on benzene in the following subsection.■

Converging CASSCF Orbitals

One unique aspect of using the CASSCF method is the necessity of approaching the target basis set via a series of intermediate steps that gradually increase the basis set size. In other words, you cannot immediately run a CASSCF calculation with the desired large basis set. Rather, you must build up gradually, starting from a minimum basis set. This process converges the orbitals with successively larger basis sets via single point energy calculation. Such an approach—referred to as the *projection* of the orbitals—allows the CASSCF procedure to proceed efficiently and reliably.

If you are a cook, it may be helpful to think of this process as analogous to making custard or crème anglaise, where the hot milk must be added very slowly to the egg mixture to avoid scrambling or curdling the eggs.

The following example will illustrate this process for a single point energy calculation on the ground state of benzene.

ADVANCED EXAMPLE 8.9 BENZENE CASSCF SINGLE POINT ENERGY CALCULATION

We will use the following approach for this calculation:

- We will begin by running an HF/STO-3G calculation. We will examine the computed orbitals and decide on the active space. We will use the results from Advanced Exercise 8.10.
- We will run a **CAS(6,6)** calculation using the same basis set. We will examine the results and verify that the orbitals remain as we expect them to be with the CASSCF method.
- Assuming that the orbitals are appropriate, we will run two more **CAS(6,6)** calculations, using the 4-31G and 6-31G(d) basis sets. For each one, the initial guess including the orbitals will be read in using the **Guess=Read** keyword.

Here is the input file for the initial job:

```
%OldChk=benzene_hf
%Chk=benzene_cas
# CASSCF(6,6)/STO-3G Guess=(Read,Alter) Geom=AllCheck

19, 17
```
blank line

Here is the input file for the second job, which can be run after the initial CAS orbitals have been examined and verified:

```
%Chk=benzene_cas
# CASSCF(6,6)/4-31G Guess=Read Geom=AllCheck

--link1--
%Chk=benzene_cas
# CASSCF(6,6)/6-31G(d) Guess=Read Geom=AllCheck
```
blank line

All of these jobs completed successfully. The key part of the CASSCF output from the final job is given in the following figure.

```
TOTAL                          -227.761333   Predicted energy
...
Final one electron symbolic density matrix:
                1              2              3
     1   0.196070D+01
     2   0.117887D-07   0.190093D+01
     3  -0.842299D-07  -0.352539D-07   0.190093D+01
     4  -0.502730D-06   0.390983D-06  -0.845607D-08
     5  -0.492211D-07   0.226906D-06   0.165016D-06
     6  -0.181446D-06   0.425582D-06   0.680802D-07

                4              5              6
     4   0.100406D+00
     5  -0.998373D-08   0.100406D+00
     6   0.346943D-07  -0.511756D-07   0.366335D-01
MCSCF converged.
```

The predicted energy is given in the first excerpted line. The values on the diagonal of the density matrix are the orbital occupancies for the active space. In this case, the values are 1.96, 1.9, 1.9, 0.1, 0.1 and 0.04. All of them are in the desired range of 0.02-1.98 (which produces the most reliable CAS procedure).■

The preceding CASSCF calculation modeled the ground state of benzene. The **CAS** keyword also supports the **NRoot** option to specify the state of interest for computing the energy and predicting properties, performing a geometry optimization, and so on. **NRoot=1** corresponds to the ground state (it is the default value), and higher numbers correspond to excited states.

Defining the Active Space via Natural Bond Orbitals (NBOs)

Natural bond orbitals provide another means of defining an active space for a CASSCF calculation. The NBO analysis is requested with the **Population=NBO** option, and an additional option is required to save the NBOs in the checkpoint file, one of:

- **Population=SaveNBO**: Saves the NBO occupied and virtual orbitals in the checkpoint file.
- **Population=SaveMix**: Saves the NBOs for the occupied orbitals and natural localized molecular orbitals for the unoccupied orbitals in the checkpoint file. This option is preferable if you plan to use the checkpoint for additional purposes beyond defining the CAS active space. We use this option in our calculations.

The NBOs are calculated using the target basis set and a DFT method. This procedure for defining the active space is illustrated in the following example.

ADVANCED EXAMPLE 8.10 BENZENE CASSCF ACTIVE SPACE REVISITED

We ran the following calculation to produce the NBOs:

```
%Chk=benz_nbo
# APFD/6-31G(d) Pop=(NBO,SaveMixed)
```

The orbitals that would comprise a six-orbital, six-electron active space by default are illustrated in the following figure:

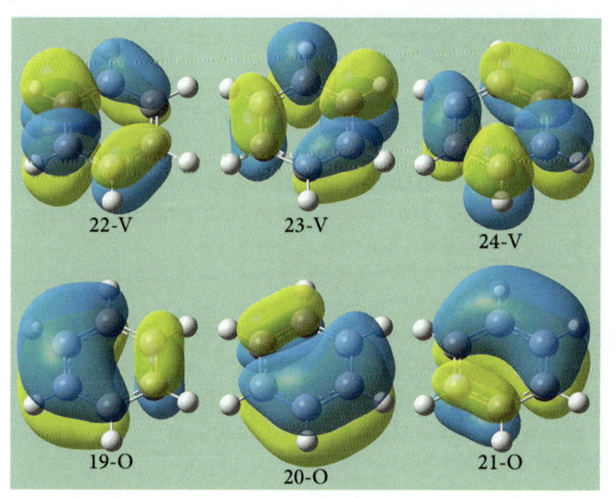

These orbitals are the appropriate ones for studying electronic excitations in benzene. We can run a single point energy calculation with **CAS(6,6)** starting from these orbitals without any need to alter the guess:

```
%OldChk=benz_nbo
%Chk=benz_cas_sp
# CASSCF(6,6)/6-31G(d) Guess=Read Geom=AllCheck
```

The job completes successfully and predicts the same total energy as the other method.■

Locating Conical Intersections

> **Conical Intersections References**
> Definition: [Teller37]
> Review articles: [Domcke11]
> Algorithms/Implementation: [Bearpark94]

A *conical intersection* is a region in which the ground state and excited state potential energy surfaces overlap (i.e., they are degenerate). As such, the lowest energy point in the region provides a pathway for radiationless decay between the excited state and the ground state.

The following diagram illustrates the main features of reactions that span both the ground state and the excited state potential energy surfaces (note that the locations of the various features are illustrative rather than literal):

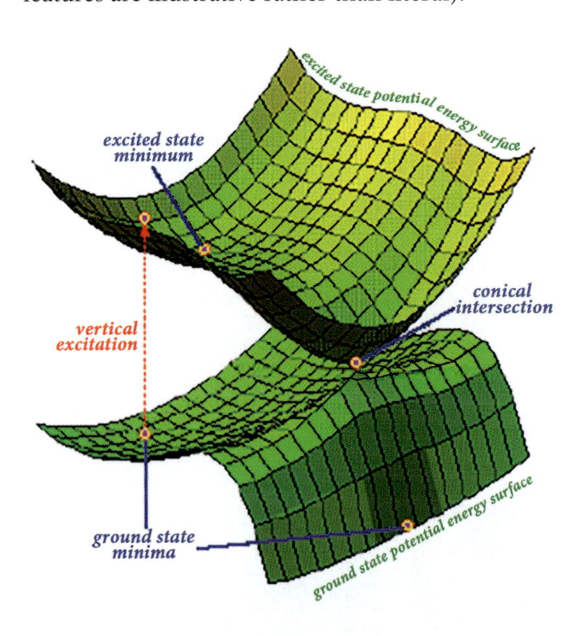

A typical reaction would begin with the reactant in its ground state (lower left portion of the ground state PES). Upon exposure to light, the molecule undergoes a vertical excitation to the excited state PES, initially retaining the same geometry. After some time, the molecular structure relaxes to the excited state minimum.

The reaction then proceeds on the excited state PES. The molecule eventually encounters the conical intersection, at which point it returns to the ground state PES. From there, it can proceed to the products and/or reactants, depending on the geometry of the specific PES.

The following exercise will guide you through the process of studying the entire photochemical reaction using the CASSCF method. See chapter 10 for more information about conical intersections (p. 500ff).

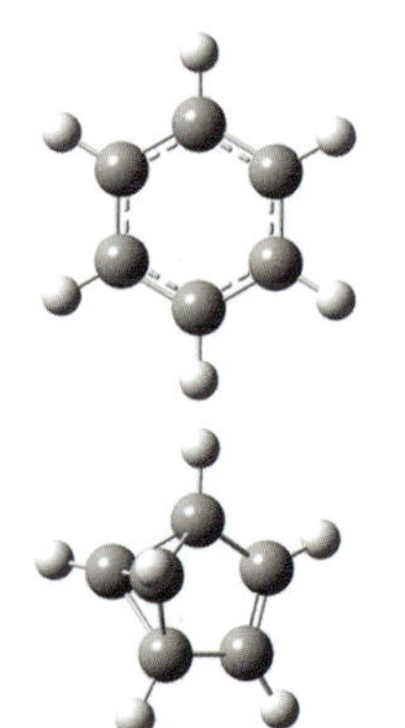

ADVANCED EXERCISE 8.12 CASSCF STUDY OF BENZENE→BENZVALENE

Illustrated in the margin are benzene and benzvalene,* two isomers of C_6H_6 that interconvert via a photochemical process. In this exercise, you will learn how to apply the CASSCF method to study all of the various species involved in this reaction. For a published CASSCF study of this reaction, see [Palmer93].

Our CASSCF calculations will use an active space of six electrons and six orbitals. This study has five phases:

* IUPAC: benzvalene: tricyclo[3.1.0.02,6]hex-3-ene.

1. Optimizing the structure of the ground state for benzene (the reactant).

2. Predicting the vertical excitation energy and then optimizing the structure of the first excited state.

3. Locating the conical intersection.

4. Following the reaction from the conical intersection on the ground state PES to the benzvalene product.

5. Optimizing the geometry of benzvalene.

The following table outlines the required steps for each phase in detail:

GROUND STATE OPTIMIZATION	EXCITED STATE OPTIMIZATION	CONICAL INTERSECTION	MOVING FROM THE CI TO THE PRODUCTS	PRODUCT OPTIMIZATION WHEN FINDING A TS
► Examine orbitals & set active space. ► Run CAS(6,6) single point energy calculation to verify active space. ► Optimize with 6-31G(d). ► Verify minimum with a frequency calculation.	► CAS(NRoot=2)/6-31G(d) to verify active space. ► CAS(NRoot=2) Opt with 6-31G(d). ► Verify minimum with a frequency calculation.	► Scan excited state PES to locate a puckered ring CI starting structure. ► Opt=Conical using CAS(NRoot=2)/ 6-31G(d).	► Starting from CI geometry, create a starting structure in the direction of benzvalene. ► Optimize and determine if a minimum or a TS. If the latter, continue to the next phase.	► Run an IRC calculation beginning from the TS. ► Opt Freq to minimum with 6-31G(d), beginning from the structure on the IRC closest to the product.

For detailed instructions for the entire process, work along with the discussion describing the solution below.

SOLUTION

Ground state optimization. Beginning from the results of either Advanced Example 8.9 or Advanced Example 8.10, we optimize the geometry with the **CAS(6,6)/6-31G(d)** model chemistry, followed by a frequency calculation. The optimization is successful, and the frequency calculation verifies that the structure is a minimum.

Excited state optimization. The excited state optimization is more straightforward than the ground state because we are starting from converged CAS orbitals using the target basis set. We run a single point energy calculation with the **CAS(NRoot=2)** option to once again verify the active space. When we are satisfied with the active space, we run an optimization plus frequency calculation, again with the **CAS(NRoot=2)** option. The optimization completes successfully, and the frequency calculation verifies that the optimized geometry corresponds to a minimum (in this case, on the potential energy surface of the first excited state).

The optimized structure is very similar to the ground state of benzene. Both the ground state and the first excited state geometries have D_{6h} symmetry. The main difference between them is an elongation of the C–C bonds in the excited state, from 1.396 Å to 1.434 Å. The energy difference between the two structures is about 4.8 eV.

Locating the conical intersection. In order to look for the conical intersection, the first step is to determine a starting structure for the optimization. We do so by running a relaxed potential energy surface scan on the excited state PES.* We begin from the optimized excited state structure, and scan the C-C-C-C dihedral angle as indicated in the margin. The input file for this calculation follows:

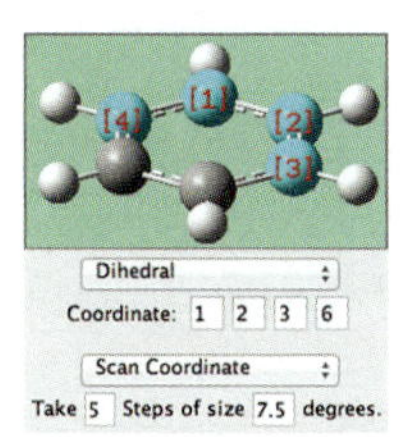

* For a review of general techniques for predicting the structures of conical intersections, see [Robb14].

```
%OldChk=benzene_s1
%Chk=benzene_scan
#P CASSCF(6,6,NRoot=2)/6-31G(d) Guess=read Geom=check
   Opt=ModRedundant NoSymm

Relaxed PES scan on benzene first singlet excited state:
move closer to the conical intersection region.

0 1

D 1 2 3 6 S 5 7.5
```

blank line

The structures located by the scan are shown in the following illustration. The numbers below each structure represent the energy difference between the ground state and the first excited state at that structure.

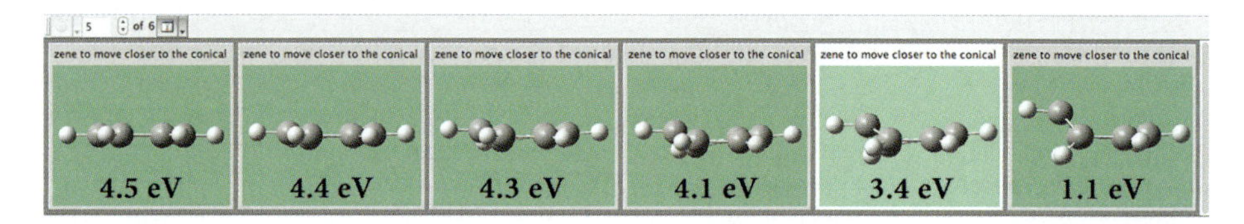

Note that the energy difference between the ground state and first excited state is included in the CASSCF section of the Gaussian output when you specify **#P** in the route section, specifically in the line labeled "EIGENVALUE" corresponding to the excited state:

NRoot		*Predicted energy (hartrees)*	*Energy difference*
(2)	EIGENVALUE	-230.5720481	1.0674 eV

Since a conical intersection is the lowest point along a seam where the ground state and excited state potential surfaces are degenerate, the energy difference between the two states will be zero at that point. Within the scan results, the lowest energy difference value corresponds to the sixth structure: the final point of this scan. It becomes our starting structure for the conical intersection optimization. We set up this job in GaussView by copying the desired structure to a new molecule group.*

We request a conical intersection optimization with the **Opt=Conical** option, in conjuction with the **NRoot=2** option to the **CAS** keyword. We perform the optimization with the 6-31G(d) basis set.† The optimization is successful.

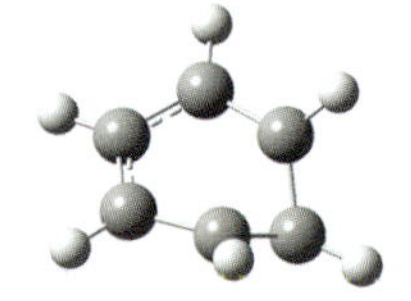

The structure of the conical intersection is illustrated in the margin. The molecule has C_s symmetry. The C–C–C–C dihedral angle including the atom below the plane of the ring is about 41.8°. The C–C bond lengths are 1.39 Å, 1.47 Å and 1.45 Å, moving from left to right in the illustration (toward the non-planar carbon atom).

* In this case, because it is the final point in the scan, the structure may also be retrieved from the checkpoint file with the **Geom=(AllCheck,NewDefinition)** options. The latter is necessary to avoid inheriting the scan job's coordinates and settings in the subsequent calculation. You could also retrieve a different structure from the scan checkpoint file with the **Geom=NGeom** option (see the *Gaussian User's Reference* for details).

† If CASSCF convergence problems are encountered when attempting to optimize a conical intersection with the desired basis set, you can begin instead with a minimum basis set, and repeat the optimization with successively larger basis sets until the desired one is reached.

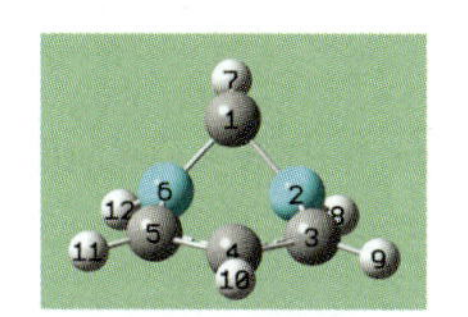

From the conical intersection toward benzvalene. From the conical intersection, one of the downhill directions should go back to benzene (corresponding to a lengthening of the C2–C6 distance), while the other downhill direction should get us closer to benzvalene (corresponding to a shortening of the C2–C6 distance). We are interested in the latter, so we begin by creating a starting structure for the upcoming optimization. We will adjust the C2–C6 bond length in the CI structure (illustrated in the margin), decreasing its value from 1.93 Å to about 1.8 Å.

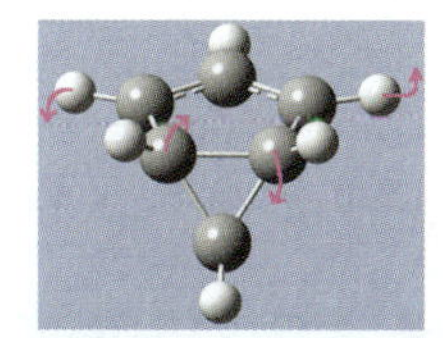

The optimization calculation using the CAS(6,6)/6-31g(d) model chemistry converges in nine steps. The subsequent frequency calculation reveals that the structure is a transition state. The imaginary frequency corresponds to a mode consisting of several CH wag motions above and below the plane of the ring, indicating a distortion of the ring from planarity.

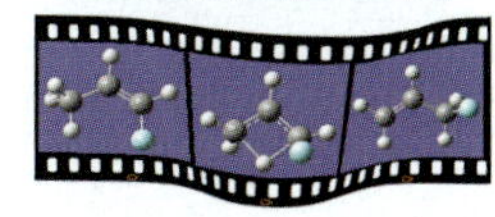

Finding a transition structure is not surprising, but one is not always present between the conical intersection and the ground state minimum. For example, when we used a similar approach to move from the conical intersection to benzene by increasing the C2–C6 bond length to ~2.1 Å, the optimization reached the benzene minimum structure.

Moving from the TS to the benzvalene product. Given that we have encountered a transition structure, we can use the usual techniques to locate the product structure: running an IRC calculation from the TS and then optimizing the minimum found at the relevant end of the reaction path. We run an IRC calculation using the options **IRC=(RCFC,ReCorrect=Never)** starting from the transition structure geometry and frequencies. We use the latter option since our interest is to get to the product as rapidly as possible.

The IRC results are shown on the left in the following illustration:

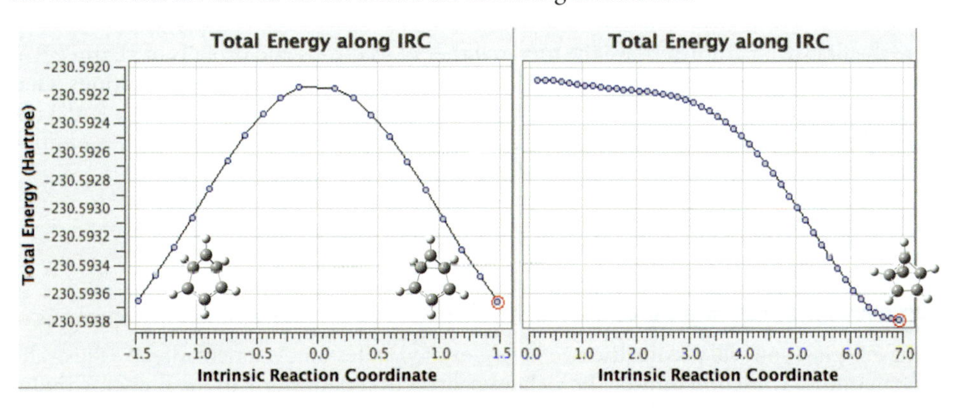

Both directions (reverse and forward) are equivalent, and they both lead to benzvalene. The two forms at opposite ends of the IRC differ only in the bonding of the individual carbon atoms. Thus, this transition structure corresponds to a benzvalene-to-benzvalene isomerization reaction. It is *not* a transition structure that connects the reactant and product on the ground state PES (although one does exist).

In order to get closer to the product, we run a one-direction IRC using the additonal options **IRC=(Forward, MaxPoints=60)**, where the latter option increases the number of IRC points from the default value of 10.

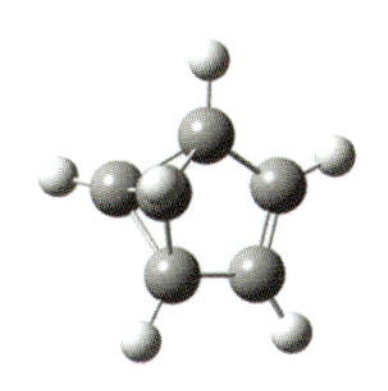

The results of this second IRC calculation are shown on the right in the preceding figure. The final point is very close to benzvalene. When we minimize the structure, we reach a stationary point that is a minimum; it is illustrated in the margin. We have successfully located benzvalene.■

RASSCF References
Definition: [Olsen88]
Algorithms/Implementation: [Klene03]

Restricted Active Space SCF (RASSCF)

The RASSCF method's goal is to improve on CASSCF results by extending the configuration expansion in a limited fashion. It does so by defining a three-part active space, with the individual components known as the RAS1, RAS2 and RAS3 spaces (in order of increasing orbital energy). RAS1 consists of occupied orbitals, RAS2 of occupied and virtual orbitals (like the CASSCF active space) and RAS3 of virtual orbitals.

All possible configurations of the electrons and orbitals within the RAS2 space are included within the multi-configuration SCF expansion (as for CASSCF). In addition, a fixed number of excitations are allowed out of RAS1 (into RAS2 or RAS3), and a fixed number of excitations are allowed into RAS3. In this way, the RAS parameters specify the minimum number of electrons occupying orbitals in RAS1 and the maximum number of electrons occupying orbitals in RAS3.

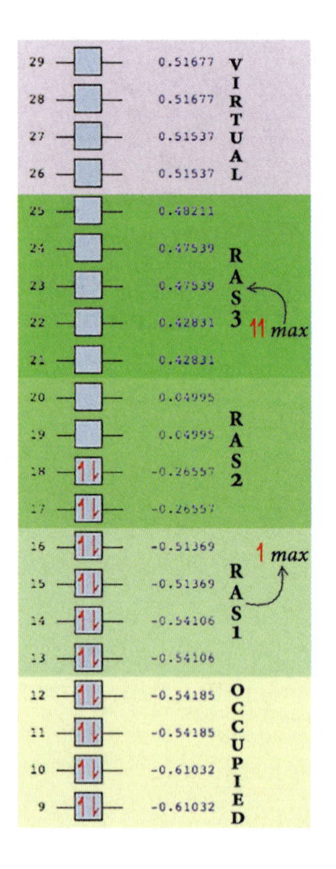

Consider the example in the margin. This active space corresponds to a RASSCF calculation involving a total of 12 electrons in 13 orbitals: eight electrons in four occupied orbitals in RAS1, four electrons in four orbitals in RAS2 and five virtual orbitals in RAS3. One excitation is allowed out of RAS1 and up to two orbitals in RAS3 may be occupied in the configurations considered. Such a calculation may be referred to in the literature as "RAS(4,4,5)," where the numeric parameters indicate the number of orbitals in the three subspaces (and the number of electrons involved is stated elsewhere).

In order to specify such a calculation to Gaussian, the **RAS** option to the **CASSCF** keyword is specified, using the following syntax:

CASSCF(*#electrons***,***#orbitals***,RAS(***maxRAS1holes***,***RAS1size***,***maxRAS3occupancies***,***RAS3size***))**

where the first two parameters are the total number of electrons and orbitals in all three RAS subspaces. The parameters to the **RAS** option are the maximum number of excitations out of RAS1, the number of orbitals in RAS1, the maximum number of occupancies in RAS3 and the number of orbitals in RAS3. Thus, the appropriate keyword and options for this example calculation would be **CASSCF(12,13,RAS(1,4,2,5))**.

The following exercise illustrates the use of the RASSCF method for modeling excited states.

ADVANCED EXERCISE 8.13 **RASSCF STUDY OF CYCLOPENTADIENE EXCITED STATES**

In this exercise, you will predict the two lowest excited states for cyclopentadiene* (illustrated in the margin). This compound's photochemical reaction path starts in the Franck-Condon (FC) region and "consists of covalent-ionic photoexcitation, an initial reaction path on an ionic state, a transition from an ionic to covalent electronic character … and finally a covalent-covalent conical intersection giving access to the ground state" [Santolinia15]. Here, we will consider the ordering of states at the FC geometry. The lowest energy excited state is a zwitterionic bright state with B_2 symmetry character. The next excited state is a covalent dark state (symmetry A_1).

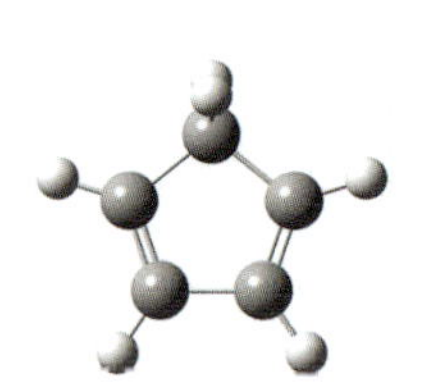

We will use the RASSCF method in order to improve on the results produced by the normal CASSCF method. For the RASSCF predictions, we will allow one excitation out of the RAS1 space and up to two excitations into the RAS3 space. The authors of the published study of this reaction [Santolinia15] describe the effect of these choices in this way:

> In a multi-reference configuration interacton expansion, if one allows only single excitations from the "closed shell sea" (RAS1 in RASSCF), then one recovers the so-called semi-internal

* IUPAC: cyclopenta-1,3-diene.

correlation of Sinanoğlu. Strictly, configurations including single excitation from RAS1 coupled with single excitation from RAS2 recover semi-internal and spin polarization effects. If one includes double excitations from the closed shell sea, then it is dynamic correlation energy that is being recovered.

The best RASSCF calculation will also include additional 3p type of π orbitals within the RAS3 subspace (see below).

Predict the energies of the ground state and first two excited states using the following procedure:

1. Optimize the structure of the ground state with the CAS(4,4)/6-31G(d) model chemistry. Verify that the optimized structure is a minimum with a frequency calculation. This structure will be used for all subsequent calculations.

2. Perform an NBO analysis at this structure using our standard theoretical method and the 6-31G(d) basis set to generate orbitals for using in the RASSCF calculations. Include the additional keyword **ExtraBasis** in the route section, which indicates that additional basis functions will be included in the input file.

   Here is the additional basis set input section:

```
4 6 8 10 0                                  Atom numbers to which additional functions apply
 P    3 0.5                                          First additional basis function
   .1727380000D+01      -.1779510000D-01
   .5729220000D+00       .2535390000D+00
   .2221920000D+00       .8006690000D+00
 P    1 0.5                                          Second additional basis function
   .7783690000D-01       .1000000000D+01
****                                                 Terminates basis function list
                                                                 Final blank line
```

   The first line of the input section indicates which atoms within the atom list should receive the additional basis functions. In this case, they correspond to the carbon atoms with only a single hydrogen atom attached. We will denote this augmented basis set as “6-31G(d)+3p.”

3. Predict the excited states using the CAS(4,4)/6-31G(d) model chemistry. Include the **Geom=ForceAbelianSymmetry** option (which aids the program in identifying the symmetries of the various states). The **CAS(NRoot=3)** option requests that the calculation solve for the ground state and two excited states. Perform this CAS as a *state-averaged* calculation using equal weighting between the two excited states.

   Normal CAS calculations focus on a single state: the ground state, the first excited state, etc. In contrast, a state-averaged calculation uses the single set of molecular orbitals to compute all multiple states. The resulting density matrix is the average for all states that were considered. This is done to ensure that the appropriate PES is followed. Note that conical intersection calculations are state-averaged by default since they locate a degeneracy in the two potential energy surfaces.

   In a state-averaged calculation, all states up to the value specified to the **CAS(NRoot)** option are included. Such calculations require a single additional input line specifying the weights to be given to the various states considered in the calculation (no following blank line is used). This input is illustrated in the input file for this calculation:

```
%OldChk=cpd_nbo
%Chk=cpd_cas
#P CASSCF(4,4,NRoot=3,StateAverage)/6-31G(d)
   Geom=Check Guess=(Read,ForceAbelianSymmetry)

Cyclopentadiene: CAS(4,4,NRoot=3)/6-31G(d) 50-50 SA

0 1

0.0 0.5 0.5
```
weights for states: ground: 0%, 1st excited: 50%, 2nd excited: 50%

4. Predict the excited states using a RASSCF calculation. Begin with a (5,4,5) RAS allowing one RAS1 hole and only one RAS3 occupancy. Use the same state averaging as in the CAS calculation. Once this has converged, increase the allowed RAS3 occupancies to two, also adding the **SCF(Conver=6,MaxCycle=128)** options, which decrease the convergence criteria slighly and double the maximum number of SCF cycles (respectively). Also include the **Population=NO** option to compute the natural orbitals for the highest excited state. These will be needed to determine the symmetry of the state.

   The orbitals selected for the RAS subspaces are as follows. Place the five occupied σ orbitals in RAS1, place the two π orbital pairs in RAS2, and place the five σ* orbitals in RAS3. These orbitals are the lowest 14 depicted in the illustration following.

5. Model the excited states again, this time using the (5,4,5) RAS with the 6-31G(d)+3p basis set. Continue to use state averaging with equal weights for the two excited states, and include the same **SCF** and **Population** options as for the previous step.

6. Finally, the RAS3 space should be expanded to include the four π 3p orbitals generated by the additional basis functions in 6-31G(d)+3p. The calculation should also use the augmented basis set.

The diagram following illustrates the active space used by the RAS calculation using the largest basis set. The orbital numbers are those from the original NBO calculation.

The desired orbitals can be moved into the active space window using **Guess=Alter** as before. Alternatively, the **Guess=Permute** option can be used. This option is preferable when more than one or two orbitals need to be reordered. It requires an additional input section that specifies the desired order for the orbitals. For example, here is the input we used to create the preceding RAS subspaces:

```
             RAS1   RAS2   RAS3              RAS3 with +3p
1-5,11-16,6-10,17-20,21-22,25,30-31,38-39,69-70
```

This example swaps the two blocks of five orbitals beginning with orbital 6 and then selects the orbitals for the RAS2 subspace, the σ orbitals for RAS3 and then the additional π orbitals for RAS3 (the latter are used only for the 6-31G(d)+3p calculation).

In general, any orbitals omitted from the list are placed at the end of the list (retaining their original order).

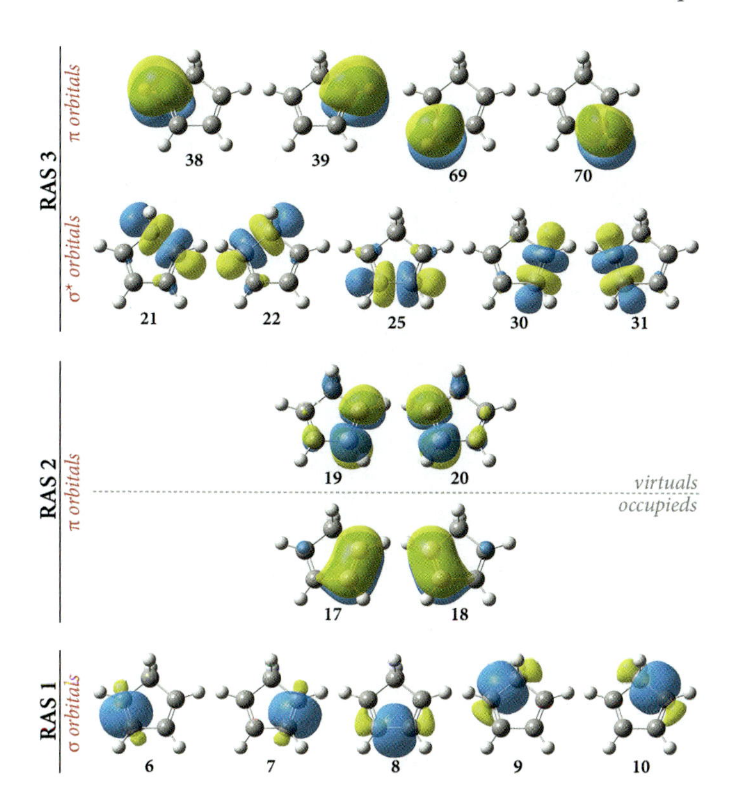

SOLUTION

Our CAS(4,4)/6-31G(d) optimization of the ground state proceeded without difficulty (without requiring any orbital alterations); the subsequent frequency calculation confirmed the structure as a minimum. The symmetry of the molecule is C_{2v}.

The APFD/6-31G(d) NBO analysis at the optimized geometry produced the natural bond orbitals for use in constructing the RAS subspaces. The selected orbitals from our calculation are indicated on the preceding page (the orbital numbers from the calculation you choose to run may differ).

Using an active space consisting of four electrons in four orbitals, we ran a CASSCF calculation to predict the excited state energies. Note that we specified **#P** in the route section to request additional detail about orbital symmetries and CAS configurations in the output file. When we run the job, it completes successfully.

Here is the key section of the output file:

```
        EIGENVALUES AND EIGENVECTORS OF CI MATRIX
ground state
  ( 1)       EIGENVALUE    -192.8352915
(    1) 0.9305296 (   10) 0.1950750 (   16)-0.1686526
excited state 1
  ( 2)       EIGENVALUE    -192.5909369          6.6492 eV
(   16) 0.5620999 (    5)-0.5282341 (    4)-0.4835352
excited state 2
  ( 3)       EIGENVALUE    -192.5546437          7.6368 eV
(    8) 0.9590367 (   14)-0.2124466 (   13)-0.1713281
```

The lines containing "EIGENVALUE"* list the energies of the computed states: here, the ground state and two excited states (the value preceding the keyword in parentheses is the **NRoot** value). As this example indicates, the total energy is reported in Hartrees, and the relative energy for states above the ground state is reported in eV.

The lines following the energy value(s) list the CAS configurations and their coefficients for that state. We will consider the first listed configuration for each excited state (indicated in green). We can consult the list of configurations appearing slightly earlier in the output to determine the symmetry of each state:

```
BOTTOM WEIGHT=   6        TOP WEIGHT= 14
Configuration             1 Symmetry 1 1100
...
Configuration             8 Symmetry 4 a10b   2nd excited state: B2
...
Configuration            15 Symmetry 4 a01b
Configuration            16 Symmetry 1 0101   1st excited state: A1
...
```

From the symmetry matrix for the two states, we can determine that the first predicted excited state has symmetry A_1 and the second excited state has symmetry B_2: the order is reversed from what is observed. The RAS calculations are an effort to correct this inaccuracy.

Here is the input file for the first calculation of the (5,4,5) RAS:

```
%OldChk=cpd_nbo
%Chk=cpd_ras
#P CASSCF(14,14,RAS(1,5,1,5),NRoot=3,StateAverage)
   6-31g(d) Geom=AllCheck
   Guess=(Read,Permute,ForceAbelianSymmetry)

1-5,11-16,6-10,17-20,21-22,25,30-31,38-39,69-70   orbital reordering

0.0 0.5 0.5                                        state-averaging weights
```

There are 14 electrons and 14 orbitals overall in all three RAS subspaces, and the latter contain five, four and five orbitals (respectively). The orbitals are reordered as we discussed above using **Guess=Permute**.

This job completed successfully, so we went on to use its results in a second RAS calculation with the same basis set, increasing the maximum number of excitations into RAS3 to two:

```
%Chk=cpd_ras
#P CASSCF(14,14,RAS(1,5,2,5),NRoot=3,StateAverage)
   6-31g(d) Geom=AllCheck SCF(Conver=6,MaxCycle=128)
   Guess=(Read,ForceAbelianSymmetry)

0.0 0.5 0.5                                        state-averaging weights
```

The job completed successfully. The energies of the various states are reported in the same manner as for a CAS calculation. This job predicted relative energies of 6.61 eV and 6.93 eV for the two excited states.

Determining the symmetries of the excited states predicted by a RASSCF calculation is not quite as straightforward as for a normal CAS because the configurations are not included in

* For CASSCF calculations. For RASSCF calculations, the string becomes "Final Eigenvalue."

the output file. However, examining the natural orbitals corresponding to the excited state(s) in question provides a method for doing so.

For each state, we generate the natural orbitals with an additional two-step job like the following (which does so for the lower energy excited state):

```
%OldChk=cpd_ras
%Chk=cpd_orbs_nroot2
#P CASSCF(14,14,RAS(1,5,2,5),NRoot=2) ChkBasis IOp(5/7=-2)
   Geom=AllCheck Guess=(Read,ForceAbelianSymmetry)

--Link1--
%Chk=cpd_orbs_nroot2
# Geom=AllCheck Guess=(Read,Only,Save,NaturalOrbitals)
  ChkBasis
```

final blank line

The first job step retrieves the converged RAS wavefunction from the checkpoint file and computes the density corresponding to the first excited state (**NRoot=2**). The second job steps computes the natural orbitals using that density and saves them in the checkpoint file.

We are interested in the orbitals for which the occupancy is partial: more than ~0 and less than ~2. These values are listed in the output file in the list of orbitals:

```
Molecular Orbital Coefficients:
...
                 16        17        18        19        20
               (B2)-O    (B1)-O    (A2)-O    (B1)-V    (A2)-V
Eigenvalues    1.9772    1.9654    1.0189    0.9790    0.0309
```

We are interested in orbitals 18-19 for the first excited state. We can use the symmetries of the relevant orbitals in order to identify the state's overall symmetry. In this case, the orbitals have symmetry A_2 and B_1, indicating that the state is symmetry B_2. We repeat the two job steps for the second excited state.

Visualizing the orbitals may also be useful. The following illustration shows the partially occupied natural orbitals for the two excited states:

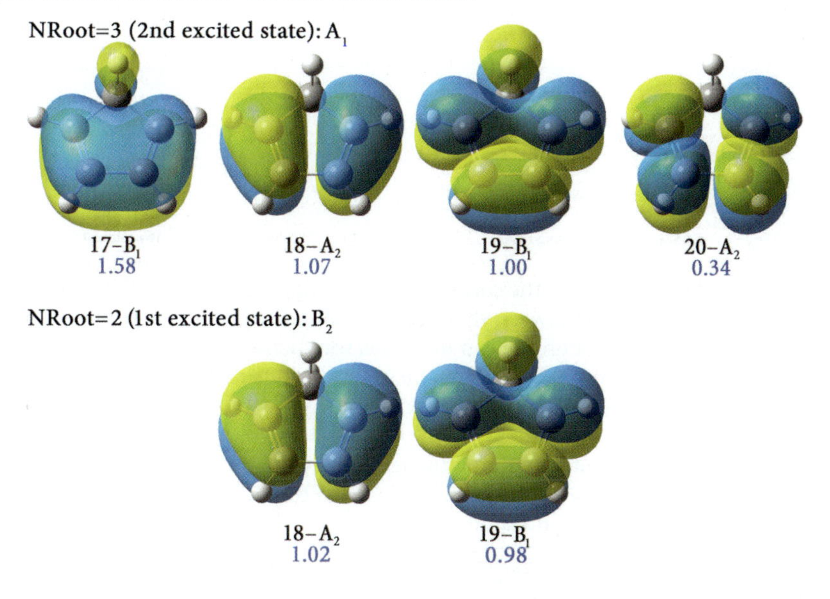

The two B_1 and two A_2 orbitals that are partially occupied in the second excited state (**NRoot=3**) yield an overall symmetry of A_1 for the higher energy excited state, indicating that this state is now higher in energy than the B_2 state.

We continue the study by running the same RAS calculation with the extra basis functions:

```
%OldChk=cpd_ras
%Chk=cpd_ras+3p
#P CASSCF(14,14,RAS(1,5,2,5),NRoot=3,StateAverage)
   6-31g(d) Geom=AllCheck Guess=(Read,ForceAbelianSymm)
   Pop=NO SCF(Conver=6,MaxCycle=128) ExtraBasis
...
```

We visualize the natural orbitals in the same manner and verify that the ordering of states is the same as for the smaller basis set. The predicted energies are 6.39 eV and 6.85 eV.

Our final calculation includes the additional four π orbitals. Here is the route section:

```
#P CASSCF(14,18,RAS(1,5,2,9),NRoot=3,StateAverage)
   6-31g(d) Geom=AllCheck Guess=(Read,ForceAbelianSymm)
   SCF(Conver=6,MaxCycle=128) ExtraBasis
```

The following plot shows the results of our calculations and experimental observations [Schalk10]:

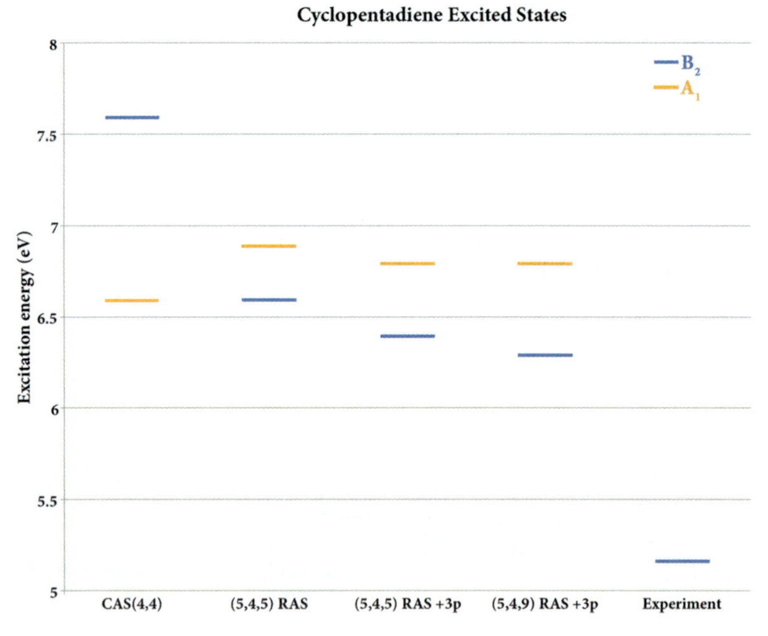

The RASSCF calculations produce the correct energetic ordering of the excited states. They also reduce the energetic distance between them compared to the ground state energy predicted by the same calculation, although the gap increases somewhat with larger RAS active spaces and additional basis functions. Nevertheless, these results illustrate how the RASSCF method may be used to recover some of the correlation effects neglected by CASSCF.

The preceding study is adequate for predicting the energy gap between the two excited states. However, the prediction of the excitation energies with respect to the ground state could be improved by predicting the ground state energy separately for each model. In the preceding calculations, the ground state energy is probably too high since the its weight is set to zero in the state-averaged calculation; the orbitals computed in this way are not optimal for the ground state. Running additional ground state calculations—which include optimizing that state's orbitals—would produce a more accurate ground state energy prediction for each model chemistry.■

9

Advanced Modeling Techniques

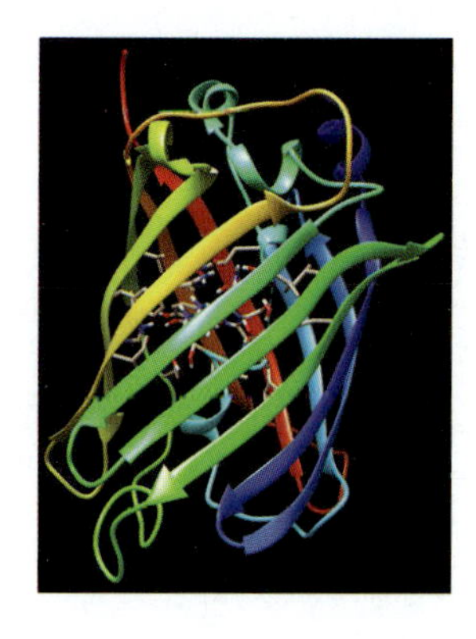

In This Chapter:

Preparing ONIOM Input for Biological Systems

Relativistic Effects

Weakly Bound Complexes:
Dispersion & Counterpoise Corrections

Electronic Spin Organization in Molecular Systems

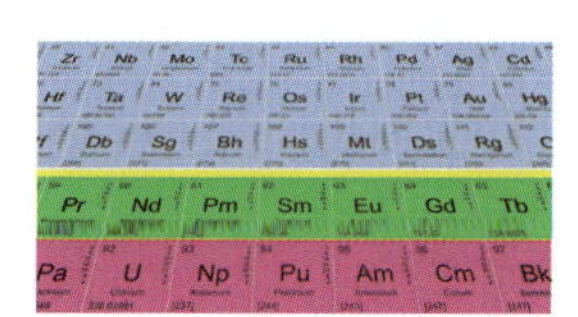

9 ◆ Advanced Modeling Techniques

Preparing ONIOM Input for Biological Systems

This section focuses on the ONIOM method and its use for modeling biological compounds. Such molecular systems typically consist of a large protein environment as well as a region of interest. The latter may be another molecule bonded to the protein (e.g., a chromophore), an active site within the protein or something else.

The challenge with using the ONIOM method for such problems lies not in the method but in preparing appropriate input for the calculation. Once set up, ONIOM jobs are not fundamentally different from other calculations, and the discussions of the previous chapters of this book apply equally to them.

Preparing a molecule specification for use with the ONIOM method in Gaussian is a rather lengthy process. While not excessively difficult, it is nevertheless challenging at times, and each part of it requires patience and care.

In general, creating a Gaussian input file for a biological molecule encompasses the following steps:

- Obtaining a starting PDB file.
- Adding hydrogen atoms to and specifying protonation for the PDB file structure.
- Specifying parameters for the molecular mechanics method (e.g., Amber): atom types and partial charges.
- Assigning atoms to ONIOM layers and defining link atoms.
- Selecting model chemistries for each ONIOM layer.
- Running the desired calculation(s).

Many of these steps can be partially or fully automated, and a variety of software packages will be useful in the process. Over time, many research groups have developed workflows for completing this process, and they differ considerably from one another. The process we will take you through here is one of several alternatives. We have chosen what seem to us to be the best available tools at the time of this writing, but they are not perfect solutions in all cases (as we'll see). We make extensive use of GaussView.*

Accordingly, keep these caveats in mind:

- Software packages are merely tools to help you prepare the molecule specification; they cannot replace human analysis. They also have their own eccentricities and imperfections, and you learn about these only through experience.
- There is no substitute for understanding the structure of the substance you are studying.
- There is no substitute for verifying the entire molecular structure manually.
- Software packages are only computer programs; they are not chemists.

* WebMO has limited features for handling PDB files. It can import them using the **Import Molecule** button on the **Build Molecule** page, and it can add hydrogen atoms to a structure via the **Add Hydrogens** item on the **Cleanup** menu on the same page. We will not be discussing WebMO further in this section.

Some useful software tools are listed in the following table, along with the relevant functions for our purposes. Note that the listed packages may also have substantially more functionality that we do not need for our current purposes.

PACKAGE	TASKS	WEBSITE	CITATION
GaussView	►import PDB (optionally remove waters) ►add H atoms ►generate and/or set Amber parameters ►define ONIOM layers ►set up Gaussian calculation	*www.gaussian.com/g_prod/gv5.htm*	[Dennington09]
ProPKA	►add H atoms ►protonate to specific pH	*propka.ki.ku.dk*	[Li05, Bas08, Olsson11, Sondergaard11]
PDB2PQR	►web interface for ProPKA	*nbcr-222.ucsd.edu/pdb2pqr_1.9.0*	[Dolinsky04]
Open Babel	►convert PQR file format to PDB	*openbabel.org*	[OBoyle11]
R.E.D.	►compute charges for Amber	*q4md-forcefieldtools.org/RED*	[Dupradeau10]
Chimera	►combine PDB files	*www.cgl.ucsf.edu/chimera*	[Pettersen04]
AmberTools	►generate & set Amber parameters ►create Amber libraries	*ambermd.org/#AmberTools*	[Case14]
Dowser	►add crystallographic waters for MD simulations	*danger.med.unc.edu/hermans/dowser/dowser.htm*	[Zhang98]
MolProbity	►analyze PDB files for problems	*molprobity.biochem.duke.edu*	[Chen10]
PISA	►generate biological assemblies from PDB files	*www.ebi.ac.uk/msd-srv/prot_int/pistart.html*	[Krissinel07]

Locating and Selecting PDB Files

Although it can be done, it is quite uncommon to build the large molecules associated with biological processes from scratch. Usually, starting structures are obtained from the X-ray diffraction data stored in the Protein Data Bank. The basic structure of a PDB file is summarized on p. 404. Additional information about PDB file structure can be found at:

- PDB file overview: *www.rcsb.org/pdb/101/static101.do?p=education_discussion/Looking-at-Structures/intro.html*
- PDB entry syntax: *www.wwpdb.org/documentation/format33/v3.3.html*

The Protein Data Bank is accessible on the Internet at *pdb.org*. The top portion of the main page is shown in the following illustration:

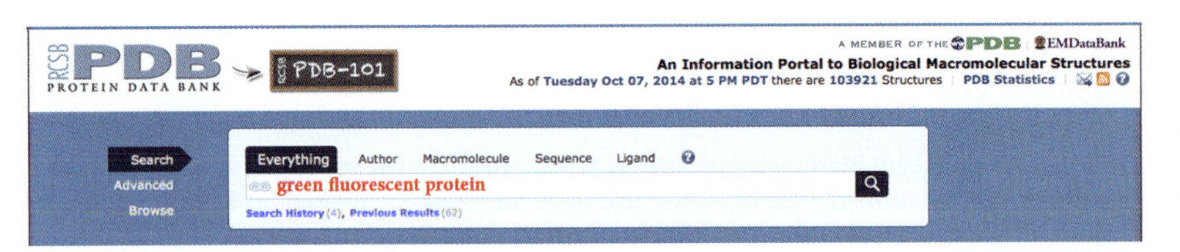

The text field can be used to perform simple search operations over the various stored structures. Here, we search for the compound we want to study: green fluorescent protein (GFP). GFP is a protein that fluoresces bright green when exposed to light in the blue to ultraviolet range. It was first isolated in the jellyfish species *Aequorea victoria*, which is native to the Pacific northwest coast of North America. Since then, variants with enhanced fluorescence properties have been engineered.

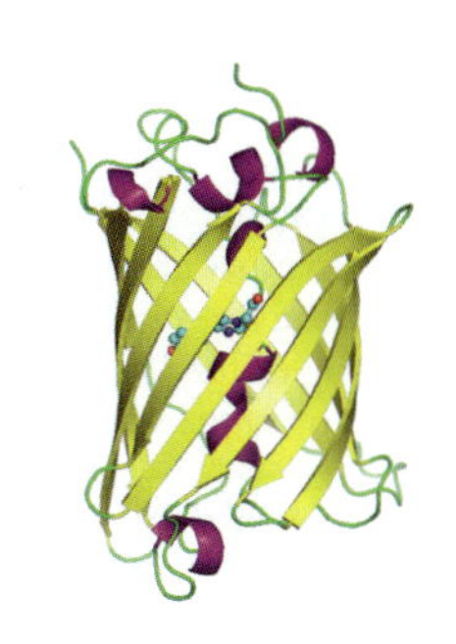

GFP consists of a chromophore within a protein chain composed of 238 amino acids. The isolated chromophore is not fluorescent. The molecule is barrel shaped (yellow portion) with alpha helices running near and through it (purple) and the chromophore located near the geometric center of the cylinder.

Simple searches in the Protein Data Bank can produce large numbers of candidates for this structure. This will be the case for almost any compound in which you are interested.

The site provides a convenient mechanism for narrowing the list via the **Query Refinements** control area:

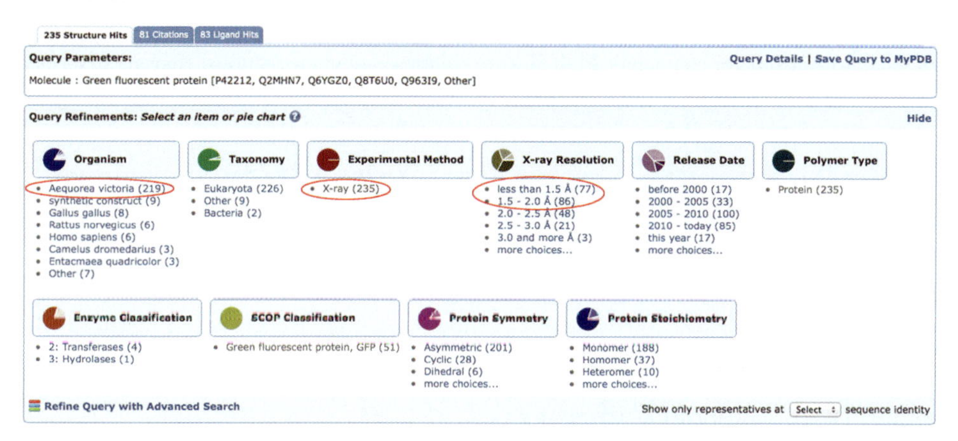

You can specify matching criteria in several categories by clicking on the desired items. Here, we will specify the source organism for GFP and the resolution of the X-ray diffraction experiment.

The resolution is directly related to the quality of the data in the PDB file. In general, smaller resolution values mean increased detail and accuracy in the structure, and you will want to choose a file with as low a resolution value as possible that is otherwise consistent with your needs. *See www.rcsb.org/pdb/101/static101.do?p=education_discussion/Looking-at-Structures/resolution.html* for an excellent discussion of resolution and PDB files.

The resolution is documented in PDB files as REMARK 2:

```
REMARK   2
REMARK   2 RESOLUTION.    1.85 ANGSTROMS.
```

The list of matching files contains entries that briefly describe the file, like this one for 1W7S. PDB:

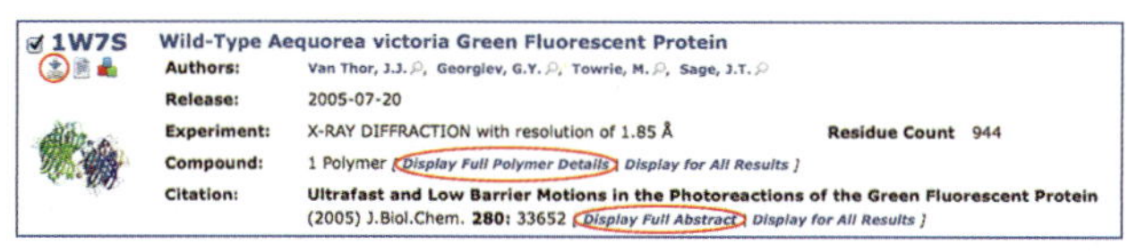

The contents and source of the file are summarized. You can click on any of the **Display Full Details** links to display more information about this entry (and optionally every entry) within the entry list. If you click on the title line in blue at the top of the entry, you will open a new page devoted to that entry with considerably more information and an integrated 3D viewer with which you can examine the structure. Finally, clicking on the leftmost icon under the PDB file name in the upper left corner of any entry will download the corresponding PDB file to your computer (the icon is circled in the preceding illustration).

Understanding PDB Files

PDB files are text files with a complicated structure that is spelled out in great detail in the official specification. However, not all programs that generate PDB files are 100% compliant with the specification (available in several formats at *www.wwpdb.org/docs.html*). The lines in a PDB file begin with standard keywords and adhere to a specific column format within each type of line; the ordering of the lines within the file is also specified. However, tools that generate PDB files do not always follow the rules to the letter, and software packages that work with these files vary substantially in how much deviation from the standard they will accept.

One consequence of all this is that it's all too easy to "break" a PDB file by modifying it manually, for example using a text editor or script/program. Nevertheless, it is sometimes necessary to perform such manual edits. When doing so, you will need to be aware of the basic format of the file, and take care to comply with the specifications such as column locations for data and ordering of lines.

Each line in a PDB file begins with a keyword indicating the type of data contained in it. The data itself follows, arranged in fixed fields and sometimes containing required separator/terminator characters. PDB files are composed of many sections. They are described in the following list. Examples come from the 1W7S.pdb file for GFP (they are considerably shortened).

Title section: Contains descriptive information about the file's contents, including documentation about the experiment:

```
HEADER    LUMINESCENT PROTEIN                     09-SEP-04   1W7S
TITLE     WILD-TYPE AEQUOREA VICTORIA GREEN FLUORESCENT PROTEIN
COMPND    MOL_ID: 1;
COMPND   2 MOLECULE: GREEN FLUORESCENT PROTEIN;              name of the compound
COMPND   3 CHAIN: A, B, C, D;                                chains present in the file
COMPND   4 ENGINEERED: YES;
COMPND   5 MUTATION: YES;
COMPND   6 OTHER_DETAILS: THE CHROMOPHORE, GYS, IS PART OF THE PEPTIDE CHAIN
COMPND   7  BETWEEN RESIDUES 64 AND 68.
SOURCE    MOL_ID: 1;
SOURCE   2 ORGANISM_SCIENTIFIC: AEQUOREA VICTORIA;           biological/chemical source of substance
SOURCE   3 ORGANISM_TAXID: 6100;
SOURCE   4 ORGAN: PHOTOGENIC ORGAN;
...
EXPDTA    X-RAY DIFFRACTION                                  type of experiment that produced the structure
JRNL        AUTH   J.J.VAN THOR,G.Y.GEORGIEV,M.TOWRIE,J.T.SAGE   corresponding reference
JRNL        TITL   ULTRAFAST AND LOW BARRIER MOTIONS IN THE PHOTOREACTIONS OF
JRNL        TITL 2 THE GREEN FLUORESCENT PROTEIN
JRNL        REF    J.BIOL.CHEM.                   V. 280 33652 2005
JRNL        DOI    10.1074/JBC.M505473200
REMARK   2                                                   remarks: details about the experiment & structure
REMARK   2 RESOLUTION.    1.85 ANGSTROMS.
REMARK 200                                                   remarks are divided into subsections by sequence number
REMARK 200 EXPERIMENTAL DETAILS
REMARK 200  EXPERIMENT TYPE                : X-RAY DIFFRACTION
REMARK 200  DATE OF DATA COLLECTION        : 08-JUL-03
REMARK 200  TEMPERATURE           (KELVIN) : 100.0
REMARK 200  PH                             : 7.80
REMARK 200  NUMBER OF CRYSTALS USED        : 1
```

Some entries take only a single line while others extend over several lines. In the latter case, the second and following lines contain a sequence number in their second field (indicated in green). This section contains a great deal of vital data about the compound contained in the file, including its name, biological or chemical source, originating article and detailed information about the experiment and the molecule. In our case, we can verify that the compound originates from the jellyfish as we expect. In the COMPND subsection, we learn that there are four protein chains present in the file, as well as the location of the chromophore within the file.

Primary structure section: Lists the sequence of residues in each chain. This section starts with the DBREF entry, but this is the most important part from a computational standpoint:

```
        chain total # ordered list of amino acids
SEQRES   1 A  236  MET SER LYS GLY GLU GLU LEU PHE THR GLY VAL VAL PRO
SEQRES   2 A  236  ILE LEU VAL GLU LEU ASP GLY ASP VAL ASN GLY HIS LYS
...
SEQRES  19 A  236  TYR LYS
```

entries for other chains follow

The SEQRES entries list the amino acids present in each chain, in order. The fourth item on each line is the total number of residues in the chain (which is the same on all of the lines).

Heterogen section: Describes any non-standard residues present in the file. In our case, this section describes the chromophore:

```
         residue  chain  residue#     # atoms
HET      GYS   A   66         21                          non-standard residue information
HET entries for chains B through D
HETNAM      GYS [(4Z)-2-(1-AMINO-2-HYDROXYETHYL)-4-(4-    chemical name of compound
HETNAM    2 GYS  HYDROXYBENZYLIDENE)-5-OXO-4,5-DIHYDRO-1H-IMIDAZOL-1-
HETNAM    3 GYS  YL]ACETIC ACID
HETSYN      GYS CHROMOPHORE (SER-TYR-GLY)                 other names (synonyms)
FORMUL    1  GYS    4(C14 H15 N3 O5)                      chemical formula
entries for chains B through D
FORMUL    5  HOH   *908(H2 O)                             water molecules
```

The HET line indicates the location of the non-standard residue; its fields are the identifier, chain and residue number (see below), as well as the number of atoms present in the structure (HETATM entries). Other lines in this subsection give the name of the compound and its chemical formula. As the last FORMUL line above indicates, water molecules may also be included in the file, typically at the end of the atom list (Here, there are 127 in each chain).

Secondary structure section: Defines helices and sheets found in the protein. Here is an example helix definition, which is defined by the starting and ending residues (specified as amino acid-chain-residue number triples):

```
       helix# ID  -starting residue-  -ending residue-  helix type      # residues
HELIX    1    1 LYS A     3   THR A     9   5                                7
```

Connectivity annotation section: Includes additional connectivity information (auxiliary to the connectivity section described below). For example, LINK records indicate additional connectivity between residues:

```
            atom 1 in residue 1            atom 2 in residue 2                      distance
LINK        C    PHE A   64                N    GYS A   66        1555    1555    1.45
```

Misc. features section: Description of the environment or essential sites within the molecule.

Crystallographic and coordinate transformation section: Documents coordinate system transformations from the experiment to that of the PDB file (e.g., change of origin, scaling).

Coordinate section: Describes the atoms in the molecule, along with their residue location and atomic coordinates:

```
       atom#  ID  residue chain residue#   -------------coordinates-------------          element
ATOM      1   N    SER A    2       13.667  22.255 106.434  1.00 19.66                    N
ATOM      2   CA   SER A    2       14.240  22.316 107.791  1.00 19.52                    C
...
ATOM   1830   NE2 HIS A 231        -1.559  17.636 103.502  1.00 14.37                    N
TER    1831       HIS A 231                 marks the end of the chain (not an actual atom)
```

The atom number is unique for each atom in the file. The alphanumeric ID distinguishes atoms of the same type within a residue (amino acid). The triplet of residue name, chain and residue number is also used to identify atoms elsewhere in the file. In some PDB files, ATOM lines are interspersed with ANISOU lines, which provide anisotropic temperature data. Due to the limitations in the resolution of X-ray crystallography, hydrogen atoms are generally not present in PDB files.

Atoms that are not part of standard residues are specified in HETATM lines, which may appear interspersed with the standard residues or following the TER line which ends the chain. These lines are quite similar to ATOM lines:

```
       atom#  ID  residue info or blank   -------------coordinates-------------           element
HETATM  473  C1   GYS A   66        17.279   1.091 105.756  1.00   8.20                   C
```

Connectivity section: Specifies bonding information. This data is typically present only for the non-standard residues defined in the heterogen section:

```
       atom# bonded to ...
CONECT  473  474  478  482
CONECT  474  473  475
```

Bookkeeping section: Summary information about the file, contained in the MASTER and END records.

PDB File Pitfalls

It would be nice if you could simply go to the protein data bank and retrieve the structure of the compound in which you are interested. Unfortunately, in real life, things are rarely this straightforward. In the first place, there may be many PDB files for the compound you are studying. Choosing among them will require careful consideration of the experimental conditions and other factors in order to choose the one most relevant to the process in which you are interested.

Once you have selected a PDB file, there are several types of pitfalls that you may encounter as you proceed:

- *Multiple copies of the molecule.* PDB files may contain more than one copy of the molecular structure (how many depends on the symmetry of the crystal used in the X-ray diffraction experiment). How many you want to include in your calculation depends on your needs.

- *Asymmetrical unit structures.* For symmetric structures, PDB files typically store only the *asymmetric unit* structure(s): the smallest unique portion(s) of the overall structure. As such, they are the components that are needed to build the functional form of the actual compound; the latter is known as the *biological assembly*. Descriptive information and transformation instructions are included in the PDB file in REMARKs 300 and 350 (respectively). See *www.rcsb.org/pdb/101/static101.do?p=education_discussion/Looking-at-Structures/bioassembly_tutorial.html* for more information.

  Note that different PDB files for the same compound may contain different asymmetric units. For example, the biological assembly for hemoglobin contains four chains in two symmetric pairs. There are PDB files for hemoglobin containing the complete molecule and others that contain half the complete molecule: specifically, the two chains that can be used to generate the entire molecule. Still others contain multiple hemoglobin molecules.

  The PISA program is capable of generating biological assemblies from PDB files. The following websites are useful:

  Main model website: *www.ccp4.ac.uk/MG/ccp4mg_help/pisa.html*
  Web service: *www.ebi.ac.uk/msd-srv/prot_int/pistart.html*
  Tutorial: *www.ebi.ac.uk/pdbe/docs/Tutorials/workshop_tutorials/PDBepisa.PDB*

  An example of creating an assembly using the PISA web service appears in the subsection "Other Useful Tools" later in this chapter (p. 434).

- *Missing residues and/or atoms.* In typical X-ray diffraction experiments, not every atom can be observed. For example, hydrogen atoms are seldom included in PDB files from these experiments. In addition, regions of the protein that are in motion during the experiment are also typically not observed. They may manifest as missing residues at the head/tail of the chain and/or breaks within the chain.

  REMARK 465 contains information about missing residues and residues with missing atoms (output is shortened):

```
REMARK 465 MISSING RESIDUES
REMARK 465 THE FOLLOWING RESIDUES WERE NOT LOCATED IN
REMARK 465 THE EXPERIMENT.
REMARK 465     residue chain residue#
REMARK 465      GLY A    232
REMARK 465      MET A    233
...
REMARK 465      LYS A    238
```

In this example, seven residues are missing from the end of the chain. You should examine the list in the PDB file you are working with. You may choose to add the missing residues by extracting them from another PDB file or even add them manually in GaussView. In the latter case, the additional atoms will not include any PDB-related data.

REMARK 470 provides information about missing atoms within standard residues:

```
REMARK 470 MISSING ATOM
REMARK 470 THE FOLLOWING RESIDUES HAVE MISSING ATOMS
REMARK 470     residue chain residue# element
REMARK 470       SER A   65      O
```

This example indicates that an oxygen atom is missing from residue 65 in chain A (the amino acid is serine). You should examine any such list in your PDB file and determine how to handle each item. Residues 65-67 in the original protein undergo a chemical transformation that results in the final chromophore. Since we will be modeling the chromophore, we can ignore the missing atom in the original residue 65.

- *Molecules encompassing multiple PDB files.* Very large molecules can be difficult or impossible to crystallize as a whole. Experimenters often opt to study manageable portions of the compound separately. In such cases, separate structures may be available for various component pieces. In order to study the entire protein, it must first be reassembled from the separate PDB files. Relevant information may be included in the REMARK section of the PDB files as well as the original publications.

  There are other cases where structures may need to be constructed from multiple PDB files. For example, if a portion of a molecule is missing from a PDB file that is otherwise desirable to use, you may be able to extract the missing piece from a different PDB file and combine it with the first PDB file to produce a structure for the entire molecule.

  An example of combining components from several PDB files appears in the subsection "Other Useful Tools" later in this chapter (p. 434).

- *Multiple conformations.* Some PDB files include multiple conformations for some portion of the structure. This is more common with PDB files derived from NMR data. You will need to choose which conformation(s) to use in your calculations.

- *Coordinates only for one atom per amino acid.* For experiments which obtained only a poor resolution image, the PDB file will be in a form known as an *alpha-carbon coordinate file*. In such files, only the coordinates for the alpha carbon in each amino acid are provided. Such files are not desirable as starting points for electronic structure calculations.

GaussView and PDB Files

GaussView can open and save PDB files. It has features for manipulating residues and secondary structures that are defined within them, via the **PDB Residues** and **PDB Secondary Structures** items on the **Edit** menu. The **Atom List Editor** can display residue and secondary structure information as well (accessed via the **P** icon in its toolbar).

Occasionally, GaussView will display the following warning when opening a PDB file:

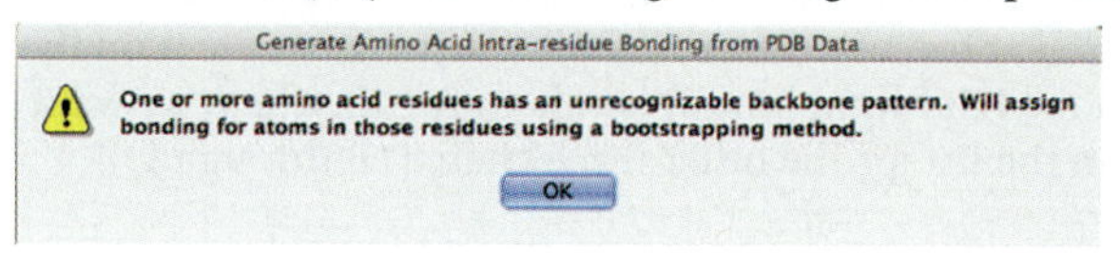

This message appears when residues have missing atoms, when the atoms are in a non-standard order and in other obscure cases. In our case, it may be safely ignored.

TUTORIAL *Preparing a Gaussian Input File for GFP*

The remainder of this section consists of an extended input file preparation example. The work is divided into several phases, indicated by numbered subheadings. We encourage you to complete the steps as you read, and we provide many intermediate files that you can use for comparison with your own work. This discussion assumes that you have read the introductory material on ONIOM and setting up ONIOM jobs earlier in this book (see pp. 15-16 and 69ff).

This example was chosen to illustrate as many tools as possible. Not all real-world examples will be this complicated or require as lengthy a process.

The example command lines can be run on any UNIX-based operating system, including Mac OS X unless otherwise stated. The command names and formats would need to be converted for Windows operating system environments.

For an ONIOM study on this structure, see [Thompson14].

T1: Processing the Initial Structure from the Protein Data Bank

Files used in this phase

starting: 1W7S.pdb
final: 1w7s_0.pdb

We have selected the file *1W7S.pdb* as the starting point for our study, and we downloaded it from the Protein Data Bank website. Examining the comments in the file reveals that the file contains four protein chains, but we need only one of them. It also contains water molecules, which we want to discard. We can accomplish both of these goals with GaussView.

When we open the PDB file in GaussView, we use the **Options** button in the lower right corner of the **Open Files** dialog to access the PDB file options:

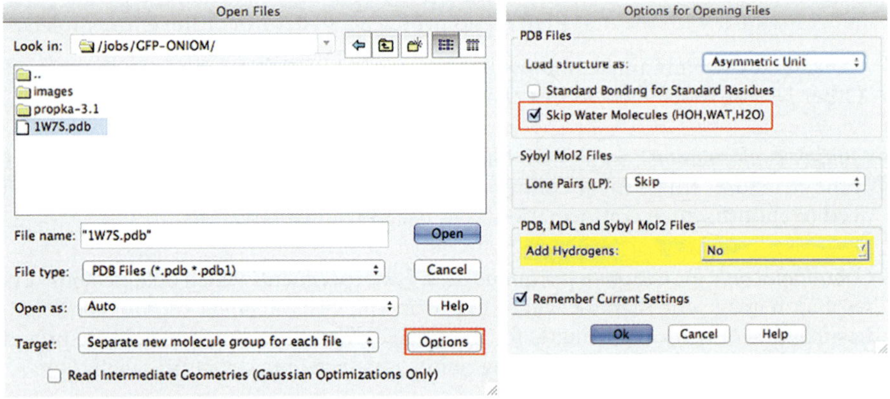

Whether or not to keep the water molecules depends on the problem you are studying. For the moment, we will discard them, but any study of GFP will ultimately need to include water molecules as they perform an important stabilization function with respect to the chromophore.

We check **Skip Water Molecules** in order to make the structure easier to work with at this point (but see the note in the margin). We leave **Add Hydrogens** set to **No** as we are not yet ready to add them. After you click **Ok** in the **Options** dialog, you must click **Open** in the **Open Files** dialog,* and then the PDB file will be opened.

Once the molecule has been imported, we easily see that there are four copies of the protein in the PDB file. We can eliminate the three unwanted chains using the **PDB Secondary Structure Editor** dialog (accessed from the similarly named item on the **Edit** menu). Here we have clicked on the **Highlight** button for chain A—the chain we want to keep—and have selected the rows in the list for the other three chains (B through D):

* In some environments, the **Open Files** dialog will be greyed out after you close the **Options** dialog, and you will need to click on it to reactivate it.

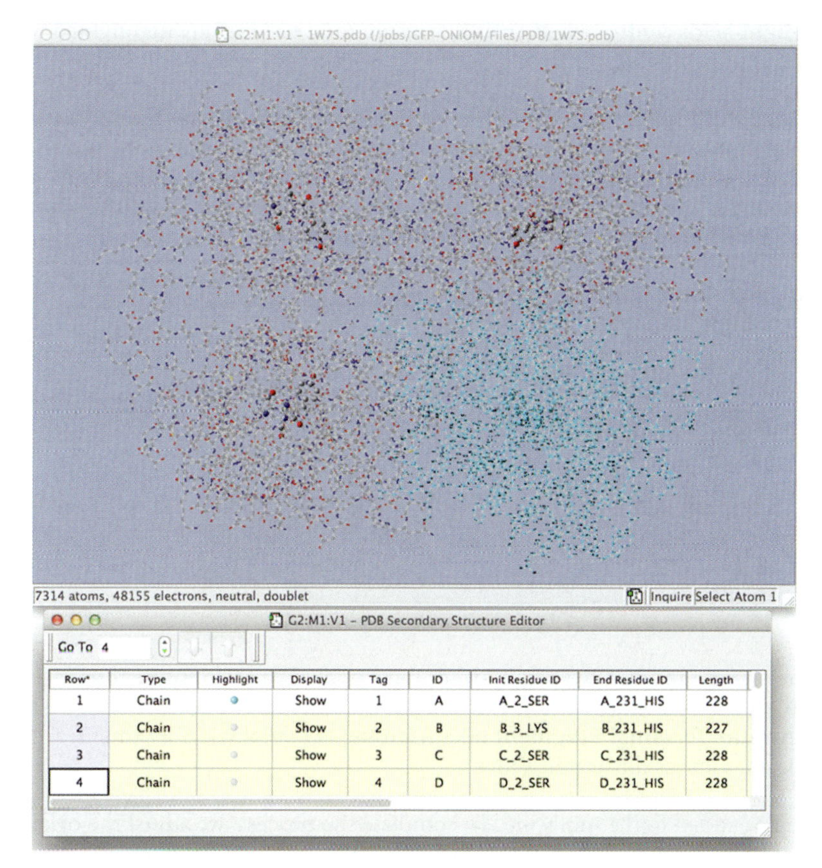

We can delete the selected chains using the **Edit→Delete→Selected Secondary Structures** menu item. We save the resulting structure as a new PDB file.

T2: Dealing with Missing Residues

Files used in this phase

starting: *1w7s_0.pdb*
after adding residue & O: *1w7s_0_add_0.pdb*
after editing: *1w7s_0_add_1.pdb*
final: *1w7s_1.pdb*

We now need to deal with the missing residues. This file is missing residues at both ends of the chain: residue 1 (alanine) and residues 232-238. Fortunately, there are no missing atoms.

We will deal with the first omission by adding the alanine amino acid manually.

In general, you can produce an isolated, close-up view of a specific portion of a structure in the following manner. Use the **PDB Residue Editor** (available from the **Edit** menu) to hide everything in the molecule other than the section of interest. Select the rows for the residue(s) you want to see, and then select **Hide Others** from the **Display** pop-up menu:

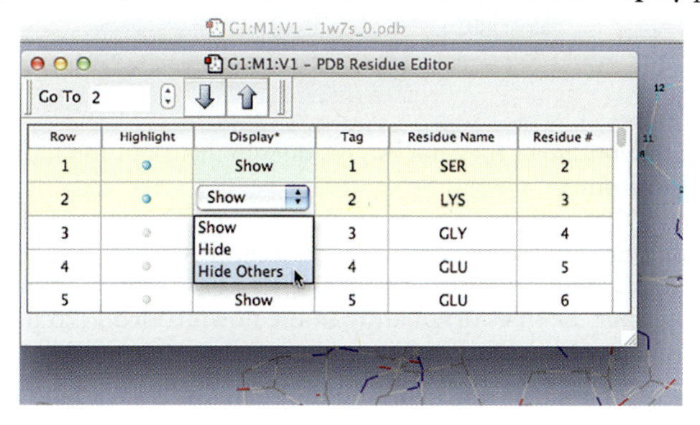

The **Highlight** fields control whether the corresponding residues are highlighted when they are visible.

We have used this technique in the following figure, which illustrates the process of adding the alanine. The larger blue window is the structure we are working on, and the smaller green window is another PDB file that does include the residue in question. We have zoomed in on the relevant residues in each structure, and we changed the **Display Format** for the **Molecule**'s **Low Layer**—the protein—to **Ball & Bond Type**:

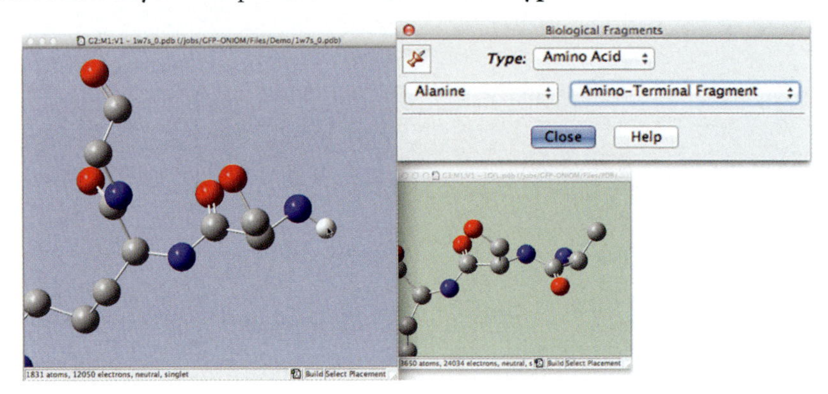

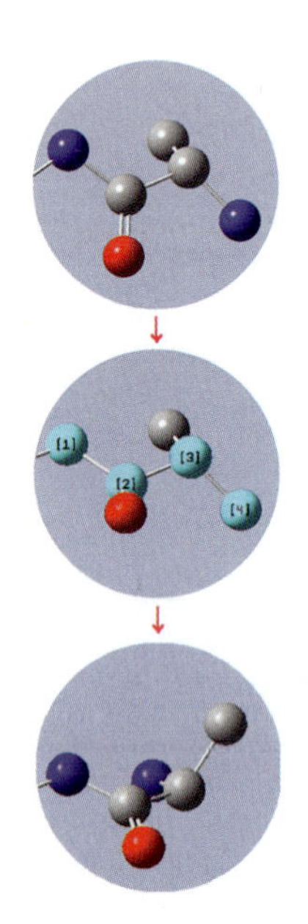

To add the alanine residue, we begin by adding a hydrogen atom to the terminal nitrogen atom in the serine residue, which will serve as an attachment point. Next, we select the **Alanine Amino-Terminal Fragment** (aka **Alanine_NT**) biological fragment, and click on the hydrogen atom (as illustrated above).

The alanine is added to the molecule. To complete the process, we adjust the orientation of a dihedral angle to match our reference structure (illustrated in the margin).

We could deal with the missing residues at the other end of the chain in a similar manner and add them manually. However, in this case, we choose to simply ignore them as they are quite far from the chromophore (~20 Å) and do not influence the chromophore's photoreactivity in any significant way.

We will need to terminate the final residue: residue 231 in the original PDB file numbering (histidine). We do so by adding an oxygen atom:

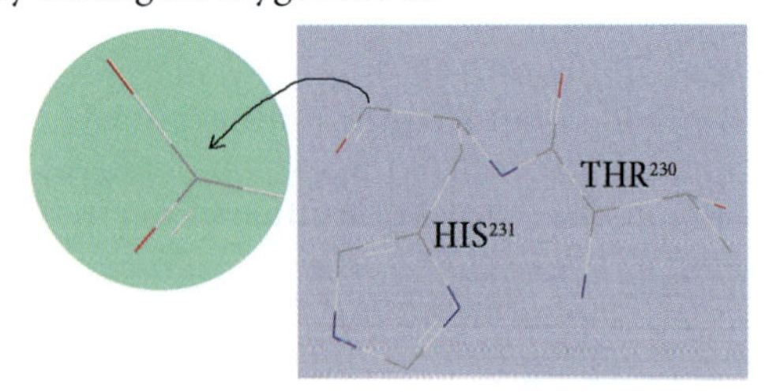

The new oxygen atom may look like this on the screen:

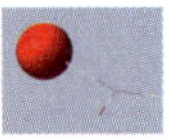

This display artifact occurs because GaussView places new atoms into the high ONIOM layer by default.

The preceding illustration shows the final two residues in the chain and indicates the location of the added oxygen atom.

We save the file as a new PDB file (*1w7s_add_0.pdb*). If we examine it, we will see that GaussView encoded all of the additional atoms as HETATMs, placing them at the end of the coordinate section, after the lines originating in the downloaded PDB file. We want them to be recognized as part of standard residues, so we will need to modify the file with a text editor. We will make the following changes:

- We will remove all hydrogen atoms present among the additional atoms.
- We will move the atoms for the alanine residue from the end of the list to their proper location at the beginning of the chain: as residue 1. The added oxygen atom in the histidine residue will also be placed within that residue.
- We will transform the HETATM lines into ATOM lines.
- We will apply standard residue atom labels to the atoms.
- We will remove unneeded CONECT lines.

Before modification, the added atoms appear as follows:

```
HETATM 1832  C           0      12.370  22.947 106.418 ...     C
HETATM 1833  C           0      11.779  23.426 105.100 ...     C
HETATM 1834  O           0      11.748  23.143 107.495 ...     O
...
HETATM 1843  H           0      10.643  22.553 103.497 ...     H
HETATM 1844  O           0      -1.070  21.901 100.869 ...     O
```

We move the lines for heavy atoms in the atom range 1832-1843 to the beginning of the atom list, before residue 2. We will then transform each line into a standard ATOM line for residue 1. The beginning of the atom list appears as follows after we have transformed the first line in the alanine residue:

```
ATOM    1832  C    ALA A   1      12.370  22.947 106.418 ...     C
HETATM 1833  C           0      11.779  23.426 105.100 ...     C
HETATM 1834  O           0      11.748  23.143 107.495 ...     O
HETATM 1835  N           0      12.802  23.995 104.246 ...     N
HETATM 1837  C           0      11.106  22.233 104.430 ...     C
ATOM       1  N    SER A   2      13.667  22.255 106.434 ...     N
```

It is not necessary to change the atom serial numbers. GaussView will clean them up for us soon.

Some programs that process PDB files require that atoms within residues be labeled in a standard way, with each atom having a unique label (the third field in each line). After we apply these labels to our alanine residue, the corresponding lines appear like this:

```
ATOM    1832  C    ALA A   1      12.370  22.947 106.418 ...     C
ATOM    1833  CA   ALA A   1      11.779  23.426 105.100 ...     C
ATOM    1834  O    ALA A   1      11.748  23.143 107.495 ...     O
ATOM    1835  N    ALA A   1      12.802  23.995 104.246 ...     N
ATOM    1837  CB   ALA A   1      11.106  22.233 104.430 ...     C
ATOM       1  N    SER A   2      13.667  22.255 106.434 ...     N
```

You can easily determine the proper atom labels by comparing the new residue to another one of the same type within the same file or a different PDB file.

The added oxygen atom in the final residue appears as follows in the edited file:

```
ATOM    1830  NE2 HIS A 231      -1.559  17.636 103.502 ...     N
ATOM    1831  OXT HIS A 231      -1.070  21.901 100.869 ...     O
TER     1831      HIS A 231
```

OXT is the atom label for a terminal oxygen atom.

The final step is to remove all CONECT lines that refer to atom numbers greater than 1831. We save the resulting file as *1w7s_0_add_1.pdb*.

The final step is to open the edited file in GaussView and then immediately save it as a PDB file, which we name *1w7s_1.pdb*. This will cause the atoms to be renumbered, and it is the file that we will use for adding hydrogen atoms in the next phase.

T3: Protonating the Protein

The next task we need to do is to add hydrogen atoms to the structure in the PDB file. For this phase, we will deal with the protein and the chromophore separately. We begin with the former.

Protonating the Protein

Files used in this phase
starting: 1w7s_1.pdb
ProPKA protonated protein: 1w7s_1_protein.pqr
after converting to PDB: 1w7s_2_protein.pdb

The tool we will use to add hydrogen atoms to the protein is call ProPKA. You can download, build and install it on your own computer or you can run it via the PDB2PQR web server* at *nbcr-222.ucsd.edu/pdb2pqr_1.9.0*. Although there are many software options for adding hydrogen atoms to a structure from a PDB file—including GaussView—ProPKA allows you to specify the desired pH and will protonate the structure accordingly. Most packages add hydrogen atoms based on templates for standard residues that do not take protonation or pH into account. They typically report the pH as 7.0.

The main window of the PDB2PQR site appears in the following figure. It consists of a form for running ProPKA jobs. You can retrieve a file directly from the Protein Data Bank or upload a file that you have prepared, as we do here.

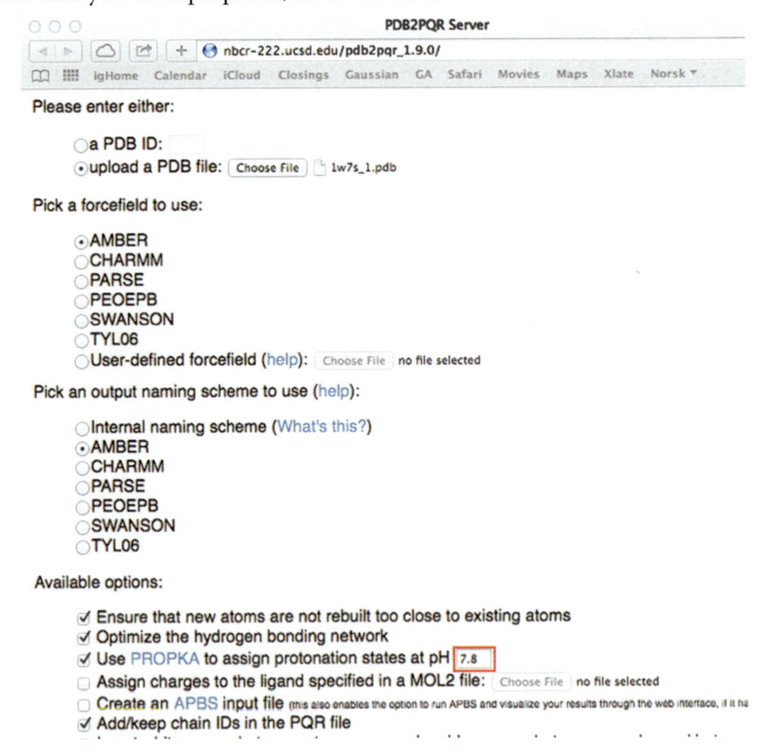

We selected the Amber force field since that is the molecular mechanics method we plan to use in our ONIOM calculations. We also set the pH to 7.8, as documented in the original PDB file, since we are modeling the same conditions as the experiment that produced the

* A few other packages also offer an interface to ProPKA, but we found the PDB2PQR service most helpful. For example, Open Babel does, but it places all of the added hydrogen atoms at the end of the PDB file atom list rather than inserting hydrogen atoms within the existing residues. Chimera can run ProPKA as well if the software is installed on the local system.

structure. Finally, we choose to retain the chain IDs (the default is to discard them). Once we have finished selecting options, we use the **Submit** button to start the calculation. The **Status** page then appears, which is updated periodically with information about the running job. When the process is finished, the final **Status** page will contain links to the report from the program and the generated output file. You will need to download the *.pqr* file, which is in PQR format (see below).

Status
Message: complete
Run time: 4 seconds
Current time: Fri Oct 17 08:51:53 2014

Here are the results:

- Input files
  - 14135610789.pdb
- Output files
  - 14135610789.propka ← *analysis*
  - 14135610789.pqr ← *output file*
- Runtime and debugging information
  - Program output (stdout)
  - Program errors and warnings (stderr)

Selecting the appropriate protonation state for your molecule is very important for accurate modeling later. For example, protonating our structure to a pH of 7.0 results in a charge on the protein of -3; a pH of 7.8 yields -6.

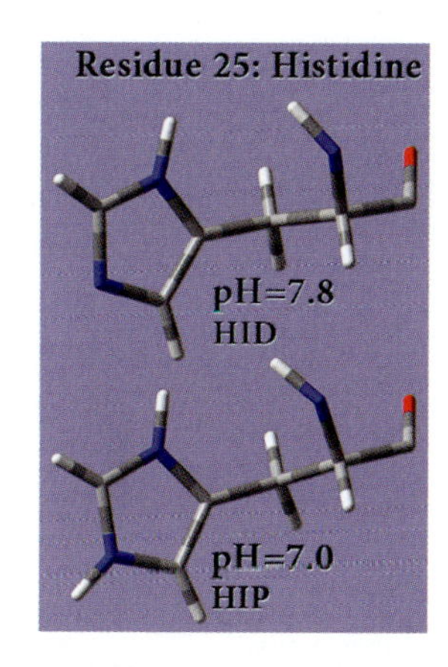

The three additional protons in the 7.0 pH structure are located on residues 25 (HIS), 222 (GLU) and 231 (HIS). The first of these is illustrated in the margin; the hydrogen atom on the ring opposite the carbon chain is not present in the 7.8 pH structure. The ProPKA program also follows the Amber naming conventions for the residues to reflect hydrogen configuration/protonation (as we requested), and the alternate names for histidine used in these two cases are noted in the figure: HID for "delta H histidine" and HIP for "protonated histidine."

ProPKA automatically discards any residues in the PDB file that it cannot recognize. This includes both non standard residues and/or HETATMs corresponding to structures other than the protein. In our case, this means that the chromophore will not be present in the generated PQR file. ProPKA reports this information in its output file as well as in the REMARKs of the generated PQR file.

HIS pKas are very difficult to predict accurately because the pKa of the isolated HIS is very close to 7. Therefore, HIS pKas are very sensitive to the environment and can shift several pKa units. Careful examination of HIS final protonation states is highly recommended, regardless of the results from PDB2PQR.

In the case of GLU, a careful examination of the final structure is needed because we ran PDB2PQR without the internal waters and the chromophore. GLU222 happens to be in the pocket near the chromophore, so the PDB2PQR run predicted a very large pKa shift for this residue as it seemed to be completely buried in a hydrophobic environment. Once you put the chromophore back, one can see that one of the OH groups in the chromophore is within hydrogen bond distance of GLU222, so obviously the PDB2PQR result is unreliable for this residue. Looking at the ProPKA output, it predicted an unreliable pKq=7.79 for this residue, just below the pH=7.8, indicating that we were lucky that the residue was not incorrectly treated as neutral (protonated).

Converting Back to PDB Format

PDB2PQR produces a PQR file as its output. This format is a slight modification of the PDB format, but it is different enough that such files must be converted before software programs will accept them.

There are also many options for converting PQR files to PDB files. Open Babel can interconvert a very wide range of data files related to chemistry. The main window of the Windows version is illustrated in the next figure (there is also a command line version). We set the formats for the input and output files via the pop-up menus indicated in green in the figure. We can navigate to the input file—the PQR file we just created—by clicking on the browse button in the input area (labeled "..." and indicated in red in the figure following). Once the file is loaded, its contents appear in the text box below these fields.

This tool works best if you enter the output file name (including directory location if you don't want it in the current directory) into the **Output file** field (indicated in red). You will also need to uncheck the **Output below only** box below this field in order to create an output file (by default, the generated file appears only in the text box in the output area).

You can initiate the conversion process by clicking the **Convert** button (indicated in yellow).

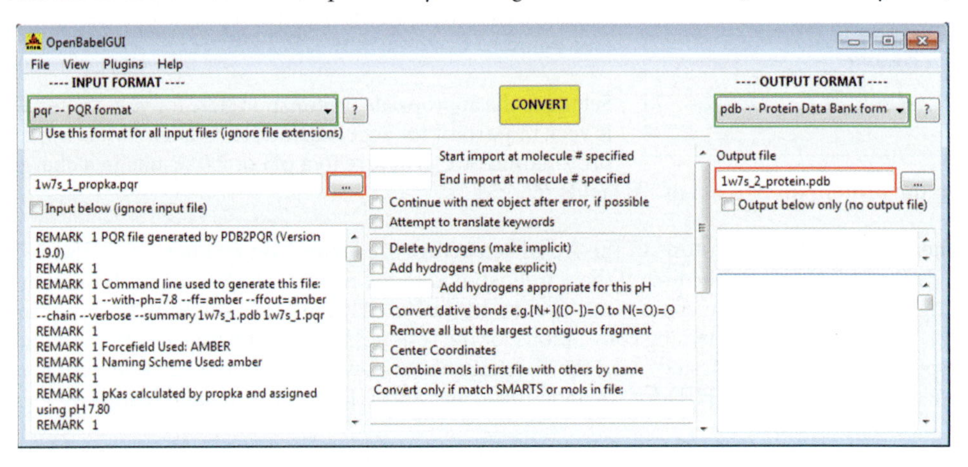

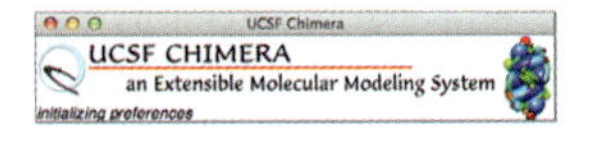

Chimera can perform the same function via a very simple process. Open the PQR file in Chimera (**File→Open**), and then select the **Save PDB** item on the **File** menu. You select the file to open and save using standard dialogs. We name the converted PDB file *1w7s_2_protein.pdb*.

Files used in this phase

starting: 1w7s_1.pdb

extracted chromophore: 1w7s_1_chromo.pdb

after adding H atoms: 1w7s_2_chromo.pdb

T4: Adding Hydrogen Atoms to the Chromophore

Now we need to add hydrogen atoms to the chromophore. In order to do so, we need to create a PDB file containing just the chromophore. This is easy to do: we simply extract the lines in the atom list corresponding to the chromophore and save them in a file with a .pdb extension:

```
TITLE      WILD-TYPE AEQUOREA VICTORIA GFP
REMARK    1 Chromophore only from 1W7S
HETATM  478  C1  GYS A  66     ...                        C
HETATM  479  N2  GYS A  66     ...                        N
...
HETATM  498  OH  GYS A  66     ...                        O
END
```

Once we have saved the file, we can open it in GaussView, this time allowing GaussView to add hydrogen atoms. Here is the resulting structure:

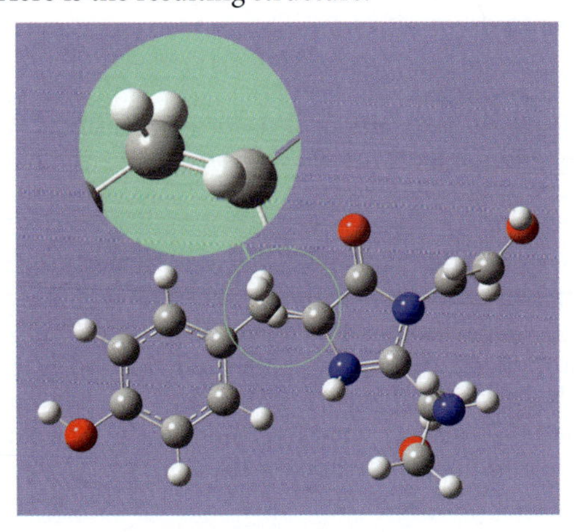

Adding hydrogen atoms to random collections of atoms—remember, computer programs don't understand chemistry—is a difficult problem due to the inherent ambiguity. GaussView has guessed where to place them, sometimes correctly, sometimes not. The highlighted area is a clear trouble spot in the resulting structure; the inset shows the results after the position of the overlapping C and H atoms has been corrected manually.

In order to know what other modifications are necessary to clean up this molecule, we need to understand the molecular structure of the chromophore. The following figure illustrates the GFP chromophore and its formation:

TYR
GLY
SER
cyclization
dehydration
oxidation
HBI

The chromophore compound is 4-(p-hydroxybenzylidene) imidazolidin-5-one (HBI).

The chromophore consists of modified residues within the protein chain:* the SER-TYR-GLY sequence conventionally numbered as residues 65-67 (as they appear in the original file in the Protein Data Bank for GFP: 1GFL). In our PDB file, they are residue 66 (label: GYS). This amino acid sequence spontaneously transforms in its natural environments into HBI through a process of cyclization and oxidation. The resulting compound fluoresces upon exposure to blue light, but only within the protein environment.

Given the known structure of the chromophore, we modify the GaussView structure, removing and repositioning hydrogen atoms as needed. Here is the final structure:

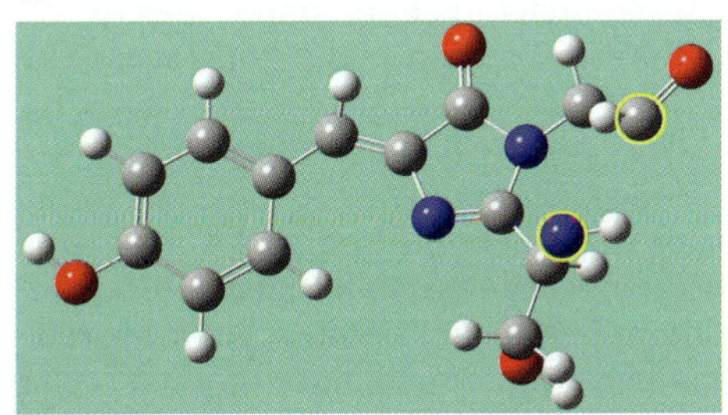

We corrected the bonds displayed by GaussView by changing the types of the bonds to match the known structure, but we have not altered any bond distances.

The highlighted atoms are the ones that are bonded to other residues in the protein chain. They remain as open valences in the structure. In addition, there are several hydrogen bonds to the protein environment, both to other residues and to water molecules. We will deal with both types of bonds later, when we recombine the chromophore and the protein into a single file.

After we finished editing the structure, we saved the file in PDB format, calling it *1w7s_2_chromo.pdb*.

Here are a couple of tips related to performing this process in GaussView:

* Specifically, within the set of six alpha helices running through the (near) center of the beta barrel.

- Be sure that the spin multiplicity reported by GaussView for the completed structure is what you expect. For example, if the structure should be a singlet but GaussView tells you it's a doublet, then you have an errant hydrogen atom somewhere.

- When you reopen the PDB for the chromophore that you've just created, be sure to prevent GaussView from adding hydrogen atoms in the **Options for Opening Files** dialog. Otherwise, GaussView will read all of the hydrogen atoms you just deleted. If you forget to change this option before opening the file, just close it again without saving it, and the added hydrogen atoms will be forgotten.

- Sometimes, the easiest and fastest way to modify the position of a hydrogen atom is to delete it and then add it again with GaussView's **Add Valence** and **Delete Valence** tools. If you do so, then the new hydrogen atoms will not be interpreted as part of the chromophore residue, but will rather appear at the end of the atom list:

```
HETATM    31 2HB1 GYS A   66     ...              H
TER       32      GYS A   66
HETATM    33  H            0     ...              H
HETATM    34  H            0     ...              H
...
```

In this case, you can edit the corresponding lines to reflect the proper residue membership:

```
HETATM    31 2HB1 GYS A   66     ...              H
HETATM    33  H   GYS A   66     ...              H
HETATM    34  H   GYS A   66     ...              H
...
```

Note that we also removed the TER line as it will not be needed when we recombine the PDB file for the two parts of the overall molecule. It is not necessary to renumber the atoms. However, if you prefer to do so, simply open the edited file with GaussView and then resave it; GaussView renumbers the atoms at each save.

T5: Computing MM Charges for the Chromophore

The next major task to perform is to prepare the molecule for modeling with molecular mechanics. We will be using the Amber method, and Amber requires that atom types and partial charges be specified for every atom in the structure. GaussView can almost always perform this task for the atoms in standard residues, but the partial charges for the atoms in any nonstandard residues—here, the chromophore—must be computed.

The R.E.D. tools were created to perform this function (*q4md-forcefieldtools.org/RED*). The name expands (somewhat confusingly) to RESP ESP charge derive. You will need to download these tools from the Internet and install them on a computer locally. Installing and running these tools requires Perl, a Fortran compiler and access to Gaussian, most conveniently all on the same computer.

We used RED* version III.52, the most recent at the time of this writing. Some of the version numbers in the discussion below may be different when you install the software.

Installation is fairly straightforward. Unpack the downloaded archive to a convenient location. We will refer to this location by the environment variable *RED*. Change to that directory's *resp-2.2* subdirectory, and compile the **resp** program from its Fortran source file using the **make** facility:†

* We will omit the periods when we refer to these tools from now on.

† On UNIX-based systems. On Windows systems, you must use the appropriate process for whatever Fortran compiler is available.

```
$ cd $RED/resp-2.2
$ make
```
version number may vary

The process of using the RED tools to compute partial charges for molecular mechanics is outlined in the following figure.

Files used in this phase

starting:
1w7s_2_chromo.pdb
added methyl groups:
1w7s_2_chromo+me.pdb
output from Ante_RED.pl:
1w7s_2_chromo+me-gau.com
1w7s_2_chromo+me-out.p2n
output from Gaussian:
1w7s_2_chromo+me-gau.log
edited log and p2n files:
Mol_red1.log, *Mol_red1.p2n*
output from RED-vIII-52.pl:
Mol_m1-o1-sm.mol2
1w7s_3_chromo_resp.mol2

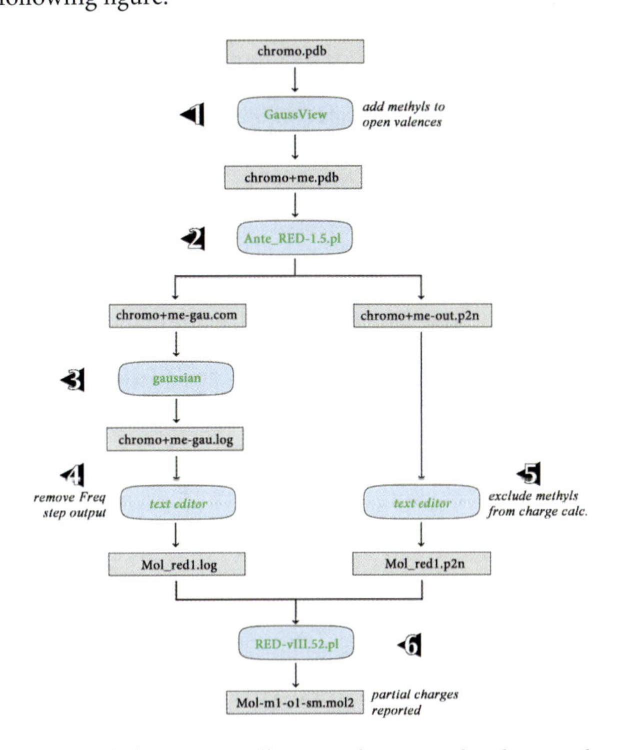

The first step is to add methyl groups to all open valences in the chromophore, and then save a new PDB file. We started with the final file from the previous phase, *1w7s_2_chromo.pdb*, and named the new file *1w7s_2_chromo+me.pdb*.

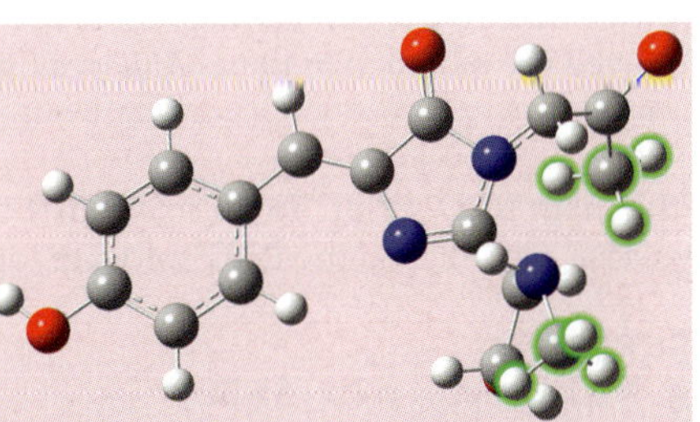

Next, we run the **Ante_RED** Perl script, specifying the PDB file for the methylated structure as its argument:

```
$ perl $RED/Ante_RED-1.5.pl 1w7s_2_chromo+me.pdb
```
produces 1w7s_2_chromo+me-gau.com, 1w7s_2_chromo+me-out.p2n and other files

This command is a preliminary one that simply sets up a series of files that will be needed as various points in the partial charges computation process. We will use two of the files it creates:

- *1w7s_2_chromo+me-gau.com*: A starting input file for a Gaussian calculation.
- *1w7s_2_chromo+me-out.p2n*: Key lines extracted from the PDB file.

We will need to edit both of these files before using them.

3 The *-gau.com* file contains a Gaussian calculation that we need to run. Before doing so, we edit it to modify the Link 0 commands in it as necessary for our environment: amount of memory, number of processors, checkpoint file location, etc. We simply removed the Link 0 commands from the generated *.com* file and let our default settings apply. If necessary, we could also have corrected the charge and/or spin multiplicity.

We ran the resulting job, a Hartree-Fock/6-31G(d) optimization+frequency calculation with tight convergence criteria and extra detail in the output file:

#P HF/6-31G* Opt=(Tight,CalcFC) Freq SCF(Conver=8) Test

The job successfully located a minimum.

4 After examining it, we edited the log file and deleted all of the output from the second job step:

```
 Leave Link 9999 at Oct 18 2014, MaxMem=536870912 cpu: 0.1
 Job cpu time:       0 days  0 hours  0 minutes 59.4 seconds.
 File lengths (MBytes):  RWF=684 Int=0 D2E=0 Chk=13 Scr=1
 Normal termination of Gaussian 09 at Oct 18 16:25:56 2014.
 (Enter //g09/l1.exe)     delete all lines starting from this line
 Link1:  Proceeding to internal job step number  2.
```

We saved the edited log file as *Mol_red1.log*; the RED tools require the use of this exact name.

5 Next, we must edit the *.p2n* file, adding a line to tell the software not to include the methyl groups when calculating the partial charges (indicated in red):

```
REMARK   1 File created by GaussView 5.0.9
REMARK INTRA-MCC 0.0 |  35  36  37  38  39  40  41  42  | R
REMARK
```

The INTRA-MCC line includes the list of atoms to skip by number between the vertical bar characters; here, we say to exclude atoms 35-42. Note that this line must be formatted in accordance with the software's inflexible specifications: there are two spaces on either side of the atoms numbers. We saved the edited *.p2n* file as *Mol_red1.p2n*, the name required by the RED software.

6 We are now ready to run the main RED script. To do so, we first clear the Gaussian scratch directory and copy the two *Mol_red1* files to the RED software top-level directory:

```
$ rm -rf $GAUSS_SCRDIR/*
$ cp Mol_red1.log Mol_red1.p2n $RED/
$ cd $RED
$ perl RED-vIII-52.pl
```

The command runs for a few minutes. It performs several operations:

- Examining the *Mol_red1.log* and *Mol_red1.p2n* files.
- Setting up and running a Gaussian **Pop=MK** electrostatic potential-derived charges calculation.*
- Retrieving results from the resulting log file

* The input file is *JOB2-gau_m1-1-1.com* in the *$RED/Data-RED* subdirectory.

- Running the **resp** program with this data to compute the partial charges

All of the intermediate and final output files are stored in a new subdirectory of *$RED* named *Data-RED.** The file *Data-RED/Mol-m1-o1-sm.mol2* lists the computed charges for each atom in the final numeric field of its line (output is condensed):

```
...
@<TRIPOS>ATOM
                                                  residue
atom# label   ------------coordinates---------- elem  #  type        partial charge
  1    C1     1.278   0.370  -0.200  C          1  GYS          0.0487  ****
  2    N2     0.008   0.398  -0.124  N          1  GYS         -0.2334  ****
...
```

We typically copy this file to the working directory for the project, giving it a more descriptive name:

```
$ cp $RED/Data-RED/Mol_m1-o1-sm.mol2 \
  /path/1w7s_3_chromo_charges.mol2
```

We will use these charges when we set up the Gaussian input file for the molecule later.

T6: Recombining the Protein and the Chromophore

The next step is to reunite the two parts of the structure that we have worked on separately: the protein and the chromophore. In some cases, there will be a third part as well: the water molecules.

There are several options for combining PDB files:

- Manually, with a text editor. This is done by appending the atom list from one PDB file to the atom list in the other PDB file. In our case, we could add the atom lines for the chromophore to the atom list in the PDB file for the protein.

  This approach will usually work, but it may be necessary to renumber the atoms that you are adding to avoid duplications in the combined atom list. In the case of the chromophore (and nonstandard residues in general) this must be done both in the HETATM lines and in the CONECT lines. You can accomplish this manually; writing a simple shell script to do this task is a more reliable approach.

- You can use a software program to combine the structures. In this case, the task will be easiest if the files use the same coordinate system (i.e., all atom coordinates share the same origin). This will be the case if you have followed the procedures we describe here, always opening and saving PDB files as you make modifications to the original structure.

We will take the latter approach and use GaussView. Once you have opened the first file in GaussView, you can open the next file, setting the **Target** field in the **Open Files** dialog to **Append all files to active molecule**. The new molecule will appear in the active window along with whatever structure(s) were there previously.

The advantages of using GaussView for this task are that it retains the original residue numbering from both files and that it is the ultimate target environment for the structures (since our goal is to set up a Gaussian job). Its disadvantage is that it does not retain all of the secondary structure information that may be present.

* If *Data-RED* already exists, the new subdirectory will be given a unique name of the form *Data-RED-n* where *n* is an integer.

Combining choices in GaussView. There are two ways to combine structures in GaussView: specifying append mode when opening the additional file(s) and cutting-and-pasting. In the first method, you open one of the files that you want to combine. Then, with that molecule group (window) active, you open the next file, setting the **Target** field in the **Open Files** dialog to **Append all files to active molecule**. This will add the molecule from the new file to the existing structure. Accomplishing the task this way will retain residue and secondary structure information from all of the combined files.

In the second method, you open the files to be combined as separate windows (molecule groups) in GaussView. You then select all of the atoms in one of the windows, and then cut-and-paste them into the other window, via **Edit→Copy** and **Edit→Paste→Append Molecule**. This is quicker than closing and reopening the file, but doing the job this way causes all of the residue information for the pasted molecules to be lost.

Files used in this phase
starting:
1w7s_2_chromo.pdb
1w7s_2_protein.pdb
combined file from Chimera:
1w7s_4.pdb
final edited structure:
1w7s_5.pdb

Chimera can also combine structures from multiple PDB files. Its advantages are that it retains all secondary structure information (and some other PDB file data as well) and that it can handle some coordinate system differences among the separate files. Its disadvantage is that residues are renumbered in the second and additional structures even when their original numbers do not conflict with those in the first file.

We will see examples of both software packages here. We begin by combining the protein and the chromophore in Chimera.

We open both files and use the **Model Panel** (opened via the **Favorites→Model Panel** menu path) to display and activate both structures (controlled by the **S** and **A** checkboxes, respectively). The display shows both molecules, color coded as indicated in the structure list: here, tan for the protein and blue for the chromophore:

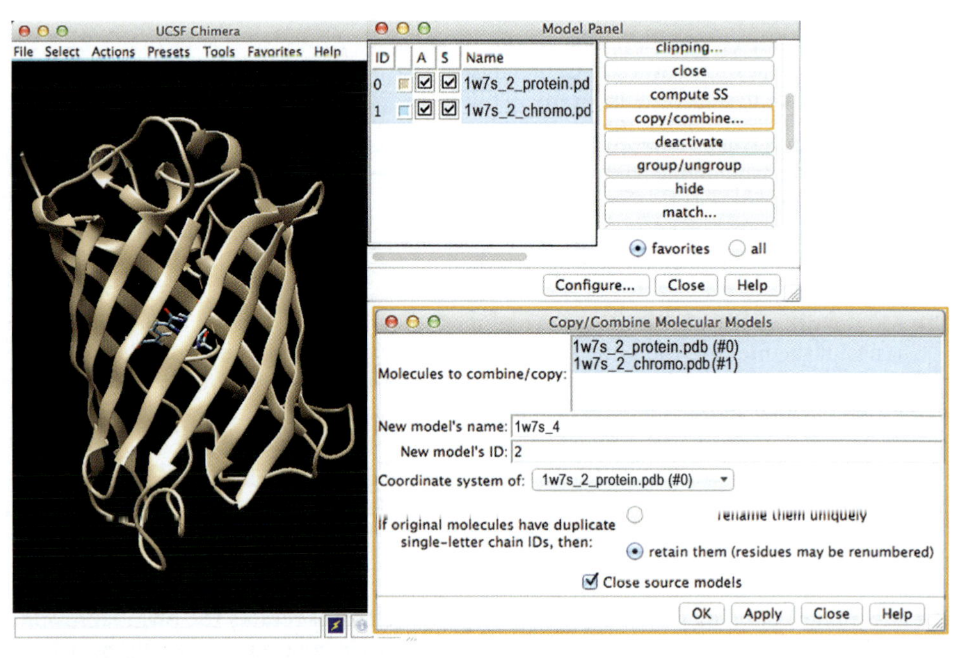

We combine the two molecules using the **copy/combine** button in the list on the right in the **Model Panel**, which opens the **Copy/Combine Molecular Models** dialog (appearing below the **Model Panel** in the figure). We select both molecules and provide a name for the new model (here "1w7s_4"). We choose to use the coordinate system of the protein; in our case,

this choice is irrelevant since the two PDB files use the same coordinate system. We choose to retain the chain IDs if they are the same residues that will reside in chain A. Clicking **OK** will perform the combination operation, resulting in the combined structures comprising a new model (Chimera's model #2). We can see that the separate molecules have been merged into a single model by the change in color coding in the display: all atoms are now the same color (as in the margin). We then save the new model as a PDB file: *1w7s_4.pdb*.

No matter how you choose to combine the molecules, you will need to verify and possibly correct the bonding between the chromophore and the protein in the combined file. The following figure shows the two bonds involving the formerly open valences in the chromophore:

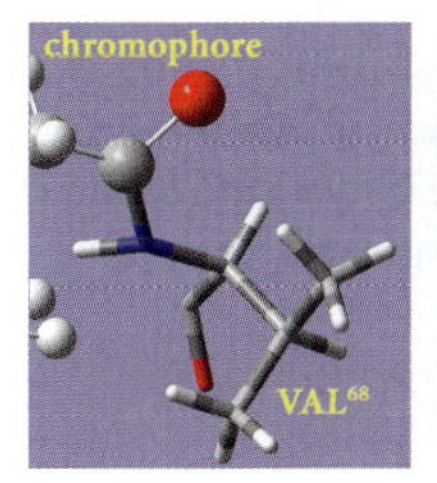

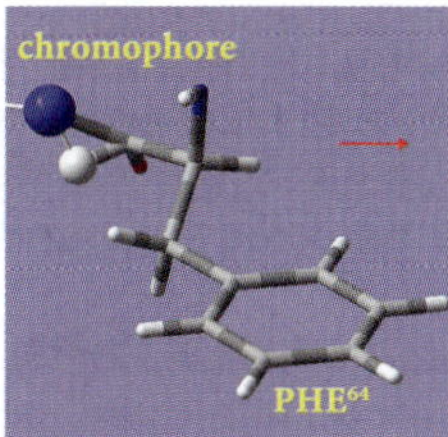

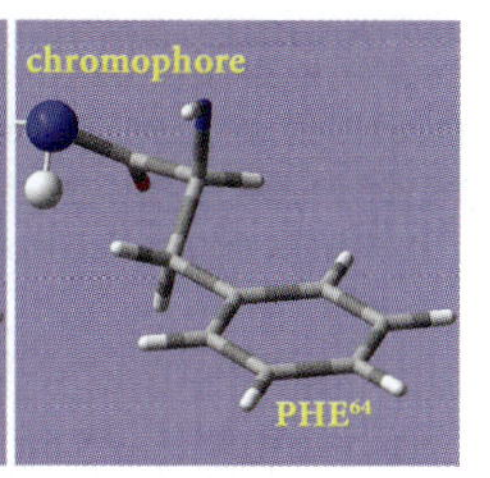

The illustration on the left shows the bond to the valine residue, which is fine. The two illustrations on the right show the bond to the phenylalanine residue, which we tweaked a bit. We saved the final structure as *1w7s_5.pdb*.

T7: Handling the Water Molecules

The final task in creating our starting structure is handling the water molecules. The specific procedure depends on the specific needs of the computational investigation:

- If we do not need the water molecules and we discarded them when opening the original PDB file, then the structure is complete, and we can move on to setting up the Gaussian input (see the next subsection).
- If we retained the water molecules from the original PDB file and are satisfied with their locations and configuration, the structure is again complete, and we can move on.
- If we stripped out the water molecules for convenience but want to use them in our calculations, then we need to add them back in at this point. We can choose to use them as specified in the original PDB file, or we can make modifications to them, depending on our needs.

Merging the Original Water Molecules

If we want to incorporate the water molecules from the original PDB file without modification, we could perform the following steps:

Files used in this phase

starting:
1W7S.pdb
water molecules file:
1w7s_0_waters.pdb

- Using a text editor, we extract the HETATM lines from original PDB file for chain A only.

```
ATOM    7317  NE2 HIS D 231  coordinates ... N
TER     7318      HIS D 231
HETATM  7319  O   HOH A2001  coordinates ... O
HETATM  7320  O   HOH A2002  coordinates ... O
...
HETATM  7571  O   HOH A2253  coordinates ... O
HETATM  7572  O   HOH B2001  coordinates ... O
```

We save them as a separate PDB file (*1w7s_0_waters.pdb*).

- We open that PDB file in GaussView, setting **Add Hydrogens** to **Yes** in the **Options for Opening Files** dialog. We save the file once the hydrogen atoms have been added and close the file.
- If necessary, we open the combined structure file we created in the preceding phase, taking care to return to the **Options for Opening Files** to set **Add Hydrogens** to **No** to ensure that no hydrogen atoms are added.
- We reopen the file containing the water molecules including hydrogen atoms, setting the **Target** field in the **Open Files** dialog to **Append all files to active molecule**. This will add the water molecules to the protein and chromophore structure.
- We set up a minimal Gaussian job using the Amber molecular mechanics method. We set the **Method** pop-ups to **Ground State**, **Mechanics** and **Amber** in the **Method** panel of the **Gaussian Calculation Setup** dialog. We also verify that the **Write PDB Data** item is selected in the **General** panel. We could then save a Gaussian input file.

A Gaussian input file prepared in this way is ready to be set up for an Amber calculation (see p. 424). In our case, however, we decide to modify the configuration of the water molecules before proceeding, as described in the following subsection.

Using Dowser to Add/Position Crystallographic Waters

Files used in this phase
starting:
1w7s_5.pdb
1w7s_0_waters.pdb
Dowser input:
1w7s_5+w.pdb
Dowser output file:
dowserwat.pdb
edited waters files:
1w7s_5_ext.pdb
1w7s_5_int.pdb
final combined file:
1w7s_6.pdb
1w7s_6.gjf

The Dowser program is designed to place water molecules at appropriate positions within the interior of large protein structures and to optimize their positions. It can do so whether or not the molecule already contains water molecules (i.e., from the originating PDB file). The program is designed to prepare structures for molecular dynamics simulations.

Dowser runs on UNIX and Linux systems (but not under Mac OS X). After downloading and unpacking the software, you must build it by running the Install script in its main directory. The build process requires the **g77** compiler. This is the name used by older versions of the GNU Fortran compiler, and simply making a symbolic link between the **gfortran** command and **g77** worked for us.*

We decided to run Dowser on the combined structure with the water molecules from the original PDB file. We created a new PDB file by adding the HETATM lines for the water molecules that we had extracted earlier to the PDB file for the combined structure, creating *1w7s_5+w.pdb*. We edited the new file and changed the HIE and HID residue names to HIS because Dowser does not recognize the Amber residue name variants.

We began by running the following commands† (the DOW environment variable points to the Dowser main directory):

```
$ . $DOW/dowserinit                              Set up environment for Dowser.
$ dowser 1w7s_4+w.pdb 2>&1 >dowser.log           Save output to a file.
```

Dowser proceeds as follows:

- It removes external water molecules from the structure (if any).
- It optimizes the positions of the existing water molecules, producing files named *xtal_hoh.pdb* and *xtal_o.pdb* (which contain/omit hydrogen atoms, respectively). We can use the latter file for comparison purposes later if we desire.

---

* **sudo ln -s /usr/bin/gfortran /usr/bin/g77**

† Assuming the bash shell.

- It identifies internal points and places water molecules at those locations.
- It optimizes and refines water molecule locations. The result of this step is stored in the file named *dowserwat_all.pdb*.
- Waters whose positions are external to the protein are removed from the set of water molecules. The result of this step is the file named *dowserwat.pdb*. This is the primary result file from Dowser, which we will use in a bit.
- The Dowser results are compared to the original set of crystallographic water molecules (if applicable). The file *dowser_report* provides a summary of this comparison.

The report from Dowser has two sections comparing the placed and optimized water molecules to the original crystallographic ones. It is similar to the following (which we've condensed):

```
* Find nearest xtal water for each dowser water
     Dowser water      ener dist  nearest xtal water
#1   13.2 -19.6 110.5 -31.2 0.3   #2140  13.4  -19.8 110.7
#2   25.1   6.8 116.1 -30.0 0.5   #2123  24.71   6.6 115.8
...
* Find nearest dowser water for each internal xtal water
   Internal xtal water  ener dist nearest Dowser water
#2001  11.1 21.9 105.8 306.0 3.0  #23  12.9  19.6 105.2
#2002  13.5 18.1 106.8  -7.3 2.3  #23  12.9  19.6 105.2
...
```

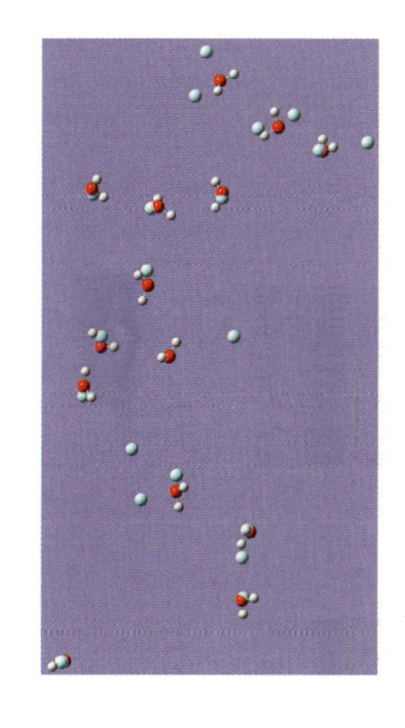

The figure in the margin shows the positions of the original water molecules (colored turquoise and sans hydrogen atoms) and the Dowser water molecules (normal appearance including hydrogen atoms). We created this figure by combining the *xtal_o.pdb* file with the *dowserwat.pdb* file in GaussView. Prior to doing so, we changed the oxygen atoms in the first file to fluorines in order to distinguish them visually.

We decide to replace the original water molecules in the interior of the protein with the ones optimized by Dowser. This will involve two steps:

- Removing the internal water molecules from the original set.
- Adding the remaining water molecules from the original PDB file and the Dowser water molecules to the protein and chromophore structure.

We perform the first task by editing the *1w7s_0_waters.pdb* file and removing all of the atoms that appeared in the "xtal water" sections of the *dowser_report* file. We saved the resulting file as *1w7s_5_ext.pdb* ("ext" for external).

We prepare Dowser's file containing the water molecules for combination with the protein and chromophore by editing *dowserwat.pdb* and changing the chain ID from X to A. We save the modified file as *1w7s_5_int.pdb* ("int" for internal).

For the second task, we first open the protein and chromophore PDB file in GaussView (*1w7s_5.pdb*). Next, we open the file containing the Dowser water molecules (*1w7s_5_int.pdb*), making sure that the **Target** field in the **Open Files** dialog is set to **Append all files to active molecule**. Then we open the third required file: the file containing the original water molecules that are external to the protein (*1w7s_5_ext.pdb*).

We initially open this file in a separate GaussView molecule group (window). The file contains only the oxygen atoms. We can add hydrogen atoms when we open the file, or we can quickly add them afterwards in this way:

- ▸Open the **Atom List Editor**: **Edit→Atom List**.
- ▸Choose the **Edit→Add Hydrogens for All Atoms** menu path.
- ▸Close the **Atom List Editor**.
- ▸Append the water molecules to the protein and chromophore structure.
- ▸We save the resulting file as a PDB file: *1w7s_6.pdb*.

We close and reopen the *1w7s_6.pdb* file (necessary for GaussView to recognize the residue information for the entire structure). We then set up a minimal Gaussian job using the Amber molecular mechanics method. We set the **Method** pop-ups to **Ground State**, **Mechanics** and **Amber** in the **Method** panel of the **Gaussian Calculation Setup** dialog. We also verify that the **Write PDB Data** item is selected in the **General** panel. We then save a Gaussian input file (*1w7s_6.gjf*).

T8: Specifying Amber Atom Types & Partial Charges

Files used in this phase

starting: *1w7s_6.gjf*

final: *1w7s_7.gjf*

The first few lines of the molecule specification section in the Gaussian input we created in the previous subsection file look something like this:

```
  -type-charge       (residue information)
N-N3-0.141400(PDBName=N,ResName=ALA,ResNum=1_A)   ...
C-CT-0.096200(PDBName=CA,ResName=ALA,ResNum=1_A) ...
C-C-0.616300(PDBName=C,ResName=ALA,ResNum=1_A)    ...
```

The usual element name is followed by two hyphen-separated fields: the Amber atom type and the partial charge on that atom. This data may also be followed by a parenthesized list of keywords and values containing residue data from the originating PDB file (as it is above). For example, the nitrogen atom has an atom type of N3 and a partial charge of ~0.14; in the PDB file, its atom label was "N," and it is in residue 1 in chain A (alanine).

Including the PDB-related data in the Gaussian input file is optional, but having it there enables the residue and secondary structure features in GaussView.

Not all atoms in the file will contain data in the Amber type and partial charge fields. This must be remedied before we can run an Amber calculation. GaussView provides tools that make doing so very convenient.

The **Atom List Editor** (**Edit→AtomList**) lets you sort the atom information based on any defined characteristic. In the following example, we sort the list of atoms according to the **AMBER Type** column. After selecting all of the rows by double clicking in the column header, we select the **Rows→Sort Selected→Ascending by AMBER Type** menu item:

We customized the **Atom List Editor** display using its toolbar buttons and **Columns** menu.

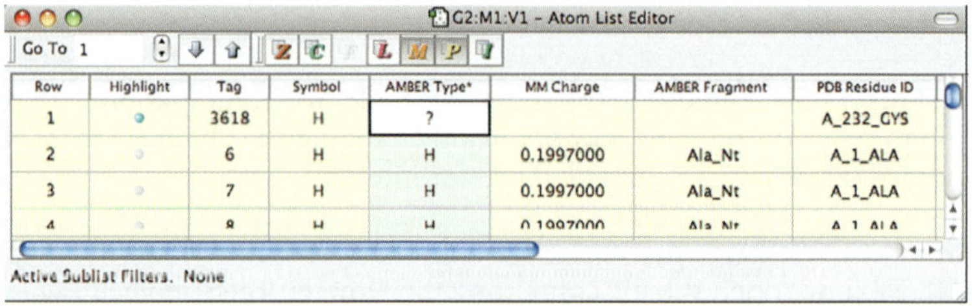

G2:M1:V1 - Atom List Editor

Go To 1

Row	Highlight	Tag	Symbol	AMBER Type*	MM Charge	AMBER Fragment	PDB Residue ID
1		3618	H	?			A_232_CYS
2		6	H	H	0.1997000	Ala_Nt	A_1_ALA
3		7	H	H	0.1997000	Ala_Nt	A_1_ALA
4		8	H	H	0.1997000	Ala_Nt	A_1_ALA

Active Sublist Filters: None

The atoms that are missing an Amber atom type relocate to the beginning of the list. In this case, there is only a single atom with an atom type of "?"; this symbol indicates that GaussView could not determine the atom type.

We will enter this atom's type—H—into the cell.

Methods for assigning Amber atom types is beyond the scope of this book, but we can provide some hints about where to look for the relevant information:

- Compare with a different file for the structure. For example, in this case, we opened the PDB file containing just the chromophore. GaussView was successful in identifying the atom type for this atom in that file, so we could use that information here.
- When the atom in question is part of a standard residue or modified from a standard residue, then you can usually apply the canonical atoms type for that environment to the atom in question. You may very well be able to locate a standard version of the residue in the same structure or another PDB file.
- You can analyze the structure and characteristics of the atom in question and then apply the appropriate Amber atom type. Lists of the defined atom types are widely available on the Internet (search for "Amber atom type list"). You can also consult the original Amber paper [Cornell95].

Of course, we can also modify the Gaussian input file directly, using a text editor. If you search for the string "-(" in the file, you will locate lines which are missing the atom type:

```
C-(PDBName=C3,ResName=GYS,ResNum=66_A) ...
```

Supplying Missing Partial Charges

The next step is to locate atoms which do not have partial charges specified. Sort the atom list by the MM Charge column in ascending order, and the atoms without charges will rise to the top of the list.

Missing charges within the protein can be obtained from the *.pqr* file created by PDB2PQR/ProPKA (*1w7s_1_protein.pqr*). For example, in order to locate the charge for this atom:

```
C-CT(PDBName=CA,ResName=PHE,ResNum=64_A) ...
```

We must find the corresponding line for the CA carbon atom in residue 64 in the *.pqr* file:

```
ATOM 939   CA   PHE A   64 20.56 -1.62 107.97 -0.0024 1.9080
```

You should verify the chain ID and residue type as well. The charge is the penultimate field in the line. We enter this value into the input file:

```
C-CT--0.0024(PDBName=CA,ResName=PHE,ResNum=64_A) ...
```

It is necessary to enter a hyphen followed by the charge value in the proper location in the corresponding line (just prior to the opening parenthesis). Note that there should not be more than a handful of protein atoms with missing charges.

The partial charges will be missing for all of the atoms in the chromophore. We will use the RED output file: *1w7s_3_chromo_resp.mol2*. Again, you can enter them in GaussView or just use a text editor. Under UNIX-compatible operating systems, you can locate lines in the input file that are missing partial charges via the following command:

```
$ grep "[^0-9](" filename                                  UNIX/Linux/Mac OS X
C-CC(PDBName=C1,ResName=GYS,ResNum=232_A) ...
```

If we're lucky, the atoms in the *.mol2* file will be in the same order as those in the Gaussian input file on which you are working. If not, then we will need to match up the atoms manually;

most often, it is the hydrogen atoms that have been reordered by the RED software. Doing so requires consulting the Gaussian input file created by the RED software (*1w7s_2_chromo+me-gau.com*), as demonstrated in the following figure:

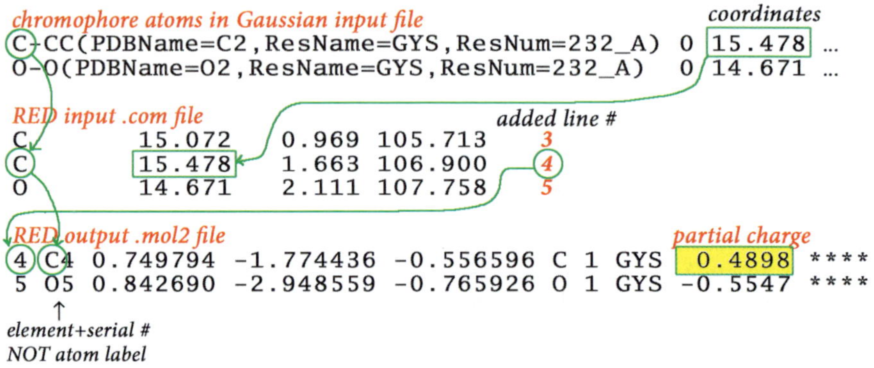

```
chromophore atoms in Gaussian input file                          coordinates
C-CC(PDBName=C2,ResName=GYS,ResNum=232_A) 0 15.478 ...
O-O(PDBName=O2,ResName=GYS,ResNum=232_A)  0 14.671 ...

RED input .com file                               added line #
C          15.072    0.969 105.713     3
C          15.478    1.663 106.900     4
O          14.671    2.111 107.758     5

RED output .mol2 file                                  partial charge
4 C4 0.749794 -1.774436 -0.556596 C 1 GYS  0.4898 ****
5 O5 0.842690 -2.948559 -0.765926 O 1 GYS -0.5547 ****
   ↑
element+serial #
NOT atom label
```

The lines at the top are the ones in the Gaussian input file to which we want to add partial charges. We note the element and x coordinate of the carbon atom (PDB label C2). We find the same atom in the RED Gaussian input file and note which line in the atom list it is (we added line numbers to a copy of the file to make this easier). Finally, we retrieve the charge from the corresponding line of the RED output *.mol2* file and enter it into our input file:

```
C-CC-0.4898(PDBName=C2,ResName=GYS,ResNum=232_A) 0 15.478 ...
```

The coordinates are different in our Gaussian input file and the RED *.mol2* file because the latter file uses the final coordinates from the Hartree-Fock geometry optimization.

The coordinate for the final hydrogen atom was different from that in the RED input file due to interaction with the methyl group, so we had to look up its original coordinate in the *1w7s_2_chromo+me.pdb* file in order to locate the corresponding line in the *.mol2* line.

Once we added all required partial charges, we saved the file as *1w7s_7.gjf*.

At some point, it is a good idea to sum the charges on each residue and verify that the values are reasonable before continuing. Gaussian will do so as part of an Amber calculation:

```
Residue 1 PDB Number 1_A ALA has 12 atoms ... charge 1.0
Residue 2 PDB Number 2_A SER has 14 atoms ... charge 0.0
Residue 3 PDB Number 3_A LYS has 25 atoms ... charge 1.0
...
```

We are not yet ready to run such a calculation, but we can do so after completing the next phase of the process. If you want to verify the partial charges prior to defining missing Amber parameters, you can do so manually or write a script for this purpose.

Files used in this phase

starting:
1w7s_7.gjf
1w7s_3_chromo_resp.mol2
Amber debug input file:
1w7s_7_amber.gjf
parmchk input file:
1w7s_7_amber.mol2
parmchk output file:
1w7s_7_amber.frcmod
final input & output files:
1w7s_8.gjf
1w7s_8.log

T9: Defining Missing Amber Parameters

We are now ready to try to run a Gaussian calculation. We will open the file from the previous phase and edit it to remove all of the ONIOM layer information at the end of each atom specification. Alternatively, we could place all of the atoms into the low ONIOM layer using the **Atom Group Editor**. After selecting all atoms, open the **Atom Group Editor**, set the **Atom Group Class** to **ONIOM Layer**, and then click on the + button for the low layer:

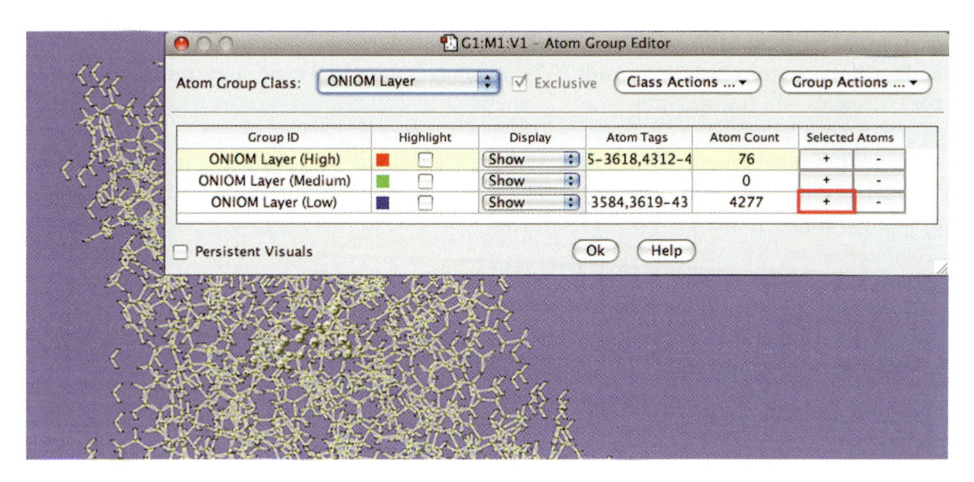

The atoms in the high layer will be moved to the low layer, and the window will be updated accordingly. We save this file as *1w7s_7_amber.gjf*, and we will use it to finish preparing the required Amber input.

We run this job in Gaussian. The job will fail due to missing Amber parameters, but Gaussian will helpfully list missing bond stretch and angle bend parameters:*

```
parameter type                              --------atom #s--------   atom types
Bondstretch undefined between atoms  940  941        CT-O
Angle bend  undefined between atoms  939  940 941  CT-CT-O
```

These sample error messages indicate the type of parameter, which atoms require it and their corresponding atom types. Missing Amber parameters messages can be caused by these conditions:

- The required parameters are not included in the set coded into Gaussian and must be added manually.
- The atom type specified for one or more atoms is incorrect, resulting in an atom sequence that is not identifiable.

The only way to distinguish between the two cases is to examine the structure. The error messages we listed above turn out to be an example of the second case: incorrect atom types. Here are the relevant lines from the input file and also the lines for the corresponding atoms in another instance of the same amino acid elsewhere in the file:

```
C-CT--0.002400(PDBName=CA,ResName=PHE,ResNum=64_A) ...
C-CT-0.597300(PDBName=C,ResName=PHE,ResNum=64_A) ...      atom 940
O-O--0.567900(PDBName=O,ResName=PHE,ResNum=64_A) ...
C-CT--0.034300(PDBName=CB,ResName=PHE,ResNum=64_A) ...
...
C-CT--0.002400(PDBName=CA,ResName=PHE,ResNum=71_A) ...
C-C-0.597300(PDBName=C,ResName=PHE,ResNum=71_A) ...
O-O--0.567900(PDBName=O,ResName=PHE,ResNum=71_A) ...
C-CT--0.034300(PDBName=CB,ResName=PHE,ResNum=71_A) ...
```

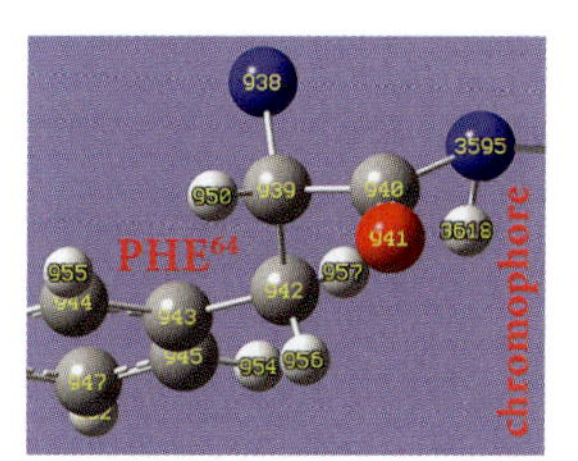

* Missing torsions, van der Waals, charges or other MM parameters will not cause an error termination. Pay careful attention to those other types of potential missing parameters and add them if necessary.

The atom in question (number 940 in the list) is bonded to the chromophore, and GaussView misidentified its atom type as a result. We can correct the atom type, and the corresponding error messages will not recur.

If there are genuinely missing parameters involving non-chromophore atoms, they can often be found or deduced from the Amber paper [Cornell95]. Note that the atom order in bond and torsion angles may be reversed from that in the error message.

The remainder of our missing Amber parameters correspond to the atoms in the chromophore. Thus, we will need to add the required parameters for them. The Amber parameters that are required describe bonds, bond angles and/or torsion angles. They are defined via a separate section in the Gaussian input file, using the following functions:

- Bonds: **HrmStr1** *Atom-type1 Atom-type2 ForceConstant* R_{eq}
- Bond angles: **HrmBnd1** *Atom-type1 Atom-type2 Atom-type3 ForceConstant* θ_{eq}
- Torsion angles: **AmbTrs** *Atom-type1 Atom-type2 Atom-type3 Atom-type4 PO1 PO2 PO3 PO4 Mag1 Mag2 Mag3 Mag4 NPaths*

Each function is defined on a single line in the Gaussian input file (the **AmbTrs** definition is wrapped solely for space reasons). The input section containing them follows all of the structure definition sections: the atom list, the variable definition list, the connectivity information, the PDB substructure data, input for the **ModRedundant** option, and so on, depending on the specific information included in the input file.

These functions begin with a keyword and then require two, three or four atom types as their initial arguments (as appropriate to the type of data). The harmonic stretch and harmonic bend functions both require two additional parameters: the force constant and the equilibrium bond length/angle value (respectively). For the Amber torsion function, the four atom types are followed by four phase offsets, four magnitudes and the number of paths.

Another important parameter type is the van der Waals parameters. For example, when adding metal centers as in metalloenzymes, one will want to add van der Waals parameters for the metal center to avoid that center collapsing with nearby atoms with opposite sign of the atomic charge. This type of parameter is not needed for our structure.

We will need to construct appropriate functions for the missing parameters before we can use the Amber method to model this molecule. The **parmchk** utility from the Amber Tools package will generate the required Amber parameters when provided with a structure including atom types and charges as input.

We need to edit the *.mol2* file produced by the RED tools (*1w7s_3_chromo_resp.mol2*) so that the atom type field contains the Amber atom type as specified in the Gaussian input file we are working on:

RED output .mol2 file

change to Amber type ↓

```
@<TRIPOS>ATOM

  1 C1 1.278752    0.370735   -0.200239  C  1 GYS  0.0487 ****
  2 N2 0.008999    0.398920   -0.124802  N  1 GYS -0.2334 ****
...
```

current Gaussian input file

```
C-CC-0.048700(PDBName=C1,ResName=GYS,ResNum=232_A) ...
N-NB--0.233400(PDBName=N2,ResName=GYS,ResNum=232_A) ...
...
```

We save the edited file as *1w7s_7_amber.mol2*.

The **parmchk** command has the following general format when used with a *.mol2* file:

parmchk -i *input-file* **-f mol2 -o** *output-file*

We ran the command as shown below. The AMBERHOME environment variable points to the top-level directory of the Amber Tools package, and defining it is required in order to run any of the utilities.

```
$ $AMBERHOME/bin/parmchk -i 1w7s_7_amber.mol2 -f mol2 \
    -o 1w7s_7_amber.frcmod
```

The final item is the output file.

The output file generated by **parmchk** contains the data required to define the missing functions. For example, here is the first missing parameter message from Gaussian for our current file:

```
Bondstretch undefined between atoms  3585  3590 CC-N
```

We will need a **HrmStr1** line like the following in order to define the required parameters:

HrmStr CC N *ForceConstant R*

We obtain the values for the force constant and the equilibrium bond length from the output of **parmchk**:

```
      force constant    R
CC-N    390.50    1.407        same as c2-n
```

Thus, the final form of the function is:

```
HrmStr CC N 390.5 1.407
```

The values required for the harmonic bend function are reported in an analogous way:

error message:
```
Angle bend undefined between atoms 3585 3586 3587 CC-NB-CM
```

parmchk output:
```
CC-NB-CM    68.600     110.190    same as cc-nc-cc
```

function definition:
```
HrmBnd1 CC NB CM 68.6 110.19
```

Things are slightly trickier for torsion angle data. Here are two examples from the **parmchk** output (preceded by the function syntax for reference):

function syntax:
AmbTrs *type1 type2 type3 type4 PO1 PO2 PO3 PO4 Mag1 Mag2 Mag3 Mag4 NPaths*

```
atom types     NPaths MagI     PO       I
CT-CT-OH-HO     1   0.160    0.000   -3.000    same as ho-oh-c3-c3
CT-CT-OH-HO     1   0.250    0.000    1.000    same as ho-oh-c3-c3
NB-CM-CC-O      1   6.650  180.000    2.000    same as X -c2-c2-X
```

The absolute value of the fifth field (I) indicates which atom type's data the line contains. When $I<0$, the parameter definition continues in the next **parmchk** ouput line. For example, the first line above specifies the values of *Mag* and *PO* for the third atom type (OH). I=-3 in the first line says that next line applies to the same paramter; it specifies the values of *Mag* and *PO* for the first atom type (CT). The third line applies to a new, different torsion angle parameter definition and gives the values for the second atom type in its list: CM. Atom types for which values are not reported have all zero values.

Thus, these lines produce the following function definitions:

```
       atom types   PO1    PO2   PO3 PO4 Mag1 Mag2 Mag3 Mag4 NP
AmbTrs CT CT OH HO  0.0    0.0  0.0 0.0 0.25 0.0  0.16 0.0  1
AmbTrs NB CM CC O   0.0  180.0  0.0 0.0 0.0  6.65 0.0  0.0  1
```

Once we have worked our way through the entire list of missing Amber parameter messages, we add the resulting functions to the end of the Gaussian input file. We also need to add an option to the **Amber** keyword to tell the program to read the additional functions from the input file; some of the relevant options are:

- **Amber=HardFirst:** Read additional parameters from the input stream, with the built-in, *hard*wired parameters having priority over the read-in—*soft*—ones. In other words, read-in parameters are used only if there is no corresponding hardwired value.
- **Amber=SoftFirst**: Read additional parameters from the input stream, with read-in parameters having priority over the hardwired values.
- **Amber=SoftOnly**: Read parameters from the input stream and use only these definitions, ignoring hardwired parameters.
- **Amber=ChkParam**: Retrieve *only* the Amber parameters from a checkpoint file. This option is useful when you want to apply Amber parameters from a previous calculation to a new structure defined in the input file. Note that Amber parameters are retrieved from the checkpoint file along with the structure when you use the **Geom=Check** or **AllCheck** option.

Since we are augmenting the built-in parameter values, we add the **HardFirst** option to the **Amber** keyword in the route section and then save the input file (*1w7s_8.gjf*). The Gaussian single point energy calculation completes successfully.

Once we successfully run a single point calculation with Amber—which serves to verify that all of the required parameters are present—we optimize the entire structure with Amber. Our optimization completes successfully. This structure will be the basis of our ONIOM study, which we will set up in the next subsection.

T10: Defining ONIOM Layers

Files used in this phase
starting:
1w7s_8.log
ONIOM=OnlyInput job:
1w7s_9.gjf
ONIOM calculations:
1w7s_10.gjf
1w7s_10_opt.gjf
Proton transfer high layer:
1w7s_10_pxfer.gjf

We are almost ready to define the high and low layers for our future ONIOM calculations. We will use the optimized structure from the previous phase as the starting point. We retrieve the structure from the Amber optimization calculation's checkpoint file. Don't make the mistake of opening the log file in GaussView, as the atom types and partial charges will not be retained.

We then check the atom types and partial charges. A few hydrogen atoms do not have atom types assigned, and the atom types for many of the atoms in the chromophore are different.* We correct them in the **Atom List Editor** (or manually with a text editor). We then place all atoms into the low ONIOM layer:

* This condition only occurs with versions of Gaussian prior to D.01.

- We select all atoms.
- We open the **Atom Group Editor**, set the **Atom Group Class** to **ONIOM Layer**, and then click on the + button for the **Low** layer.

We now need to define the high layer. Before we consider the mechanics of doing so, we need to decide which atoms will become part of the high layer. [Clemente10] discusses guidelines for making this decision.

- Only single bonds should be cut.
- The best sort of bonds to cut are nonpolar carbon-carbon bonds. However, other nonpolar single bonds may also be considered.
- Separations should not occur within cyclic structures.
- Any bond breaking/formation should take place at least three bonds away from the MM region.

In our initial study, we want to define the high layer—the model system—as the chromophore. Here is the procedure we followed:

- In the **PDB Residue Editor**, we select the row for residue 232 (GYS).
- We select the **Rows→Select Atoms of Selected Rows** menu path and then close the dialog.
- We open the **Atom Group Editor**, set the **Atom Group Class** to **ONIOM Layer** and then click on the + button for the high layer. The display format for the selected atom changes to ball-and-bond type in the molecule window, and the **Atom Tags** and **Atom Count** columns in the **Atom Group Editor** are updated.

Using a similar process, we will also add the CO and NH groups from residues PHE[64] and VAL[68] (respectively), as illustrated below. We do so to avoid a layer partitioning that cuts through peptide bonds. We display the relevant residues, select the desired atoms manually, and then click the + button again:

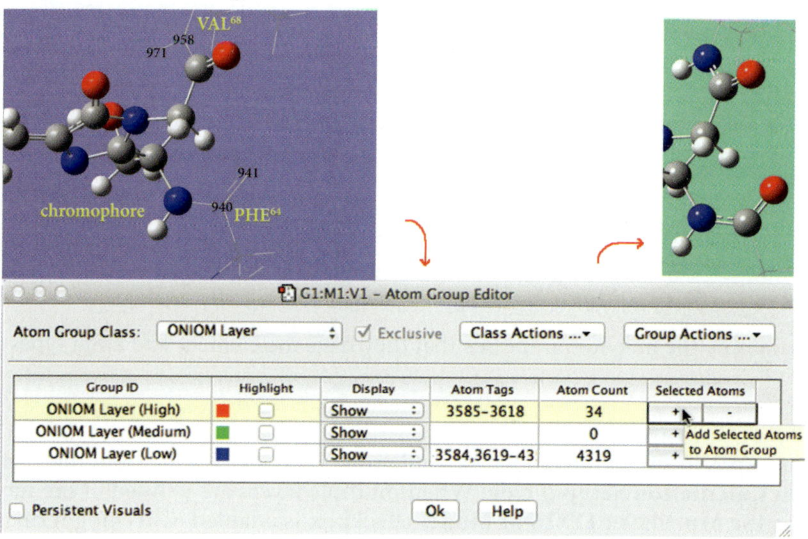

The molecule view at the extreme right in the illustration shows the display change that occurs following the atoms' assignment to the high layer.

ONIOM Terminology Review

real system: entire molecule (high & low layers).

model system: high layer plus link atoms.

MM & QM regions: usually synonymous with the low and high layers (respectively).

link atom: replacement atom for an MM region heavy atom bonded to an atom in the QM region.

electronic embedding: the incorporation of partial charges from the MM region into the QM part of the calculation in order to improve the description of the electrostatic interaction between the regions. The **ONIOM=Embed** option requests this.

In this definition of the high layer, we place a boundary at a C-N bond, a bond type that tends to be slightly polar. We could also choose to include the carbon atom and its hydrogen atom bonded to nitrogen atom 940 (illustrated in the margin).

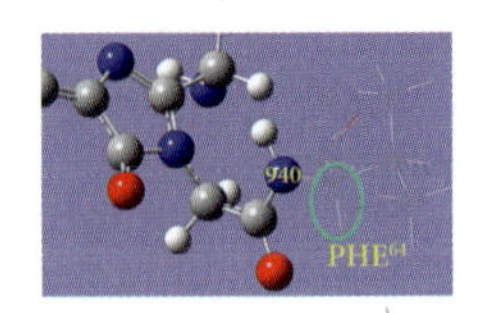

We will now check on the link atom definitions added by GaussView and verify that they are what we want. The easiest way to do this is via the sublist filters feature of the **Atom List Editor**.

- Select the **Rows→Sublist Filters** menu item. The top dialog in the illustration following will open.
- Specify the **Criteria** as **ONIOM**.
- Click the **New Filter** button to create a new filter.

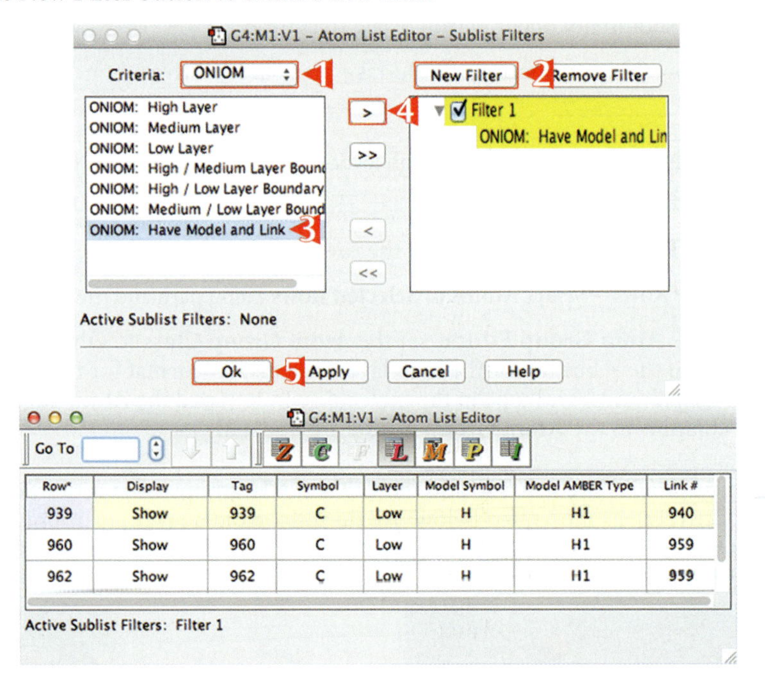

- Select the indicated item from the list of criteria in the left list box, and then click the > button to add it to the filter.
- The filter defined in this way will limit the items in the atom list to those atoms that have a link atom defined.
- When we click **Ok**, the dialog closes, and the atom list is displayed (as shown in the lower portion of the illustration).

We can examine the link atoms and see that they have the element and atom types that we expect.

Setting up the First ONIOM Calculation

Our next step is to set up the ONIOM calculation. This can be done from the GaussView **Gaussian Calculation Setup** dialog. When multiple layers are defined in the molecular structure, the **Multilayer ONIOM Model** checkbox is enabled. Checking it divides the **Method** panel into subpanels for the various defined layers.

There's a bit of a trick needed to include an option to the **Amber** keyword when it is used for one of the ONIOM component methods. The ONIOM keyword in the input section will have a format like the following:

high-level method *low-level method* *options to ONIOM keyword*
ONIOM(APFD/6-311+G(2d,p):Amber=HardFirst)=OnlyInput

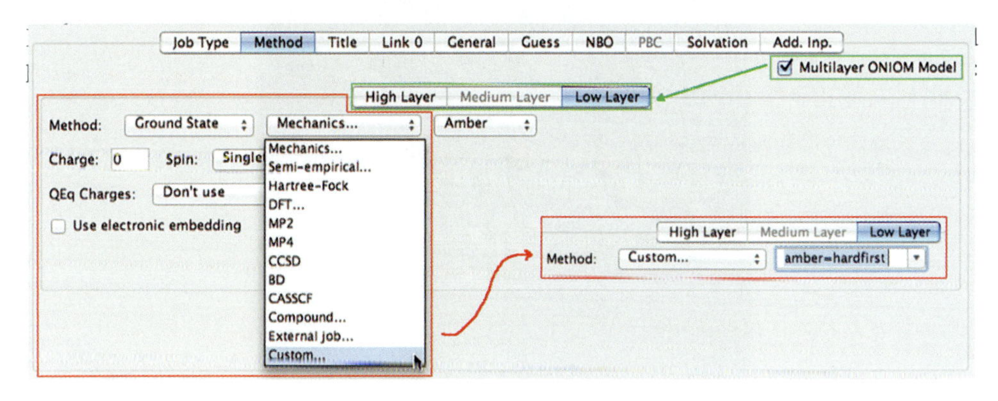

You can select an item from the resulting menu or enter anything you want into the text field.

For our first calculation, we will include the **ONIOM=OnlyInput** option, which simply checks the ONIOM layering and other information and reports on any errors. We then save the file (*1w7s_9.gjf*).

When we run the job, the output shows the following errors:

```
Bondstretch undefined atoms  939  940 H1-C [L,H]
Bondstretch undefined atoms 3597 3617 OH-HO [H,H] *
Angle bend undefined atoms  939  940  941 H1-C-O [L,H,H]
Angle bend undefined atoms  939  940 3595 H1-C-N [L,H,H]
Angle bend undef atoms 3596 3597 3617 CT-OH-HO [H,H,H] *
* These undefined terms cancel in the ONIOM expression.
```

These missing parameters arise from bonds and angles involving atoms in the chromophore and link atoms. They were not identified earlier because these connections do not exist except in the context of an ONIOM calculation

We can ignore the starred items in the list because, as the output indicates, they cancel in the actual ONIOM calculation. We add additional Amber functions for the other required parameters.

We are now ready to try an actual ONIOM calculation. We remove the **OnlyInput** option from the **ONIOM** keyword and then save the resulting input file (*1w7s_10.gjf*). The ONIOM single point energy calculation is successful! ■

Continuing from Here

It's been a long road, but we've finally reached the end of the beginning. We are now ready to model actual molecular structures and processes. We can offer you a couple of tips:

- The natural next step is to perform a geometry optimization. Although we ultimately want to study this system using electronic embedding, we will begin by optimizing using mechanical embedding (this is the default). We can then reoptimize the resulting minimum with the **ONIOM=Embed** option added.

- The high layer we defined is a good one for an initial optimization and property prediction study. However, it will need to be modified in order to investigate other processes. For example, the following high layer might be used to study a proton transfer reaction:

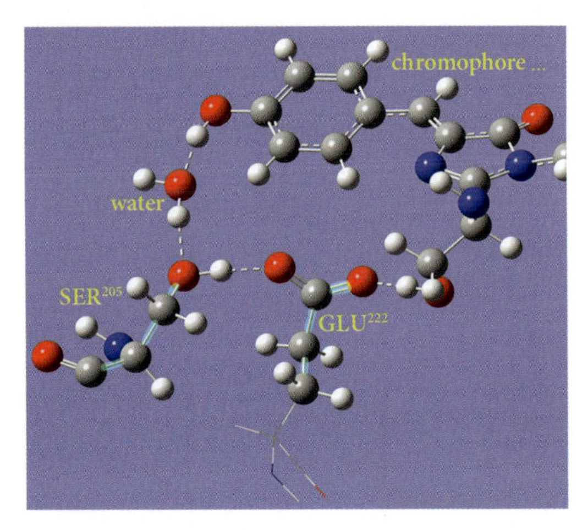

We retain the complete high layer from our first calculation and incorporate additional atoms from the SER[205] and GLU[222] residues as well as a water molecule (which has been repositioned). The illustration highlights the four bonds from the active bond formation and breaking in these two residues, which determined where we divided the high and low layers. We also included the oxygen atom double bonded to the lowest oxygen atom in the serine residue in accordance with the guidelines. Note that the hydrogen atom bonded to the oxygen atom attached to the chromophore's six-membered ring has also been repositioned.

T11: Other Useful Tools

We conclude this tutorial by briefly mentioning a couple of other useful tools that were not needed in our case.

Using PISA to Generate Biological Assemblies

The PISA program is useful for generating biological assemblies from PDB files. It is available via a web service at *www.ebi.ac.uk/msd-srv/prot_int/pistart.html*. Once you launch PDBePISA, the following page appears:

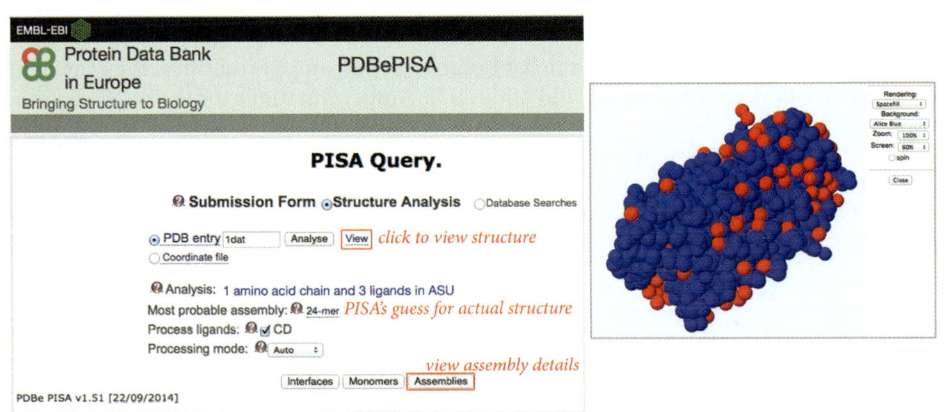

The **PDB entry field** can be used to enter the desired starting structure within the Protein Data Bank. Here, we have chosen the entry 1DAT.

The results of the initial analysis of the PDB file structure are presented in the **Analysis** area. In this case, PISA identifies the most probable assembly as containing 24 instances of the basis unit. PISA bases its analysis on PDB file information and structural analysis. Click the **View** link to visualize the PDB file contents (requires a compatible browser). A sample **View** page is shown above on the right. The various controls in its upper right corner control the display format and orientation, and there are additional controls on the right click menu. Closing the page will return to PISA.

Clicking the **Assemblies** button on the **PISA Query** page opens the **PISA Assembly List**:

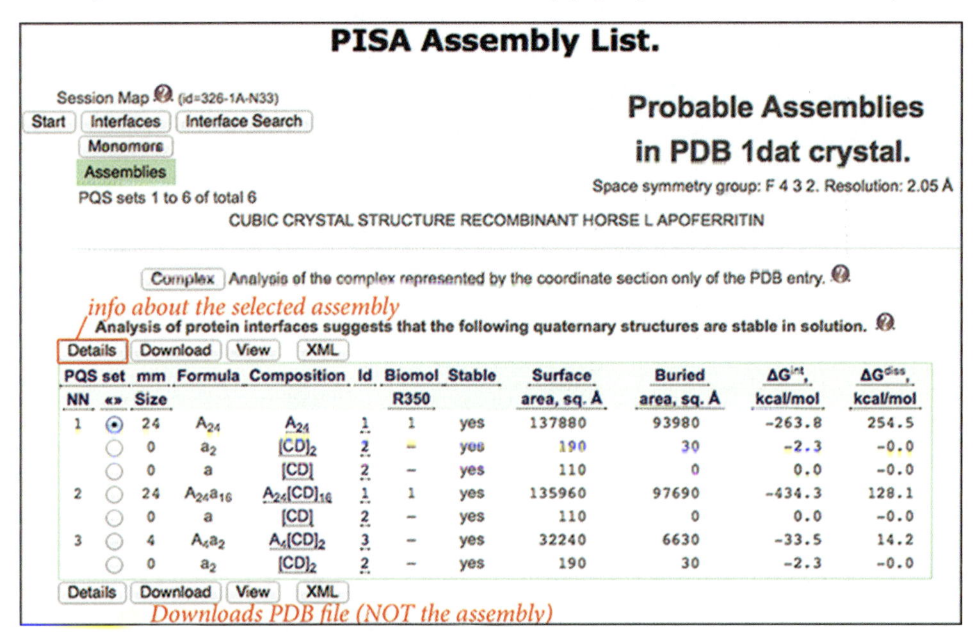

PQS set NN	mm «»	Formula Size	Composition	Id	Biomol R350	Stable	Surface area, sq. Å	Buried area, sq. Å	ΔG^int, kcal/mol	ΔG^diss, kcal/mol
1	24	A_{24}	A_{24}	1	1	yes	137880	93980	-263.8	254.5
	0	a_2	$[CD]_2$	2	-	yes	190	30	-2.3	-0.0
	0	a	[CD]	2	-	yes	110	0	0.0	-0.0
2	24	$A_{24}a_{16}$	$A_{24}[CD]_{16}$	1	1	yes	135960	97690	-434.3	128.1
	0	a	[CD]	2	-	yes	110	0	0.0	-0.0
3	4	A_4a_2	$A_4[CD]_2$	3	-	yes	32240	6630	-33.5	14.2
	0	a_2	$[CD]_2$	2	-	yes	190	30	-2.3	-0.0

The list of potential assemblies that PISA has identified appear on this page. You can select an item from the list (via the corresponding radio button) and then click the **Details** button above the list in order to open another page devoted to that assembly. Note that the **Download** button below the list downloads only the original PDB file, not the file corresponding to the selected assembly. For the latter, continue on to its details page.

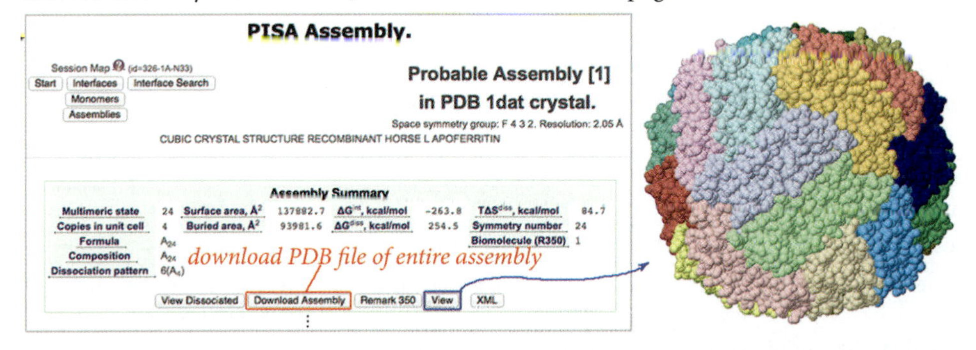

A great deal of information is provided about the assembly. If you want to view it graphically, then the **View** button may be used to access the visualization software. The image for the 24-mer assembly is shown on the right in the preceding illustration.

You can access a PDB file of the complete assembly using the **Download Assembly** button. Depending on your browser and its configuration, this will either download the file or open it in a new window in text format. In the latter case, you can cut-and-paste the text to a local file and save it with an appropriate name.

The PISA software has many more capabilities, but this introduction will enable you to use it to generate starting PDB files for ONIOM studies.

Structural Analysis with MolProbity

MolProbity is an integrated suite of structural validation programs from the Richardson group at Duke University. There is a web interface for MolProbity, available at *molprobity.biochem.duke.edu.*

Here is an example of the reports that MolProbity produces. This one is for the PDB file that we used for this tutorial:

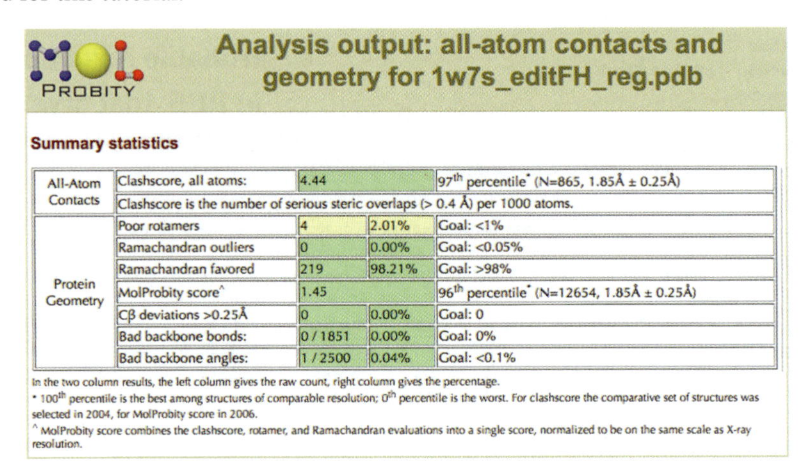

Analysis output: all-atom contacts and geometry for 1w7s_editFH_reg.pdb

Summary statistics

All-Atom Contacts	Clashscore, all atoms:	4.44		97th percentile* (N=865, 1.85Å ± 0.25Å)
	Clashscore is the number of serious steric overlaps (> 0.4 Å) per 1000 atoms.			
Protein Geometry	Poor rotamers	4	2.01%	Goal: <1%
	Ramachandran outliers	0	0.00%	Goal: <0.05%
	Ramachandran favored	219	98.21%	Goal: >98%
	MolProbity score^	1.45		96th percentile* (N=12654, 1.85Å ± 0.25Å)
	Cβ deviations >0.25Å	0	0.00%	Goal: 0
	Bad backbone bonds:	0 / 1851	0.00%	Goal: 0%
	Bad backbone angles:	1 / 2500	0.04%	Goal: <0.1%

In the two column results, the left column gives the raw count, right column gives the percentage.

* 100th percentile is the best among structures of comparable resolution; 0th percentile is the worst. For clashscore the comparative set of structures was selected in 2004, for MolProbity score in 2006.

^ MolProbity score combines the clashscore, rotamer, and Ramachandran evaluations into a single score, normalized to be on the same scale as X-ray resolution.

The program found a few minor problems with this structure (indicated by the yellow shading above). You can get full details about them by examining the details of the report.

The following table summarizes the steps required to produce this report:

MOLPROBITY PAGE	ACTION	BUTTON TO GO TO NEXT STEP
Main page	Enter "1w7s" in the **PDB code** field.	carriage return
Uploaded PDB file as ...	View summary information.	**Continue**
Main page		**Edit PDB file**
Edit PDB file	Verify desired file is selected.	**Choose editing options**
Edit PDB file	Select chains B, C, D in **Remove unwanted chains** area.	**Continue**
Created PDB file ...	View summary information.	**Continue**
Main page		**Add hydrogens**
Add hydrogens	We used the default settings.	**Start adding H**
Review flips	We accepted all suggested flips.	**Regenerate H, applying only selected flips**
Added H with -build to get ...	Read summary information.	**Continue**
Main page		**Analyze all-atom contacts and geometry**
Analyze all-atom contact and geometry	We used the default settings.	**Run program to perform these analyses**
Analysis output	View the summary table. The **View in KiNG** link for the Multi-criterion chart and the **View** link for the Multi-criterion chart display graphical and tabular details of the results.	**Continue**

As indicated in the final row of the table, details about the final results can be viewed in tabular format or graphically. Here is an example of the latter:

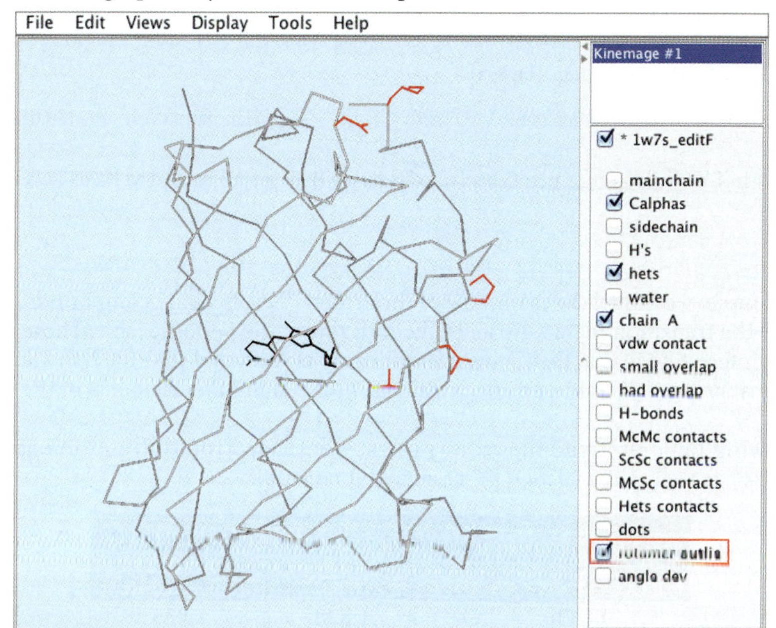

The rotamer outliers—identified as potential problems in the report—are shown in red. The controls in the right column of the window control the data that is highlighted in the structure.

This brief introduction to MolProbity should give you insight into the available analyses and enable you to explore it further on your own.

Relativistic Effects

We introduced some of the considerations that arise for elements in the fourth and higher rows of the periodic table in the first chapter of this book (see p. 29). Here we demonstrate techniques for performing calculations on molecules containing them by using effective core potentials (ECPs).

Gaussian offers a wide variety of ECPs, including the following:

ECP	BASIS SET	ELEMENTS	KEYWORD(S)	DESCRIPTION
Los Alamos	D95V[a]	H,Li-La,Hf-Bi	**LANL2DZ**	The Los Alamos ECPs plus the DZ basis set for the second row and beyond in combination with the D95V basis set for the first row.
Stevens/Basch/Krauss	31G	H-Rn	**CEP-31G**	The CEP ECPs in combination with a modest split valence basis set.
Stuttgart/Dresden	D95[a]	all but Fr,Ra	**SDD**	Stuttgart/Dresden ECPs for the 4th row and beyond in combination with the D95 basis set for the first 3 rows.
	TZVPP QZVPP	H-La,Hf-Rn	**def2TZVPP** **def2QZVPP**	Stuttgart/Dresden ECPs in combination with the triple/quadruple zeta basis sets[†] of Ahlrichs and coworkers designed for use with them.

[a][Dunning76]
[†]Second formulation: version with additional polarization functions [Weigend05, Weigend06].

When citing references for ECPs, you need to cite the main references as well as any specific ones corresponding to the elements in your calculations. These are the references for the ECPs we will use in this section:

- LANL2DZ: [Hay85, Wadt85, Hay85a].
- SDD: [Fuentealba82, Fuentealba83, Wedig86]; Cl, Br, I: [IgelMann88]; W, Ir, Pt: [Andrae90].

The *Gaussian User's Reference* provides full reference lists for all included basis sets and ECPs.

EXAMPLE 9.1 THE GEOMETRY OF METAL HEXAFLUORIDE COMPOUNDS

In this example, we model the geometry of three metal hexafluoride compounds: WF_6, IrF_6 and PtF_6.* The transition metals are all in the fifth row of the periodic table. These particular metals are interesting in that they show a lengthening of the metal-fluorine bond length as we move across the periodic table, an effect which is not seen in other elements in the same row.

The following table presents the results of gas phase electron diffraction experiments [Richardson00] as well as the CCSD(T)† results of [Craciun10]:

	SYMMETRY		BOND LENGTH (Å)	
METAL	EXP.	CCSD(T)	EXP. ±0.002	CCSD(T) (axial, equatorial)
W	O_h	O_h	1.828	1.835
Ir	O_h	O_h	1.838	1.832
Pt	O_h	D_{4h}	1.851	1.823, 1.856

The results from the coupled cluster theory calcuations are excellent, but the calculations are extremely expensive.

We want to explore how well various ECPs perform in combination with density functional theory for this problem. We will perform geometry optimizations using the following ECPs with our standard theoretical model:

- SDD with its default basis set for fluorine.
- SDD in combination with 6-311+G(2d,p) for fluorine.
- TZVPP.

The first and third calculations will use the **SDD** and **def2TZVPP** keywords (respectively) as the basis set part of the model chemistry. The second job will use the **GenECP** basis set keyword, which will allow us to specify the basis set and ECP in the input file (see below). We will request tightened convergence criteria for the optimization with the **Opt=Tight** option.

When you specify the **GenECP** keyword in the route section of a Gaussian input file, the basis set and ECP specifications are given in two separate input sections following the molecule specification. Here is the input file for the iridium job:

```
# APFD/GenECP Opt=Tight Freq

IrF6 Opt Freq: standard model
```

* IUPAC: hexafluorotungsten, hexafluoroiridium and hexafluoroplatinum (respectively).

† The basis set used was aug-cc-pVTZ with added pseudopotentials.

molecule specification

```
F 0
6-311+G(2d,p)
****
Ir 0
MWB60
****

Ir 0
MWB60

```

basis set specifications by element
standard basis set keywords may be used
required separator between elements

blank line ends basis set input
ECP specifications by element

blank line ends ECP input

Don't forget to include the ECP section in your input! It's easy to accidentally omit it when the specification is the same as for the basis set section.

The extra input sections specify the basis functions and ECPs for each element present in the molecule. This input can be explicit basis functions/ECPs or the names of standard basis sets/ECPs from which to draw predefined items. We choose the latter approach here as we simply want to pair the basis functions for fluorine from our standard basis set with the basis set and ECP for the metal element defined in the Stuttgart-Dresden set. The latter are specified via keywords defined on the **Pseudo** page of the *Gaussian User's Reference*; the ones corresponding to our elements is **MWB60**.

All of our calculations located structures with O_h symmetry for the tungsten and iridium compounds, but the symmetry of the platinum compound was D_{4h} (as was also true for coupled cluster theory).

The following table presents the results of our calculations:

	BOND LENGTH DIFFERENCE FROM EXPERIMENT (Å)			
METAL	CCSD(T)	SDD	SDD with 6-311+G(2d,p)	def2TZVPP
W	0.007	0.045	0.024	0.005
Ir	-0.006	0.068	0.043	0.024
Pt (eq.)	0.005	0.078	0.053	0.026

Although none of these DFT-based models is as accurate as coupled cluster theory, they are all in at least reasonable agreement with experiment, and the **def2TZVPP** results are quite good. Substituting our standard basis set for the very small basis set included by default with the Stuttgart-Dresden ECPs almost halves the errors in the predicted values. The basis set from the Ahlrichs group shows the best performance across the three elements as its errors for iridium and platinum are consistent (rather than increasing, as occurs with the other two models).■

ECPs are useful for predicting the properties of compounds containing heavy atoms as well as energies and minima, as is illustrated in the following example.

EXAMPLE 9.2 ^{17}O NMR CHEMICAL SHIFTS IN TRANSITION METAL-OXO COMPLEXES

In this example, we will compute ^{17}O NMR chemical shifts with respect to water vapor for some MO_4 transition metal complexes, for M=Cr, Mo and W.* We will use the **def2TZVPP** basis set/ECPs along with our standard theoretical method. We will thus run four **Opt Freq NMR** calculations: CrO_4^{2-}, MoO_4^{2-}, WO_4^{2-} and H_2O.

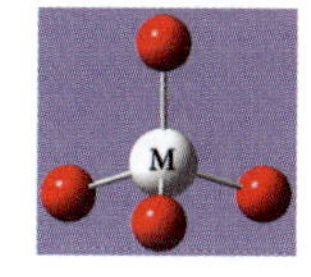

* IUPAC: dioxido(dioxo)chromium, dioxido(dioxo)molybdenum and dioxido(dioxo)tungsten (respectively).

The following table lists the results of our calculations and compares the predicted chemical shifts to experiment [Kaupp95, Kaupp95a]:

COMPOUND	SHIELDING	SHIFT	EXP.
H_2O	333.0926		
CrO_4^{2-}	-551.6738	885	871
MoO_4^{2-}	-245.0050	578	576
WO_4^{2-}	-125.9680	459	456

These results are in excellent agreement with experiment. ■

Weakly Bound Complexes: Dispersion & Counterpoise Corrections

Weakly bound complexes present special challenges to electronic structure modeling due to the nature of the weak interatomic forces that hold them together (e.g., van der Waals forces, hydrogen bonding). In this section, we consider two topics of relevance to such molecular systems:

- *Dispersion* is the weak interaction arising from the fluctuation of the charge distribution around a molecular system due to the movement of electrons. In a weakly bound complex, the movement of electrons within one region can create a temporary, instantaneous dipole, that will in turn induce a dipole in another region, resulting in a weak interaction between them. Many widely-used DFT functionals (e.g., B3LYP) do not take these effects into account.

- *Basis set superposition error* [Jansen69, Liu73] occurs when basis functions present in one part of a molecule or complex overlap with other atoms and are used to describe the latter's electron distribution, resulting in an artificial—and erroneous—stabilization of the molecular structure. For example, in a dimer at close intermolecular distances, basis functions from one molecule can be used to describe the other.

  This effect is most significant with small basis sets and is usually minor with one the size of our standard basis set. A *counterpoise correction* [Boys70, Simon96] can be computed to estimate and correct for basis set superposition error.

We will consider these effects via examples involving dimers. The first of them will illustrate the importance of treating dispersion when modeling weakly bound compounds, and the second will demonstrate basis set superposition error and counterpoise corrections.

EXAMPLE 9.3 MODELING METHANE DIMER

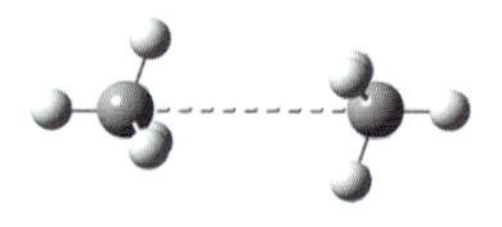

In this example, we consider methane dimer in its lowest energy conformation (illustrated in the margin). We will perform a relaxed PES scan, varying the C-C distance from 2.6 Å to 6.0 Å, in steps of 0.2 Å. We will also optimize the structures of the dimer and methane monomer. We will perform both calculations using the APF functional and our standard APFD theoretical method. In this way, we can view the effect of including dispersion in the calculation.

The following figure reports the binding of two molecules. The energy of the dimer is plotted relative to the energy of the isolated monomers: twice the predicted monomer energy. Thus, zero on the y-axis corresponds to molecules at infinite separation. We use the computed total energy at each scan point without zero point or thermal corrections.

The plot includes the results of our scans, along with the results of a CCSD(T) study [Hellmann08]* and with results extrapolated from experimental data [Matthews76]. The energy is plotted as a function of the C-C distance. The y-axis on the left is in units of K while the one on the right is in kcal/mol:

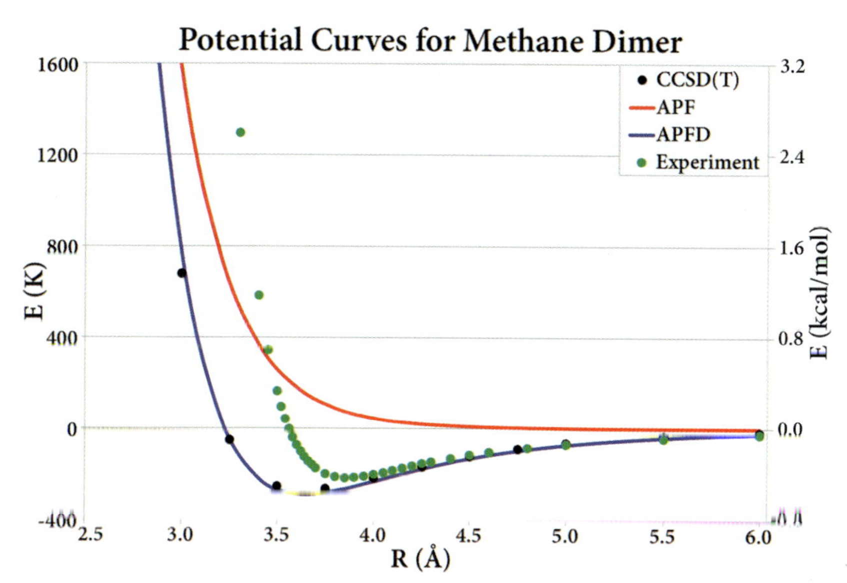

The results from our standard model chemistry (blue line) match those from the high accuracy CCSD(T) method very closely. These results match experiment as the molecules approach one another until reaching a distance of ~3.8 Å, the point at which the repulsive part of the potential begins. The theoretical predictions have the repulsive part starting at ~3.6 Å instead.

The APF results (red line) clearly indicate the deficiencies of functionals not including dispersion for studying this type of problem. The results are qualitatively incorrect in that the model does not predict any attraction; the potential is repulsive at all distances.

The following table lists the minima found in each case (indicating the equilibrium distance):

MODEL	C-C (Å)	E (K)
Experiment	3.85	-217
CCSD(T)/CBS	3.62	-286
APFD/6-311+G(2d,p)	3.64	-288

APFD and CCSD(T) in very good agreement with experiment. The APF functional tails off to 0 at C-C distances above ~4 Å. ■

* This work uses the CCSD(T) method in conjunction with the aug-cc-pVTZ and aug-cc-pVQZ basis sets and then extrapolated the results to the basis set limit, a model denoted as CCSD(T)/CBS. This study includes counterpoise corrections.

EXAMPLE 9.4 MODELING PHENOL DIMER

In this example, we will examine the effect of basis set superposition error on the predicted structure of phenol dimer. The following illustration shows two views of this complex:

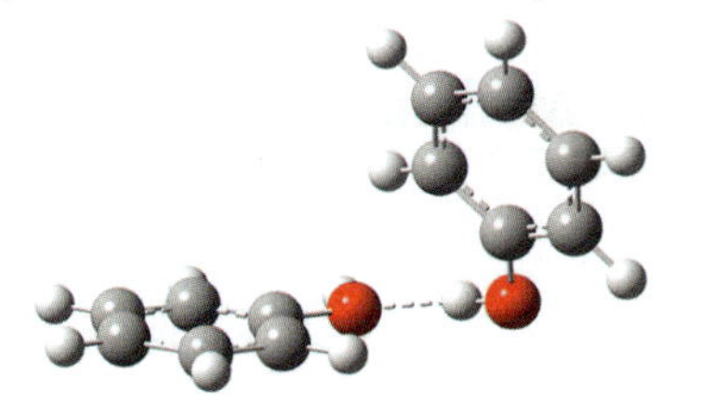
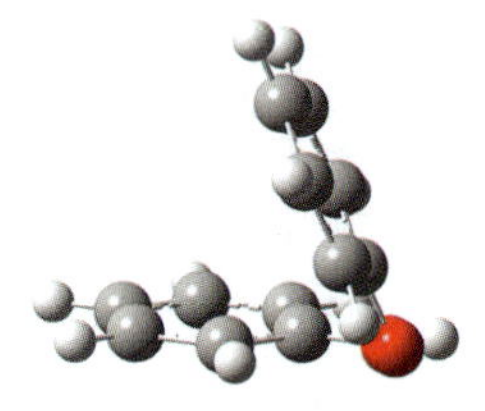

We will optimize the structure of this complex, both with and without counterpoise correction, using our standard APFD method and three different basis sets: 3-21G, 6-31G(d) and 6-31+G(d,p).

The **Counterpoise** keyword is used to request a calculation including counterpoise corrections. It requires an argument specifying the number of monomers or fragments in the molecular structure. In our case, this parameter will be 2.

In order to run a counterpoise correction calculation, it is necessary to define the separate fragments within the molecular system. In this case, each phenol ring will be defined as a separate fragment.

You can perform this task using GaussView's **Atom Group Editor** (**Edit→Atom Groups**):

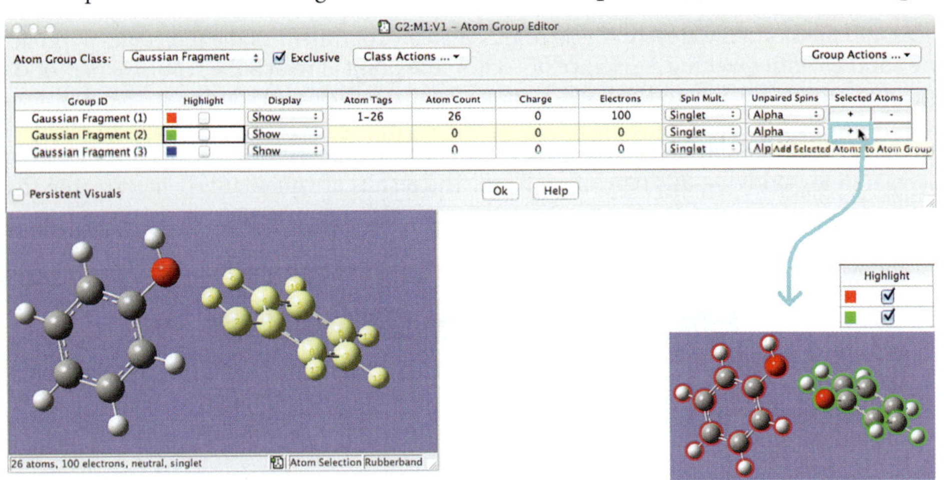

After setting the **Atom Group Class** to **Gaussian Fragment**, all of the atoms will be placed into the first fragment. To define the second fragment, select all of the atoms in one of the phenol groups, and then click on the + button corresponding to the second fragment: **Gaussian Fragment (2)**. Once you have done so, and enabled highlighting for both defined fragments, the molecule display will change as depicted in the lower right corner of the figure.

Fragment definitions are indicated in the molecule specification within the input file as follows:

```
# APFD/3-21G Opt Freq Counterpoise=2

Phenol dimer optimization with CC

0 1 0 1 0 1
O(Fragment=1)       1.86728800    2.11908000    0.00115000
```

charge & spin: overall, fragment 1, fragment 2

```
H(Fragment=1)       2.51377500     2.32519600    -0.68289400
...
O(Fragment=2)      -1.27801600     2.40484100     0.00693400
H(Fragment=2)      -0.40527500     2.61690300    -0.34963600
...
```

The fragment definition is included in parentheses following each element symbol. Note also that the charge and spin multiplicity line has additional information: the charge and spin for the molecular system as a whole is followed by the same data for each fragment.

The counterpoise-corrected energy appears in the Gaussian output as follows:

```
Counterpoise corrected energy =    -611.048772807729
                  BSSE energy =       0.009602424730
```

The predicted total energy including the counterpoise correction appears in the first line of the preceding output, while the correction itself appears in the second line.

[Schmitt06] reports the structure of phenol dimer, finding a center of mass-to-center of mass distance of 5.251 Å and an angle of 63° between the two rings. We can determine the predicted values of the latter from the C-O-O-C dihedral angle of the optimized structures. In order to determine the former, however, we will need to run an additional single point energy calculation on each monomer using the following route section:

```
# UFF Symmetry=COM IOp(2/33=2)
```

The molecular structures are taken from the optimized dimer (discussed below).

The purpose of these jobs is to translate each monomer structure to a coordinate system centered at the molecule's center of mass. The **UFF** molecular mechanics method is used because it is very fast. The **Symmetry=COM** option says to use the center of mass as the origin for the standard orientation, and the **IOp** causes the translation vector to be reported in the output file:

```
PtGrp-- Translation vector: -4.89586  0.00846 -0.06091
```

The distance between the centers of mass for the two monomers is the length of the difference vector between the two translations vectors. This result will be in units of bohr, so it will be necessary to convert to angstroms.

1 Å ≈ 0.52918 bohr

In order to set up these jobs, we will complete the following steps for each optimized structure:

1. Open the output file from the optimization calculation in GaussView.
2. Select all of the atoms in one of the monomers.
3. Cut-and-paste the selected atoms as a structure within the molecule group: **Edit→Cut** followed by **Edit→Paste→Add to Molecule Group**.
4. Set up and run a Gaussian job as specified above for each monomer.

The following table summarizes the results of our six sets of calculations:

GaussView Note:

The Builder's context menu includes a feature that adds an atom at the "centroid" of a group of selected atoms. It may be tempting to use it to add dummy atoms to each monomer in the optimized dimer structure. It would then be a simple task to measure the distance between them in inquire mode. However, the centroid position is determined spatially and is not equivalent to the center of mass.

BASIS SET	COM-COM DISTANCE (Å) default	with CC	RING ANGLE (°) default	with CC	COUNTERPOISE CORRECTION (au)
3-21G	5.17	5.21	95.6	67.6	0.01054
6-31G(d)	4.86	4.99	60.4	53.5	0.00362
6-31+G(d,p)	4.83	5.06	49.0	58.4	0.00157
Experiment	5.251		63		

In most cases, including the counterpoise correction brings significant improvement to the predicted geometry (although the 3-21G basis set without counterpoise correction achieves fortuitously good results for the distance parameter). The counterpoise corrections to the total energy decrease substantially as the basis set size increases.

In Exercise 9.4 (p. 455), we will study the same system with our standard model chemistry. ■

Electronic Spin Organization in Molecular Systems

Although closed shell singlet systems comprise a large fraction of the systems studied with electronic structure methods, there are also many systems with interest which have one or more unpaired electrons. Such systems exhibit a non-zero *spin angular momentum.*

Spin angular momentum is a vector quantity whose z-component is the simple addition of the z-components from each electron. As we've seen, one must specify the spin multiplicity as part of the molecule specification. This quantity, computed as the number of spin-unpaired electrons plus 1, is related to the number of degenerate eigenstates of $\langle \hat{\mathbf{S}}_z \rangle$, denoted S_z. For further discussion and a table of values, see pp. 478-479.

Various spin states are possible in a multi-electron system:

- Closed shell systems have the simplest organization of electronic spin in that all electrons are located in doubly occupied orbitals. For these systems, S_z=0.
- There are other electronic configurations for which S_z=0. For example, an equal number of electrons could be in spin up and spin down singly occupied orbitals. As long as there are the same number of each type, the total spin remains zero.
- When the numbers of singly occupied orbitals of spin up and spin down differ—as in radicals—then S_z will be non-zero: i.e., $S_z = ½, 1, ³⁄₂, ...$

In this section, we will illustrate some of the different approaches necessary for handling molecules with complex spin states:

- We begin with an example comparing related singlet and triplet species, followed by a consideration of some simple radical species.
- Next, we consider antiferromagnetic coupling, which is found in species with multiple electrons in singly occupied orbitals organized into a stable arrangement via a regular pattern of opposite spins. The spin state of such species are poorly described by a single spin multiplicity parameter.
- We close the section by discussing a compound for which the spin state varies as we move across its potential energy surface.

Two Examples of Open Shell Species

EXAMPLE 9.5 NITROGEN MOLECULE AND NITROGEN DIANION

In this first example, we will predict the equilibrium bond lengths and spin states for nitrogen molecule and nitrogen dianion: N_2 and N_2^{2-}.* We build both molecules and perform optimization calculations on each one's singlet and triplet electronic states. *Tip:* Be sure that your starting structures have reasonable bond lengths that correspond to the type of bond in each molecule (i.e., double vs. triple).

The following table presents the results of our calculations:

COMPOUND	SPIN STATE	ENERGY (hartrees)	BOND LENGTH (Å)	<S²>
N_2	singlet	-109.46814	1.091	0.0
	triplet	-109.20529	1.198	2.0077
N_2^{2-}	singlet	-109.08700	1.202	0.0
	triplet	-109.12639	1.170	2.0021

Although both species have an even number of electrons, their ground states have different electronic spin states. The singlet state of the neutral species is lower in energy than the triplet state whereas this is reversed for the dianion. N_2^{2-} is therefore predicted to be a paramagnetic molecule with two unpaired electrons of parallel spin. Spin contamination, as measured by the deviation from 2.0 for the $<S^2>$ value of the triplet states, is small for both molecules (see p. 479 for a discussion of this concept).

Note that the observed differences in bond length and magnetic character are consistent with qualitative MO theory for these molecules. The two additional electrons of the dianion will occupy two degenerate π* antibonding orbitals, meaning that the bond will weaken (lengthen) and the electrons will have parallel spin.

By default, Gaussian will perform unrestricted calculations on the triplet systems; such calculations can also be requested explicitly by prepending a **U** to the method keyword: **UAPFD**. As we've seen previously, a stability calculation—**RAPFD Stable**—for each singlet species would have indicated that the neutral N_2 wavefunction is stable and that the dianion wavefunction has a singlet-to-triplet instability.■

In the next example, we consider some radical species.

EXAMPLE 9.6 A REACTION INVOLVING RADICAL SPECIES

Reactions that are initiated by radical species are very important in chemistry (see [Wong98] for a detailed discussion). We are interested in the enthalpy change for the reaction of methyl radical and ethene forming propyl radical, specifically 1-dehydropropane:

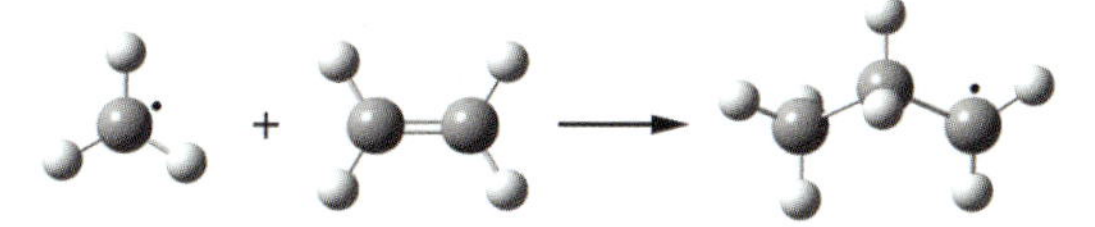

We will use our standard model chemistry to compute both the reaction enthalpy (ΔH) and the enthalpy barrier at 298° for this reaction and compare them to the experimental

* IUPAC: dinitrogen and diazenediide (respectively).

quantities. This will require unrestricted calculations on two different radical species, both of which are doublets, and for locating a radical transition structure.

We chose to do an **Opt=TS** calculation for the transition structure, beginning with the structure on the right in the following illustration. We constructed the starting transition structure by adding the optimized methyl radical molecule to the optimized ethene molecule, placing the former at a distance of ~2.4 Å with a C-C-C angle of ~110°. We then added the half bond, resulting in the structure on the left. Using the clean function resulted in the structure on the right.

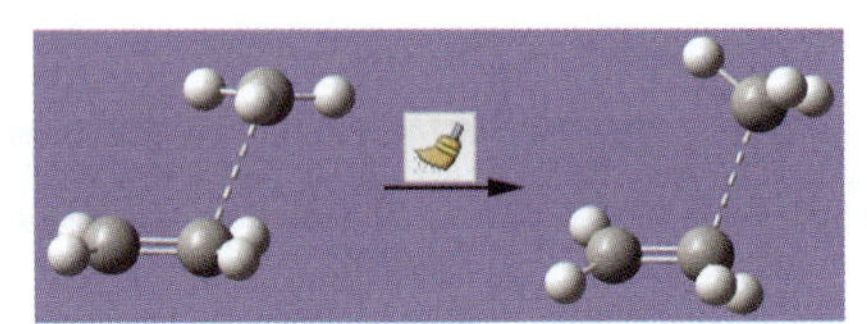

The transition structure search was done in two phases: an initial constrained optimization that froze the C-C bond coordinate between the two fragments, followed by a full optimization beginning from the stationary point located by the first optimization.

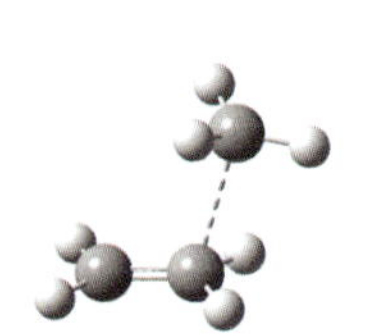

This strategy successfully located the transition state (illustrated in the margin.)

The following table presents the results of our calculations (all values are in kJ/mol) in comparison with the observed results [Zytowski96]:

QUANTITY	CALCULATED		EXPERIMENT[a]
	STD. MODEL	W1BD	
activation barrier	18.4	31.3	23±5
reaction enthalpy	-116.2	-98.6	-98±5

[a]Experimental Arrhenius activation energy converted using $\Delta H^{\dagger}=E_a+(1-\Delta n)$ RT where Δn is the change in the number of moles for the process (here -1).

Our standard model chemistry provides an estimate within the error bars of the experiment for the activation barrier but results in a reaction enthalpy that is a bit too exothermic.

In an attempt to improve on these results, we ran W1BD calculations [Barnes09] for all four species. The W1BD method is a member of the Weizmann-*n* family of very high accuracy compound model chemistries (see p. 156) [Martin99, Parthiban01] and is distinguished by its use of a Brueckner Doubles (BD) calculation instead of a coupled cluster calculation as the basis to which later corrections are applied.

In contrast to DFT, the W1BD results are very close to the experimental reaction enthalpy change while the activation barrier is predicted to be somewhat higher. This difference between the two theoretical models could be due to the approximate nature of the APFD functional, its inability to treat spin correctly, or both. Calibration with a more sophisticated method in this case causes us to question the results of our standard model (and the experiment!) for the activation energy for the reaction.

Antiferromagnetic Coupling

Ferromagnetic substances—of which iron is the canonical example—are strongly attracted to magnets. In such substances, unpaired electron spins are held in alignment by *ferromagnetic coupling*. *Antiferromagnetic* materials are typically repelled by a magnet at room temperature (although their interactions vary with temperature and are quite complex). These substances have equal numbers of unpaired electrons of each spin type arranged in a pattern so that neighboring spins have opposite polarity. This alignment is stabilized by *antiferromagnetic coupling*. The latter effect is the subject of our next example.

The complexities of spin and sometimes charge in such systems are handled by the fragment feature of Gaussian. We introduced this earlier in this chapter in the context of counterpoise corrections (see p. 442).

EXAMPLE 9.7 MODELING ANTIFERROMAGNETISM IN FERREDOXINS

Ferredoxins are a class of proteins that facilitate electron transfer reactions in organisms. We will study them using the following simple structure, the dimer of $FeS(SCH_3)_2^-$, which has an overall charge of -2. The four colors of shading indicate the division of the molecule into fragments.

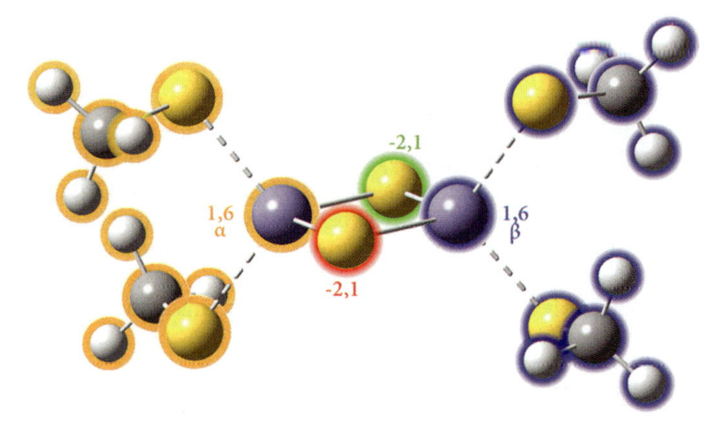

This system has an even number of electrons, and its ground state wavefunction is a singlet. However, a closed shell optimization+frequency calculation on the molecule results in a second order saddle point, characterized by a distorted structure and an unstable wavefunction (see the table on the following page). Merely using an unrestricted method will also fail. Since the system has equal numbers of spin up and spin down electrons, the calculation collapses to the restricted wavefunction solution and exhibits an internal instability.

The challenge with this system is to create a wavefunction that accurately describes its very complex electronic configuration. We will provide additional detail about the charge and spin distribution within the molecule by defining regions—fragments—for which these items can be specified individually.

In this system, each of the Fe(III) atoms has five unpaired electrons: one iron atom has five α electrons, and the other iron atom has five β electrons.

We used GaussView's **Atom Group Editor (Edit→Atom Groups)** to define the four fragments. After assigning the atoms, we defined the charge and spin multiplicity for each one as indicated in the dialog following:

Atom Group Class: Gaussian Fragment ☑ Exclusive Class Actions ... Group Actions ...

Group ID	Highlight	Atom Tags	Atom Count	Charge	Electrons	Spin Mult.	Unpaired Spins	Selected Atoms
Gaussian Fragment (1)		4	1	-2	18	Singlet	Alpha	+ -
Gaussian Fragment (2)		3	1	-2	18	Singlet	Alpha	+ -
Gaussian Fragment (3)		2,5-6,17-24	11	1	75	Sextet	Alpha	+ -
Gaussian Fragment (4)		1,7-16	11	1	75	Sextet	Beta	+ -

The two central sulfur atoms have a charge of -2 and a spin multiplicity of 1, while the two fragments containing the iron atoms have charges of +1 and spin multiplicities of 6.

The charge/spin multiplicity line from the Gaussian input file is below:

```
overall  frag.1  frag.2  frag.3  frag.4
-2,1     -2,1    -2,1    1,6     1,-6
```

The negative value for the spin multiplicity for fragment 4 indicates that unpaired electrons should be assigned β spin (a positive value implies α spin).

In order to generate a wavefunction corresponding to these characteristics, we use the **Guess=Fragment=***n* option, which says to use the fragment information within the molecule specification when generating the wavefunction (where *n* is the number of fragments). We run a preliminary **Stable** calculation to verify that the resulting wavefunction is stable.

We use the same wavefunction for our **Opt Freq** calculation—accomplished by including the **Guess=(Fragment=4,Only)** options in the first job step—followed by the optimization+frequency calculation as the second job step. **Guess=Only** says to limit the calculation to calculating the initial guess. The resulting wavefunction can then be retrieved from the checkpoint file in a subsequent calculation using the **Guess=Read** option.

Here is the general structure of our input file for modeling this compound:

```
%Chk=antiferro
# UAPFD/6-311+G(2d,p) Guess=(Fragment=4,Only)

Generate wavefunction using fragment guess approach

-2,1  -2,1  -2,1  1,6  1,-6
 Fe(Fragment=4)    -1.42690200    0.00000000    0.00000000
 Fe(Fragment=3)     1.42690200    0.00000000    0.00000000
 S(Fragment=2)      0.00000000    0.00000000   -1.75214400
 ...

--link1--
%Chk=antiferro
# UAPFD/6-311+G(2d,p) Opt Freq Guess=Read Geom=Check

Opt+Freq model for ferredoxin dianion: [Fe2S2(SCH3)4]2-

-2 1
```

blank line

The following table presents the results of our calculations. Neglecting the complexity of spin by using a spin-restricted model chemistry leads to a higher energy, an unstable wavefunction and much shorter bond lengths.

PREDICTION	RAPFD	UAPFD	UAPFD FRAGMENT GUESS
Fe-Fe (Å)	2.32	2.32	2.42
Fe-S (Å)	2.02	2.02	2.31
Fe-$S(CH_3)_2$ (Å)	2.36	2.36	2.39
E (hartrees)	-5075.53045	-5075.53053	-5075.72539
Stable calculation results	RHF-to-UHF instability	internal instability	stable
structure of central atoms			
optimization result	2nd order saddle point	2nd order saddle point	minimum

The following figure plots the spin density for the fragment guess wavefunction:

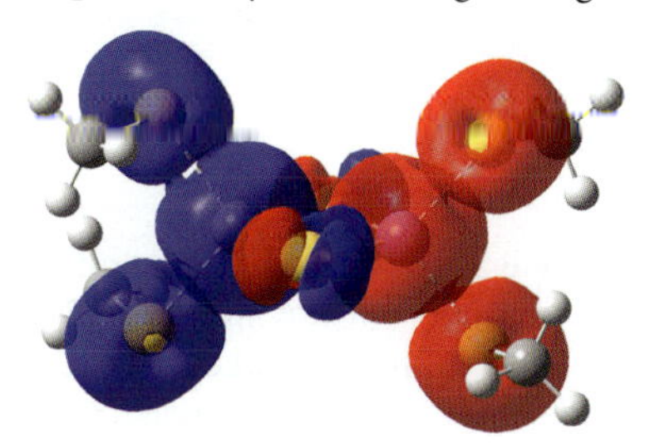

The unpaired electrons in each monomer have opposite spin. The spin density is mainly on the iron atoms, with a lesser amount on the methylated sulfur atoms. ■

Spin States on the Potential Energy Surface

When your goal is to study more than simple species, then finding the appropriate spin state for the relevant molecule(s) may not be sufficient. For example, the spin state may vary in different regions of the potential energy surface. Our next example will illustrate this phenomenon.

EXAMPLE 9.8 SCANNING THE POTENTIAL ENERGY SURFACE OF 2,6-PYRIDYNE

In this example, we consider the 2,6-isomer of didehydropyridine (2,6-pyridyne). The equilibrium geometry of 2,6-pyridyne is not completely certain [Debbert00, Li08, Prochnow09, Chattopadhyay10, Jagau12, Saito12]. It may have a monocyclic or bicyclic structure (both forms have C_{2v} symmetry), where the former is a diradical, and the latter is a closed-shell system:

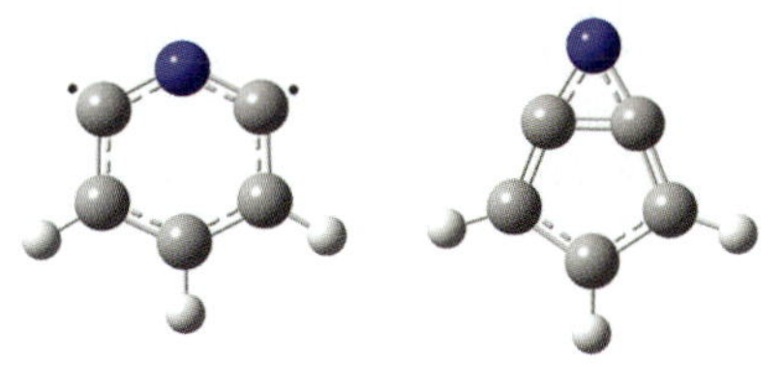

We ran stability calculations on the singlet and triplet states of both forms:

CALCULATION	MONOCYCLIC STABLE?	MONOCYCLIC E (hartrees)	BICYCLIC STABLE?	BICYCLIC E (hartrees)
RAPFD singlet	RHF-to-UHF instability	-246.72046	stable	-246.75776
UAPFD singlet with **Guess=Mix**	stable	-246.76277	stable	-246.75776
UAPFD triplet	stable	-246.76685	internal instability	-246.65436

We include calculations using the **Guess=Mix** option as well. This option ensures that the wavefunction is an open shell singlet that breaks the α-β and spatial symmetries of the initial guess. It is useful in producing stable UHF wavefunctions for singlet states, and it succeeds in doing so in these cases.* The lowest energy wavefunctions are the monocyclic triplet and the bicyclic singlet (the RHF and UHF solutions are the same for the latter molecule).

When we optimized these two structures, using our standard theoretical method and the 6-311G(2d,p) basis set, we obtained the following results:

FORM & STATE	C–C (Å)	C–N–C (°)	E (hartrees)
monocyclic triplet	2.22	118.2	-246.77862
bicyclic singlet	1.48	67.1	-246.76459

The C–C bond length for the carbon atoms adjacent to the nitrogen atom for the bicyclic form is in good agreement with the best theoretical results (1.483 Å); these were obtained using the state-specific MRCC method of Mukherjee and co-workers [Mahapatra98], specifically the Mk-MRCCSD/cc-pCVTZ model chemistry [Prochnow09]. The corresponding result for the monocyclic form is 2.017 Å; our optimized triplet structure deviates significantly from this.

The PES surface corresponding to the lengthening of the C–C bond contains two minima, corresponding to these two forms of the molecule. As it is traversed, not only does the energy rise and fall, but the electronic state also changes. We ran three relaxed PES scans for this molecule using the following model chemistries:

- Closed shell singlet: **rAPFD/6-311G(2d,p)**
- Open shell singlet: **uAPFD/6-311G(2d,p) Guess=Mix**
- Open shell triplet: **uAPFD/6-311G(2d,p)**

We scanned the C–C bond distance from 1.35 Å to 2.35 Å, in steps of 0.05 Å, moving from the bicyclic to the monocyclic form. We include the **Guess=Mix** option for the unrestricted singlet scan to ensure that the wavefunction for the starting structure is an open shell singlet, with unpaired electrons of opposite spin on the two carbon atoms of interest.

This approach to producing the unrestricted method scan may allow the state to change smoothly from the closed shell wavefunction at small distances to the broken symmetry solution at longer distances. This scan follows the *lowest energy* wavefunction across the range of C–C bond distances; its nature changes from closed shell to open shell at a distance of about 1.85 Å.

* There is no guarantee that a broken symmetry solution will be found. Thus, you should inspect the results of such calculations carefully (the value of S^2 and the spin densities) to determine whether such a solution was produced.

The figure on the next page presents the results of these scans. Our unrestricted method scan is plotted in blue.

We note the following important conclusions:

- All three curves contain two minima, roughly corresponding to the monocyclic and bicyclic forms. However, the predicted energy varies considerably by method.

- The abrupt jump in the closed shell curve (red) at 2.19 Å is due to the breaking of one C–N bond and the formation of a different molecular structure.

- The blue curve represents the open shell, unrestricted calculation and is equivalent to the red curve where the bond distance ≤ 1.8 Å. Above that, the broken-symmetry solution is more stable.

- The atomic spin densities for the open shell singlet structure at 2.2 Å—which is approximately the minimum for this model—indicate that there are two unpaired electrons of opposite spin, located primarily on the two carbon atoms adjacent to the nitrogen atom (shown in the margin). Thus, the broken-symmetry solution is not a pure spin description of the biradical species; there is contamination from the triplet state.

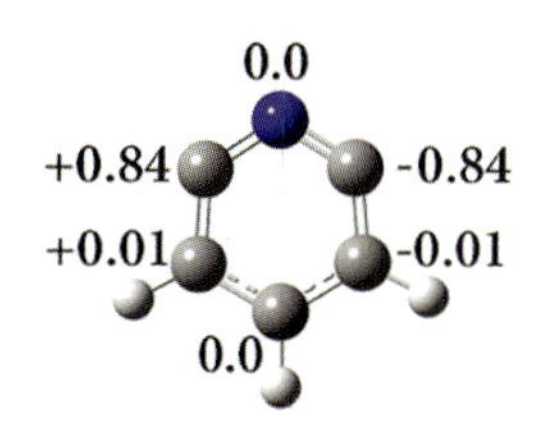

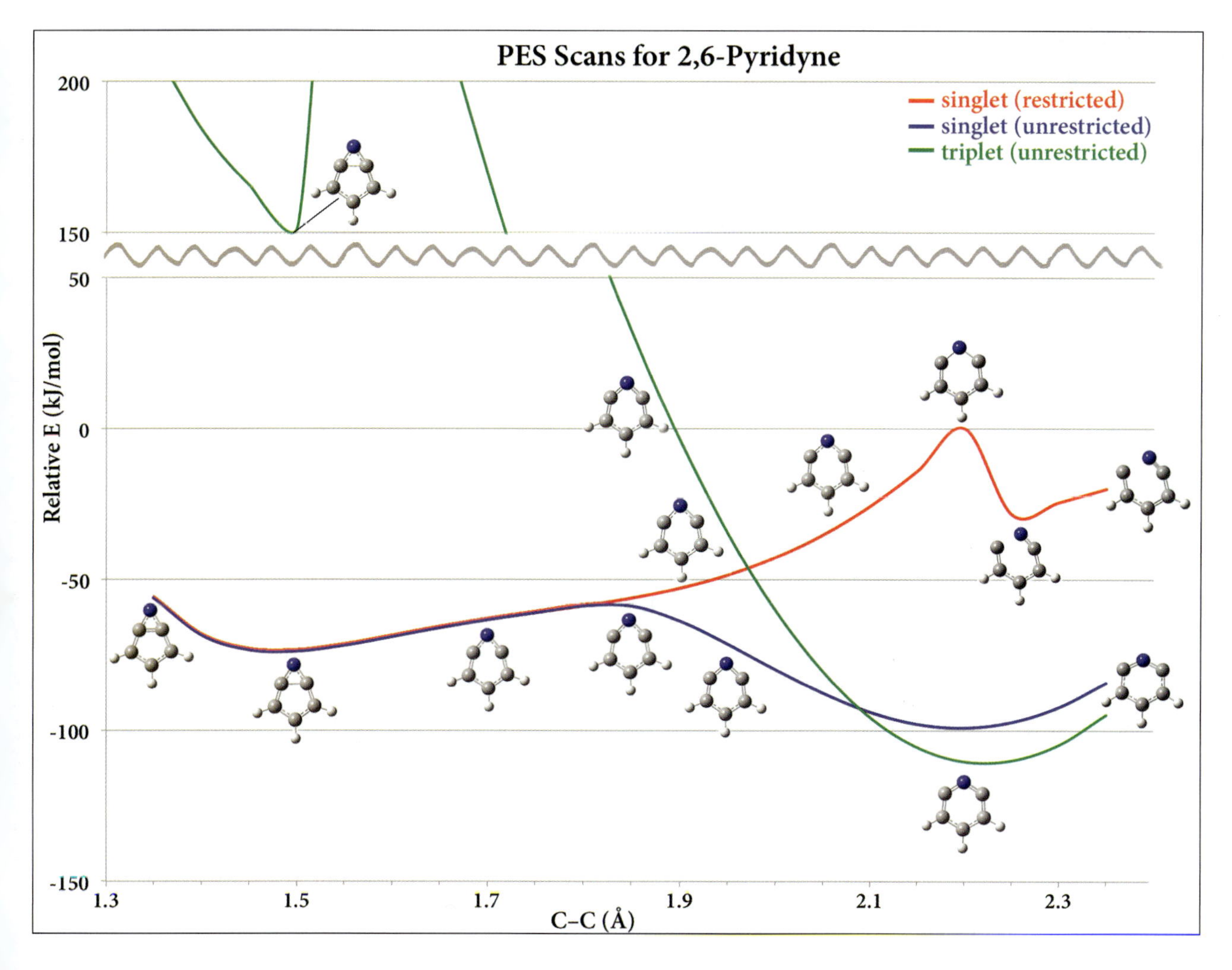

This is also evident if we examine the values of <S^2> for the structures in this range of bond distances. In the following illustration, we reproduce the relevant portion of the scan plot which we have annotated with the value of <S^2> for each point on the open shell singlet curve:

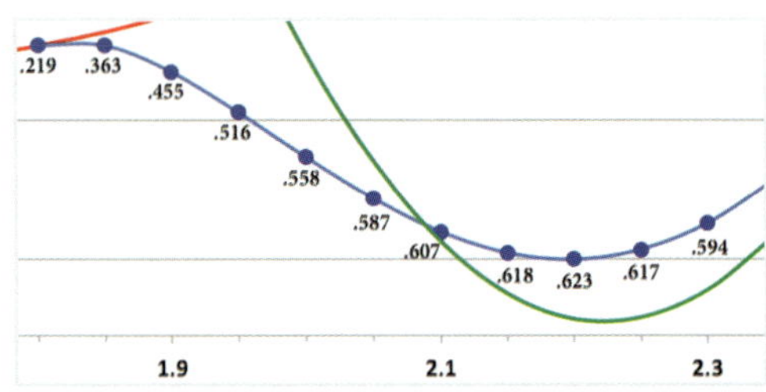

At smaller bond distances, <S^2> is ~0.0. It increases to its maximum value at a bond distance of 2.2 Å: the predicted minimum for the monocyclic structure. Notice how the curve for the open shell singlet (M_s=0) state is influenced by the triplet state.

- The open shell singlet curve collapses to the closed shell wavefunction at short C–C distances, predicting a second minimum corresponding to a bicyclic structure.

- The triplet curve attains reasonable energy values for bond distances ≥ ~2.1 Å. The spin density at the optimized structure is shown in the margin.

In summary, closed shell calculations can model the bicyclic system, and a reasonable structure for the minimum is predicted. Open shell calculations can be used to predict a monocyclic, diradical species for 2,6-pyridyne. However, the minimum occurs at much too long a distance as compared to the accepted value of 2.02 Å [Prochnow09]. This is due to the large amount of spin contamination from the triplet state, whose minimum occurs at the longer distance.

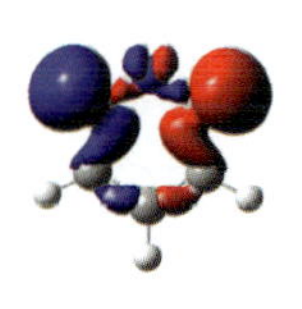

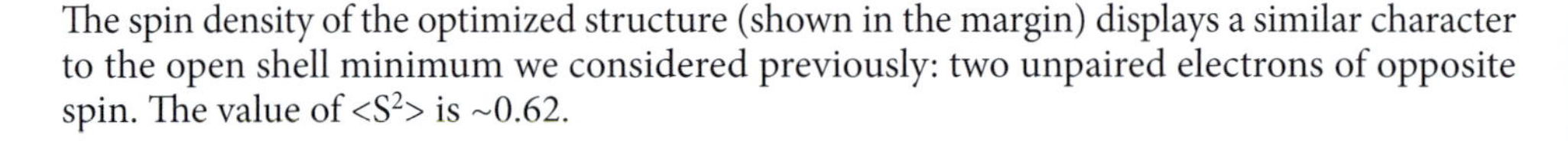

The spin density of the optimized structure (shown in the margin) displays a similar character to the open shell minimum we considered previously: two unpaired electrons of opposite spin. The value of <S^2> is ~0.62.

We will return to this problem in Exercise 9.8 (p. 460), where we will compare these results to methods that do not suffer from spin contamination.

Exercises

EXERCISE 9.1 M-F BOND LENGTHS IN METAL HEXAFLUORIDE COMPOUNDS

Optimize the geometries of the three metal hexafluoride compounds from Example 9.1 (p. 438), using the following ECPs in combination with our standard theoretical method:

- LANL2DZ (keyword: **LANL2DZ**)
- QZVPP (keyword: **def2QZVPP**)

How do the Los Alamos ECPs perform compared to the Stuttgart-Dresden ones? Does the larger basis set from the Ahlrichs group improve upon the results from the triple-zeta version?

SOLUTION

Our additional calculations found minima with the same symmetries as the previous ones: O_h for WF_6 and IrF_6 and D_{4h} for PtF_6. [David08] reports that four-component treatments of relativistic effects are required in order to locate an O_h minimum for PtF_6 (such methods are not currently available in Gaussian).

The following figure plots the experimental and predicted bond lengths for the various models we are considering (the equatorial bond lengths are plotted for PtF_6):

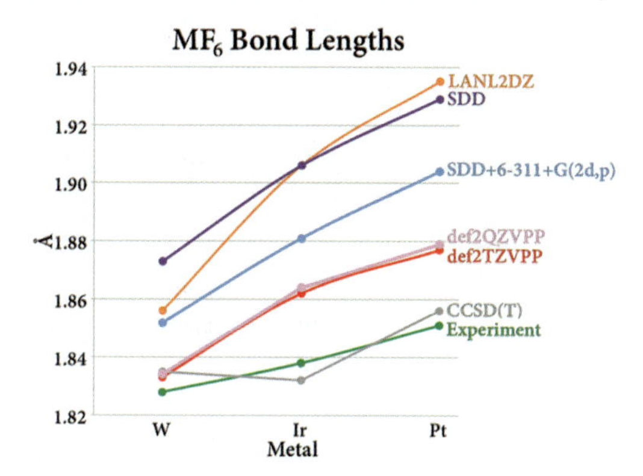

The LANL2DZ results are similar to SDD although its predicted W-F bond length is better. The additional basis functions in the quadruple-zeta basis set bring no advantage, as the triple-zeta basis set seems to represent the basis set limit with this method.

The SDD-6-311+G(2d,p) and def2TZVPP basis sets do best at reproducing the trend across the set of elements although both overestimate the lengthening in IrF_6 compared to WF_6. Interestingly, CCSD(T) underestimates the bond length for the same compound.■

EXERCISE 9.2 ^{17}O NMR CHEMICAL SHIFTS IN TRANSITION METAL-OXO COMPLEXES

Continue the study from Example 9.2 (p. 439) for these additional transition metal complexes: MnO_4^-, TcO_4^-, ReO_4^-, FeO_4, RuO_4 and OsO_4.* Compare the predicted metal-oxygen bond lengths and ^{17}O chemical shifts with the following experimental results [Kaupp95, Kaupp95a]:

COMPOUND	BOND LENGTH	SHIFT
MnO_4^-	1.629	1255
TcO_4^-	1.711	786
ReO_4^-	1.730	605
FeO_4		
RuO_4	1.705	1142
OsO_4	1.711	832

What periodic trends do you observe in the results?

* IUPAC: MnO_4^-: permanganate; TcO_4^-: technetate; ReO_4^-: rhenate; FeO_4: iron tetraoxide; RuO_4: tetraoxoruthenium; OsO_4: tetraoxoosmium.

SOLUTION

The following table presents the results of our calculations (including those from Example 9.2):

		BOND LENGTH		^{17}O SHIFT	
COMPOUND	CALC. SYMM.	CALC.	EXP.	CALC.	EXP.
CrO_4^{2-}	D_{2d}	1.64	1.65	885	871
MoO_4^{2-}	D_{2d}	1.77	1.76	578	576
WO_4^{2-}	D_{2d}	1.79	1.79	459	456
MnO_4^-	C_{3v}	1.59	1.63	1250	1255
TcO_4^-	C_{3v}	1.70	1.71	778	786
ReO_4^-	C_{3v}	1.73	1.73	613	605
FeO_4	T_d	1.56		1995	
RuO_4	T_d	1.66	1.71	1135	1142
OsO_4	C_{3v}	1.69	1.71	856	832

The predicted chemical shifts for the latter six compounds are based on the averaged shielding value.

As we move across the rows of the periodic table, the bond length decreases, while it lengthens as we move down the various columns. The largest differences are between corresponding elements in the third and fourth rows (e.g., Cr vs. Mo).

The positive chemical shift values indicate that the oxygen atoms in all of the compounds are less shielded than the one in water. Chemical shifts increase from left to right within the rows, indicating greater difference in shielding with respect to water, while they decrease as we move down each row.

All in all, these results are in excellent agreement with experiment. ECPs succeed in treating the relativistic effects in these compounds.■

EXERCISE 9.3 COUNTERPOISE CORRECTIONS: METHANE DIMER

Perform the scan and optimization calculations on methane dimer from Example 9.3 (p. 440), this time including counterpoise corrections. How do the results change? Is neglecting dispersion more or less significant than basis set superposition error?

SOLUTION

The following figure shows the results of our calculations including counterpoise correction along with the previous results:

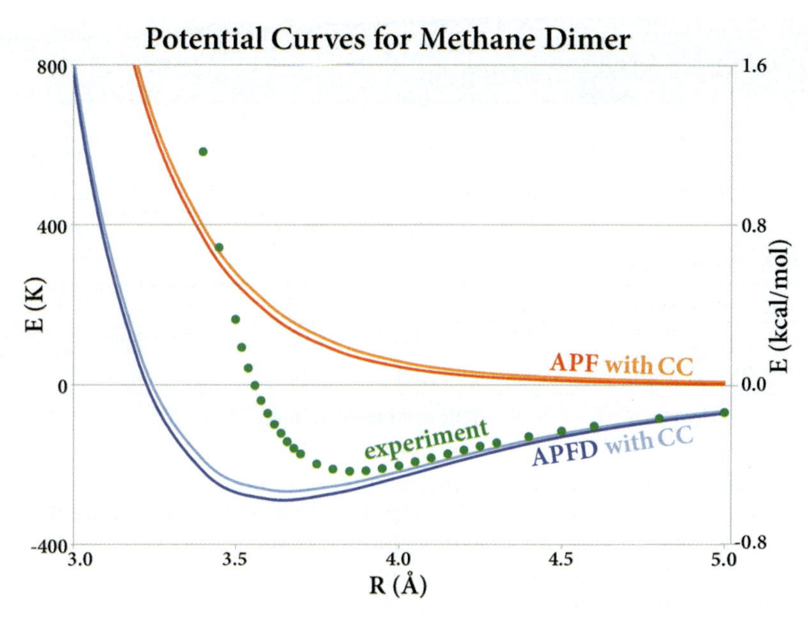

For this system, basis set superposition error is minimal and treating dispersion appropriately is much more important for achieving accurate results.■

GOING DEEPER: ADDITIONAL CALCULATIONS

Methane dimer exists in a variety of conformations. Two others of interest are shown below:

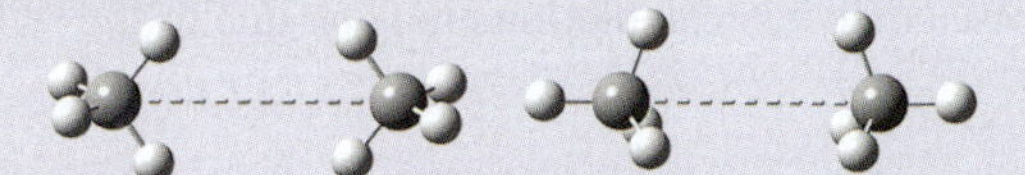

Calculations could be performed for other forms of this complex for comparison purposes:

EXERCISE 9.4 STUDYING PHENOL DIMER WITH 6-311+G(2d,p)

Optimize the structure of phenol dimer with our standard model chemistry, with and without counterpoise correction. Repeat the calculations using the APF functional instead of APFD. How do the results compare with one another and with those we obtained previously in Example 9.4 (p. 442)? What are the relative importances of counterpoise corrections and dispersion?

SOLUTION

The following table includes the results of the calculations using our standard model chemistry as well as additional results from the literature:

MODEL CHEMISTRY	COM-COM DISTANCE (Å) default	with CC	RING ANGLE (°) default	with CC	ENERGY CORRECTION
Experiment[a]	5.251		63		
APFD/3-21G	5.17	5.21	95.6	67.6	0.01054
APFD/6-31G(d)	4.86	5.00	60.4	53.5	0.00362
APFD/6-31+G(d,p)	4.83	5.06	49.0	58.4	0.00157
APFD/6-311+G(2d,p)	4.76	4.89	47.0	50.0	0.00121
APF/6-311+G(2d,p)	5.78	6.13	70.3	105.4	0.00087
B3LYP/cc-pVTZ[b]	6.24	5.86	119	-113.1[c]	
MP2/cc-PVTZ[b]	5.33	5.04	40	47	

[a][Schmitt06]; [b][Kolar07]; [c]Our calc. for value not reported in ref.

Using our standard theoretical method, the counterpoise correction produces somewhat to slightly more accurate geometries than the default method, with the difference decreasing as the basis set size increases. At the same time, the correction to the energy decreases to less than 1 kcal/mol with the largest basis set. Interestingly, the larger basis set does not improve the geometry compared to the 6-31+G(d) basis set, as we might expect.

Comparing the results for the APFD and APF functional shows the importance of treating dispersion properly for problems like these. The counterpoise corrections for the APF functional are in the wrong direction with respect to experiment.

The geometry calculated by the B3LYP/cc-pVTZ model chemistry deviates slightly more from experiment than that of our standard model chemistry, and the counterpoise correction has a more dramatic impact with this level of theory. The large differences from experiment in the non-counterpoise corrected calculations may be due to the lack of diffuse functions in the basis set.

Thus, for this problem, including dispersion is at least as important as counterpoise corrections, especially when using basis sets with multiple polarization functions and diffuse functions like our standard one.■

EXERCISE 9.5 OXYGEN MOLECULE AND OXYGEN DICATION

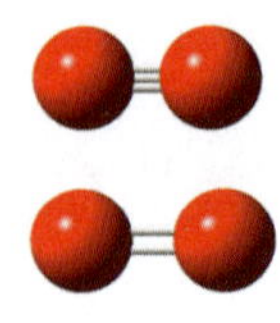

In Chapter 2 (see Advanced Exercise 2.8, p. 72), we found that the ground state of O_2 is a triplet. Now examine the dication of molecular oxygen.* Use our standard model chemistry to predict the bond lengths for both the open shell singlet and triplet states. Is the ground state diamagnetic or paramagnetic? Do your predictions agree with qualitative MO theory?

SOLUTION

The following table presents the results of our calculations:

COMPOUND	SPIN STATE	ENERGY (hartrees)	BOND LENGTH (Å)	<S²>
O_2	singlet	-150.24180	1.198	0.0
	triplet	-150.25841	1.199	2.0093
O_2^{2+}	singlet	-148.89691	1.032	0.0
	triplet	-148.63149	1.161	2.0086

* IUPAC: O_2: dioxygen; O_2^{2+}: dioxide dication.

Oxygen dication's electronic ground state is a singlet. Removing two electrons from oxygen molecule results in a significant shortening (strengthening) of the O-O bond.

Note that O_2^{2+} is isoelectronic with neutral N_2. Qualitative MO theory predicts a shorter bond length and an electronic ground state with all electrons paired—a closed-shell singlet—results which our calculations confirm. Notice the unrestricted calculations for the triplet species are again only slightly spin contaminated.■

EXERCISE 9.6 METHYL RADICAL ADDITION TO CYANOETHENE

Using the techniques introduced in Example 9.6 (p. 445), model the following reaction involving radicals:*

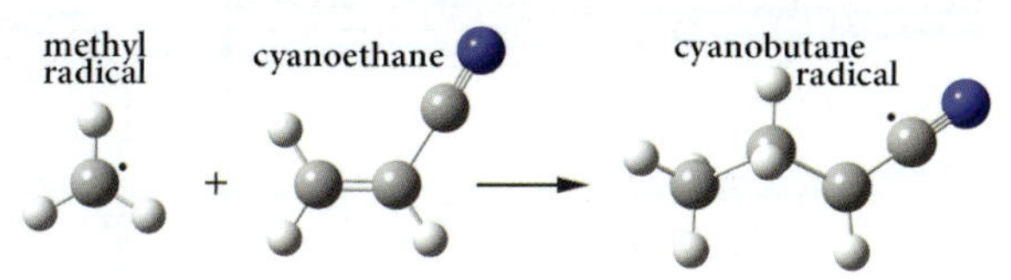

Compare your results to the observed values [Zytowski96]†:

- Activation barrier: 10±5 kJ/mol
- Reaction enthalpy: -139±5 kJ/mol

SOLUTION

Calculations using our standard model chemistry showed spin contamination: 0.7686 for butanenitrile radical and 0.7768 for the transition state. We also modeled each compound using the W1BD compound model in an effort to improve on these results. The TS optimization was performed in two phases: a constrained optimization freezing the intermolecular C-C bond coordinate followed by a full optimization.

The following table presents the results of our calculations as well as the observed values (all values are in kJ/mol):

QUANTITY	APFD/ 6-311+G(2d,p)	W1BD	EXPERIMENT
activation barrier	4.9	17.8	10±5
reaction enthalpy	-151.2	-129.3	-139±5

The results indicate that the reaction is a challenging one to model. Nevertheless, the W1BD model produces results which are in fair agreement with experiment.■

EXERCISE 9.7 CARBON NANOTUBES

Molecular nanostructures have received much attention for their potential biomedical and molecular electronic applications. Single-walled carbon nanotubes (SWCNTs) are now commercially available for experiments. They have a variety of lengths, widths, and chiralities. One interesting case is the so-called zigzag SWCNT in which adjacent carbon atoms form planes, potentially resulting in an antiferromagnetic ground state in which carbons in adjacent planes have unpaired electrons of opposite spin (with M_s=0 for the molecule overall).

* IUPAC: cyanoethane: propanenitrile; cyanobutane: pentanenitrile.

† Corrected as before using $\Delta H^{\dagger}=E_a+(1-\Delta n)RT$.

Building the molecular structures of carbon nanotubes can be quite challenging. However, we can fortunately use a program called TubeGen [Frey11], which is available on the Internet: *turin.nss.udel.edu/research/tubegenonline.html*. It will generate Gaussian input files for carbon nanotube structures that you specify via its web form, as illustrated in the figure following. Fields which we have modified from the default are highlighted in the illustration. Use the indicated settings to create an input file for a carbon nanotube with (7,0) Nanotube (1,1,4) replication. A front and side view of the resulting structure are included in the figure.

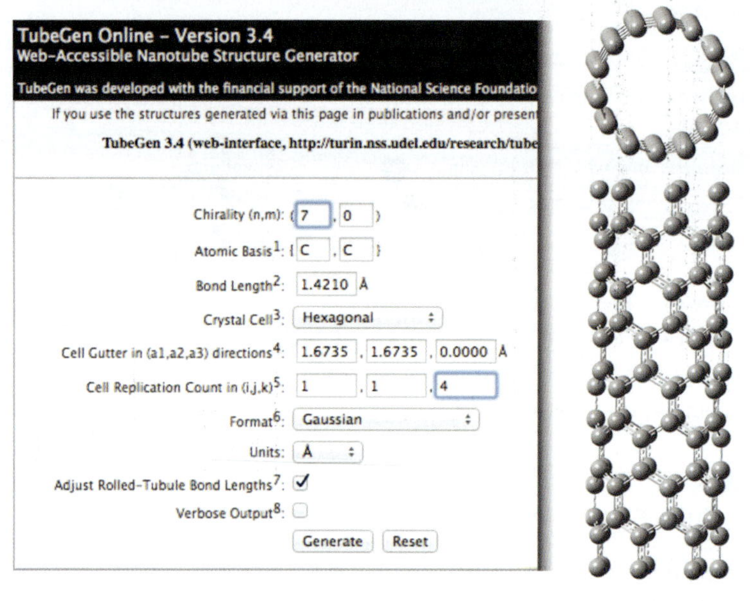

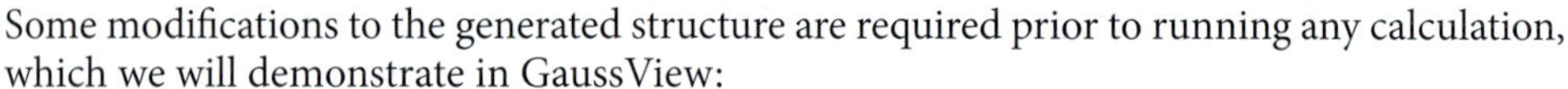

Some modifications to the generated structure are required prior to running any calculation, which we will demonstrate in GaussView:

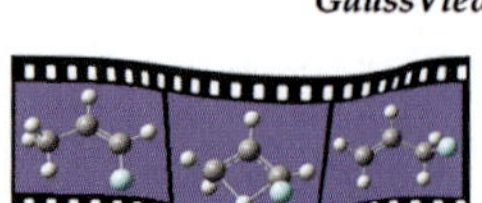

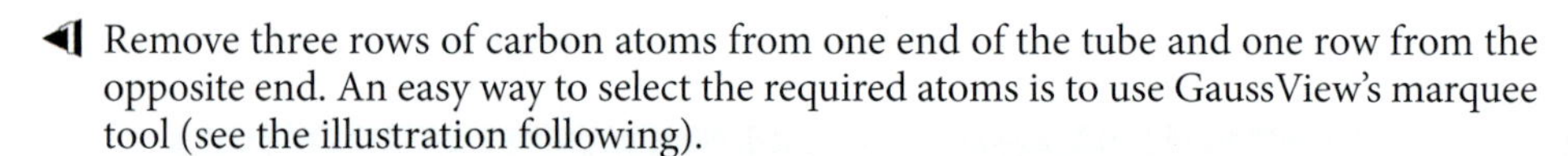

1. Remove three rows of carbon atoms from one end of the tube and one row from the opposite end. An easy way to select the required atoms is to use GaussView's marquee tool (see the illustration following).
2. Add hydrogen atoms to the outermost remaining carbon atoms.
3. Impose D_{7h} symmetry on the molecule, but do not use the **Clean** function.

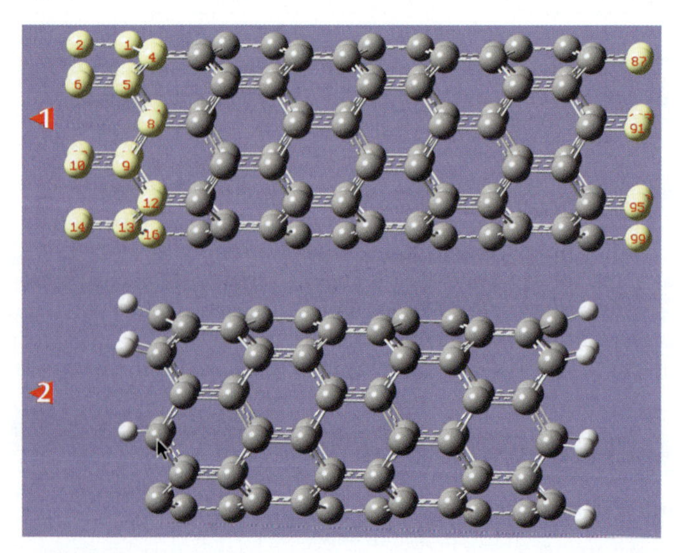

Run the appropriate calculations to determine whether an antiferromagnetic state exists for this molecular system. Use the 6-31G(d) basis set and our standard theoretical method.

If you need help in setting up the fragments for this molecule, consult the beginning portion of the solution discussion on the next page.

SOLUTION

We modeled this molecule using a variety of wavefunctions, including an open-shell antiferromagnetic state. We used the following fragment definitions for our antiferromagnetic wavefunction:

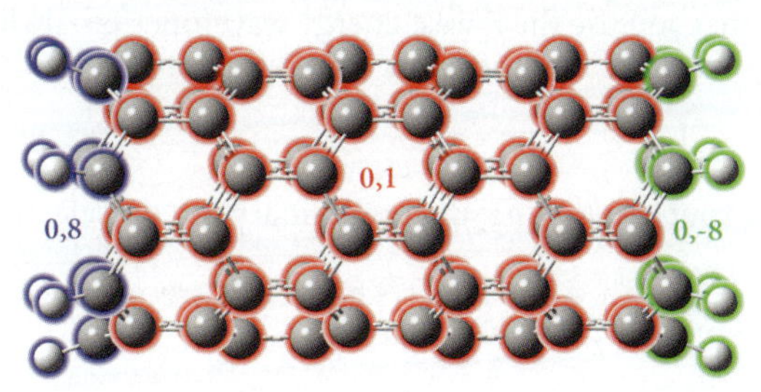

The following table presents the results of our calculations:

WAVEFUNCTION	E (hartrees)	REL. E (kJ/mol)
closed shell singlet	-3206.70918	209.4
antiferromagnetic singlet	-3206.78892	0.0
open shell triplet	-3206.74257	121.7
open shell quintet	-3206.78500	10.3
open shell septet	-3206.73413	143.9

The antiferromagnetic singlet structure has the lowest energy. Here is the spin density corresponding to it:

The pattern of adjacent electrons of opposite spin is very clear.

See [Hod08] for more information about this interesting compound.■

EXERCISE 9.8 MODELING THE BIRADICAL 2,6-PYRIDYNE

In Example 9.8 (p. 449), we discovered that the UAPFD/6-311G(2d,p) model predicted a monocyclic structure with a C-C bond distance of 2.22 Å. Repeat the relaxed scan using the CASSCF method with two electrons and two orbitals. Finally, perform single point **BD(T)** calculations at each geometry located by the scan. Plot the BD, BD(T), and CAS(2,2) energies. How do these results compare with those we obtained earlier?

The CAS(2,2)/6-311G(2d,p) scan calculation should be started using the natural orbitals from a broken symmetry UHF calculation. Natural orbitals can be generated and saved in the checkpoint file by specifying the **Pop=(NaturalOrbital,SaveNatural)** options in the job's route section. Be sure that you use the lowest energy wavefunction. Include the **UNO** option to the **CAS** keyword to indicate that the guess employs natural orbitals.

SOLUTION

We used the following route section for the preliminary job for selecting the CAS orbitals:

```
# UHF/6-311g(2d,p) Guess=Mix Stable=Opt
  Pop(NaturalOrbital,SaveNatural)
```

The following figure shows the orbitals that were used to construct the active space:

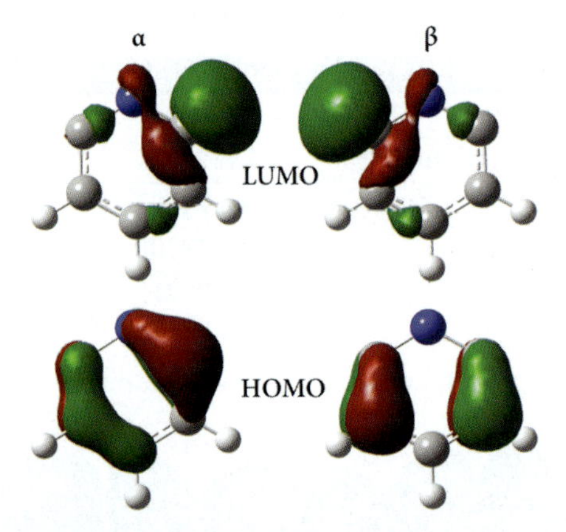

This wavefunction will allow the molecule to adopt a pure* M_s=0 biradical spin state with one electron occupying each of the MOs shown at the left above.

Next, we examined the active space to be used in the CASSCF scan by running a **CAS(2,2,UNO)** single point calculation, reading in the guess from the checkpoint file of the preceding job. When the job completed, we visualized the orbitals corresponding to the converged CAS wavefunction:

These orbitals again allow the molecule to attain a pure biradical spin state, so we verified that we have an appropriate active space.

* Non-spin contaminated.

We then ran the CASSCF scan calculation, once again starting from the saved wavefunction. Once it completed, we extracted the set of optimized geometries and performed a single point calculation using the **BD(T)/6-311G(2d,p)** model chemistry for each one.

The following plot shows the results of our calculations, along with the relevant portions of the singlet and triplet curves computed previously with our standard model chemistry for reference. Each energy value has been plotted relative to the lowest energy for that model chemistry. As compared to unrestricted DFT, the use of the CASSCF or BD theoretical method results in a more consistent picture for the PES of this molecule.

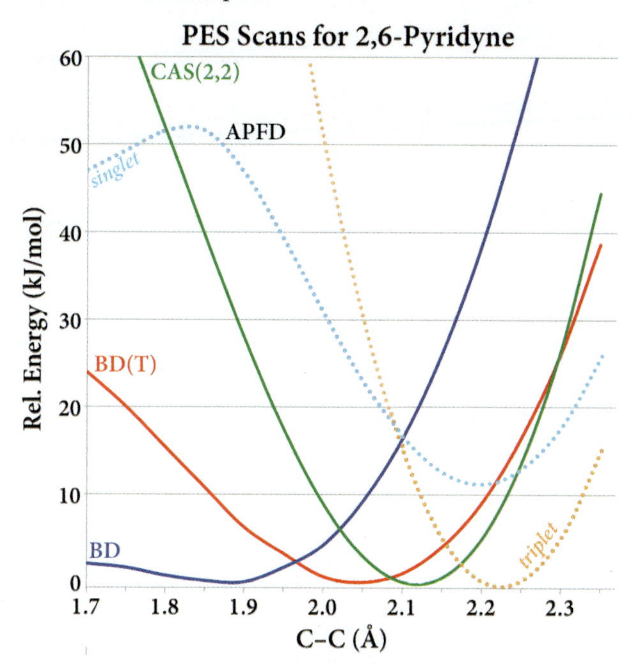

The triples correction to BD is essential in reproducing the energetics. The **CAS(2,2)** treatment predicts the minimum for the monocyclic structure at a C–C bond length of 2.15 Å, as well as a large barrier to conversion to the bicyclic structure. The Brueckner Doubles method without the triples correction underestimates the bond length at the minimum (~1.89 Å). The **BD(T)** minimum is close to the accepted value of 2.02 obtained from multiconfigurational-CI calculations [Prochnow09].

Another method for studying a system like this has been recently proposed. This involves projecting out the spin contamination from a double-hybrid DFT calculation [Thompson14].■

10

The Theoretical Background

$$\mathbf{H}\Psi = E\Psi$$

In This Chapter:

10 ◆ The Theoretical Background

Electronic structure theory is based on applying the fundamental laws of quantum mechanics to atoms and molecules in order to predict their energies and properties. It necessarily uses a variety of approximation methods to solve the challenging mathematical equations that are involved.

The goal of this chapter is to provide an overview of these procedures.

- We begin with a detailed look at Hartree-Fock theory for molecular ground states.
- Next, we provide high-level descriptions of electron correlation methods, density functional theory and complete basis set extrapolation.
- We conclude with a brief summary of excited state methods.

We devote the greatest level of detail to Hartree-Fock theory because it provides a pedagogically effective entrée to electronic structure theory and also serves as the foundation for other theoretical methods. Space constraints limit the depth of our discussions of the other methods to an overview level with substantially less mathematical details. They also exclude consideration of many related topics of interest.

Accordingly, this chapter cannot replace full-scale treatments of quantum mechanics and electronic structure theory. For general introductions to quantum mechanics from a physics and chemistry point of view, we recommend [Schiff68] and [Levine13] (respectively). For details about electronic structure theory, helpful books include [Cramer04], [Jensen07] and [Szabo96].

This chapter assumes familiarity with the basics of linear algebra and differential equations and with quantum mechanics notation (e.g., bra and ket).

Mathematics & Quantum Mechanics

We begin, perhaps somewhat surprisingly, with mathematics rather than physics. An equation in which a mathematical operator $\hat{\mathbf{O}}$ applied to a function φ yields that same function multiplied by a constant is known as an *eigenvalue equation*, and φ is an *eigenfunction* of the operator $\hat{\mathbf{O}}$. In fact, every operator has an associated set of eigenfunctions satisfying the equation:

$$\hat{\mathbf{O}}\varphi_i = c_i\varphi_i \tag{10.1}$$

Notation Conventions
operator: bold+circumflex ($\hat{\mathbf{H}}$)
function: Greek letter (φ,Ψ)
scalar: italic (c,E)

where the set of functions φ_i is the set of eigenfunctions, and the scalars c_i are the corresponding eigenvalues.

Eigenvectors were first studied intensively by Lagrange in the eighteenth century (following on some work by Euler). Cauchy was the first mathematician to give a name to the eigenvalue—he called it a *racine caractéristique* (characteristic root)—and, according to one account, eigenvalues were given their modern name by Hilbert: in German, *eigen* is an adjective meaning *own,* as in "my own idea." Hilbert spaces figure prominantly in the mathematics of quantum mechanics.

In the same general era as Hilbert (the early twentieth century), physicists noticed that some natural observables took on only certain values even though classical mechanics predicted that the associated functions would be continuous (could potentially take on any value). Put another way, they observed that certain natural phenomena could be described by equations whose solutions were eigenfunctions and their associated eigenvalues. Quantum mechanics is the theory that was developed to describe this (non-classical) behavior. All quantum mechanical theories have this basic construct of "allowed" values for observables, even if they use different mathematical constructs (e.g., matrices for Heisenberg) or incorporate the theory of relativity (as with Dirac).

Quantum mechanics rests on a set of fundamental postulates. Different descriptions of quantum mechanics will state them quite differently, especially if you compare ones from, say, pure mathematics,* physics and chemistry. They can be summarized in this way:

- A particle—really, a quantum mechanical system—moving in a field has a *wavefunction* associated with it, generally denoted as Ψ, which is a function of its position in space and of time. For example, the wavefunction for a molecule depends on the coordinates of its nuclei and electrons (as well as time).

  The wavefunction is a superposition of all of the available states of the system; it contains all information which is knowable about the system.

- Every *observable* of a quantum mechanical system—call it S—has an associated operator, and measurements on S always yield an eigenstate of the operator. The act of measuring is said to "collapse" the wavefunction into one of its possible states.

- Measurements on S have a *probabilistic* outcome; exact results cannot be foreseen. However, immediately repeated measurements of the same observable will yield identical results.

- Different observables can be *compatible* or *incompatible*. Identical measurements of compatible observables can be taken one after another without changing the state of the quantum mechanical system; this is not true for incompatible observables.

  Consider two observables Q_1 and Q_2, and suppose we measure the current value of Q_1 for S; we'll call this q_a. Suppose that we then measure the value of the other observable Q_2 (its value is not important here), and then immediately measure the value of Q_1 again, which we will call q_b.

  If Q_1 and Q_2 are compatible observables, measuring the value of Q_2 won't have affected S, and q_b will be equal to q_a. On the other hand, if $q_b \neq q_a$, then Q_1 and Q_2 are incompatible since measuring Q_2 altered the state of S and changed the value yielded by measuring Q_1.

- The *evolution* of the quantum mechanical system over time is described by the time-dependent Schrödinger equation: $i\hbar\,\partial/\partial t\,\Psi = \hat{\mathbf{H}}\Psi$. Electronic structure theory does not need to solve this most general form (see below).

The following points deserve special emphasis:

- Only quantities for which there is a quantum mechanical operator—e.g., the total energy, angular momentum, square of the spin angular momentum—can be observed. Certain chemical concepts—partial atomic charge, electronegativity, hardness—do not have corresponding operators and accordingly cannot be measured.

- Operators can have eigenfunctions in common. When they do, the observables associated with them can be known simultaneously to the same precision. Conversely, operators that do not share eigenfunctions (e.g., position and momentum) cannot be known to the same precision simultaneously.

- Eigenfunctions themselves cannot be observed; they are simply mathematical constructs.

* For an excellent description of quantum mechanics from a mathematical point of view, see [Moretti13].

- Being a superposition of states, the wavefunction itself is not an eigenfunction of any operator.

According to the *average value theorem*, we can calculate the average value of an observable Q by integration over all space:

$$\langle Q \rangle = \frac{\int \Psi^*(\bar{r})\hat{Q}\Psi(\bar{r})d\bar{r}}{\int \Psi^*(\bar{r})\Psi(\bar{r})d\bar{r}} \quad (10.2)$$

Notation Conventions
vector/set of vectors: arrow ($\bar{r}$)
indiv. vector in set: subscripted ($\bar{r}_i$)

where $\hat{Q}$ is the operator associated with Q, and Ψ^* is the complex conjugate of Ψ. Time drops out here as we are computing an instantaneous average value.

The wavefunction can be *normalized* so that the denominator is one. Once we have done so, we can then write the following expression for the probability (*P*) that the quantum mechanical system lies within certain bounds of the coordinates, denoted as the volume τ:

$$P = \int^{volume\,\tau} \Psi^*(\bar{r})\Psi(\bar{r})d\tau \quad (10.3)$$

The squared modulus of a normalized wavefunction—the integrand of the above expression—can be interpreted as a probability distribution. Evaluating this quantity over a grid of points will yield data that can be visualized to find the most likely places the system exists. One such plot is given at the start of Chapter 8 (see p. 329) for the radial probabilities of the hydrogen atom (places to find the electron). As it applies to the electrons of a molecule, the wavefunction can be normalized to the total number of electrons, in which case the probability is referred to as the *total electron density*, and it can be visualized in a variety of ways using graphical interfaces to Gaussian.

$\Psi^*\Psi$ *can be written* $|\Psi|^2$

Schrödinger's Equation and the Molecular Hamiltonian

Of particular interest to us is the eigenvalue equation associated with the energy of a molecular system. The *Schrödinger equation* [Schrodinger26] describes how a quantum mechanical system changes over time. For molecular systems, however, time is not variable, and the Schrödinger equation can be simplified mathematically to yield the *time-independent* form that is a function of only the coordinates of the nuclei and the electrons. Here is its familiar, compact form:

$$\hat{\mathbf{H}}\Psi(\bar{r}) = E\Psi(\bar{r}) \quad (10.4)$$

The operator $\hat{\mathbf{H}}$ is known as the *Hamiltonian*, and its corresponding observable is the total energy of the system (*E*). The Hamiltonian operator for the time-independent, non-relativistic Schrödinger equation is:

$$\hat{\mathbf{H}} = -\sum_i^{electrons} \frac{\hbar^2}{2m_e}\nabla_i^2 - \sum_A^{nuclei} \frac{\hbar^2}{2m_A}\nabla_A^2 - \sum_i^{electrons}\sum_A^{nuclei} \frac{e^2 Z_A}{r_{iA}} + \sum_{i>j}^{electrons} \frac{e^2}{r_{ij}} + \sum_{A>B}^{nuclei} \frac{e^2 Z_A Z_B}{r_{AB}} \quad (10.5)$$

$$\nabla^2 = \frac{\partial^2}{\partial x^2} + \frac{\partial^2}{\partial y^2} + \frac{\partial^2}{\partial z^2}$$

where r_{ln} represents the distance between *l* and *n*. Conventionally, *i* is the subscript for electrons, and *A* is the subscript for nuclei. m_e and m_A are the masses of the electron and nucleon, respectively, and $\hbar$ is Planck's constant divided by 2π.

The first two terms in the Hamiltonian represent the kinetic energy of the electrons and nuclei, respectively. The remaining terms involve the potential energy due to the attraction of the electrons for the nuclei, the electron-electron repulsions, and the nuclear-nuclear repulsions (terms three, four and five, respectively).

Since the nuclei are much heavier and move much more slowly than the electrons, we can freeze the nuclear positions and solve for an energy and wavefunction for the molecular system that involves only the coordinates of the electrons. This will provide a route to a uniquely defined energy that depends on a set of atomic coordinates. When we do this, we are making the *Born-Oppenheimer approximation* [Born27], and the consequences are:

- The second term in the Hamiltonian (which involves only the nuclei) can be ignored.
- The final term in the Hamiltonian is trivial to evaluate since we only need to know the charges on the nuclei and their distances from one another. This is the nuclear-repulsion energy E_{NRE} (which is reported in the Gaussian log file, labeled as "nuclear repulsion energy"). Accordingly, it can be evaluated separately.
- The remaining terms—one, three and four—remain in the eigenvalue problem, and the energy that emerges from solving it can be added to the nuclear-repulsion energy to yield a total energy, E_{total}, for the molecular system. Note that this energy has a zero value for all nuclei and electrons at infinite separation.

Atomic Units

When working with expressions that have many small-valued constants—such as the electron mass and Planck's constant—it may be easier to simply set a constant or group of constants equal to the value 1. If we apply this thinking to the equations that emerge from quantum mechanics, we can derive the set of atomic units, summarized in the following table:

QUANTITY SET EQUAL TO 1	CONSTANT EXPRESSION	LHS SI VALUE
mass of an electron	$m_e = 1$	9.109382×10^{-31} kg
charge of an electron	$e = 1$	$1.6021765 \times 10^{-19}$ C
reduced Planck's constant	$\hbar = \frac{h}{2\pi} = 1$	$1.0545716 \times 10^{-34}$ J·sec
bohr: *distance* which is the most probable radius of a 1s electron in hydrogen atom	$a_o = \frac{4\pi\varepsilon_o \hbar^2}{m_e e^2} = 1$	$5.2917721 \times 10^{-11}$ m
hartree: amount of *energy* equal to twice the ionization energy of hydrogen atom	$\frac{m_e e^4}{(4\pi\varepsilon_o \hbar)^2} = 1$	4.359744×10^{-18} J
time it takes for an electron to travel one period in the first Bohr orbit	$\frac{(4\pi\varepsilon_o)^2 \hbar^3}{m_e e^4} = 1$	$2.4188843 \times 10^{-17}$ s
permittivity of a vacuum (ε_0) times 4π	$4\pi\varepsilon_0 = 1$	1.11265×10^{-10} $C^2/(J{\cdot}m)$

We will use these units in all future expressions in this chapter.

The Schrödinger Equation in Atomic Units

Here is the time-independent, non-relativistic Born-Oppenheimer Schrödinger equation in atomic units:

$$\hat{\mathbf{H}}_{BO}\Psi(\vec{r}_e) = \left(\sum_i^{electrons} \frac{-\nabla_i^2}{2} + \sum_i^{electrons} \sum_A^{nuclei} \frac{-Z_A}{r_{iA}} + \sum_{i>j}^{electrons} \frac{1}{r_{ij}} \right) \Psi(\vec{r}_e) = E_{BO}\Psi(\vec{r}_e) \tag{10.6}$$

We solve for E_{BO} and add this quantity to the nuclear repulsion energy to obtain the total electronic energy: $E_{total} = E_{BO} + E_{NRE}$.

E_{total} is the energy due to the movement of the electrons within the molecule and electrostatic energies of all of the component particles, including the fixed nuclei. Real molecules possess additional energy: vibrational energy at 0 K and thermal energy due to translations, rotations and vibrations at temperatures above absolute zero.

The Schrödinger equation can be solved exactly only for one-electron systems. In the case of the hydrogen atom, the energy eigenvalues that will emerge are given by:

$$E_{\text{elec}}^{\text{H atom}} = -\frac{1}{2n^2},\ n = 1,2,3,... \tag{10.7}$$

The kinetic and potential energy operators can be used with the average value theorem, equation (10.2), to derive the expectation values for the corresponding energies of the hydrogen atom:

$$\left\langle E_{\text{kinetic}}^{\text{H atom}} \right\rangle = +\frac{1}{2n^2} \qquad \left\langle E_{\text{potential}}^{\text{H atom}} \right\rangle = -\frac{1}{n^2} \tag{10.8}$$

This demonstrates that quantum mechanics is consistent with the *Virial theorem*:

$$\left\langle E_{\text{potential}} \right\rangle = -2\left\langle E_{\text{kinetic}} \right\rangle \tag{10.9}$$

The Virial theorem was derived by Clausius in the late 1800s using only classical mechanics.

Be aware that approximate solutions to Schrödinger's equation may lead to energies that do not precisely satisfy the Virial theorem. Gaussian prints the virial ratio, $\langle E_{\text{potential}} \rangle / \langle E_{\text{kinetic}} \rangle$, labeled as "-V/T":

```
SCF Done:  E(RHF) =  -1122.37211556  A.U. after  2 cycles
     NFock= 19  Conv=0.70D-14     -V/T= 2.0061
```

Restriction on the Wavefunction

The *Pauli principle* requires that wavefunctions representing fermions—quantum particles with half-integer spin—be antisymmetric with respect to exchanging any two particles. Since electrons are fermions, for a two electron system then we insist that:

$$\Psi(\bar{r}_1,\bar{r}_2) = -\Psi(\bar{r}_2,\bar{r}_1) \tag{10.10}$$

A consequence of applying this restriction is that electrons can each be associated with a unique set of quantum numbers and yet remain indistinguishable from one another in the overall wavefunction. The probability distribution (the square of the wavefunction) will remain the same no matter how we label the electrons of the system.

What we have presented so far was understood by physicists in the 1920s. As Dirac put it in 1929:

> The underlying physical laws necessary for the mathematical theory of a large part of physics and the whole of chemistry are thus completely known, and the difficulty is only that the exact application of these laws leads to equations much too complicated to be soluble.

It would take both advances in computer technology and the willingness to try approximate (inexact) approaches to solving the fundamental equations in order to change the pessimistic state of affairs described by Dirac and make quantum mechanics useable for studying non-trivial molecular systems.

Oscar Blanco Rodriguez

Hartree-Fock Theory

Even though it is seldom used for production calculations, understanding how Hartree-Fock theory approximates a solution to the Schrödinger equation is still the best introduction to the methods of electronic structure theory.

The Hartree-Fock approach [Hartree28, Fock30] can be described as an attempt to solve Schrödinger's equation for a multi-electron system by reducing the problem to a set of one-electron problems. Each electron exists in the field of the nuclei and experiences the influence of all the other electrons only in an averaged way, neglecting how an electron of opposite spin might specifically influence its motion. This is a severe approximation, but one that historically allowed chemists to get started modeling molecular systems. Today, Hartree-Fock theory still has its place as the basis for more accurate calculations that correct for its deficiencies.

Molecular Orbitals and Basis Sets

In order to solve equation (10.6), we first consider how to represent the wavefunction. Based on the success of applying the theory to the hydrogen atom, we can build the wavefunction from functions centered on each nucleus, which resemble the eigenfunctions of the hydrogen atom Hamiltonian. The set of functions, including how many of each type and their orbital exponents, are parameters in the Hartree-Fock calculation. We refer to this construct as the *basis set*.

It is common to define these atomic functions, which we denote χ_μ, using a linear combination of gaussian functions, chosen because they allow integrals and integral derivatives to be calculated efficiently. Gaussian functions can be added together to mimic the radial hydrogen orbitals, which are exponential.* Typically, the number of component gaussian functions

* The description following will assume Pople-type basis sets composed of component gaussian functions. Other types of basis sets are constructed differently.

$g_p(\zeta,r)$, known as *primitives,* that are required to construct an atomic function ranges from six in the core region to one in the outer valence region.

Notation Conventions
AO basis function: chi (χ)
molecular orbital: phi (ϕ)
Gaussian function: g
orbital exponent: zeta (ζ)
orbital radius: r

Thus,

$$\chi_\mu = \sum_p d_{\mu,p} g_p(\zeta,r)$$
$$\text{where } g_p(\zeta,r) = cx^n y^m z^l e^{-\zeta r^2} \qquad (10.11)$$

where $d_{\mu,p}$ is the fixed coefficient for the gaussian primitive g_p for atomic orbital χ_μ. The orbital exponent, ζ, controls the tightness or diffuseness of each primitive function, while the exponents on *x*, *y* and *z* will dictate its shape (s, p, d, and so on). The coefficient *c* serves to normalize the function such that the integral over all space of its square is 1: $\int g^2=1$.* Note that separate basis functions are defined for each element.

In Gaussian, you can view all the parameters for any basis set by including the **GFPrint** keyword in the route section for a job like this one:

```
%Kjob l301
# basis-set-name GFPrint
```

final blank line

A *molecular orbital,* ϕ_i, is defined as the linear combination of the applicable functions in the basis set:

$$\phi_i = \sum_\mu c_{i\mu} \chi_\mu \qquad (10.12)$$

where the $c_{i\mu}$ are the molecular orbital coefficients (unrelated to the *c* used in the construction of the gaussian primitive above). A given molecular orbital (MO) will have contributions from atomic orbitals distributed throughout the molecular system.

Keep in mind that the MOs are simply mathematical conveniences; they are not real and cannot be observed. In fact, there may be different sets of orbitals/coefficients that yield equivalent results:

- Canonical orbitals: delocalized orbitals emerging from the Hartree-Fock procedure (described below).
- Localized orbitals, found by a unitary transformation of the canonical orbitals.
- Natural or Löwdin orbitals: eigenvectors of the reduced one-particle density matrix.
- Natural bond orbitals, located by an analysis procedure constrained to find MOs that resemble those in a Lewis structure description.

Incorporating the property of electron spin into the description will allow us to create an antisymmetric wavefunction.

Electron Spin Operators

Electrons possess an intrinsic characteristic known as *spin*. This quantum mechanical property can be observed by passing a beam of neutral hydrogen atoms through an inhomogeneous magnetic field and recording where they appear on a detector screen. Instead of being continuously scattered, they are found clustered equally in two precise regions, above and below the axis of propagation. For this reason, we refer to electrons as being *spin up* (alpha) or *spin down* (beta), and we have an equal chance of detecting one or the other.

* The contracted functions are usually normalized as well.

Focusing on just the z-component of the spin, we can write spin operator eigenvalue equations:

$$\hat{S}_z\,\alpha = +\tfrac{1}{2}\alpha \qquad \hat{S}_z\,\beta = +\tfrac{1}{2}\beta \tag{10.13}$$

In atomic units, the only eigenvalues possible for the z-component of the spin operator are +0.5 and -0.5; they are associated with the eigenfunctions α and β, respectively.

Constructing the Hartree-Fock Wavefunction

In order to build the complete wavefunction, $\Psi(\bar{r})$, we need to combine the molecular orbitals in such a way that the Pauli principle of antisymmetry is obeyed. Put another way, we want to write the many electron wavefunction as an antisymmetric combination of one electron wavefunctions.

The simplest possible combination is the product of the various molecular orbitals, known as their *Hartree product*. However, this is not antisymmetric. Instead, we will combine the spin functions with the spatial part of the wavefunction—the part that depends only on the coordinates of the electrons—into an antisymmetric wavefunction by employing a mathematical tool from linear algebra: the determinant of a matrix.

For a closed shell* molecular system having *n* electrons that are all spin-paired, we can write the wavefunction as follows:

This formulation for closed shell systems is termed "restricted" Hartree-Fock (RHF) since the spatial orbitals are constrained to be doubly occupied with electrons of opposite spin.

$$\Psi(\bar{r}) = \frac{1}{\sqrt{n!}}\begin{vmatrix} \phi_1(\bar{r}_1)\alpha(1) & \phi_1(\bar{r}_1)\beta(1) & \phi_2(\bar{r}_1)\alpha(1) & \phi_2(\bar{r}_1)\beta(1) & \cdots \\ \phi_1(\bar{r}_2)\alpha(2) & \phi_1(\bar{r}_2)\beta(2) & \phi_2(\bar{r}_2)\alpha(2) & \phi_2(\bar{r}_2)\beta(2) & \cdots \\ \phi_1(\bar{r}_3)\alpha(3) & \phi_1(\bar{r}_3)\beta(3) & \phi_2(\bar{r}_3)\alpha(3) & \phi_2(\bar{r}_3)\beta(3) & \cdots \\ \phi_1(\bar{r}_4)\alpha(4) & \phi_1(\bar{r}_4)\beta(4) & \phi_2(\bar{r}_4)\alpha(4) & \phi_2(\bar{r}_4)\beta(4) & \cdots \\ \vdots & \vdots & \vdots & \vdots & \ddots \end{vmatrix} \tag{10.14}$$

where the subscripts on the MOs run from 1 to $n/2$, and the subscripts on the position vectors run from 1 to n. In this notation, $\alpha(m)$ and $\beta(m)$ are the alpha and beta spin functions for electron m, and their arguments run from 1 to n. Each MO appears twice for a given electron since each orbital is doubly occupied. Thus, there will be as many rows and columns in the determinant as there are electrons in the system. The factor in front of the determinant is for normalization.

Each element in the matrix is referred to as a *spin orbital*, and the set of these can be represented as the set $\{\Phi_i(m)\}$. If there are $n/2$ spatial orbitals, there are n spin orbitals available to be occupied by electron m:

$$\begin{aligned} \Phi_i(m) &= \phi_i(\bar{r}_m)\alpha(m) \\ \Phi_{\frac{n}{2}+i}(m) &= \phi_i(\bar{r}_m)\beta(m) \qquad \text{where } i = 1,2,\ldots,\tfrac{n}{2} \end{aligned} \tag{10.15}$$

Each row of the determinant is formed by representing all possible assignments of electron i in all the spin orbitals of the system. Interchanging two electrons corresponds to swapping two rows of the matrix, which will have the effect of changing the sign of the determinant as required. The algebraic expansion of the determinant will generate products of spin orbitals such that each term has each of the electrons appearing once, meaning that each electron exists in its own unique spin orbital as required by quantum mechanics. The total wavefunction

* See p. 478 for a discussion of modeling open shell systems.

is then a sum of all the possible ways of distributing the indistinguishable electrons among the set of spin orbitals.*

The Variational Principle and the Self-Consistent Field (SCF) Approach

We return now to solving equation (10.6) using our properly defined wavefunction. The wavefunction still is not precisely known—merely its functional form—because we do not know the molecular orbital coefficients ($c_{i\mu}$) of equation (10.12). However, we can create a trial wavefunction by making a guess at these and solving for the energy, using the average value theorem, equation (10.2). We can use the information learned in each trial to proceed in an iterative matter (see below).

There are several methods of estimating the initial MO coefficients (see the discussion of the **Guess** keyword). The most common approach is to use eigenvectors from a truncated version of the actual Hamiltonian.

The *variational theorem* states that the energy derived in this manner is an upper bound to the actual energy:

$$\frac{\int \Psi^* \hat{\mathbf{H}} \Psi d\tau}{\int \Psi^* \Psi d\tau} = E_{\text{variational}} \geq E_{\text{exact}} \qquad (10.16)$$

where $d\tau$ now indicates integration over spatial and spin coordinates.

The problem reduces to finding the set of orthonormal spin orbitals that minimize the energy:

$$\frac{\partial E_{\text{var}}}{\partial \Phi_i} = 0 \quad \text{where} \quad \int \Phi_i^* \Phi_j d\tau = \delta_{ij} = \begin{matrix} 0 \;\; \mathit{if} \;\; i \neq j \\ 1 \;\; \mathit{if} \;\; i = j \end{matrix} \qquad (10.17)$$

Using the method of Lagrange multipliers, D.R. Hartree and V. Fock each developed an operator equation in which the eigenvectors are the "best" molecular spin orbitals possible for a wavefunction that is a single determinant:

$$\hat{\mathbf{F}} \Phi_i = \varepsilon_i \Phi_i \qquad (10.18)$$

This equation is known as the *Fock equation*. The *Fock operator*, $\hat{\mathbf{F}}$, is an effective one-electron Hamiltonian for an orbital. The eigenvalues, ε_i, are referred to as *orbital energies*. In this formulation, each orbital sees the average distribution of all the other electrons.

The Fock equation can be expanded as follows (using atomic units):

$$\hat{\mathbf{F}} \Phi_i = \hat{\mathbf{T}} \Phi_i + \hat{\mathbf{V}}_{\text{NE}} + \hat{\mathbf{J}} \Phi_i - \hat{\mathbf{K}} \Phi_i = \varepsilon_i \Phi_i \qquad (10.19)$$

$$\hat{\mathbf{T}} = -\tfrac{1}{2} \nabla^2 \qquad \textit{kinetic energy operator}$$

$$\hat{\mathbf{V}}_{\text{NE}} = \sum_{A}^{\mathit{nuclei}} \frac{-Z_A}{r_{iA}} \qquad \textit{nuclear - electron attraction operator}$$

$$\hat{\mathbf{J}} \Phi_i(1) = \left\{ \sum_{j}^{\mathit{occupied}} \int \Phi_j(2) \tfrac{1}{r_{12}} \Phi_j(2) d\tau_2 \right\} \Phi_i(1) \qquad \textit{Coulomb operator}$$

$$\hat{\mathbf{K}} \Phi_i(1) = \left\{ \sum_{j}^{\mathit{occupied}} \int \Phi_i(2) \tfrac{1}{r_{12}} \Phi_j(2) d\tau_2 \right\} \Phi_j(1) \qquad \textit{exchange operator}$$

* What we have done here is to graft the concept of spin into the wavefunction associated with the non-relativistic, time-independent Schrödinger equation. Dirac's version of quantum mechanics incorporating relativity yields spin and its associated eigenfunctions as a natural consequence.

Here $\hat{\mathbf{T}}$ accounts for the kinetic energy of an electron, and $\hat{\mathbf{V}}_{NE}$ accounts for the attraction it has for each of the nuclei. The Coulomb operator $\hat{\mathbf{J}}$ represents the coulombic repulsion between an electron in Φ_i and the charge distribution created by another electron in Φ_j. The $\hat{\mathbf{K}}$ operator represents the *exchange term:* a non-local quantity present as a result of the antisymmetry requirement of the wavefunction. Notice that the electrons are labeled 1 and 2 to identify the coordinates of the integration. They could be any two electrons of the system.

The summations are over all the occupied spin orbitals. Note that self-interaction and self-exchange are included and cancel one another out since the two integrals are equivalent when $i=j$.

It is clear from the two integral quantities that the Fock operator itself depends on the orbitals in order to be evaluated. This unique feature of the equations requires us to solve them iteratively, using a process known as the self-consistent field (SCF) approach. The computation process is illustrated on the left below:

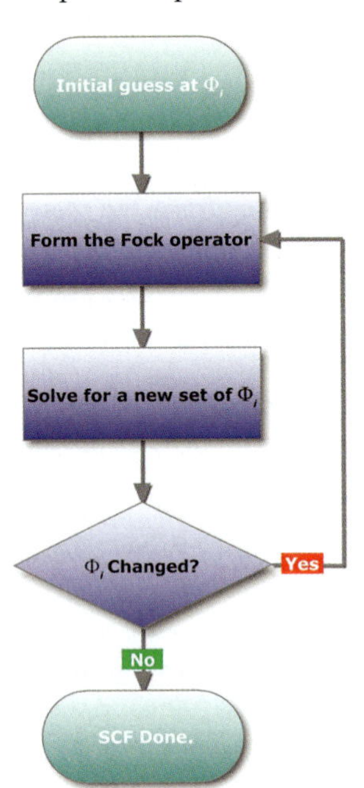

The test condition in the loop checks whether the new orbitals are the same as the ones from the previous cycle within some tolerance. When they are, the SCF procedure has *converged.* The specific criteria used to determine convergence will vary according to the calculation type.

Gaussian reports the number of cycles taken by the completed SCF procedure after reporting the energy:

```
SCF Done: E(RHF)= ... after 13 cycles
```

The item in parentheses indicates the method used for the calculation: here, restricted Hartree-Fock theory.

The Roothaan-Hall Equations

Since the molecular orbitals are represented by a basis set, we can expand the Fock operator in terms of the basis functions χ_μ to form a set of matrix equations. These equations were originally developed independently by Roothaan and Hall (each of whom was a graduate student at the time!) [Roothaan51, Hall51]. Using N to represent the number of basis functions and n to represent the number of electrons, the matrix equation for the closed-shell case (i.e., an even number of electrons with each MO doubly occupied) is:

Notation Conventions
matrix: boldface (**F**)
matrix elements: subscripts ($\mathbf{F}_{\mu\nu}$)

$$\mathbf{F}_{\mu\nu} = \mathbf{H}_{\mu\nu}^{\text{core}} + \sum_{\lambda}^{N}\sum_{\sigma}^{N} \mathbf{P}_{\lambda\sigma}\left[(\mu\nu|\lambda\sigma) - \tfrac{1}{2}(\mu\lambda|\nu\sigma)\right] \tag{10.20}$$

F is the Fock matrix. The Fock matrix includes *one-electron integrals* involving kinetic and nuclear-electron attraction operators and *two-electron integrals* involving electron-electron interaction. Note that we have simplified the integral quantities in the equation so that they only involve the spatial coordinates of the electrons, making the factor of one half necessary to account for the terms that are annihilated by integration over spin coordinates.

In the literature, the core Hamiltonian operator is frequently denoted by a lowercase h: $\hat{\mathbf{h}} = \hat{\mathbf{T}} + \hat{\mathbf{V}}_{NE}$.

The elements of matrix $\mathbf{H}^{core}$—known as the *core Hamiltonian*—are the one-electron integrals, accounting for the movement of individual electrons in the field of the bare nuclei:

$$\mathbf{H}^{core}_{\mu\nu} = \int \chi_\mu (\hat{\mathbf{T}} + \hat{\mathbf{V}}_{NE}) \chi_\nu d\tau \qquad (10.21)$$

The electron-electron interactions—both coulomb and exchange—are found in the four-suffixed arrays of the two-electron integrals: the double summation terms from equation (10.20). The elements of the density matrix (**P**) are calculated as a sum over the occupied MOs:

$$P_{\lambda\sigma} = 2\sum_{i}^{n/2} c_{i,\lambda} c_{i,\sigma} \qquad (10.22)$$

We obtain the following expression for the two-electron integrals:

$$\langle \mu\lambda | \nu\sigma \rangle = (\mu\nu | \lambda\sigma) = \int \chi_\mu^*(\bar{r}_1)\chi_\nu(\bar{r}_1)\frac{1}{r_{12}}\chi_\lambda^*(\bar{r}_2)\chi_\sigma(\bar{r}_2) d\bar{r}_1 d\bar{r}_2 \qquad (10.23)$$

Chemist's Notation

Chemists replace the bra-ket notation for the two-electron integrals with a parenthesized one having both functions depending on electron 1 on the left and those depending on electron 2 on the right, with functions indicated solely by their indices

The one-electron integrals are relatively few in number—$\mathcal{O}(N^2)$ where N is the number of basis functions—and they are easy to evaluate computationally. In contrast, the two-electron integrals are so great in number $\mathcal{O}(N^4)$ that they may not fit in memory. However, it turns out that storing them to disk and later retrieving them frequently takes more time and resources than simply recomputing them as needed. Intelligent algorithms exist for doing this, including the capability of ignoring integrals whose contribution to the Fock matrix is not significantly different from the previous cycle.

In some cases, the amount of memory required is reported in the Gaussian output file:

```
Keep R1 ints in mem in symm-blocked form, NReq=1162840
```

The reported value is in 4-byte words. Thus, this example job required only about 8.8 MB.

We now arrive at the actual Roothaan-Hall matrix equation that is solved self-consistently by the SCF procedure:

$$\mathbf{FC} = \boldsymbol{\varepsilon}\mathbf{SC} \qquad (10.24)$$

Each component is an NxN matrix:

- **F** is the Fock matrix.
- **C** is the matrix of MO coefficients.
- **ε** is the diagonal matrix of orbital energies.
- **S** is the *overlap matrix*. Since basis functions are not orthogonal to one another, their overlap must be accounted for: $\mathbf{S}_{\mu\nu} = \int \chi_\mu \chi_\nu d\tau$.

Gaussian will print these matrices in the output file if you include **IOp(3/33=1, 5/33=3)** in the job's route section. Be warned that this produces lots of output for all but the smallest jobs!

In all, there will be N orbitals emerging from the solution of this matrix equation; only the lowest $N/2$ will be occupied with electrons. The remaining ones are referred to as *virtual* or *unoccupied* orbitals.

It is important to recognize that the Hartree-Fock energy of the entire molecular system (the energy that is minimized in the SCF process) is not simply a sum of the orbital eigenvalues.

Using the total single-determinant wavefunction in the average value theorem, we can derive the following expression for the Hartree-Fock energy:

$$E^{HF}_{total} = E_{NRE} + \tfrac{1}{2}\sum_{\mu\nu} \mathbf{P}_{\mu\nu}\left(\mathbf{H}^{core}_{\mu\nu} + \mathbf{F}_{\mu\nu}\right) \tag{10.25}$$

The SCF procedure used to obtain the Hartree-Fock orbitals and energies can be applied more generally. For example, *semi-empirical methods* start with the minimum basis set Hartree-Fock formulation and make various approximations in the interest of calculation speed, typically by replacing some ab initio matrix elements with empirical expressions fitted to experimental data. An example is the PM6 method, which uses experimental heats of formation. Semi-empirical methods also neglect some matrix elements entirely.

EXERCISE 10.1 CALCULATION OF THE HARTREE-FOCK ENERGY

Perform a single-point restricted Hartree-Fock (RHF) calculation on the water molecule using O-H bond distances of 1.0 Angstroms and an H-O-H angle of 104.5 degrees. Specify the STO-3G basis set. Request printing of all matrices as mentioned above, and also include the **GFPrint** and **Pop=Full** keywords in your job.

Using the Gaussian output, manually compute the SCF energy using equation (10.25). Using a spreadsheet is one way to accomplish this. Compare your result to the energy reported in the output file. In addition, compute the orbital energies by solving equation (10.24) for ε:

$$\varepsilon = (\mathbf{SC})^{-1}\,\mathbf{FC} \tag{10.26}$$

The eigenvalues will be the diagonal entries in the result matrix.

For this calcuation, there are 7 basis functions, so each matrix will be 7x7.

SOLUTION

We used a spreadsheet to gather all the matrices. Equation (10.25) can be evaluated using functions that perform dot products, and equation (10.26) requires functions for matrix multiplication and matrix inversion.*

d_1	a	b	c	d
a	d_2	e	f	g
b	e	d_3	h	i
c	f	h	d_4	j
d	g	i	j	d_5

In the Gaussian output file, matrices are reported in lower triangular form: the matrix is expressed in a symmetric form reflected across the diagonal. Only the shaded elements in the diagram at the left are included in the output (the example is a 5x5 matrix). The remaining elements of the matrix—the ones above the diagonal—are the same as those below the diagonal, as indicated by the corresponding letters in the diagram. Of course, for this exercise, the entire matrix must be built in the spreadsheet program.

We needed to gather information from various parts of the Gaussian output file:

* In Excel and compatible spreadsheet applications, the functions are SUMPRODUCT, MMULT and MINVERSE (respectively).

ITEM	IDENTIFYING OUTPUT HEADER
list of basis functions	AO basis set (Overlap normalization):
nuclear repulsion energy	nuclear repulsion energy *value* Hartrees.
overlap matrix	*** Overlap ***
core Hamiltonian matrix	****** Core Hamiltonian ******
Fock matrix†	Fock matrix (alpha): (FINAL CYCLE)
MO coefficients†	Alpha MO coefficients at cycle *n*:
density matrix†	Total density matrix:

†Taken from/after the final SCF iteration.

With the exception of the Fock matrix, all of the matrices have their rows and columns labeled by basis function number. For the former, the rows are again basis functions, and the columns are molecular orbitals.

The basis functions are identified by number early in the output file:

```
AO basis set (Overlap normalization):
  element                   AO type                 basis functions
Atom O1         Shell        1 S        3     bf      1 - 1
Atom O1         Shell        2 SP       3     bf      2 - 5
Atom H2         Shell        3 S        3     bf      6 - 6
Atom H3         Shell        4 S        3     bf      7 - 7
```

Basis function 1 is the 1s function on the oxygen atom, functions 2 through 5 are the 2s and 2p on the same atom, and functions 6 and 7 are the 1s on the two hydrogen atoms.

Knowing this makes interpreting the matrix output easier. Here is the section of the output corresponding to the overlap matrix:

```
*** Overlap ***  The row & column numbers refer to basis functions, as numbered earlier in the output.
                  O 1s            O 2s            O 2px           O 2py           O 2pz
                   1               2               3               4               5
O 1s    1  0.100000D+01
O 2s    2  0.236704D+00   0.100000D+01
O 2px   3  0.000000D+00   0.000000D+00   0.100000D+01
O 2py   4  0.000000D+00   0.000000D+00   0.000000D+00   0.100000D+01
O 2pz   5  0.000000D+00   0.000000D+00   0.000000D+00   0.000000D+00   0.100000D+01
H1 1s   6  0.487518D-01   0.447147D+00   0.000000D+00   0.299042D+00  -0.231543D+00
H2 1s   7  0.487518D-01   0.447147D+00   0.000000D+00  -0.299042D+00  -0.231543D+00
                  H1 1s           H2 1s
                   6               7
H1 1s   6  0.100000D+01
H2 1s   7  0.228239D+00   0.100000D+01
```

The matrix is printed in sections for historical reasons: it used to have to fit on the available printer paper.

Our computed Hartree-Fock energy is -74.9647 (decimals beyond this are not possible, given the limited precision of the numbers reported in the output file). The value predicted by Gaussian agrees with this one to the same number of digits.

The computed orbital energies are -20.247, -1.248, -0.596, -0.448, -0.389, +0.564, +0.693. The negative values correspond to occupied orbital and the positive values to unoccupied ones (in the ground state). The predicted orbital energies are found within the population analysis section of the output file (look for the "Alpha MOs" header); the energy value for each orbital appears directly under the orbital number, in the row labeled "Eigenvalues." All of our orbital energies agree with the corresponding values reported in the Gaussian output.■

Open Shell Systems & Unrestricted Wavefunctions

Modeling molecular systems in which the electrons of the system are not spin-paired present additional challenges. Examples of such systems include the following:

- Any system with an odd number of electrons: organic radicals, transition metal complexes, and so on.
- Systems with an even number of electrons where the lowest energy configuration has a total spin greater than zero: triplet ground states such as molecular oxygen, transition metal complexes, and so on.
- Systems with an even number of electrons and a total spin of zero, but in which pairs of electrons of opposite spin do not all reside in the same orbital: e.g., antiferromagnetic substances.

A variety of approaches are available for modeling these systems, and each has both strengths and weaknesses. For example, *unrestricted Hartree-Fock* theory (UHF), first proposed by Pople and Nesbet [Pople54, McWeeny68], constructs a single determinant wavefunction that optimizes separate spatial orbitals for the electrons of each spin type (α and β):

$$\Phi_i^\alpha(m) = \phi_i^\alpha(\bar{r}_m)\alpha(m) \qquad \Phi_i^\beta(m) = \phi_i^\beta(\bar{r}_m)\beta(m) \tag{10.27}$$

This will translate into separate Fock matrices for α and β spin:

$$\mathbf{F}^\alpha\mathbf{C}^\alpha = \varepsilon^\alpha\,\mathbf{S}^\alpha\mathbf{C}^\alpha \qquad \mathbf{F}^\beta\mathbf{C}^\beta = \varepsilon^\beta\,\mathbf{S}^\beta\mathbf{C}^\beta \tag{10.28}$$

as well as a unique set of molecular orbital coefficients for each. The total electron density is the sum of the α electron and β electron densities, and these components are allowed to be different if doing so results in a lower total energy.

This different-orbitals-for-different-spin model—also referred to as unrestricted Hartree-Fock theory (UHF)—has the ability to treat many open shell problems.* For example, it can model bond dissociation and predicts the correct products. However, the eigenfunctions it computes are not pure spin states (e.g., doublets have some quartet character mixed in).

Other approaches to open shell systems at the Hartree-Fock level include:

- *Restricted open shell Hartree-Fock* theory (ROHF), which restricts all but one electron to be spin-paired in a doublet, all but two electrons to be spin-paired in a triplet, and so on.
- *Multiconfiguration SCF* (MCSCF) allows the wavefunction to be a linear combination of multiple determinants so that each degenerate component of a non-zero spin system can be represented on equal grounds (see p. 499ff).

Modeling Spin

Spin is an intrinsic property of electrons first demonstrated by Stern and Gerlach in their famous experiment in which they passed silver atoms through a magnetic field with a detector

* This unrestricted approach can also be applied to other methods. For example, unrestricted Kohn-Sham theory (UKS) employs a density functional model to calculate the energy in a manner that is analogous to UHF theory. In general, prepending a **U** to a method keyword will request an unrestricted calculation: **UB3LYP**, **UAPFD**, **UCCSD**.

collecting them in two distinct regions above and below the beam axis [Gerlach22]. Measuring any one cartesian component of this quantity can have one of two possible outcomes: $\pm\frac{1}{2}\hbar$.

In our earlier discussion of spin, we used atomic units, which set $\hbar$=1.

These findings can be expressed as operator equations:

$$\hat{S}_z\,\alpha=+\tfrac{1}{2}\hbar\alpha \qquad \hat{S}_z\,\beta=+\tfrac{1}{2}\hbar\beta \tag{10.29}$$

We chose the z-component as an example and refer to the eigenfunctions of this measurement as being in a spin up (α) or spin down (β) arrangement. Regardless of which spin state the electron is in, the total spin angular momentum of an electron is the same. Written in the form of the spin-squared operator, we can express this as:

$$\hat{S}^2\,\alpha=\tfrac{1}{2}(\tfrac{1}{2}+1)\hbar^2\alpha \qquad \hat{S}^2\,\beta=\tfrac{1}{2}(\tfrac{1}{2}+1)\hbar^2\beta \tag{10.30}$$

For the spin of a multi-electron system, the spin-squared operator equations can be generalized to:

$$\hat{S}^2\,\Psi = s(s+1)\hbar^2\Psi \tag{10.31}$$

where s is the maximum z-component over all electrons.* The following table summarizes the lowest few possible spin states for molecules:

MULTIPLICITY	# UNPAIRED ELECTRONS	S	EIGENVALUES OF $\hat{S}^2$
1 (singlet)	0	0	0
2 (doublet)	1	½	¾
3 (triplet)	2	1	2
4 (quartet)	3	3/2	3¾
5 (quintet)	4	2	6

When performing a calculation, one specifies the appropriate spin state for the molecular system (Gaussian cannot assign this). You can evaluate the energy of different spin states in order to predict which one is most stable via **Stable** calculations on each state.

Spin Density

As we noted, the α and β electron densities can differ in the results of an unrestricted calculation. By subtracting the electron density of the β orbitals from that of the α orbitals, one derives a difference density that would indicate probable regions where the unpaired electron(s) exist. This *spin density* is available for visualization by graphics packages. For an example, see Advanced Example 4.10 (p. 196).

A related quantity is the *atomic spin density*, which is the spin density partitioned among the atoms of the system. The atomic spin density is printed in the population section of the Gaussian output file.

Spin Contamination

Spin contamination is an artifact that arises from the articifical mixing of multiple electronic states within the wavefunction. In general, it may be present in an unrestricted calculation since the wavefunction is not constrained to be an eigenvector of the spin-squared operator.

* Assuming also that Φ is an eigenfunction of S^2.

Gaussian will use the wavefunction to calculate the expectation value of $\hat{\mathbf{S}}^2$ and report it. Here is the relevant output from a calculation on methyl radical:

```
SCF Done: E(UHF) = -39.6063713661 A.U. after 11 cycles
          NFock= 11  Conv=0.45D-08    -V/T= 2.0186
<Sx>= 0.0 <Sy>= 0.0 <Sz>= 0.50 <S**2>= 0.7549 S= 0.6025
```

To the extent that the reported value differs from the corresponding value in the table, we say the wavefunction is *contaminated*. This is an unavoidable result of higher spin states mixing in with the one of interest during the variational part of the calculation. In this example, the reported value is very close to the theoretical value of 0.75, indicating that spin contamination is not a problem.

In contrast, for a calculation on cobalt hydride, Gaussian reports a value of ~1.0 for S_z, indicating significant contribution from the triplet state to this singlet system. The calculation predicted a relatively inaccurate value for the equilibrium bond length, differing by 0.044 Å from experiment.

In general, spin contamination is more of a problem for inorganic systems, especially ones containing metal elements, than for organic systems. Note that, even when present, spin contamination is not always indicative of unreliable results. There are many cases where spin-contaminated wavefunctions still predict meaningful results in terms of energy. This is particularly true when using density functional models.

*Interpreting <S**2>*
S_z is the sum of the weighted contributions from each of the individual states:
$S_z = c_1 S_{z\text{-state1}} + c_2 S_{z\text{-state2}}$
subject to the constaint that $c_1 + c_2 = 1$. Thus, $c_1 \leq 0.70$ indicates a potential problem.

How much spin contamination is too much? Opinions differ on this point. We use the general guideline that the situation requires further exploration whenever the secondary state contributes more than about 30% to S_z.

When it is deemed necessary to correct for spin contamination, there are several options, including the following:

- To the extent that these methods approach exact eigenfunctions (full CI), very accurate electron correlation methods, such as BD(T) and CCSD(T), can have less spin contamination associated with their wavefunctions.
- Compound models that approximate high accuracy electron correlation methods may be useful in treating difficult cases (e.g., CBS-QB3, W1BD).
- Projected methods: these approaches start from a UHF wavefunciton and project out the higher order spin states. The Hartree-Fock version is known as PUHF. The resulting wavefunction can also be used as a starting point for electron correlation methods to produce the PUMP2, PUMP3, etc. model chemistries, which have proven useful for exploring potential energy surfaces where spin contamination can lead to incorrect features. These methods are beyond the scope of this book.
- Some systems will require a multireference approach (i.e., CASSCF) since they are not adequately described by a single determinant wavefunction.

For more information, see [Sonnenberg09]. For example applications to inorganic hydrides and their cations, see [Goel08].

Restricted open shell methods like ROHF are often proposed as an approach to systems with spin contamination since, unlike for unrestricted methods, their wavefuncitons are constrained to be eigenfunctions of $\hat{\mathbf{S}}^2$. ROHF is a variation of Hartree-Fock that allows for some orbitals to be singly occupied. However, such wavefunctions do not account for spin polarization in doubly occupied orbitals; in this respect, they are physically unrealistic. For example, the following illustration compares the predicted spin density in methyl radical for ROHF (left) and UHF (right) wavefunctions:

The RHF description finds no spin density in the plane of the molecule. In contrast, the UHF shows a small amount of negative spin density (green) in that location. This spin density arises from electron correlation effects that ROHF cannot model. For an excellent treatment of spin in open shell systems, see section 6.3.3 in [Cramer04] (and passim).

Electron Correlation Methods

When Hartree-Fock theory fulfills the requirement that Ψ be invariant with respect to the exchange of any two electrons by antisymmetrizing the wavefunction, it automatically includes the major correlation effects arising from pairs of electrons with the same spin. This correlation is termed *Pauli exchange*. However, the single determinant wavefunction does not include correlation between electrons of opposite spin (which turns out to be much larger than same-spin correlation).

Traditionally, methods that go beyond SCF in attempting to treat this phenomenon properly are known as an *electron correlation methods* or *post-SCF methods*. Many of them center on expressing the wavefunction as a combination of multiple determinants corresponding to different electronic configurations with respect to Hartree-Fock.

The following figure shows a visual representation of three excitations with respect to the Hartree-Fock wavefunction:

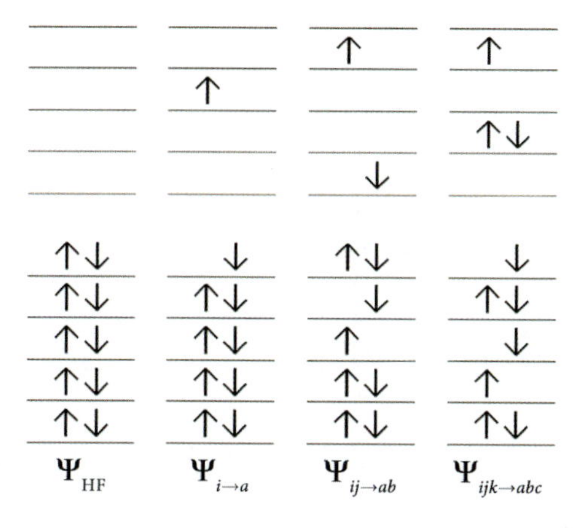

It is conventional to use $\{i, j, k \ldots\}$ as subscripts to represent occupied orbitals and $\{a, b, c \ldots\}$ for unoccupied orbitals from a Hartree-Fock calculation. The figure shows one particular singly excited configuration ($\Psi_{i \to a}$) along with a doubly ($\Psi_{ij \to ab}$) and a triply ($\Psi_{ijk \to abc}$) excited configuration:

$$\Psi_0 = |\phi_1 \cdots \phi_n|$$

$$\Psi_{i \to a} = |\phi_1 \cdots \phi_{i-1}\phi_a\phi_{i+1} \cdots \phi_{a-1}\phi_i\phi_{a+1} \cdots \phi_n|$$

$$\Psi_{ij \to ab} = |\phi_1 \cdots \phi_{i-1}\phi_a \cdots \phi_{j-1}\phi_b \cdots \phi_n|$$

$$\Psi_{ijk \to abc} = |\phi_1 \cdots \phi_{i-1}\phi_a \cdots \phi_{j-1}\phi_b \cdots \phi_{k-1}\phi_c \cdots \phi_n|$$

These other determinants can be mixed with the Hartree-Fock determinant to provide a more complete wavefunction to use in Schrödinger's equation. Different electron correlation methods are defined by specifying which excited configurations are included and how their interaction is treated. In addition, you can perform each of these calculations with or without including determinants with substitutions out of the core electrons. The latter is referred to as a frozen core (FC) calculation, and it is the default in Gaussian.

Allowing electrons to move into these configurations allows electrons of opposite spin to avoid one another (a physical feature that is missing from the Hartree-Fock energy). We define the correlation energy (E_{corr}) as the difference between Hartree-Fock theory at an infinite basis set* and the exact solution to the Schrödinger equation:

$$\hat{\mathbf{H}}\Psi = E_{exact}\Psi$$
$$E_{corr} = E_{exact} - E_{HF}^{\infty} \tag{10.32}$$

We will look briefly at some different approaches to the electron correlation problem in this section:

- Configuration interaction (CI) methods work by expressing the wavefunction as an expansion starting with the Hartree-Fock wavefunction and systematically adding in determinants for various electronic excitations. This is the most intuitively "direct" approach to improving on Hartree-Fock theory. However, truncations of the expansion short of full CI are problematic.
- Coupled cluster (CC) methods also include electronic excitation determinants via a *cluster operator* that is an exponential operator. This method has the same general goal as full CI but is formulated differently. Truncations of the expansion at various levels provide useful computational methods, avoiding full CI's prohibitive computational cost.
- Møller-Plesset perturbation theory, derived from many body perturbation theory of mathematical physics, adds higher excitations to Hartree-Fock theory as a non-iterative correction.

Technically, CC is somewhat more expensive than CI at the same truncation level (e.g., comparing CCSD to CISD). However, truncations of the CI expansion do not produce acceptable computational methods (as we'll see shortly), and full CI is combinational in cost and very expensive.

In general, the following are desirable qualities for an electron correlation method. We will use them as we discuss and evaluate the various approaches:

- Applicable to a wide range of molecular systems.
- Variational: the predicted energy is an upper bound to the exact energy.
- Well-defined, requiring no ad-hoc choices, and thus useful for defining a model chemistry.
- Size-consistent/size-extensive (defined in the next subsection).

We hope that the introductory overviews here will give you the flavor of each electron correlation approach. For full mathematical details, consult the references cited at the beginning of the chapter.

Configuration Interaction

Configuration interaction [Pople77, Raghavachari80a] mixes excited state determinants into the Hartree-Fock wavefunction using the variational principle:

$$\Psi_{CI} = \Psi_{HF} + \sum_{ia} t_i^a \Psi_{i\to a} + \sum_{ijab} t_{ij}^{ab} \Psi_{ij\to ab} + \cdots \tag{10.33}$$

The various t amplitudes are parameters that can be optimized using the variational theorem such that the CI energy is an upper bound to the true energy.

If this expansion is taken as far as possible, we obtain the *full CI* wavefunction [Knowles84, Bauschlicher86, Bytautas04]. However, this level of theory is practical only for extremely small systems. In practice, the CI expansion is truncated at some point. For instance, the CISD model will include just the terms shown explicitly in the preceding equation: all single and double excitations. Similarly, the CID model will include only double excitations, and it

* Technically, RHF with all symmetries included in the traditional Löwdin definition.

would omit the second term in the previous equation. The Gaussian keywords for these methods have the same names.

The CI approach has two serious drawbacks. First, the full CI method is too expensive to be useful. Second, methods using truncated CI wavefunctions are not size-consistent/size-extensive:

- A calculation done on two molecules at infinite separation is not equivalent to the sum of their energies in isolation, a characteristic known as *size consistency*. For example, the energy of two helium atoms at long distance would need to include a quadruple excitation in order to reproduce the two doubly excited determinants simultaneously present in the composite system.
- The energy resulting from them does not scale linearly with the size of the system. This means that the methods are not *size-extensive* (a related concept to size consistency).

This lack is what makes CISD unsuitable for studying chemical reactions. Moreover, any truncation of CI with higher excitations fails to solve this fundamental deficiency.

Coupled Cluster Theory

The coupled cluster method [Bartlett78] uses an operator approach to constructing a multi-determinant wavefunction, expanding the exponential cluster operator as a Taylor series:

$$\Psi = e^{\hat{\mathrm{T}}}\Psi_{HF} = (1+\hat{\mathrm{T}}+\tfrac{1}{2}\hat{\mathrm{T}}^2+...)\Psi_{HF} \tag{10.34}$$

In this formulation, the CID wavefunction would be:

$$\Psi_{CID} = (1+\ddot{\mathrm{T}})\Psi_{HF}$$

where in practice the operator is truncated at some level of excitation. For example, the CCSD method [Cizek69, Scuseria88] includes both single and double excitations (the Gaussian keyword is also **CCSD**).

The $\hat{\mathrm{T}}$ operator generates the excitation configurations. For example, in the case of CCSD:

$$\hat{\mathrm{T}}\Psi_{HF} = \sum_{ia} t_i^a \Psi_{i\to a} + \sum_{ijab} t_{ij}^{ab}\Psi_{ij\to ab} \tag{10.35}$$

The $\{t_i^a, t_{ij}^{ab}, ...\}$ *amplitudes* and the CCSD energy, E_{CCSD}, are solved for iteratively using a set of non-linear equations derived by projecting the total wavefunction onto the Hartree-Fock wavefunction as well as each of the singly and doubly substituted determinants. The resulting theory is no longer variational, but does yield energies that preserve size consistency/extensivity at any level of truncation (provided the reference determinant is allowed to break all Hamiltonian symmetries).

The CCSD method can be augmented to account for most of the contributions of triple excitations [Purvis82]. This technique is discussed following the next exercise.

EXERCISE 10.2 SIZE CONSISTENCY: HELIUM ATOM CLUSTER

Compute both the CISD and CCSD energies for the helium atom using the 6-31G basis set. Do the same calculations for a cluster of seven helium atoms* formed by placing six atoms in an octahedral arrangement around a central He atom at a distance of 10 Å. Compare the Hartree-Fock, CISD and CCSD energies to see if they are size-consistent.

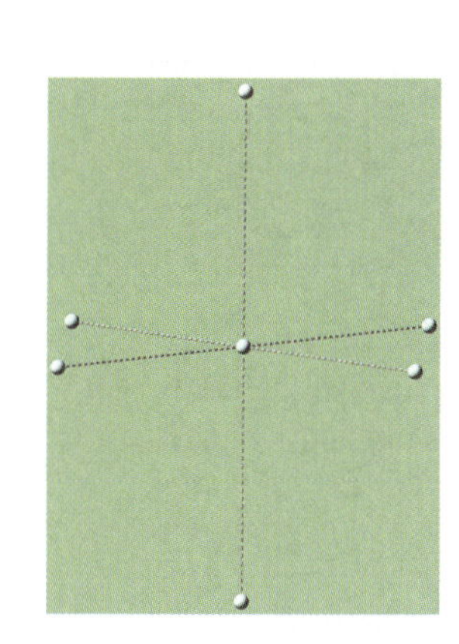

Note that the Hartree-Fock energy is reported in the output for **CISD** and **CCSD** calculations.

* IUPAC: septahelium.

SOLUTION

The following table presents the results we obtained:

	ENERGY (hartrees)		
	HF	**CISD**	**CCSD**
He atom	-2.85516	-2.87016	-2.87016
7 × He atom	-19.98612	-20.09113	-20.09113
octahedral He_7	-19.98612	-20.08854	-20.09113
E_{corr}(6-31G)		-0.10242	-0.10501

The CISD and CCSD energies for the helium atom are equivalent, since for this system each procedure would represent the equivalent of a full-CI calculation. Both the Hartree-Fock and CCSD theories are size-consistent since the energy of seven non-interacting atoms is the same as seven times the energy of one atom. For this basis set, CISD underestimates the correlation energy by ~2.5%.■

The Triples Correction to CCSD

For systems containing more than two electrons, we can improve upon the calculation of correlation energies by including additional determinants in the expansion. For instance, CCSD does not include the interactions from triply excited determinants, which have been shown to be important for systems with multiple bonds. To include these explicitly would greatly increase the time and resources needed, since the CCSDT method [Lee84] formally scales as mO^3V^5 where m is the number of iterations required, and O and V are the number of occupied and virtual orbitals (respectively).

Scaling formulas indicate that in general the CPU time required is proportional to the result of the formula for a given problem. Naturally, the proportionality constant would have to be determined by running and timing a reference calculation.

A more practical approach is to include these effects in a non-iterative fashion, using a formula developed from perturbation theory that recovers nearly all the important energetic contributions arising from the triple excitations:

$$E_{\text{CCSD(T)}} = E_{\text{CCSD}} - \tfrac{1}{36}\sum \frac{\left|\mathbf{u}_{ijk}^{abc}\right|^2}{\varepsilon_a + \varepsilon_b + \varepsilon_c - \varepsilon_i - \varepsilon_j - \varepsilon_k} \tag{10.36}$$

Here, the $\mathbf{u}_{ijk}^{abc}$ are the six-way indexed matrix elements of the Hamiltonian between the CCSD single and double excitation amplitudes and the triply excited determinant. In simple terms, this formulation takes advantage of some computational machinery from the MP4 method (see below), using a form of the amplitudes from CCSD, in order to compute the triples contribution to the energy far more rapidly than possible in the usual iterative approach.

While expensive, computing this energy correction does not require an iterative approach. The CPU cost of the one-time triples correction in CCSD(T) scales as O^3V^4, making it a much more practical alternative to CCSDT. For systems for which it is appropriate, the CCSD(T) method [Raghavachari89] is the "gold standard" for computing electron correlation energies. However, large basis sets must be used to take full advantage of its accuracy (see Exercise 10.5). The Gaussian keyword for this method is **CCSD(T)**.

Brueckner Doubles Theory

Finally, there is a version of Coupled-Cluster Theory known as Brueckner Doubles (BD) [Dykstra77, Handy89, Kobayashi91] in which the Hartree-Fock reference orbitals are transformed such that their singles amplitudes ($\{t_i^a\}$) are all zero in a CCSD calculation. A version incorporating the triples correction is known as BD(T) theory. This theory showed initial promise in modeling molecular systems with unpaired electrons. More recently, its usefulness has been demonstrated as part of the compound model chemistry known as W1BD [Barnes09].

Møller-Plesset Perturbation Theory

Another approach to solving the electron correlation problem is to apply Rayleigh-Schrödinger perturbation theory to the Hamiltonian. In general, perturbation theory approximates an exact solution to a problem by starting from a related problem for which an exact solution is known—known as the *zeroth-order* or *unperturbed* problem—and treating whatever remains from the real problem as a perturbation to this solution: in general, $Q = Q_0 + V$, where Q_0 is solvable exactly. The perturbation is conventionally denoted as V. Note that it is unrelated to the potential energy, to the nuclear-electron attraction operator and to every other chemical usage of "V."

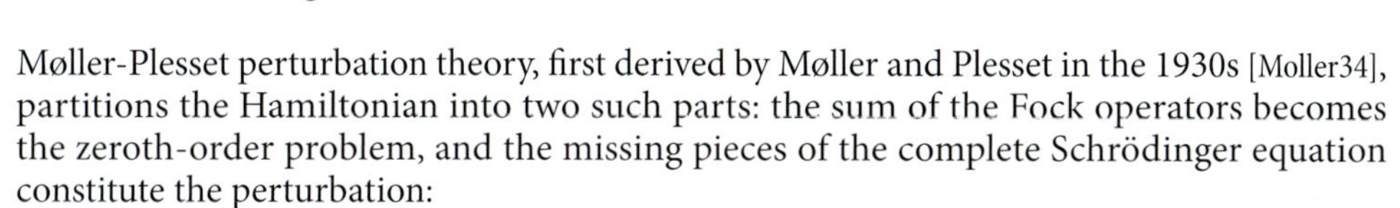

Møller-Plesset perturbation theory, first derived by Møller and Plesset in the 1930s [Moller34], partitions the Hamiltonian into two such parts: the sum of the Fock operators becomes the zeroth-order problem, and the missing pieces of the complete Schrödinger equation constitute the perturbation:

$$\hat{\mathbf{H}}_{MP} = \hat{\mathbf{H}}_0 + \hat{\mathbf{V}} \qquad \hat{\mathbf{H}}_0 = \sum \hat{\mathbf{F}}_v \tag{10.37}$$

Møller-Plesset perturbation theory methods define a series of numeric *orders*—first order, second order, etc.—with higher numbered orders incorporating more types of excitations. They are referred to by acronyms of the form MP*n*, where *n* indicates the perturbation order: MP2 for second order, MP3 for third order, through MP5 for fifth order [Pople76, Pople77, Raghavachari78, Raghavachari80, Raghavachari90]. The Gaussian keywords are the same as these acronyms.

The following table summarizes the equations for the energies for the zeroth-order reference as well as the first four orders of MP theory:

Zeroth-order: the sum of the orbital energies.

$$E^{(0)} = \langle \Psi_0 | \hat{\mathbf{H}}_0 | \Psi_0 \rangle = \sum_i \varepsilon_i$$

First-order: yields the Hartree-Fock energy.

$$E^{(1)} = \langle \Psi_0 | \hat{\mathbf{H}} | \Psi_0 \rangle \qquad E^{\mathrm{HF}} = E^{(0)} + E^{(1)}$$

Second-order: the single excitation terms are 0, so the final expression contains only double excitations.

$$E^{(2)} = E^{\mathrm{HF}} - \sum_{s \neq 0} \frac{\left| \langle \Psi_0 | \hat{\mathbf{V}} | \Psi_s \rangle \right|^2}{E_s - E_0} = E^{\mathrm{HF}} - \tfrac{1}{4} \sum_{ijab} \frac{\left| \langle ij \| ab \rangle \right|^2}{\varepsilon_a + \varepsilon_b - \varepsilon_i - \varepsilon_j}$$

Third-order: s and t run over double excitations.

$$E^{(3)} = E^{(2)} + \sum_{st} \frac{\langle \Psi_0 | \hat{\mathbf{V}} | \Psi_s \rangle \langle \Psi_s | \hat{\mathbf{V}} | \Psi_t \rangle \langle \Psi_t | \hat{\mathbf{V}} | \Psi_0 \rangle}{(E_s - E_0)(E_t - E_0)}$$

Fourth-order: s and u run over double excitations; t runs over singles, doubles, triples and quadruples. The numerator of the final expression is also written: $t_s \hat{\mathbf{V}}_{st} \hat{\mathbf{V}}_{tu} t_u = |\mathbf{u}_t|^2$.

$$E^{(4)} = E^{(3)} + \sum_{stu} \frac{\langle \Psi_0 | \hat{\mathbf{V}} | \Psi_s \rangle \langle \Psi_s | \hat{\mathbf{V}} | \Psi_t \rangle \langle \Psi_t | \hat{\mathbf{V}} | \Psi_u \rangle \langle \Psi_u | \hat{\mathbf{V}} | \Psi_0 \rangle}{(E_s - E_0)(E_t - E_0)(E_u - E_0)}$$

$$= E^{(3)} + \sum_{stu} \frac{t_s \hat{\mathbf{V}}_{st} \hat{\mathbf{V}}_{tu} t_u}{\varepsilon_a + \varepsilon_b + \varepsilon_c - \varepsilon_i - \varepsilon_j - \varepsilon_k}$$

Notation Notes

- *In the second-order equation, i and j are occupied, and a and b are virtual.*
- *In general, the s,t,u subscripts run over possible excitations: single, double, triple, ...*
- t_s *and* t_u *in the final equation for* $E^{(4)}$ *are amplitudes (analogous to what we saw earlier in the formulation of coupled cluster theory).*
- *Double-bar notation:* $\langle ij\|ab\rangle = \langle ij|ab\rangle - \langle ij|ba\rangle$, $(ia\|jb) = (ia|jb) - (ij|ab)$

MP theory is size-extensive. Like coupled-cluster theory, it is not variational, and therefore the energy obtained at any particular order is not necessarily higher than the Schrödinger energy. Nevertheless, the approach—especially at the MP2 level—offers a model that is much less computationally expensive than coupled-cluster theory and thus can be applied to a much wider range of chemical systems. Higher orders of MP theory are more complicated and progressively more expensive, and the entire series is very slow to converge, as we will see in the following exercise.

EXERCISE 10.3 CORRELATION ENERGIES OF A WATER MOLECULE

Compute the total electronic energy of the water molecule using the HF, CISD, CCSD, CCSD(T), BD, BD(T) and MP2 through MP5 methods and the cc-pVTZ basis set. Compare the predicted energies to the full CI results of -76.32971 hartrees reported in [Bytautas04]. Use the same geometry as Exercise 10.1 and the frozen core approximation in order to compare to the literature. The latter can be requested explicitly with the **FC** option to the method keywords.

What percent of the electron correlation energy is recovered by each method?

Hint: Only four calculations are needed: **CISD**, **CCSD(T)**, **BD(T)**, and **MP5**. The other results can be obtained within these output files.

SOLUTION

The following table presents the results of our calculations:

	ENERGY (hartrees)		
METHOD	E_{elec}	E_{corr}	**%CORR**
HF	-76.05099	0.0000	0.00
CISD	-76.30994	-0.2589	92.9
MP2	-76.31559	-0.2646	94.9
MP3	-76.31902	-0.2680	96.2
MP4	-76.33016	-0.2792	100.2
MP5	-76.32851	-0.2775	99.6
CCSD	-76.32123	-0.2702	97.0
CCSD(T)	-76.32924	-0.2782	99.8
BD	-76.32091	-0.2699	96.8
BD(T)	-76.32912	-0.2781	99.8

Notice that the MP series is far from converged at MP5. It will continue to oscillate about the full CI result at higher orders. The MP4 result reinforces the fact that Møller-Plesset perturbation theory is not variational.

The results also indicate that there is a significant component missing in CCSD that is almost completely accounted for in the CCSD(T) treatment. Finally, since this is a closed-shell system at or near the equilibrium geometry, there is little difference between BD(T) and CCSD(T).■

Density Functional Theory

Electron correlation methods can lead to highly accurate energies. However, this accuracy comes at significant computational cost, limiting the application of these methods to relatively small molecules. If we want to answer questions that can only be addressed by modeling large numbers of atoms, another approach will need to be taken.

Density Functional Theory (DFT) provides this alternative [Hohenberg64, Kohn65]. The rudiments of this theory predate most of the methods discussed in the previous section,* but it was not until the 1990s that they became widely used. Although the theory itself is quite precise and even elegant, its implementation has not been. Researchers continue to work to refine and improve the models, and each year many new density functional approximations (DFAs) are offered to the computational chemistry community.

Hohenberg and Kohn demonstrated that:

- *The exact energy is a functional of the density: E(ρ).*
- *This functional is independent of any particular system.*
- *The exact density minimizes the energy.*

The basic premise behind DFT is that the energy and associated properties of any system containing electrons are calculable from the probability distribution that is the total electron density, $\rho(\bar{r})$. Using an orbital basis set, this quantity can be easily obtained as:

$$\rho(\bar{r}_g) = \sum_{i}^{\text{occupied}} \phi_i^2(\bar{r}_g) \tag{10.38}$$

Here the position vector, $\bar{r}_g$, represents all space coordinates for the system as defined by a grid g. The probability of observing an electron in the volume element at $\bar{r}_g$ is $\rho(\bar{r}_g)$. While it may be intuitively useful to picture the volume surrounding a molecule as divided into a three-dimensional rectangular grid of small cubic elements, the grids typically used in calculations employ spherical and radial coordinates centered on each atom.

Although it generally produces much more accurate results, typical DFT implementations use much of the same computational infrastructure as Hartree-Fock, which accounts for its speed and efficiency. In its "pure" form, DFT uses the following expression for the energy as a functional of the density:

$$E_{\text{DFT}}[\rho(\bar{r})] = T[\rho(\bar{r})] + V_{\text{NE}} + J[\rho(\bar{r})] + E_{\text{XC}}[\rho(\bar{r})] \tag{10.39}$$

Compare this to equation (10.19) on p. 473.

Hopefully, the first three terms are recognizable from our discussion of Hartree-Fock theory. The final term is known as the *exchange-correlation* term. It models the parts of the electron-electron interactions that are neglected by Hartree-Fock theory. What is missing from this formulation is any term corresponding to Hartree-Fock exchange (K). Pure DFT functionals eliminate it entirely; hybrid functionals (see below) include some portion of it in the energy expression.

The preceding equation yields a modified Fock matrix:

$$\mathbf{F}_{\mu\nu}^{\text{DFT}} = \mathbf{H}_{\mu\nu}^{\text{core}} + \mathbf{J}_{\mu\nu}(\mathbf{P}_{\lambda\sigma}) + \sum_{g} w_g \mathbf{F}_{\mu\nu}^{\text{XC}}\left[\rho(\bar{r}_g)\right] \tag{10.40}$$

$\mathbf{H}^{\text{core}}$ is obtained from the one-electron integrals (kinetic plus nuclear-electronic attraction) just as before. The second term ($\mathbf{J}_{\mu\nu}$) can be found by considering the classical electron-electron repulsions and depends on the molecular orbitals through the density matrix $\mathbf{P}_{\lambda\sigma}$, which will again require the calculation of two-electron integrals.

The final term is the new quantity introduced by DFT that accounts for all electron correlations: both exchange and correlation. Unlike the coulombic repulsions, which are evaluated using integrals involving basis functions, this new quantity is evaluated by summing over a grid of points, with each volume element multiplied by a weighting factor, w_g. This technique was suggested by Kohn and Sham in 1965. The additional work required scales more or less the

* They originate in the Thomas-Fermi-Dirac model of the 1920s and from Slater's work in the 1950s. The Hohenberg-Kohn theorem, which proves the existence of a unique functional that determines the ground state energy and density exactly, was published in 1964 [Hohenberg64].

same as a Hartree-Fock calculation, allowing DFT models to study a wide range of systems, including ones with very large numbers of atoms.

Unfortunately, the exact specification of $\mathbf{F}^{XC}$ is not known. Its form has been postulated from the equations describing a uniform electron gas, from the exact solution of the hydrogen atom, by parameterization with experiment, and in other fashions, giving rise to a multitude of DFAs.

Hybrid Functionals

The B in the B3LYP functional's name is for Becke.

While the early application of pure DFT functionals showed promise, their results were nevertheless decidedly mixed, and many significant failures accompanied the successes. In the early 1990s, Axel Becke [Becke93a] demonstrated that more accurate energies could be obtained if some part of the Hartree-Fock exchange was included within the functional. This crucial insight is what made DFT methods accurate and consistent and enabled them to be applied to the full range of chemical problems, resulting in the widespread use they see today.

Functionals that mix exchange-correlation functionals with Hartree-Fock exchange are termed *hybrid* functionals. Conceptually, a hybrid functional defines the DFT exchange-correlation term as a sum of Hartree-Fock exchange and pure DFT exchange-correlation, mixed in some fixed percentages:

$$E_{XC}^{\text{hybrid}}[\rho(\bar{r})] = c_{HF}\,\mathbf{K} + c_{DFT}\,E_{XC}^{\text{pure}}[\rho(\bar{r})] \qquad (10.41)$$

The constants (c) specify how much of each term is included.

The preceding equation is merely descriptive in that it does not specify the form of the pure exchange-correlation functional. Becke's actual formulation is more specific. It separates the DFT exchange-correlation into distinct exchange and correlation parts. It further distinguishes the *local* and *non-local* portions of each. These terms refer to the part of the functional that has either a direct dependence on the density (local) or a dependence on its mathematical derivatives (non-local).

Here is the generalization of Becke's formulation in which each of the resulting five terms has its own weighting factor:

$$E_{XC}[\rho(\bar{r})] = c_1 E_X^{HF} + c_2 E_X^{\text{local}} + c_3 E_X^{\text{non-local}} + c_4 E_C^{\text{local}} + c_5 E_C^{\text{non-local}} \qquad (10.42)$$

Three or Five?

B3LYP defines c_1=0.2, c_2=1.0-c_1, c_3=0.72, c_4=1.0 and c_5=0.81. Thus, three parameters.

The original specification for the B3LYP functional is the following:

$$E_{XC}^{\text{Becke3}}[\rho(\bar{r})] = 0.2E_X^{HF} + 0.8E_X^{LDA} + 0.72E_X^{B88} + E_C^{\text{local PW91}} + 0.81E_C^{\text{non-local PW91}} \qquad (10.43)$$

The components come from three different functionals:

- local exchange: LDA [Slater74]
- non-local exchange: B88 [Becke88b]
- local correlation and non-local correlation: PW91 [Perdew91, Perdew92, Perdew93a]*

In Gaussian, it is possible to define hybrid functionals by choosing any exchange functional(s) partnered with any correlation functional(s) and specifying the values of the five parameters (see the discussion of the DFT methods in the *Gaussian User's Reference* for details). However,

* In the B3LYP implementation in Gaussian, local correlation is provided by the VWN functional and non-local correlation by the difference between the VWN [Vosko80] and LYP [Lee88] functionals, with Becke's coefficient: $0.81(E_C^{LYP} - E_C^{VWN})$. The subtraction is due to the fact that LYP includes both local and non-local parts. The fact that the same coefficients work well with different functionals reflects the underlying physical justification for using such a mixture of Hartree-Fock and DFT exchange first pointed out by Becke.

selecting one of the available hybrid functionals—i.e., B3LYP or APFD—is usually a better choice.

The following table compares several hybrid functions with respect to the weighting used for each component. Be aware, there are many other differences among the definitions of the actual functionals, sometimes including additional parameterization:

FUNCTIONAL	c_1	c_2	c_3	c_4	c_5
B3PW91 [Becke93]	0.20	0.80	0.72	1.00	0.81
PBE1PBE†	0.25	0.75	0.75	1.00	1.00
TPSSh [Staroverov03]	0.10	0.90	0.90	1.00	1.00
BMk [Boese04]	0.42	1.00	1.00	1.00	1.00

†The "PBE hybrid" of [Ernzerhof99] and PBE0 of [Adamo99a].

B3PW91 uses the same parameters as B3LYP. As their names suggest, they both use the B88 exchange functional but different correlation functionals.

The PBE1PBE hybrid functional is a "one parameter" hybrid since c_2=c_3=1.0-c_1. The parameter for this functional was not optimized but rather was chosen based on theoretical considerations. For this reason, some claim this functional is parameter-free [Adamo99a].

The TPSSh functional uses a smaller amount of Hartree-Fock exchange, since the functional itself contains additional physics (through the use of the kinetic energy density) that purportedly improves the DFA exchange contribution.

Finally, BMk is an example of a functional using a large amount of Hartree-Fock exchange. It was constructed with the goal of producing better energies for transition structures (here, "k" stands for kinetics).

With extra printing enabled by **#P** in the route section, each of the parameters are reported in the Gaussian output file:

```
                                                      c1
IExCor=   408 DFT=T Ex+Corr=B3PW91 ExCW=0 ScaHFX=   0.200
ScaDFX=   0.800000  0.720000  1.000000  0.810000 ...
            c2          c3          c4          c5
```

EXERCISE 10.4 PROTON AFFINITY OF METHYL ANION

Using a 6-311+G(2d,p) basis set, estimate the electronic energy change for the gas-phase reaction $CH_4 \rightarrow H^+ + CH_3^-$: the electronic energy difference between methyl anion and methane (the hydrogen cation has no electrons and therefore an electronic energy of zero). Optimize the molecular geometries, and predict their ground state energies using the following theoretical methods:

- Use the Becke-Roussel pure exchange-correlation functional (keyword **BRxBRc**).
- Define a one-parameter hybrid functional that mixes Hartree-Fock exchange with the BRxBRc functional. Find the value of the c_1 parameter that best reproduces the W1BD predicted value of 1785 kJ/mol for this problem (consistent with the best experimental estimates).

SOLUTION

We varied c_1 from 5 to 40 in steps of 5 (with c_2=c_3=1-c_1 and c_4=c_5=1). We found a linear correlation between the error and the c_1 parameter:

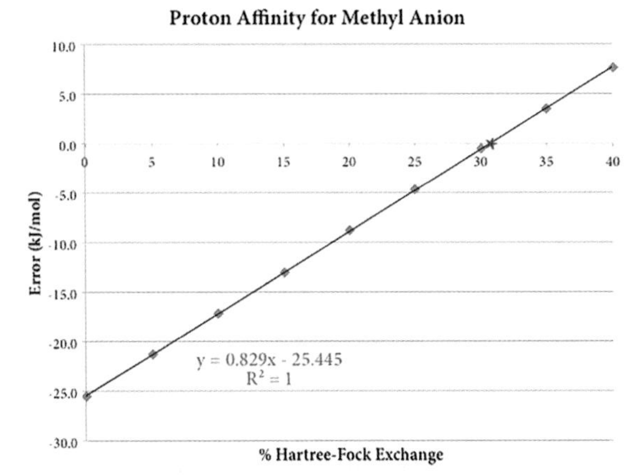

This relationship should have been obvious since the energy is linear in A according to equation (10.42), and we could have determined c_1=30.7 using just two calculations.

Congratulations, you have just invented a new density functional approximate model—although it's not one you'd want to use for actual production calculations! Many researchers have gone before you (although they generally use much more substantial training sets). Notice that the basis set chosen would have an effect on the parameterization, so many authors have suggested particular basis sets to use in conjunction with their functionals. Hybrid functionals also differ in mathematical complexity.

The following table presents the results for the proton affinity of methyl anion as well as the optimized C-H bond length in methane for some hybrid functionals and the Hartree-Fock and MP2 methods:

METHOD	PA (kJ/mol)	R (Å)
HF	1814	1.0832
pure functionals:		
BLYP	1761	1.0963
BRxBRc	1759	1.0960
TPSSTPSS	1786	1.0925
PBEPBE	1765	1.0972
hybrid functionals:		
BRxBRc, c_1=0.307	1785	1.0879
B1B95	1787	1.0871
PBE1PBE	1785	1.0901
TPSSh	1792	1.0904
B3LYP	1778	1.0896
BMk	1788	1.0908
B98	1786	1.0908
MP2	1780	1.0888
accepted value	1785[a]	1.0859[b]

[a]Estimated from W1BD calculations [b][Landolt76]

Hartree-Fock theory greatly overestimates the proton affinity of methyl anion, and all the DFA methods correct it in the direction of the accepted value. In general, pure DFA models slightly underestimate it (with the exception of TPSS) while overestimating the C-H bond length of methane. The use of hybrid DFAs improves both quantities, and their results are in line with MP2 for this problem.■

As we have seen, the performance of a given DFA will depend on basis set. However, increased accuracy does not always follow by increasing the basis set. We will investigate this in the next exercise.

EXERCISE 10.5 HCN GEOMETRY AND FREQUENCIES

Optimize the geometry of the linear HCN* molecule and determine its vibrational frequencies. Use both the B3LYP and CCSD(T) models with basis sets of increasing size: STO-3G, cc-pVDZ, cc-pVTZ, cc-pVQZ and cc-pV5Z. Compare your results to the following experimental values taken from [Kobayashi91]:

- Bond lengths (Å): C-H: 1.065 (R_1), C-N: 1.153 (R_2) [Winnewisser71]
- Frequencies: C-H wag: 727 (ω_1, doubly degenerate), C-N stretch: 2128 (ω_3), C-H stretch: 3440 (ω_4) [Quapp87]

How does increasing the basis set affect the results for each of these models?

You will need to use a Z-matrix in order to optimize the molecular geometry with the CCSD(T) model chemistries since there are no analytical derivatives available for this method. For a linear molecule, a dummy atom is required:

```
0 1
C
X 1 1.0
H 1 R1 2 90.0
N 1 R2 2 90.0 3 180.0
  Variables:
R1=1.08
R2=1.15
```

SOLUTION

The following table presents our results:

BASIS SET	METHOD	R_1 (Å)	R_2 (Å)	ω_1	ω_3	ω_4
STO-3G	B3LYP	1.088	1.193	794	2203	3614
	CCSD(T)	1.095	1.205	752	2115	3580
cc-pVDZ	B3LYP	1.077	1.158	773	2200	3464
	CCSD(T)	1.082	1.175	706	2099	3449
cc-pVTZ	B3LYP	1.065	1.146	759	2201	3455
	CCSD(T)	1.067	1.160	716	2112	3444
cc-pVQZ	B3LYP	1.065	1.145	758	2202	3444
	CCSD(T)	1.067	1.156	722	2124	3436
cc-pV5Z	B3LYP	1.065	1.145	755	2201	3444
	CCSD(T)	1.067	1.156	725	2125	3438
Experiment		1.065	1.153	727	2128	3440

* IUPAC: hydrogen cyanide.

For the B3LYP model chemistries, the results improve when we move from the DZ to the TZ basis set, but they do not significantly change beyond that. In contrast, the CCSD(T) model yields bond lengths and frequencies that steadily improve as you increase the basis set, with very high accuracy associated with the cc-pVQZ and cc-pV5Z basis sets. Unfortunately, this basis set is much too demanding to use with larger molecular systems. Using CCSD(T) with small basis sets, such as cc-pVDZ, does not offer any improvement for bond lengths over B3LYP theory.■

Long-Range Corrections

According to the laws of physics, the exchange potential in molecules should properly decay as $-1/r$, where r is the distance between two electrons. One reason that hybrid DFAs are more accurate than pure DFAs is that the Hartree-Fock exchange does exactly that, while pure DFAs exhibit exponential decay in their exchange potentials. Hybrid DFAs only partially overcome this problem; their decay is proportional to the amount of HF exchange they include ($-c_1/r$). This deficiency introduces spurious interactions that are not physical. There have been several attempts to correct for this.

One approach is to partition the exchange operator into two regions, computing the exchange correlation by using exact Hartree-Fock at long electron-electron distances and pure DFT exchange at short electron-electron distances:

$$E_{XC}[\rho(\bar{r})] = E_X^{HF:long} + E_X^{DFT:short} + E_C^{DFT} \tag{10.44}$$

The LC-ωPBE [Vydrov06a] is an example of this type of functional. It uses an error function with one parameter to define the short range and long range regions.

Other approaches include:

- *Long-range corrected* versions of some pure functionals have been constructed (e.g., LC-BLYP) via various formulas [Iikura01].
- The CAM-B3LYP functional [Yanai04] uses the concept of coulomb-attenuation, treating the short-range exchange at 19% Hartree-Fock and the long-range at 65%.

Double Hybrid Functionals

Double hybrid functionals hybridize the correlation contribution by incorporating the MP2 energy computed using the Kohn-Sham orbitals (in addition to including some Hartree-Fock exchange):

$$E_{XC}[\rho(\bar{r})] = d_1 E_X^{HF} + d_2 E_X^{DFT} + d_3 E_X^{DFT} + d_4 E_C^{MP2} \tag{10.45}$$

Grimme and coworkers were the first to suggest this type of DFA in 2006 [Grimme06a], and he derived the parameters 0.53, 0.47, 0.73 and 0.27 for the four coefficients when constructing the B2PLYP functional. Double hybrid functionals have performed fairly well to predict thermochemical quantities, with a computational cost similar to a conventional MP2 calculation.

Dispersion

Dispersion refers to the weak interaction arising from the fact that the charge distribution around a molecular system isn't constant; rather, it fluctuates due to the movement of electrons. Dispersion forces—both intra- and intermolecular—are present in any chemical system.

They are often dwarfed by the other covalent and dipole-dipole type interactions, which are of much larger magnitude. However, dispersion forces dominate in the absence of other types of intermolecular forces, and they can be important in other situations when highly accurate energies are desired.

Both pure and hybrid DFA models fail to describe dispersion adequately. Most functionals exhibit spurious long-range repulsion or short range attraction. For instance, if you were to use B3LYP to study the attraction of two noble gas atoms (or even two methane molecules), you would find that the theory predicts repulsion at all distances. MP2 theory can recover some of the effects of dispersion, but it often underestimates them.

Since approximate density functionals already have parameterization built into them, it seems reasonable to add one more empirical term, E_{disp}, to account for dispersion. The APF functional [Austin12] is based upon a combination of 41.1% B3PW91 and 58.9% PBE1PBE, a combination that was shown to cancel the spurious interactions present in the individual hybrid functionals, making it a better candidate to augment with an empirical dispersion term. Next, the dispersion term—formulated as a spherical atom model—is parameterized using a very small dataset of true dispersion cases along with the use of computed ionization potentials and atomic polarizabilities of the elements. Proceeding in this way makes it less likely that the dispersion correction is merely accounting for random errors in the functional and more likely to be achieving the goal of treating dispersion.

Note that the simplest empirical term that might be added to a functional involves pairwise contributions from the atoms:

$$E_{disp} = -k_1 \sum_{\text{atom pairs}} \frac{\sqrt{C_i C_j}}{R_{ij}^6} \times \frac{1}{1 + exp\left\{-k_2\left(\frac{R_{ij}}{R_i^{VDW} + R_j^{VDW}} - 1\right)\right\}} \tag{10.46}$$

where k_1 and k_2 are adjustable parameters that vary according to the functional to which the dispersion correction is applied. The $\{C\}$ and $\{R^{VDW}\}$ are atomic parameters determined by fitting to experimental data on non-covalently bonded complexes. Using this approach, Grimme added dispersion to the B3LYP functional [Grimme06] and to the B2PLYP double hybrid functional [Goerigk11].

There is a troublesome aspect to this. Since experimental data is being used to train the dispersion correction, it is difficult to know whether the parameterization is truly providing dispersion that is missing in the functional or if it is rather repairing some other missing physics (inadequate correlation energy, for instance). For this reason, we prefer the approach of APFD.

EXERCISE 10.6 ARGON DIMER BINDING ENERGY

Calculate the energy of diargon, Ar_2, relative to the energy of two isolated Ar atoms at interatomic distances of 3.6 Å to 6.4 Å, in increments of 0.4 Å. Use the 6-311+G(2d,p) basis set with the following DFAs: B3LYP, B97d, APF and APFD.

Compare your results to the CCSD(T)/aug-cc-pV5Z results given below (ΔE values are in millihartrees):

$\mathbf{R}^{Ar\text{-}Ar}$ (Å)	3.6	4.0	4.4	4.8	5.2	5.6	6.0	6.4
$\mathbf{\Delta E}^{Ar_2\text{-}2Ar}$	-0.4032	-0.4167	-0.2744	-0.1654	-0.0999	-0.0628	-0.0410	-0.0274

SOLUTION

The results of our calculations are plotted in the following figure (which also includes some other model chemistries for reference):

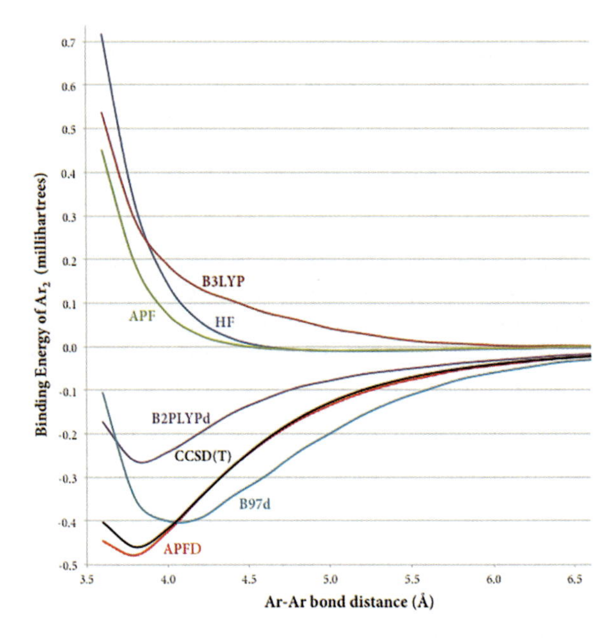

Hartree Fock, APF (without dispersion) and B3LYP all fail to show any substantial binding. The models with dispersion corrections all show binding, but B97d locates the minimum at a very long distance. The APFD results closely follow those of CCSD(T) using a large basis set.■

Integration Grids and DFT Calculations

As we have noted, DFT calculations evaluate the terms added to Hartree-Fock theory via numerical integration. Such calculations employ a grid of points in space in order to perform the numerical integration. Grids are specified as a number of radial shells around each atom, each of which contains a set number of integration points. For example, in the (75,302) grid, there are 75 radial shells each containing 302 points, resulting in a total of 22,650 integration points per atom.

Uniform and pruned versions of many grids have been defined. Uniform grids contain the same number of angular points at each radial distance, while pruned grids are reduced from their full form so that fewer points are used on the shells near the core and far from the nucleus, where less density is needed for a given level of computational accuracy. Put another way, pruned grids are designed to be densest in the region of the atom where its properties are changing most rapidly. For example, the pruned (75,302) grid, denoted "(75,302)p," contains about 7,500 integration points per atom. In general, pruning reduces the size of a uniform grid by about 66%.

In addition to producing more accurate results, larger grids also have better rotational invariance properties and are thus much more suitable for molecular systems involving transition metals and calculations using pseudopotentials.

As of this writing, (75,302)p—known as **FineGrid**—is the default grid in Gaussian 09. The SG1 grid, a pruned (50,194) grid containing about 3,600 points per atom is used for lower-accuracy single point calculations. However, we recommend using the larger **UltraFine** grid, (99,590)p for all production calculations, and we have done so throughout this book.

This grid not only produces more accurate results, but it also increases the stability of all numerical computation processes.

The following exercise illustrates the differences in accuracy that can result from employing different grids.

EXERCISE 10.7 COMPARING INTEGRATION GRIDS

Run a series of single point calculations for Si_5H_{12} and Al_4P_4,* using our standard model chemistry. Run calculations specifying the following grids available in Gaussian:

- **Integral=SG1**: (50,194)p
- **Integral=FineGrid**: (75,302)p
- **Integral=UltraFine**: (99,590)p
- **Integral=SuperFine**: (150,974)p for the first two rows of the periodic table, (225,974)p for the third row and beyond.

Repeat the calculations for Al_4P_4, adding the **NoSymm** keyword to the route section. Doing so will allow you to explore the rotational invariance behavior of the various grids.

SOLUTION

The following table presents our results on the various grid sizes for the two compounds:

	$\Delta E^{superfine}$ (kJ/mol)	
GRID	Si_5H_{12}	Al_4P_4
SG1	-0.16	0.56
FineGrid	0.17	0.25
UltraFine	0.03	0.02

SG1 seems to have more trouble with second-row atoms than first-row atoms. The energy differences between the SG1 and the large UltraFine and SuperFine grids are modest but significant. The accuracy of the FineGrid is comparable to SG1 for Si_5H_{12} but substantially better for Al_4P_4. There is no substantive difference between the two largest grids for these small problems.

SG1 also suffers from substantial deviations from rotation invariance: changing the orientation of the molecule can substantially alter the predicted energy. All DFT methods using finite grids will exhibit some degree of deviation from rotational invariance, but SG1 is more sensitive than most grids—the effect is generally more pronounced with smaller grids—as the results on Al_4P_4 indicate:

GRID	ΔE^{NoSymm} (kJ/mol)
SG1	1.18
FineGrid	0.31
UltraFine	0.02
SuperFine	0.001

* IUPAC: silicon hydride and aluminum phosphide (respectively).

Although a great improvement on SG1, FineGrid nevertheless shows a small deviation from rotational invariance. The problem is essentially eliminated for the two largest grids.■

Petersson's Complete Basis Set Extrapolation

As we noted earlier in this book, the methods in the Complete Basis Set (CBS) family all include a component that extrapolates from calculations using a finite basis set to the estimated complete basis set limit. In this section, we very briefly introduce this procedure.

The extrapolation to the complete basis set energy limit is based upon the Møller-Plesset expansion of the energy $E^{MP} = E^{(0)} + E^{(1)} + E^{(2)} + E^{(3)} + E^{(4)} + ...$, as described earlier. Recall that $E^{(0)} + E^{(1)}$ is the Hartree-Fock energy. We will denote $E^{(3)}$ and all higher terms as $E^{(3\rightarrow\infty)}$, resulting in this expression for E:

$$E^{CBS} = E^{HF} + E^{(2)} + E^{(3\rightarrow\infty)} \tag{10.47}$$

CBS extrapolation computes the second and third terms in this equation—the second-order and infinite-order corrections to the energy—via an extrapolation procedure [Nyden81, Petersson00]. No explicit extrapolation of the SCF energy is included because the CBS models begin with a large enough SCF calculation to obtain the desired level of accuracy.

Perhaps the most obvious way to extrapolate the MP2 energy term would be to calculate it using a set of increasingly large basis sets and then projecting the resulting series of values to the basis set limit. This would be very costly in terms of computation resources, depending as it does on large MP2 calculations. Moreover, careful analysis of the problem reveals the components of the Møller-Plesset second-order term (MP2) converge at different rates, which would cause the most lengthy parts to dominate the time required for any monolithic extrapolation scheme.

The CBS extrapolation process for the second-order correlation energy proceeds by decomposing it into a sum of pair energies, each of which represents the correlation energy for two of the electrons in the molecular system. This work draws on that of Schwartz [Schwartz62], who derived an expression for how the energy converges as successive s functions, p functions, d functions, f functions, and so on, are added to the description for a helium-like ion. Petersson and coworkers extended this two-electron formulation to many-electron atoms by writing the MP2 correlation energy as a sum of *pair energies*, each describing the energetic effect of the electron correlation between that pair of electrons.

In general, the process proceeds in this way:

- Perform an MP2 calculation. The basis set used depends on the specific CBS method and its target accuracy level. For example, the CBS-4 method uses the 6-31+G(d,p) basis set [Ochterski96, Montgomery00], and the CBS-QB3 method uses 6-311G(3d2f,2p) basis set [Montgomery99, Montgomery00] (with one fewer d and f function for first row atoms).
- Transform the orbitals and the MP2 amplitudes to the localized pair orbital formalism.
- For each electron pair:
  - Compute the base pair energy (the contribution of that pair to the SCF energy).
  - Calculate the second-order energy for each pair by adding in successive shells of virtual orbitals and extrapolating to the basis set limit.
- The sum of the extrapolated pair energies is the second-order correlation energy.

CBS extrapolation is illustrated in the following figure:

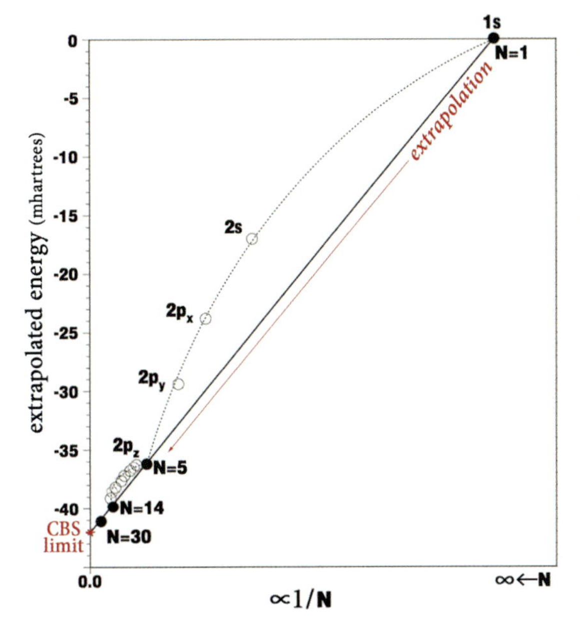

N is the number of orbitals in the expansion of the pair energy.

The extrapolation begins at the top right, where the point $N=1$ corresponds to the uncorrelated pair energy and finishes in the lower left. The x-axis plots a quantity that is proportional to N^{-1}, moving from right to left from 0 added virtual orbitals to the basis set limit: $N \to \infty$.

The circles indicate the contributions of each successive natural orbital to the pair energy; filled circles indicate complete shells. The extrapolation is linear in *N* when it corresponds to complete shells, and only these points are required for extrapolating to the complete basis set limit (indicated in red on the y-axis).

Computing the second-order energy by decomposition into pair energies is not only much faster than extrapolating the MP2 energy *in toto*, but it also produces a more accurate value for the basis set limit for a given basis set.

The $E^{(3 \to \infty)}$ term in equation (10.47) also requires only the extrapolated second-order energy because the ratio of the total full CI basis set truncation error to this term is known to be described by the following expression for incorporating the *interference factor*:

$$\frac{\sum_{ij} \left[\sum_{\mu} \mathbf{C}_{\mu_{ij}} \right]^2 \Delta e_{ij}^{(2)}}{\sum_{ij} \Delta e_{ij}^{(2)}} \tag{10.48}$$

The denominator is the extrapolated second-order energy (the $\Delta e_{ij}^{(2)}$ are the pair orbital energies). **C** holds the coefficients of the virtual orbitals in the first-order perturbation wavefunction. The *i* and *j* indices run over the electron pairs, and the μ index runs over the virtual orbitals. In this way, when the MP2 CBS limit is known for a given basis set, the CCSD(T) CBS limit can be computed from a single CCSD(T) calculation with the same basis set. This is what is done in the CBS-QB3 compound model chemistry to obtain very accurate thermochemistry predictions at reasonable computational cost.

Excited State Methods

The methods we have considered so far treat the molecule in its ground state: the lowest energy electronic configuration. The mathematics built into ground state methods has the feature of trying to optimize parameters and minimize the energy. They will require additional constraints in order to solve for other stable electronic states that are not the ground state. In this brief section, we introduce you to the main excited state methods in wide use.

The first approach is to express the excited state as a sum of singly excited determinants, which is known as the CI-Singles method:

$$\Psi_{\text{CIS}} = \sum_{ia} t_i^a \Psi_{i \to a} \tag{10.49}$$

In minimizing the energy with respect to the $\{t_i^a\}$, we are assured that we will not collapse to the ground state since the Hamiltonian integral relating each of the singly excited determinants to the ground state is zero (a result known as *Brillouin's theorem* [Brillouin34]). One can solve for any of the roots of this CI expansion, allowing a large number of excited states to be found at once, all treated at a consistent level.

This level of theory is variational (it will always lead to an energy above the true energy), and, unlike regular CI, it is size-extensive. It has been efficiently implemented in Gaussian so that one can use the theory to study systems with very large numbers of atoms. In addition, the availability of first and second derivatives for the CIS method eases the exploration of the excited state surface. The method is similar in accuracy to Hartree-Fock theory for ground states, meaning that excitation energies are often overestimated due to lack of flexibility (i.e., neglect of electron correlation) in the wavefunction.

One way to repair CIS without adding substantially to the cost of the calculation is to allow the CIS wavefunction to partially relax by introducing de-excitation amplitudes into the expansion. Stated as an operator equation in matrix form, we can express this as:

Notation Notes
Double-bar notation:
(ia||jb)=(ia|jb)-(ij|ab)

$$\begin{pmatrix} \mathbf{A} & \mathbf{B} \\ \mathbf{B} & \mathbf{A} \end{pmatrix} \begin{pmatrix} \mathbf{X} \\ \mathbf{Y} \end{pmatrix} = \omega \begin{pmatrix} 1 & 0 \\ 0 & -1 \end{pmatrix} \begin{pmatrix} \mathbf{X} \\ \mathbf{Y} \end{pmatrix}$$
$$\mathbf{A}_{ia,jb} = \delta_{ij}\delta_{ab}(\varepsilon_a - \varepsilon_i) + (ai \| jb)$$
$$\mathbf{B}_{ia,jb} = (ai \| bj) \tag{10.50}$$

A is the matrix corresponding to the original CIS Hamiltonian. The **B** matrix is the same size and contains Hamiltonian terms relating determinants formed by a single excitation on one side and a single de-excitation on the other side. Solving this eigenvalue equation leads to the excitation energy, here denoted ω, and a vector of excitation (**X**) and de-excitation (**Y**) amplitudes.

CI-Singles is equivalent to TDHF with B=0.

The theory is known as *time-dependent* Hartree-Fock (TDHF) [Dirac30],* and it recovers some of the electron correlation missing in the CIS approach at basically the same computational cost. The "time dependence" in the name comes from the fact that the fundamental equations of the method are ultimately derived from the time-dependent Schrödinger equation; the derivation relies on linear response theory.

A much more accurate approach is to incorporate density functional theory into the time-dependent formalism. The Runge-Gross theorem [Runge84] describes the connection between the time-dependent density of an evolving system and the time-dependent potential in which the system evolves; it is analogous to the Hohenberg-Kohn theorem for a system that evolves over time starting from a specific initial state.

A time-dependent DFT theory (TD-DFT) can be derived in an analogous way to TDHF using a density functional reference [Bauernschmitt96a, Casida98]. The **A** and **B** matrices can then be evaluated using Kohn-Sham orbitals from a DFT calculation. The excited states

* The method is also referred to as RPA for *random phase approximation* (primarily in the physics community).

that emerge will include exchange correlation effects as well as the flexibility coming from the time-dependent approach. This theory is implemented in Gaussian [Stratmann98] for any of the available pure and hybrid functionals. First derivatives are also available [VanCaillie99, VanCaillie00, Furche02, Scalmani06], allowing geometry optimizations to be performed.

For very high-accuracy excited state calculations, the *equations of motion coupled cluster* (EOM-CCSD) method is available [Stanton93, Bartlett07]. EOM-CCSD is an extension of CCSD for modeling excited states, consisting of a CI expansion of a CCSD reference state. It has comparable accuracy and cost to CCSD for one-electron excitations.

The closely-related SAC-CI method [Nakatsuji78, Nakatsuji79, Nakatsuji79a] is formally the same as EOM-CC; its implementation neglects various higher order terms and selects configurations based on estimates from perturbation theory. SAC-CI is capable of treating states that are primarily higher-order excitations. For review articles, see [Nakatsuji97] and [Ehara02].

Complete Active Space SCF

CASSCF is a method that takes multiple electronic configurations into account as it models a molecular system. It is particularly useful for molecular systems for which wavefunctions based on a single determinant are inadequate (e.g., Hartree-Fock or DFT wavefunctions); such systems include ground states which are quasi-degenerate with low-lying excited states, intermediate geometries during bond breaking processes, some excited states and reactions involving both ground and excited state potential energy surfaces.

Conceptually, a CASSCF calculation is a combination of an SCF computation on the entire molecule and a full CI procedure involving a subset of the molecular orbitals, with this subset referred to as the active space. The total orbital space of the molecular system is thus decomposed into three subspaces: core/inactive orbitals, active orbitals and virtual orbitals (denoted C, A and V, respectively).

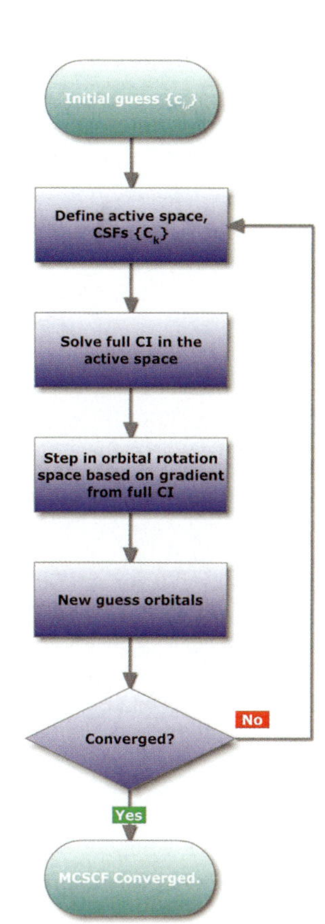

The CASSCF orbital optimization process is similar to that of Hartree-Fock but includes the additional complexity of optimizing the mixing between these subspaces: C-A, A-V and C-V. This mixing will change the predicted total energy of the system, and finding the optimal mix is necessary for locating the wavefunction with the lowest total energy. Note that the CASSCF energy is invariant to the mixing of orbitals within a single subspace.

The CASSCF method proceeds by optimizing a linear combination of configuration state functions (CSFs) that comprise all possible occupations of the orbitals in the active space:

CASSCF wavefunction: $\Psi^{\text{CASSCF}} = \sum_k C_k \Phi_k$

configuration state functions: $\Phi_k = \left| \phi_{k_1} \ \phi_{k_2} \ \phi_{k_3} \ \ldots \ \phi_{k_m} \right|$

molecular orbitals: $\phi_i = \sum_\mu c_{i\mu} \chi_\mu$

In the expression for the CASSCF wavefunction, the $\{C_k\}$ are configuration mixing coefficients for the CSFs. The CSFs are determinants corresponding to the configurations of m electrons in k orbitals (i.e., antisymmetrized). The molecular orbitals are defined in terms of spin functions as we saw previously, and $\{c_{i\mu}\}$ are the MO coefficients.

During the CASSCF procedure, both the $\{C_k\}$ and $\{c_{i\mu}\}$ are optimized. The process proceeds as indicated in the diagram at right. In the final converged wavefunction, the optimized active space orbitals will have fractional occupations, and all other orbitals remain empty or full (as appropriate), but they too will have been optimized.

Gaussian lists the configurations in the output file from CASSCF jobs:

```
no. active orbitals (n) 6
no. active ELECTRONS (N) 6
                          IRREPS TO BE RETAINED = 1 2
                          GROUP IRREP. MULT. TABLE
                              1    2
                              2    1
                          IRREP. LABELS FOR ORBITALS
                           2 2 2 2 2 2
BOTTOM WEIGHT= 12  TOP WEIGHT= 30
Configuration 1 Symmetry 1 111000
Configuration 2 Symmetry 1 11ab00
Configuration 3 Symmetry 1 110100
...
```

Symmetries are numbered in Cotton ordering.

1: doubly occupied
a,b: alpha,beta
0: unoccupied

In addition to modeling molecules in their ground and excited states, CASSCF has the ability to treat reactions involving both the ground and excited state potential energy surfaces, which are typical of many photochemical processes.

The following figure depicts the general features of such reactions [SerranoPerez13]:

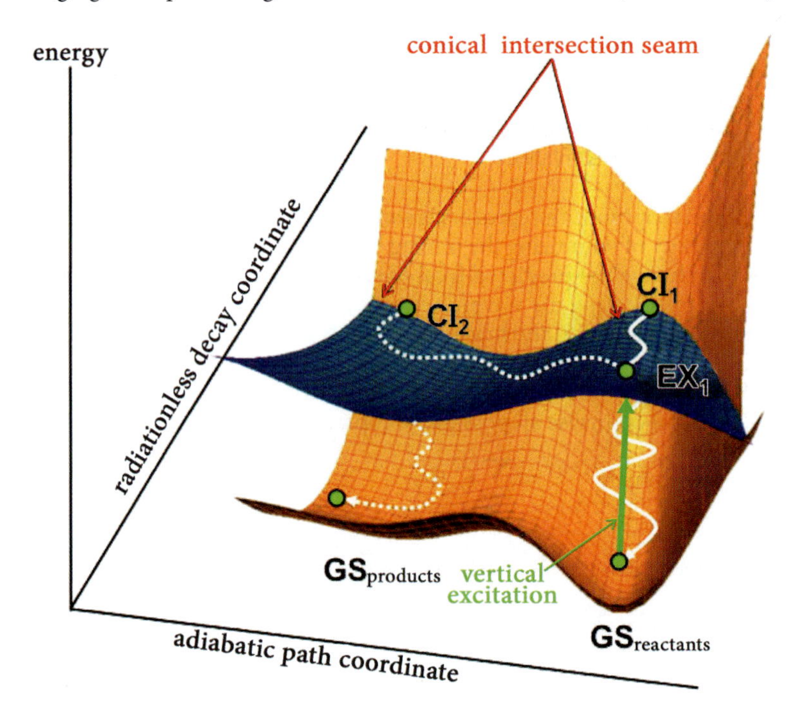

The ground state (orange) and excited state (blue) PESs intersect. The PESs are represented as 2D surfaces, but in reality, CASSCF treats all of the molecular degrees of freedom. In general, a photochemical reaction path has two branches, one on each PES, which are joined by a *conical intersection*: a point common to both surfaces through which radiationless decay can occur.

In this example, a vertical excitation occurs from GS_1 to XS_1. If we focus on reactions starting from XS_1, there are multiple possible reaction paths from the excited state to structures on the ground state PES: through CI_1 back to $GS_{reactants}$, through CI_2 to the structure $GS_{products}$.

The two conical intersections lie on a *conical intersection seam* (indicated in red). The seam is parallel to the adiabatic path coordinate (see below) and perpendicular to the radiationless decay coordinate. Photochemistry can occur at the minimum of a CI seam as well as at higher energy points on it. Conical intersection seams can be studied with geometry optimization techniques in order to locate minima, maxima and saddle points and to follow minimum energy paths along it. For an excellent example of study incorporating this approach, see [SerranoPerez13].

As we noted earlier, the Born-Oppenheimer approximation allows the nuclei to be treated as essentially fixed with respect to the electrons. Situations where this holds are referred to as *adiabatic*, indicating that the wavefunction is an equilibrium state with respect to nuclear motion.

In regions where the ground state and excited state PESs overlap—i.e., where the ground state and excited state energies are equal—*non-adiabatic coupling* becomes significant; there is interaction between the nuclear kinetic energy and the electronic wavefunction, and the Born-Oppenheimer approximation does not apply.

We can conceptually separate out the degree of freedom corresponding to the non-adiabatic coupling from the others describing the molecular state. In the following diagram, this coordinate is one of the primary axes (labeled x_2); the other axis (x_1) in the conventional x-y plane is the gradient difference with respect to the nuclear coordinates ($\vec{R}$). All of the other degrees of freedom form the axis perpendicular to this plane in the conventional z direction (x_3 through x_n):

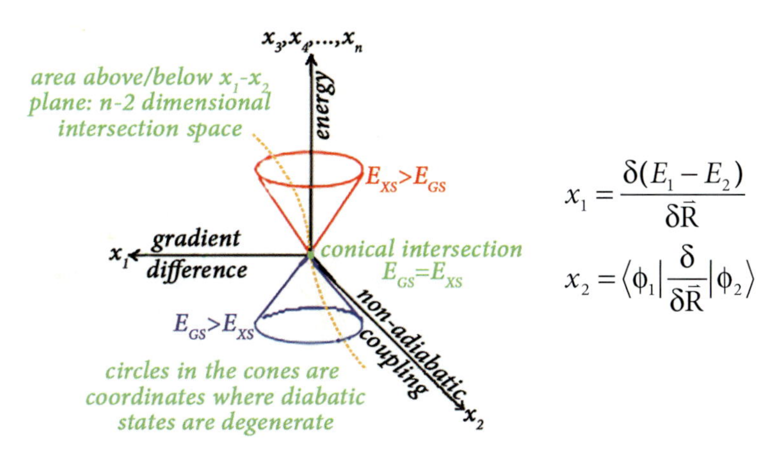

$$x_1 = \frac{\delta(E_1 - E_2)}{\delta\vec{R}}$$

$$x_2 = \langle \phi_1 | \frac{\delta}{\delta\vec{R}} | \phi_2 \rangle$$

Thus, the area perpendicular to the x_1-x_2 plane is an n-2 dimensional intersection space in which the energy is minimized. Points at which the energies of the ground and excited states are equal correspond to two diabatic states in which the ground state and excited adiabatic state wavefunctions mix. Each circle in the cones in the diagram corresponds to the range of coordinates in the n-2 dimensional space where the diabatic states are degenerate. The circles decrease in size as we approach the conical intersection (from either PES); at the latter minimum, the circle contracts to a single point.

The orange line in this figure (which includes the apex of the cone) corresponds to the conical intersection seam in the previous figure.

In a search for a conical intersection (**Opt=Conical**), the CASSCF procedure uses *state averaging*: this technique uses a single set of MOs to compute all the states of the required spatial and spin degrees of freedom rather than one(s) optimized for a single state. State averaging prevents the variational collapse of the wavefunction that can occur when the potential energy surface of a higher-order root passes through that of a lower one (sometimes called *eigenvector sag*).

References

[Adamo13] Adamo, C.; Jacquemin, D., "The calculations of excited-state properties with Time-Dependent Density Functional Theory," *Chemical Society Reviews*, 2013, **42**, 845-56, DOI: 10.1039/C2CS35394F.

[Adamo90] Adamo, C.; Cossi, M.; Rega, N.; Barone, V., "New Computational Strategies for the Quantum Mechanical Study of Biological Systems in Condensed Phases" in *Theoretical Biochemistry: Processes and Properties of Biological Systems*; ed. Eriksson, L. A., *Theoretical and Computational Chemistry*, vol. 9, Elsevier: New York, 1990

[Adamo99a] Adamo, C.; Barone, V., "Toward reliable density functional methods without adjustable parameters: The PBE0 model," *J. Chem. Phys.*, 1999, **110**, 6158-69, DOI: 10.1063/1.478522.

[Adcock74] Adcock, W.; Gupta, B. D.; Khor, T. C.; Doddrell, D.; Jordan, D.; Kitching, W., "Concerning the Carbon-13 chemical shifts of benzocycloalkenes," *J. Am. Chem. Soc.*, 1974, **96**, 1595, DOI: 10.1021/ja00812a054.

[Ali08] Ali, E. M. A.; Edwards, H. G. M.; Hargreaves, M. D.; Scowen, I. J., "In-situ detection of drugs-of-abuse on clothing using confocal Raman microscopy," *Anal. Chim. Acta*, 2008, **615**, 6372, DOI: 10.1016/j.aca.2008.03.051.

[Ali08a] Ali, E. M. A.; Edwards, H. G. M.; Hargreaves, M. D.; Scowen, I. J., "Raman spectroscopic investigation of cocain hydrochloride on human nail in a forensic context," *Anal. Bioanal. Chem.*, 2008, **390**, 1159-66, DOI: 10.1007/s00216-007-1776-z.

[Almlof78] Almlöf, J.; Roos, B. O.; Siegbahn, P. E. M., "An MC-SCF computation scheme for large-scale calculations on polyatomic systems," *Comput. Chem.*, 1978, **2**, 89, DOI: 10.1016/0097-8485(78)87007-7.

[Amovilli98] Amovilli, C.; Mennucci, B.; Floris, F. M., "MCSCF Study of the S_N2 Menshutkin Reaction in Aqueous Solution within the Polarizable Continuum Model," *J. Phys. Chem. B*, 1998, **102**, 3023, DOI: 10.1021/jp9803945.

[Anderson01u] Anderson, W. P.; Foresman, J. B., "Use of Ab Initio Calculations to Help Interpret the UV-Visible Spectra of Aquavanadium Complexes: A New Look at an Old Experiment," 2001, unpublished.

[Andrae90] Andrae, D.; Häussermann, U.; Dolg, M.; Stoll, H.; Preuss, H., "Energy-adjusted ab initio pseudopotentials for the 2nd and 3rd row transition-elements," *Theor. Chim. Acta*, 1990, **77**, 123-41, DOI: 10.1007/BF01114537.

[Aparicio07] Aparicio, S., "Computational study on the properties and structure of methyl lactate," *J. Phys. Chem. A*, 2007, **111**, 4671, DOI: 10.1021/jp070841t.

[Austin12] Austin, A.; Petersson, G. A.; Frisch, M. J.; Dobek, F. J.; Scalmani, G.; Throssell, K., "A density functional with spherical atom dispersion terms," *J. Chem. Theory and Comput.*, 2012, **8**, 4989-5007, DOI: 10.1021/ct300778e.

[Autschbach02] Autschbach, J.; Ziegler, T.; van Gisbergen, S. J. A.; Baerends, E. J., "Chiroptical properties from time-dependent density functional theory. I. Circular dichroism spectra of organic molecules," *J. Chem. Phys.*, 2002, **116**, 6930-40, DOI: 10.1063/1.1436466.

[Baboul99] Baboul, A. G.; Curtiss, L. A.; Redfern, P. C.; Raghavachari, K., "Gaussian-3 theory using density functional geometries and zero-point energies," *J. Chem. Phys.*, 1999, **110**, 7650-57, DOI: 10.1063/1.478676.

[Bak95] Bak, K. L.; Hansen, A. E.; Ruud, K.; Helgaker, T.; Olsen, J.; Jørgensen, P., "Ab Initio Calculation of Electronic Circular-Dichroism for trans-Cyclooctene Using London Atomic Orbitals," *Theoretical Chemistry Accounts*, 1995, **90**, 441-58, DOI: 10.1007/BF01113546.

[Balci08] Balci, K.; Akyuz, S., "A vibrational spectroscopic investigation on benzocaine molecule," *Vibr. Spect.*, 2008, **48 (2)**, 215-28, DOI: 10.1016/j.vibspec.2008.02.001.

[Barnes09] Barnes, E. C.; Petersson, G. A.; Montgomery Jr., J. A.; Frisch, M. J.; Martin, J. M. L., "Unrestricted Coupled Cluster and Brueckner Doubles Variations of W1 Theory," *J. Chem. Theory and Comput.*, 2009, **5**, 2687-93, DOI: 10.1021/ct900260g.

[Barone02] Barone, V.; Peralta, J. E.; Contreras, R. H.; Snyder, J. P., "DFT Calculation of NMR JFF Spin-Spin Coupling Constants in Fluorinated Pyridines," *J. Phys. Chem. A*, 2002, **106**, 5607-12, DOI: 10.1021/jp020212d.

[Barone03] Barone, V.; Cossi, M.; Rega, N.; Scalmani, G., "Energies, structures, and electronic properties of molecules in solution with the C-PCM solvation model," *J. Comp. Chem.*, 2003, **24**, 669-81, DOI: 10.1002/jcc.10189.

[Barone04] Barone, V., "Vibrational zero-point energies and thermodynamic functions beyond the harmonic approximation," *J. Chem. Phys.*, 2004, **120**, 3059-65, DOI: 10.1063/1.1637580.

[Barone05] Barone, V., "Anharmonic vibrational properties by a fully automated second-order perturbative approach," *J. Chem. Phys.*, 2005, **122**, 014108: 1-10, DOI: 10.1063/1.1824881.

[Barone09] Barone, V.; Bloino, J.; Biczysko, M.; Santoro, F., "Fully integrated approach to compute vibrationally resolved optical spectra: From small molecules to macrosystems," *J. Chem. Theory and Comput.*, 2009, **5**, 540-54, DOI: 10.1021/ct8004744.

[Barone95a] Barone, V.; Adamo, C., "Density functional study of intrinsic and environmental effects in the tautomeric equilibrium of 2-pyridone," *J. Phys. Chem.*, 1995, **99**, 15062-68, DOI: 10.1021/j100041a022.

[Barron04] Barron, L. D., *Molecular Light Scattering and Optical Activity*; 2nd ed.; Cambridge University Press: Cambridge, UK, 2004.

[Barron73] Barron, L. D.; Bogaard, M. P.; Buckingham, A. D., "Raman scattering of circularly polarized light by optically active molecules," *J. Am. Chem. Soc.*, 1973, **95**, 603-05, DOI: 10.1021/ja00783a058.

[Bartlett07] Bartlett, R. J.; Musial, M., "Coupled-cluster theory in quantum chemistry," *Rev. Mod. Phys.*, 2007, **79**, 291, DOI: 10.1103/RevModPhys.79.291.

[Bartlett78] Bartlett, R. J.; Purvis III, G. D., "Many-body perturbation-theory, coupled-pair many-electron theory, and importance of quadruple excitations for correlation problem," *Int. J. Quantum Chem.*, 1978, **14**, 561-81, DOI: 10.1002/qua.560140504.

[Bas08] Bas, D. C.; Rogers, D. M.; Jensen, J., "Very Fast Prediction and Rationalization of pKa Values for Protein-Ligand Complexes," *Proteins*, 2008, **73**, 765-83, DOI: 10.1002/prot.22102.

[Bauernschmitt96a] Bauernschmitt, R.; Ahlrichs, R., "Treatment of electronic excitations within the adiabatic approximation of time dependent density functional theory," *Chem. Phys. Lett.*, 1996, **256**, 454-64, DOI: 10.1016/0009-2614(96)00440-X.

[Baughcum81] Baughcum, S. I.; Duerst, R. W.; Rowe, W. F.; Smith, Z.; Wilson, E. B., "Microwave Spectroscopic Study of Malonadlehyde (3-Hydroxy-2-propenal). 2. Structure, Dipole Moment, and Tunneling," *J. Am. Chem. Soc.*, 1981, **103**, 6296-303, DOI: 10.1021/ja00411a005.

[Baughcum84] Baughcum, S. I.; Smith, Z.; Wilson, E. B.; Duerst, R. W., "Microwave Spectroscopic Study of Malonaldehyde. 3. Vibration-Rotation Interaction and One-Dimensional Model for Proton Tunneling," *J. Am. Chem. Soc.*, 1984, **106**, 2260-65, DOI: 10.1021/ja00320a007.

[Bauschlicher86] Bauschlicher Jr., C. W.; Taylor, P. R., "Benchmark full configuration-interaction calculations on H_2O, F, and F^-," *J. Chem. Phys.*, 1986, **85**, 2779, DOI: 10.1063/1.451034.

[Bauschlicher95] Bauschlicher Jr., C. W.; Partridge, H., "Modification of the Gaussian-2 approach using density functional theory," *J. Chem. Phys.*, 1995, **103**, 1788-91, DOI: 10.1063/1.469752.

[Bearpark94] Bearpark, M. J.; Robb, M. A.; Schlegel, H. B., "A Direct Method for the Location of the Lowest Energy Point on a Potential Surface Crossing," *Chem. Phys. Lett.*, 1994, **223**, 269-74, DOI: 10.1016/0009-2614(94)00433-1.

[Becke14] Becke, A. D., "Perspective: Fifty years of density-functional theory in chemical physics," *The Journal of Chemical Physics*, 2014, **140**, 18A301, DOI: 10.1063/1.4869598.

[Becke88b] Becke, A. D., "Density-functional exchange-energy approximation with correct asymptotic-behavior," *Phys. Rev. A*, 1988, **38**, 3098-100, DOI: 10.1103/PhysRevA.38.3098.

[Becke93] Becke, A. D., "A new mixing of Hartree-Fock and local density-functional theories," *J. Chem. Phys.*, 1993, **98**, 1372-77, DOI: 10.1063/1.464304.

[Becke93a] Becke, A. D., "Density-functional thermochemistry. III. The role of exact exchange," *J. Chem. Phys.*, 1993, **98**, 5648-52, DOI: 10.1063/1.464913.

[Bender67] Bender, M. L.; Heck, H. d. A., "Carbonyl Oxygen Exchange in General Base Catalyzed Ester Hydrolysis," *J. Am. Chem. Soc.*, 1967, **89**, 1211, DOI: 10.1021/ja00981a030.

[Berkowitz69] Berkowitz, J.; Chupka, W. A.; Walter, T. A., "Photoionization of HCN: the electron affinity and heat of formation of CN," *J. Chem. Phys.*, 1969, **50**, 1497, DOI: 10.1063/1.1671233.

[Bernardi84] Bernardi, F.; Bottini, A.; McDougall, J. J. W.; Robb, M. A.; Schlegel, H. B., "MCSCF gradient calculation of transition structures in organic reactions," *Far. Symp. Chem. Soc.*, 1984, **19**, 137-47, DOI: 10.1039/FS9841900137.

[Bernardi96] Bernardi, F.; Olivucci, M.; Robb, M. A., "Potential energy surface crossings in organic photochemistry," *Chem. Soc. Reviews*, 1996, **25**, 321, DOI: 10.1039/CS9962500321

[Biedermann99] Biedermann, P. U.; Cheeseman, J. R.; Frisch, M. J.; Schurig, V.; Gutman, I.; Agranat, I., "Conformational Spaces and Absolute Configurations of Chiral Fluorinated Inhalation Anaesthetics. A Theoretical Study," *J. Org. Chem.*, 1999, **64**, 3878-84, DOI: 10.1021/jo9821325.

[Binning90] Binning Jr., R. C.; Curtiss, L. A., "Compact contracted basis-sets for 3rd-row atoms - Ga-Kr," *J. Comp. Chem.*, 1990, **11**, 1206-16, DOI: 10.1002/jcc.540111013.

[Blanksby03] Blanksby, S. J.; Ellison, G. B., "Bond Dissociation Energies of Organic Molecules," *Acc. Chem. Res.*, 2003, **36**, 255-63, DOI: 10.1021/ar020230d.

[Blaudeau97] Blaudeau, J.-P.; McGrath, M. P.; Curtiss, L. A.; Radom, L., "Extension of Gaussian-2 (G2) theory to molecules containing third-row atoms K and Ca," *J. Chem. Phys.*, 1997, **107**, 5016-21, DOI: 10.1063/1.474865.

[Bloino12] Bloino, J.; Barone, V., "A second-order perturbation theory route to vibrational averages and transition properties of molecules: General formulation and application to infrared and vibrational circular dichroism spectroscopies," *J. Chem. Phys.*, 2012, **136**, 124108, DOI: 10.1063/1.3695210.

[Bock81a] Bock, H.; Solouki, B., "Photoelectron spectra and electron properties: Real-time gas analysis in flow systems," *Angew. Chem. Int. Ed.*, 1981, **20**, 427, DOI: 10.1002/anie.198104271.

[Bock83] Bock, H.; Dammel, R.; Aygen, S., "Gas-phase reactions. 36. Pyrolysis of vinyl azide," *J. Am. Chem. Soc.*, 1983, **105**, 7681-85, DOI: 10.1021/ja00364a037.

[Bock88] Bock, H.; Dammel, R., "Gas-Phase Reactions. 66. Gas-Phase Pyrolyses of Alkyl Azides: Experimental Evidence for Chemical Activation," *J. Am. Chem. Soc.*, 1988, **110**, 5261, DOI: 10.1021/ja00224a004.

[Boese04] Boese, A. D.; Martin, J. M. L., "Development of Density Functionals for Thermochemical Kinetics," *J. Chem. Phys.*, 2004, **121**, 3405-16, DOI: 10.1063/1.1774975.

[Born27] Born, M.; Oppenheimer, J. R., "Zur Quantentheorie der Molekeln," *Phys. Rev. précis (22 pgs):*, 1927, **389**, 0457-84, DOI: 10.1002/andp.19273892002.

[Bouwens96] Bouwens, R. J.; Hammerschmidt, J. A.; Grzeskowiak, M. M.; Stegink, T. A.; Yorba, P. M.; Polik, W. F., "Pure vibrational spectroscopy of S_0 formaldehyde by dispersed fluorescence," *J. Chem. Phys.*, 1996, **104**, 460, DOI: 10.1063/1.470844.

[Bowen83] Bowen, K. H.; Liesegang, G. W.; Sanders, R. A.; Herschbach, D. W., "Electron Attachment to Molecular Clusters by Collisional Charge Transfer," *J. Chem. Phys.*, 1983, **87**, 557-65, DOI: 10.1021/j100227a009.

[Boys70] Boys, S. F.; Bernardi, F., "Calculation of Small Molecular Interactions by Differences of Separate Total Energies - Some Procedures with Reduced Errors," *Molecular Physics*, 1970, **19**, 553, DOI: 10.1080/00268977000101561.

[Bradforth93] Bradforth, S. E.; Kim, E. H.; Arnold, D. W.; Neumark, D. M., "Photoelectron spectroscopy of CN^-, NCO^-, and NCS^-," *J. Chem. Phys.*, 1993, **98**, 800, DOI: 10.1063/1.464244.

[Brancato06] Brancato, G.; Rega, N.; Barone, V., "A quantum mechanical/molecular dynamics/mean field study of acrolein in aqueous solution: Analysis of H bonding and bulk effects on spectroscopic properties," *J. Chem. Phys.*, 2006, **125**, 164515, DOI: 10.1063/1.2359723.

[Brillouin34] Brillouin, L., "Les champs self-consistents de Hartree et de Fock," *Actualités Scientifiques et Industrielles*, 1934, **159**.

[Buckingham67] "Permanent & Induced Molecular Moments and Long-Range Intermolecular Forces" in Buckingham, A. D., *Advances in Chemical Physics*, Wiley Interscience: Hoboken, NJ, 1967, Vol. 12, pp. 107.

[Buckingham67a] Buckingham, A. D.; Orr, B. J., "Molecular hyperpolarisabilities," *Q. Rev. Chem. Soc.*, 1967, **21**, 195-212, DOI: 10.1039/QR9672100195.

[Buckingham87] Buckingham, A. D.; Fowler, P. W.; Galwas, P. A., "Velocity-dependent property surfaces and the theory of vibrational circular dichroism," *Chem. Phys.*, 1987, **112**, 1-14, DOI: 10.1016/0301-0104(87)85017-6.

[Bulliard99] Bulliard, C.; Allan, M.; Wirtz, G.; Haselbach, E.; Zachariasse, K. A.; Detzer, N.; Grimme, S., "Electron Energy Loss and DFT/SCI Study of the Singlet and Triplet Excited States of Aminobenzonitriles and Benzoquinuclidines: Role of the Amino Group Twist Angle," *J. Phys. Chem. A*, 1999, **103**, 7766, DOI: 10.1021/jp990922s.

[Burk07] Burk, R., "Paramagnetism of Liquid Oxygen" [video], 2007; *www.youtube.com/watch?v=Isd9IEnR4bw*.

[Bytautas04] Bytautas, L.; Ruedenberg, K., "Correlation energy extrapolation by intrinsic scaling. II. The water and the nitrogen molecule," *J. Chem. Phys.*, 2004, **121**, 10919, DOI: 10.1063/1.1811604.

[Califano76] Califano, S., *Vibrational States*; Wiley: London, 1976.

[Cami10] Cami, J.; Bernard-Salas, J.; Peeters, E.; Malek, S. E., "Detection of C_{60} and C_{70} in a Young Planetary Nebula," *Science*, 2010, **329**, 1180, DOI: 10.1126/science.1192035

[Cances97] Cancès, E.; Mennucci, B.; Tomasi, J., "A new integral equation formalism for the polarizable continuum model: Theoretical background and applications to isotropic and anisotropic dielectrics," *J. Chem. Phys.*, 1997, **107**, 3032-41, DOI: 10.1063/1.474659.

[Cardini94] Cardini, G.; Bini, R.; Salvi, P. R.; Schettino, V., "Infrared spectrum of two fullerene derivatives: $C_{60}O$ and $C_{61}H_2$," *The Journal of Chemical Physics*, 1994, **98**, 9966-71, DOI: 10.1021/j100091a006.

[Caricato12b] Caricato, M., "Absorption and Emission Spectra of Solvated Molecules with the EOM-CCSD-PCM Method," *J. Chem. Theory and Comput.*, 2012, **8**, 4494, DOI: 10.1021/ct3006997.

[Caricato13] Caricato, M.; Lipparini, F.; Scalmani, G.; Cappelli, C.; Barone, V., "Vertical electronic excitations in solution with the EOM-CCSD method combined with a polarizable explicit/implicit solvent model," *J. Chem. Theory and Comput.*, 2013, **9** , 3035, DOI: 10.1021/ct40032881.

[Caricato13b] Caricato, M., "Implementation of the CCSD-PCM linear response function for frequency dependent properties in solution: Application to polarizability and specific rotation," *J. Chem. Phys.*, 2013, **139**, 114103 1-6, DOI: 10.1063/1.4821087.

[Caricato14] Caricato, M., "A corrected-linear response formalism for the calculation of electronic excitation energies of solvated molecules with the CCSD-PCM method," *Comput. Theoret. Chem.*, 2014, **1040-1041**, 99-105, DOI: 10.1016/j.comptc.2014.02.001.

[Carroll82] Carroll, F. I.; Coleman, M. L.; Lewin, A. H., "Syntheses and conformational analyses of isomeric cocaines: a proton and carbon-13 NMR study," *J. Org. Chem.*, 1982, **47**, 13-19, DOI: 10.1021/jo00340a004.

[Case14] Case, D. A.; Babin, V.; Berryman, R. M.; Cai, Q.; Cerutti, D. S.; Cheatham III, T. E.; Darden, T. A.; Duke, R. E.; Gohlke, H.; Goetz, A. W.; Gusarov, S.; Homeyer, N.; Janowski, P.; Kaus, J.; Kolossváry, I.; Kovalenko, A.; Lee, T. S.; LeGrand, S.; Luchko, T.; Luo, R.; Madej, B.; Merz, K. M.; Paesani, F.; Roe, D. R.; Roitberg, A.; Sagui, C.; Salomon-Ferrer, R.; Seabra, G.; Simmerling, C. L.; Smith, W.; Swails, J.; Walker, R. C.; Wang, J.; Wolf, R. M.; Wu, X.; Kollman, P. A., *AMBER14*, University of California, San Francisco, 2014.

[Casida98] Casida, M. E.; Jamorski, C.; Casida, K. C.; Salahub, D. R., "Molecular excitation energies to high-lying bound states from time-dependent density-functional response theory: Characterization and correction of the time-dependent local density approximation ionization threshold," *J. Chem. Phys.*, 1998, **108**, 4439-49, DOI: 10.1063/1.475855

[Castejon99] Castejon, H.; Wiberg, K. B., "Solvent Effects on Methyl Transfer Reactions. 1. The Menshutkin Reaction," *J. Am. Chem. Soc.*, 1999, **121**, 2139, DOI: 10.1021/ja983736t.

[Chai08] Chai, J.-D.; Head-Gordon, M., "Systematic optimization of long-range corrected hybrid density functionals," *J. Chem. Phys.*, 2008, **128**, 084106, DOI: 10.1063/1.2834918.

[Champagne98] Champagne, B.; Perpète, E. A.; van Gisbergen, S. J. A.; Baerends, E.-J.; Snijders, J. G.; Soubra-Ghaoui, C.; Robins, K. A.; Kirtman, B., "Assessment of conventional density functional schemes for computing the polarizabilities and hyperpolarizabilities of conjugated oligomers: An ab initio investigation of polyacetylene chains," *J. Chem. Phys.*, 1998, **109**, 10489, DOI: 10.1063/1.477731.

[Chase98] Chase Jr., M. W., *NIST-JANAF Thermochemical Tables*; 4th ed.; American Inst. of Physics: New York, 1998; Vol. 1-2.

[Chattopadhyay10] Chattopadhyay, S.; Chaudyhuri, R. K.; Mahapatra, U. S., "Studies on m-benzyne and phenol via improved virtual orbital-complete active space configuration interaction (IVO-CASCI) analytical gradient method," *Chem. Phys. Lett.*, 2010, **491**, 102-08, DOI: 10.1016/j.cplett.2010.04.005.

[Cheeseman11] Cheeseman, J. R.; Shaik, M. S.; Popelier, P. L. A.; Blanch, E. W., "Calculation of Raman optical activity spectra of methyl-β-D-glucose incorporating a full molecular dynamics simulation of hydration effects," *J. Am. Chem. Soc.*, 2011, **133**, 4991-97, DOI: 10.1021/ja110825z.

[Cheeseman11a] Cheeseman, J. R.; Frisch, M. J., "Basis set dependence of vibrational raman and raman optical activity intensities," *J. Chem. Theory and Comput.*, 2011, **7**, 3323-34, DOI: 10.1021/ct200507e.

[Cheeseman96] Cheeseman, J. R.; Trucks, G. W.; Keith, T. A.; Frisch, M. J., "A Comparison of Models for Calculating Nuclear Magnetic Resonance Shielding Tensors," *J. Chem. Phys.*, 1996, **104**, 5497-509, DOI: 10.1063/1.471789.

[Cheeseman96a] Cheeseman, J. R.; Frisch, M. J.; Devlin, F. J.; Stephens, P. J., "Ab Initio Calculation of Atomic Axial Tensors and Vibrational Rotational Strengths Using Density Functional Theory," *Chem. Phys. Lett.*, 1996, **252**, 211-20, DOI: 10.1016/0009-2614(96)00154-6.

[Chen10] Chen, V. B.; Arendall III, W. B.; Headd, J. J.; Keedy, D. A.; Immormino, R. M.; Kapral, G. J.; Murray, L. W.; Richardson, J. S.; Richardson, D. C., "MolProbity: all-atom structure validation for macromolecular crystallography," *Acta Crystallogr.*, 2010, **D66**, 12-21, DOI: 10.1107/S0907444909042073.

[Chiara92] Chiara, J. L.; Gómez-Sánchez, A.; Bellanato, J. J., "Spectral Properties and Isomerism of Nitroenamines 3," *J. Chem. Soc., Perkin Trans. 2*, 1992, **5**, 787, DOI: 10.1039/P29880001691

[Cho07] Cho, L.-L., "Identification of textile fiber by Raman microspectroscopy," *Forensic Science J.*, 2007, **6**, 55-62.

[Chung15] Chung, L. W.; Sameera, W. M. C.; Ramozzi, R.; Page, A. J.; Hatanaka, M.; Petrova, G. P.; Harris, T. V.; Li, X.; Ke, Z.; Liu, F., "The ONIOM Method and Its Applications," *Chemical reviews*, 2015.

[Cizek69] Čížek, J., "On the Use of the Cluster Expansion and the Technique of Diagrams in Calculations of Correlation Effects in Atoms And Molecules" in *Advances in Chemical Physics: Correlation Effects in Atoms and Molecules*; ed. Hariharan, P. C., Wiley Interscience: New York, 1969, pp. 35-89, DOI: 10.1002/9780470143599.

[Clabo88] Clabo, D. A.; Allen, W. D.; Remington, R. B.; Yamaguchi, Y.; Schaefer III, H. F., "A systematic study of molecular vibrational anharmonicity and vibration-rotation interaction by self-consistent-field higher-derivative methods - asymmetric-top molecules," *Chem. Phys.*, 1988, **123**, 187-239, DOI: 10.1016/0301-0104(88)87271-9.

[Clemente10] Clemente, F.; Vreven, T.; Frisch, M. J., "Getting the Most out of ONIOM: Guidelines and Pitfalls" in *Quantum Biochemistry*; ed. Matta, C., Wiley VCH: Weinheim, 2010, pp. 61-84, DOI: 10.1002/9783527629213.ch2.

[Condon26] Condon, E. U., "A Theory of Intensity Distribution in Band Systems," *Phys. Rev.*, 1926, **28**, 1182, DOI: 10.1103/PhysRev.28.1182.

[Condon37] Condon, E. U., "Theories of optical rotatory power," *Rev. Mod. Phys.*, 1937, **9**, 432-57, DOI: 10.1103/RevModPhys.9.432

[Cook76] Cook, M. J.; Katritzky, A. R.; Hepler, L. G.; Matsui, T., "Heats of solution and tautomeric equilibrium constants. The 2-pyridone:2-hydroxypyridine equilibrium in non-aqueous media," *Tetrahedron Lett.*, 1976, **17**, 2685, DOI: 10.1016/S0040-4039(00)77795-1.

[Cooper80] Cooper, J. W., *Spectroscopic Techniques for Organic Chemists*; Wiley-Interscience: New York, 1980, pp. 376.

[Cornell95] Cornell, W. D.; Cieplak, P.; Bayly, C. I.; Gould, I. R.; Merz Jr., K. M.; Ferguson, D. M.; Spellmeyer, D. C.; Fox, T.; Caldwell, J. W.; Kollman, P. A., "A second generation force-field for the simulation of proteins, nucleic-acids, and organic-molecules," *J. Am. Chem. Soc.*, 1995, **117**, 5179-97, DOI: 10.1021/ja00124a002.

[Cossi00] Cossi, M.; Barone, V., "Solvent effect on vertical electronic transitions by the polarizable continuum model," *J. Chem. Phys.*, 2000, **112**, 2427-35, DOI: 10.1063/1.480808.

[Cossi96] Cossi, M.; Barone, V.; Cammi, R.; Tomasi, J., "Ab initio study of solvated molecules: A new implementation of the polarizable continuum model," *Chem. Phys. Lett.*, 1996, **255**, 327-35, DOI: 10.1016/0009-2614(96)00349-1.

[Costentin06] Costentin, C.; Robert, M.; Savéant, J.-M., "Electrochemical and Homogeneous Proton-Coupled Electron Transfers: Concerted Pathways in the One-Electron Oxidation of a Phenol Coupled with an Intramolecular Amine-Driven Proton Transfer," *J. Am. Chem. Soc.*, 2006, **128**, 4552, DOI: 10.1021/ja060527x.

[Coto11] Coto, P. B.; Serrano-Andrés, L.; Gustavsson, T.; Fujiwara, T.; Lim, E. C., "Intramolecular charge transfer and dual fluorescence of 4-(dimethylamino)benzonitrile: ultrafast branching followed by a two-fold decay mechanism," *Phys. Chem. Chem. Phys.*, 2011, **13**, 15183-88, DOI: 10.1039/c1cp21089k.

[Cox84] *CODATA Key Values for Thermodynamics*; Cox, J. D.; Wagnam, D. D.; Medvedev, V. A., Eds.; Hemisphere: New York, 1984.

[Craciun10] Craciun, R.; Picone, D.; Long, R. T.; Li, S.; Dixon, D. A.; Peterson, K. A.; Christe, K. O., "Third Row Transition Metal Hexafluorides, Extraordinary Oxidizers, and Lewis Acids: Electron Affinities, Fluoride Affinities, and Heats of Formation of WF_6, ReF_6, OsF_6, IrF_6, PtF_6, and AuF_6," *Inorg. Chem.*, 2010, **49**, 1956-70, DOI: 10.1021/ic901967h.

[Craig06] Craig, N. C.; Groner, P.; McKean, D. C., "Equilibrium structures for butadiene and ethylene: Compelling evidence for pi-electron delocalization in butadiene," *J. Phys. Chem. A*, 2006, **110**, 7461, DOI: 10.1021/jp060695b.

[Cramer04] Cramer, C. J., *Essentials of Computational Chemistry: Theories and Models*; 2nd ed.; Wiley & Sons, 2004.

[CRC00] *CRC Handbook of Chemistry and Physics*; 81st ed.; Lide, D. R., Ed.; CRC Press: Boca Raton, FL, 2000.

[Curtiss00a] Curtiss, L. A.; Raghavachari, K.; Redfern, P. C.; Pople, J. A., "Assessment of Gaussian-3 and density functional theories for a larger experimental test set," *J. Chem. Phys.*, 2000, **112**, 7374-83.

[Curtiss02a] Curtiss, L. A.; Raghavachari, K., "Gaussian-3 and related methods for accurate thermochemistry," *Theoretical Chemistry Accounts*, 2002, **108**, 61-70, DOI: 10.1007/s00214-002-0355-9.

[Curtiss05] Curtiss, L. A.; Redfern, P. C.; Raghavachari, K., "Assessment of Gaussian-3 and density-functional theories on the G3/05 test set of experimental energies," *J. Chem. Phys.*, 2005, **123**, 124107 1-12, DOI: 10.1063/1.2039080.

[Curtiss07] Curtiss, L. A.; Redfern, P. C.; Raghavachari, K., "Gaussian-4 theory," *J. Chem. Phys.*, 2007, **126**, 084108, DOI: 10.1063/1.2436888.

[Curtiss07a] Curtiss, L. A.; Redfern, P. C.; Raghavachari, K., "Gaussian-4 theory using reduced order perturbation theory," *J. Chem. Phys.*, 2007, **127**, 124105, DOI: 10.1063/1.2770701.

[Curtiss90] Curtiss, L. A.; Jones, C.; Trucks, G. W.; Raghavachari, K.; Pople, J. A., "Gaussian-1 theory of molecular energies for second-row compounds," *J. Chem. Phys.*, 1990, **93**, 2537-45, DOI: 10.1063/1.458892.

[Curtiss95] Curtiss, L. A.; McGrath, M. P.; Blaudeau, J.-P.; Davis, N. E.; Binning Jr., R. C.; Radom, L., "Extension of Gaussian-2 theory to molecules containing third-row atoms Ga-Kr," *J. Chem. Phys.*, 1995, **103**, 6104-13, DOI: 10.1063/1.470438.

[Curtiss97] Curtiss, L. A.; Raghavachari, K.; Redfern, P. C.; Pople, J. A., "Assessment of Gaussian-2 and density functional theories for the computation of enthalpies of formation," *J. Chem. Phys.*, 1997, **106**, 1063-79.

[Curtiss98] Curtiss, L. A.; Raghavachari, K.; Redfern, P. C.; Rassolov, V.; Pople, J. A., "Gaussian-3 (G3) theory for molecules containing first and second-row atoms," *J. Chem. Phys.*, 1998, **109**, 7764-76, DOI: 10.1063/1.477422.

[Cybulski07] Cybulski, H.; Sadlej, J., "On the calculations of the vibrational Raman spectra of small water clusters," *Chem. Phys.*, 2007, **342**, 163, DOI: 10.1016/j.chemphys.2007.09.058.

[Dapprich99] Dapprich, S.; Komáromi, I.; Byun, K. S.; Morokuma, K.; Frisch, M. J., "A New ONIOM Implementation in Gaussian 98. 1. The Calculation of Energies, Gradients and Vibrational Frequencies and Electric Field Derivatives," *J. Mol. Struct. (Theochem)*, 1999, **462**, 1-21, DOI: 10.1016/S0166-1280(98)00475-8.

[Das08] Das, U.; Raghavachari, K., "Al_5SO_4: A Superatom with Potential for New Materials Design," *J. Chem. Theory and Comput.*, 2008, **4**, 2011-19, DOI: 10.1021/ct800232b.

[David08] David, J.; Fuentealba, P.; Restrepo, A., "Relativistic effects on the hexafluorides of group 10 metals," *Chem. Phys. Lett.*, 2008, **457**, 42-44, DOI: 10.1016/j.cplett.2008.04.003.

[Davidson96] Davidson, E. R., "Comment on 'Comment on Dunning's correlation-consistent basis sets,'" *Chem. Phys. Lett.*, 1996, **260**, 514-18, DOI: 10.1016/0009-2614(96)00917-7.

[Debbert00] Debbert, S. L.; Cramer, C. J., "Systematic comparison of the benzynes, pyridynes, and pyridynium cations and characterization of the Bergman cyclization of Z-but-1-en-3-yn-1-yl isonitrile to the meta diradical 2,4-pyridyne," *Int. J. Mass Spectrom.*, 2000, **201**, 1-15, DOI: 10.1016/S1387-3806(00)00160-3.

[Debie08] Debie, E.; Jaspers, L.; Bultinck, P.; Herrebout, W.; Veken, B. V. D., "Induced solvent chirality: A VCD study of camphor in $CDCl_3$," *Chem. Phys. Lett.*, 2008, **450**, 426-30, DOI: 10.1016/j.cplett.2007.11.064.

[DelRio08] del Rio, D.; Resa, I.; Rodríguez, A.; Sánchez, L.; Köppe, R.; Downs, A. J.; Tang, C. Y.; Carmona, E., "IR and Raman Characterization of the Zincocenes $(\eta^5\text{-}C_5Me_5)_2Zn_2$ and $(\eta^5\text{-}C_5Me_5)(\eta^1\text{-}C_5Me_5)Zn$," *J. Phys. Chem. A*, 2008, **112**, 10516-25, DOI: 10.1021/jp805291e.

[Deng06] Deng, W.; Cheeseman, J. R.; Frisch, M. J., "Calculation of Nuclear Spin-Spin Coupling Constants of Molecules with First and Second Row Atoms in Study of Basis Set Dependence," *J. Chem. Theory and Comput.*, 2006, **2**, 1028-37, DOI: 10.1021/ct600110u.

[Dennington09] Dennington, R.; Keith, T. A.; Millam, J., *GaussView 5*, Semichem, Inc., Shawnee Mission, KS, 2009.

[Devlin96] Devlin, F. J.; Stephens, P. J.; Cheeseman, J. R.; Frisch, M. J., "Prediction of vibrational circular dichroism spectra using density functional theory: Camphor and fenchone," *J. Am. Chem. Soc.*, 1996, **118**, 6327-28.

[Devlin97] Devlin, F. J.; Stephens, P. J.; Cheeseman, J. R.; Frisch, M. J., "Ab initio prediction of vibrational absorption and circular dichroism spectra of chiral natural products using density functional theory: Camphor and fenchone," *J. Phys. Chem. A*, 1997, **101**, 6322-33, DOI: 10.1021/jp9712359.

[Devlin97a] Devlin, F. J.; Stephens, P. J.; Cheeseman, J. R.; Frisch, M. J., "Ab initio prediction of vibrational absorption and circular dichroism spectra of chiral natural products using density functional theory: α-pinene," *J. Phys. Chem. A*, 1997, **101**, 9912-24, DOI: 10.1021/jp971905A.

[Devlin99] Devlin, F. J.; Stephens, P. J., "Conformational Analysis Using ab initio Vibrational Spectroscopy: 3-Methylcyclohexanone," *J. Am. Chem. Soc.*, 1999, **121**, 7413-14, DOI: 10.1021/ja9910513.

[Diels28] Diels, O.; Alder, K., "Synthesen in der hydroaromatischen Reihe," *Annalen der Chemie*, 1928, **460**, 98, DOI: 10.1002/jlac.19284600106.

[Dirac30] Dirac, P. A. M., "Note on Exchange Phenomena in the Thomas Atom," *Math. Proc. Camb. Phil. Soc.*, 1930, **26**, 376, DOI: 10.1017/S0305004100016108.

[Ditchfield71] Ditchfield, R.; Hehre, W. J.; Pople, J. A., "Self-Consistent Molecular Orbital Methods. 9. Extended Gaussian-type basis for molecular-orbital studies of organic molecules," *J. Chem. Phys.*, 1971, **54**, 724, DOI: 10.1063/1.1674902

[Ditchfield74] Ditchfield, R., "Self-consistent perturbation theory of diamagnetism. 1. Gauge-invariant LCAO method for N.M.R. chemical shifts," *Molecular Physics*, 1974, **27**, 789-807, DOI: 10.1080/00268977400100711.

[Doering69] Doering, J. P., "Low-Energy Electron-Impact Study of the First, Second, and Third Triplet States of Benzene," *J. Chem. Phys.*, 1969, **51**, 2866, DOI: 10.1063/1.1672424.

[Dolinsky04] Dolinsky, T. J.; Nielsen, J. E.; McCammon, J. A.; Baker, N. A., "PDB2PQR: an automated pipeline for the setup of Poisson-Boltzmann electrostatics calculations," *Nucleic Acids Res.*, 2004, **32 (suppl. 2)**, W665-67, DOI: 10.1093/nar/gkh381.

[Domcke11] *Conical Intersections: Theory, Computation and Experiment*; Domcke, W.; Yarkony, D. R.; Köppel, H., Eds.; World Scientific Publishing, 2011; Vol. 17.

[Drakenberg76] Drakenberg, T.; Sommer, J. M.; Jost, R., "^{13}C nuclear magnetic resonance studies on acetophenones: Barriers to internal rotation," *Magn. Res. Chem.*, 1976, **8**, 579-81, DOI: 10.1002/mrc.1270081110

[Duchovic82] Duchovic, R. J.; Hase, W. L.; Schlegel, H. B.; Frisch, M. J.; Raghavachari, K., "Ab initio potential energy curve for CH bond dissociation in methane," *Chem. Phys. Lett.*, 1982, **89**, 120-25, DOI: 10.1016/0009-2614(82)83386-1.

[Duncan97] Duncan, T. M., *Principal Components of Chemical Shift Tensors: A compilation*; 2nd ed.; Farragut Press: Madison, WI, 1997.

[Dunlap00] Dunlap, B. I., "Robust and variational fitting: Removing the four-center integrals from center stage in quantum chemistry," *J. Mol. Struct. (Theochem)*, 2000, **529**, 37-40, DOI: 10.1016/S0166-1280(00)00528-5.

[Dunlap83] Dunlap, B. I., "Fitting the Coulomb Potential Variationally in Xα Molecular Calculations," *J. Chem. Phys.*, 1983, **78**, 3140-42, DOI: 10.1063/1.445228.

[Dunning76] Dunning Jr., T. H.; Hay, P. J., "Gaussian Basis Sets for Molecular Calculations" in *Methods of Electronic Structure Theory*; ed. Schaefer III, H. F., *Modern Theoretical Chemistry*, vol. 3, Plenum: New York, 1976, pp. 1-28, DOI: 10.1007/978-1-4757-0887-5_1.

[Dunning89] Dunning Jr., T. H., "Gaussian basis sets for use in correlated molecular calculations. I. The atoms boron through neon and hydrogen," *J. Chem. Phys.*, 1989, **90**, 1007-23, DOI: 10.1063/1.456153.

[Dupradeau10] Dupradeau, F.-Y.; Pigache, A.; Zaffran, T.; Savineau, C.; Lelong, R.; Grivel, N.; Lelong, D.; Rosanski, W.; Cieplak, P., "The R.E.D. tools: Advances in RESP and ESP charge derivation and force field library building," *Phys. Chem. Chem. Phys.*, 2010, **12**, 7821-39, DOI: 10.1039/c0cp00111b

[Dykstra77] Dykstra, C. E., "Examination of Brueckner condition for selection of molecular-orbitals in correlated wavefunctions," *Chem. Phys. Lett.*, 1977, **45**, 466-69, DOI: 10.1016/0009-2614(77)80065-1

[Ehara02] Ehara, M.; Ishida, M.; Toyota, K.; Nakatsuji, H., "SAC-CI General-R Method: Theory and Applications to the Multi-Electron Processes" in *Reviews of Modern Quantum Chemistry*; ed. Sen, K. D., World Scientific: Singapore, 2002, pp. 293, DOI: 10.1142/9789812775702_0011.

[Ernzerhof99] Ernzerhof, M.; Scuseria, G. E., "Assessment of the Perdew-Burke-Ernzerhof exchange-correlation functional," *The Journal of Chemical Physics*, 1999, **110**, 5029-36, DOI: 10.1063/1.478401.

[Eyring31] Eyring, H.; Polanyi, M., "On Simple Gas Reactions," *Zeitschrift für Physikalische Chemie*, 1931, **227**, 1221-46, DOI: 10.1524/zpch.2013.9023.

[Eyring35] Eyring, H., "The activated complex in chemical reactions," *J. Chem. Phys.*, 1935, **3**, 107-15, DOI: 10.1063/1.1749604.

[Field90] Field, M. J.; Bash, P. A.; Karplus, M., "A Combined Quantum-Mechanical and Molecular Mechanical Potential for Molecular-Dynamics Simulations," *J. Comp. Chem.*, 1990, **11**, 700-33, DOI: 10.1002/jcc.540110605.

[Fillet56] Fillet, P.; Letort, M., *Journal de Chimie Physique et de Physico-Chimie Biologique*, 1956, **53**, 8.

[Finkelmeier78] Finkelmeier, H.; Luettke, W., "Carbon-13-carbon-13 and carbon-13-hydrogen coupling constants in 2,2,4,4-tetramethylbicyclo[1.1.0]butane," *J. Am. Chem. Soc.*, 1978, **100**, 6261-62, DOI: 10.1021/ja00487a065.

[Finnerty10] Finnerty, J.; Koch, R., "Accurate Calculated Optical Properties of Substituted Quaterphenylene Nanofibers," *J. Phys. Chem. A*, 2010, **114**, 474-80, DOI: 10.1021/jp906233d.

[Fischer89] Fischer, B.; Wijkens, P.; Boersma, J.; van Koten, G.; Smeets, W. J. J.; Spek, A. L.; Budzelaar, P. H. M., "The unusual solid state structures of the penta-substituted bis(cyclopentadienyl)zinc compounds bis(pentamethylcyclopentadienyl)zinc and bis(tetramethylpenylcyclopentadienyl)zinc," *J. Organomet. Chem.*, 1989, **376**, 223-33, DOI: 10.1016/0022-328X(89)85132-0.

[Fock30] Fock, V., "'Self-consistent field' with interchange for sodium," *Zeitschrift für Physik*, 1930, **62**, 795.

[Fogarasi92] Fogarasi, G.; Zhou, X.; Taylor, P.; Pulay, P., "The calculation of ab initio molecular geometries: Efficient optimization by natural internal coordinates and empirical correction by offset forces," *J. Am. Chem. Soc.*, 1992, **114**, 8191-201, DOI: 10.1021/ja00047a032.

[Foresman13] Foresman, J. B.; Clarke, D. C., "Substituent Interactions in Aromatic Rings: Student Exercises Using FT-NMR And Electronic Structure Calculations" in *NMR Spectroscopy in the Undergraduate Curriculum*; ed. Soulsby, D., Anna, L., Wallner, T., *ACS Symposium Series*, ACS: Washington, D.C., 2013, DOI: 10.1021/bk-2013-1128.ch012.

[Foresman92] Foresman, J. B.; Head-Gordon, M.; Pople, J. A.; Frisch, M. J., "Toward a Systematic Molecular Orbital Theory for Excited States," *J. Phys. Chem.*, 1992, **96**, 135-49, DOI: 10.1021/j100180a030.

[Foresman96] Foresman, J. B.; Keith, T. A.; Wiberg, K. B.; Snoonian, J.; Frisch, M. J., "Solvent Effects. 5. The Influence of Cavity Shape, Truncation of Electrostatics, and Electron Correlation on ab initio Reaction Field Calculations," *J. Phys. Chem.*, 1996, **100**, 16098-104, DOI: 10.1021/jp960488j.

[Foster80] Foster, J. P.; Weinhold, F., "Natural hybrid orbitals," *J. Am. Chem. Soc.*, 1980, **102**, 7211-18, DOI: 10.1021/ja00544a007.

[Franck26] Franck, J.; Dymond, E. G., "Elementary processes of photochemical reactions," *Trans. Faraday Soc.*, 1926, **21**, 536, DOI: 10.1039/TF9262100536

[Frey11] Frey, J. T.; Doren, D. J., *TubeGen 3.4*, University of Delaware, Newark DE, 2011.

[Frisch85c] Frisch, M. J.; Scheiner, A. C.; Schaefer III, H. F.; Binkley, J. S., "Malonaldehyde equilibrium geometry: A major structural shift due to the effects of electron correlation," *J. Chem. Phys.*, 1985, **82**, 4194, DOI: 10.1063/1.448861.

[Frisch92] Frisch, M. J.; Ragazos, I. N.; Robb, M. A.; Schlegel, H. B., "An Evaluation of 3 Direct MC-SCF Procedures," *Chem. Phys. Lett.*, 1992, **189**, 524-28, DOI: 10.1016/0009-2614(92)85244-5.

[Frosch52] Frosch, R. A.; Foley, H. M., "Magnetic Hyperfine Structure in Diatomic Molecules," *Physical Review*, 1952, **88**, 1337-49, DOI: 10.1103/PhysRev.88.1337.

[Frum91] Frum, C. I.; Rolf Engleman, J.; Hedderich, H. G.; Bernath, P. F., "The infrared emission spectrum of gas-phase C_{60} (Buckminsterfullerene)," *Chem. Phys. Lett.*, 1991, **176**, 504-08, DOI: 10.1016/0009-2614(91)90245-5.

[Fuentealba82] Fuentealba, P.; Preuss, H.; Stoll, H.; Szentpály, L. v., "A Proper Account of Core-polarization with Pseudopotentials - Single Valence-Electron Alkali Compounds," *Chem. Phys. Lett.*, 1982, **89**, 418-22, DOI: 10.1016/0009-2614(82)80012-2.

[Fuentealba83] Fuentealba, P.; Stoll, H.; Szentpály, L. v.; Schwerdtfeger, P.; Preuss, H., "On the reliability of semi-empirical pseudopotentials: Simulation of Hartree-Fock and Dirac-Fock results," *Journal of Physics B: Atomic and Molecular Physics*, 1983, **16**, L323-28, DOI: 10.1088/0022-3700/16/11/001.

[Fukui81] Fukui, K., "The path of chemical-reactions: The IRC approach," *Acc. Chem. Res.*, 1981, **14**, 363-68, DOI: 10.1021/ar00072a001.

[Furche02] Furche, F.; Ahlrichs, R., "Adiabatic time-dependent density functional methods for excited state properties," *J. Chem. Phys.*, 2002, **117**, 7433-47, DOI: 10.1063/1.1508368.

[G09] Frisch, M. J.; Trucks, G. W.; Schlegel, H. B.; Scuseria, G. E.; Robb, M. A.; Cheeseman, J. R.; Scalmani, G.; Barone, V.; Mennucci, B.; Petersson, G. A.; Nakatsuji, H.; Caricato, M.; Li, X.; Hratchian, H. P.; Izmaylov, A. F.; Bloino, J.; Zheng, G.; Sonnenberg, J. L.; Hada, M.; Ehara, M.; Toyota, K.; Fukuda, R.; Hasegawa, J.; Ishida, M.; Nakajima, T.; Honda, Y.; Kitao, O.; Nakai, H.; Vreven, T.; Montgomery Jr., J. A.; Peralta, J. E.; Ogliaro, F.; Bearpark, M. J.; Heyd, J.; Brothers, E. N.; Kudin, K. N.; Staroverov, V. N.; Kobayashi, R.; Normand, J.; Raghavachari, K.; Rendell, A. P.; Burant, J. C.; Iyengar, S. S.; Tomasi, J.; Cossi, M.; Rega, N.; Millam, N. J.; Klene, M.; Knox, J. E.; Cross, J. B.; Bakken, V.; Adamo, C.; Jaramillo, J.; Gomperts, R.; Stratmann, R. E.; Yazyev, O.; Austin, A. J.; Cammi, R.; Pomelli, C.; Ochterski, J. W.; Martin, R. L.; Morokuma, K.; Zakrzewski, V. G.; Voth, G. A.; Salvador, P.; Dannenberg, J. J.; Dapprich, S.; Daniels, A. D.; Farkas, Ö.; Foresman, J. B.; Ortiz, J. V.; Cioslowski, J.; Fox, D. J., *Gaussian 09*, Gaussian, Inc., Wallingford, CT, 2009.

[Galwas83] Galwas, P. A., *On the Distribution of Optical Polarization in Molecules* [PhD Thesis], University of Cambridge (Cambridge, UK), 1983.

[Gamot85] Gamot, A. P.; Vercoten, G.; Fleury, G., "Étude par spectroscopie Raman du chlorhydrate de cocaine," *Talanta*, 1985, **32**, 363-72, DOI: 10.1016/0039-9140(85)80100-4.

[Gao91] Gao, J., "A priori Computation of a Solvent-enhanced S_N2 Reaction Profile in Water: The Menshutkin Reaction," *J. Am. Chem. Soc.*, 1991, **113**, 7796-97, DOI: 10.1021/ja00020a070.

[Gauss92] Gauss, J., "Calculation of NMR chemical shifts at second-order many-body perturbation theory using gauge-including atomic orbitals," *Chem. Phys. Lett.*, 1992, **191**, 614-20, DOI: 10.1016/0009-2614(92)85598-5.

[Gauss95] Gauss, J., "Accurate Calculation of NMR Chemical-Shifts," *Phys. Chem. Chem. Phys.*, 1995, **99**, 1001-08, DOI: 10.1002/bbpc.199500022.

[Gavroglu12] Gavroglu, K.; Simões, A., *Neither Chemistry nor Physics: A History of Quantum Chemistry*; MIT Press: Cambridge, MA, 2012.

[Gelbart80] Gelbart, W. M.; Elert, M. L.; Heller, D. F., "Photodissociation of the formaldehyde molecule: Does it or doesn't it?," *Chem. Rev.*, 1980, **80**, 403-16, DOI: 10.1021/cr60327a002.

[George00] George, W. O.; Jones, B. F.; Lewis, R.; Price, J. M., "Ab initio computations on simple carbonyl compounds," *J. Mol. Struct.*, 2000, **550-551**, 281, DOI: 10.1016/S0022-2860(00)00391-4.

[Gerlach22] Gerlach, W.; Stern, O., "Das magnetische Moment des Silberatoms," *Zeitschrift für Physik*, 1922, **9**, 353-55, DOI: 10.1007/BF01326984.

[Gilbert14] Gilbert, K. E., *Pcmodel 10.0*, Serena Software, Bloomington, IN, 2014.

[Goel08] Goel, S.; Masunov, A. E., "Potential energy curves and electronic structure of 3d transition metal hydrides and their cations. ," *The Journal of Chemical Physics* 2008, **129**, 214302-15, DOI: 10.1063/1.2996347.

[Goerigk11] Goerigk, L.; Grimme, S., "Efficient and Accurate Double-Hybrid-Meta-GGA Density Functionals - Evaluation with the Extended GMTKN30 Database for General Main Group Thermochemistry, Kinetics, and Noncovalent Interactions," *J. Chem. Theory and Comput.*, 2011, **7**, 291-309, DOI: 10.1021/ct100466k.

[Good69] Good, W. D.; Smith, N. K., "Enthalpies of Combustion of Toluene, Benzene, Cyclohexane, Cyclohexene, Methylcyclopentane, 1-Methylcyclopentene, and n-Hexane," *J. Chem. & Eng. Data*, 1969, **14**, 102-06, DOI: 10.1021/je60040a036.

[Grimme06] Grimme, S., "Semiempirical GGA-type density functional constructed with a long-range dispersion correction," *J. Comp. Chem.*, 2006, **27**, 1787-99, DOI: 10.1002/jcc.20495.

[Grimme06a] Grimme, S., "Semiempirical hybrid density functional with perturbative second-order correlation," *J. Chem. Phys.*, 2006, **124**, 034108, DOI: 10.1063/1.2148954.

[Grirrane08] Grirrane, A.; Resa, I.; Rodríguez, A.; Carmona, E., "Synthesis and Structural Characterization of Dizincocenes $Zn_2(\eta^5\text{-}C_5Me_5)_2$ and $Zn_2(\eta^5\text{-}C_5Me_4Et)_2$," *Coord. Chem. Rev.*, 2008, **252**, 1532-39, DOI: 10.1016/j.ccr.2008.01.014.

[Gronert06] Gronert, S.; Keefe, J. R., "Primary Semiclassical Kinetic Hydrogen Isotope Effects in Identity Carbon-to-Carbon Proton- and Hydride-Transfer Reactions, an ab Initio and DFT Computational Study," *J. Org. Chem.*, 2006, **71**, 5959-68, DOI: 10.1021/jo0606296.

[Gronert07] Gronert, S.; Keefe, J. R., "The Protenation of Allene and Some Heteroallenes, a Computational Study," *J. Org. Chem.*, 2007, **72**, 6343-52, DOI: 10.1021/jo0704107.

[Gunther96] Günther, H., *NMR Spectroscopy: Basic principles, concepts and applications in chemistry*; 2nd ed.; John Wiley & Sons, Inc.: Chichester, 1996.

[Guthrie73] Guthrie, J. P., "Hydration of carboxylic acids and esters. Evaluation of the free energy change for addition of water to acetic and formic acids and their methyl esters," *J. Am. Chem. Soc.*, 1973, **95**, 6999-7003, DOI: 10.1021/ja00802a021.

[Guzman12] Guzmán, V.; Roueff, E.; Gauss, J.; Pety, J.; Gratier, P.; Goicoechea, J. R.; Gerin, M.; Teyssier, D., "The hyperfine structure in the rotational spectrum of CF^+," *A & A*, 2012, **548**, A94, DOI: 10.1051/0004-6361/201220174.

[Guzman12a] Guzmán, V.; Pety, J.; Gratier, P.; Goicoechea, J. R.; Gerin, M.; Roueff, E.; Teyssier, D., "The IRAM-30m line survey of the Horsehead PDR. I. CF^+ as a tracer of C^+ and as a measure of the fluorine abundance," *A & A*, 2012, **543**, L1, DOI: 10.1051/0004-6361/201219449.

[Haaland03] Haaland, A.; Samdal, S.; Tverdova, N. V.; Girichev, G. V.; Giricheva, N. I.; Shlykov, S. A.; Garkusha, O. G., "The molecular structure of dicyclopentadienylzinc (zincocene) determined by gas electron diffraction and density functional theory calculations: η^5, η^5, η^3, η^3 or η^5, η^1 coordination of the ligand rings?," *J. Organomet. Chem.*, 2003, **684**, 351-58, DOI: 10.1016/S0022-328X(03)00770-8.

[Haeffner99] Hæffner, F.; Hu, C.-H.; Brinck, T.; Norin, T., "The catalytic effect of water in basic hydrolysis of methyl acetate: a theoretical study," *J. Mol. Struct. (Theochem)*, 1999, **459**, 85-93, DOI: 10.1016/S0166-1280(98)00251-6.

[Hall51] Hall, G. G., "The Molecular Orbital Theory of Chemical Valency. VIII. A Method of Calculating Ionization Potentials," *Royal Soc. Proc. A*, 1951, **205**, 541-52, DOI: 10.1098/rspa.1951.0048

[Haloui10] Haloui, A.; Arfaoui, Y., "A DFT study of the conformational behavior of para-substituted acetophenones in vacuum and in various solvents," *J. Mol. Struct. (Theochem)*, 2010, **950**, 1319, DOI: 10.1016/j.theochem.2010.03.012.

[Handy89] Handy, N. C.; Pople, J. A.; Head-Gordon, M.; Raghavachari, K.; Trucks, G. W., "Size-consistent Brueckner theory limited to double substitutions," *Chem. Phys. Lett.*, 1989, **164**, 185-92, DOI: 10.1016/0009-2614(89)85013-4.

[Hansen99] Hansen, A. E.; Bak, K. L., "Ab initio calculations of electronic circular dichroism," *Enantiomer*, 1999, **4**, 455-76.

[Harding80] Harding, L. B.; Schlegel, H. B.; Krishnan, R.; Pople, J. A., "Møller-Plesset study of the H_4CO potential energy surface," *J. Phys. Chem.*, 1980, **84**, 3394-401, DOI: 10.1021/j100462a017.

[Hariharan73] Hariharan, P. C.; Pople, J. A., "Influence of polarization functions on molecular-orbital hydrogenation energies," *Theoretical Chemistry Accounts*, 1973, **28**, 213-22, DOI: 10.1007/BF00533485.

[Hartree28] Hartree, D. R., "The wave mechanics of an atom with a non-coulomb central field," *Math. Proc. Camb. Phil. Soc.*, 1928, **24**, 89, DOI: 10.1017/S0305004100011919.

[Hay77] Hay, P. J., "Gaussian basis sets for molecular calculations: Representation of 3d orbitals in transition-metal atoms," *J. Chem. Phys.*, 1977, **66**, 4377-84, DOI: 10.1063/1.433731.

[Hay85] Hay, P. J.; Wadt, W. R., "Ab initio effective core potentials for molecular calculations: Potentials for the transition-metal atoms Sc to Hg," *J. Chem. Phys.*, 1985, **82**, 270-83, DOI: 10.1063/1.448799.

[Hay85a] Hay, P. J.; Wadt, W. R., "Ab initio effective core potentials for molecular calculations: Potentials for K to Au including the outermost core orbitals," *J. Chem. Phys.*, 1985, **82**, 299-310, DOI: 10.1063/1.448975.

[He11] He, Y.; Bo, W.; Dukor, R.; Nafie, L. A., "Determination of Absolute Configuration of Chiral Molecules Using Vibrational Optical Activity: A Review," *Appl. Spectrosc.*, 2011, **65**, 194A-212A, DOI: 10.1366/11-06321.

[Hehre72] Hehre, W. J.; Ditchfield, R.; Pople, J. A., "Self-Consistent Molecular Orbital Methods. 12. Further extensions of Gaussian-type basis sets for use in molecular-orbital studies of organic-molecules," *J. Chem. Phys.*, 1972, **56**, 2257, DOI: 10.1063/1.1677527.

[Heine94] Heine, A.; Herbst-Irmer, R.; Stalke, D.; Kühnle, W.; Zachariasse, K. A., "Structure and crystal packing of 4-aminobenzonitriles and 4-amino-3,5-dimethylbenzonitriles at various temperatures," *Acta Crystallogr. B: Struct. Snc.*, 1994, **50**, 363, DOI: 10.1107/S0108768193008523.

[Helgaker91] Helgaker, T.; Jørgensen, P., "An Electronic Hamiltonian for Origin Independent Calculations of Magnetic-Properties," *J. Chem. Phys.*, 1991, **95**, 2595-601, DOI: 10.1063/1.460912.

[Helgaker94] Helgaker, T.; Ruud, K.; Bak, K. L.; Jørgensen, P.; Olsen, J., "Vibrational Raman Optical-Activity Calculations Using London Atomic Orbitals," *Faraday Discussions*, 1994, **99**, 165-80, DOI: 10.1039/FD9949900165.

[Helgaker99] Helgaker, T.; Jaszuński, M.; Ruud, K., "Ab Initio Methods for the Calculation of NMR Shielding and Indirect Spin-Spin Coupling Constants," *Chem. Rev.*, 1999, **99**, 293-352, DOI: 10.1021/cr960017t.

[Hellmann08] Hellmann, R.; Bich, E.; Vogel, E., "Ab initio intermolecular potential energy surface and second pressure virial coefficients of methane," *J. Chem. Phys.*, 2008, **128**, 214303, DOI: 10.1063/1.2932103.

[Hellmann35] Hellmann, H., "A New Approximation Method in the Problem of Many Electrons," *J. Chem. Phys.*, 1935, **3**, 61, DOI: 10.1063/1.1749559.

[Herzberg91] Herzberg, G., *Infrared and Raman Spectra of Polyatomic Molecules*; Reprint ed.; Krieger Publishing Company: Malabar, FL, 1991; Vol. 2.

[Hirota92] Hirota, E., "Microwave and infrared spectra of free radicals and molecular ions," *Chem. Rev.*, 1992, **92**, 141-73, DOI: 10.1021/cr00009a006.

[Ho82] Ho, P.; Bamford, D. J.; Buss, R. J.; Lee, Y. T.; Moore, C. B., "Photodissociation of Formaldehyde in a Molecular Beam," *J. Chem. Phys.*, 1982, **76**, 3630-36, DOI: 10.1063/1.443400.

[Hod08] Hod, O.; Scuseria, G. E., "Half-metallic zigzag carbon nanotube dots," *ACS Nano*, 2008, **2**, 2243-49, DOI: 10.1021/nn8004069.

[Hohenberg64] Hohenberg, P.; Kohn, W., "Inhomogeneous Electron Gas," *Phys. Rev.*, 1964, **136**, B864-71, DOI: 10.1103/PhysRev.136.B864

[Holzwarth72] Holzwarth, G.; Chabay, I., "Optical activity of vibrational transitions. Coupled oscillator model," *J. Chem. Phys.*, 1972, **57**, 1632-35, DOI: 10.1063/1.1678447.

[Hopmann11] Hopmann, K. H.; Ruud, K.; Pecul, M.; Kudelski, A ; Dračínský, M.; Bouř, P., "Explicit versus Implicit Solvent Modeling of Raman Optical Activity Spectra," *J. Phys. Chem. B*, 2011, **115**, 4128-37, DOI: 10.1021/jp110662w.

[Horng95] Horng, M. L.; Gardecki, J. A.; Papazyan, A.; Maroncelli, M., "Subpicosecond Measurements of Polar Solvation Dynamics: Coumarin 153 Revisited," *J. Phys. Chem.*, 1995, **99**, 17311, DOI: 10.1021/j100048a004.

[Horowitz78] Horowitz, A.; Calvert, J. G., "The quantum efficiency of the primary processes in formaldehyde photolysis at 3130 Å and 25°C," *Int. J. Chem. Kinet.*, 1978, **10**, 713-32, DOI: 10.1002/kin.550100706.

[Hratchian04] Hratchian, H. P.; Chowdhury, S. K.; Gutiérrez-García, V. M.; Amarasinghe, K. K. D.; Heeg, M. J.; Schlegel, H. B.; Montgomery Jr., J. A., "Combined experimental and computational investigation of the mechanism of nickel-catalyzed three-component addition processes," *Organometallics*, 2004, **23** , 4636-46, DOI: 10.1021/om049471a.

[Hratchian04corr] Hratchian, H. P.; Chowdhury, S. K.; Gutiérrez-García, V. M.; Amarasinghe, K. K. D.; Heeg, M. J.; Schlegel, H. B.; Montgomery Jr., J. A., "Combined experimental and computational investigation of the mechanism of nickel-catalyzed three-component addition processes: Additions/Corrections," *Organometallics*, 2004, **23** , 5652, DOI: 10.1021/om0492799.

[Hratchian05a] Hratchian, H. P.; Schlegel, H. B., "Finding minima, transition states, and following reaction pathways on ab initio potential energy surfaces" in *Theory and Applications of Computational Chemistry: The First 40 Years*; ed. Dykstra, C. E., Frenking, G., Kim, K. S., Scuseria, G., Elsevier: Amsterdam, 2005, pp. 195-249

[Hratchian05b] Hratchian, H. P.; Schlegel, H. B., "Using Hessian updating to increase the efficiency of a Hessian based predictor-corrector reaction path following method," *J. Chem. Theory and Comput.*, 2005, **1**, 61-69, DOI: 10.1021/ct0499783.

[Hratchian12] Hratchian, H. P.; Li, X., "Thirty years of geometry optimization in quantum chemistry and beyond: A tribute to Berny Schlegel," *J. Chem. Theory and Comput.*, 2012, **8**, 4853-55, DOI: 10.1021/ct300950r.

[Hughes00] Hughes, R. A.; Brown, J. M.; Evenson, K. M., "Rotational Spectrum of the AsH_2 Radical in Its Ground State, Studied by Far-Infrared Laser Magnetic Resonance " *J. Mol. Spectrosc.*, 2000, **200**, 210-28, DOI: 10.1006/jmsp.1999.8037.

[Hunter98] Hunter, E. P.; Lias, S. G., "Evaluated Gas Phase Basicities and Proton Affinities of Molecules: An Update," *J. Phys. Chem. Ref. Data*, 1998, **27**, 413, DOI: 10.1063/1.556018.

[Huron72] Huron, M. J.; Claverie, P., "Calculation of the interaction energy of one molecule with its whole surrounding. I. Method and application to pure nonpolar compounds," *The Journal of Physical Chemistry*, 1972, **76**, 2123-33, DOI: 10.1021/j100659a011.

[IgelMann88] Igel-Mann, G.; Stoll, H.; Preuss, H., "Pseudopotentials for main group elements (IIIA through VIIA)," *Molecular Physics*, 1988, **65**, 1321-28, DOI: 10.1080/00268978800101811.

[Iikura01] Iikura, H.; Tsuneda, T.; Yanai, T.; Hirao, K., "Long-range correction scheme for generalized-gradient-approximation exchange functionals," *J. Chem. Phys.*, 2001, **115**, 3540-44, DOI: 10.1063/1.1383587.

[Improta06] Improta, R.; Barone, V.; Scalmani, G.; Frisch, M. J., "A state-specific polarizable continuum model time dependent density functional method for excited state calculations in solution," *J. Chem. Phys.*, 2006, **125**, 054103: 1-9, DOI: 10.1063/1.2222364.

[Improta07] Improta, R.; Scalmani, G.; Frisch, M. J.; Barone, V., "Toward effective and reliable fluorescence energies in solution by a new State Specific Polarizable Continuum Model Time Dependent Density Functional Theory Approach," *J. Chem. Phys.*, 2007, **127**, 074504: 1-9, DOI: 10.1063/1.2757168.

[Ingold53] Ingold, C. K., *Structure and mechanism in organic chemistry*; Cornell Univ. Press: Ithaca, NY, 1953.

[Isaacs87] Isaacs, N. S., *Physical Organic Chemistry*; 1st ed.; Longman Scientific and Technical: Essex, UK 1987.

[Ishikawa07] Ishikawa, Y.-i.; Kawakami, K., "Structure and Infrared Spectroscopy of Group 6 Transition-Metal Carbonyls in the Gas Phase: DFT Studies on M(CO)n (M = Cr, Mo, and W; n = 6, 5, 4, and 3)," *J. Phys. Chem. A*, 2007, **111**, 9940-44, DOI: 10.1021/jp071509k.

[Jacquemin07] Jacquemin, D.; Perpète, E. A.; Scalmani, G.; Frisch, M. J.; Kobayashi, R.; Adamo, C., "Assessment of the Efficiency of Long-Range Corrected Functionals for Some Properties of Large Compounds," *J. Chem. Phys.*, 2007, **126**, 144105: 1-12, DOI: 10.1063/1.2715573.

[Jagau12] Jagau, T.-C.; Gauss, J., "Ground and excited state geometries via Mukherjee's multireference coupled-cluster method," *J. Chem. Phys.*, 2012, **401**, 73, DOI: 10.1016/j.chemphys.2011.10.016.

[Jansen69] Jansen, H. B.; Ros, P., "Non-empirical molecular orbital calculations on the protonation of carbon monoxide," *Chem. Phys. Lett.*, 1969, **3**, 140, DOI: 10.1016/0009-2614(69)80118-1.

[Jensen07] Jensen, F., *Introduction to Computational Chemistry*; 2nd ed.; Wiley & Sons: West Sussex, England, 2007, pp. 620.

[Jensen69] Jensen, F. R.; Bushweller, C. H., "Separation of Conformers. II. Axial and Equatorial Isomers of Chlorocyclohexane and Trideuteriomethoxycyclohexane," *J. Am. Chem. Soc.*, 1969, **91**, 3223-25, DOI: 10.1021/ja01040a022.

[Johnson04] Johnson, E. R.; Wolkow, R. A.; DiLabio, G. A., "Application of 25 density functionals to dispersion-bound homomolecular dimers," *Chem. Phys. Lett.*, 2004, **394**, 334-38, DOI: 10.1016/j.cplett.2004.07.029.

[Jorgensen62] Jørgensen, C. K., *Absorption Spectra and Chemical Bonding in Complexes*; Pergamon Press: Oxford, UK, 1962, pp. 352.

[Jorgensen93] Jorgensen, W. L.; Lim, D.; Blake, J. F., "Ab Initio Study of Diels-Alder Reactions of Cyclopentadiene with Ethylene, Isoprene, Cyclopentadiene, Acrylonitrile, and Methyl Vinyl Ketone," *J. Am. Chem. Soc.*, 1993, **115**, 2936-42, DOI: 10.1021/ja00060a048.

[Jost75] Jost, A.; Rees, B.; Yelon, W. B., "Electronic Structure of Chromium Hexacarbonyl at 78 K. 1. Neutron Diffraction Study," *Acta Crystallogr.*, 1975, **B31**, 2649-58, DOI: 10.1107/S0567740875008394.

[Kajimoto91] Kajimoto, O.; Yokohama, H.; Ooshima, Y.; Endo, Y., "The structure of 4-(N,N-dimethylamino)benzonitrile and its van der Waals complexes," *Chem. Phys. Lett.*, 1991, **179**, 455, DOI: 10.1016/0009-2614(91)87085-P.

[Kallies01] Kallies, B.; Meier, R., "Electronic Structure of 3d $[M(H_2O)_6]^{3+}$ Ions from Sc^{III} to Fe^{III}: A Quantum Mechanical Study Based on DFT Computations and Natural Bond Orbital Analyses," *Inorg. Chem.*, 2001, **40**, 3101, DOI: 10.1021/ic001258t.

[Kang06] Kang, L.; Novick, S. E., "The microwave spectrum of the 1,1-difluoroprop-2-ynyl radical," *J. Chem. Phys.*, 2006, **125**, 054309, DOI: 10.1063/1.2215599.

[Karna91] Karna, S. P.; Dupuis, M., "Frequency-Dependent Nonlinear Optical-Properties of Molecules: Formulation and Implementation in the HONDO Program," *J. Comp. Chem.*, 1991, **12**, 487-504, DOI: 10.1002/jcc.540120409.

[Karton06] Karton, A.; Rabinovich, E.; Martin, J. M. L.; Ruscic, B., "W4 theory for computational thermochemistry: In pursuit of confident sub-kJ/mol predictions," *J. Chem. Phys.*, 2006, **125**, 144108, DOI: 10.1063/1.2348881.

[Kaupp95] Kaupp, M.; Malkin, V.; Malkina, O.; Salahub, D. R., "Scalar Relativistic Effects on ^{17}O NMR Chemical Shifts in Transition-Metal Oxo Complexes. An ab Initio ECP/DFT Study," *J. Am. Chem. Soc.*, 1995, **117**, 1851-2, DOI: 10.1021/ja00111a032.

[Kaupp95a] Kaupp, M.; Malkin, V.; Malkina, O.; Salahub, D. R., "Erratum: Transition-Metal Oxo Complexes. An ab Initio ECP/DFT Study.," *J. Am. Chem. Soc.*, 1995, **117**, 8492, DOI: 10.1021/ja00137a036.

[Kealy51] Kealy, T. J.; Pauson, P. L., "A New Type of Organo-Iron Compound," *Nature*, 1951, **168**, 1039, DOI: 10.1038/1681039b0.

[Keegstra96] Keegstra, M. A.; De Feyter, S.; De Schryver, F. C.; Müllen, K., "Hexaterphenylyl- und Hexaquaterphenylylbenzol: das Verhalten von Chromophoren und Elektrophoren auf engem Raum," *Angew. Chem.*, 1996, **108**, 830, DOI: 10.1002/ange.19961080721 (original in German); 10.1002/anie.199607741 (in English).

[Kendall92] Kendall, R. A.; Dunning Jr., T. H.; Harrison, R. J., "Electron affinities of the first-row atoms revisited. Systematic basis sets and wave functions," *J. Chem. Phys.*, 1992, **96**, 6796-806, DOI: 10.1063/1.462569.

[Kim12] Kim, D.; Sambasivan, S.; Nam, H.; Kim, K. H.; Kim, J. Y.; Joo, T.; Lee, K.-H.; Kim, K.-T.; Ahn, K. H., "Reaction-based two-photon probes for in vitro analysis and cellular imaging of monoamine oxidase activity," *Chem. Commun.*, 2012, **48**, 6833, DOI: 10.1039/C2CC32424E

[Kirkwood34] Kirkwood, J. G., "Theory of Solutions of Molecules Containing Widely Separated Charges with Special Application to Zwitterions," *J. Chem. Phys.*, 1934, **2**, 351, DOI: 10.1063/1.1749489.

[Klene00] Klene, M.; Robb, M. A.; Frisch, M. J.; Celani, P., "Parallel implementation of the CI-vector evaluation in full CI/CAS-SCF," *J. Chem. Phys.*, 2000, **113**, 5653-65, DOI: 10.1063/1.1290014.

[Klene03] Klene, M.; Robb, M. A.; Blancafort, L.; Frisch, M. J., "A New Efficient Approach to the Direct restricted active space self-consistent field Method," *J. Chem. Phys.*, 2003, **119**, 713-28, DOI: 10.1063/1.1578620

[Knight95] Lon B. Knight, J.; Jones, G. C.; King, G. M.; Babb, R. M.; McKinley, A. J., "Electron spin resonance and theoretical studies of the PO_2 and AsO_2 radicals in neon matrices at 4 K: Laser vaporization and x-irradiation radical generation techniques," *J. Chem. Phys.*, 1995, **103**, 493, DOI: 10.1063/1.470135.

[Knowles84] Knowles, P. J.; Handy, N., "A new determinant-based full configuration interaction method," *Chem. Phys. Lett.*, 1984, **111**, 315-21, DOI: 10.1016/0009-2614(84)85513-X.

[Kobayashi91] Kobayashi, R.; Handy, N. C.; Amos, R. D.; Trucks, G. W.; Frisch, M. J.; Pople, J. A., "Gradient theory applied to the Brueckner doubles method," *J. Chem. Phys.*, 1991, **95**, 6723-33, DOI: 10.1063/1.461544.

[Kohn65] Kohn, W.; Sham, L. J., "Self-Consistent Equations Including Exchange and Correlation Effects," *Phys. Rev.*, 1965, **140**, A1133-38, DOI: 10.1103/PhysRev.140.A1133

[Kolar07] Kolář, M.; Hobza, P., "Accurate Theoretical Determination of the Structure of Aromatic Complexes Is Complicated: The Phenol Dimer and Phenol···Methanol Cases," *The Journal of Physical Chemistry*, 2007, **111**, 5851-54, DOI: 10.1021/jp071486.

[Kondru98] Kondru, R. K.; Wipf, P.; Beratan, D. N., "Theory-assisted determination of absolute stereochemistry for complex natural products via computation of molar rotation angles," *J. Am. Chem. Soc.*, 1998, **120**, 2204-05, DOI: 10.1021/ja973690o.

[Krissinel07] Krissinel, E.; Henrick, K., "Inference of Macromolecular Assemblies from Crystalline State," *J. Mol. Biol.*, 2007, **372**, 774-97, DOI: 10.1016/j.jmb.2007.05.022.

[Krivdin03] Krivdin, L. B.; Kuznetsova, T. A., "Spin-spin coupling constants ^{13}C - ^{13}C in structural studies. Part 34 - Nonempirical calculations: Small heterocycles," *Russ. J. Org. Chem.*, 2003, **39**, 1618-28, DOI: 10.1023/B:RUJO.0000013137.36900.82.

[Krowczynski83] Krówczyński, A.; Kozerski, L., "A General Approach to Aliphatic 2-Nitroenamines," *Synthesis*, 1983, **6**, 489-91, DOI: 10.1055/s-1983-30397.

[Kwiatkowski92] Kwiatkowski, J. S.; Leszczynski, J., "Ab initio post-Hartree-Fock studies on molecular structure and vibrational IR spectrum of formaldehyde," *Int. J. Quantum Chem., Quantum Chem. Symp.*, 1992, **44**, 421, DOI: 10.1002/qua.560440837.

[Landolt76] *Landolt-Börnstein: Numerical Data and Functional Relationships in Science and Technology, Group II*; Springer Verlag: Berlin, 1979; Vol. 7.

[Lassettre68] Lassettre, E. N.; Skerbele, A.; Dillon, M. A.; Ross, K. J., "High-Resolution Study of Electron-Impact Spectra at Kinetic Energies between 33 and 100 eV and Scattering Angles to 16°," *J. Chem. Phys.*, 1968, **48**, 5066, DOI: 10.1063/1.1668178.

[Lee84] Lee, Y. S.; Kucharski, S. A.; Bartlett, R. J., "Coupled cluster approach with triple excitations," *J. Chem. Phys.*, 1984, **81**, 5906-12.

[Lee88] Lee, C.; Yang, W.; Parr, R. G., "Development of the Colle-Salvetti correlation-energy formula into a functional of the electron density," *Phys. Rev. B*, 1988, **37**, 785-89, DOI: 10.1103/PhysRevB.37.785

[Levine13] Levine, I. N., *Quantum Chemistry*; 7th ed.; Prentice Hall: Englewood Cliffs, NJ, 2013.

[Lewars11] "The Concept of the Potential Energy Surface" in Lewars, E. G., *Computational Chemistry: Introduction to the Theory and Applications of Molecular and Quantum Mechanics*, 2nd ed., Kluwer Acad. Pub.: Boston, 2011, pp. 9-41.

[Li05] Li, H.; Robertson, A.; Jensen, J., "Very Fast Empirical Prediction and Interpretation of Protein pKa Values," *Proteins*, 2005, **61**, 704-21, DOI: 10.1002/prot.20660.

[Li06] Li, X.; Frisch, M. J., "Energy-represented DIIS within a hybrid geometry optimization method," *J. Chem. Theory and Comput.*, 2006, **2**, 835-39, DOI: 10.1021/ct050275a.

[Li08] Li, X.; Josef, P., "Electronic structure of organic diradicals. Evaluation of the performance of coupled-cluster methods," *J. Chem. Phys.*, 2008, **129**, 174101, DOI: 10.1063/1.2999560.

[Lias88] Lias, S. G.; Bartmess, J. E.; Liebman, J. F.; Holmes, J. L.; Levin, R. D.; Mallard, W. G., "Gas-phase Ion and Neutral Thermochemistry," *J. Phys. Chem. Ref. Data*, 1988, **17 (S1)**, 861.

[Limacher09] Limacher, P. A.; Mikkelsen, K. V.; Lüthi, H.-P., "On the accurate calculation of polarizabilities and second hyperpolarizabilities of polyacetylene oligomer chains using the CAM-B3LYP density functional," *J. Chem. Phys.*, 2009, **130**, 194114, DOI: 10.1063/1.3139023.

[Lin09] Lin, W.; Ganguly, A.; Minei, A. J.; Lindeke, G. L.; Pringle, W. C.; Novick, S. E.; Durig, J. R., "Microwave spectra and structural parameters of equatorial-trans cyclobutanol," *J. Mol. Struct.*, 2009, **922**, 83-7, DOI: 10.1016/j.molstruc.2009.01.040.

[Linke67] Linke, S.; Tisue, G. T.; Lwowski, W., "Curtius and Lossen Rearrangements. II. Pivaloyl Azide," *J. Am. Chem. Soc.*, 1967, **89**, 6308, DOI: 10.1021/ja01000a057.

[Lipinski80] Lipiński, J.; Chojnacki, H.; Grabowski, Z. R.; Rotkiewicz, K., "Theoretical model for the double fluorescence of p-cyano-N,N-dimethylaniline in polar solvents," *Chem. Phys. Lett.*, 1980, **70**, 449, DOI: 10.1016/0009-2614(80)80102-3.

[Lipparini11a] Lipparini, F.; Scalmani, G.; Mennucci, B.; Frisch, M. J., "Self-consistent field and polarizable continuum model: A new strategy of solution for the coupled equations," *J. Chem. Theory and Comput.*, 2011, **7**, 610-17, DOI: 10.1021/ct1005906.

[Little89] Little, T. S.; Qiu, J.; Durig, J. R., "Asymmetric torsional potential function and conformational analysis of furfural by far infrared and Raman spectroscopy," *Spectrochim. Acta, Part A*, 1989, **45**, 789-94, DOI: 10.1016/0584-8539(89)80215-6.

[Liu73] Liu, B.; McLean, A. D., "Accurate calculation of the attractive interaction of two ground state helium atoms," *J. Chem. Phys.*, 1973, **59**, 4557-58, DOI: 10.1063/1.1680654.

[Lohr83] Lohr Jr., L. L.; Hanamura, M.; Morokuma, K., "The 1,2 hydrogen shift as an accompaniment to ring closure and opening: ab initio MO study of thermal rearrangements on the C_2H_3N potential energy hypersurface," *J. Am. Chem. Soc.*, 1983, **105**, 5541-47, DOI: 10.1021/ja00355a003.

[London37] London, F., "The quantic theory of inter-atomic currents in aromatic combinations," *Journal de Physique et le Radium*, 1937, **8**, 397-409, DOI: 10.1051/jphysrad:01937008010039700.

[Ludlow05] Ludlow, M. K.; Foresman, J. B., "Computational analysis of zincocene, decamethylzincocene, and decamethyldizincocene," presented at the 229th National Meeting of the ACS, March 13-17, 2005, San Diego, CA, Session: 910-CHED.

[Mahapatra98] Mahapatra, U. S.; Datta, B.; Mukherjee, D., "A state-specific multi-reference coupled cluster formalism with molecular applications," *Molecular Physics*, 1998, **94**, 157, DOI: 10.1080/002689798168448.

[Marcelin1915] Marcelin, R., "Physicochemical kinetics," *Ann. Phys*, 1915, **3**, 120-84.

[Marenich09] Marenich, A. V.; Cramer, C. J.; Truhlar, D. G., "Universal Solvation Model based on Solute Electron Density and on a Continuum Model of the Solvent Defined by the Bulk Dielectric Constant and Atomic Surface Tensions," *J. Phys. Chem. B*, 2009, **113**, 6378-96, DOI: 10.1021/jp810292n.

[Marenich09a] "Minnesota Solvation Database - version 2009," 2009, *comp.chem.umn.edu/mnsol/*.

[Maris02] Maris, A.; Melandri, S.; Caminati, W.; Favero, P. G., "The proton donor/acceptor double role of the peptidic group: Free jet rotational spectrum and computational study of lactamide," *Chem. Phys.*, 2002, **283**, 111, DOI: 10.1016/S0301-0104(02)00499-8.

[Martin03] Martin, R. L., "Natural Transition Orbitals," *J. Chem. Phys.*, 2003, **118**, 4775, DOI: 10.1063/1.1558471.

[Martin99] Martin, J. M. L.; de Oliveira, G., "Towards standard methods for benchmark quality ab initio thermochemistry - W1 and W2 theory," *J. Chem. Phys.*, 1999, **111**, 1843-56, DOI: 10.1063/1.479454.

[Matthews76] Matthews, G. P.; Smith, E. B., "An intermolecular pair potential energy function for methane," *Molecular Physics*, 1976, **32**, 1719-29, DOI: 10.1080/00268977600103031.

[McGrath91] McGrath, M. P.; Radom, L., "Extension of Gaussian-1 (G1) theory to bromine-containing molecules," *J. Chem. Phys.*, 1991, **94**, 511-16, DOI: 10.1063/1.460367.

[McIntyre71] McIntyre, N. S.; Thompson, K. R.; Weltner Jr., W., "Spectroscopy of titanium oxide and titanium dioxide molecules in inert matrices at 4° K," *J. Phys. Chem.*, 1971, **75**, 3243, DOI: 10.1021/j100690a008.

[McLean80] McLean, A. D.; Chandler, G. S., "Contracted Gaussian-basis sets for molecular calculations. 1. 2nd row atoms, Z=11-18," *J. Chem. Phys.*, 1980, **72**, 5639-48, DOI: 10.1063/1.438980

[McNeill77] McNeill, E. A.; Scholer, F. R., "Molecular structure of the gaseous metal carbonyl hydrides of manganese, iron and cobalt," *J. Am. Chem. Soc.*, 1977, **99**, 6243-49, DOI: 10.1021/ja00461a011.

[McWeeny62] McWeeny, R., "Perturbation Theory for Fock-Dirac Density Matrix," *Phys. Rev.*, 1962, **126**, 1028, DOI: 10.1103/PhysRev.126.1028

[McWeeny68] McWeeny, R.; Dierksen, G., "Self-consistent perturbation theory. 2. Extension to open shells," *J. Chem. Phys.*, 1968, **49**, 4852, DOI: 10.1063/1.1669970.

[Mennucci00] Mennucci, B.; Toniolo, A.; Tomasi, J., "Ab Initio Study of the Electronic Excited States in 4-(N,N-Dimethylamino)benzonitrile with Inclusion of Solvent Effects: The Internal Charge Transfer Process," *J. Am. Chem. Soc.*, 2000, **122**, 10621-30, DOI: 10.1021/ja000814f.

[Mennucci02] Mennucci, B.; Tomasi, J.; Cammi, R.; Cheeseman, J. R.; Frisch, M. J.; Devlin, F. J.; Gabriel, S.; Stephens, P. J., "Polarizable continuum model (PCM) calculations of solvent effects on optical rotations of chiral molecules," *J. Phys. Chem. A*, 2002, **106**, 6102-13, DOI: 10.1021/jp020124t.

[Mennucci11] Mennucci, B.; Cappelli, C.; Cammi, R.; Tomasi, J., "Modeling solvent effects on chiroptical properties," *Chirality*, 2011, **23**, 717-29, DOI: 10.1002/chir.20984.

[Menshutkin1890] Menshutkin, N., "Beiträgen zur Kenntnis der Affinitätskoeffizienten der Alkylhaloide und der Organischen Amine," *Z. Physik. Chem.*, 1890, **5**, 589-600.

[Menshutkin1890a] Menshutkin, N., "Über die Affinitätskoeffizienten der Alkylhaloide und der Amine," *Z. Physik. Chem.*, 1890, **6**, 41-57.

[Miertus81] Miertuš, S.; Scrocco, E.; Tomasi, J., "Electrostatic Interaction of a Solute with a Continuum. A Direct Utilization of ab initio Molecular Potentials for the Prevision of Solvent Effects," *Chem. Phys.*, 1981, **55**, 117-29, DOI: 10.1016/0301-0104(81)85090-2.

[Miller52] Miller, S. A.; Tebboth, J. A.; Tremaine, J. F., "Dicyclopentadienyliron," *J. Chem. Soc. (Resumed)*, 1952, **0**, 632-35, DOI: 10.1039/JR9520000632

[Miller77] Miller, C. K.; Ward, J. F., "Measurements of nonlinear optical polarizabilities for some halogenated methanes: The role of bond-bond interactions," *Phys. Rev. A*, 1977, **16**, 1179, DOI: 10.1103/PhysRevA.16.1179

[Miller80] Miller, W. H.; Handy, N. C.; Adams, J. E., "Reaction-path Hamiltonian for polyatomic-molecules," *J. Chem. Phys.*, 1980, **72**, 99-112, DOI: 10.1063/1.438959.

[Miller90] Miller, W. H.; Hernandez, R.; Handy, N. C.; Jayatilaka, D.; Willets, A., "Ab initio calculation of anharmonic constants for a transition-state, with application to semiclassical transition-state tunneling probabilities," *Chem. Phys. Lett.*, 1990, **172**, 62-68, DOI: 10.1016/0009-2614(90)87217-F.

[Mills30] Mills, W. H.; Nixon, I. G., "CCCXXXII -- Stereochemical influences on aromatic substitution. Substitution derivatives of 5-hydroxyhydrindene," *J. Chem. Soc.*, 1930, **0**, 2510, DOI: 10.1039/JR9300002510

[Moller34] Møller, C.; Plesset, M. S., "Note on an approximation treatment for many-electron systems," *Phys. Rev.*, 1934, **46**, 0618-22, DOI: 10.1103/PhysRev.46.618

[Montgomery00] Montgomery Jr., J. A.; Frisch, M. J.; Ochterski, J. W.; Petersson, G. A., "A complete basis set model chemistry. VII. Use of the minimum population localization method," *J. Chem. Phys.*, 2000, **112**, 6532-42, DOI: 10.1063/1.481224.

[Montgomery99] Montgomery Jr., J. A.; Frisch, M. J.; Ochterski, J. W.; Petersson, G. A., "A complete basis set model chemistry. VI. Use of density functional geometries and frequencies," *J. Chem. Phys.*, 1999, **110**, 2822-27, DOI: 10.1063/1.477924.

[Moretti13] Moretti, V., *Spectral Theory and Quantum Mechanics*; Springer-Verlag: Milan, 2013, pp. 735.

[Mulliken55] Mulliken, R. S., "Electronic Population Analysis on LCAO-MO Molecular Wave Functions I," *The Journal of Chemical Physics*, 1955, **23**, 1833-40, DOI: 10.1063/1.1740588.

[Nafie11] Nafie, L. A., *Vibrational Optical Activity: Principles and Applications*; John Wiley & Sons, Ltd., 2011.

[Nafie83] Nafie, L. A., "Adiabatic molecular properties beyond the Born-Oppenheimer approximation: Complete adiabatic wave functions and vibrationally induced electronic current density," *J. Chem. Phys.*, 1983, **79**, 4950-57, DOI: 10.1063/1.445588.

[Nafie83a] Nafie, L. A.; Freedman, T. B., "Vibronic coupling theory of infrared vibrational transitions," *J. Chem. Phys.*, 1983, **78**, 7108-16, DOI: 10.1063/1.444741.

[Nafie92] Nafie, L. A., "Velocity-gauge formalism in the theory of vibrational circular dichroism and infrared absorption," *J. Chem. Phys.*, 1992, **96**, 5687-702, DOI: 10.1063/1.462668.

[Nafie94] Nafie, L. A.; Che, D., "Theory and Measurement of Raman Optical Activity" in *Modern Nonlinear Optics, Part 3*; ed. Evans, M., Kielich, S., *Advances in Chemical Physics*, vol. 85, Wiley: New York, 1994, pp. 105-49

[Nakashima80] Nakashima, N.; Sumitani, M.; Ohmine, I.; Yoshihara, K., "Nanosecond laser photolysis of the benzene monomer and eximer," *J. Chem. Phys.*, 1980, **72**, 2226, DOI: 10.1063/1.439465.

[Nakashima80a] Nakashima, N.; Inoue, H.; Sumitani, M.; Yoshihara, K., "Laser flash photolysis of benzene. III. S_n <-- S_1 absorption of gaseous benzene," *J. Chem. Phys.*, 1980, **73**, 5976, DOI: 10.1063/1.440131.

[Nakatsuji78] Nakatsuji, H.; Hirao, K., "Cluster expansion of the wavefunction: Symmetry-adapted-cluster expansion, its variational determination, and extension of open-shell orbital theory," *J. Chem. Phys.*, 1978, **68**, 2053-65, DOI: 10.1063/1.436028.

[Nakatsuji79] Nakatsuji, H., "Cluster expansion of the wavefunction: Calculation of electron correlations in ground and excited states by SAC and SAC-CI theories," *Chem. Phys. Lett.*, 1979, **67**, 334-42, DOI: 10.1016/0009-2614(79)85173-8.

[Nakatsuji79a] Nakatsuji, H., "Cluster expansion of the wavefunction: Electron correlations in ground and excited states by SAC (Symmetry-Adapted-Cluster) and SAC-CI theories," *Chem. Phys. Lett.*, 1979, **67**, 329-33, DOI: 10.1016/0009-2614(79)85172-6.

[Nakatsuji97] Nakatsuji, H., "SAC-CI Method: Theoretical aspects and some recent topics" in *Computational Chemistry: Reviews of Current Trends*; ed. Leszczynski, J., vol. 2, World Scientific: Singapore, 1997, pp. 62-124, DOI: 10.1142/9789812812148_0002.

[Nash68] Nash, L. K., *Elements of Statistical Thermodynamics*; Addison-Wesley: Reading, MA, 1968.

[NBO3.1] Glendening, E. D.; Reed, A. E.; Carpenter, J. E.; Weinhold, F., *NBO Version 3.1* Theoretical Chemistry Institute, Univ. WI, Madison, c.1988.

[NietoOrtega11] Nieto-Ortega, B.; Casado, J.; Blanch, E. W.; López Navarrete, J. T.; Quesada, A. R.; Ramírez, F. J., "Raman Optical Activity Spectra and Conformational Elucidation of Chiral Drugs. The Case of the Antiangiogenic Aeroplysinin-1," *J. Phys. Chem. A*, 2011, **115**, 2752-55, DOI: 10.1021/jp2009397.

[NIST08] *NIST Chemistry WebBook*; Linstrom, P. J.; Mallard, W. G., Eds.; National Institute of Standards and Technology: Gaithersburg, MD, 2008.

[Nobes87] Nobes, R. H.; Pople, J. A.; Radom, L.; Handy, N. C.; Knowles, P. J., "Slow Convergence of the Møller-Plesset Perturbation Series: The Dissociation-Energy of Hydrogen Cyanide and the Electron-Affinity of the Cyano Radical," *Chem. Phys. Lett.*, 1987, **138**, 481-85, DOI: 10.1016/0009-2614(87)80545-6.

[Noonan05] Noonan, K. Y.; Beshire, M.; Darnell, J.; Frederick, K. A., "Qualitative and Quantitative Analysis of Illicit Drug Mixtures on Paper Currency Using Raman Microspectroscopy," *Appl. Spectrosc.*, 2005, **59**, 1493, DOI: 10.1366/000370205775142610.

[Nyden81] Nyden, M. R.; Petersson, G. A., "Complete basis set correlation energies. I. The asymptotic convergence of pair natural orbital expansions," *J. Chem. Phys.*, 1981, **75**, 1843-62, DOI: 10.1063/1.442208.

[OBoyle11] O'Boyle, N. M.; Banck, M.; James, C. A.; Morley, C.; Vandermeersch, T.; Hutchison, G. R., "Open Babel: An open chemical toolbox," *J. Cheminform.*, 2011, **3**, 7 Oct. 2011, DOI: 10.1186/1758-2946-3-33.

[Ochterski00] Ochterski, J. W., "Thermochemistry in Gaussian," 2000.

[Ochterski93] Ochterski, J. W., *Complete Basis Set Model Chemistries* [PhD Thesis], Wesleyan University (Middletown, CT), 1993.

[Ochterski95] Ochterski, J. W.; Petersson, G. A.; Wiberg, K. B., "A Comparison of Model Chemistries," *J. Am. Chem. Soc.*, 1995, **117**, 11299-308, DOI: 10.1021/ja00150a030.

[Ochterski96] Ochterski, J. W.; Petersson, G. A.; Montgomery Jr., J. A., "A complete basis set model chemistry. V. Extensions to six or more heavy atoms," *J. Chem. Phys.*, 1996, **104**, 2598-619, DOI: 10.1063/1.470985.

[Odell70] O'Dell Jr., M. S.; Darwent, B. d. B., "Thermal decomposition of methyl azide," *Can. J. Chem.*, 1970, **48**, 1140, DOI: 10.1139/v70-187.

[Okamoto67] Okamoto, K.; Fukui, S.; Shingu, H., "Kinetic Studies of Bimolecular Nucleophilic Substitution. 6. Rates of Menschtkin Reaction of Methyl Iodide with Methylamines and Ammonia in Aqueous Solutions," *Bull. Chem. Soc. Jpn.*, 1967, **40**, 1920, DOI: 10.1246/bcsj.40.1920.

[Olivucci93] Olivucci, M.; Ragazos, I. N.; Bernardi, F.; Robb, M. A., "Conical intersection mechanism for the photochemistry of butadiene: an MC-SCF study," *J. Am. Chem. Soc.*, 1993, **115**, 3710-21, DOI: 10.1021/ja00062a042.

[Olsen85] Olsen, J.; Jørgensen, P., "Linear and Nonlinear Response Functions for an Exact State and for an MCSCF State," *J. Chem. Phys.*, 1985, **82**, 3235-64, DOI: 10.1063/1.448223.

[Olsen88] Olsen, J.; Roos, B. O.; Jørgensen, P.; Jensen, H. J. A., "Determinant Based Configuration-Interaction Algorithms for Complete and Restricted Configuration-Interaction Spaces," *J. Chem. Phys.*, 1988, **89**, 2185-92, DOI: 10.1063/1.455063.

[Olsen95] Olsen, J.; Bak, K. L.; Ruud, K.; Helgaker, T.; Jørgensen, P., "Orbital Connections for Perturbation-Dependent Basis-Sets," *Theoretical Chemistry Accounts*, 1995, **90**, 421-39, DOI: 10.1007/BF01113545.

[Olsson11] Olsson, M. H. M.; Søndergaard, C. R.; Rostkowski, M.; Jensen, J. H., "PROPKA3: Consistent Treatment of Internal and Surface Residues in Empirical pKa predictions," *J. Chem. Theory and Comput.*, 2011, **7**, 525-37, DOI: 10.1021/ct100578z.

[Onsager36] Onsager, L., "Electric Moments of Molecules in Liquids," *J. Am. Chem. Soc.*, 1936, **58**, 1486-93, DOI: 10.1021/ja01299a050.

[Ophardt84] Ophardt, C. E., "Synthesis and spectra of vanadium complexes," *J. Chem. Educ.*, 1984, **61**, 1102, DOI: 10.1021/ed061p1102.

[ORegan91] O'Regan, B.; Grätzel, M., "A low-cost, high-efficiency solar cell based on dye-sensitized colloidal TiO_2 films," *Nature*, 1991, **353**, 737-40, DOI: 10.1038/353737a0.

[Orendt85] Orendt, A. M.; Facelli, J. C.; Grant, D. M.; Michl, J.; Walker, F. H.; Dailey, W. P.; Waddell, S. T.; Wiberg, K. B.; Schindler, M.; Kutzelnigg, W., "Low temperature ^{13}C NMR magnetic resonance in solids. 4. Cyclopropane, bicyclo[1.1.0]butane, and [1.1.1]propellane," *Theor. Chim. Acta*, 1985, **68**, 421, DOI: 10.1007/BF00527667.

[Oyler96] Oyler, J.; Darwin, W. D.; Cone, E. J., "Cocaine Contamination of United States Paper Currency," *J. Anal. Toxicol.*, 1996, **20**, 213-16, DOI: 10.1093/jat/20.4.213.

[Page88] Page, M.; McIver Jr., J. W., "On evaluating the reaction path Hamiltonian," *J. Chem. Phys.*, 1988, **88**, 922-35, DOI: 10.1063/1.454172.

[Page90] Page, M.; Doubleday Jr., C.; McIver Jr., J. W., "Following steepest descent reaction paths: the use of higher energy derivatives with ab initio electronic-structure methods," *J. Chem. Phys.*, 1990, **93**, 5634-42, DOI: 10.1063/1.459634.

[Palafox89] Palafox, M. A., "Raman spectra and vibrational analysis for benzocaine," *J. Raman Spectrosc.*, 1989, **20**, 765-71, DOI: 10.1002/jrs.1250201203.

[Palafox93] Palafox, M. A., "Infrared and Raman Study of Benzocaine Hydrochloride," *Spectroscopy Lett.*, 1993, **26**, 1395-415, DOI: 10.1080/00387019308011618

[Palmer93] Palmer, I. J.; Ragazos, I. N.; Bernardi, F.; Olivucci, M.; Robb, M. A., "An MC-SCF study of the S_1 and S_2 photochemical reactions of benzene," *J. Am. Chem. Soc.*, 1993, **115**, 673-82, DOI: 10.1021/ja00055a042.

[Papousek82] Papoušek, D.; Aliev, M. R., *Molecular Vibrational-Rotational Spectra, Studies in Physical and Theoretical Chemistry*, vol. 17, ed. Durig, J. R., Elsevier: New York, 1982

[Pappalardo93] Pappalardo, R. R.; Marcos, E. S.; Ruiz-López, M. F.; Rinaldi, D.; Rivail, J.-L., "Solvent effects on molecular geometries and isomerization processes: A study of push-pull ethylenes in solution," *J. Am. Chem. Soc.*, 1993, **115**, 3722-30, DOI: 10.1021/ja00062a043.

[Parmenter63] Parmenter, C. S.; Noyes, W. A., "Energy Dissipation from Excited Acetaldehyde Molecules," *J. Am. Chem. Soc.*, 1963, **85**, 416, DOI: 10.1021/ja00887a010.

[Parr89] Parr, R. G.; Yang, W., *Density-functional theory of atoms and molecules*; Oxford Univ. Press: Oxford, 1989, pp. 352.

[Parthiban01] Parthiban, S.; Martin, J. M. L., "Assessment of W1 and W2 theories for the computation of electron affinities, ionization potentials, heats of formation, and proton affinities," *J. Chem. Phys.*, 2001, **114**, 6014-29, DOI: 10.1063/1.1356014.

[Parusel98] Parusel, A. B. J.; Köhler, G.; Grimme, S., "Density Functional Study of Excited Charge Transfer State Formation in 4-(N,N-Dimethylamino)benzonitrile," *J. Phys. Chem. A*, 1998, **102**, 6297, DOI: 10.1021/jp9800867.

[Paulisse00] Paulisse, K. W.; Friday, T. O.; Graske, M. L.; Polik, W. F., "Vibronic spectroscopy and lifetime of S_1 acrolein," *J. Chem. Phys.*, 2000, **113**, 184, DOI: 10.1063/1.481785.

[Pecul06] Pecul, M.; Lamparska, E.; Cappelli, C.; Frediani, L.; Ruud, K., "Solvent Effects on Raman Optical Activity Spectra Calculated Using the Polarizable Continuum Model," *J. Phys. Chem. A*, 2006, **110**, 2807-15, DOI: 10.1021/jp056443c.

[Pedersen95] Pedersen, T. B.; Hansen, A. E., "Ab initio calculation and display of the rotatory strength tensor in the random phase approximation. Method and model studies," *Chem. Phys. Lett.*, 1995, **246**, 1-8, DOI: 10.1016/0009-2614(95)01036-9.

[Pedley94] Pedley, J. B., *Thermochemical Data and Structures of Organic Compounds*; CRC Press: TX, 1994; Vol. 1.

[Pell65] Pell, A. S.; Pilcher, G., "Measurements of Heats of Combustion by Flame Calorimetry. 3. Ethylene Oxide, Trimethylene Oxide, Tetrahydrofuran and Tetrahydropy," *Trans. Faraday Soc.*, 1965, **61**, 71-77, DOI: 10.1039/TF9656100071

[Peng10] Peng, B.; Yang, S.; Li, L.; Cheng, F.; Chen, J., "A density functional theory and time-dependent density functional theory investigation on the anchor comparison of triarylamine-based dyes," *J. Chem. Phys.*, 2010, **132**, 034305, DOI: 10.1063/1.3292639.

[Peng93] Peng, C.; Schlegel, H. B., "Combining Synchronous Transit and Quasi-Newton Methods for Finding Transition States," *Israel J. Chem.*, 1993, **33**, 449-54, DOI: 10.1002/ijch.199300051.

[Peng96] Peng, C.; Ayala, P. Y.; Schlegel, H. B.; Frisch, M. J., "Using redundant internal coordinates to optimize equilibrium geometries and transition states," *J. Comp. Chem.*, 1996, **17**, 49-56, DOI: 10.1002/(SICI)1096-987X(19960115)17:1<49::AID-JCC5>3.0.CO;2-0.

[Peralta03] Peralta, J. E.; Scuseria, G. E.; Cheeseman, J. R.; Frisch, M. J., "Basis set dependence of NMR Spin-Spin Couplings in Density Functional Theory Calculations: First row and hydrogen atoms," *Chem. Phys. Lett.*, 2003, **375**, 452-58, DOI: 10.1016/S0009-2614(03)00886-8.

[Perdew91] Perdew, J. P., "Unified Theory of Exchange and Correlation Beyond the Local Density Approximation" in *Electronic Structure of Solids '91. Proceedings of the 75th WE-Heraeus-Seminar and 21st Annual International Symposium on Electronic Structure of Solids held in Gaussig (Germany)*; ed. Ziesche, P., Eschrig, H., Akademie Verlag: Berlin, 1991, pp. 11-20

[Perdew92] Perdew, J. P.; Chevary, J. A.; Vosko, S. H.; Jackson, K. A.; Pederson, M. R.; Singh, D. J.; Fiolhais, C., "Atoms, molecules, solids, and surfaces: Applications of the generalized gradient approximation for exchange and correlation," *Phys. Rev. B*, 1992, **46**, 6671-87, DOI: 10.1103/PhysRevB.46.6671.

[Perdew92a] Perdew, J. P.; Wang, Y., "Accurate and Simple Analytic Representation of the Electron Gas Correlation Energy," *Phys. Rev. B*, 1992, **45**, 13244-49, DOI: 10.1103/PhysRevB.45.13244

[Perdew93a] Perdew, J. P.; Chevary, J. A.; Vosko, S. H.; Jackson, K. A.; Pederson, M. R.; Singh, D. J.; Fiolhais, C., "Erratum: Atoms, molecules, solids, and surfaces: Applications of the generalized gradient approximation for exchange and correlation," *Phys. Rev. B*, 1993, **48**, 4978, DOI: 10.1103/PhysRevB.48.4978.2.

[Petersson00] Petersson, G. A.; Frisch, M. J., "A journey from generalized valence bond theory to the full CI complete basis set limit," *J. Phys. Chem. A*, 2000, **104**, 2183-90, DOI: 10.1021/jp991947u.

[Petersson81] Petersson, G. A.; Nyden, M. R., "Interference effects in pair correlation energies: Helium L-limit energies," *J. Chem. Phys.*, 1981, **75**, 3423-25, DOI: 10.1063/1.442450.

[Petersson85] Petersson, G. A.; Yee, A. K.; Bennett, A., "Complete basis set correlation energies. III. The total correlation energy of the neon atom," *J. Chem. Phys.*, 1985, **83**, 5105-28, DOI: 10.1063/1.449724.

[Petersson88] Petersson, G. A.; Bennett, A.; Tensfeldt, T. G.; Al-Laham, M. A.; Shirley, W. A.; Mantzaris, J., "A complete basis set model chemistry. I. The total energies of closed-shell atoms and hydrides of the first-row elements," *J. Chem. Phys.*, 1988, **89**, 2193-218, DOI: 10.1063/1.455064.

[Petersson91] Petersson, G. A.; Al-Laham, M. A., "A complete basis set model chemistry. II. Open-shell systems and the total energies of the first-row atoms," *J. Chem. Phys.*, 1991, **94**, 6081-90, DOI: 10.1063/1.460447.

[Pettersen04] Pettersen, E. F.; Goddard, T. D.; Huang, C. C.; Couch, G. S.; Greenblatt, D. M.; Meng, E. C.; Ferrin, T. E., "UCSF Chimera: a visualization system for exploratory research and analysis," *J. Comp. Chem.*, 2004, **25**, 1605-12, DOI: 10.1002/jcc.20084.

[Phillips59] Phillips, J. C.; Kleinman, L., "New Method for Calculating Wave Functions in Crystals and Molecules," *Phys. Rev.*, 1959, **116**, 287-94, DOI: 10.1103/PhysRev.116.287.

[Pickett91] Pickett, H. M., "The Fitting and Prediction of Vibration-Rotation Spectra with Spin Interactions," *J. Mol. Spectrosc.*, 1991, **148**, 371-77, DOI: 10.1016/0022-2852(91)90393-O.

[Pisarenko03] Pisarenko, A.; Foresman, J. B.; Clarke, D. D., "Toward more accurate computational methods to predict C-13 chemical shifts: A study of 2,2,4-trimethylpentane-1,3-diol," 2003, unpublished.

[Pittam72] Pittam, D. A.; Pilcher, G., "Measurements of Heats of Combustion by Flame Calorimetry. 8. Methane, ethane, propane, n-butane and 2-methylpropane," *J. Chem. Soc., Faraday Trans. 1*, 1972, **68**, 2224-29, DOI: 10.1039/F19726802224

[Plass98] Plass, R.; Egan, K.; Collazo-Davila, C.; Grozea, D.; Landree, E.; Marks, L. D.; Gajdardziska-Josifovska, M., "Cyclic Ozone Identified in Magnesium Oxide (111) Surface Reconstructions," *Phys. Rev. Lett.*, 1998, **81**, 4891-94, DOI: 10.1103/PhysRevLett.81.4891.

[Polavarapu99] Polavarapu, P. L.; Zhao, C.; Cholli, A. L.; Vernice, G. G., "Vibrational Circular Dichroism, Absolute Configuration, and Predominant Conformations of Volatile Anesthetics: Desflurane," *J. Phys. Chem. B*, 1999, **103**, 6127-32, DOI: 10.1021/jp990550n.

[Pople54] Pople, J. A.; Nesbet, R. K., "Self-Consistent Orbitals for Radicals," *J. Chem. Phys.*, 1954, **22**, 571-72, DOI: 10.1063/1.1740120.

[Pople73] Pople, J. A., "Theoretical Models for Chemistry" in *Energy, Structure, and Reactivity: Proceedings of the 1972 Boulder Summer Research Conference on Theoretical Chemistry*; ed. Smith, D. W., Wiley & Sons: New York, 1973, pp. 399

[Pople76] Pople, J. A.; Binkley, J. S.; Seeger, R., "Theoretical Models Incorporating Electron Correlation," *Int. J. Quantum Chem.*, 1976, **10**, 1-19, DOI: 10.1002/qua.560100802.

[Pople77] Pople, J. A.; Seeger, R.; Krishnan, R., "Variational Configuration Interaction Methods and Comparison with Perturbation Theory," *Int. J. Quantum Chem.*, 1977, **12**, 149-63, DOI: 10.1002/qua.560120820.

[Pople89] Pople, J. A.; Head-Gordon, M.; Fox, D. J.; Raghavachari, K.; Curtiss, L. A., "Gaussian-1 theory: A general procedure for prediction of molecular energies," *J. Chem. Phys.*, 1989, **90**, 5622-29, DOI: 10.1063/1.456415.

[Pople93] Pople, J. A.; Scott, A. P.; Wong, M. W.; Radom, L., "Scaling Factors for Obtaining Fundamental Vibrational Frequencies and Zero-Point Energies from HF/6-31G* and MP2/6-31G* Harmonic Frequencies," *Israel J. Chem.*, 1993, **33**, 345-50, DOI: 10.1002/ijch.199300041.

[Porezag00] Porezag, D.; Pederson, M. R.; Liu, A. Y., "The Accuracy of the Pseudopotential Approximation with Density-Functional Theory," *Phys. Status Solidi B*, 2000, **217**, 219-30, DOI: 10.1002/(SICI)1521-3951(200001)217:1<219::AID-PSSB219>3.0.CO;2-V.

[Pranata94] Pranata, J., "Ab Initio Study of the Base-Catalyzed Hydrolysis of Methyl Formate," *J. Phys. Chem.*, 1994, **98**, 1180-84, DOI: 10.1021/j100055a023.

[Prochnow09] Prochnow, E.; Evangelista, F. A.; Schaefer III, H. F.; Allen, W. D.; Gauss, J., "Analytic gradients for the state-specific multireference coupled cluster singles and doubles model," *J. Chem. Phys.*, 2009, **131**, 064109, DOI: 10.1063/1.3204017.

[Pulay69] Pulay, P., "Ab initio calculation of force constants and equilibrium geometries in polyatomic molecules. I. Theory," *Molecular Physics*, 1969, **17**, 197-204.

[Pulay79] Pulay, P.; Fogarasi, G.; Pang, F.; Boggs, J. E., "Systematic ab initio gradient calculation of molecular geometries, force constants, and dipole-moment derivatives," *J. Am. Chem. Soc.*, 1979, **101**, 2550-60, DOI: 10.1021/ja00504a009.

[Pulay92] Pulay, P.; Fogarasi, G., "Geometry optimization in redundant internal coordinates," *J. Chem. Phys.*, 1992, **96**, 2856-60, DOI: 10.1063/1.462844.

[Purvis82] Purvis III, G. D.; J., B. R., "A full coupled-cluster singles and doubles model: the inclusion of disconnected triples," *J. Chem. Phys.*, 1982, **76**, 1910-18, DOI: 10.1063/1.443164.

[Qiu10] Qiu, S.; Li, G.; Liu, P.; Wang, C.; Feng, Z.; Li, C., "Chirality transition in the epoxidation of (-)-α-pinene and successive hydrolysis studied by Raman optical activity and DFT," *Phys. Chem. Chem. Phys.*, 2010, **12**, 3005-13, DOI: 10.1039/B919993D

[Quapp87] Quapp, W., "A redefined anharmonic potential energy surface of HCN," *J. Mol. Spectrosc.*, 1987, **125**, 122, DOI: 10.1016/0022-2852(87)90198-6.

[Raghavachari78] Raghavachari, K.; Pople, J. A., "Approximate 4th-order perturbation-theory of electron correlation energy," *Int. J. Quantum Chem.*, 1978, **14**, 91-100, DOI: 10.1002/qua.560140109.

[Raghavachari80] Raghavachari, K.; Frisch, M. J.; Pople, J. A., "Contribution of triple substitutions to the electron correlation energy in fourth-order perturbation theory," *J. Chem. Phys.*, 1980, **72**, 4244-45, DOI: 10.1063/1.439657.

[Raghavachari80a] Raghavachari, K.; Schlegel, H. B.; Pople, J. A., "Derivative studies in configuration-interaction theory," *J. Chem. Phys.*, 1980, **72**, 4654-55, DOI: 10.1063/1.439708.

[Raghavachari80b] Raghavachari, K.; Binkley, J. S.; Seeger, R.; Pople, J. A., "Self-Consistent Molecular Orbital Methods. 20. Basis set for correlated wave-functions," *J. Chem. Phys.*, 1980, **72**, 650-54, DOI: 10.1063/1.438955.

[Raghavachari88] Raghavachari, K., "Sequential Clustering Reactions of Si^+ with Silane: A Theoretical Study of the Reaction Mechanisms," *J. Chem. Phys.*, 1988, **88**, 1688-702, DOI: 10.1063/1.454147.

[Raghavachari89] Raghavachari, K.; Trucks, G. W., "Highly correlated systems: Excitation energies of first row transition metals Sc-Cu," *J. Chem. Phys.*, 1989, **91**, 1062-65, DOI: 10.1063/1.457230.

[Raghavachari90] Raghavachari, K.; Pople, J. A.; Replogle, E. S.; Head-Gordon, M., "Fifth Order Møller-Plesset Perturbation Theory: Comparison of Existing Correlation Methods and Implementation of New Methods Correct to Fifth Order," *J. Phys. Chem.*, 1990, **94**, 5579-86, DOI: 10.1021/j100377a033.

[Raghavachari92] Raghavachari, K., "Ground state of C_{84}: Two almost isoenergetic isomers," *Chem. Phys. Lett.*, 1992, **190**, 397-400, DOI: 10.1016/0009-2614(92)85162-4.

[Rappe92] Rappé, A. K.; Casewit, C. J.; Colwell, K. S.; III, W. A. G.; Skiff, W. M., "UFF, a full periodic-table force-field for molecular mechanics and molecular-dynamics simulations," *J. Am. Chem. Soc.*, 1992, **114**, 10024-35, DOI: 10.1021/ja00051a040.

[Reed85] Reed, A. E.; Weinstock, R. B.; Weinhold, F., "Natural-population analysis," *J. Chem. Phys.*, 1985, **83**, 735-46, DOI: 10.1063/1.449486.

[Reed85a] Reed, A. E.; Weinhold, F., "Natural Localized Molecular Orbitals," *J. Chem. Phys.*, 1985, **83**, 1736-40, DOI: 10.1063/1.449360.

[Resa04] Resa, I.; Carmona, E.; Gutierrez-Puebla, E.; Monge, A., "Decamethyldizincocene, a Stable Compound of Zn(I) with a Zn-Zn Bond," *Science*, 2004, **305**, 1136-38, DOI: 10.1126/science.1101356

[Rettig99] Rettig, W.; Bliss, B.; Dirnberger, K., "Pseudo-Jahn-Teller and TICT-models: a photophysical comparison of meta- and para-DMABN derivatives," *Chem. Phys. Lett.*, 1999, **305**, 8-14, DOI: 10.1016/S0009-2614(99)00316-4.

[Reusch13] Reusch, W. H., "Aromatic Substitution Reactions," 2013, *www2.chemistry.msu.edu/faculty/reusch/VirtTxtJml/benzrx1.htm*.

[Rhile06] Rhile, I. J.; Markle, T. F.; Nagao, H.; DiPasquale, A. G.; Lam, O. P.; Lockwood, M. A.; Rotter, K.; Mayer, J. M., "Concerted Proton-Electron Transfer in the Oxidation of Hydrogen-Bonded Phenols," *J. Am. Chem. Soc.*, 2006, **128**, 6075, DOI: 10.1021/ja054167+.

[Rhinehart12] Rhinehardt, J. M.; Challa, J. R.; McCamant, D. W., "Multimode Charge-Transfer Dynamics of 4-(Dimethylamino)benzonitrile Probed with Ultraviolet Femtosecond Stimulated Raman Spectroscopy," *J. Phys. Chem. B*, 2012, **116**, 10522-34, DOI: 10.1021/jp3020645.

[Ribeiro11] Ribeiro, R. F.; Marenich, A. V.; Cramer, C. J.; Truhlar, D. G., "Use of Solution-Phase Vibrational Frequencies in Continuum Models for the Free Energy of Solvation," *J. Phys. Chem. B*, 2011, **115**, 14556, DOI: 10.1021/jp205508z.

[Rice90] Rice, J. E.; Amos, R. D.; Colwell, S. M.; Handy, N. C.; Sanz, J., "Frequency-Dependent Hyperpolarizabilities with Application to Formaldehyde and Methyl-Fluoride," *J. Chem. Phys.*, 1990, **93**, 8828-39, DOI: 10.1063/1.459221.

[Rice91] Rice, J. E.; Handy, N. C., "The Calculation of Frequency-Dependent Polarizabilities as Pseudo-Energy Derivatives," *J. Chem. Phys.*, 1991, **94**, 4959-71, DOI: 10.1063/1.460558.

[Rice92] Rice, J. E.; Handy, N. C., "The Calculation of Frequency-Dependent Hyperpolarizabilities Including Electron Correlation-Effects," *Int. J. Quantum Chem.*, 1992, **43**, 91-118, DOI: 10.1002/qua.560430110.

[Richardson00] Richardson, A. D.; Hedberg, K.; Lucier, G. M., "Gas-Phase Molecular Structures of Third Row Transition-Metal Hexafluorides WF_6, ReF_6, OsF_6, IrF_6, and PtF_6. An Electron-Diffraction and ab Initio Study," *Inorg. Chem.*, 2000, **39**, 2787-93, DOI: 10.1021/ic000003c.

[Rinaldi73] Rinaldi, D.; Rivail, J.-L., "Polarisabilites moléculaires et effet diélectrique de milieu à l'état liquide," *Theoretica chimica acta*, 1973, **32**, 57-70, DOI: 10.1007/BF00527479.

[Robb14] Robb, M. A., "In This Molecule There Must be a Conical Intersection" in *Advances in Physical Organic Chemistry*, vol. 48, ed. Williams, I. H., Williams, N. H., Elsevier: 2014, pp. 189-228, DOI: 10.1016/B978-0-12-800256-8.00003-5.

[Robb90] Robb, M. A.; Niazi, U., "The Unitary Group Approach to Electronic Structure Computations" in *Reports in Molecular Theory*; ed. Weinstein, H., Náray-Szabó, G., vol. 1, CRC Press: 1990, pp. 23-55

[Roos80] Roos, B. O.; Taylor, P. R.; Siegbahn, P. E. M., "A complete active space SCF method (CASSCF) using a density matrix formulated super-CI approach," *Chem. Phys.*, 1980, **48**, 157-73, DOI: 10.1016/0301-0104(80)80045-0.

[Roothaan51] Roothaan, C. C. J., "New Developments in Molecular Orbital Theory," *Rev. Mod. Phys.*, 1951, **23**, 69, DOI: 10.1103/RevModPhys.23.69

[Rosenau02] Rosenau, T.; Potthast, A.; Elder, T.; Kosma, P., "Stabilization and First Direct Spectroscopic Evidence of the o-Quinone Methide Derived from Vitamin E," *Org. Lett.*, 2002, **4**, 4285, DOI: 10.1021/ol026917f.

[Rosenau04] Rosenau, T.; Ebner, G.; Stanger, A.; Perl, S.; Nuri, L., "From a Theoretical Concept to Biochemical Reactions: Strain-Induced Bond Localization (SIBL) in Oxidation of Vitamin E," *Chem. - Eur. J.*, 2004, **11**, 280-87, DOI: 10.1002/chem.200400265.

[Rosenau07] Rosenau, T.; Kloser, E.; Gille, L.; Mazzani, F.; Netscher, T., "Vitamin E Chemistry. Studies into Initial Oxidation Intermediates of α-Tocopherol: Disproving the Involvement of 5a-C-Centered 'Chromanol Methide' Radicals," *J. Org. Chem.*, 2007, **72**, 3268-81, DOI: 10.1021/jo062553j.

[Rosenfeld28] Rosenfeld, L., "Quantum-mechanical theory of the natural optical activity of liquids and gases," *Zeitschrift fuer Physik*, 1928, **52**, 161-74.

[Roux08] Roux, M. V.; Temprado, M.; Chickos, J. S.; Nagano, Y., "Critically Evaluated Thermochemical Properties of Polycyclic Aromatic Hydrocarbons," *J. Phys. Chem. Ref. Data*, 2008, **37**, 1855-996, DOI: 10.1063/1.2955570.

[Runge84] Runge, E.; Gross, E. K. U., "Density-functional theory for time-dependent systems," *Phys. Rev. Lett.*, 1984, **52**, 997-1000, DOI: 10.1103/PhysRevLett.52.997.

[Rusakov13] Rusakov, A. A.; Krivdin, L. B., "Modern quantum chemical methods for calculating spin-spin coupling constants: Theoretical basis and structural applications in chemistry," *Russ. Chem. Rev.*, 2013, **82**, 99-130, DOI: 10.1070/RC2013v082n02ABEH004350.

[Ruud02] Ruud, K.; Helgaker, T., "Optical rotation studied by density-functional and coupled-cluster methods," *Chem. Phys. Lett.*, 2002, **352**, 533-39, DOI: 10.1016/S0009-2614(01)01492-0.

[Ruud02a] Ruud, K.; Helgaker, T.; Bouř, P., "Gauge-origin independent density-functional theory calculations of vibrational Raman optical activity," *J. Phys. Chem. A*, 2002, **106**, 7448-55, DOI: 10.1021/jp026037i.

[Sadlej88] Sadlej, A. J., "Medium-sized polarized basis sets for high-level correlated calculations of molecular electric properties," *Collect. Czech. Chem. Commun.*, 1988, **53**, DOI: 10.1135/cccc19881995

[Saito12] Saito, T.; Thiel, W., "Analytical Gradients for Density Functional Calculations with Approximate Spin Projection," *J. Phys. Chem. A*, 2012, **116**, 10864-69, DOI: 10.1021/jp308916s.

[Sala14] Sala, M.; Kirkby, O. M.; Guérin, S.; Fielding, H. H., "New insight into the potential energy landscape and relaxation pathways of photoexcited aniline from CASSCF and XMCQDPT2 electronic structure calculations," *Phys. Chem. Chem. Phys.*, 2014, **16**, 3122-33, DOI: 10.1039/c3cp54418d.

[Salahub89] *The Challenge of d and f Electrons*; Salahub, D. R.; Zerner, M. C., Eds.; ACS: Washington, D.C., 1989; Vol. 394.

[Salter98] Salter, C.; Foresman, J. B., "Naphthalene and azulene. I. Semimicro bomb calorimetry and quantum mechanical calculations," *J. Chem. Educ.*, 1998, **75**, 1341-45, DOI: 10.1021/ed075p1341.

[Santolinia15] Santolinia, V.; Malhadoa, J. P.; Robb, M. A.; Garavelli, M.; Bearpark, M. J., "Photochemical reaction paths of cis-dienes studied with RASSCF: the changing balance between ionic and covalent excited states," *Molecular Physics*, submitted (2015), **113**, DOI: 10.1080/00268976.2015.1025880.

[Santoro07] Santoro, F.; Improta, R.; Lami, A.; Bloino, J.; Barone, V., "Effective method to compute Franck-Condon integrals for optical spectra of large molecules in solution," *J. Chem. Phys.*, 2007, **126**, 084509: 1-13, DOI: 10.1063/1.2437197.

[Scalmani06] Scalmani, G.; Frisch, M. J.; Mennucci, B.; Tomasi, J.; Cammi, R.; Barone, V., "Geometries and properties of excited states in the gas phase and in solution: Theory and application of a time-dependent density functional theory polarizable continuum model," *J. Chem. Phys.*, 2006, **124**, 094107: 1-15, DOI: 10.1063/1.2173258.

[Scalmani10] Scalmani, G.; Frisch, M. J., "Continuous surface charge polarizable continuum models of solvation: I. General formalism," *J. Chem. Phys.*, 2010, **132**, 114110: 1-15, DOI: 10.1063/1.3359469.

[Schalk10] Schalk, O.; Boguslavskiy, A.; Stolow, A., "Substituent Effects on Dynamics at Conical Intersections: Cyclopentadienes," *J. Phys. Chem. A*, 2010, **114**, 4058-64, DOI: 10.1021/jp911286s.

[Schellman73] Schellman, J. A., "Vibrational optical activity," *J. Chem. Phys.*, 1973, **58**, 2882-86, DOI: 10.1063/1.1679592.

[Schiek07] Schiek, M.; Al-Shamery, K.; Lützen, A., "Synthesis of Symmetrically and Unsymmetrically para-Functionalized p-Quaterphenylenes," *Synthesis*, 2007, **4**, 613, DOI: 10.1055/s-2007-965891

[Schiek07a] Schiek, M., *Organic Molecular Nanotechnology* [PhD Thesis], University of Oldenburg, Institute of Pure and Applied Chemistry (Germany), 2007.

[Schiff68] Schiff, L. I., *Quantum Mechanics*; 3rd ed.; McGraw-Hill: New York, 1968.

[Schlegel82] Schlegel, H. B., "Optimization of Equilibrium Geometries and Transition Structures," *J. Comp. Chem.*, 1982, **3**, 214-18, DOI: 10.1002/jcc.540030212.

[Schlegel82a] Schlegel, H. B.; Robb, M. A., "MC SCF gradient optimization of the $_2$CO --> H2 + CO transition structure," *Chem. Phys. Lett.*, 1982, **93**, 43-46, DOI: 10.1016/0009-2614(82)85052-5.

[Schlegel84a] Schlegel, H. B., "Estimating the Hessian for gradient-type geometry optimizations," *Theoretical Chemistry Accounts*, 1984, **66**, 333-40, DOI: 10.1007/BF00554788.

[Schlegel88] Schlegel, H. B., "Møller-Plesset perturbation theory with spin projection," *J. Phys. Chem.*, 1988, **92**, 3075-78, DOI: 10.1021/j100322a014.

[Schlegel91] Schlegel, H. B.; Frisch, M. J., "Computational Bottlenecks in Molecular Orbital Calculations" in *Theoretical and Computational Models for Organic Chemistry*; ed. Formosinho, S. J., Csizmadia, I. G., Arnaut, L. G., *NATO-ASI Series C*, vol. 339, Kluwer Academic: The Netherlands, 1991, pp. 5-33, DOI: 10.1007/978-94-011-3584-9_2.

[Schlegel91a] Schlegel, H. B.; McDouall, J. J., "Do you have SCF Stability and Convergence Problems?" in *Computational Advances in Organic Chemistry*; ed. Ögretir, C., Csizmadia, I. G., Kluwer Academic: The Netherlands, 1991, pp. 167-85, DOI: 10.1007/978-94-011-3262-6_2.

[Schmitt06] Schmitt, M.; Böhm, M.; Ratzer, C.; Krügler, D.; Kleinermanns, K.; Kalkman, I.; Berden, G.; Meetz, W. L., "Determining the intermolecular structure in the S_0 and S_1 states of the phenol dimer by rotationally resolved electronic spectroscopy," *ChemPhysChem*, 2006, **7**, DOI: 10.1002/cphc.200500670.

[Schnatter14] Schnatter, W. F. K.; Rogers, D. W.; Zavitsas, A. A., "Teaching Electrophilic Aromatic Substitution: Enthalpies of Hydrogenation of the Rings of C6H5X Predict Relative Reactivities; ^{13}C NMR Shifts Predict Directing Effects of X," *J. Chem. Educ.*, 2014, DOI: 10.1021/ed3007742.

[Schrodinger26] Schrödinger, E., "Quantisierung als Eigenwertproblem," *Phys. Rev. précis (22 pgs):*, 1926, **79**, 361, DOI: 10.1002/andp.19263840404.

[Schultz93] Schultz, G.; Hargittai, I., "Molecular-Structure of N,N-Dimethylformamide from Gas-Phase Electron-Diffraction," *J. Phys. Chem.*, 1993, **97**, 4966-69, DOI: 10.1021/j100121a018.

[Schwartz62] Schwartz, C., "Importance of Angular Correlations Between Atomic Electrons," *Phys. Rev.*, 1962, **126**, 1015, DOI: 10.1103/PhysRev.126.1015

[Schwartz63] Schwartz, C., *Methods in Computational Physics*; ed. Alder, B. J., Fernback, S., Rotenberg, M., vol. 2, Academic Press: New York, 1963, pp. 271

[Schwerdtfeger11a] Schwerdtfeger, P., "The Pseudopotential Approximation in Electronic Structure Theory," *ChemPhysChem*, 2011, **12**, 3143-55, DOI: 10.1002/cphc.201100387.

[Scuseria88] Scuseria, G. E.; Janssen, C. L.; Schaefer III, H. F., "An efficient reformulation of the closed-shell coupled cluster single and double excitation (CCSD) equations," *J. Chem. Phys.*, 1988, **89**, 7382-87, DOI: 10.1063/1.455269.

[Scuseria91] Scuseria, G. E., "Analytic evaluation of energy gradients for the singles and doubles coupled cluster method including perturbative triple excitations: Theory and Applications to FOOF and Cr_2," *J. Chem. Phys.*, 1991, **94**, 442-47, DOI: 10.1063/1.460359.

[SDBS2336] "#2336 (o-nitroaniline)," Spectral Database for Organic Compounds (SDBS), National Institute of Advanced Industrial Science and Technology (AIST), *sdbs.db.aist.go.jp*.

[Seeger77] Seeger, R.; Pople, J. A., "Self-Consistent Molecular Orbital Methods. 28. Constraints and Stability in Hartree-Fock Theory," *J. Chem. Phys.*, 1977, **66**, 3045-50, DOI: 10.1063/1.434318.

[Sekino07] Sekino, H.; Maeda, Y.; Kamiya, M.; Hirao, K., "Polarizability and second hyperpolarizability evaluation of long molecules by the density functional theory with long-range correction," *J. Chem. Phys.*, 2007, **126**, 014107, DOI: 10.1063/1.2428291.

[Sekino86] Sekino, H.; Bartlett, R. J., "Frequency-Dependent Nonlinear Optical-Properties of Molecules," *J. Chem. Phys.*, 1986, **85**, 976-89, DOI: 10.1063/1.451255.

[Selander71] Selander, H.; Nilsson, J. L. G., "Tritium Exchange in Specifically Labelled Xylenols, Indanols and Tetrahydronaphthols and Their Methyl Ethers," *Acta Chem. Scand.*, 1971, **25**, 1182-84, DOI: 10.3891/acta.chem.scand.25-1182.

[Sellgren10] Sellgren, K.; Werner, M. W.; Ingalls, J. G.; Smith, J. D. T.; Carleton, T. M.; Joblin, C., "C_{60} in Reflection Nebulae," *Astrophys. J. Lett.*, 2010, **722**, L54-L57, DOI: 10.1088/2041-8205/722/1/L54.

[Senthilkumar13] Senthilkumar, L.; Umadevi, P.; Nithya, K. N. S.; Kolandaivel, P., "Density functional theory investigation of cocaine water complexes," *J. Mol. Model*, 2013, **19**, 3411, DOI: 10.1007/s00894-013-1866-0.

[SerranoPerez13] Serrano-Pérez, J. J.; de Vleeschouwer, F.; de Proft, F.; Mendive-Tapia, D.; Bearpark, M. J.; Robb, M. A., "How the Conical Intersection Seam Controls Chemical Selectivity in the Photocycloaddition of Ethylene and Benzene," *J. Org. Chem.*, 2013, **78**, 1874-86, DOI: 10.1021/jo3017549

[Shaik92] Shaik, S.; Schlegel, H. B.; Wolfe, S., *Theoretical Aspects of Physical Organic Chemistry: The S_N2 mechanism*; John Wiley & Sons: New York, 1992.

[Shelton82] Shelton, D. P.; Buckingham, A. D., "Optical second-harmonic generation in gases with a low-power laser," *Phys. Rev. A*, 1982, **26**, 2787, DOI: 10.1103/PhysRevA.26.2787

[Shelton94] Shelton, D. P.; Rice, J. E., "Measurements and Calculations of the Hyperpolarizabilities of Atoms and Small Molecules in the Gas Phase," *Chem. Rev.*, 1994, **94**, 3-29, DOI: 10.1021/cr00025a001.

[Shi89] Shi, Z.; Boyd, R. J., "Transition State Electronic Structures in S_N2 Reactions," *J. Am. Chem. Soc.*, 1989, **111**, 1575-79, DOI: 10.1021/ja00187a007.

[Shuai91] Shuai, Z.; Brédas, J. L., "Static and Dynamic Third-Harmonic Generation in Long Polyacetylene and Polyparaphenylene Vinylene Chains," *Phys. Rev. B*, 1991, **44**, 5962, DOI: 10.1103/PhysRevB.44.5962

[Siegbahn81] Siegbahn, P. E. M.; Almlöf, J.; Heiberg, A.; Roos, B. O., "The complete active space SCF (CASSCF) method in a Newton-Raphson formulation with application to the HNO molecule," *J. Chem. Phys.*, 1981, **74**, 2384-96, DOI: 10.1063/1.441359.

[Siegbahn84] Siegbahn, P. E. M., "A new direct CI method for large CI expansions in a small orbital space," *Chem. Phys. Lett.*, 1984, **109**, 417-23, DOI: 10.1016/0009-2614(84)80336-X.

[Silverstein91] Silverstein, R. M.; Bassler, G. C.; Morrill, T. C., *Spectrometric Identification of Organic Compounds*, 5th, Wiley & Sons, Inc.: New York, 1991, pp. 236-39.

[Simon96] Simon, S.; Duran, M.; Dannenberg, J. J., "How does basis set superposition error change the potential surfaces for hydrogen bonded dimers?," *J. Chem. Phys.*, 1996, **105**, 11024-31, DOI: 10.1063/1.472902.

[Singh84] Singh, U. C.; Kollman, P. A., "An approach to computing electrostatic charges for molecules," *J. Comp. Chem.*, 1984, **5**, 129-45, DOI: 10.1002/jcc.540050204.

[Singh86] Singh, U. C.; Kollman, P. A., "A Combined ab initio Quantum-Mechanical and Molecular Mechanical Method for Carrying out Simulations on Complex Molecular Systems: Applications to the CH_3Cl + Cl- Exchange-Reaction and Gas-Phase Protonation of Polyethers," *J. Comp. Chem.*, 1986, **7**, 718-30, DOI: 10.1002/jcc.540070604.

[Slater74] Slater, J. C., *The Self-Consistent Field for Molecules and Solids*; McGraw-Hill: New York, 1974; Vol. 4.

[Sohn10] Sohn, W. Y.; Kim, T. W.; Lee, J. S., "Structure and Energetics of $C_{60}O$: A Theoretical Study," *J. Phys. Chem.*, 2010, **114**, 1939-43, DOI: 10.1021/jp9093386.

[Sondergaard11] Søndergaard, C. R.; Olsson, M. H. M.; Rostkowski, M.; Jensen, J. H., "Improved Treatment of Ligands and Coupling Effects in Empirical Calculation and Rationalization of pKa Values," *J. Chem. Theory and Comput.*, 2011, **7**, 2284-95, DOI: 10.1021/ct200133y.

[Sonnenberg09] Sonnenberg, J. L.; Schlegel, H. B.; Hratchian, H. P., "Spin Contamination in Inorganic Chemistry Calculations" in *Computational Inorganic and Bioinorganic Chemistry*; ed. Soloman, E. I., Scott, R. A., B., K. R., John Wiley & Sons, Ltd: Chichester, 2009, pp. 173-86, DOI: 10.1002/0470862106.ia617.

[Sonnenberg09a] Sonnenberg, J. L.; Wong, K. F.; Voth, G. A.; Schlegel, H. B., "Distributed Gaussian Valence Bond Surface Derived from Ab Initio Calculations," *J. Chem. Theory and Comput.*, 2009, **5**, 949-61, DOI: 10.1021/ct800477y.

[Stahelin93] Stähelin, M.; Moylan, C. R.; Burland, D. M.; Willetts, A.; Rice, J. E.; Shelton, D. P.; Donley, E. A., "A comparison of calculated and experimental hyperpolarizabilities for acetonitrile in gas and liquid phases," *J. Chem. Phys.*, 1993, **98**, 5595, DOI: 10.1063/1.464904.

[Staley84] Staley, S. W.; Norden, T. D., "Synthesis and Direct Observation of Methylenecyclopropene," *J. Am. Chem. Soc.*, 1984, **106**, 3699-700, DOI: 10.1021/ja00324a065.

[Stanton93] Stanton, J. F.; Bartlett, R. J., "Equation of motion coupled-cluster method: A systematic biorthogonal approach to molecular excitation energies, transition probabilities, and excited state properties," *J. Chem. Phys.*, 1993, **98**, 7029-39, DOI: 10.1063/1.464746.

[Staroverov03] Staroverov, V. N.; Scuseria, G. E.; Tao, J.; Perdew, J. P., "Comparative assessment of a new nonempirical density functional: Molecules and hydrogen-bonded complexes," *J. Chem. Phys.*, 2003, **119**, 12129, DOI: 10.1063/1.1626543.

[Steel09] Steel, W. H.; Foresman, J. B.; Burden, D. K.; Lau, Y. Y.; Walker, R. A., "Solvation of Nitrophenol Isomers: Consequences for Solute Electronic Structure and Alkane/Water Partitioning," *J. Phys. Chem.*, 2009, **113**, 759-66, DOI: 10.1021/jp805184w.

[Stephens01] Stephens, P. J.; Devlin, F. J.; Cheeseman, J. R.; Frisch, M. J., "Calculation of optical rotation using Density Functional Theory," *J. Phys. Chem. A*, 2001, **105**, 5356-71, DOI: 10.1021/jp0105138.

[Stephens03] Stephens, P. J.; Devlin, F. J.; Cheeseman, J. R.; Frisch, M. J.; Bortolini, O.; Besse, P., "Determination of Absolute Configuration Using Ab Initio Calculation of Optical Rotation," *Chirality*, 2003, **15**, S57-64, DOI: 10.1002/chir.10270.

[Stephens05] Stephens, P. J.; McCann, D. M.; Cheeseman, J. R.; Frisch, M. J., "Determination of absolute configurations of chiral molecules using ab initio time-dependent Density Functional Theory calculations of optical rotation: How reliable are absolute configurations obtained for molecules with small rotations?," *Chirality*, 2005, **17**, S52-64, DOI: 10.1002/chir.20109.

[Stephens05a] Stephens, P. J.; McCann, D. M.; Devlin, F. J.; Flood, T. C.; Butkus, E.; Stončius, S.; Cheeseman, J. R., "Determination of molecular structure using vibrational circular dichroism spectroscopy: The keto-lactone product of Baeyer-Villiger oxidation of (+)-(1R,5S)-bicyclo[3.3.1]nonane-2,7-dione," *J. Org. Chem.*, 2005, **70**, 3903-13, DOI: 10.1021/jo047906y.

[Stephens07] Stephens, P. J.; Pan, J. J.; Devlin, F. J.; Krohn, K.; Kurtán, T., "Determination of the Absolute Configurations of Natural Products via Density Functional Theory Calculations of Vibrational Circular Dichroism, Electronic Circular Dichroism, and Optical Rotation: The Iridoids Plumericin and Isoplumericin," *J. Org. Chem.*, 2007, **72**, 3521-36, DOI: 10.1021/jo070155q.

[Stephens12] Stephens, P. J.; Devlin, F. J.; Cheeseman, J. R., *VCD Spectroscopy for Organic Chemists*; CRC Press: Boca Raton, FL, 2012.

[Stephens85] Stephens, P. J., "Theory of vibrational circular dichroism," *J. Phys. Chem.*, 1985, **89**, 748-52, DOI: 10.1021/j100251a006

[Stewart07] Stewart, J. J. P., "Optimization of parameters for semiempirical methods. V. Modification of NDDO approximations and application to 70 elements," *J. Mol. Model.*, 2007, **13**, 1173-213, DOI: 10.1007/s00894-007-0233-4.

[Stewart89] Stewart, J. J. P., "Optimization of parameters for semiempirical methods. I. Method," *J. Comp. Chem.*, 1989, **10**, 209-20, DOI: 10.1002/jcc.540100208.

[Stieglitz1896] Stieglitz, J., "On the 'Beckmann Rearrangement' I. Chlorimido-ethers," *Am. Chem. J.*, 1896, **18**, 751-61.

[Stojanovic12] Stojanović, L., "Theoretical Study of Hyperfine Interactions in Small Arsenic-Containing Radicals," *J. Phys. Chem. A*, 2012, **116**, 8624-33, DOI: 10.1021/jp304786r.

[Stratmann98] Stratmann, R. E.; Scuseria, G. E.; Frisch, M. J., "An efficient implementation of time-dependent density-functional theory for the calculation of excitation energies of large molecules," *J. Chem. Phys.*, 1998, **109**, 8218-24, DOI: 10.1063/1.477483.

[Sustmann96] Sustmann, R.; Tappanchai, S.; Bandmann, H., "a(E)-1-Methoxy-1,3-butadiene and 1,1-Dimethoxy-1,3-butadiene in (4 + 2) Cycloadditions. A Mechanistic Comparison," *J. Am. Chem. Soc.*, 1996, **118**, 12555, DOI: 10.1021/ja961390l.

[Sychrovsky00] Sychrovsky, V.; Gräfenstein, J.; Cremer, D., "Nuclear magnetic resonance spin-spin coupling constants from coupled perturbed density functional theory," *J. Chem. Phys.*, 2000, **113**, 3530-47, DOI: 10.1063/1.1286806.

[Szabo96] Szabo, A.; Ostlund, N. S., *Modern Quantum Chemistry: Introduction to Advanced Electronic Structure Theory* ; Dover, 1996.

[Takano05] Takano, Y.; Houk, K. N., "Benchmarking the Conductor-like Polarizable Continuum Model (CPCM) for Aqueous Solvation Free Energies of Neutral and Ionic Organic Molecules," *J. Chem. Theory and Comput.*, 2005, **1**, 70-77, DOI: 10.1021/ct049977a.

[Takashima78] Takashima, K.; Riveros, J., "Gas-phase pathways for ester hydrolysis," *J. Am. Chem. Soc.*, 1978, **100**, 6128-32, DOI: 10.1021/ja00487a027.

[Tanaka97] Tanaka, K.; Sumiyoshi, Y.; Ohshima, Y.; Endo, Y.; Kawaguchi, K., "Pulsed discharge nozzle Fourier transform microwave spectroscopy of the propargyl radical (H2CCCH)," *J. Chem. Phys.*, 1997, **197**, 2728, DOI: 10.1063/1.474631.

[Tapia75] Tapia, O.; Goscinski, O., "Self-consistent reaction field theory of solvent effects," *Molecular Physics*, 1975, **29**, 1653-61, DOI: 10.1080/00268977500101461.

[Taylor10] Taylor, D. J.; Paterson, M. J., "Calculations of the low-lying excited states of the TiO_2 molecule," *J. Chem. Phys.*, 2010, **133**, 204302, DOI: 10.1063/1.3515477.

[Teale13] Teale, A. M.; Lutnæs, O. B.; Helgaker, T.; Tozer, D. J.; Gauss, J., "Benchmarking density-functional theory calculations of NMR shielding constants and spin-rotation constants using accurate coupled-cluster calculations," *The Journal of Chemical Physics*, 2013, **138**, 024111, DOI: 10.1063/1.4773016.

[Teller37] Teller, E., "The Crossing of Potential Surfaces," *J. Chem. Phys.*, 1937, **41**, 109-16, DOI: 10.1021/j150379a010.

[Thompson14] Thompson, L. M.; Hratchian, H. P., "Spin projection with double hybrid density functional theory," *J. Chem. Phys.*, 2014, **141**, 034108, DOI: 10.1063/1.4887361

[Thorvaldsen08] Thorvaldsen, A. J.; Ruud, K.; Kristensen, K.; Jørgensen, P.; Coriani, S., "A density matrix-based quasienergy formulation of the Kohn-Sham density functional response theory using perturbation- and time-dependent basis sets," *J. Chem. Phys.*, 2008, **129**, 214108, DOI: 10.1063/1.2996351.

[Tomasi05] Tomasi, J.; Mennucci, B.; Cammi, R., "Quantum mechanical continuum solvation models," *Chem. Rev.*, 2005, **105**, 2999-3093, DOI: 10.1021/cr9904009.

[Tondo05] Tondo, D. W.; Pliego Jr., J. R., "Modeling Protic to Dipolar Aprotic Solvent Rate Acceleration and Leaving Group Effects in S_N2 Reactions: A theoretical study of the reaction of acetate ion with ethyl halides in aqueous and dimethyl sulfoxide solutions," *J. Phys. Chem. A*, 2005, **109**, 507-11, DOI: 10.1021/jp047386a.

[Torrent00] Torrent, M.; Solà, M.; Frenking, G., "Theoretical Studies of Some Transition-Metal-Mediated Reactions of Industrial and Synthetic Importance," *Chem. Rev.*, 2000, **100**, 439-93, DOI: 10.1021/cr980452i.

[Townsend04] Townsend, D.; Lahankar, S. A.; Chambreau, S. D.; Suits, A. G.; Zhang, X.; Rheinecker, J.; Harding, L. B.; Bowman, J. M., "The Roaming Atom: Straying from the reaction path in formaldehyde decomposition," *Science*, 2004, **306**, 1158-61, DOI: 10.1126/science.1104386.

[Truong97] Truong, T. N.; Truong, T. T.; Stefanovich, E. V., "A general methodology for quantum modeling of free-energy profile of reactions in solution: An application to the Menshutkin $NH_3 + CH_3Cl$ reaction in water," *J. Chem. Phys.*, 1997, **107**, 1881, DOI: 10.1063/1.474538.

[VanCaillie00] Van Caillie, C.; Amos, R. D., "Geometric derivatives of density functional theory excitation energies using gradient-corrected functionals," *Chem. Phys. Lett.*, 2000, **317**, 159-64, DOI: 10.1016/S0009-2614(99)01346-9.

[VanCaillie99] Van Caillie, C.; Amos, R. D., "Geometric derivatives of excitation energies using SCF and DFT," *Chem. Phys. Lett.*, 1999, **308**, 249-55, DOI: 10.1016/S0009-2614(99)00646-6.

[VanLonkhuyzen84] Lonkhuyzen, H. v.; Lange, C. A. d., "High-resolution UV photoelectron spectroscopy of diatomic halogens," *Chem. Phys.*, 1984, **89**, 313-22, DOI: 10.1016/0301-0104(84)85319-7.

[Vogtle91] Vögtle, F.; Kadei, K., "Großflächige Makrocyclen aus p-Quaterphenyl-Bauteilen (Flat Macrocycles Based on p-Quaterphenyl Units)," *Chem. Ber.*, 1991, **124**, 903, DOI: 10.1002/cber.19911240434.

[Vosko80] Vosko, S. H.; Wilk, L.; Nusair, M., "Accurate spin-dependent electron liquid correlation energies for local spin density calculations: A critical analysis," *Can. J. Phys.*, 1980, **58**, 1200-11, DOI: 10.1139/p80-159.

[Vreven06] Vreven, T.; Byun, K. S.; Komáromi, I.; Dapprich, S.; Montgomery Jr., J. A.; Morokuma, K.; Frisch, M. J., "Combining quantum mechanics methods with molecular mechanics methods in ONIOM," *J. Chem. Theory and Comput.*, 2006, **2**, 815-26, DOI: 10.1021/ct050289g.

[Vreven06b] "Hybrid Methods: ONIOM (QM:MM) and QM/MM" in Vreven, T.; Morokuma, K., *Annual Reports in Computational Chemistry*, Elsevier: 2006, Vol. 2, pp. 35-51.

[Vreven08] "The ONIOM Method for Layered Calculations" in Vreven, T.; Morokuma, K., *Continuum Solvation Models in Chemical Physics: From Theory to Applications*, Wiley: 2008

[Vydrov06a] Vydrov, O. A.; Heyd, J.; Krukau, A.; Scuseria, G. E., "Importance of short-range versus long-range Hartree-Fock exchange for the performance of hybrid density functionals," *J. Chem. Phys.*, 2006, **125**, 074106, DOI: 10.1063/1.2244560.

[Wachters70] Wachters, A. J. H., "Gaussian basis set for molecular wavefunctions containing third-row atoms," *J. Chem. Phys.*, 1970, **52**, 1033, DOI: 10.1063/1.1673095

[Wadt85] Wadt, W. R.; Hay, P. J., "Ab initio effective core potentials for molecular calculations: potentials for main group elements Na to Bi," *J. Chem. Phys.*, 1985, **82**, 284-98, DOI: 10.1063/1.448800.

[Wallmann08] Wallmann, I.; Schiek, M.; Koch, R.; Lützen, A., "Synthesis of Monofunctionalized p-Quaterphenyls," *Synthesis*, 2008, **15**, 2446, DOI: 10.1055/s-2008-1067163.

[Ward78] Ward, J. F.; Elliott, D. S., "Measurements of molecular hyperpolarizabilities for ethylene, butadiene, hexatriene, and benzene," *J. Chem. Phys.*, 1978, **69**, 5438, DOI: 10.1063/1.436534.

[Warshel76] Warshel, A.; Levitt, M., "Theoretical Studies of Enzymatic Reactions: Dielectric, Electrostatic and Steric Stabilization of Carbonium Ion in the Reaction of Lysozyme," *J. Mol. Biol.*, 1976, **103**, 227-49, DOI: 10.1016/0022-2836(76)90311-9.

[Webster07] Webster, R. D., "New Insights into the Oxidative Electrochemistry of Vitamin E," *Acc. Chem. Res.*, 2007, **40**, 251, DOI: 10.1021/ar068182a.

[Wedig86] Wedig, U.; Dolg, M.; Stoll, H.; Preuss, H., "Energy-Adjusted Pseudopotentials For Transition-Metal Elements" in *Quantum Chemistry: The Challenge of Transition Metals and Coordination Chemistry. [Proceesings of the NATO Advanced Research Workshop and 40th International Meeting of the Societe de Chimie Physique, Strasbourg, France, 16-20 Sept. 1985]*; ed. Veillard, A., D. Reidel Pub. Co.: Dordrecht, The Netherlands, 1986, pp. 79-90

[Weigend05] Weigend, F.; Ahlrichs, R., "Balanced basis sets of split valence, triple zeta valence and quadruple zeta valence quality for H to Rn: Design and assessment of accuracy," *Phys. Chem. Chem. Phys.*, 2005, **7**, 3297-305, DOI: 10.1039/b508541a

[Weigend06] Weigend, F., "Accurate Coulomb-fitting basis sets for H to Rn," *Phys. Chem. Chem. Phys.*, 2006, **8**, 1057-65, DOI: 10.1039/B515623H

[Weinhold88] Weinhold, F.; Carpenter, J. E., "The Natural Bond Orbital Lewis Structure Concept for Molecules, Radicals, and Radical Ions" in *The Structure of Small Molecules and Ions*; ed. Naaman, R., Vager, Z., Plenum: 1988, pp. 227-36

[Werner04] Werner, M. W.; Uchida, K. I.; Sellgren, K.; Marengo, M.; Gordon, K. D.; Morris, P. W.; Houck, J. R.; Stansberry, J. A., "New Infrared Emission Features and Spectral Variations in NGC 7023," *Astrophys. J., Suppl. Ser.*, 2004, **154**, 309-14, DOI: 10.1086/422413.

[West09] West, M. J.; Went, M. J., "The spectroscopic detection of drugs of abuse in fingerprints after development with powders and recovery with adhesive lifters," *Spectrochim. Acta, Part A*, 2009, **71**, 1984-88, DOI: 10.1016/j.saa.2008.07.024.

[Wiberg04c] Wiberg, K. B.; Hammer, J. D.; Zilm, K. W.; Keith, T. A.; Cheeseman, J. R.; Duchamp, J. C., "NMR chemical shifts: Substituted acetylenes," *J. Org. Chem.*, 2004, **69**, 1086-96, DOI: 10.1021/jo030258i.

[Wiberg07] Wiberg, K. B.; Wilson, S. M.; Wang, Y.-g.; Vaccaro, P. H.; Cheeseman, J. R.; Luderer, M. R., "Effect of Substituents and Conformations on the Optical Rotations of Cyclic Oxides and Related Compounds. Relationship between the Anomeric Effect and Optical Rotation," *J. Org. Chem.*, 2007, **72**, 6206-14, DOI: 10.1021/jo070816j.

[Wiberg82] Wiberg, K. B.; Walker, F. H., "[1.1. 1] Propellane," *Journal of the American Chemical Society*, 1982, **104**, 5239-40, DOI: 10.1021/ja00383a046.

[Wiberg85] Wiberg, K. B.; Dailey, W. P.; Walker, F. H.; Waddell, S. T.; Crocker, L. S.; Newton, M., "Vibrational Spectrum, Structure, and Energy of [1.1.1]propellane," *J. Am. Chem. Soc.*, 1985, **107**, 7247-57, DOI: 10.1021/ja00311a003.

[Wiberg86] Wiberg, K. B.; Matturro, M. G.; Okarma, P. J.; Jason, M. E.; Dailey, W. P.; Burgmaier, G. J.; Bailey, W. F.; Warner, P., "Bicyclo[2.2.0]hex-1(4)-ene," *Tetrahedron*, 1986, **42**, 1895-902, DOI: 10.1016/S0040-4020(01)87609-2

[Wiberg87] Wiberg, K. B.; Murcko, M. A., "Rotational Barriers. 1. 1,2-Dihaloethanes," *J. Phys. Chem.*, 1987, **91**, 3616-20, DOI: 10.1021/j100297a030.

[Wiberg91] Wiberg, K. B.; Crocker, L. S.; Morgan, K. M., "Thermochemical Studies of Carbonyl Compounds. 5. Enthalpies of Reduction of Carbonyl Groups," *J. Am. Chem. Soc.*, 1991, **113**, 3447-50, DOI: 10.1021/ja00009a033.

[Wiberg92] Wiberg, K. B.; Hadad, C. M.; LePage, T. J.; Breneman, C. M.; Frisch, M. J., "An Analysis of the Effect of Electron Correlation on Charge Density Distributions," *J. Phys. Chem.*, 1992, **96**, 671-79, DOI: 10.1021/j100181a030.

[Wiberg92b] Wiberg, K. B.; Rosenberg, R. E., "Infrared Intensities: Cyclobutene: A Normal-Coordinate Analysis and Comparison with Cyclopropene," *J. Phys. Chem.*, 1992, **96**, 8282-92, DOI: 10.1021/j100200a016.

[Wiberg92c] Wiberg, K. B.; Rosenberg, R. E.; Waddell, S. T., "Infrared Intensities: Bicyclo[1.1.1]pentane: A Normal-Cöordinate Analysis and Comparison with [1.1.1]propellane," *J. Phys. Chem.*, 1992, **96**, 8293-303, DOI: 10.1021/j100200a017.

[Wiberg92d] Wiberg, K. B.; Hadad, C. M.; Rablen, P. R.; Cioslowski, J., "Substituent Effects. 4. Nature of Substituent Effects at Carbonyl Groups," *J. Am. Chem. Soc.*, 1992, **114**, 8644-54, DOI: 10.1021/ja00048a044.

[Wiberg95] Wiberg, K. B.; Cheeseman, J. R.; Ochterski, J. W.; Frisch, M. J., "Substituent Effects. 6. Heterosubstituted allyl radicals: Comparison with substituted allyl cations and anions," *J. Am. Chem. Soc.*, 1995, **117**, 6535-43, DOI: 10.1021/ja00129a018.

[Wiberg95c] Wiberg, K. B.; Thiel, Y.; Goodman, L.; Leszczynski, J., "Acetaldehyde: Harmonic Frequencies, Force Field, and Infrared Intensities," *J. Phys. Chem.*, 1995, **99**, 13850, DOI: 10.1021/j100038a016.

[Wilson05] Wilson, S. M.; Wiberg, K. B.; Cheeseman, J. R.; Frisch, M. J.; Vaccaro, P. H., "Nonresonant optical activity of isolated organic molecules," *J. Phys. Chem. A*, 2005, **109**, 11752-64, DOI: 10.1021/jp054283z.

[Winnewisser71] Winnewisser, G.; Maki, A. G.; Johnson, D. R., "Rotational constants for HCN and DCN," *J. Mol. Spectrosc.*, 1971, **39**, 149, DOI: 10.1016/0022-2852(71)90286-4.

[Wohar91] Wohar, M. M.; Jagodzinski, P. W., "Infrared Spectra of $_2CO$, $_2{}^{13}CO$, $_2CO$, and $_2{}^{13}CO$ and Anomalous Values in Vibrational Force Fields," *J. Mol. Spectrosc.*, 1991, **148**, 13-19, DOI: 10.1016/0022-2852(91)90030-E.

[Wolinski90] Wolinski, K.; Hilton, J. F.; Pulay, P., "Efficient Implementation of the Gauge-Independent Atomic Orbital Method for NMR Chemical Shift Calculations," *J. Am. Chem. Soc.*, 1990, **112**, 8251-60, DOI: 10.1021/ja00179a005.

[Wong91a] Wong, M. W.; Wiberg, K. B.; Frisch, M. J., "Hartree-Fock Second Derivatives and Electric Field Properties in a Solvent Reaction Field: Theory and Application," *J. Chem. Phys.*, 1991, **95**, 8991-98, DOI: 10.1063/1.461230.

[Wong96] Wong, M. W., "Vibrational frequency prediction using density functional theory," *Chem. Phys. Lett.*, 1996, **256**, 391-99, DOI: 10.1016/0009-2614(96)00483-6.

[Wong98] Wong, M. W.; Radom, L., "Radical Addition to Alkenes: Further Assessment of Theoretical Procedures," *J. Phys. Chem. A*, 1998, **102**, 2237-45, DOI: 10.1021/jp973427+.

[Woon93] Woon, D. E.; Dunning Jr., T. H., "Gaussian-basis sets for use in correlated molecular calculations. 3. The atoms aluminum through argon," *J. Chem. Phys.*, 1993, **98**, 1358-71, DOI: 10.1063/1.464303.

[Xu08] Xu, W.; Peng, B.; Chen, J.; Liang, M.; Cai, F., "New Triphenylamine-Based Dyes for Dye-Sensitized Solar Cells," *J. Phys. Chem. C*, 2008, **112**, 874-80, DOI: 10.1021/jp076992d.

[Yamamoto96] Yamamoto, N.; Vreven, T.; Robb, M. A.; Frisch, M. J.; Schlegel, H. B., "A Direct Derivative MC-SCF Procedure," *Chem. Phys. Lett.*, 1996, **250**, 373-78, DOI: 10.1016/0009-2614(96)00027-9.

[Yanai04] Yanai, T.; Tew, D.; Handy, N., "A new hybrid exchange-correlation functional using the Coulomb-attenuating method (CAM-B3LYP)," *Chem. Phys. Lett.*, 2004, **393**, 51-57, DOI: 10.1016/j.cplett.2004.06.011.

[Yang14] Yang, Y.; Ratner, M. A.; Schatz, G. C., "Multireference Ab Initio Study of Ligand Field d-d Transitions in Octahedral Transition-Metal Oxide Clusters," *J. Phys. Chem. C*, 2014, **(in press)**, DOI: 10.1021/jp5052672.

[Yao09] Yao, W. W.; Peng, H. M.; Webster, R. D., "Electrochemistry of α-Tocopherol (Vitamin E) and α-Tocopherol Quinone Films Deposited on Electrode Surfaces in the Presence and Absence of Lipid Multilayers," *J. Phys. Chem. C*, 2009, **113**, 21805, DOI: 10.1021/jp9079124.

[Yencha95] Yencha, A. J.; Hopkirk, A.; Hiraya, A.; Donovan, R. J.; Goode, J. G.; Maier, R. R. J.; King, G. C.; Kvaran, A., "Threshold Photoelectron Spectroscopy of Cl_2 and Br_2 up to 35 eV," *J. Phys. Chem.*, 1995, **99**, 7231, DOI: 10.1021/j100019a004.

[York99] York, D. M.; Karplus, M., "Smooth solvation potential based on the conductor-like screening model," *J. Phys. Chem. A*, 1999, **103**, 11060-79, DOI: 10.1021/jp992097l.

[Zeng04] Zeng, Y.; Sun, Q.; Meng, L.; Zheng, S.; Wang, D., "Theoretical calculational studies on the mechanism of thermal dissociations for RN_3 (R=CH_3, CH_3CH_2, $(CH_3)2CH$, $(CH_3)_3C$)," *Chem. Phys. Lett.*, 2004, **390**, 362, DOI: 10.1016/j.cplett.2004.04.045.

[Zerner91] "Semi Empirical Molecular Orbital Methods" in Zerner, M. C., *Reviews of Computational Chemistry*, VCH Publishing: New York, 1991, Vol. 2, pp. 313-66.

[Zhang98] Zhang, L.; Hermans, J., "Hydrophilicity of cavities in proteins," *Proteins*, 1998, **24**, 433-38, DOI: 10.1002/(SICI)1097-0134(199604)24:4<433::AID-PROT3>3.0.CO;2-F.

[Zytowski96] Zytowski, T.; Fischer, H., "Absolute Rate Constants for the Addition of Methyl Radicals to Alkenes in Solution: New Evidence for Polar Interactions," *J. Am. Chem. Soc.*, 1996, **118**, 437-39, DOI: 10.1021/ja953085q.

Molecule Index

The following index lists the molecules studied in this book, ordered by common name. A sortable and searchable version of this list is available on the book's website: *www.expchem3.com/molecules*. Common names are simplified in some cases, most chirality indicators are omitted, and single atoms are not listed. In general, neutral forms are shown except when doing so would introduce confusion among multiple compounds. Initial numbers and other prefixes are ignored in sorting. IUPAC names are omitted when identical with the common name.

Index